CIVIL AIRCRAFT MARKINGS 2005

Alan J. Wright & Dave Peel

Ian Allan PUBLISHING

Contents

Introduction 4
International Civil Aircraft Markings 6
Aircraft Type Designations & Abbreviations 10
British Civil Registrations 13
Military to Civil Cross-Reference 285
Republic of Ireland Civil Registrations 307
Overseas Airliner Registrations. 319
Radio Frequencies 402
Airline Flight Codes. 403
BAPC Register 405
Future Allocations Logs. 410
Future Allocation Groups. 411
Overseas Airlines Registration Log. 412
Addenda 413

This fifty sixth edition published 2005

ISBN 0 7110 3051 0

Published by Ian Allan Publishing

an imprint of Ian Allan Publishing Ltd, Hersham, Surrey KT12 4RG.
Printed in England by Ian Allan Printing Ltd, Hersham, Surrey KT12 4RG.

Code: 0503/G

Front cover: Boeing 747-200, MK Airlines.
Richard Cooper

All photographs by Alan J. Wright (AJW) or Dave Peel (DP) unless otherwise indicated.

3

Introduction

The familiar 'G' prefixed four letter registration system was adopted in 1919 after a short-lived spell with serial numbers commencing at K-100. Until July 1928 the UK allocations were issued in the G-Exxx range, but as a result of further International agreements, this series ended at G-EBZZ, the replacement being G-Axxx. From this point registrations were issued in a reasonably orderly manner through to G-AZZZ, the position reached in July 1972. There were, however, two exceptions. In order to prevent possible confusion with signal codes, the G-AQxx sequence was omitted, while G-AUxx was reserved for Australian use originally. In recent years however, individual requests for a mark in the latter range have been granted by the Authorities.

Although the next logical sequence was started at G-Bxxx, it was not long before the strictly applied rules relating to aircraft registration began to be relaxed. Permission was readily given for personalised marks to be issued incorporating virtually any four-letter combination, while re-registration also became a common feature, a practice almost unheard of in the past. In this book, where this has taken place at some time, all previous UK identities carried appear in parenthesis after the operator's/owner's name. For example, during its career Jetstream 31 G-PLAH has also carried the identity G-LOVA, G-OAKA, G-BUFM and G-LAKH.

Some aircraft have also been allowed to wear military markings without displaying their civil identity. In this case the serial number actually carried is shown in parenthesis after the type's name. For example Auster 6A G-ARRX flies in military colours as VF512, its genuine previous identity. As an aid to the identification of such machines, a conversion list is provided.

Other factors caused a sudden acceleration in the number of registrations allocated by the Civil Aviation Authority in the early 1980s. The first surge followed the discovery that it was possible to register plastic bags and other items even less likely to fly, on payment of the standard fee. This erosion of the main register was checked in early 1982 by the issue of a special sequence for such devices commencing with G-FYAA. Powered hang-gliders provided the second glut of allocations as a result of the decision that these types should be officially registered. Although a few of the early examples penetrated the current in-sequence register, in due course all new applicants were given marks in special ranges, this time G-MBxx, G-MGxx, G-MJxx, G-MMxx, G-MNxx, G-MTxx, G-MVxx, G-MWxx, G-MYxx and G-MZxx. It took some time before all microlights displayed an official mark but gradually the registration was carried, the size and position depending on the dimensions of the component to which it was applied.

There was news of a further change in mid-1998 when the CAA announced that with immediate effect microlights would be issued with registrations in the normal sequence alongside aircraft in other classes. In addition, it meant that owners could also apply for a personalised identity upon payment of the then current fee of £170 from April 1999, a low price for those wishing to display their status symbol. These various changes played their part in exhausting the current G-Bxxx range after some 26 years, with G-BZxx coming into use before the end of 1999. As this batch approached completion the next series to be used began at G-CBxx instead of the anticipated G-CAxx. The reason for this step was to avoid the re-use of marks issued in Canada during the 1920s, although a few have appeared more recently as personalised UK registrations.

The various alterations have resulted in the introduction of some changes of format for this edition of Civil Aircraft Markings in which all the UK registrations appear in alphabetical order including the sections containing microlights, toy balloons and the ever expanding out-of-sequence listing. Hopefully the new layout will be of help.

Throughout the UK section of the book, there are instances when the probable base of the aircraft has been included. This is positioned at the end of the owner/operator details preceded by an oblique stroke. It must of course be borne in mind that changes do take place and that no attempt has been made to record the residents at the many private strips. The base of airline equipment has been given as the company's headquarters airport, although frequently aircraft are outstationed for long periods.

Acknowledgements

Once again thanks are extended to the Registration Department of the Civil Aviation Authority for its assistance and allowing access to its files. As always the comments and amendments flowing from the ever-observant Wal Gandy have proved of considerable value, while similarly George Pennick has also provided much useful material on a regular basis. Sincere thanks are extended, not only to both of these contributors, but to all who have supplied items for possible use. Special thanks must go to Sara Wright who has used her considerable experience to resolve computer problems before they became disasters. Her keyboard speed also proved beneficial for the project which was otherwise a much slower operation. AJW.

We welcome Dave Peel to the editorial team who is to look after the Overseas Airliner Registartions section of Civil Aircraft Markings.

International Civil Aircraft Markings

A2-	Botswana	N-	United States of America
A3-	Tonga	OB-	Peru
A4O-	Oman	OD-	Lebanon
A5-	Bhutan	OE-	Austria
A6-	United Arab Emirates	OH-	Finland
A7-	Qatar	OK-	Czech Republic
A9C-	Bahrain	OM-	Slovakia
AP-	Pakistan	OO-	Belgium
B-	China/Taiwan/Hong Kong	OY-	Denmark
C-F/C-G-	Canada	P-	Korea (North)
C2-	Nauru	P2-	Papua New Guinea
C3	Andora	P4-	Aruba
C5-	Gambia	PH-	Netherlands
C6-	Bahamas	PJ-	Netherlands Antilles
C9-	Mozambique	PK-	Indonesia and West Irian
CC-	Chile	PP-,PR-,PT-	Brazil
CN-	Morocco	PZ-	Surinam
CP-	Bolivia	RA-	Russia
CS-	Portugal	RDPL-	Laos
CU-	Cuba	RP-	Philippines
CX-	Uruguay	S2-	Bangladesh
D-	Germany	S5-	Slovenia
D2-	Angola	S7-	Seychelles
D4-	Cape Verde Islands	S9-	São Tomé
D6-	Comores Islands	SE-	Sweden
DQ-	Fiji	SP-	Poland
E3-	Eritrea	ST-	Sudan
EC-	Spain	SU-	Egypt
EI-	Republic of Ireland	SU-Y	Palestine
EK-	Armenia	SX-	Greece
EL-	Liberia	T2-	Tuvalu
EP-	Iran	T3-	Kiribati
ER-	Moldova	T7-	San Marino
ES-	Estonia	T8A	Palau
ET-	Ethiopia	T9-	Bosnia-Herzegovina
EW-	Belarus	TC-	Turkey
EX-	Kyrgyzstan	TF-	Iceland
EY-	Tajikistan	TG-	Guatemala
EZ-	Turkmenistan	TI-	Costa Rica
F-	France, Colonies and Protectorates	TJ-	United Republic of Cameroon
G-	United Kingdom	TL-	Central African Republic
H4-	Solomon Islands	TN-	Republic of Congo (Brazzaville)
HA-	Hungary	TR-	Gabon
HB-	Switzerland and Liechtenstein	TS-	Tunisia
HC-	Ecuador	TT-	Tchad
HH-	Haiti	TU-	Ivory Coast
HI-	Dominican Republic	TY-	Benin
HK-	Colombia	TZ-	Mali
HL-	Korea (South)	UK-	Uzbekistan
HP-	Panama	UN-	Kazakhstan
HR-	Honduras	UR-	Ukraine
HS-	Thailand	V2-	Antigua
HV-	The Vatican	V3-	Belize
HZ-	Saudi Arabia	V4	St Kitts & Nevis
I-	Italy	V5-	Namibia
J2-	Djibouti	V6	Micronesia
J3-	Grenada	V7-	Marshall Islands
J5-	Guinea Bissau	V8-	Brunei
J6-	St Lucia	VH-	Australia
J7-	Dominica	VN-	Vietnam
J8-	St Vincent	VP-B	Bermuda
JA-	Japan	VP-C	Cayman Islands
JU	Mongolia	VP-F	Falkland Islands
JY-	Jordan	VP-G	Gibraltar
LN-	Norway	VP-LA	Anguilla
LV-	Argentina	VP-LM	Montserrat
LX-	Luxembourg	VP-LV	Virgin Islands
LY-	Lithuania	VQ-T	Turks & Caicos Islands
LZ-	Bulgaria	VT-	India
MT	Mongolia	XA-,XB-,XC-	Mexico

THESE AIRLINER CLASSICS NEED A PROFESSIONAL!

Just Flight are proud to present two classic airliners which you can fly in Microsoft Flight Simulator - can you handle these big birds and live the dream, in full realistic detail!

CONCORDE PROFESSIONAL

The best commercial aircraft ever built? We think so, and now that the real Concorde has stopped flying it's time for flight simulation to step up to the crease and carry on batting for this remarkable aviation achievement that used to cross the Atlantic faster than the sun! Concorde Professional is the most detailed re-creation of this great aircraft developed for the consumer market and its unparalleled accuracy will make flying a challenge similar to that provided by the real Concorde.

Thanks to the development team at Phoenix Simulation Software (PSS), you are able to operate this supersonic marvel in Flight Simulator 2004 – it will be your most rewarding and exciting flying, bar none!

A340 PROFESSIONAL

Make the professional choice with two of Europe's most advanced airliners developed for Microsoft's Flight Simulator 2002 & 2004.

The Airbus A330 and A340 are spearheading Europe's domination of the airline market and now these two high-tech heroes are brought to life by the experts in high-quality flight simulation aircraft, Just Flight and Phoenix Simulations.

Eight variants of the A330 and A340 in over 40 airline liveries are painstakingly reproduced to a life-like level of detail with accurate animations of the flight surfaces, engines, undercarriage and passenger and cargo doors - even special light and engine effects are included. Incredible 2D and 3D cockpit interiors give you the feeling of 'being there' and flight testing by real Airbus pilots has ensured that true professional experience.

Available from all good computer games and aviation stores or direct from Just Flight www.justflight.com

XT-	Burkina Faso	5R-	Malagasy Republic (Madagascar)
XU-	Cambodia	5T-	Mauritania
XY-	Myanmar	5U-	Niger
YA-	Afghanistan	5V-	Togo
YI-	Iraq	5W-	Western Samoa (Polynesia)
YJ-	Vanuatu	5X-	Uganda
YK-	Syria	5Y-	Kenya
YL-	Latvia	6O-	Somalia
YN-	Nicaragua	6V-	Senegal
YR-	Romania	6Y-	Jamaica
YS-	El Salvador	7O-	Yemen
YU-	Yugoslavia	7P-	Lesotho
YV-	Venezuela	7Q-	Malawi
Z-	Zimbabwe	7T-	Algeria
Z3-	Macedonia	8P-	Barbados
ZA-	Albania	8Q-	Maldives
ZK-	New Zealand	8R-	Guyana
ZP-	Paraguay	9A-	Croatia
ZS-, ZU	South Africa	9G-	Ghana
3A-	Monaco	9H-	Malta
3B-	Mauritius	9J-	Zambia
3C-	Equatorial Guinea	9K-	Kuwait
3D-	Swaziland	9L-	Sierra Leone
3X-	Guinea	9M-	Malaysia
4K-	Azerbaijan	9N-	Nepal
4L-	Georgia	9Q-	Congo Kinshasa
4R-	Sri Lanka	9U-	Burundi
4X-	Israel	9V-	Singapore
5A-	Libya	9XR-	Rwanda
5B-	Cyprus	9Y-	Trinidad and Tobago
5H-	Tanzania		
5N-	Nigeria		

Aircraft Type Designations & Abbreviations

(eg PA-28 Piper Type 28)

A.	Beagle, Auster, Airbus
AAC	Army Air Corps
AA-	American Aviation, Grumman American
AB	Agusta-Bell
AG	American General
ANEC	Air Navigation & Engineering Co.
ANG	Air National Guard
AS	Aérospatiale
A.S.	Airspeed
A.W.	Armstrong Whitworth
B.	Blackburn, Bristol, Boeing, Beagle
BA	British Airways
BAC	British Aircraft Company
BAC	British Aircraft Corporation
BAe	British Aerospace
BAPC	British Aviation Preservation Council
BAT	British Aerial Transport
B.K.	British Klemm
BN	Britten-Norman
Bo	Bolkow
Bu	Bücker
CAARP	Co-operative des Ateliers A de la Région Parisienne
CAC	Commonwealth Aircraft Corporation
CAF	Canadian Air Force
C.A.S.A.	Construcciones Aeronautics SA
CCF	Canadian Car & Foundry Co
CEA	Centre-Est Aviation
C.H.	Chrislea
CHABA	Cambridge Hot-Air Ballooning Association
CLA	Comper
CP.	Piel
CUAS	Cambridge University Air Squadron
Cycl	Cyclone
D.	Druine
DC-	Douglas Commercial
D.H.	de Havilland
D.H.A.	de Havilland Australia
D.H.C.	de Havilland Canada
DR.	Jodel (Robin-built)
EE	English Electric
EAA	Experimental Aircraft Association
EMB	Embraer Empresa Brasileira de Aeronautica SA
EoN	Elliotts of Newbury
EP	Edgar Percival
ETPS	Empire Test Pilots School
F.	Fairchild, Fokker
F.A.A.	Fleet Air Arm
FFA	Flug und Fahrzeugwerke AG
FH	Fairchild-Hiller
FrAF	French Air Force
FRED	Flying Runabout Experimental Design
Fw	Focke-Wulf
G.	Grumman
GA	Gulfstream American
G.A.L.	General Aircraft
G.C.	Globe
GECAS	General Electric Capital Aviation Services
GY	Gardan
H	Helio
HM.	Henri Mignet

HP.	Handley Page
HPR	Handley Page Reading
HR.	Robin
H.S.	Hawker Siddeley
ICA	Intreprinderea de Constructii
IHM	International Helicopter Museum
I.I.I.	Iniziative Industriali Italiane
IL	Ilyushin
ILFC	International Lease Finance Corporation
IMCO	Intermountain Manufacturing Co
IWM	Imperial War Museum
J.	Auster
JT	John Taylor
KR	Rand-Robinson
L.	Lockheed
L.A.	Luton, Lake
L.V.G.	Luft-Verkehrs Gesellschaft
M.	Miles, Mooney
MBA	Micro Biplane Aviation
MBB	Messerschmitt-Bölkow-Blohm
McD	McDonnell
MDH	McDonnell Douglas Helicopters
MH	Max Holste
MHCA	Manhole Cover
MJ	Jurca
M.S.	Morane-Saulnier
NA	North American
NC	Nord
NE	North East
P.	Hunting (formerly Percival), Piaggio
PA-	Piper
PC.	Pilatus
PZL	Panstwowe Zaklady Lotnicze
QAC	Quickie Aircraft Co
R.	Rockwell
RAF	Rotary Air Force
RAAF	Royal Australian Air Force
RAFGSA	Royal Air Force Gliding & Soaring Association
RCAF	Royal Canadian Air Force
RF	Fournier
R.N.	Royal Navy
S.	Short, Sikorsky
SA, SE, SO	Sud-Aviation, Aérospatiale, Scottish Aviation
SAAB	Svenska Aeroplan Aktieboleg
SC	Short
SCD	Side Cargo Door
SNCAN	Societe Nationale de Constructions Aeronautiques du Nord
SOCATA	Societe de Construction d'Avions de Tourisme et d'Affaires
Soc	Society
SpA	Societa per Azioni
SPP	Strojirny Prvni Petilesky
S.R.	Saunders-Roe, Stinson
SS	Special Shape
ST	SOCATA
SW	Solar Wings
T.	Tipsy
TB	SOCATA
Tu	Tupolev
UH.	United Helicopters (Hiller)
UK	United Kingdom
USAF	United States Air Force
USAAC	United States Army Air Corps
USN	United States Navy
V.	Vickers-Armstrongs
V.L.M.	Vlaamse Luchttransportmaatschappij
V.S.	Vickers-Supermarine
WA	Wassmer
WAR	War Aircraft Replicas
WHE	W.H.Ekin
W.S.	Westland
Z.	Zlin

Reg.	Type (†False registration)	Owner or Operator	Notes
G-AAAH†	D.H.60G Moth (replica) (BAPC 168) ★	Yorkshire Air Museum/Elvington	
G-AAAH	D.H.60G Moth ★	Science Museum Jason/S. Kensington	
G-AACA†	Avro 504K (BAPC 177) ★	Brooklands Museum of Aviation/Weybridge	
G-AACN	H.P.39 Gugnunc ★	Science Museum/Wroughton	
G-AADR	D.H.60GM Moth	E. V. Moffatt	
G-AAEG	D.H.60G Gipsy Moth	I. B. Grace	
G-AAHI	D.H.60G Moth	N. J. W. Reid	
G-AAHY	D.H.60M Moth	D. J. Elliott	
G-AAIN	Parnall Elf II	The Shuttleworth Collection/O. Warden	
G-AALY	D.H.60G Moth	K. M. Fresson	
G-AAMX	D.H.60GM Moth ★	Aerospace Museum/Cosford	
G-AAMY	D.H.60GMW Moth	Totalsure Ltd	
G-AANG	Blériot XI	The Shuttleworth Collection/O. Warden	
G-AANH	Deperdussin Monoplane	The Shuttleworth Collection/O. Warden	
G-AANI	Blackburn Monoplane	The Shuttleworth Collection/O. Warden	
G-AANJ	L.V.G.-C VI (7198/18)	RAF Museum/Hendon	
G-AANL	D.H.60M Moth	A. L. Berry	
G-AANM	Bristol 96A F.2B (D7889)	Aero Vintage Ltd (BAPC166)	
G-AANO	D.H.60GMW Moth	A. W. & M. E. Jenkins	
G-AANV	D.H.60G Moth	R. A. Seeley	
G-AAOK	Curtiss Wright Travel Air 12Q	Shipping & Airlines Ltd/Biggin Hill	
G-AAOR	D.H.60G Moth	B. R. Cox & N. J. Stagg	
G-AAPZ	Desoutter I (mod.)	The Shuttleworth Collection/O. Warden	
G-AAUP	Klemm L.25-1A	J. I. Cooper	
G-AAWO	D.H.60G Moth	N. J. W. Reid	
G-AAXK	Klemm L.25-1A ★	C. C. Russell-Vick (stored)	
G-AAYT	D.H.60G Moth	P. Groves	
G-AAYX	Southern Martlet	The Shuttleworth Collection/O. Warden	
G-AAZG	D.H.60G Moth	J. A. Pothecary & ptnrs	
G-AAZP	D.H.80A Puss Moth	R. P. Williams	
G-ABAA	Avro 504K ★	Manchester Museum of Science & Industry	
G-ABAG	D.H.60G Moth	A. & P. A. Wood	
G-ABBB	B.105A Bulldog IIA (K2227) ★	RAF Museum/Hendon	
G-ABDW	D.H.80A Puss Moth (VH-UQB) ★	Museum of Flight/E. Fortune	
G-ABDX	D.H.60G Moth	M. D. Souch	
G-ABEV	D.H.60G Moth	S. L. G. Darch	
G-ABLM	Cierva C.24 ★	De Havilland Heritage Museum	
G-ABLS	D.H.80A Puss Moth	R. I. & D. E. Souch	
G-ABMR	Hart 2 (J9941) ★	RAF Museum	
G-ABNT	Civilian C.A.C.1 Coupe	Shipping & Airlines Ltd/Biggin Hill	
G-ABNX	Redwing 2	Redwing Syndicate/Redhill	
G-ABOI	Wheeler Slymph ★	Midland Air Museum/Coventry	
G-ABOX	Sopwith Pup (N5195)	C. M. D. & A. P. St. Cyrien/Middle Wallop	
G-ABSD	D.H.A.60G Moth	M. E. Vaisey	
G-ABTC	CLA.7 Swift	P. Channon (stored)	
G-ABUL†	D.H.82A Tiger Moth ★	F.A.A. Museum (G-AOXG)/Yeovilton	
G-ABUS	CLA.7 Swift	R. C. F. Bailey	
G-ABVE	Arrow Active 2	Real Aircraft Co./Breighton	
G-ABWP	Spartan Arrow	R. E. Blain/Barton	
G-ABXL	Granger Archaeopteryx ★	J. R. Granger	
G-ABYA	D.H.60G Gipsy Moth	J. F. Moore & D. A. Hay/Biggin Hill	
G-ABZB	D.H.60G-III Moth Major	R. Earl & B. Morris	
G-ACAA	Bristol 96A F.2B (D8084†)	Patina Ltd/Duxford	
G-ACBH	Blackburn B.2 ★	–/West Hanningfield, Essex	
G-ACCB	D.H.83 Fox Moth	E. A. Gautrey	
G-ACDA	D.H.82A Tiger Moth	B. D. Hughes	
G-ACDC	D.H.82A Tiger Moth	Tiger Club Ltd/Headcorn	
G-ACDI	D.H.82A Tiger Moth	J. A. Pothecary/Shoreham	
G-ACDJ	D.H.82A Tiger Moth	de Havilland School of Flying Ltd	
G-ACEJ	D.H.83 Fox Moth	Newbury Aeroplane Co	
G-ACET	D.H.84 Dragon	M. D. Souch	
G-ACGT	Avro 594 Avian IIIA ★	Yorkshire Light Aircraft Ltd/Leeds	
G-ACGZ	D.H.60G-III Moth Major	N. H. Lemon	
G-ACIT	D.H.84 Dragon ★	Science Museum/Wroughton	
G-ACLL	D.H.85 Leopard Moth	V. M & D. C. M. Stiles	

Notes	Reg.	Type	Owner or Operator
	G-ACMA	D.H.85 Leopard Moth	S. J. Filhol/Sherburn
	G-ACMD	D.H.82A Tiger Moth	M. J. Bonnick
	G-ACMN	D.H.85 Leopard Moth	M. R. & K. E. Slack
	G-ACNS	D.H.60G-III Moth Major	R. I. & D. Souch
	G-ACOJ	D.H.85 Leopard Moth	Norman Aeroplane Trust/Rendcomb
	G-ACSP	D.H.88 Comet ★	T. M., M. L., D. A. & P. M. Jones
	G-ACSS	D.H.88 Comet ★	The Shuttleworth Collection *Grosvenor House*/ O. Warden
	G-ACSS†	D.H.88 Comet (replica)	G. Gayward (BAPC216) ★
	G-ACSS†	D.H.88 Comet (replica)	The Galleria/Hatfield (BAPC257) ★
	G-ACTF	CLA.7 Swift ★	The Shuttleworth Collection/O. Warden
	G-ACUS	D.H.85 Leopard Moth	R. A. & V. A. Gammons
	G-ACUU	Cierva C.30A (HM580) ★	G. S. Baker/Duxford
	G-ACUX	S.16 Scion (VH-UUP) ★	Ulster Folk & Transport Museum
	G-ACVA	Kay Gyroplane ★	Glasgow Museum of Transport
	G-ACWM	Cierva C.30A (AP506) ★	IHM/Weston-s-Mare
	G-ACWP	Cierva C.30A (AP507) ★	Science Museum/S. Kensington
	G-ACXB	D.H.60G-III Moth Major	D. F. Hodgkinson
	G-ACXE	B.K. L-25C Swallow	J. G. Wakeford
	G-ACYK	Spartan Cruiser III ★	Museum of Flight *(front fuselage)*/E. Fortune
	G-ACZE	D.H.89A Dragon Rapide	Wessex Aviation & Transport Ltd G-AJGS/
	G-ADAH	D.H.89A Dragon Rapide ★	Manchester Museum of Science & Industry Pioneer
	G-ADEV	Avro 504K (H5199)	The Shuttleworth Collection (G-ACNB)/ O. Warden
	G-ADFV	Blackburn B-2 ★	Lincolnshire Aviation Heritage Centre
	G-ADGP	M.2L Hawk Speed Six	R. A. Mills
	G-ADGT	D.H.82A Tiger Moth	M. F. Dalton
	G-ADGV	D.H.82A Tiger Moth	K. J. Whitehead (G-BACW)
	G-ADHD	D.H.60G-III Moth Major	M. E. Vaisey
	G-ADIA	D.H.82A Tiger Moth	S. J. Beaty
	G-ADJJ	D.H.82A Tiger Moth	J. M. Preston
	G-ADKC	D.H.87B Hornet Moth	A. J. Davy/Carlisle
	G-ADKK	D.H.87B Hornet Moth	R. M. Lee
	G-ADKL	D.H.87B Hornet Moth	P. R. & M. J. F. Gould
	G-ADKM	D.H.87B Hornet Moth	L. V. Mayhead
	G-ADLY	D.H.87B Hornet Moth	Totalsure Ltd
	G-ADMT	D.H.87B Hornet Moth	S. & H. Roberts
	G-ADMW	M.2H Hawk Major (DG590) ★	Museum of Army Flying/Middle Wallop
	G-ADND	D.H.87B Hornet Moth (W9385)	The Shuttleworth Collection/O. Warden
	G-ADNE	D.H.87B Hornet Moth	G-ADNE Group
	G-ADNL	M.5 Sparrowhawk ★	A. P. Pearson
	G-ADNZ	D.H.82A Tiger Moth (DE673)	D. C. Wall
	G-ADOT	D.H.87B Hornet Moth ★	De Havilland Heritage Museum
	G-ADPC	D.H.82A Tiger Moth	D. J. Marshall
	G-ADPJ	B.A.C. Drone ★	N. H. Ponsford/Brighton
	G-ADPS	B.A. Swallow 2	J. F. Hopkins
	G-ADRA	Pietenpol Air Camper	A. J. Mason
	G-ADRG†	Mignet HM.14 (replica)	Lower Stondon Transport Museum (BAPC77) ★
	G-ADRH	D.H.87B Hornet Moth	R. G. Grocott/Switzerland
	G-ADRR	Aeronca C.3	S. J. Rudkin
	G-ADRX†	Mignet HM.14 (replica)	S. Copeland Aviation Group (BAPC231) ★
	G-ADRY†	Mignet HM.14 (replica) (BAPC29) ★	Brooklands Museum of Aviation/Weybridge
	G-ADUR	D.H.87B Hornet Moth	R. A. Seeley
	G-ADVU†	Mignet HM.14 (replica)	N.E. Aircraft Museum/Usworth (BAPC211) ★
	G-ADWJ	D.H.82A Tiger Moth	C. Adams
	G-ADWO	D.H.82A Tiger Moth (BB807) ★	Southampton Hall of Aviation
	G-ADWT	M.2W Hawk Trainer	R. Earl & B. Morris
	G-ADXS	Mignet HM.14 ★	Thameside Aviation Museum/Shoreham
	G-ADXT	D.H.82A Tiger Moth	J. R, Hanauer
	G-ADYS	Aeronca C.3	J. I. Cooper/Rendcomb
	G-ADYV†	Mignet HM.14 (replica)	P. Ward (BAPC243) ★
	G-ADZW†	Mignet HM.14 (replica)	H. Shore/Sandown (BAPC253) ★
	G-AEAJ†	D.H.89A Dragon Rapide (replica) ★	Marriott Liverpool South Hotel
	G-AEBB	Mignet HM.14 ★	The Shuttleworth Collection/O. Warden
	G-AEBJ	Blackburn B-2	BAE Systems (Operations) Ltd/Warton
	G-AEDB	B.A.C. Drone 2	R. E. Nerou & P. L. Kirk
	G-AEDU	D.H.90 Dragonfly	Norman Aeroplane Trust/Rendcomb
	G-AEEG	M.3A Falcon Skysport	P. R. Holloway/O.Warden
	G-AEEH	Mignet HM.14 ★	Aerospace Museum/Cosford
	G-AEFG	Mignet HM.14 (BAPC75) ★	N. H. Ponsford/Brighton

Reg.	Type	Owner or Operator	Notes
G-AEFT	Aeronca C.3	N. S. Chittenden	
G-AEGV	Mignet HM.14 ★	Midland Air Museum/Coventry	
G-AEHM	Mignet HM.14 ★	Science Museum/Wroughton	
G-AEXZ	Piper J-2 Cub	Mrs M. & J. R. Dowson/Leicester	
G-AEJZ	Mignet HM.14 (BAPC120) ★	Bomber County Museum/Hemswell	
G-AEKR	Mignet HM.14 (BAPC121) ★	S. Yorks Aviation Soc	
G-AEKV	Kronfeld Drone ★	Brooklands Museum of Aviation/Weybridge	
G-AEKW	M.12 Mohawk ★	RAF Museum	
G-AELO	D.H.87B Hornet Moth	M. J. Miller	
G-AEML	D.H.89 Dragon Rapide	Amanda Investments Ltd	
G-AENP	Hawker Hind (K5414) (BAPC78)	The Shuttleworth Collection/O. Warden	
G-AEOA	D.H.80A Puss Moth	P. & A. Wood/O. Warden	
G-AEOF†	Mignet HM.14 (BAPC22) ★	Aviodome/Schiphol, Netherlands	
G-AEOF	Rearwin 8500	Shipping & Airlines Ltd/Biggin Hill	
G-AEOH	Mignet HM.14 ★	Midland Air Museum	
G-AEPH	Bristol F.2B (D8096)	The Shuttleworth Collection/O. Warden	
G-AERV	M.11A Whitney Straight	R. A. Seeley	
G-AESB	Aeronca C.3	R. J. M. Turnbull	
G-AESE	D.H.87B Hornet Moth	J. G. Green/Redhill	
G-AESZ	Chilton D.W.1	R. E. Nerou	
G-AETA	Caudron G.3 (3066) ★	RAF Museum/Hendon	
G-AEUJ	M.11A Whitney Straight	R. E. Mitchell	
G-AEVS	Aeronca 100	A. M. Lindsay & N. H. Ponsford/Breighton	
G-AEXD	Aeronca 100	Mrs M. A. & R. W. Mills	
G-AEXF	P.6 Mew Gull	Real Aircraft Co/Breighton	
G-AEXT	Dart Kitten II	A. J. Hartfield	
G-AEZF	S.16 Scion 2 ★	Acebell Aviation/Redhill	
G-AEZJ	P.10 Vega Gull	D. P. H. Hulme	
G-AEZX	Bücker Bü133C Jungmeister (LG+03)	T. J. Reeve	
G-AFAP†	C.A.S.A. C.352L ★	Aerospace Museum/Cosford	
G-AFAX	B. A. Eagle 2	J. G. Green & M. J. Miller	
G-AFBS	M.14A Hawk Trainer 3 ★	G. D. Durbridge-Freeman (G-AKKU)/ Duxford	
G-AFCL	B. A. Swallow 2	C. P. Bloxham	
G-AFDO	Piper J-3F-60 Cub	R. Wald	
G-AFDX	Hanriot HD.1 (HD-75) ★	RAF Museum/Hendon	
G-AFEL	Monocoupe 90A	M. Rieser	
G-AFFD	Percival Q-6 ★	B. D. Greenwood	
G-AFFH	Piper J-2 Cub	M. J. Honeychurch	
G-AFFI†	Mignet HM.14 (replica) (BAPC76) ★	Yorkshire Air Museum/Elvington	
G-AFGC	B. A. Swallow 2	G. E. Arden	
G-AFGD	B. A. Swallow 2	A. T. Williams & ptnrs/Shobdon	
G-AFGE	B. A. Swallow 2	G. R. French	
G-AFGH	Chilton D.W.1.	M. L. & G. L. Joseph	
G-AFGI	Chilton D.W.1.	J. E. & K. A. A. McDonald	
G-AFGM	Piper J-4A Cub Coupé	P. H. Wilkinson/Carlisle	
G-AFGZ	D.H.82A Tiger Moth	M. R. Paul & P. A. Shaw (G-AMHI)	
G-AFHA	Mosscraft MA.1. ★	C. V. Butler	
G-AFIN	Chrislea LC.1 Airguard (BAPC203) ★	N. Wright	
G-AFIR	Luton LA-4 Minor	A. J. Mason	
G-AFJA	Watkinson Dingbat ★	K. Woolley	
G-AFJB	Foster-Wikner G.M.1. Wicko (DR613) ★	J. Dibble	
G-AFJR	Tipsy Trainer 1	M. E. Vaisey *(stored)*	
G-AFJU	M.17 Monarch ★	Museum of Flight/E. Fortune	
G-AFJV	Mosscraft MA.2 ★	C. V. Butler	
G-AFNG	D.H.94 Moth Minor	The Gullwing Trust	
G-AFNI	D.H.94 Moth Minor	J. Jennings	
G-AFOB	D.H.94 Moth Minor	K. Cantwell	
G-AFOJ	D.H.94 Moth Minor ★	De Havilland Heritage Museum	
G-AMHJ	Douglas C-47A ★	Assault Glider Association/Shawbury	
G-AFPN	D.H.94 Moth Minor	J. W. & A. R. Davy/Carlisle	
G-AFRZ	M.17 Monarch	R. E. Mitchell (G-AIDE)	
G-AFSC	Tipsy Trainer 1	D. M. Forshaw	
G-AFSV	Chilton D.W.1A	R. E. Nerou	
G-AFSW	Chilton D.W.2 ★	R. I. Souch	
G-AFTA	Hawker Tomtit (K1786)	The Shuttleworth Collection/O. Warden	
G-AFTN	Taylorcraft Plus C2 ★	Leicestershire County Council Museums	
G-AFUP	Luscombe 8A Silvaire	R. Dispain	
G-AFVE	D.H.82A Tiger Moth (T7230)	Tigerfly/Booker	
G-AFVN	Tipsy Trainer 1	D. F. Lingard	
G-AFWH	Piper J-4A Cub Coupé	C. W. Stearn & R. D. W. Norton	
G-AFWI	D.H.82A Tiger Moth	E. Newbigin	

Notes	Reg.	Type	Owner or Operator
	G-AFWT	Tipsy Trainer 1	N. Parkhouse
	G-AFYD	Luscombe 8F Silvaire	J. D. Iliffe
	G-AFYO	Stinson H.W.75	M. Lodge
	G-AFZA	Piper J-4A Cub Coupe	R. A. Benson
	G-AFZK	Luscombe 8A Silvaire	M. G. Byrnes
	G-AFZL	Porterfield CP.50	P. G. Lucas & S. H. Sharpe/White Waltham
	G-AFZN	Luscombe 8A Silvaire	A. L. Young/Henstridge
	G-AGAT	Piper J-3F-50 Cub	O. T. Taylor & C. J. Marshall
	G-AGBN	G.A.L.42 Cygnet 2 ★	Museum of Flight/E. Fortune
	G-AGEG	D.H.82A Tiger Moth	Norman Aeroplane Trust/Rendcomb
	G-AGFT	Avia FL.3 (W7)	K. Joynson & K. Cracknell
	G-AGHY	D.H.82A Tiger Moth	P. Groves
	G-AGIV	Piper J-3C-65 Cub	P. C. & F. M. Gill
	G-AGJG	D.H.89A Dragon Rapide	M. J. & D. J. T. Miller/Duxford
	G-AGLK	Auster 5D	C. R. Harris/Biggin Hill
	G-AGMI	Luscombe 8A Silvaire	P. R. Bush
	G-AGNJ	D.H.82A Tiger Moth	B. P. Borsberry & ptnrs
	G-AGNV	Avro 685 York 1 (TS798) ★	Aerospace Museum/Cosford
	G-AGOH	J/1 Autocrat ★	Newark Air Museum
	G-AGOS	R.S.4 Desford Trainer (VZ728) ★	Museum of Flight/E. Fortune
	G-AGOY	M.48 Messenger 3 (U-0247)	P. A. Brook
	G-AGPG	Avro 19 Srs 2 ★	Avro Heritage Soc/Woodford
	G-AGPK	D.H.82A Tiger Moth	Delta Aviation Ltd
	G-AGRU	V.498 Viking 1A ★	Brooklands Museum of Aviation/Weybridge
	G-AGSH	D.H.89A Dragon Rapide 6	Venom Jet Promotions Ltd/Bournemouth
	G-AGTM	D.H.89A Dragon Rapide 6	Air Atlantique Ltd/Coventry
	G-AGTO	J/1 Autocrat	M. J. Barnett & D. J. T. Miller/Duxford
	G-AGTT	J/1 Autocrat	R. Farrer
	G-AGVG	J/1 Autocrat (modified)	S. J. Riddington/Leicester
	G-AGVN	J/1 Autocrat	G. H. Farrar
	G-AGVV	Piper J-3C-65 Cub	M. Molina-Ruano/Spain
	G-AGXN	J/1N Alpha	Gentleman's Aerial Touring Carriage Group
	G-AGXT	J/1N Alpha ★	Nene Valley Aircraft Museum
	G-AGXU	J/1N Alpha	B. H. Austen
	G-AGXV	J/1 Autocrat	B. S. Dowsett & I. M. Oliver
	G-AGYD	J/1N Alpha	P. D. Hodson
	G-AGYK	J/1 Autocrat	Autocrat Syndicate
	G-AGYL	J/1 Autocrat ★	Military Vehicle Conservation Group
	G-AGYT	J/1N Alpha	P. J. Barrett
	G-AGYU	D.H.82A Tiger Moth (DE208)	P. L. Jones
	G-AGZZ	D.H.82A Tiger Moth	R. C. Mercer
	G-AHAG	D.H.89A Rapide	Pelham Ltd/Membury
	G-AHAL	J/1N Alpha	Wickenby Aviation
	G-AHAM	J/1 Autocrat	C. P. L. Jenkin
	G-AHAN	D.H.82A Tiger Moth	Tiger Associates Ltd
	G-AHAP	J/1 Autocrat	W. D. Hill
	G-AHAT	J/1N Alpha ★	Dumfries & Galloway Aviation Museum
	G-AHAU	J/1 Autocrat	Andreas Auster Group
	G-AHAV	J/1 Autocrat	J. S. Antrum
	G-AHBL	D.H.87B Hornet Moth	H. D. Labouchere
	G-AHBM	D.H.87B Hornet Moth	P. A. & E. P. Gliddon
	G-AHCK	J/1N Alpha	Skegness Air Taxi Service Ltd
	G-AHCL	J/1N Alpha	Electronic Precision Ltd (G-OJVC)
	G-AHCN	J/1N Alpha	G. L. Brown & E. Martinsen
	G-AHCR	Gould-Taylorcraft Plus D Special	D. E. H. Balmford & D. R. Shepherd/Dunkeswell
	G-AHEC	Luscombe 8A Silvaire	P. G. Baxter
	G-AHED	D.H.89A Dragon Rapide (RL962) ★	RAF Museum Storage & Restoration Centre/ RAF Wyton
	G-AHGD	D.H.89A Dragon Rapide	R. Jones
	G-AHGW	Taylorcraft Plus D (LB375)	C. V. Butler/Coventry
	G-AHGZ	Taylorcraft Plus D (LB367)	M. Pocock
	G-AHHH	J/1 Autocrat	H. A. Jones/Norwich
	G-AHHT	J/1N Alpha	A. C. Barber & N. J. Hudson
	G-AHHU	J/1N Alpha ★	L. A. Groves & I. R. F. Hammond
	G-AHIP	Piper J-3C-65 Cub	A. R. Mangham
	G-AHIZ	D.H.82A Tiger Moth	C.F.G. Flying Ltd/Cambridge
	G-AHKX	Avro 19 Srs 2	BAe PLC/Avro Heritage Soc/Woodford
	G-AHKY	Miles M.18 Series 2 ★	Museum of Flight/E. Fortune
	G-AHLK	Auster 3	E. T. Brackenbury/Leicester
	G-AHLT	D.H.82A Tiger Moth	M. P. Waring

Reg.	Type	Owner or Operator	Notes
G-AHMN	D.H.82A Tiger Moth (N6985)	Museum of Army Flying/Middle Wallop	
G-AHNR	Taylorcraft BC-12D	T. P. Hancock	
G-AHOO	D.H.82A Tiger Moth	J. T. & A. D. Milsom	
G-AHPZ	D.H.82A Tiger Moth	N. J. Wareing	
G-AHRI	D.H.104 Dove 1 ★	Newark Air Museum	
G-AHRO	Cessna 140	R. H. Screen/Kidlington	
G-AHSA	Avro 621 Tutor (K3215)	The Shuttleworth Collection/O. Warden	
G-AHSD	Taylorcraft Plus D	A. L. Hall-Carpenter	
G-AHSO	J/1N Alpha	W. P. Miller	
G-AHSP	J/1 Autocrat	R. M. Weeks	
G-AHSS	J/1N Alpha	A. M. Roche	
G-AHST	J/1N Alpha	A. C. Frost	
G-AHTE	P.44 Proctor V	D. K. Tregilgas	
G-AHTW	A.S.40 Oxford (V3388) ★	Skyfame Collection/Duxford	
G-AHUF	D.H.Tiger Moth	Dream Ventures Ltd	
G-AHUG	Taylorcraft Plus D	D. Nieman	
G-AHUI	M.38 Messenger 2A ★	Museum of Berkshire Aviation/Woodley	
G-AHUJ	M.14A Hawk Trainer 3 (R1914) ★	Strathallan Aircraft Collection	
G-AHUN	Globe GC-1B Swift	R. J. Hamlett	
G-AHUV	D.H.82A Tiger Moth	A. D. Gordon	
G-AHVU	D.H.82A Tiger Moth (T6313)	Foley Farm Flying Group	
G-AHVV	D.H.82A Tiger Moth	B. M. Pullen	
G-AHWJ	Taylorcraft Plus D (LB294)	M. Pocock	
G-AHXE	Taylorcraft Plus D (LB312)	J. M. C. Pothecary/Shoreham	
G-AIBE	Fulmar II (N1854) ★	F.A.A. Museum/Yeovilton	
G-AIBH	J/1N Alpha	M. J. Bonnick	
G-AIBM	J/1 Autocrat	D. G. Greatrex	
G-AIBR	J/1 Autocrat	R. H. & J. A. Cooper	
G-AIBW	J/1N Alpha	W. E. Bateson/Blackpool	
G-AIBX	J/1 Autocrat	Wasp Flying Group	
G-AIBY	J/1 Autocrat	D. Morris/Sherburn	
G-AICX	Luscombe 8A Silvaire	R. V. Smith/Henstridge	
G-AIDL	D.H.89A Dragon Rapide 6	Atlantic Air Transport Ltd/Coventry	
G-AIDS	D.H.82A Tiger Moth	K. D. Pogmore & T. Dann	
G-AIEK	M.38 Messenger 2A (RG333)	J. Buckingham	
G-AIFZ	J/1N Alpha	M. D. Ansley	
G-AIGD	J/1 Autocrat	R. B. Webber	
G-AIGF	J/1N Alpha	A. R. C. Mathie	
G-AIGT	J/1N Alpha	R. R. Harris	
G-AIGU	J/1N Alpha	N. K. Geddes	
G-AIIH	Piper J-3C-65 Cub	J. A. de Salis	
G-AIJI	J/1N Alpha ★	C. J. Baker	
G-AIJM	Auster J/4	N. Huxtable	
G-AIJS	Auster J/4 ★	(stored)	
G-AIJT	Auster J/4 Srs 100	Aberdeen Auster Flying Group	
G-AIJZ	J/1 Autocrat	A. A. Marshall (stored)	
G-AIKE	Auster 5	C. J. Baker	
G-AIPR	Auster J/4	MPM Flying Group/Booker	
G-AIPV	J/1 Autocrat	W. P. Miller	
G-AIRC	J/1 Autocrat	R. C. Tebbett/Shobdon	
G-AIRI	D.H.82A Tiger Moth	E. R. Goodwin (stored)	
G-AIRK	D.H.82A Tiger Moth	R. C. Teverson & ptnrs	
G-AISA	Tipsy B Srs 1	A. A. M. Huke	
G-AISC	Tipsy B Srs 1	Wagtail Flying Group	
G-AISS	Piper J-3C-65 Cub	K. W. Wood & F. Watson/Insch	
G-AIST	V.S.300 Spitfire 1A (AR213/PR-D)	Sheringham Aviation UK Ltd	
G-AISX	Piper J-3C-65 Cub	Cubfly	
G-AITB	A.S.10 Oxford (MP425) ★	RAF Museum Store/Cardington	
G-AIUA	M.14A Hawk Trainer 3 ★	K. P. & D. L. Hunt	
G-AIUL	D.H.89A Dragon Rapide 6	I. Jones/Chirk	
G-AIXA	Taylorcraft Plus D (LB264)★	Aerospace Museum/Cosford	
G-AIXJ	D.H.82A Tiger Moth	D. Green/Goodwood	
G-AIXN	Benes-Mraz M.1C Sokol	A. J. Wood	
G-AIYG	SNCAN Stampe SV-4B	R. Lageirse & H. waut//Belgium	
G-AIYR	D.H.89A Dragon Rapide	Spectrum Leisure Ltd/Duxford/Clacton	
G-AIYS	D.H.85 Leopard Moth	Wessex Aviation & Transport Ltd	
G-AIZE	F.24W Argus 2 ★	Aerospace Museum/Cosford	
G-AIZF	D.H.82A Tiger Moth ★	(stored)	
G-AIZG	V.S. Walrus 1 (L2301) ★	F.A.A. Museum/Yeovilton	
G-AIZU	J/1 Autocrat	C. J. & J. G. B. Morley	

Notes	Reg.	Type	Owner or Operator
	G-AJAD	Piper J-3C-65 Cub	C. R. Shipley
	G-AJAE	J/1N Alpha	Litchfield Auster Group
	G-AJAJ	J/1N Alpha	R. B. Lawrence
	G-AJAM	J/2 Arrow	D. A. Porter
	G-AJAP	Luscombe 8A Silvaire	R. J. Thomas
	G-AJAS	J/1N Alpha	C. J. Baker
	G-AJCP	D.31 Turbulent	B. R. Pearson
	G-AJDW	J/1 Autocrat	D. R. Hunt
	G-AJEB	J/1N Alpha ★	Manchester Museum of Science & Industry
	G-AJEH	J/1N Alpha	J. T. Powell-Tuck
	G-AJEI	J/1N Alpha	J. Siddall
	G-AJEM	J/1 Autocrat	C. D. Wilkinson
	G-AJES	Piper J-3C-65 Cub (330485)	G. W. Jarvis
	G-AJGJ	Auster 5 (RT486)	British Classic Aircraft Restoration Flying Group
	G-AJHJ	Auster 5 ★	stored
	G-AJHS	D.H.82A Tiger Moth	Vliegend Museum/Netherlands
	G-AJHU	D.H.82A Tiger Moth	G. Valenti/Italy
	G-AJIH	J/1 Autocrat	D. G. Curran
	G-AJIS	J/1N Alpha	Husthwaite Auster Group
	G-AJIT	J/1 Kingsland Autocrat	A. J. Kay
	G-AJIU	J/1 Autocrat	M. D. Greenhalgh/Netherthorpe
	G-AJIW	J/1N Alpha	Truman Aviation Ltd/Tollerton
	G-AJJP	Jet Gyrodyne (XJ389) ★	Museum of Berkshire Aviation/Woodley
	G-AJJS	Cessna 120	Robhurst Flying Group
	G-AJJT	Cessna 120	J. S. Robson
	G-AJJU	Luscombe 8E Silvaire	S. C. Weston & R. J. Hopcraft
	G-AJKB	Luscombe 8E Silvaire	A. F. Hall & S. P. Collins/Tibenham
	G-AJOC	M.38 Messenger 2A ★	Ulster Folk & Transport Museum
	G-AJOE	M.38 Messenger 2A	P. W. Bishop
	G-AJON	Aeronca 7AC Champion	A. Biggs & D. S. Moores
	G-AJOV†	Sikorsky S-51 ★	Aerospace Museum/Cosford
	G-AJOZ	F.24W Argus 2 ★	Thorpe Camp Preservation Group/Woodhall Spa
	G-AJPI	F.24R-41a Argus 3 (314887)	K. A. Doornbos/Netherlands
	G-AJPZ	J/1 Autocrat ★	Wessex Aviation Soc
	G-AJRB	J/1 Autocrat	G-AJRB Flying Group
	G-AJRC	J/1 Autocrat	M. Baker
	G-AJRE	J/1 Autocrat (Lycoming)	Air Tech Spares
	G-AJRH	J/1N Alpha ★	Charnwood Museum/Loughborough
	G-AJRS	M.14A Hawk Trainer 3 (P6382)	The Shuttleworth Collection/O. Warden
	G-AJTW	D.H.82A Tiger Moth (N6965)	J. A. Barker/Tibenham
	G-AJUE	J/1 Autocrat	P. H. B. Cole
	G-AJUL	J/1N Alpha	M. J. Crees
	G-AJVE	D.H.82A Tiger Moth	R. A. Gammons
	G-AJWB	M.38 Messenger 2A	G. E. J. Spooner
	G-AJXC	Auster 5	J. E. Graves
	G-AJXV	Auster 4 (NJ695)	B. A. Farries/Leicester
	G-AJXY	Auster 4	D. A. Hall
	G-AJYB	J/1N Alpha	P. J. Shotbolt
	G-AKAT	M.14A Hawk Trainer 3 (T9738)	R. A. Fleming/Brighton
	G-AKAZ	Piper J-3C-65 Cub (57-H)	Frazerblades Ltd/Duxford
	G-AKBM	M.38 Messenger 2A ★	Bristol Plane Preservation Unit
	G-AKBO	M.38 Messenger 2A	P. R. Holloway
	G-AKDW	D.H.89A Dragon Rapide ★	De Havilland Heritage Museum
	G-AKEL	M.65 Gemini 1A ★	Ulster Folk & Transport Museum
	G-AKER	M.65 Gemini 1A ★	Berkshire Aviation Group/Woodley
	G-AKGD	M.65 Gemini 1A ★	Berkshire Aviation Group/Woodley
	G-AKGE	M.65 Gemini 3C ★	Ulster Folk & Transport Museum
	G-AKHP	M.65 Gemini 1A	P. A. Brook/Shoreham
	G-AKHZ	M.65 Gemini 7 ★	Museum of Berkshire Aviation/Woodley
	G-AKIB	Piper J-3C-90 Cub (480015)	M. C. Bennett
	G-AKIF	D.H.89A Dragon Rapide	Airborne Taxi Services Ltd/Booker
	G-AKIN	M.38 Messenger 2A	Sywell Messenger Group
	G-AKIU	P.44 Proctor V	Air Atlantique Ltd/Coventry
	G-AKKB	M.65 Gemini 1A	J. Buckingham
	G-AKKH	M.65 Gemini 1A	J. S. Allison
	G-AKKR	M.14A Magister (T9707) ★	Manchester Museum of Science & Industry
	G-AKKY	M.14A Hawk Trainer 3 (L6906) ★ (BAPC44)	Museum of Berkshire Aviation/Woodley
	G-AKLW	SA.6 Sealand 1 ★	Ulster Folk & Transport Museum
	G-AKOT	Auster 5 ★	C. J. Baker
	G-AKOW	Auster 5 (TJ569) ★	Museum of Army Flying/Middle Wallop

Reg.	Type	Owner or Operator
G-AKPF	M.14A Hawk Trainer 3 (V1075)	P. R. Holloway
G-AKRA	Piper J-3C-65 Cub	W. R. Savin
G-AKRP	D.H.89A Dragon Rapide 4	Fordaire Aviation Ltd/Sywell
G-AKSY	Auster 5	A, Brier
G-AKSZ	Auster 5C	P. W. Yates & R. G. Darbyshire
G-AKTH	Piper J-3C-65 Cub	G. H. Harry & Viscount Goschen
G-AKTI	Luscombe 8A Silvaire	M. W. Olliver
G-AKTK	Aeronca 11AC Chief	Aeronca Tango Kilo Group
G-AKTN	Luscombe 8A Silvaire	M. G. Rummey
G-AKTO	Aeronca 7BCM Champion	D. C. Murray
G-AKTP	PA-17 Vagabond	Golf Tango Papa Group
G-AKTR	Aeronca 7AC Champion	C. Fielder
G-AKTS	Cessna 120	M. Isterling
G-AKTT	Luscombe 8A Silvaire	S. J. Charters
G-AKUE	D.H.82A Tiger Moth	D. F. Hodgkinson
G-AKUG	Luscombe 8A Silvaire	J. C. Holland
G-AKUH	Luscombe 8E Silvaire	E. J. Lloyd
G-AKUI	Luscombe 8E Silvaire	D. A. Sims
G-AKUJ	Luscombe 8E Silvaire	R. C. Green
G-AKUK	Luscombe 8A Silvaire	Leckhampstead Flying Group
G-AKUL	Luscombe 8A Silvaire	E. A. Taylor
G-AKUM	Luscombe 8F Silvaire	D. A. Young
G-AKUN	Piper J-3F-65 Cub	W. R. Savin
G-AKUO	Aeronca 11AC Chief	L. W. Richardson
G-AKUP	Luscombe 8E Silvaire	D. A. Young
G-AKUR	Cessna 140	J. Greenaway & C. A. Davies/Popham
G-AKUW	Chrislea C.H.3 Super Ace 2	J. & S. Rickett
G-AKVF	Chrislea C.H.3 Super Ace 2	B. Metters
G-AKVM	Cessna 120	N. Wise & S. Walker
G-AKVN	Aeronca 11AC Chief	Breckland Aeronca Group
G-AKVO	Taylorcraft BC-12D	M. Gibson
G-AKVP	Luscombe 8A Silvaire	J. M. Edis
G-AKVR	Chrislea C.H.3 Skyjeep 4	C. L. Needham
G-AKVZ	M.38 Messenger 4B	Shipping & Airlines Ltd/Biggin Hill
G-AKWS	Auster 5A-160 (RT610)	The Interesting Aircraft Co
G-AKWT	Auster 5 ★	C. Baker
G-AKXP	Auster 5	M. Pocock
G-AKXS	D.H.82A Tiger Moth	J. & G. J. Eagles
G-AKZN	P.34A Proctor 3 (Z7197) ★	RAF Museum/Hendon
G-ALAH	M.38 Messenger 4A (RH377) ★	RAF Museum/Henlow
G-ALAX	D.H.89A Dragon Rapide ★	Durney Aeronautical Collection/Andover
G-ALBJ	Auster 5	P. N. Elkington
G-ALBK	Auster 5	S. J. Wright
G-ALBN	Bristol 173 (XF785) ★	RAF Museum Storage & Restoration Centre/ Cardington
G-ALCK	P.34A Proctor 3 (LZ766) ★	Skyfame Collection/Duxfor?d
G-ALCS	M.65 Gemini 3C ★	*(stored)*
G-ALCU	D.H.104 Dove 2 ★	Midland Air Museum/Coventry
G-ALDG	HP.81 Hermes 4 ★	Duxford Aviation Soc *(fuselage only)*
G-ALEH	PA-17 Vagabond	A. D. Pearce/White Waltham
G-ALFA	Auster 5	S. P. Barrett
G-ALFT	D.H.104 Dove 6 ★	Caernarfon Air World
G-ALFU	D.H.104 Dove 6 ★	Duxford Aviation Soc
G-ALGA	PA-15 Vagabond	G. A. Brady
G-ALGT	V.S.379 Spitfire F.XIV★	Rolls-Royce PLC
G-ALIJ	PA-17 Vagabond	Popham Flying Group/Popham
G-ALIW	D.H.82A Tiger Moth	D. I. M. Geddes & F. Curry/Booker
G-ALJF	P.34A Proctor 3	J. F. Moore/Biggin Hill
G-ALJL	D.H.82A Tiger Moth	R. I. & D. Souch
G-ALLF	Slingsby T.30A Prefect (ARK)	J. F. Hopkins & K. M. Fresson/Parham Park
G-ALNA	D.H.82A Tiger Moth	R. J. Doughton
G-ALND	D.H.82A Tiger Moth (N9191)	J. T. Powell-Tuck
G-ALNV	Auster 5 ★	C. J. Baker (stored)
G-ALOD	Cessna 140	J. R. Stainer
G-ALRI	D.H.82A Tiger Moth (T5672)	Mark Squared Ltd
G-ALSP	Bristol 171 Sycamore (WV783)★	R.N. Fleetlands Museum
G-ALSS	Bristol 171 Sycamore (WA576)★	Dumfries & Galloway Aviation Museum
G-ALST	Bristol 171 Sycamore (WA577)★	N.E. Aircraft Museum/Usworth
G-ALSW	Bristol 171 Sycamore (WT933)★	Newark Air Museum
G-ALSX	Bristol 171 Sycamore (G-48-1)★	IHM/Weston-s-Mare
G-ALTO	Cessna 140	J. M. Edis

Notes	Reg.	Type	Owner or Operator
	G-ALTW	D.H.82A Tiger Moth ★	A. Mangham
	G-ALUC	D.H.82A Tiger Moth	D. R. & M. Wood
	G-ALVP	D.H.82A Tiger Moth ★	V. & R. Wheele (stored)
	G-ALWB	D.H.C.1 Chipmunk 22A	D. M. Neville
	G-ALWF	V.701 Viscount ★	Duxford Aviation Soc
			RMA Sir John Franklin/Duxford
	G-ALWS	D.H.82A Tiger Moth	A. P. Benyon/Welshpool
	G-ALWW	D.H.82A Tiger Moth	D. E. Findon
	G-ALXT	D.H.89A Dragon Rapide ★	Science Museum/Wroughton
	G-ALXZ	Auster 5-150	M. F. Cuming
	G-ALYB	Auster 5 (RT520) ★	S. Yorks Aircraft Preservation Soc
	G-ALYG	Auster 5D	A. L. Young/Henstridge
	G-ALYW	D.H.106 Comet 1 ★	RAF Exhibition Flight
			(fuselage converte to Nimrod)
	G-ALZE	BN-1F ★	M. R. Short/Southampton Hall of Aviation
	G-ALZO	A.S.57 Ambassador ★	Duxford Aviation Soc
	G-AMAW	Luton LA-4 Minor	R. H. Coates
	G-AMBB	D.H.82A Tiger Moth	J. Eagles
	G-AMCK	D.H.82A Tiger Moth	Avia Special Ltd
	G-AMCM	D.H.82A Tiger Moth	A. K. & J. I. Cooper
	G-AMDA	Avro 652A Anson 1 (N4877) ★	Skyfame Collection/Duxford
	G-AMEN	PA-18 Super Cub 95	A. Lovejoy & W. Cook
	G-AMHF	D.H.82A Tiger Moth	Wavendon Social Housing Ltd
	G-AMIU	D.H.82A Tiger Moth	M. D. Souch
	G-AMKU	J/1B Aiglet	P. G. Lipman
	G-AMLZ	P.50 Prince 6E ★	Caernarfon Air World Museum
	G-AMMS	J/5K Aiglet Trainer	R. B. Webber
	G-AMNN	D.H.82A Tiger Moth	M. Thrower Spirit of Pashley/Shoreham
	G-AMOG	V.701 Viscount ★	Aerospace Museum/Cosford
	G-AMPG	PA-12 Super Cruiser	N.Sutton
	G-AMPI	SNCAN Stampe SV-4C	T. W. Harris
	G-AMPO	Douglas C-47B (FZ626/YS-DH) ★	Gate Guardian/RAF Lyneham
	G-AMPY	Douglas C-47B	Atlantic Air Ltd/Thales/Coventry
	G-AMPZ	Douglas C-47B★	Air Service Berlin GmbH/Templehof
	G-AMRA	Douglas C-47B	Atlantic Air Transport Ltd/Coventry
	G-AMRF	J/5F Aiglet Trainer	A. L. Tuttle
	G-AMRK	G.37 Gladiator I (423/427)	The Shuttleworth Collection/O. Warden
	G-AMSG	SIPA 903	S. W. Markham
	G-AMSN	Douglas C-47B ★	Aceball Aviation/Redhill
	G-AMSV	Douglas C-47B	Atlantic Air Transport Ltd/Coventry
	G-AMTA	J/5F Aiglet Trainer	J. D. Manson
	G-AMTK	D.H.82A Tiger Moth	S. W. McKay & M. E. Vaisey
	G-AMTM	J/1 Autocrat	R. J. Stobo (G-AJUJ)
	G-AMTV	D.H.82A Tiger Moth	Tango Victor Flying Group
	G-AMUF	D.H.C.1 Chipmunk 21	Redhill Tailwheel Flying Club Ltd
	G-AMUI	J/5F Aiglet Trainer	R. B. Webber
	G-AMVD	Auster 5	M.Hammond
	G-AMVP	Tipsy Junior	A. R. Wershat
	G-AMVS	D.H.82A Tiger Moth	J. T. Powell-Tuck
	G-AMXA	D.H.106 Comet 2 (nose only) ★	Spectators' Terrace/Gatwick
	G-AMYA	Zlin Z.381	D. M. Fenton
	G-AMYD	J/5L Aiglet Trainer	G. H. Maskell
	G-AMYJ	Douglas C-47B ★	Yorkshire Air Museum/Elvington
	G-AMYL	PA-17 Vagabond	P. J. Penn-Sayers/Shoreham
	G-AMZI	J/5F Aiglet Trainer	J. F. Moore/Biggin Hill
	G-AMZT	J/5F Aiglet Trainer	D. Hyde & ptnrs/Cranfield
	G-AMZU	J/5F Aiglet Trainer	J. A. Longworth & ptnrs
	G-ANAF	Douglas C-47B	Air Atlantique Ltd/Thales/Coventry
	G-ANAP	D.H.104 Dove 6 ★	Brunel Technical College/Lulsgate
	G-ANCF	B.175 Britannia 308 ★	Bristol Aero Collection (stored)/Kemble
	G-ANCS	D.H.82A Tiger Moth	C. E. Edwards & E. A. Higgins
	G-ANCX	D.H.82A Tiger Moth	D. R. Wood/Biggin Hill
	G-ANDM	D.H.82A Tiger Moth	N. J. Stagg
	G-ANDP	D.H.82A Tiger Moth	A. H. Diver
	G-ANEC	D.H.82A Tiger Moth ★	(stored)
	G-ANEH	D.H.82A Tiger Moth (N6797)	G. J. Wells/Goodwood
	G-ANEL	D.H.82A Tiger Moth	Totalsure Ltd
	G-ANEM	D.H.82A Tiger Moth	P. J. Benest
	G-ANEN	D.H.82A Tiger Moth	A. J. D. Douglas-Hamilton
	G-ANEW	D.H.82A Tiger Moth	A. L. Young

Reg.	Type	Owner or Operator	Notes
G-ANEZ	D.H.82A Tiger Moth	C. D. J. Bland/Sandown	
G-ANFC	D.H.82A Tiger Moth	H. J. E. Pierce/Chirk	
G-ANFH	Westland S-55 ★	IHM/Weston-s-Mare	
G-ANFI	D.H.82A Tiger Moth (DE623)	G. P. Graham	
G-ANFL	D.H.82A Tiger Moth	Felthorpe Tiger Group Ltd	
G-ANFM	D.H.82A Tiger Moth	Reading Flying Group/White Waltham	
G-ANFP	D.H.82A Tiger Moth	G. D. Horn	
G-ANFU	Auster 5 (NJ719) ★	N.E. Aircraft Museum/Usworth	
G-ANFV	D.H.82A Tiger Moth (DF155)	R. A. L. Falconer	
G-ANGK	Cessna 140A	G. A. Copeland	
G-ANHK	D.H.82A Tiger Moth	J. D. Iliffe	
G-ANHR	Auster 5	H. L. Swallow	
G-ANHS	Auster 4	Tango Uniform Group	
G-ANHU	Auster 4	D. J. Baker (stored)	
G-ANHX	Auster 5D	D. J. Baker	
G-ANIE	Auster 5 (TW467)	M. J. Whitwell	
G-ANIJ	Auster 5D (TJ672)	M. Pocock	
G-ANIS	Auster 5	J. Clarke-Cockburn	
G-ANJA	D.H.82A Tiger Moth (N9389)	P. Auckland	
G-ANJD	D.H.82A Tiger Moth	I. Laws	
G-ANKK	D.H.82A Tiger Moth (T5854)	Halfpenny Green Tiger Group	
G-ANKT	D.H.82A Tiger Moth (T6818)	The Shuttleworth Collection/O. Warden	
G-ANKV	D.H.82A Tiger Moth (T7793) ★	Westmead Business Group/Croydon Airport	
G-ANKZ	D.H.82A Tiger Moth	D. W. Grahamt	
G-ANLD	D.H.82A Tiger Moth	K. Peters	
G-ANLS	D.H.82A Tiger Moth	P. A. Gliddon	
G-ANLW	W.B.1. Widgeon (MD497) ★	Norfolk & Suffolk Museum/Flixton	
G-ANMO	D.H.82A Tiger Moth (K4259)	E. & K. M. Lay	
G-ANMV	D.H.82A Tiger Moth (T7404)	J. W. Davy/Cardiff	
G-ANMY	D.H.82A Tiger Moth (DE470)	Lotmead Flying Group	
G-ANNB	D.H.82A Tiger Moth	G. M. Bradley	
G-ANNE	D.H.82A Tiger Moth	C. R. Hardiman	
G-ANNG	D.H.82A Tiger Moth	P. F. Walter	
G-ANNI	D.H.82A Tiger Moth	C. E. Ponsford & ptnrs	
G-ANNK	D.H.82A Tiger Moth	D. R. Wilcox	
G-ANOA	Hiller UH-12A ★	Redhill Technical College	
G-ANOH	D.H.82A Tiger Moth	N. Parkhouse/White Waltham	
G-ANOK	SAAB S.91C Safir ★	A. F. Galt & Co (stored)	
G-ANOM	D.H.82A Tiger Moth	A. L. Creer	
G-ANON	D.H.82A Tiger Moth (T7909)	Hields Aviation/Sherburn	
G-ANOO	D.H.82A Tiger Moth	R. K. Packman/Shoreham	
G-ANOV	D.H.104 Dove 6 ★	Museum of Flight/E. Fortune	
G-ANPE	D.H.82A Tiger Moth	I. E. S. Huddleston (G-IESH)/Clacton	
G-ANPP	P.34A Proctor 3	C. P. A. & J. Jeffrey	
G-ANRF	D.H.82A Tiger Moth	C. D. Cyster	
G-ANRM	D.H.82A Tiger Moth (DF112)	Spectrum Leisure Ltd/Duxford/Clacton	
G-ANRN	D.H.82A Tiger Moth	J. J. V. Elwes/Rush Green	
G-ANRP	Auster 5 (TW439)	I. C. Naylor	
G-ANRX	D.H.82A Tiger Moth ★	De Havilland Heritage Museum	
G-ANSM	D.H.82A Tiger Moth	Northamptonshire School of Flying Ltd	
G-ANTE	D.H.82A Tiger Moth (T6562)	P. Reading	
G-ANTK	Avro 685 York ★	Duxford Aviation Soc	
G-ANTS	D.H.82A Tiger Moth (N6532)	J. G. Green	
G-ANUO	D.H.114 Heron 2D (G-AOXL) ★	Westmead Business Group/Croydon Airport	
G-ANUW	D.H.104 Dove 6 ★	Jet Aviation Preservation Group	
G-ANWB	D.H.C.1 Chipmunk 21	G. Briggs/Blackpool	
G-ANWO	M.14A Hawk Trainer 3 ★	A. G. Dunkerley	
G-ANXB	D.H.114 Heron 1B ★	Newark Air Museum	
G-ANXC	J/5R Alpine	Alpine Group	
G-ANXR	P.31C Proctor 4 (RM221)	L. H. Oakins/Biggin Hill	
G-ANZJ	P.31C Proctor 4 (NP303) ★	A. Hillyard	
G-ANZT	Thruxton Jackaroo	D. J. Neville & P. A. Dear	
G-ANZU	D.H.82A Tiger Moth	P. A. Jackson	
G-ANZZ	D.H.82A Tiger Moth	J. I. B. Bennett & P. P. Amershi	
G-AOAA	D.H.82A Tiger Moth	R. C. P. Brookhouse	
G-AOBG	Somers-Kendall SK.1 ★	(stored)/Breighton	
G-AOBH	D.H.82A Tiger Moth (NL750)	P. Nutley/Thruxton	
G-AOBO	D.H.82A Tiger Moth	J. S. & S. V. Shaw	
G-AOBU	P.84 Jet Provost T.1 (XD693)	T. J. Manna/North Weald	
G-AOBV	J/5P Autocar	P. E. Champney (stored)	
G-AOBX	D.H.82A Tiger Moth	David Ross Flying Group	

Notes	Reg.	Type	Owner or Operator
	G-AOCP	Auster 5 ★	C. J. Baker *(stored)*
	G-AOCR	Auster 5D (NJ673)	G. J. McDill
	G-AOCU	Auster 5	S. J. Ball/Leicester
	G-AODA	Westland S-55 Srs 3 ★	IHM/Weston-s-Mare
	G-AODR	D.H.82A Tiger Moth	L. A. Groves (G-ISIS)
	G-AODT	D.H.82A Tiger Moth (R5250)	R. A. Harrowven
	G-AOEH	Aeronca 7AC Champion	R. A & S. P. Smith
	G-AOEI	D.H.82A Tiger Moth	C.F.G. Flying Ltd/Cambridge
	G-AOEL	D.H.82A Tiger Moth ★	Museum of Flight/E. Fortune
	G-AOES	D.H.82A Tiger Moth	K. A. & A. J. Broomfield
	G-AOET	D.H.82A Tiger Moth	Venom Jet Promotions Ltd/Bournemouth
	G-AOEX	Thruxton Jackaroo	A. T. Christian
	G-AOFE	D.H.C.1 Chipmunk 22A (WB702)	W. J. Quinn
	G-AOFJ	J/1N Alpha	N. Fraser
	G-AOFM	J/5P Autocar	S. J. Cooper
	G-AOFS	J/5L Aiglet Trainer	P. N. A. Whitehead
	G-AOGA	M.75 Aries ★	Irish Aviation Museum *(stored)*
	G-AOGE	P.34A Proctor 3 ★	N. I. Dalziel *(stored)*/Biggin Hill
	G-AOGI	D.H.82A Tiger Moth	W. J. Taylor
	G-AOGR	D.H.82A Tiger Moth (XL714)	M. I. Edwards
	G-AOGV	J/5R Alpine	R. E. Heading
	G-AOHY	D.H.82A Tiger Moth (N6537)	R. H. & S. J. Cooper
	G-AOHZ	J/5P Autocar	A. D. Hodgkinson
	G-AOIL	D.H.82A Tiger Moth (XL716)	C. D. Davidson
	G-AOIM	D.H.82A Tiger Moth	C. R. Hardiman
	G-AOIR	Thruxton Jackaroo	K. A. & A. J. Broomfield
	G-AOIS	D.H.82A Tiger Moth	J. K. Ellwood
	G-AOIY	J/5G Autocar 160	R. A. Benson
	G-AOJD	V.802 Viscount ★	Jersey Airport Fire Service
	G-AOJH	D.H.83C Fox Moth	Connect Properties Ltd/Kemble
	G-AOJJ	D.H.82A Tiger Moth (DF128)	E. & K. M. Lay
	G-AOJK	D.H.82A Tiger Moth	R. J. Willies
	G-AOJR	D.H.C.1 Chipmunk 22	G. J-H. Caubergs & N. Marien/Belgium
	G-AOJT	D.H.106 Comet 1 (F-BGNX) ★	De Havilland Heritage Museum *(fuselage only)*
	G-AOKH	P.40 Prentice 1	J. F. Moore/Biggin Hill
	G-AOKL	P.40 Prentice 1 (VS610)	The Shuttleworth Collection/O. Warden
	G-AOKO	P.40 Prentice 1 ★	Airport Fire Section/Coventry
	G-AOKZ	P.40 Prentice 1 (VS623) ★	Midland Air Museum/Coventry
	G-AOLK	P.40 Prentice 1	Hilton Aviation Ltd/Southend
	G-AOLU	P.40 Prentice 1 (VS356)	N. J. Butler
	G-AORB	Cessna 170B	Eaglescott Parachute Centre
	G-AORG	D.H.114 Heron 2	Duchess of Brittany (Jersey) Ltd
	G-AORW	D.H.C.1 Chipmunk 22A	Skylark Aviation Ltd
	G-AOSF	D.H.C.1 Chipmunk 22 (WB571)	T. S. Olsen
	G-AOSK	D.H.C.1 Chipmunk 22 (WB726)	L. J. Irvine
	G-AOSU	D.H.C.1 Chipmunk 22 (Lycoming)	Fulmar Gliding Club/Easterton
	G-AOSY	D.H.C.1 Chipmunk 22 (WB585)	WFG Chipmunk Group
	G-AOTD	D.H.C.1 Chipmunk 22 (WB588)	S. Piech
	G-AOTF	D.H.C.1 Chipmunk 23 (Lycoming)	Clevelands Gliding Club/Dishforth
	G-AOTI	D.H.114 Heron 2D ★	De Havilland Heritage Museum
	G-AOTK	D.53 Turbi	T. J. Adams
	G-AOTR	D.H.C.1 Chipmunk 22	M. R. Woodgate/Aldergrove
	G-AOTY	D.H.C.1 Chipmunk 22A (WG472)	A. A. Hodgson
	G-AOUJ	Fairey Ultra-Light ★	IHM/Weston-s-Mare
	G-AOUO	D.H.C.1 Chipmunk 22 (Lycoming)	Wrekin/Cosford
	G-AOUP	D.H.C.1 Chipmunk 22	A. R. Harding
	G-AOUR	D.H.82A Tiger Moth ★	Ulster Folk & Transport Museum
	G-AOVF	B.175 Britannia 312F ★	Aerospace Museum/Cosford
	G-AOVS	B.175 Britannia 312F ★	Airport Fire Section/Luton
	G-AOVT	B.175 Britannia 312F ★	Duxford Aviation Soc
	G-AOVW	Auster 5	B. Marriott/Cranwell
	G-AOXN	D.H.82A Tiger Moth	S. L. G. Darch
	G-AOZH	D.H.82A Tiger Moth (K2572)	M. H. Blois-Brooke
	G-AOZL	J/5Q Alpine	R. M. Weeks/Stapleford
	G-AOZP	D.H.C.1 Chipmunk 22	S. J. Davies
	G-APAA	J/5R Alpine ★	L. A. Groves *(stored)*
	G-APAF	Auster 5 (TW511)	J. J. J. Mostyn (G-CMAL)
	G-APAH	Auster 5	T. J. Goodwin
	G-APAJ	Thruxton Jackaroo	J. T. H. Page
	G-APAL	D.H.82A Tiger Moth (N6847)	Avia Special Ltd
	G-APAM	D.H.82A Tiger Moth	R. P. Williams

Reg.	Type	Owner or Operator	Notes
G-APAO	D.H.82A Tiger Moth	H. J. Maguire	
G-APAP	D.H.82A Tiger Moth	J. C. Wright	
G-APAS	D.H.106 Comet 1XB ★	Aerospace Museum/Cosford	
G-APBE	Auster 5	R. J. Napp	
G-APBI	D.H.82A Tiger Moth	A. Wood	
G-APBO	D.53 Turbi	R. C. Hibberd	
G-APBW	Auster 5	N. Huxtable	
G-APCB	J/5Q Alpine	A. A. Beswick	
G-APCC	D.H.82A Tiger Moth	L. J. Rice/Henstridge	
G-APDB	D.H.106 Comet 4 ★	Duxford Aviation Soc	
G-APEP	V.953C Merchantman ★	Brooklands Museum of Aviation/ Weybridge	
G-APFA	D.54 Turbi	F. J. Keitch	
G-APFG	Boeing 707-436 ★	Cabin water spray tests/Cardington	
G-APFJ	Boeing 707-436 ★	Aerospace Museum/Cosford	
G-APFU	D.H.82A Tiger Moth	Leisure Assets Ltd	
G-APGL	D.H.82A Tiger Moth	K. A. Broomfield	
G-APHV	Avro 19 Srs 2 (VM360) ★	Museum of Flight/E. Fortune	
G-APIE	Tipsy Belfair B	D. Beale	
G-APIH	D.H.82A Tiger Moth	K. Stewering	
G-APIK	J/1N Alpha	J. H. Powell-Tuck	
G-APIM	V.806 Viscount ★	Brooklands Museum of Aviation/ Weybridge	
G-APIT	P.40 Prentice 1 (VR192) ★	WWII Aircraft Preservation Soc/Lasham	
G-APIY	P.40 Prentice 1 (VR249) ★	Newark Air Museum	
G-APIZ	D.31 Turbulent	E. J. I. Musty/White Waltham	
G-APJB	P.40 Prentice 1 (VR259) ★	Atlantic Air Transport Ltd/Coventry	
G-APJJ	Fairey Ultra-light ★	Midland Aircraft Preservation Soc	
G-APJO	D.H.82A Tiger Moth	D. R. & M. Wood	
G-APJZ	J/1N Alpha	P. G. Lipman	
G-APKM	J/1N Alpha	D. E. A. Huggins (stored)	
G-APLG	J/5L Aiglet Trainer ★	Solway Aviation Soc	
G-APLO	D.H.C.1 Chipmunk 22A (WD379)	Lindholme Aircraft Ltd/Jersey	
G-APLU	D.H.82A Tiger Moth	R. A. Bishop & M. E. Vaisey	
G-APMB	D.H.106 Comet 4B ★	Gatwick Handling Ltd (ground trainer)	
G-APMH	J/1U Workmaster	J. L. Thorogood	
G-APMX	D.H.82A Tiger Moth	M. A. Boughton	
G-APMY	PA-23 Apache 160 ★	South Yorkshire Aviation Museum	
G-APNJ	Cessna 310 ★	Chelsea College/Shoreham	
G-APNS	Garland-Bianchi Linnet	P. M. Busaidy	
G-APNT	Currie Wot	B. J. Dunford	
G-APNZ	D.31 Turbulent	J. Knight	
G-APOI	Saro Skeeter Srs 8	B. Chamberlain	
G-APPA	D.H.C.1 Chipmunk 22	D. M. Squires	
G-APPL	P.40 Prentice 1	S. J. Saggers/Biggin Hill	
G-APPM	D.H.C.1 Chipmunk 22 (WB711)	S. D. Wilch	
G-APPN	D.H.82A Tiger Moth (T7328)	John Colours SRL/Spain	
G-APRF	Auster 5	W. B. Bateson/Blackpool	
G-APRJ	Avro 694 Lincoln B.2 ★	D. Copley/Sandtoft	
G-APRL	AW.650 Argosy 101 ★	Midland Air Museum/Coventry	
G-APRR	Super Aero 45	R. H. Jowett	
G-APRS	SA Twin Pioneer Srs 3	Atlantic Air Transport Ltd (G-BCWF)/Coventry	
G-APRT	Taylor JT.1 Monoplane	D. A. Slater	
G-APSA	Douglas DC-6A	Atlantic Air Transport Ltd/Coventry	
G-APSR	J/1U Workmaster	D. & K. Aero Services Ltd/Shobdon	
G-APTR	J/1N Alpha	C. J. & D. J. Baker	
G-APTU	Auster 5	G-APTU Flying Group	
G-APTW	W.B.1 Widgeon ★	N.E. Aircraft Museum/Usworth	
G-APTY	Beech G.35 Bonanza	G. E. Brennand	
G-APTZ	D.31 Turbulent	The Tiger Club (1990) Ltd/Headcorn	
G-APUD	Bensen B.7M (modified) ★	Manchester Museum of Science & Industry	
G-APUE	L.40 Meta Sokol	S. E. & M. J. Aherne	
G-APUP	Sopwith Pup (replica) (N5182) ★	RAF Museum/Hendon	
G-APUR	PA-22 Tri-Pacer 160	L. F. Miller	
G-APUW	J/5V-160 Autocar	E. A. J. Hibbard	
G-APUY	D.31 Turbulent	C. Jones/Barton	
G-APUZ	PA-24 Comanche 250	Tatenhill Aviation	
G-APVF	Putzer Elster B (97+04)	A. & E. A. Wiseman	
G-APVG	J/5L Aiglet Trainer	R. Farrer/Cranfield	
G-APVL	Saro P.531-2	R. E. Dagless	
G-APVN	D.31 Turbulent	R. Sherwin/Shoreham	
G-APVS	Cessna 170B	N. Simpson Stormin' Norman	
G-APVU	L.40 Meta Sokol	S. E. & M. J. Aherne	
G-APVZ	D.31 Turbulent	The Tiger Club 1990 Ltd	

Notes	Reg.	Type	Owner or Operator
	G-APWA	HPR-7 Herald 101 ★	Museum of Berkshire Aviation/Woodley
	G-APWJ	HPR-7 Herald 201 ★	Duxford Aviation Soc
	G-APWL	EoN AP.10 460 Srs 1A	D. G. Andrew
	G-APWN	WS-55 Whirlwind 3 ★	Midland Air Museum/Coventry
	G-APWY	Piaggio P.166 ★	Science Museum/Wroughton
	G-APXJ	PA-24 Comanche 250	T. Wildsmith/Netherthorpe
	G-APXR	PA-22 Tri-Pacer 160	A. Troughton
	G-APXT	PA-22 Tri-Pacer 150 (modified)	A. E. Cuttler
	G-APXU	PA-22 Tri-Pacer 125 (modified)	The Scottish Aero Club Ltd/Perth
	G-APXW	EP.9 Prospector (XM819) ★	Museum of Army Flying/Middle Wallop
	G-APXX	D.H.A.3 Drover 2 (VH-FDT) ★	WWII Aircraft Preservation Soc/Lasham
	G-APXY	Cessna 150	Merlin Flying Club Ltd/Hucknall
	G-APYB	Tipsy T.66 Nipper 3	B. O. Smith
	G-APYD	D.H.106 Comet 4B ★	Science Museum/Wroughton
	G-APYG	D.H.C.1 Chipmunk 22	PYGS Flying Group
	G-APYI	PA-22 Tri-Pacer 135	B. T. & J. Cullen
	G-APYN	PA-22 Tri-Pacer 160	S. J. Raw
	G-APYT	Champion 7FC Tri-Traveller	B. J. Anning
	G-APZJ	PA-18 Super Cub 150	Southern Sailplanes Ltd/Membury
	G-APZL	PA-22 Tri-Pacer 160	B. Robins
	G-APZR	Cessna 150 ★	*Engine test-bed*/Biggin Hill
	G-APZX	PA-22 Tri-Pacer 150	Applied Signs Ltd
	G-ARAI	PA-22 Tri-Pacer 160	S. T. A. Hutchinson
	G-ARAM	PA-18 Super Cub 150	Spectrum Leisure Ltd
	G-ARAN	PA-18 Super Cub 150	A. P. Docherty/Redhill
	G-ARAO	PA-18 Super Cub 95 (607327)	R. G. Manton
	G-ARAS	Champion 7FC Tri-Traveller	Alpha Sierra Flying Group
	G-ARAT	Cessna 180C	S. D. Pryke
	G-ARAW	Cessna 182C Skylane	Ximango UK/Rufforth
	G-ARAX	PA-22 Tri-Pacer 150	J. J. Bywater
	G-ARAZ	D.H.82A Tiger Moth	D. A. Porter
	G-ARBE	D.H.104 Dove 8	M. Whale & M. W. A. Lunn/Old Sarum
	G-ARBG	Tipsy T.66 Nipper 2	D. Shrimpton
	G-ARBO	PA-24 Comanche 250	Arrow Aviation Services Ltd
	G-ARBP	Tipsy T.66 Nipper 2	F. W. Kirk
	G-ARBS	PA-22 Tri-Pacer 160 (tailwheel)	S. D. Rowell
	G-ARBV	PA-22 Tri-Pacer 160	Oaksey Pacers
	G-ARBZ	D.31 Turbulent	G. Richards
	G-ARCC	PA-22 Tri-Pacer 150	Popham Flying Group/Popham
	G-ARCF	PA-22 Tri-Pacer 150	M. J. Speakman
	G-ARCI	Cessna 310D ★	*(stored)*/Blackpool
	G-ARCS	Auster D6/180	E. A. Matty/Shobdon
	G-ARCT	PA-18 Super Cub 95	C. F. O'Neill
	G-ARCV	Cessna 175A	R. Francis & C. Campbell
	G-ARCW	PA-23 Apache 160	F. W. Ellis
	G-ARCX	A.W. Meteor 14 ★	Museum of Flight/E. Fortune
	G-ARDB	PA-24 Comanche 250	P. Crook
	G-ARDD	CP.301C1 Emeraude	M. W. Bodger & M. H. Hoffman
	G-ARDE	D.H.104 Dove 6 ★	T. E. Evans
	G-ARDJ	Auster D.6/180	RN Aviation (Leicester Airport) Ltd
	G-ARDO	Jodel D.112	W. R. Prescott
	G-ARDP	PA-22 Tri-Pacer 150	G. M. Jones
	G-ARDS	PA-22 Caribbean 150	N. P. McGowan
	G-ARDT	PA-22 Tri-Pacer 160	B. W. Haston
	G-ARDV	PA-22 Tri-Pacer 160	R. W. Christie
	G-ARDY	Tipsy T.66 Nipper 2	D. House
	G-ARDZ	Jodel D.140A	M. J. Wright
	G-AREA	D.H.104 Dove 8 ★	De Havilland Heritage Museum
	G-AREF	PA-23 Aztec 250 ★	Southall College of Technology
	G-AREH	D.H.82A Tiger Moth	C. D. Cyster & A. J. Hastings
	G-AREI	Auster 3 (MT438)	G. M. Bauer
	G-AREL	PA-22 Caribbean 150	H. H. Cousins/Fenland
	G-AREO	PA-18 Super Cub 150	Vale of the White Horse Gliding Club
	G-ARET	PA-22 Tri-Pacer 160	I. S. Runnalls
	G-AREV	PA-22 Tri-Pacer 160	D. J. Ash
	G-AREX	Aeronca 15AC Sedan	R. J. M. Turnbull
	G-ARFB	PA-22 Caribbean 150	D. Shaw
	G-ARFD	PA-22 Tri-Pacer 160	J. R. Dunnett
	G-ARFG	Cessna 175A Skylark	Foxtrot Golf Group
	G-ARFH	PA-24 Comanche 250	A. B. W. Taylor
	G-ARFI	Cessna 150A	A. R. Abrey

Reg.	Type	Owner or Operator	Notes
G-ARFO	Cessna 150A	Breakthrough Aviation Ltd	
G-ARFT	Jodel DR. 1050	R. Shaw	
G-ARFV	Tipsy T.66 Nipper 2	C. J. Pidler	
G-ARGB	Auster 6A ★	C. J. Baker (stored)	
G-ARGG	D.H.C.1 Chipmunk 22 (WD305)	D. Curtis	
G-ARGO	PA-22 Colt 108	D. R. Smith	
G-ARGV	PA-18 Super Cub 180	Wolds Gliding Club Ltd	
G-ARGY	PA-22 Tri-Pacer 160	D. H.& R. T. Tanner (G-JEST)	
G-ARGZ	D.31 Turbulent	The Tiger Club (1990) Ltd/Headcorn	
G-ARHB	Forney F-1A Aircoupe	K. J. Peacock & S. F. Turner/Earls Colne	
G-ARHC	Forney F-1A Aircoupe	A. P. Gardner/Elstree	
G-ARHI	PA-24 Comanche 180	D. D. Smith	
G-ARHL	PA-23 Aztec 250	C. J. Freeman/Headcorn	
G-ARHM	Auster 6A	R. C. P. Brookhouse	
G-ARHN	PA-22 Caribbean 150	I. S. Hodge & S. Haughton	
G-ARHP	PA-22 Tri-Pacer 160	R. N. Morgan	
G-ARHR	PA-22 Caribbean 150	A. R. Wyatt	
G-ARHT	PA-22 Caribbean 150 ★	Moston Technical College	
G-ARHW	D.H.104 Dove 8	Pacelink Ltd	
G-ARHX	D.H.104 Dove 8 ★	N.E. Aircraft Museum/Usworth	
G-ARHZ	D.62 Condor	E. Shouler	
G-ARID	Cessna 172B	L. M. Edwards	
G-ARIF	Ord-Hume O-H.7 Minor Coupé ★	N. H. Ponsford (stored)	
G-ARIH	Auster 6A (TW591)	R. J. Griffin	
G-ARIK	PA-22 Caribbean 150	H. E. Barraclough	
G-ARIL	PA-22 Caribbean 150	A. Fergusson & A. W. McBlain	
G-ARIM	D.31 Turbulent	R. M. White	
G-ARJB	D.H.104 Dove 8	M. Whale & M. W. A. Lunn	
G-ARJE	PA-22 Colt 108	C. I. Fray	
G-ARJF	PA-22 Colt 108	Tandycel Co Ltd	
G-ARJH	PA-22 Colt 108	A. Vine	
G-ARJR	PA-23 Apache 160G ★	Instructional airframe/Kidlington	
G-ARJS	PA-23 Apache 160G	Bencray Ltd/Blackpool	
G-ARJT	PA-23 Apache 160G	J. A. Cole	
G-ARJU	PA-23 Apache 160G	G. R. Manley	
G-ARJV	PA-23 Apache 160G	Metham Aviation Ltd/Blackbushe	
G-ARJW	PA-23 Apache 160G	(stored)/Bristol	
G-ARKG	J/5G Autocar	B. C. C. Harrison	
G-ARKJ	Beech N35 Bonanza	P. D. & J. L. Jenkins	
G-ARKK	PA-22 Colt 108	R. D. Welfare	
G-ARKM	PA-22 Colt 108	D. Dytch & J. Moffat	
G-ARKN	PA-22 Colt 108	R. Redfern	
G-ARKP	PA-22 Colt 108	J. P. A. Freeman/Headcorn	
G-ARKS	PA-22 Colt 108	R. A. Nesbitt-Dufort	
G-ARLB	PA-24 Comanche 250	D. Heater (G-BUTL)	
G-ARLG	Auster D.4/108	Auster D4 Group	
G-ARLK	PA-24 Comanche 250	I. Kazi	
G-ARLO	A.61 Terrier 1 ★	(stored)	
G-ARLP	A.61 Terrier 1	Gemini Flying Group	
G-ARLR	A.61 Terrier 2	M. Palfreman	
G-ARLU	Cessna 172B Skyhawk ★	Instructional airframe/Irish AC	
G-ARLW	Cessna 172B Skyhawk ★	(spares source)/Barton	
G-ARLX	Jodel D.140B	J. S. & S. V. Shaw	
G-ARLZ	D.31A Turbulent	Little Bear Ltd	
G-ARMC	D.H.C.1 Chipmunk 22A (WB703)	John Henderson Children's Trust	
G-ARMD	D.H.C.1 Chipmunk 22A (WD297)	D. M. Squires/Wellesbourne	
G-ARMF	D.H.C.1 Chipmunk 22A (WZ868)	D. M. Squires	
G-ARMG	D.H.C.1 Chipmunk 22A	MG Group	
G-ARML	Cessna 175B Skylark	R. W. Boote	
G-ARMN	Cessna 175B Skylark	B. R. Nash	
G-ARMO	Cessna 172B Skyhawk	I. M. Latiff	
G-ARMR	Cessna 172B Skyhawk	Sunsaver Ltd/Barton	
G-ARMZ	D.31 Turbulent	J. Mickleburgh & D. Clark	
G-ARNB	J/5G Autocar	R. F. Tolhurst	
G-ARND	PA-22 Colt 108	E. J. Clarke	
G-ARNE	PA-22 Colt 108	T. D. L. Bowden/Shipdham	
G-ARNG	PA-22 Colt 108	F. B. Rothera	
G-ARNH	PA-22 Colt 108 ★	Fenland Aircraft Preservation Soc	
G-ARNJ	PA-22 Colt 108	R. A. Keech	
G-ARNK	PA-22 Colt 108 (tailwheel)	A. T. J. & D. Hyatt/France	
G-ARNL	PA-22 Colt 108	J. A. Dodsworth/White Waltham	
G-ARNO	A.61 Terrier 1 ★	~/Sywell	

Notes	Reg.	Type	Owner or Operator
	G-ARNP	A.109 Airedale	S. W. & M. Isbister
	G-ARNY	Jodel D.117	D. P. Jenkins
	G-ARNZ	D.31 Turbulent	The Tiger Club (1990) Ltd/Headcorn
	G-AROA	Cessna 172B Skyhawk	D. E. Partridge
	G-AROC	Cessna 175B	A. J. Symes (G-OTOW)
	G-AROJ	A.109 Airedale ★	D. J. Shaw (stored)
	G-ARON	PA-22 Colt 108	R. W. Curtis
	G-AROO	Forney F-1A Aircoupe	W. J. McMeekan/Newtownards
	G-AROW	Jodel D.140B	A. R. Crome
	G-AROY	Boeing Stearman A.75N.1	Abbey Security Services Ltd
	G-ARPH	H.S.121 Trident 1C ★	Aerospace Museum/Cosford
	G-ARPK	H.S.121 Trident 1C ★	Manchester Airport Authority
	G-ARPO	H.S.121 Trident 1C ★	CAA Fire School/Teesside
	G-ARRD	Jodel DR.1050	D. J. Taylor & J. P. Brady
	G-ARRE	Jodel DR.1050	G. & R. Ward
	G-ARRI	Cessna 175B	R. D. Fowden
	G-ARRL	J/1N Alpha	A. C. Ladd
	G-ARRM	Beagle B.206-X ★	Bristol Aero Collection (stored)
	G-ARRO	A.109 Airedale	M. & S. W. Isbister
	G-ARRS	CP.301A Emeraude	J. P. Drake/Sturgate
	G-ARRT	Wallis WA-116-1	K. H. Wallis
	G-ARRU	D.31 Turbulent	D. G. Huck
	G-ARRX	Auster 6A (VF512)	J. E. D. Mackie
	G-ARRY	Jodel D.140B	Fictionview Ltd
	G-ARRZ	D.31 Turbulent	C. I. Jefferson
	G-ARSG	Roe Triplane Type IV (replica)	The Shuttleworth Collection/O. Warden
	G-ARSL	A.61 Terrier 1	D. J. Colclough
	G-ARSU	PA-22 Colt 108	D. P. Owen
	G-ARTH	PA-12 Super Cruiser	R. I. Souch & B. J. Dunford
	G-ARTJ	Bensen B.8M ★	Museum of Flight/E. Fortune
	G-ARTL	D.H.82A Tiger Moth (T7281)	F. G. Clacherty
	G-ARTT	MS.880B Rallye Club	R. N. Scott
	G-ARTZ	McCandless M.4 gyroplane	W. R. Partridge
	G-ARUG	J/5G Autocar	D. P. H. Hulme/Biggin Hill
	G-ARUH	Jodel DR.1050	PFA Group/Denham
	G-ARUI	A.61 Terrier	T. W. J. Dann
	G-ARUL	LeVier Cosmic Wind	P. G. Kynsey/Headcorn
	G-ARUV	CP.301A Emeraude	P. O'Fee
	G-ARUY	J/1N Alpha	D. Burnham
	G-ARUZ	Cessna 175C	Cardiff Skylark Group
	G-ARVM	V.1101 VC10 ★	Aerospace Museum/Cosford
	G-ARVO	PA-18 Super Cub 95	Northamptonshire School of Flying Ltd/Sywell
	G-ARVT	PA-28 Cherokee 160	Red Rose Aviation Ltd/Liverpool
	G-ARVU	PA-28 Cherokee 160	Barton Mudwing Ltd
	G-ARVV	PA-28 Cherokee 160	G. E. Hopkins/Shobdon
	G-ARVZ	D.62B Condor	A. A. M. Huke
	G-ARWB	D.H.C.1 Chipmunk 22 (WK611)	Thruxton Chipmunk Flying Club
	G-ARWH	Cessna 172C ★	(stored)
	G-ARWO	Cessna 172C	J. J. Sheeran
	G-ARWR	Cessna 172C	Devanha Flying Group/Insch
	G-ARWS	Cessna 175C	M. D. Fage
	G-ARXB	A.109 Airedale	S. W. & M. Isbister
	G-ARXD	A.109 Airedale	D. Howden
	G-ARXG	PA-24 Comanche 250	Fairoaks Comanche
	G-ARXH	Bell 47G	A. B. Searle
	G-ARXP	Luton LA-4 Minor	E. Evans
	G-ARXT	Jodel DR.1050	CJM Flying Group
	G-ARXU	Auster AOP.6A (VF526)	E. C. Tait & M. Pocock
	G-ARXW	M.S.885 Super Rallye	M. J. Kirk
	G-ARYB	H.S.125 Srs 1 ★	Midland Air Museum/Coventry
	G-ARYC	H.S.125 Srs 1 ★	De Havilland Heritage Museum
	G-ARYD	Auster AOP.6 (WJ358) ★	Museum of Army Flying/Middle Wallop
	G-ARYF	PA-23 Aztec 250B	D. A. Hitchcockl
	G-ARYH	PA-22 Tri-Pacer 160	C. Watt
	G-ARYI	Cessna 172C	J. Rhodes
	G-ARYK	Cessna 172C	G. W. Goodban
	G-ARYR	PA-28 Cherokee 180	G-ARYR Flying Group
	G-ARYS	Cessna 172C	Squires Gear & Engineering Ltd
	G-ARYV	PA-24 Comanche 250	D. C. Hanss
	G-ARYZ	A.109 Airedale	C. W. Tomkins
	G-ARZB	Wallis WA-116 Srs 1	K. H. Wallis
	G-ARZE	Cessna 172C ★	Parachute jump trainer/Cockerham

Reg.	Type	Owner or Operator	Notes
G-ARZM	D.31 Turbulent ★	The Tiger Club (1990) Ltd/Headcorn	
G-ARZN	Beech N35 Bonanza	D. W. Mickleburgh/Leicester	
G-ARZS	A.109 Airedale	M. & S. W. Isbister	
G-ARZW	Currie Wot	B. R. Pearson/Eaglescott	
G-ASAA	Luton LA-4 Minor	M. J. Aubrey (stored)/Netherthorpe	
G-ASAI	A.109 Airedale	K. R. Howden	
G-ASAJ	A.61 Terrier 2 (WE569)	G-ASAJ Flying Group	
G-ASAK	A.61 Terrier 2	I. Frank	
G-ASAL	SA Bulldog Srs 120/124	Pioneer Flying Co. Ltd/Prestwick	
G-ASAM	D.31 Turbulent ★	The Tiger Club (1990) Ltd/Headcorn	
G-ASAT	M.S.880B Rallye Club	M. Cutovic	
G-ASAU	M.S.880B Rallye Club	L. O. Queen	
G-ASAX	A.61 Terrier 2	P. G. Morris	
G-ASAZ	Hiller UH-12E4 (XS165)	Hields Aviation	
G-ASBA	Phoenix Currie Wot	J. C. Lister	
G-ASBH	A.109 Airedale	D. T. Smollett	
G-ASBY	A.109 Airedale	F. A. Forster	
G-ASCC	Beagle E3 Mk 11 (XP254)	P. T. Bolton	
G-ASCD	A.61 Terrier 2 (TJ704) ★	Yorkshire Air Museum/Elvington	
G-ASCM	Isaacs Fury II (K2050)	E. C. & P. King	
G-ASCU	PA-18A Super Cub 150	D. J. O'Mahony/Ireland	
G-ASCZ	CP.301A Emeraude	I. Denham-Brown	
G-ASDF	Edwards Gyrocopter ★	B. King	
G-ASDK	A.61 Terrier 2	J. Swallow (G-ARLM)	
G-ASDL	A.61 Terrier 2	G-ASDL Group (G-ARLN)	
G-ASDY	Wallis WA-116/F	K. H. Wallis	
G-ASEA	Luton LA-4A Minor	C. Willmott	
G-ASEB	Luton LA-4A Minor	S. R. P. Harper	
G-ASEO	PA-24 Comanche 250	M. Scott	
G-ASEP	PA-23 Apache 235	Arrowstate Ltd/Denham	
G-ASEU	D.62A Condor	W. M. Grant	
G-ASFA	Cessna 172D	D. Halfpenny	
G-ASFD	L-200A Moravia	M. Emery	
G-ASFK	J/5G Autocar	T. D. G. Lancaster	
G-ASFL	PA-28 Cherokee 180	J. Simpson & D. Kennedy	
G-ASFR	Bo 208A1 Junior	S. T. Dauncey	
G-ASFX	D.31 Turbulent	E. F. Clapham & W. B. S. Dobie	
G-ASGC	V.1151 Super VC10 ★	Duxford Aviation Soc	
G-ASHD	Brantly B-2A ★	IHM/Weston-s-Mare	
G-ASHH	PA-23 Aztec 250	C. Fordham & L. Barr	
G-ASHS	SNCAN Stampe SV-4C	D. G. Girling	
G-ASHT	D.31 Turbulent	C. W. N. Huke	
G-ASHU	PA-15 Vagabond (modified)	The Calybe Flying Group	
G-ASHV	PA-23 Aztec 250B	R. J. Ashley & G. O'Gorman	
G-ASHX	PA-28 Cherokee 180	Powertheme Ltd/Barton	
G-ASIB	Cessna F.172D	G-ASIB Flying Group	
G-ASII	PA-28 Cherokee 180	T. R. Hart & R. W. S. Matthews	
G-ASIJ	PA-28 Cherokee 180	G-ASIJ Group	
G-ASIL	PA-28 Cherokee 180	C. D. Powell	
G-ASIS	Jodel D.112	W. R. Prescott	
G-ASIT	Cessna 180	R. A. Seeley	
G-ASIY	PA-25 Pawnee 235	RAFGSA/Bicester	
G-ASJL	Beech H.35 Bonanza	A. J. Orchard	
G-ASJM	PA-30 Twin Comanche 160 ★	Via Nova Ltd (stored)	
G-ASJO	Beech B.23 Musketeer	R. M. Wilson/Sandown	
G-ASJV	V.S.361 Spitfire IX (MH434/PK-K)	Merlin Aviation Ltd/Duxford	
G-ASJY	GY-80 Horizon 160	No.6 Group Aviation Ltd	
G-ASJZ	Jodel D.117A	W. J. Siertsema	
G-ASKC	D.H.98 Mosquito 35 (TA719) ★	Skyfame Collection/Duxford	
G-ASKK	HPR-7 Herald 211 ★	Norwich Aviation Museum	
G-ASKL	Jodel 150	J. M. Graty	
G-ASKP	D.H.82A Tiger Moth	Tiger Club (1990) Ltd/Headcorn	
G-ASKT	PA-28 Cherokee 180	T. J. Herbert	
G-ASLH	Cessna 182F	A. L. Brown & A. L. Butcher	
G-ASLL	Cessna 336 ★	(stored)/Bournemouth	
G-ASLR	Agusta-Bell 47J-2	N. M. G. Pearson	
G-ASLV	PA-28 Cherokee 235	Sackville Flying Group/Riseley	
G-ASLX	CP.301A Emeraude	P. White	
G-ASMA	PA-30 Twin Comanche 160 C/R	K. Cooper	
G-ASME	Bensen B.8M	R. M. Harris	
G-ASMF	Beech D.95A Travel Air	M. J. A. Hornblower	

Notes	Reg.	Type	Owner or Operator
	G-ASMJ	Cessna F.172E	Aeroscene Ltd
	G-ASML	Luton LA-4A Minor	S. Slater
	G-ASMM	D.31 Tubulent	W. J. Browning
	G-ASMO	PA-23 Apache 160G ★	Aviation Enterprises/Fairoaks
	G-ASMS	Cessna 150A	P. J. Husband
	G-ASMT	Fairtravel Linnet 2	A. F. Cashin
	G-ASMW	Cessna 150D	Aviation Business Centres Ltd
	G-ASMY	PA-23 Apache 160H	R. D. Forster
	G-ASMZ	A.61 Terrier 2 (VF516)	B. Andrews
	G-ASNB	Auster 6A (VX118)	B. C. C. Harrison
	G-ASNC	Beagle D.5/180 Husky	Peterborough & Spalding Gliding Club Ltd/ Crowland
	G-ASNI	CP.1310-C3 Super Emeraude	D. Chapman
	G-ASNK	Cessna 205	Justgold Ltd
	G-ASNN	Cessna 182F ★	*Parachute jump trainer*/Tilstock
	G-ASNW	Cessna F.172E	G-ASNW Group
	G-ASNY	Campbell-Bensen B.8M gyroplane ★	R. Light & T. Smith
	G-ASOC	Auster 6A	M. J. Kirk
	G-ASOH	Beech 95-B55A Baron	GMD Group
	G-ASOI	A.61 Terrier 2	G.D.B. Delmege
	G-ASOK	Cessna F.172E	D. W. Disney
	G-ASOM	A.61 Terrier 2	D. Humphries (G-JETS)
	G-ASPF	Jodel D.120	T. J. Bates
	G-ASPK	PA-28 Cherokee 140	Westward Airways (Lands End) Ltd/St Just
	G-ASPP	Bristol Boxkite (replica)	The Shuttleworth Collection/O. Warden
	G-ASPS	Piper J-3C-90 Cub	A. J. Chalkley/Blackbushe
	G-ASPU	D.31 Turbulent	M. K. Field
	G-ASPV	D.H.82A Tiger Moth	Z. J. Rockey
	G-ASRB	D.62B Condor	R. J. Bentley/Ireland
	G-ASRC	D.62C Condor	C. R. Isbell
	G-ASRI	PA-23 Aztec 250B ★	Graham Collins Associates Ltd
	G-ASRK	A.109 Airedale	Bio Pathica Ltd/Lydd
	G-ASRO	PA-30 Twin Comanche 160	D. W. Blake
	G-ASRT	Jodel 150	P. Turton
	G-ASRW	PA-28 Cherokee 180	Alliance Aerolink Ltd
	G-ASSF	Cessna 182G Skylane	J. D. Bingham
	G-ASSM	H.S.125 Srs 1/522 ★	Science Museum/S. Kensington
	G-ASSP	PA-30 Twin Comanche 160	P. H. Tavener
	G-ASSS	Cessna 172E	D. H. N. Squires & P. R. March/Filton
	G-ASST	Cessna 150D	F. R. H. Parker
	G-ASSU	CP.301A Emeraude	R. W. Millward *(stored)*/Redhill
	G-ASSV	Kensinger KF	C. I. Jefferson
	G-ASSW	PA-28 Cherokee 140	C. G. Stone
	G-ASSY	D.31 Turbulent	M. N. King & J. A. Thomas
	G-ASTA	D.31 Turbulent	P. A. Cooke
	G-ASTH	Mooney M.20C ★	E. L. Martin *(stored)*/Guernsey
	G-ASTI	Auster 6A	C. C. Burton
	G-ASTL	Fairey Firefly I (Z2033) ★	F.A. A. Museum/Yeovilton
	G-ASTP	Hiller UH-12C ★	IHM/Weston-s-Mare
	G-ASTV	Cessna 150D (tailwheel) ★	*(stored)*
	G-ASUB	Mooney M.20E Super 21	S. C. Coulbeck
	G-ASUD	PA-28 Cherokee 180	S. J. Rogers & M. N. Petchey
	G-ASUE	Cessna 150D	D. Huckle
	G-ASUG	Beech E18S ★	Museum of Flight/E. Fortune
	G-ASUI	A.61 Terrier 2	K. W. Chigwell & D. R. Lee
	G-ASUP	Cessna F.172E	GASUP Air/Cardiff
	G-ASUR	Dornier Do 28A-1	Sheffair Ltd
	G-ASUS	Jurca MJ.2B Tempete	D. G. Jones/Coventry
	G-ASVG	CP.301B Emeraude	K. S. Woodard
	G-ASVM	Cessna F.172E	R. Seckington
	G-ASVN	Cessna U.206 Super Skywagon	British Skysports Paracentre
	G-ASVO	HPR-7 Herald 214 ★	Archive Visitor Centre/Shoreham *(cockpit section)*
	G-ASVP	PA-25 Pawnee 235	Aquila Gliding Club Ltd/Hinton-in-the-Hedges
	G-ASVZ	PA-28 Cherokee 140	J. S. Garvey
	G-ASWH	Luton LA-5A Major	J. T. Powell-Tuck
	G-ASWJ	Beagle 206 Srs 1 (8449M) ★	Brunel Technical College/Bristol
	G-ASWL	Cessna F.172F	Ensiform Aviation Ltd
	G-ASWN	Bensen B.8M	D. R. Shepherd
	G-ASWW	PA-30 Twin Comanche 160	R. J. Motors
	G-ASWX	PA-28 Cherokee 180	A. F. Dadds
	G-ASXD	Brantly B.2B	Lousada PLC
	G-ASXI	Tipsy T.66 Nipper 3	B. Dixon

Reg.	Type	Owner or Operator	Notes
G-ASXJ	Luton LA-4A Minor	P. N. Stacey	
G-ASXR	Cessna 210	A. Schofield	
G-ASXS	Jodel DR.1050	R. A. Hunter	
G-ASXU	Jodel D.120A	G-ASXU Group	
G-ASXX	Avro 683 Lancaster 7 (NX611) ★	Lincolnshire Aviation Heritage Centre/E. Kirkby	
G-ASXY	Jodel D.117A	P. A. Davies & ptnrs/Cardiff	
G-ASXZ	Cessna 182G Skylane	Last Refuge Ltd	
G-ASYD	BAC One-Eleven 475 ★	Brooklands Museum of Aviation/Weybridge	
G-ASYG	A.61 Terrier 2	R. H. & S. J. Cooper	
G-ASYJ	Beech D.95A Travel Air	Crosby Aviation (Jersey) Ltd	
G-ASYP	Cessna 150E	Henlow Flying Group	
G-ASYZ	Victa Airtourer 100	N. C. Grayson	
G-ASZB	Cessna 150E	R. J. Scott	
G-ASZD	Bo 208A2 Junior	M. J. Ayers	
G-ASZE	A.61 Terrier 2	P. J. Moore	
G-ASZR	Fairtravel Linnet 2	K. H. Bunt & R. Palmer	
G-ASZS	GY.80 Horizon 160	ZS Group	
G-ASZU	Cessna 150E	T. H. Milburn	
G-ASZV	Tipsy T.66 Nipper 2	J. M. Gough	
G-ASZX	A.61 Terrier 1	R. B. Webber	
G-ATAF	Cessna F.172F	Summit Media Ltd	
G-ATAG	Jodel DR.1050	T. M. Dawes-Gamble	
G-ATAS	PA-28 Cherokee 180	ATAS Group	
G-ATAT	Cessna 150E	The Derek Pointon Group (stored)	
G-ATAU	D.62B Condor	M. A. Peare/Redhill	
G-ATAV	D.62C Condor	J. R. Hornby	
G-ATBG	Nord 1002 (NJ+C11)	T. W. Harris/Little Snoring	
G-ATBH	Aero 145	P. D. Aberbach	
G-ATBI	Beech A.23 Musketeer	Three Musketeers Flying Group	
G-ATBJ	Sikorsky S-61N	British International	
G-ATBL	D.H.60G Moth	J. M. Greenland	
G-ATBP	Fournier RF-3	D. McNicholl	
G-ATBS	D.31 Turbulent	J. A. Lear	
G-ATBU	A.61 Terrier 2	K9 Flying Group	
G-ATBW	Tipsy T.66 Nipper 2	Stapleford Nipper Group	
G-ATBX	PA-20 Pacer 135	G. D. & P. M. Thomson	
G-ATBZ	W.S.58 Wessex 60 ★	IHM/Weston-s-Mare	
G-ATCC	A.109 Airedale	J. R. Bowden	
G-ATCD	Beagle D.5/180 Husky	T. C. O'Gorman	
G-ATCE	Cessna U.206	C. M. J. Fitzmaurice	
G-ATCJ	Luton LA-4A Minor	P. R. Diffey	
G-ATCL	Victa Airtourer 100	A. D. Goodall/Cardiff	
G-ATCR	Cessna 310 ★	ITD Aviation Ltd/Denham	
G-ATCU	Cessna 337	University of Cambridge	
G-ATCX	Cessna 182H	Craft Associates Worldwide Ltd	
G-ATDA	PA-28 Cherokee 160	Portway Aviation Ltd/Shobdon	
G-ATDB	Nord 1101 Noralpha	J. W. Hardie	
G-ATDN	A.61 Terrier 2 (TW641)	S. J. Saggers/Biggin Hill	
G-ATDO	Bo 208C1 Junior	P. Thompson/Crosland Moor	
G-ATEF	Cessna 150E	Swans Aviation/Blackbushe	
G-ATEM	PA-28 Cherokee 180	G. D. Wyles	
G-ATEP	EAA Biplane ★	E. L. Martin (stored)/Guernsey	
G-ATES	PA-32 Cherokee Six 260 ★	Parachute jump trainer/Stirling	
G-ATEV	Jodel DR.1050	J. C. Carter & J. L. Altrip	
G-ATEW	PA-30 Twin Comanche 160	Air Northumbria (Woolsington) Ltd	
G-ATEX	Victa Airtourer 100	Halton Victa Group	
G-ATEZ	PA-28 Cherokee 140	EFI Aviation Ltd	
G-ATFD	Jodel DR.1050	V. Usher	
G-ATFF	PA-23 Aztec 250C	T. J. Wassell	
G-ATFG	Brantly B.2B ★	Museum of Flight/E. Fortune	
G-ATFK	PA-30 Twin Comanche 160	D. J. Crinnon/White Waltham	
G-ATFM	Sikorsky S-61N	British International Ltd	
G-ATFR	PA-25 Pawnee 150	Borders (Milfield) Gliding Club Ltd	
G-ATFV	Agusta-Bell 47J-2A ★	Caernarfon Air World	
G-ATFW	Luton LA-4A Minor	P. A. Rose	
G-ATFY	Cessna F.172G	Jaguar Aviation Ltd	
G-ATGE	Jodel DR.1050	J. Turner	
G-ATGN	Thorn Coal Gas balloon	British Balloon Museum/Newbury	
G-ATGO	Cessna F.172G	Poetpilot Ltd	
G-ATGP	Jodel DR.1050	Madley Flying Group/Shobdon	
G-ATGY	GY.80 Horizon	D. Cowen	

Notes	Reg.	Type	Owner or Operator
	G-ATGZ	Griffiths GH-4 Gyroplane	R. W. J. Cripps
	G-ATHA	PA-23 Apache 235 ★	Brunel Technical College/Bristol
	G-ATHD	D.H.C.1 Chipmunk 22 (WP971)	Spartan Flying Group Ltd/Denham
	G-ATHF	Cessna 150F ★	Lincolnshire Aviation Heritage Centre/E. Kirkby
	G-ATHK	Aeronca 7AC Champion	T. P. McDonald & T. Crawley
	G-ATHM	Wallis WA-116 Srs 1	Wallis Autogyros Ltd
	G-ATHN	Nord 1101 Noralpha ★	E. L. Martin (stored)/Guernsey
	G-ATHR	PA-28 Cherokee 180	Britannia Airways Ltd/Luton
	G-ATHT	Victa Airtourer 115	D. A. Breeze
	G-ATHU	A.61 Terrier 1	J. A. L. Irwin
	G-ATHV	Cessna 150F	Cessna Hotel Victor Group
	G-ATHZ	Cessna 150F	R. D. Forster
	G-ATIA	PA-24 Comanche 260	L. A. Brown
	G-ATIC	Jodel DR.1050	R. J. Major
	G-ATIE	Cessna 150F ★	Parachute jump trainer/Chetwynd
	G-ATIN	Jodel D.117	A. Ayre
	G-ATIR	AIA Stampe SV-4C	Austin Trueman Ltd
	G-ATIS	PA-28 Cherokee 160	M. J. Barton
	G-ATIZ	Jodel D.117	D. K. Shipton/Leicester
	G-ATJA	Jodel DR.1050	Bicester Flying Group
	G-ATJC	Victa Airtourer 100 (modfied)	Aviation West Ltd/Cumbernauld
	G-ATJG	PA-28 Cherokee 140	C. A. McGee & L. K. G. Manning
	G-ATJL	PA-24 Comanche 260	M. J. Berry & T. R. Quinn/Blackbushe
	G-ATJM	Fokker Dr.1 (replica) (152/17)	R. Lamplough/North Weald
	G-ATJN	Jodel D.119	Oxenhope Flying Group
	G-ATJT	GY.80 Horizon 160	N. Huxtable
	G-ATJV	PA-32 Cherokee Six 260	Wingglider Ltd/Hibaldstow
	G-ATKF	Cessna 150F	P. Ashbridge
	G-ATKH	Luton LA-4A Minor	H. E. Jenner
	G-ATKI	Piper J-3C-65 Cub	KI Group
	G-ATKT	Cessna F.172G	P. J. Megson
	G-ATKX	Jodel D.140C	Kilo XraySyndicate
	G-ATLA	Cessna 182J Skylane	J. W. & J. T. Whicher
	G-ATLB	Jodel DR.1050/M1	Le Syndicate du Petit Oiseau/Brighton
	G-ATLC	PA-23 Aztec 250C ★	Alderney Air Charter Ltd (stored)
	G-ATLG	Hiller UH-12B	Bristow Helicopters Ltd
	G-ATLM	Cessna F.172G	Air Fotos Aviation Ltd/Newcastle
	G-ATLP	Bensen B.8M	R. F. G. Moyle
	G-ATLT	Cessna U.206A	A. I. M. & A. J. Guest
	G-ATLV	Jodel D.120	L. S. Thorne
	G-ATMC	Cessna F.150F	G. H. Farrah & D. Cunnane
	G-ATMH	Beagle D.5/180 Husky	Dorset Gliding Club Ltd
	G-ATMJ	H.S.748 Srs 2A	Emerald Airways Ltd/Liverpool
	G-ATML	Cessna F.150F	G. I. Smith
	G-ATMM	Cessna F.150F	Skytrax Aviation Ltd
	G-ATMT	PA-30 Twin Comanche 160	Montagu-Smith & Co Ltd
	G-ATMW	PA-28 Cherokee 140	Bencray Ltd/Blackpool
	G-ATMY	Cessna 150F	A. Dobson
	G-ATNB	PA-28 Cherokee 180	Bravo-180 Group
	G-ATNE	Cessna F.150F	A. D. Revill
	G-ATNK	Cessna F.150F	Pegasus Aviation Ltd
	G-ATNL	Cessna F.150F	G. A. Lauf
	G-ATNV	PA-24 Comanche 260	A. Heydn & K. Powell
	G-ATOA	PA-23 Apache 160G	Oscar Alpha Ltd/Stapleford
	G-ATOD	Cessna F.150F	D. Lugg
	G-ATOH	D.62B Condor	Three Spires Flying Group
	G-ATOI	PA-28 Cherokee 140	R. W. Nash
	G-ATOJ	PA-28 Cherokee 140	A Flight Aviation Ltd
	G-ATOK	PA-28 Cherokee 140	ILC Flying Group
	G-ATOL	PA-28 Cherokee 140	L. J. Nation & G. Alford
	G-ATOM	PA-28 Cherokee 140	A. Flight Aviation Ltd
	G-ATON	PA-28 Cherokee 140	R. G. Walters/Shobdon
	G-ATOO	PA-28 Cherokee 140	A. K. Komosa
	G-ATOP	PA-28 Cherokee 140	P. R. Coombs/Blackbushe
	G-ATOR	PA-28 Cherokee 140	Aligator Group
	G-ATOT	PA-28 Cherokee 180	Totair Ltd
	G-ATOU	Mooney M.20E Super 21	M20 Flying Group/Sherburn
	G-ATOY	PA-24 Comanche 260 ★	Museum of Flight/E. Fortune
	G-ATOZ	Bensen B.8M	N. C. White
	G-ATPN	PA-28 Cherokee 140	M. F. Hatt & ptnrs/Southend
	G-ATPT	Cessna 182J Skylane	G. B. Scholes
	G-ATPV	JB.01 Minicab	C. F. O'Neill

Reg.	Type	Owner or Operator	Notes
G-ATRA	LET L.13 Blanik (BXV)	Blanik Syndicate/Husbands Bosworth	
G-ATRB	LET L.13 Blanik (BXW)	Avon Soaring Centre/Bickmarsh	
G-ATRG	PA-18 Super Cub 150	Lasham Gliding Soc Ltd	
G-ATRI	Bolkow Bo. 208C1 Junior	H. P. Brooks	
G-ATRK	Cessna F.150F	Armstrong Aviation	
G-ATRL	Cessna F.150F	A. A. W. Stevens	
G-ATRM	Cessna F.150F	J. Redfearn	
G-ATRO	PA-28 Cherokee 140	M. A. Smith	
G-ATRR	PA-28 Cherokee 140	Marnham Investments Ltd	
G-ATRW	PA-32 Cherokee Six 260	Pringle Brandon Architects	
G-ATRX	PA-32 Cherokee Six 260	A. M. Harrhy & ptnrs	
G-ATSI	Bo 208C1 Junior	R. S. Jordan	
G-ATSL	Cessna F.172G	Alpha Aviation	
G-ATSM	Cessna 337A	Landscape & Ground Maintenance	
G-ATSR	Beech M.35 Bonanza	G-ATSR Group	
G-ATSX	Bo 208C1 Junior	Little Bear Ltd	
G-ATSY	Wassmer WA41 Super Baladou IV	McLean Aviation	
G-ATSZ	PA-30 Twin Comanche 160B	Sierra Zulu Aviation Ltd	
G-ATTB	Wallis WA-116-1 (XR944)	D. A. Wallis	
G-ATTD	Cessna 182J	Batesons Hotels (1958))td	
G-ATTI	PA-28 Cherokee 140	G-ATTI Flying Group	
G-ATTK	PA-28 Cherokee 140	G-ATTK Flying Group/Southend	
G-ATTM	Jodel DR.250-160	R. W. Tomkinson	
G-ATTN	Piccard balloon ★	Science Museum/S. Kensington	
G-ATTR	Bo 208C1 Junior	S. Luck	
G-ATTV	PA-28 Cherokee 140	D. B. & M. E. Meeks	
G-ATTX	PA-28 Cherokee 180	IPAC Aviation Ltd	
G-ATUB	PA-28 Cherokee 140	R. H. Partington & M. J. Porter	
G-ATUD	PA-28 Cherokee 140	J. J. Ferguson	
G-ATUF	Cessna F.150F	D. P. Williams	
G-ATUG	D.62B Condor	R. Crosby	
G-ATUH	Tipsy T.66 Nipper 1	M. D. Barnard & C. Voelger	
G-ATUI	Bo 208C1 Junior	M. J. Grundy	
G-ATUL	PA-28 Cherokee 180	Barry Fielding Aviation Ltd	
G-ATVF	D.H.C.1 Chipmunk 22	Four Counties Fling Club/Syerston	
G-ATVK	PA-28 Cherokee 140	Broadland Flyers Ltd	
G-ATVO	PA-28 Cherokee 140	G. R. Bright	
G-ATVP	F.B.5 Gunbus (2345) ★	RAF Museum/Hendon	
G-ATVS	PA-28 Cherokee 180	T. A. Buckley	
G-ATVW	D.62B Condor	G. G. Roberts	
G-ATVX	Bo 208C1 Junior	A. V. Hurley & ptnrs	
G-ATWA	Jodel DR.1050	One Twenty Group	
G-ATWB	Jodel D.117	Andrewsfield Whiskey Bravo Group	
G-ATWJ	Cessna F.172F	J. P. A. Freeman/Headcorn	
G-ATWR	PA-30 Twin Comanche 160B	Lubair (Transport Services) Ltd/E. Midlands	
G-ATXA	PA-22 Tri-Pacer 150	S. Hildrop	
G-ATXD	PA-30 Twin Comanche 160B	P. A. Brook	
G-ATXJ	H.P.137 Jetstream 300 ★	Fire Service Training Airframe/Cardiff	
G-ATXM	PA-28 Cherokee 180	G-ATXM Flying Group	
G-ATXN	Mitchell-Proctor Kittiwake 1	R. G. Day/Biggin Hill	
G-ATXO	SIPA 903	D. F. Hurn	
G-ATXZ	Bolkow Bo 208C1 Junior	G-ATXZ Group	
G-ATYM	Cessna F.150G	P. D' Costa	
G-ATYN	Cessna F.150G	J. S. Grant	
G-ATYS	PA-28 Cherokee 180	G-ATYS Flying Group	
G-ATZG	AFB2 gas balloon	S. Cameron	
G-ATZK	PA-28 Cherokee 180	G-ZK Group	
G-ATZM	Piper J-3C-90 Cub	R. W. Davison	
G-ATZS	Wassmer WA41 / Super Baladou IV	Little Bear Ltd	
G-ATZY	Cessna F.150G	Prestwick Flight Centre Ltd	
G-AVAK	M.S.893A Rallye Commodore 180	W. K. Anderson *(stored)*/Perth	
G-AVAR	Cessna F.150G	J. A. Rees	
G-AVAU	PA-30 Twin Comanche 160B	Enrico Ermano Ltd	
G-AVAW	D.62B Condor	Condor Aircraft Group	
G-AVAX	PA-28 Cherokee 180	J. J. Parkes	
G-AVBG	PA-28 Cherokee 180	G-AVBG Flying Group/White Waltham	
G-AVBH	PA-28 Cherokee 180	T. R. Smith (Agricultural Machinery) Ltd	
G-AVBS	PA-28 Cherokee 180	A. G. Arthur	
G-AVBT	PA-28 Cherokee 180	J. F. Mitchell	
G-AVCM	PA-24 Comanche 260	R. F. Smith/Stapleford	

Notes	Reg.	Type	Owner or Operator
	G-AVCV	Cessna 182J Skylane	University of Manchester, School of Earth Atmospheric and Environmental Sciences
	G-AVDA	Cessna 182K Skylane	F. W. Ellis
	G-AVDF	Beagle Pup 100 ★	Beagle Owners Club
	G-AVDG	Wallis WA-116 Srs 1	K. H. Wallis
	G-AVDS	Beech 65-B80 Queen Air ★	Airport Fire Service/Filton
	G-AVDT	Aeronca 7AC Champion	D. Cheney & G. Moore
	G-AVDV	PA-22 Tri-Pacer 150 (tailwheel)	S. C. Brooks/Slinfold
	G-AVDY	Luton LA-4A Minor	R. Targonski
	G-AVEC	Cessna F.172H	Quick Flight Images LLP
	G-AVEF	Jodel 150	Heavy Install Ltd
	G-AVEH	SIAI-Marchetti S.205	R. E. Gretton & ptnrs
	G-AVEM	Cessna F.150G	T. D. & J. A. Warren
	G-AVEN	Cessna F.150G	R. A. Lambert
	G-AVER	Cessna F.150G	LAC (Enterprises) Ltd
	G-AVEU	Wassmer WA.41 Baladou IV	S. Roberts
	G-AVEX	D.62B Condor	C. A. Macleod
	G-AVEY	Currie Super Wot	B. J. Anning
	G-AVEZ	HPR-7 Herald 210 ★	*Rescue trainer*/Norwich
	G-AVFB	H.S.121 Trident 2E ★	Duxford Aviation Soc
	G-AVFE	H.S.121 Trident 2E ★	Belfast Airport Authority
	G-AVFH	H.S.121 Trident 2E ★	De Havilland Heritage Museum *(fuselage only)*/ London Colney
	G-AVFM	H.S.121 Trident 2E ★	Brunel Technical College/Bristol
	G-AVFP	PA-28 Cherokee 140	R. L. Howells/Barton
	G-AVFR	PA-28 Cherokee 140	R. R. Orr
	G-AVFU	PA-32 Cherokee Six 300	Trixstar Farms Ltd
	G-AVFX	PA-28 Cherokee 140	J. Watson
	G-AVFZ	PA-28 Cherokee 140	G-AVFZ Flying Group
	G-AVGA	PA-24 Comanche 260	Conram Aviation/Biggin Hill
	G-AVGC	PA-28 Cherokee 140	R. Dagg-Heston
	G-AVGD	PA-28 Cherokee 140	T. Akeroyd
	G-AVGE	PA-28 Cherokee 140	A. J. Cutler
	G-AVGI	PA-28 Cherokee 140	GI Group
	G-AVGK	PA-28 Cherokee 180	M. A. Bush
	G-AVGU	Cessna F.150G	Coulson Flying Services Ltd
	G-AVGY	Cessna 182K Skylane	R. M. C. Sears
	G-AVGZ	Jodel DR.1050	D. C. Webb
	G-AVHH	Cessna F.172H	The Bristol & Wessex Aeroplane Club
	G-AVHL	Jodel DR.105A	P. J. McMahon
	G-AVHM	Cessna F.150G	M & N Flying Group
	G-AVHT	Auster AOP.9 (WZ711)	J. Pyatt
	G-AVHY	Fournier RF.4D	I. G. K. Mitchell
	G-AVIA	Cessna F.150G	Cheshire Air Training Services Ltd/Liverpool
	G-AVIB	Cessna F.150G	Far North Aviation
	G-AVIC	Cessna F.172H	Leeside Flying Ltd
	G-AVID	Cessna 182K	Jaguar Aviation Ltd
	G-AVII	AB-206A JetRanger	Bristow Helicopters Ltd
	G-AVIL	Alon A.2 Aircoupe (VX147)	D. J. Hulks
	G-AVIN	M.S.880B Rallye Club	B. Bunce
	G-AVIP	Brantly B.2B	W. G. B. Yard
	G-AVIS	Cessna F.172H	J. P. A. Freeman
	G-AVIT	Cessna F.150G	P. J. Mitchell
	G-AVIZ	Scheibe SF.25A Motorfalke	Spilsby Gliding Trust
	G-AVJE	Cessna F.150G	G-AVJE Syndicate
	G-AVJF	Cessna F.172H	J. A. & G. M. Rees
	G-AVJI	Cessna F.172H ★	Northbrook College/Shoreham
	G-AVJJ	PA-30 Twin Comanche 160B	A. H. Manser
	G-AVJK	Jodel DR.1050/M1	A. A. Robertson & D. S. Spillane
	G-AVJO	Fokker E.III (replica) (422-15)	Bianchi Aviation Film Services Ltd/Booker
	G-AVJV	Wallis WA-117 Srs 1	K. H. Wallis (G-ATCV)
	G-AVJW	Wallis WA-118 Srs 2	K. H. Wallis (G-ATPW)
	G-AVKB	MB.50 Pipistrelle	W. B. Cooper
	G-AVKD	Fournier RF-4D	Lasham RF4 Group
	G-AVKE	Gadfly HDW.1 ★	IHM/Weston-s-Mare
	G-AVKG	Cessna F.172H	Springbank Aviation Ltd
	G-AVKI	Slingsby T.66 Nipper 3	J. M. Greenway
	G-AVKK	Slingsby T.66 Nipper 3	C. Watson
	G-AVKL	PA-30 Twin Comanche 160B	Bravo Aviation Ltd/Jersey
	G-AVKN	Cessna 401	Law Leasing Ltd
	G-AVKP	A.109 Airedale	D. R. Williams
	G-AVKR	Bolkow Bo 208C1 Junior	C. H. Morris

Reg.	Type	Owner or Operator	Notes
G-AVLB	PA-28 Cherokee 140	M. Wilson	
G-AVLC	PA-28 Cherokee 140	C. M. Tyers	
G-AVLE	PA-28 Cherokee 140	Video Security Services/Tollerton	
G-AVLF	PA-28 Cherokee 140	G. H. Hughesdon	
G-AVLG	PA-28 Cherokee 140	C. H. R. Hewitt	
G-AVLI	PA-28 Cherokee 140	Lima India Aviation Group	
G-AVLJ	PA-28 Cherokee 140	Cherokee Aviation Holdings Jersey Ltd	
G-AVLM	B.121 Pup 3	T. M. & D. A. Jones/Egginton	
G-AVLN	B.121 Pup 2	Dogs Flying Group	
G-AVLO	Bo 208C1 Junior	P. J. Swain	
G-AVLT	PA-28 Cherokee 140	G-AVLT Flying Group (G-KELC)	
G-AVLW	Fournier RF-4D	J. C. A. C. da Silva	
G-AVLY	Jodel D.120A	N. V. de Candole	
G-AVMA	GY-80 Horizon 180	B. R. Hildick	
G-AVMB	D.62B Condor	L. J. Dray	
G-AVMD	Cessna 150G	Bagby Aviation Flying Group	
G-AVMF	Cessna F. 150G	J. F. Marsh	
G-AVMJ	BAC One-Eleven 510ED ★	European Aviation Ltd *(cabin trainer)*	
G-AVMK	BAC One-Eleven 510ED ★	Gravesend College *(fuselage only)*	
G-AVMN	BAC One-Eleven 510ED ★	Aviation Museum/Bournemouth	
G-AVMO	BAC One-Eleven 510ED ★	Aerospace Museum/Cosford	
G-AVMU	BAC One-Eleven 510ED ★	Duxford Aviation Soc	
G-AVNC	Cessna F.150G	J. R. Alderson	
G-AVNE	W.S.58 Wessex Mk 60 Srs 1 ★	IHM/Weston-s-Mare	
G-AVNN	PA-28 Cherokee 180	G-AVNN Flying Group	
G-AVNO	PA-28 Cherokee 180	November Oscar Flying Group	
G-AVNS	PA-28 Cherokee 180	J. G. O'Brien	
G-AVNU	PA-28 Cherokee 180	D. Durrant	
G-AVNW	PA-28 Cherokee 180	Len Smith's School of Sports Ltd	
G-AVNZ	Fournier RF-4D	C. D. Pidler	
G-AVOA	Jodel DR.1050	D. A. Willies/Cranwell	
G-AVOC	CEA Jodel DR.221	Alpha One Flying Group	
G-AVOH	D.62B Condor	Halegreen Associates Ltd	
G-AVOM	CEA Jodel DR.221	Avon Flying Group	
G-AVOO	PA-18 Super Cub 150	Dublin Gliding Club Ltd	
G-AVOZ	PA-28 Cherokee 180	Oscar Zulu Flying Group	
G-AVPD	Jodel D.9 Bebe ★	S. W. McKay *(stored)*	
G-AVPJ	D.H.82A Tiger Moth	C. C. Silk	
G-AVPM	Jodel D.117	J. C. Haynes/Breighton	
G-AVPN	HPR-7 Herald 213 ★	Yorkshire Air Museum/Elvington	
G-AVPO	Hindustan HAL-26 Pushpak	M. B. Johns	
G-AVPS	PA-30 Twin Comanche 160B	J. M. Bisco/Staverton	
G-AVPV	PA-18 Cherokee 180	K. A. Passmore	
G-AVPY	PA-25 Pawnee 235C	Southdown Gliding Club Ltd	
G-AVRK	PA-28 Cherokee 180	J. Gama	
G-AVRP	PA-28 Cherokee 140	Trent 199 F/Group	
G-AVRS	GY-80 Horizon 180	C. Clark-Monks & B. A. Heath	
G-AVRU	PA-28 Cherokee 180	G-AVRU Partnership/Clacton	
G-AVRW	GY-20 Minicab	Kestrel Flying Group/Tollerton	
G-AVRY	PA-28 Cherokee 180	Brigfast Ltd/Blackbushe	
G-AVRZ	PA-28 Cherokee 180	Mantavia Group Ltd	
G-AVSA	PA-28 Cherokee 180	G-AVSA Flying Group	
G-AVSB	PA-28 Cherokee 180	D. L. Macdonald	
G-AVSC	PA-28 Cherokee 180	MCS019 Ltd	
G-AVSD	PA-28 Cherokee 180	Landmate Ltd	
G-AVSE	PA-28 Cherokee 180	F. Glendon/Ireland	
G-AVSF	PA-28 Cherokee 180	Monday Club/Blackbushe	
G-AVSI	PA-28 Cherokee 140	G-AVSI Flying Group	
G-AVSP	PA-28 Cherokee 180	Airways Flight Training (Exeter) Ltd	
G-AVSR	Beagle D.5/180 Husky	A. L. Young	
G-AVSZ	AB-206B JetRanger	Patriot Aviation Ltd/Cranfield	
G-AVTC	Slingsby Nipper T.66 RA.45 Srs 3	M. W. Bodger	
G-AVTP	Cessna F.172H	Tango Papa Group/White Waltham	
G-AVTT	Ercoupe 415D	Wright's Farm Eggs Ltd/Andrewsfield	
G-AVTV	M.S.893A Rallye Commodore	Staffordshire Gliding Club Ltd	
G-AVUD	PA-30 Twin Comanche 160B	F.M.Aviation/Biggin Hill	
G-AVUG	Cessna F.150H	Skyways Flying Club	
G-AVUH	Cessna F.150H	C. M. Chinn	
G-AVUS	PA-28 Cherokee 140	D. J. Hunter	
G-AVUT	PA-28 Cherokee 140	Bencray Ltd/Blackpool	
G-AVUU	PA-28 Cherokee 140	A. Jahanfar & ptnrs/Southend	
G-AVUZ	PA-32 Cherokee Six 300	Ceesix Ltd/Jersey	

BRITISH CIVIL REGISTRATIONS

Notes	Reg.	Type	Owner or Operator
	G-AVVC	Cessna F.172H	M. Turnbull
	G-AVVE	Cessna F.150H ★	R. Windley (stored)
	G-AVVJ	M.S.893A Rallye Commodore	M. Powell
	G-AVVL	Cessna F.150H	International Aerospace Engineering Ltd/Cranfield
	G-AVVO	Avro 652A Anson 19 (VL348) ★	Newark Air Museum
	G-AVWA	PA-28 Cherokee 140	SFG Ltd
	G-AVWD	PA-28 Cherokee 140	Evelyn Air
	G-AVWI	PA-28 Cherokee 140	L. M. Veitch
	G-AVWJ	PA-28 Cherokee 140	A. M. Harrhy
	G-AVWL	PA-28 Cherokee 140	S.H. & C.L. Maynard
	G-AVWM	PA-28 Cherokee 140	P. E. Preston & ptnrs/Southend
	G-AVWN	PA-28R Cherokee Arrow 180	Vawn Air Ltd/Jersey
	G-AVWO	PA-28R Cherokee Arrow 180	The Whisky Oscar Group
	G-AVWR	PA-28R Cherokee Arrow 180	S. J. French & ptnrs/Dunkeswell
	G-AVWT	PA-28R Cherokee Arrow 180	Cloudbase Aviation Ltd
	G-AVWU	PA-28R Cherokee Arrow 180	A. M. Alam/Elstree
	G-AVWV	PA-28R Cherokee Arrow 180	Strathtay Flying Group
	G-AVWY	Fournier RF-4D	P. Turner
	G-AVXA	PA-25 Pawnee 235	S. Wales Gliding Club Ltd/Usk
	G-AVXC	Slingsby T.66 Nipper 3	P. A. Gibbs
	G-AVXD	Slingsby T.66 Nipper 3	Tayside Nipper Group
	G-AVXF	PA-28R Cherokee Arrow 180	G-AVXF Group
	G-AVXW	D.62B Condor	A. J. Cooper/Rochester
	G-AVXY	Auster AOP.9	G. J. Siddall
	G-AVXZ	PA-28 Cherokee 140 ★	ATC Hayle (instructional airframe)
	G-AVYB	H.S.121 Trident 1E-140 ★	SAS training airframe/Hereford
	G-AVYK	A.61 Terrier 3	J. P. Roland/Aboyne
	G-AVYL	PA-28 Cherokee 180	N. E. Binner
	G-AVYM	PA-28 Cherokee 180	Carlisle Aviation (1985) Ltd/Crosby
	G-AVYR	PA-28 Cherokee 140	SAS Flying Group/Thruxton
	G-AVYS	PA-28R Cherokee Arrow 180	Musicbank Ltd
	G-AVYT	PA-28R Cherokee Arrow 180	J. R. Tindale
	G-AVYV	Jodel D.120	A. J. Sephton
	G-AVZB	Aero Z-37 Cmelak ★	Science Museum/Wroughton
	G-AVZI	Bo 208C1 Junior	C. F. Rogers
	G-AVZM	B.121 Pup 1	ARAZ Group/Elstree
	G-AVZN	B.121 Pup 1	Shipdham Aviators Flying Group
	G-AVZO	B.121 Pup 1 ★	Thamesside Aviation Museum/E. Tilbury
	G-AVZP	B.121 Pup 1	T. A. White
	G-AVZR	PA-28 Cherokee 180	Lincoln Aero Club Ltd/Sturgate
	G-AVZU	Cessna F.150H	R. D. Forster
	G-AVZV	Cessna F.172H	E. M. & D. S. Lightbown/Crosland Moor
	G-AVZW	EAA Biplane Model P	R. G. Maidment & G. R. Edmundson/Goodwood
	G-AVZX	M.S.880B Rallye Club	J. Nugent
	G-AWAA	M.S.880B Rallye Club ★	P. A. Cairns (stored)/St Just
	G-AWAC	GY-80 Horizon 180	P. B. Hodgson
	G-AWAH	Beech 95-D55 Baron	B. J. S. Grey/Duxford
	G-AWAJ	Beech 95-D55 Baron	Standard Hose Ltd/Leeds
	G-AWAT	D.62B Condor	B. Woolford
	G-AWAU	Vickers F.B.27A Vimy (replica) (F8614) ★	Bomber Command Museum/Hendon
	G-AWAW	Cessna F.150F ★	Science Museum/S. Kensington
	G-AWAX	Cessna 150D	J. Haunch
	G-AWAZ	PA-28R Cherokee Arrow 180	Aero Angia Ltd
	G-AWBA	PA-28R Cherokee Arrow 180	March Flying Group/Stapleford
	G-AWBB	PA-28R Cherokee Arrow 180	D. L. Claydon
	G-AWBC	PA-28R Cherokee Arrow 180	Anglo Aviation (UK) Ltd
	G-AWBE	PA-28 Cherokee 140	B. E. Boyle
	G-AWBG	PA-28 Cherokee 140	G. D. Cooper
	G-AWBH	PA-28 Cherokee 140	Mainstreet Aviation
	G-AWBJ	Fournier RF-4D	J. M. Adams
	G-AWBM	D.31 Turbulent	A. D. Pratt
	G-AWBN	PA-30 Twin Comanche 160B	Stourfield Investments Ltd/Jersey
	G-AWBS	PA-28 Cherokee 140	M. A. English & T. M. Brown
	G-AWBT	PA-30 Twin Comanche 160B ★	Instructional airframe/Cranfield
	G-AWBU	Morane-Saulnier N (replica) (MS.824)	Bianchi Aviation Film Services Ltd
	G-AWBW	Cessna F.172H ★	Brunel Technical College/Bristol
	G-AWBX	Cessna F.150H	J. Meddings/Tatenhill
	G-AWCM	Cessna F.150H	R. Garbett
	G-AWCN	Cessna FR.172E	R. C Lunnon & A. J. Speight.
	G-AWCP	Cessna F.150H (tailwheel)	C. E. Mason/Shobdon
	G-AWDA	Slingsby T.66 Nipper 3	J. A. Cheesebrough

Reg.	Type	Owner or Operator	Notes
G-AWDI	PA-23 Aztec 250C ★	Queens Head/Willington, Beds	
G-AWDO	D.31 Turbulent	R. N. Crosland	
G-AWDP	PA-28 Cherokee 180	B. H. & P. M. Illston/Shipdham	
G-AWDR	Cessna FR.172E	B. A. Wallace	
G-AWDU	Brantly B.2B	B. M. Freeman	
G-AWEF	SNCAN Stampe SV-4B	RAF Buchanan	
G-AWEI	D.62B Condor	J. M. C. Coyle	
G-AWEK	Fournier RF-4D	P. Barrett	
G-AWEL	Fournier RF-4D	A. B. Clymo/Halfpenny Green	
G-AWEM	Fournier RF-4D	B. J. Griffin/Wickenby	
G-AWEP	Barritault JB-01 Minicab	A. Louth	
G-AWES	Cessna 150H	P. Montgomery-Stuart	
G-AWET	PA-28 Cherokee 180D	Broadland Flying Group Ltd/Shipdham	
G-AWEV	PA-28 Cherokee 140	Norflight Ltd	
G-AWEX	PA-28 Cherokee 140	Sir W. G. Armstrong Whitworth Flying Group/ Coventry	
G-AWEZ	PA-28R Cherokee Arrow 180	T. R. Leighton & ptnrs	
G-AWFB	PA-28R Cherokee Arrow 180	J. C. Luke/Filton	
G-AWFC	PA-28R Cherokee Arrow 180	B. J. Hines	
G-AWFD	PA-28R Cherokee Arrow 180	D. J. Hill	
G-AWFF	Cessna F.150H	Westflight Aviation Ltd	
G-AWFJ	PA-28R Cherokee Arrow 180	Parplon Ltd	
G-AWFN	D.62B Condor	P. B. Lowry	
G-AWFO	D.62B Condor	T. A. Major	
G-AWFP	D.62B Condor	Blackbushe Flying Club	
G-AWFT	Jodel D.9 Bebe	W. H. Cole	
G-AWFW	Jodel D.117	C. J. Rodwell	
G-AWFZ	Beech A23 Musketeer	Bob Crowe Aircraft Sales Ltd/Cranfield	
G-AWGA	A.109 Airedale ★	(stored)/Sevenoaks	
G-AWGD	Cessna F.172H	R. P. Vincent	
G-AWGJ	Cessna F.172H	J. & C. J. Freeman/Headcorn	
G-AWGK	Cessna F.150H	G. E. Allen	
G-AWGN	Fournier RF-4D	M. P. Barley	
G-AWGZ	Taylor JT.1 Monoplane	R. L. Sambell	
G-AWHB	C.A.S.A. 2-111D (6J+PR) ★	Aces High Ltd/North Weald	
G-AWHX	Rollason Beta B.2	S. G. Jones	
G-AWHY	Falconar F.11-3	Why Fly Group (G-BDPB)	
G-AWIF	Brookland Mosquito 2	C. A. Reeves/Gloucester	
G-AWII	V.S.349 Spitfire VC (AR501)	The Shuttleworth Collection/O. Warden	
G-AWIP	Luton LA-4A Minor	J. Houghton	
G-AWIR	Midget Mustang	K. E. Sword/Leicester	
G-AWIT	PA-28 Cherokee 180	Cherry Orchard Aparthotel Ltd	
G-AWIV	Airmark TSR.3	F. R. Hutchings	
G-AWIW	SNCAN Stampe SV-4B	R. E. Mitchell	
G-AWJE	Slingsby T.66 Nipper 3	K G. G. Howe	
G-AWJV	D.H.98 Mosquito TT Mk 35 (TA634) ★	De Havilland Heritage Museum	
G-AWJX	Zlin Z.526 Trener Master	P. A. Colman	
G-AWJY	Zlin Z.526 Trener Master	M. Gainza	
G-AWKD	PA-17 Vagabond	A. T. & M. R. Dowie/ White Waltham	
G-AWKO	B.121 Pup 1	S. E. Ford	
G-AWKP	Jodel DR.253	G-AWKP Group	
G-AWKT	M.S.880B Rallye Club	A. Ringland & P. Keating	
G-AWKX	Beech A65 Queen Air ★	(instructional airframe)/Shoreham	
G-AWLA	Cessna F.150H	Bagby Aviation	
G-AWLF	Cessna F.172H	Gannet Aviation	
G-AWLG	SIPA 903	S. W. Markham	
G-AWLI	PA-22 Tri-Pacer 150	J. S. Lewery/Shoreham	
G-AWLO	Boeing Stearman E.75	N. D. Pickard/Shoreham	
G-AWLP	Mooney M.20F	I. C. Lomax	
G-AWLR	Slingsby T.66 Nipper 3	T. D. Reid	
G-AWLS	Slingsby T.66 Nipper 3	G. A. Dunster & B. Gallagher	
G-AWLZ	Fournier RF-4D	Nympsfield RF-4 Group	
G-AWMD	Jodel D.11	R. M. Worth	
G-AWMF	PA-18 Super Cub 150 (modified)	Booker Gliding Club Ltd	
G-AWMI	Glos-Airtourer 115	M. Furse/Cardiff/Wales	
G-AWMM	M.S.893A Rallye Commodore 180	D. P. & S. White	
G-AWMN	Luton LA-4A Minor	B. J. Douglas	
G-AWMP	Cessna F.172H	R. J. D. Blois	
G-AWMR	D.31 Turbulent	M. J. Freeman	
G-AWMT	Cessna F.150H	M. Paisley	
G-AWMZ	Cessna F.172H ★	Parachute jump trainer/Cark	
G-AWNT	BN-2A Islander	Sterling Helicopters Ltd/Norwich	

Notes	Reg.	Type	Owner or Operator
	G-AWOA	M.S.880B Rallye Club	J. A. Rimmer
	G-AWOE	Aero Commander 680E	J. M. Houlder/Elstree
	G-AWOF	PA-15 Vagabond	C. M. Hicks
	G-AWOH	PA-17 Vagabond	I. M. Callier
	G-AWOT	Cessna F.150H	M. J. Willoughby
	G-AWOU	Cessna 170B	S. Billington/Denham
	G-AWOX	W.S.58 Wessex 60 (150225) ★	IHM/Weston-s-Mare
	G-AWPH	P.56 Provost T.1	J. A. D. Bradshaw
	G-AWPJ	Cessna F.150H	W. J. Greenfield
	G-AWPN	Shield Xyla	K. R. Snell
	G-AWPS	PA-28 Cherokee 140	A. R. Matthews
	G-AWPU	Cessna F.150J	LAC (Enterprises) Ltd/Barton
	G-AWPW	PA-12 Super Cruiser	AK Leasing (Jersey) Ltd
	G-AWPY	Bensen B.8M	J. Jordan
	G-AWPZ	Andreasson BA-4B	J. M. Vening
	G-AWRK	Cessna F.150J	Systemroute Ltd/Shoreham
	G-AWRP	Cierva Rotorcraft ★	IHM/Weston-s-Mare
	G-AWRS	Avro 19 Srs. 2 ★	N. E. Aircraft Museum/Usworth
	G-AWRY	P.56 Provost T.1 (XF836)	Slymar Aviation & Services Ltd
	G-AWSA	Avro 652A Anson 19 (VL349) ★	Norfolk & Suffolk Aviation Museum
	G-AWSH	Zlin Z.526 Trener Master	Avia Special Ltd
	G-AWSL	PA-28 Cherokee 180D	Fascia Services Ltd/Southend
	G-AWSM	PA-28 Cherokee 235	Aviation Projects
	G-AWSN	D.62B Condor	M. K. A. Blyth
	G-AWSP	D.62B Condor	R. Q. & A. S. Bond/Wellesbourne
	G-AWSS	D.62A Condor	N. J. & D. Butler
	G-AWST	D.62B Condor	T. P. Lowe
	G-AWSV	Skeeter 12 (XM553)	Maj. M. Somerton-Rayner/Middle Wallop
	G-AWSW	D.5/180 Husky (XW635)	Windmill Aviation/Spanhoe
	G-AWTJ	Cessna F.150J	P. J. Jameson
	G-AWTL	PA-28 Cherokee 180D	I. R. Chaplin
	G-AWTS	Beech A.23 Musketeer	Cinque Ports Flying Group Ltd
	G-AWTV	Beech 19A Musketeer Sport	J. Whittaker
	G-AWTX	Cessna F.150J	R. D. Forster
	G-AWUB	GY-201 Minicab	R. A. Hand
	G-AWUE	Jodel DR.1050	K. W. Wood & F. M. Watson
	G-AWUG	Cessna F.150H	Prestwick Flight Centre Ltd
	G-AWUJ	Cessna F.150H	S. R. Hughes
	G-AWUL	Cessna F.150H	C. A. & L. P. Green
	G-AWUN	Cessna F.150H	C150 Group
	G-AWUO	Cessna F.150H	SAS Flying Group
	G-AWUT	Cessna F.150J	S. J. Black/Leeds
	G-AWUU	Cessna F.150J	A. L. Grey
	G-AWUX	Cessna F.172H	G-AWUX Group/St.Just
	G-AWUZ	Cessna F.172H	I. R. Judge
	G-AWVA	Cessna F.172H	Barton Air Ltd
	G-AWVB	Jodel D.117	H. Davies
	G-AWVC	B.121 Pup 1	J. J. West/Sturgate
	G-AWVE	Jodel DR.1050/M1	E. A. Taylor/Southend
	G-AWVF	P.56 Provost T.1 (XF877)	J. H. Powell-Tuck
	G-AWVG	AESL Airtourer T.2	C. J. Schofield
	G-AWVN	Aeronca 7AC Champion	Champ Flying Group
	G-AWVZ	Jodel D.112	D. C. Stokes
	G-AWWE	B.121 Pup 2	J. M. Randle/Coventry
	G-AWWI	Jodel D.117	W. J. Evans
	G-AWWM	GY-201 Minicab	P. J. Brayshaw
	G-AWWN	Jodel DR.1051	R. A. J. Hurst
	G-AWWO	Jodel DR.1050	Whiskey Oscar Group/Barton
	G-AWWP	Aerosport Woody Pusher III	M. S. Bird & R. D. Bird
	G-AWWT	D.31 Turbulent	E. L. Phillips
	G-AWWU	Cessna FR.172F	Westward Airways (Lands End) Ltd
	G-AWWW	Cessna 401	Treble Whisky Aviation Ltd
	G-AWXR	PA-28 Cherokee 180D	Aero Club de Portugal
	G-AWXS	PA-28 Cherokee 180D	J. A. Hardiman/Shobdon
	G-AWXY	M.S.885 Super Rallye	K. Henderson/Hibaldstow
	G-AWXZ	SNCAN Stampe SV-4C	Bianchi Aviation Film Services Ltd
	G-AWYB	Cessna FR.172F	J. R. Sharpe/Southend
	G-AWYJ	B.121 Pup 2	H. C. Taylor
	G-AWYL	Jodel DR.253B	K. Gillham
	G-AWYO	B.121 Pup 1	B. R. C. Wild/Popham
	G-AWYX	M.S.880B Rallye Club	M. J. Edwards/Henstridge
	G-AWYY	T.57 Camel replica (B6401) ★	F.A.A. Museum/Yeovilton

Reg.	Type	Owner or Operator	Notes
G-AWZI	H.S.121 Trident 3B ★	A. Lee/FAST Museum (nose only)/Farnborough	
G-AWZJ	H.S.121 Trident 3B ★	Dumfries & Galloway Museum	
G-AWZK	H.S.121 Trident 3B ★	Trident Preservation Soc./Manchester	
G-AWZM	H.S.121 Trident 3B ★	Science Museum/Wroughton	
G-AWZP	H.S.121 Trident 3B ★	Manchester Museum of Science & Industry (nose only)	
G-AWZX	H.S.121 Trident 3B ★	BAA Airport Fire Services/Gatwick	
G-AXAB	PA-28 Cherokee 140	D. M. Loughlin	
G-AXAN	D.H.82A Tiger Moth (EM720)	Leading Edge Marketing Ltd	
G-AXAS	Wallis WA-116T	K. H. Wallis (G-AVDH)	
G-AXAT	Jodel D.117A	P. S. Wilkinson	
G-AXBF	Beagle D.5/180 Husky	J. H. Powell-Tuck	
G-AXBH	Cessna F.172H	D. F. Ranger	
G-AXBJ	Cessna F.172H	BJ Flying Group/Leicester	
G-AXBW	D.H.82A Tiger Moth (T5879)	Hunter Wing Ltd/Bournemouth	
G-AXBZ	D.H.82A Tiger Moth	W. J. de Jong Cleyndert	
G-AXCA	PA-28R Cherokee Arrow 200	W. H. Nelson	
G-AXCG	Jodel D.117	Charlie Golf Group/Andrewsfield	
G-AXCI	Bensen B.8M	N. Martin (stored)	
G-AXCM	M.S.880B Rallye Club	D. C. Manifold	
G-AXCX	B.121 Pup 2	L. A. Pink	
G-AXCY	Jodel D.117A	R. S. Marom	
G-AXCZ	SNCAN Stampe SV-4C	J. Price	
G-AXDC	PA-23 Aztec 250D	N. J. Lilley	
G-AXDI	Cessna F.172H	M. F. & J. R. Leusby/Conington	
G-AXDK	Jodel DR.315	Delta Kilo Flying Group/Sywell	
G-AXDN	BAC-Sud Concorde 01 ★	Duxford Aviation Soc	
G-AXDV	B.121 Pup 1	T. A. White	
G-AXDW	B.121 Pup 1	Cranfield Delta Whiskey Group	
G-AXDY	Falconar F-11	J. Nunn	
G-AXDZ	Cassutt Racer IIIM	A. Chadwick/Little Staughton	
G-AXEB	Cassutt Racer IIIM	G. E. Horder/Redhill	
G-AXED	PA-25 Pawnee 235	Wolds Gliding Club Ltd/Pocklington	
G-AXEH	B.125 Bulldog 1 ★	Museum of Flight/E. Fortune	
G-AXEI	Ward Gnome ★	Real Aeroplane Club/Breighton	
G-AXEO	Scheibe SF.25B Falke	The Borders (Milfield) Gliding Club Ltd	
G-AXEV	B.121 Pup 2	D. S. Russell & D. G. Benson	
G-AXFG	Cessna 337D	County Garage (Cheltenham) Ltd	
G-AXFN	Jodel D.119	B. M. Jackson	
G-AXGE	M.S.880B Rallye Club	R. P. Loxton	
G-AXGG	Cessna F.150J	S. G. Moores & A. J. Simpson	
G-AXGP	Piper J-3C-90 Cub	G-AXGP Group	
G-AXGR	Luton LA-4A Minor	B. A. Schlussler	
G-AXGS	D.62B Condor	G-AXGS Condor Group	
G-AXGV	D.62B Condor	R. J. Wrixon	
G-AXGZ	D.62B Condor	A. J. Cooper	
G-AXHA	Cessna 337A	I. M. Latiff	
G-AXHC	SNCAN Stampe SV-4C	D. L. Webley	
G-AXHE	BN-2A Islander ★	(parachute trainer)/Strathallan	
G-AXHO	B.121 Pup 2	L. W. Grundy/Stapleford	
G-AXHP	Piper J-3C-65 Cub (480636)	Witham (Specialist) Vehicles Ltd	
G-AXHR	Piper J-3C-65 Cub (329601/D-44)	K. B. Raven & E. Cundy	
G-AXHS	M.S.880B Rallye Club	B. & A. Swales	
G-AXHT	M.S.880B Rallye Club	Hotel Tango Group	
G-AXHV	Jodel D.117A	Derwent Flying Group/Hucknall	
G-AXIA	B.121 Pup 1	P. S. Shuttleworth	
G-AXIE	B.121 Pup 2	G. McD. Moir	
G-AXIF	B.121 Pup 2	J. R. Faulkner	
G-AXIG	B.125 Bulldog 104	A. A. A. Hamilton	
G-AXIO	PA-28 Cherokee 140B	Just Plane Trading Ltd	
G-AXIR	PA-28 Cherokee 140B	A. G. Birch	
G-AXIW	Scheibe SF.25B Falke	M. B. Hill	
G-AXIX	Glos-Airtourer 150	J. C. Wood	
G-AXJB	Omega 84 balloon	Southern Balloon Group	
G-AXJH	B.121 Pup 2	The Henry Flying Group	
G-AXJI	B.121 Pup 2	J. J. Sanders	
G-AXJJ	B.121 Pup 2	M. L. Jones & ptnrs	
G-AXJO	B.121 Pup 2	J. A. D. Bradshaw	
G-AXJR	Scheibe SF.25B Falke	Falke Syndicate	
G-AXJV	PA-28 Cherokee 140B	ATC (Lasham) Ltd	
G-AXJX	PA-28 Cherokee 140B	Patrolwatch Ltd/Sleap	

G-AXKH – G-AXVM

Notes	Reg.	Type	Owner or Operator
	G-AXKH	Luton LA-4A Minor	M. E. Vaisey
	G-AXKJ	Jodel D.9	P. D. Smalley
	G-AXKO	Westland-Bell 47G-4A	G. P. Hinkley
	G-AXKS	Westland Bell 47G-4A ★	Museum of Army Flying/Middle Wallop
	G-AXKW	Westland-Bell 47G-4A	Eyre Spier Associates Ltd
	G-AXKX	Westland Bell 47G-4A	South Yorkshire Aviation Ltd
	G-AXLG	Cessna 310K	Smiths (Outdrives) Ltd
	G-AXLI	Slingsby T.66 Nipper 3	D. & M. Shrimpton
	G-AXLS	Jodel DR.105A	Axle Flying Club
	G-AXLZ	PA-18 Super Cub 95	R. J. Quantrell
	G-AXMA	PA-24 Comanche 180	J. A. & S. M. Fletcher
	G-AXMD	Omega O-56 balloon ★	British Balloon Museum/Newbury
	G-AXMN	J/5B Autocar	C. D. Wilkinson
	G-AXMT	Bücker Bü133 Jungmeister	R. A. Fleming/Breighton
	G-AXMW	B.121 Pup 1	DJP Engineering (Knebworth) Ltd
	G-AXMX	B.121 Pup 2	Susan A. Jones/Cannes
	G-AXNJ	Wassmer Jodel D.120	Clive Flying Group/Sleap
	G-AXNL	B.121 Pup 1	Appleton Aviation Ltd
	G-AXNM	B.121 Pup 1	T. W. Anderson/Sandown
	G-AXNN	B.121 Pup 2	Gabrielle Aviation Ltd/Shoreham
	G-AXNP	B.121 Pup 2	J. W. Ellis & R. J. Hemmings
	G-AXNR	B.121 Pup 2	The November Romeo Group
	G-AXNS	B.121 Pup 2	Derwent Aero Group/Gamston
	G-AXNW	SNCAN Stampe SV-4C	C. S. Grace
	G-AXNX	Cessna 182M	D. B. Harper
	G-AXNZ	Pitts S.1C Special	W. A. Jordan
	G-AXOG	PA-E23 Aztec 250D	G. H. Nolan
	G-AXOH	M.S.894 Rallye Minerva	Bristol Cars Ltd/White Waltham
	G-AXOJ	B.121 Pup 2	Pup Flying Group
	G-AXOR	PA-28 Cherokee 180D	Oscar Romeo Aviation Ltd
	G-AXOS	M.S.894A Rallye Minerva	M. N. Stevens
	G-AXOT	M.S.893 Rallye Commodore 180	P. Evans & J. C. Graves
	G-AXOZ	B.121 Pup 1	R. J. Ogborn/Liverpool
	G-AXPA	B.121 Pup 1	D. G. Lewendon
	G-AXPB	B.121 Pup 1	M. J. K. Seary & R. T. Austin
	G-AXPC	B.121 Pup 2	T. A. White/Bagby
	G-AXPF	Cessna F.150K	D. R. Marks/Denham
	G-AXPG	Mignet HM-293	W. H. Cole (stored)
	G-AXPM	B.121 Pup 1	S. C. Stanton
	G-AXPN	B.121 Pup 2	A. Richardson
	G-AXPZ	Campbell Cricket	W. R. Partridge
	G-AXRC	Campbell Cricket	L. R. Morris
	G-AXRK	Practavia Pilot Sprite 115 ★	M. Oliver
	G-AXRP	SNCAN Stampe SV-4A	Skysport Engineering (G-BLOL) (stored)/Hatch
	G-AXRR	Auster AOP.9 (XR241)	R. B. Webber
	G-AXRT	Cessna FA.150K (tailwheel)	C. C. Walley
	G-AXSC	B.121 Pup 1	R. J. MacCarthy/Swansea
	G-AXSD	B.121 Pup 1	AURS Aviation Ltd
	G-AXSF	Nash Petrel	Nash Aircraft Ltd/Lasham
	G-AXSG	PA-28 Cherokee 180	Admiral Property Ltd
	G-AXSI	Cessna F.172H	R. Collins (G-SNIP)
	G-AXSM	Jodel DR.1051	T. R. G. Barnby & M. S. Regendanz
	G-AXSW	Cessna FA.150K	R. Mitchell
	G-AXSZ	PA-28 Cherokee 140B	The White Wings Flying Group/White Waltham
	G-AXTA	PA-28 Cherokee 140B	G-AXTA Aircraft Group
	G-AXTC	PA-28 Cherokee 140B	G-AXTC Group
	G-AXTJ	PA-28 Cherokee 140B	K. Patel/Elstree
	G-AXTL	PA-28 Cherokee 140B	Pegasus Aviation (Midlands) Ltd
	G-AXTO	PA-24 Comanche 260	J. L. Wright
	G-AXTP	PA-28 Cherokee 180	M. Whyte/Ireland
	G-AXTX	Jodel D.112	C. Sawford
	G-AXUA	B.121 Pup 1	P. Wood
	G-AXUB	BN-2A Islander	Headcorn Parachute Club Ltd
	G-AXUC	PA-12 Super Cruiser	J. J. Bunton
	G-AXUF	Cessna FA.150K	W. B. Bateson/Blackpool
	G-AXUJ	J/1 Autocrat	C. J. Harrison & S. J. McKenna (G-OSTA)
	G-AXUK	Jodel DR.1050	Downland Flying Group
	G-AXUM	H.P.137 Jetstream 1 ★	Sodeteg Formation/France
	G-AXUW	Cessna FA.150K	Coventry Air Training School
	G-AXVB	Cessna F.172H	R. & J. Turner
	G-AXVK	Campbell Cricket	B. Jones
	G-AXVM	Campbell Cricket	D. M. Organ

Reg.	Type	Owner or Operator	Notes
G-AXVN	McCandless M.4	W. R. Partridge	
G-AXWA	Auster AOP.9 (XN437)	C. M. Edwards	
G-AXWT	Jodel D.11	R. C. Owen	
G-AXWV	Jodel DR.253	R. Friedlander & D. C. Ray	
G-AXWZ	PA-28R Cherokee Arrow 200	P. Walkley	
G-AXXC	CP.301B Emeraude	L. F. Clayton	
G-AXXV	D.H.82A Tiger Moth (DE992)	C. N. Wookey	
G-AXXW	Jodel D.117	D. F. Chamberlain & M. A. Hughes	
G-AXYK	Taylor JT.1 Monoplane	G. D. Bailey	
G-AXYU	Jodel D.9 Bebe	P. Turton & H. C. Peake-Jones	
G-AXYZ	WHE Airbuggy	B. Gunn	
G-AXZB	WHE Airbuggy	B. Gunn	
G-AXZD	PA-28 Cherokee 180E	G. M. Whitmore	
G-AXZF	PA-28 Cherokee 180E	E. P. C. & W. R. Rabson/Southampton	
G-AXZK	BN-2A-26 Islander	P. Johnson	
G-AXZM	Slingsby T.66 Nipper 3	G. R. Harlow	
G-AXZO	Cessna 180	Bourne Park Flyers	
G-AXZP	PA-E23 Aztec 250D	D. M. Harbottle	
G-AXZT	Jodel D.117	N. Batty	
G-AXZU	Cessna 182N	C. D. Williams	
G-AYAB	PA-28 Cherokee 180E	Films Ltd	
G-AYAC	PA-28R Cherokee Arrow 200	Fersfield Flying Group	
G-AYAJ	Cameron O-84 balloon	E. T. Hall	
G-AYAL	Omega 56 balloon ★	British Balloon Museum/Newbury	
G-AYAN	Slingsby Motor Cadet III	D. C. Pattison	
G-AYAR	PA-28 Cherokee 180E	A. Jahanfar/Southend	
G-AYAT	PA-28 Cherokee 180E	G-AYAT Flying Group	
G-AYAW	PA-28 Cherokee 180E	G-AYAW Group	
G-AYBD	Cessna F.150K	Apollo Aviation Advisory Ltd/Shoreham	
G-AYBG	Scheibe SF.25B Falke	H. H. T. Wolf	
G-AYBO	PA-23 Aztec 250D	Forster and Hales Ltd	
G-AYBP	Jodel D.112	G. J. Langston	
G-AYBR	Jodel D.112	I. S. Parker	
G-AYCC	Campbell Cricket	D. J. M. Charity	
G-AYCE	CP.301C Emeraude	S. D. Glover	
G-AYCF	Cessna FA.150K	E. J. Atkins/Popham	
G-AYCG	SNCAN Stampe SV-4C	N. Bignall/Booker	
G-AYCJ	Cessna TP.206D	White Knuckle Airways Ltd	
G-AYCK	AIA Stampe SV-4C	The Real Flying Co. Ltd (G-BUNT)/Shoreham	
G-AYCN	Piper J-3C-65 Cub	W. R. & B. M. Young	
G-AYCO	CEA DR.360	Charlie Oscar Club	
G-AYCP	Jodel D.112	D. J. Nunn	
G-AYCT	Cessna F.172H	Haimoss Ltd	
G-AYDI	D.H.82A Tiger Moth	R. B. Woods & ptnrs	
G-AYDR	SNCAN Stampe SV-4C	A. J. McLuskie	
G-AYDV	Coates SA.II-1 Swalesong	J. R. Coates	
G-AYDW	A.61 Terrier 2	A. S. Topen	
G-AYDX	A.61 Terrier 2	R. A. Kirby/Barton	
G-AYDY	Luton LA-4A Minor	T. Littlefair	
G-AYDZ	Jodel DR.200	Zero One Group	
G-AYEB	Jodel D.112	R. A. Durance	
G-AYEC	CP.301A Emeraude	Redwing Flying Group	
G-AYEE	PA-28 Cherokee 180E	Halegreen Associates Ltd & Demero Ltd	
G-AYEF	PA-28 Cherokee 180E	G-AYEF Group	
G-AYEG	Falconar F-9	A. L. Smith	
G-AYEH	Jodel DR.1050	John Scott Jodel Group	
G-AYEJ	Jodel DR.1050	J. M. Newbold	
G-AYEN	Piper J-3C-65 Cub	P. Warde & C. F. Morris	
G-AYET	M.S.892A Rallye Commodore 150	A. T. R. Bingley	
G-AYEV	Jodel DR.1050	L. G. Evans/Headcorn	
G-AYEW	Jodel DR.1051	J. M. Gale & J. R. Hope	
G-AYFA	SA Twin Pioneer Srs 3 ★	Solway Aviation Soc/Carlisle	
G-AYFC	D.62B Condor	A. R. Chadwick/Breighton	
G-AYFD	D.62B Condor	B. G. Manning	
G-AYFF	D.62B Condor	D. Ellis & A. W. Maycock	
G-AYFG	D.62C Condor	W. A. Braim	
G-AYFJ	M.S.880B Rallye Club	Rallye FJ Group	
G-AYFP	Jodel D.140	A. R. Wood	
G-AYFV	Crosby BA-4B	A. R. C. Mathie/Norwich	
G-AYGA	Jodel D.117	M. F. Sedgwick	

Notes	Reg.	Type	Owner or Operator
	G-AYGB	Cessna 310Q ★	Instructional airframe/Perth
	G-AYGC	Cessna F.150K	Alpha Aviation Group/Barton
	G-AYGD	Jodel DR.1051	Jodel Flying Group
	G-AYGE	SNCAN Stampe SV-4C	L. J. Proudfoot & ptnrs/Booker
	G-AYGG	Jodel D.120	J. M. Dean
	G-AYGX	Cessna FR.172G	Reims Rocket Group/Barton
	G-AYHA	AA-1 Yankee	S. J. Carr
	G-AYHX	Jodel D.117A	L. J. E. Goldfinch
	G-AYIA	Hughes 369HS ★	G. D. E. Bilton/Sywell
	G-AYIG	PA-28 Cherokee 140C	Biggles Ltd
	G-AYII	PA-28R Cherokee Arrow 200	Double India Group/Exeter
	G-AYIJ	SNCAN Stampe SV-4B	T. C. Beadle/Headcorn
	G-AYIM	H.S.748 Srs 2A	Emerald Airways Ltd/Liverpool
	G-AYIT	D.H.82A Tiger Moth	Ulster Tiger Group/Newtownards
	G-AYJA	Jodel DR.1050	G. Connell
	G-AYJB	SNCAN Stampe SV-4C	F. J. M. & J. P. Esson/Middle Wallop
	G-AYJD	Alpavia-Fournier RF-3	I. O. Bull
	G-AYJP	PA-28 Cherokee 140C	RAF Brize Norton Flying Club Ltd
	G-AYJR	PA-28 Cherokee 140C	RAF Brize Norton Flying Club Ltd
	G-AYJW	Cessna FR.172G	Sir W. G. Armstrong-Whitworth Flying Group
	G-AYJY	Isaacs Fury II	M. F. Newman
	G-AYKA	Beech 95-B55A Baron	Walsh Bros (Tunnelling) Ltd/Elstree
	G-AYKD	Jodel DR.1050	I. M. D. L. Weston
	G-AYKJ	Jodel D.117A	J. M. Alexander
	G-AYKK	Jodel D.117	J. M. Whitham
	G-AYKL	Cessna F.150L	M. A. Judge
	G-AYKS	Leopoldoff L-7	W. B. Cooper
	G-AYKT	Jodel D.117	D. I. Walker
	G-AYKW	PA-28 Cherokee 140C	S. P. Rooney & D. Griffiths
	G-AYKX	PA-28 Cherokee 140C	Robin Flying Group/Woodford
	G-AYKZ	SAI KZ-8	R. E. Mitchell/Cosford
	G-AYLA	Glos-Airtourer 115	D. S. P. Disney
	G-AYLC	Jodel DR.1051	E. W. B. Trollope
	G-AYLF	Jodel DR.1051 (modified)	Sicile Flying
	G-AYLL	Jodel DR.1050	C. Joly
	G-AYLP	AA-1 Yankee	D. Nairn
	G-AYLV	Jodel D.120	M. R. Henham
	G-AYLZ	SPP Super Aero 45 Srs 04	M. J. Cobb
	G-AYME	Fournier RF-5	R. D. Goodger/Biggin Hill
	G-AYMK	PA-28 Cherokee 140C	The Piper Flying Group/Newcastle
	G-AYMO	PA-23 Aztec 250C	A. Wardle
	G-AYMP	Currie Wot	N. S. Chittenden
	G-AYMR	Lederlin 380L	P. J. Brayshaw
	G-AYMU	Jodel D.112	M. R. Baker
	G-AYMV	Western 20 balloon	R. G. Turnbull
	G-AYMW	Bell 206A JetRanger 2	PLM Dollar Group Ltd
	G-AYNA	Phoenix Currie Wot	I. D. Folland
	G-AYND	Cessna 310Q	Source Group Ltd/Bournemouth
	G-AYNF	PA-28 Cherokee 140C	BW Aviation
	G-AYNJ	PA-28 Cherokee 140C	R. H. Ribbons
	G-AYNN	Cessna 185B Skywagon	Bencray Ltd/Blackpool
	G-AYNP	W.S.55 Whirlwi?nd Srs 3 ★	IHM/Weston-s-Mare
	G-AYOW	Cessna 182N Skylane	D. W. Parfrey
	G-AYOY	Sikorsky S-61N Mk 2	British International Ltd
	G-AYOZ	Cessna FA.150L	S. A. Hughes
	G-AYPE	MBB Bo 209 Monsun	Papa Echo Ltd/Biggin Hill
	G-AYPG	Cessna F.177RG	D. P. McDermott
	G-AYPH	Cessna F.177RG	M. R. & K. E. Slack
	G-AYPJ	PA-28 Cherokee 180	R. B. Petrie
	G-AYPM	PA-18 Super Cub 95	R. Horner
	G-AYPO	PA-18 Super Cub 95	A. W. Knowles
	G-AYPR	PA-18 Super Cub 95	R. G. Manton
	G-AYPS	PA-18 Super Cub 95	R. J. Hamlett & ptnrs
	G-AYPT	PA-18 Super Cub 95	B. L. Proctor & T. F. Lyddon
	G-AYPU	PA-28R Cherokee Arrow 200	Monalto Investments Ltd
	G-AYPV	PA-28 Cherokee 140D	Ashley Gardner Flying Club Ltd
	G-AYPZ	Campbell Cricket	A. Melody
	G-AYRF	Cessna F.150L	D. T. A. Rees
	G-AYRH	M.S.892A Rallye Commodore 150	S. O'Ceallaigh & J. Barry
	G-AYRI	PA-28R Cherokee Arrow 200	A. E. Thompson & Delta Motor Co (Windsor) Sales Ltd/White Waltham
	G-AYRM	PA-28 Cherokee 140D	M. J. Saggers/Biggin Hill

Reg.	Type	Owner or Operator	Notes
G-AYRO	Cessna FA.150L Aerobat	Fat Boys Flying Club	
G-AYRS	Jodel D.120A	L. R. H. D'Eath	
G-AYRT	Cessna F.172K	P. E. Crees	
G-AYRU	BN-2A-6 Islander	Army Parachute Association/Netheravon	
G-AYSB	PA-30 Twin Comanche 160C	N. J. Goff	
G-AYSD	Slingsby T.61A Falke	P. W. Hextall	
G-AYSH	Taylor JT.1 Monoplane	C. J. Lodge	
G-AYSJ	Bücker Bü133C Jungmeister (LG+01)	Patina Ltd/Duxford	
G-AYSK	Luton LA-4A Minor	Luton Minor Group	
G-AYSX	Cessna F.177RG	A. P. R. Dean	
G-AYSY	Cessna F.177RG	S. A. Tuer	
G-AZSZ	PA-23 Aztec 250D	A. A. Mattacks	
G-AYTA	SOCATA M.S.880B Rallye Club ★	Manchester Museum of Science & Industry	
G-AYTR	CP.301A Emeraude	Croft Farm Flying Group	
G-AYTT	Phoenix PM-3 Duet	H. E. Jenner	
G-AYTV	MJ.2A Tempete	Shoestring Flying Group	
G-AYUA	Auster AOP.9 (XK416)	De Havilland Aviation Ltd/Swansea	
G-AYUB	CEA DR.253B	D. J. Clark	
G-AYUH	PA-28 Cherokee 180F	Starpress Ltd	
G-AYUJ	Evans VP-1	T. N. Howard	
G-AYUM	Slingsby T.61A Falke	M. H. Simms	
G-AYUN	Slingsby T.61A Falke	G-AYUN Group	
G-AYUP	Slingsby T.61A Falke	P. R. Williams	
G-AYUR	Slingsby T.61A Falke	R. Hanningan & R. Lingard	
G-AYUS	Taylor JT.1 Monoplane	A. M. Sutton	
G-AYUT	Jodel DR.1050	M. L. Robinson	
G-AYUV	Cessna F.172H	Justgold Ltd	
G-AYVO	Wallis WA-120 Srs 1	K. H. Wallis	
G-AYVP	Woody Pusher	J. R. Wraight	
G-AYVT	Brochet MB.84 ★	Dunelm Flying Group (stored)	
G-AYWA	Avro 19 Srs 2 ★	N. K. Geddes	
G-AYWD	Cessna 182N	Wild Dreams Group	
G-AYWE	PA-28 Cherokee 140	Intelcomm (UK) Ltd	
G-AYWH	Jodel D.117A	D. Kynaston/Cambridge	
G-AYWM	Glos-Airtourer Super 150	The Star Flying Group/Staverton	
G-AYWT	AIA Stampe SV-4C	Meridian Aviation Group Ltd	
G-AYXP	Jodel D.117A	G. N. Davies	
G-AYXS	SIAI-Marchetti S205-18R	P. J. Bloore & J. M. Biles	
G-AYXT	W.S. 55 Whirlwind Srs 2 (XK940) ★	IHM/Weston-s-Mare	
G-AYXU	Champion 7KCAB Citabria	E. T. & P. A. Wild	
G-AYYK	Slingsby T.61A Falke	Cornish Gliding & Flying Club Ltd/Perranporth	
G-AYYL	Slingsby T.61A Falke	C. Wood	
G-AYYO	Jodel DR.1050/M1	Bustard Flying Club Ltd	
G-AYYT	Jodel DR.1050/M1	Yankee Tango Group	
G-AYYU	Beech C23 Musketeer	G-AYYU Group	
G-AYYW	BN-2A-21 Islander	Secretary of State for Foreign & Commonwealth Affairs	
G-AYYX	M.S.880B Ralle Club	J. G. MacDonald	
G-AYZE	PA-39 Twin Comanche 160 C/R	J. E. Palmer/Staverton	
G-AYZI	SNCAN Stampe SV-4C	D. M. & P. A. Fenton	
G-AYZJ	W.S.55 Whirlwind HAS.7	Newark Air Museum (XM685)★	
G-AYZK	Jodel DR.1050/M1	D. G. Hesketh	
G-AYZS	D.62B Condor	M. N. Thrush	
G-AYZU	Slingsby T.61A Falke	The Falcon Gliding Group/Elstree	
G-AYZW	Slingsby T.61A Falke	Y-ZW Group	
G-AZAB	PA-30 Twin Comanche 160B	Bickertons Aerodromes Ltd	
G-AZAJ	PA-28R Cherokee Arrow 200B	J. McHugh & P. Woulfe/Stapleford	
G-AZAW	GY-80 Horizon 160	J. W. Foley	
G-AZAZ	Bensen B.8M ★	F.A.A. Museum/Yeovilton	
G-AZBA	T.66 Nipper 3	L. A. Brown	
G-AZBB	MBB Bo 209 Monsun 160FV	G. N. Richardson/Staverton	
G-AZBE	Glos-Airtourer Super 150	BE Flying Group/Staverton	
G-AZBI	Jodel 150	F. M. Ward	
G-AZBL	Jodel D.9 Bebe	J. Hill	
G-AZBN	AT-16 Harvard IIB (FT391)	Swaygate Ltd/Shoreham	
G-AZBU	Auster AOP.9 (XR246)	Auster Nine Group	
G-AZBY	W.S.58 Wessex 60 Srs 1 ★	IHM/Weston-s-Mare	
G-AZBZ	W.S.58 Wessex 60 Srs 1 ★	IHM/Weston-s-Mare	
G-AZCB	SNCAN Stampe SV-4C	M. L. Martin	
G-AZCK	B.121 Pup 2	D. R. Newell	
G-AZCL	B.121 Pup 2	J. J. Watts & D. Fletcher	

Notes	Reg.	Type	Owner or Operator
	G-AZCN	B.121 Pup 2	D. M. Callaghan & ptnrs
	G-AZCP	B.121 Pup 1	T. J. Watson/Elstree
	G-AZCT	B.121 Pup 1	J. Coleman
	G-AZCU	B.121 Pup 1	A. A. Harris/Shobdon
	G-AZCV	B.121 Pup 2	N. R. W. Long/Elstree
	G-AZCZ	B.121 Pup 2	L. & J. M. Northover/Cardiff-Wales
	G-AZDA	B.121 Pup 1	B. D. Deubelbeiss
	G-AZDD	MBB Bo 209 Monsun 150FF	Double Delta Flying Group/Elstree
	G-AZDE	PA-28R Cherokee Arrow 200B	C. Wilson
	G-AZDG	B.121 Pup 2	D. J. Sage & J. R. Heaps
	G-AZDJ	PA-32 Cherokee Six 300	K. J. Mansbridge & D. C. Gibbs/Car
	G-AZDX	PA-28 Cherokee 180F	M. Cowan
	G-AZDY	D.H.82A Tiger Moth	J. B. Mills
	G-AZEE	M.S.880B Rallye Club	J. Shelton
	G-AZEF	Jodel D.120	D. Street
	G-AZEG	PA-28 Cherokee 140D	Ashley Gardner Flying Club Ltd
	G-AZEU	B.121 Pup 2	G. M. Moir/Egginton
	G-AZEV	B.121 Pup 2	C. J. Partridge
	G-AZEW	B.121 Pup 2	Dukeries
	G-AZEY	B.121 Pup 2	M. E. Reynolds
	G-AZFA	B.121 Pup 2	J. Smith/Sandown
	G-AZFC	PA-28 Cherokee 140D	WLS Flying Group
	G-AZFF	Jodel D.112	T. Mackey/Ireland
	G-AZFI	PA-28R Cherokee Arrow 200B	G-AZFI Ltd/Sherburn
	G-AZFM	PA-28R Cherokee Arrow 200B	P. J. Jenness
	G-AZFR	Cessna 401B	Harding Wragg/Blackpool
	G-AZGA	Jodel D.120	A. F. Vizoso
	G-AZGE	SNCAN Stampe SV-4A	M. R. L. Astor/Booker
	G-AZGF	B.121 Pup 2	K. Singh
	G-AZGJ	M.S.880B Rallye Club	P. Rose
	G-AZGL	M.S.894A Rallye Minerva	The Cambridge Aero Club Ltd
	G-AZGY	CP.301B Emeraude	D. J. Gibson
	G-AZGZ	D.H.82A Tiger Moth (NM181)	R. J. King
	G-AZHB	Robin HR.100/200B	P. Fenwick
	G-AZHC	Jodel D.112	Aerodel Flying Group
	G-AZHD	Slingsby T.61A Falke	N. J. Orchard-Armitage
	G-AZHE	Slingsby T.61B Falke	M. R. Shelton/Tatenhill
	G-AZHH	SA 102.5 Cavalier	D. W. Buckle
	G-AZHI	Glos-Airtourer Super 150	Flying Grasshoppers Ltd
	G-AZHJ	SA Twin Pioneer Srs 3 ★	Air Atlantique Ltd/Coventry
	G-AZHK	Robin HR.100/200B	D. J. Sage (G-ILEG)
	G-AZHR	Piccard Ax6 balloon	C. Fisher
	G-AZHT	AESL Airtourer (modified)	Aviation West Ltd/Glasgow
	G-AZHU	Luton LA-4A Minor	W. Cawrey/Netherthorpe
	G-AZIB	ST-10 Diplomate	W. B. Bateson/Blackpool
	G-AZID	Cessna FA.150L	Aerobat Ltd
	G-AZII	Jodel D.117A	J. S. Brayshaw
	G-AZIJ	Jodel DR.360	K. J. Fleming/Liverpool
	G-AZIK	PA-34-200 Seneca II	Walkbury Aviation Ltd
	G-AZIL	Slingsby T.61A Falke	D. W. Savage/Portmoak
	G-AZIO	SNCAN Stampe SV-4C	–/Booker (Lycoming) ★
	G-AZIP	Cameron O-65 balloon	Dante Balloon Group
	G-AZJC	Fournier RF-5	W. St. G. V. Stoney/Italy
	G-AZJE	Ord-Hume JB-01 Minicab	J. B. Evans/Sandown
	G-AZJN	Robin DR.300/140	Wright Farm Eggs Ltd
	G-AZJV	Cessna F.172L	R. P. Smith
	G-AZJY	Cessna FRA.150L	P. J. McCartney
	G-AZKC	M.S.880B Rallye Club	L. J. Martin/Redhill
	G-AZKE	M.S.880B Rallye Club	D. A. Thompson & J. D. Headlam/Germany
	G-AZKK	Cameron O-56 balloon	Gemini Balloon Group Gemini
	G-AZKO	Cessna F.337F	Willpower Garage Ltd
	G-AZKP	Jodel D.117	A. M. & J. L. Moar
	G-AZKR	PA-24 Comanche 180	J. Van Der Kwast
	G-AZKS	AA-1A Trainer	M. D. Henson
	G-AZKW	Cessna F.172L	J. C. C. Wright
	G-AZKZ	Cessna F.172L	R. D. & E. Forster/Swanton Morley
	G-AZLE	Boeing N2S-5 Kaydet (1102/102)	A. E. Paulson
	G-AZLF	Jodel D.120	M. S. C. Ball
	G-AZLH	Cessna F.150L	L. Papatheocharis & I. Buck
	G-AZLN	PA-28 Cherokee 180F	Liteflite Ltd/Kidlington
	G-AZLV	Cessna 172K	G-AZLV Flying Group
	G-AZLY	Cessna F.150L	M. Stewart

Reg.	Type	Owner or Operator	Notes
G-AZLZ	Cessna F.150L	A. G. Martlew	
G-AZMC	Slingsby T.61A Falke	G-AZMC Group	
G-AZMD	Slingsby T.61C Falke	R. A. Rice/Wellesbourne	
G-AZMJ	AA-5 Traveler	R. T. Love/Bodmin	
G-AZMN	Glos-Airtourer T.5	W. Crozier & I. Young	
G-AZMX	PA-28 Cherokee 140 ★	NE Wales Institute of Higher Education	
		(Instructional airframe)/Flintshire	
G-AZMZ	M.S.893A Rallye Commodore 150	D. R. Wilcox	
G-AZNK	SNCAN Stampe SV-4A	November Kilo Group	
G-AZNL	PA-28R Cherokee Arrow 200D	B. P. Liversidge	
G-AZNO	Cessna 182P	S. Turton	
G-AZOA	MBB Bo 209 Monsun 150FF	M. W. Hurst	
G-AZOB	MBB Bo 209 Monsun 150FF	G. N. Richardson/Staverton	
G-AZOE	Glos-Airtourer 115	G-AZOE 607 Group/Newcastle	
G-AZOF	Glos-Airtourer Super 150	Cirrus Flying Group/Denham	
G-AZOG	PA-28R Cherokee Arrow 200D	Southend Flying Club	
G-AZOL	PA-34-200 Seneca II	Stapleford Flying Club Ltd	
G-AZOS	Jurca MJ.5-H1 Sirocco	P. J. Tanulak	
G-AZOT	PA-34-200 Seneca II	Northern Aviation Ltd	
G-AZOU	Jodel DR.1050	Horsham Flying Group/Slinfold	
G-AZOZ	Cessna FRA.150L	Seawing Flying Club Ltd/Southend	
G-AZPA	PA-25 Pawnee 235	Black Mountains Gliding Club Ltd/Talgarth	
G-AZPC	Slingsby T.61C Falke	The Surrey Hills Gliding Club Ltd/Kenley	
G-AZPF	Fournier RF-5	R. Pye/Blackpool	
G-AZPH	Craft-Pitts S-1S Special ★	Science Museum/S. Kensington	
G-AZPV	Luton LA-4A Minor	J. R. Faulkner	
G-AZRA	MBB Bo 209 Monsun 150FF	Alpha Flying Ltd/Denham	
G-AZRD	Cessna 401B	Romeo Delta Group	
G-AZRH	PA-28 Cherokee 140D	Trust Flying Group	
G-AZRK	Fournier RF-5	A. B. Clymo & J. F. Rogers	
G-AZRL	PA-18 Super Cub 95	B. J. Stead	
G-AZRM	Fournier RF-5	Romeo Mike Group	
G-AZRN	Cameron O-84 balloon	C. J. Desmet/Belgium	
G-AZRP	Glos-Airtourer 115	B. F. Strawford/Shobdon	
G-AZRR	Cessna 310Q	M. A. Rooney	
G-AZRS	PA-22 Tri-Pacer 150	R. H. Hulls	
G-AZRZ	Cessna U.206F	Hinton Skydiving Centre	
G-AZSA	Stampe et Renard SV-4B	J. K. Faulkner/Biggin Hill	
G-AZSC	AT-16 Harvard IIB (43SC)	Machine Music Ltd/Duxford	
G-AZSD	Slingsby T.29B Motor Tutor	Essex Aviation	
G-AZSF	PA-28R Cherokee Arrow 200D	Wellesbourne Aviation	
G-AZSW	B.121 Pup 1	J. R. Parry	
G-AZSZ	PA-23 Aztec250D	A. A. Mattacks	
G-AZTA	MBB Bo 209 Monsun 150FF	K. T. Pierce	
G-AZTF	Cessna F.177RG	R. Burgun	
G-AZTK	Cessna F.172F	S. O'Ceallaigh	
G-AZTS	Cessna F.172L	C. E. Stringer	
G-AZTV	Stolp SA.500 Starlet	G. R. Rowland	
G-AZTW	Cessna F.177RG	I. M. Richmond	
G-AZUM	Cessna F.172L	Fowlmere Flyers	
G-AZUP	Cameron O-65 balloon	R. S. Bailey & A. B. Simpson	
G-AZUT	M.S.893A Rallye Commodore 180	J. Palethorpe	
G-AZUV	Cameron O-65 balloon ★	British Balloon Museum/Newbury	
G-AZUY	Cessna E.310L	W. B. Bateson/Blackpool	
G-AZUZ	Cessna FRA.150L	D. J. Parker/Netherthorpe	
G-AZVA	MBB Bo 209 Monsun 150FF	C. Elder	
G-AZVB	MBB Bo 209 Monsun 150FF	E. & P. M. L. Cliffe	
G-AZVF	M.S.894A Rallye Minerva	Minerva Flying Group	
G-AZVG	AA-5 Traveler	Whelan Building & Development Ltd	
G-AZVH	M.S.894A Rallye Minerva	P. L. Jubb	
G-AZVI	M.S.892A Rallye Commodore	R. E. Knapton	
G-AZVJ	PA-34-200 Seneca II	Andrews Professional Colour Laboratories Ltd/Lydd	
G-AZVL	Jodel D.119	Forest Flying Group/Stapleford	
G-AZVM	Hughes 369HS	GTS Engineering (Coventry) Ltd	
G-AZVP	Cessna F.177RG	Cardinal Flyers Ltd	
G-AZWB	PA-28 Cherokee 140	B. N. Rides & L. Connor	
G-AZWD	PA-28 Cherokee 140	Solent School of Flying	
G-AZWE	PA-28 Cherokee 140	G-AZWE Flying Group	
G-AZWF	SAN Jodel DR.1050	Cawdor Flying Group	
G-AZWS	PA-28R Cherokee Arrow 180	Arrow 88 Flying Group/Newcastle	
G-AZWT	Westland Lysander IIIA (V9367)	The Shuttleworth Collection/O. Warden	
G-AZWY	PA-24 Comanche 260	Keymer Son & Co Ltd/Biggin Hill	

Notes	Reg.	Type	Owner or Operator
	G-AZXB	Cameron O-65 balloon	R. J. Mitchener & P. F. Smart
	G-AZXC	Cessna F.150L	D. C. Bonsall
	G-AZXD	Cessna F.172L	Birdlake Ltd/Wellesbourne
	G-AZXG	PA-23 Aztec 250D ★	Instructional airframe/Cranfield
	G-AZYA	GY-80 Horizon 160	P. J. Fahie
	G-AZYB	Bell 47H-1 ★	IHM/Weston-s-Mare
	G-AZYD	M.S.893A Rallye Commodore	Storey Aviation Services
	G-AZYS	CP.301C-1 Emeraude	C. G. Ferguson & D. Drew
	G-AZYU	PA-23 Aztec 250E	L. J. Martin/Biggin Hill
	G-AZYY	Slingsby T.61A Falke	J. A. Towers
	G-AZYZ	WA.51A Pacific	C. R. Buxton/France
	G-AZZH	Practavia Pilot Sprite	A. Moore
	G-AZZO	PA-28 Cherokee 140	R. J. Hind/Elstree
	G-AZZR	Cessna F.150L	A. J. Hobbs
	G-AZZS	PA-34-200 Seneca II	Robin Cook Aviation/Shoreham
	G-AZZT	PA-28 Cherokee 180 ★	Ground instruction airframe/Cranfield
	G-AZZV	Cessna F.172L	Cristal Air Ltd
	G-AZZZ	D.H.82A Tiger Moth	S. W. McKay
	G-BAAD	Evans Super VP-1	Breighton VP-1 Group
	G-BAAF	Manning-Flanders MF1 (replica)	Aviation Film Services Ltd/Booker
	G-BAAI	M.S.893A Rallye Commodore	R. D. Taylor/Thruxton
	G-BAAT	Cessna 182P	British Skysports Paracentre
	G-BAAU	Enstrom F-28A-UK	G. Firbank
	G-BAAW	Jodel D.119	Alpha Whiskey Flying Group
	G-BABB	Cessna F.150L	Seawing Flying Club Ltd/Southend
	G-BABC	Cessna F.150L	Fordaire Ltd/Sywell
	G-BABD	Cessna FRA.150L (modified)	Anglia Flight
	G-BABE	Taylor JT.2 Titch	M. Bonsall/Netherthorpe
	G-BABG	PA-28 Cherokee 180	Mendip Flying Group/Bristol
	G-BABH	Cessna F.150L	D. B. Ryder & Co Ltd
	G-BABK	PA-34-200 Seneca II	D. F. J. Flashman/Biggin Hill
	G-BACB	PA-34-200 Seneca II	Milbrooke Motors
	G-BACC	Cessna FRA.150L	P. J. Dalby & P. Huckle
	G-BACE	Fournier RF-5	W. B. Hosie
	G-BACJ	Jodel D.120	Wearside Flying Association/Newcastle
	G-BACL	Jodel 150	M. L. Sargeant/Biggin Hill
	G-BACN	Cessna FRA.150L	F. Bundy
	G-BACO	Cessna FRA.150L	M. A. McLoughlin
	G-BACP	Cessna FRA.150L	M. Markwick
	G-BADC	Rollason Beta B.2A	D. H. Greenwood
	G-BADH	Slingsby T.61A Falke	A. P. Askwith
	G-BADJ	PA-E23 Aztec 250E	K. A. W. Ashcroft
	G-BADM	D.62B Condor	D. J. Wilson
	G-BADW	Pitts S-2A Special	R. E. Mitchell/Cosford
	G-BAEB	Robin DR.400/160	R. Hatton
	G-BAEC	Robin HR.100/210	Datacorp Enterprises Pty Ltd
	G-BAEE	Jodel DR.1050/M1	R. Little
	G-BAEM	Robin DR.400/125	M. A. Webb/Booker
	G-BAEN	Robin DR.400/180	European Soaring Club Ltd
	G-BAEO	Cessna F.172M	L. W. Scattergood
	G-BAEP	Cessna FRA.150L (modified)	A. M. Lynn
	G-BAER	Cosmic Wind	R. S. Voice/Redhill
	G-BAET	Piper J-3C-65 Cub	C. J. Rees
	G-BAEU	Cessna F.150L	L. W. Scattergood
	G-BAEV	Cessna FRA.L150L	T. J. Richardson
	G-BAEW	Cessna F.172M ★	Westley Aircraft/Cranfield
	G-BAEY	Cessna F.172M	Skytrax Aviation Ltd
	G-BAEZ	Cessna FRA.150L	Donair Flying Club Ltd/E. Midlands
	G-BAFA	AA-5 Traveler	C. F. Mackley/Stapleford
	G-BAFG	D.H.82A Tiger Moth	D. Watt
	G-BAFL	Cessna 192P	M. Langhammer
	G-BAFP	Robin DR.400/160	A. S. Langdale & J. Bevis-Lawson
	G-BAFT	PA-18 Super Cub 150	T. J. Wilkinson/Riseley
	G-BAFU	PA-28 Cherokee 140	D. Matthews
	G-BAFV	PA-18 Super Cub 95	T. F. & S. J. Thorpe
	G-BAFW	PA-28 Cherokee 140	R. S. Chance
	G-BAFX	Robin DR.400/140	R. Foster
	G-BAGB	SIAI-Marchetti SF.260	British Midland Airways Ltd/E. Midlands
	G-BAGC	Robin DR.400/140	W. P. Nutt
	G-BAGE	Cessna T.210L ★	Aeroplane Collection Ltd
	G-BAGF	Jodel D.92 Bebe	E. Evans

Reg.	Type	Owner or Operator	Notes
G-BAGG	PA-32 Cherokee Six 300E	Channel Islands Aero Club (Jersey) Ltd	
G-BAGL	SA.341G Gazelle Srs 1	Foremans Aviation Ltd	
G-BAGN	Cessna F.177RG	R. W. J. Andrews	
G-BAGO	Cessna 421B	Golden Aviation Ltd	
G-BAGR	Robin DR.400/140	J. D. Last	
G-BAGS	Robin DR.400/180 2+2	M. Whale & M. W. A. Lunn	
G-BAGT	Helio H.295 Courier	B. J. C. Woodall Ltd	
G-BAGV	Cessna U.206F	Scottish Parachute Club/Strathallan	
G-BAGX	PA-28 Cherokee 140	Golf X-Ray Group	
G-BAGY	Cameron O-84 balloon	P. G. Dunnington	
G-BAHD	Cessna 182P Skylane	Lambley Flying Group	
G-BAHE	PA-28 Cherokee 140	M. W. Kilvert & A. O. Jones	
G-BAHF	PA-28 Cherokee 140	BJ Services (Midlands) Ltd	
G-BAHH	Wallis WA-121	K. H. Wallis	
G-BAHI	Cessna F.150H	MJP Aviation & Sales	
G-BAHJ	PA-24 Comanche 250	K. Cooper	
G-BAHL	Robin DR.400/160	Robin Group	
G-BAHO	Beech C.23 Sundowner	G-BAHO Group	
G-BAHP	Volmer VJ.22 Sportsman	Seaplane Group	
G-BAHS	PA-28R Cherokee Arrow 200-II	A. R. N. Morris	
G-BAHX	Cessna 182P	Dupost Group	
G-BAIG	PA-34-200-2 Seneca	Mid-Anglia School of Flying	
G-BAIH	PA-28R Cherokee Arrow 200-II	M. G. West	
G-BAII	Cessna FRA.150L	Cornwall Flying Club Ltd/Bodmin	
G-BAIK	Cessna F.150L	Wickenby Aviation Ltd	
G-BAIN	Cessna FRA.150L	S. J. Windle/Bodmin	
G-BAIP	Cessna F.150L	G. & S. A. Jones	
G-BAIS	Cessna F.177RG	Cardinal Syndicate	
G-BAIW	Cessna F.172M	W. J. Greenfield/Humberside	
G-BAIX	Cessna F.172M	R. A. Nichols/Elstree	
G-BAIZ	Slingsby T.61A Falke	Falke Syndicate/Hinton-in-the-Hedges	
G-BAJA	Cessna F.177RG	D. W. Ward	
G-BAJB	Cessna F.177RG	LDJ Ltd	
G-BAJC	Evans VP-1	S. J. Greer	
G-BAJE	Cessna 177	D. M. Dawson	
G-BAJN	AA-5 Traveler	I. M. Snelson	
G-BAJO	AA-5 Traveler	S. Bradshaw	
G-BAJR	PA-28 Cherokee 180	Belfast Flying Club Ltd	
G-BAJY	Robin DR.400/180	L. J. Murray	
G-BAJZ	Robin DR.400/125	Rochester Aviation Ltd	
G-BAKD	PA-34-200 Seneca II	Andrews Professional Colour Laboratories/Elstree	
G-BAKH	PA-28 Cherokee 140	Marnham Investments Ltd	
G-BAKJ	PA-30 Twin Comanche 160B	G. D. Colover & ptnrs	
G-BAKK	Cessna F.172H ★	*Parachute jump trainer*/Hinton-in-the-Hedges	
G-BAKM	Robin DR.400/140	D. V. Pieri	
G-BAKN	SNCAN Stampe SV-4C	M. Holloway	
G-BAKR	Jodel D.117	R. W. Brown	
G-BAKV	PA-18 Super Cub 150	Western Air (Thruxton) Ltd & F. Taylor	
G-BAKW	B.121 Pup 2	H. Beavan	
G-BAKY	Slingsby T.61C Falke	Buckminster Gliding Club Ltd/Saltby	
G-BALF	Robin DR.400/140	G. & D. A. Wasey	
G-BALG	Robin DR.400/180	R. Jones	
G-BALH	Robin DR.400/140B	G-BALH Flying Group	
G-BALI	Robin DR.400 2+2	A. Brinkley	
G-BALJ	Robin DR.400/180	D. A. Bett & D. de Lacey-Rowe	
G-BALN	Cessna T.310Q	O'Brien Properties Ltd/Shoreham	
G-BALZ	Bell 212	Bristow Helicopters Ltd	
G-BAMB	Slingsby T.61C Falke	Flying Group G-BAMB	
G-BAMC	Cessna F.150L	K. Evans	
G-BAMF	MBB Bo 105D	Bond Air Services/Aberdeen	
G-BAMJ	Cessna 182P	A. E. Kedros	
G-BAMK	Cameron D-96 airship ★	British Balloon Museum	
G-BAMM	PA-28 Cherokee 235	Group 235	
G-BAMR	PA-16 Clipper	H. Royce	
G-BAMS	Robin DR.400/160	G-BAMS Ltd/Headcorn	
G-BAMU	Robin DR.400/160	The Alternative Flying Group	
G-BAMV	Robin DR.400/180	K. Jones & E. A. Anderson/Booker	
G-BAMY	PA-28R Cherokee Arrow 200-II	G-BAMY Group/Birmingham	
G-BANA	Robin DR.221	G. T. Pryor	
G-BANB	Robin DR.400/180	D. R. L. Jones	
G-BANC	GY-201 Minicab	C. R. Shipley	
G-BANU	Wassmer Jodel D.120	W. M. & C. H. Kilner	

Notes	Reg.	Type	Owner or Operator
	G-BANV	Phoenix Currie Wot	K. Knight
	G-BANW	CP.1330 Super Emeraude	P. S. Milner
	G-BANX	Cessna F.172M	Oakfleet 2000 Ltd
	G-BAOB	Cessna F.172M	S. O. Smith & R. H. Taylor/Andrewsfield
	G-BAOH	M.S.880B Rallye Club	A. P. Swain
	G-BAOJ	M.S.880B Rallye Club	R. E. Jones
	G-BAOM	M.S.880B Rallye Club	P. J. D. Feehan
	G-BAOP	Cessna FRA.150L	R. D. Forster
	G-BAOS	Cessna F.172M	Wingtask 1995 Ltd
	G-BAOU	AA-5 Traveler	R. C. Mark
	G-BAPB	D.H.C.1 Chipmunk 22	G. V. Bunyan
	G-BAPI	Cessna FRA.150L	Industrial Supplies (Peterborough) Ltd/Sibson
	G-BAPJ	Cessna FRA.150L	M. D. Page/Manston
	G-BAPL	PA-23 Turbo Aztec 250E	Donington Aviation Ltd/E. Midlands
	G-BAPR	Jodel D.11	J. B. Liber & J. F. M. Bartlett
	G-BAPS	Campbell Cougar ★	IHM/Weston-s-Mare
	G-BAPV	Robin DR.400/160	J. D. & M. Millne/Newcastle
	G-BAPW	PA-28R Cherokee Arrow 180	IA. G. Bourne
	G-BAPX	Robin DR.400/160	G-BAPX Group
	G-BAPY	Robin HR.100/210	D. M. Hansell
	G-BARC	Cessna FR.172J	Severn Valley Aviation Group
	G-BARF	Jodel D.112 Club	G. P. Jewell
	G-BARG	Cessna E.310Q	Tibus Aviation Ltd
	G-BARH	Beech C.23 Sundowner	G. Moorby & J. Hinchcliffe
	G-BARN	Taylor JT.2 Titch	R. G. W. Newton
	G-BARP	Bell 206B JetRanger 2	S.W. Electricity Board/Bristol
	G-BARS	D.H.C.1 Chipmunk 22 (1377)	J. Beattie/Yeovilton
	G-BARV	Cessna 310Q	Old England Watches Ltd/Elstree
	G-BARZ	Scheibe SF.28A Tandem Falke	K. Kiely
	G-BASH	AA-5 Traveler	BASH Flying Group
	G-BASJ	PA-28 Cherokee 180	Bristol Aero Club/Filton
	G-BASL	PA-28 Cherokee 140	Justgold Ltd
	G-BASM	PA-34-200 Seneca II	M. Gipps & J. R. Whetlor
	G-BASN	Beech C.23 Sundowner	J. Greenwood
	G-BASO	Lake LA-4 Amphibian	C. J. A. Macauley
	G-BASP	B.121 Pup 1	B. J. Coutts/Sywell
	G-BASX	PA-34-200 Seneca II	Air Consul SL/Spain
	G-BATC	MBB Bo 105D	Bond Air Services/Aberdeen
	G-BATJ	Jodel D.119	D. J. & K. S. Thomas
	G-BATN	PA-23 Aztec 250E	Marshall of Cambridge Ltd
	G-BATR	PA-34-200 Seneca II	Falcon Flying Services/Biggin Hill
	G-BATV	PA-28 Cherokee 180D	J. N. Rudsdale
	G-BATW	PA-28 Cherokee 140	C. D. Sainsbury
	G-BAUC	PA-25 Pawnee 235	Southdown Gliding Club Ltd/Parham Park
	G-BAUH	Jodel D.112	G. A. & D. Shepherd
	G-BAUZ	SNCAN NC.854S	W. A. Ashley & D. Horne
	G-BAVB	Cessna F.172M	Taylor Aviation Ltd
	G-BAVH	D.H.C.1 Chipmunk 22	Portsmouth Naval Gliding Club/Lee-on-Solent
	G-BAVL	PA-23 Aztec 250E	S. P. & A. V. Chillott
	G-BAVO	Boeing Stearman N2S (26)	(Stored)
	G-BAVR	AA-5 Traveler	G. E. Murray
	G-BAWG	PA-28R Cherokee Arrow 200-II	Solent Air Ltd
	G-BAWK	PA-28 Cherokee 140	Newcastle-upon-Tyne Aero Club Ltd
	G-BAWR	Robin HR.100/210	T. Taylor
	G-BAXE	Hughes 269A	Reethorpe Engineering Ltd
	G-BAXJ	PA-32 Cherokee Six 300B	UK Parachute Services/Stirling
	G-BAXK	Thunder Ax7-77 balloon ★	A. R. Snook
	G-BAXS	Bell 47G-5	RK Helicopters
	G-BAXU	Cessna F.150L	M. W. Sheppardson
	G-BAXV	Cessna F.150L	G. & S. A. Jones
	G-BAXY	Cessna F.172M	Eaglesoar Ltd
	G-BAXZ	PA-28 Cherokee 140	G-BAXZ (87) Syndicate
	G-BAYL	SNCAN Nord 1101 Norecrin ★	(stored)/Chirk
	G-BAYO	Cessna 150L	Messrs Rees of Poyston West
	G-BAYP	Cessna 150L	Yankee Papa Flying Group
	G-BAYR	Robin HR.100/210	P. Chamberlain
	G-BAYV	SNCAN 1101 Noralpha (3+) ★	Macclesfield Historical Aviation Soc/Barton
	G-BAZC	Robin DR.400/160	Southern Sailplanes Ltd/Membury
	G-BAZJ	HPR-7 Herald 209 ★	Guernsey Airport Fire Services
	G-BAZM	Jodel D.11	A. F. Simpson
	G-BAZS	Cessna F.150L	L. W. Scattergood
	G-BAZT	Cessna F.172M	Exeter Flying Club Ltd

Reg.	Type	Owner or Operator	Notes
G-BBAW	Robin HR.100/210	J. R. Williams	
G-BBAX	Robin DR.400/140	G. J. Bissex & P. H. Garbutt	
G-BBAY	Robin DR.400/140	Rothwell Group	
G-BBBC	Cessna F.150L	W. J. Greenfield	
G-BBBI	AA-5 Traveler	Go Baby Aviation Group	
G-BBBN	PA-28 Cherokee 180	Estuary Aviation Ltd	
G-BBBO	SIPA 903	G. K. Brothwood/Liverpool	
G-BBBW	FRED Srs 2	M. Palfreman	
G-BBBX	Cessna 310L	Atlantic Air Transport Ltd/Coventry	
G-BBBY	PA-28 Cherokee 140	S. J. Mount	
G-BBCA	Bell 206B JetRanger 2	Heliflight (UK) Ltd/Wolverhampton	
G-BBCC	PA-23 Aztec 250D	County Garage (Cheltenham) Ltd	
G-BBCH	Robin DR.400/2+2	Charlie Hotel Syndicate	
G-BBCI	Cessna 150H	A. M. & F. Alam	
G-BBCK	Cameron O-77 balloon	W. R. Teasdale	
G-BBCN	Robin HR.100/210	Gloucestershire Flying Club	
G-BBCS	Robin DR.400/140	A. N. Kaschevski	
G-BBCY	Luton LA-4A Minor	Wingnuts	
G-BBCZ	AA-5 Traveler	A. D. Massey	
G-BBDC	PA-28 Cherokee 140	G-BBDC Group/Andrewsfield	
G-BBDE	PA-28R Cherokee Arrow 200-II	R. L. Coleman & ptnrs	
G-BBDG	Concorde 100 ★	BAE Systems. *Stored* Filton	
G-BBDH	Cessna F.172M	J. D. Woodward	
G-BBDJ	Thunder Ax6-56 balloon	Balloon Preservation Flying Group	
G-BBDL	AA-5 Traveler	Delta Lima Flying Group	
G-BBDM	AA-5 Traveler	Jackeroo Aviation Group	
G-BBDO	PA-23 Turbo Aztec 250E	J. W. Anstee/Bristol	
G-BBDP	Robin DR.400/160	Robin Lance Aviation Associates Ltd	
G-BBDS	PA-31 Turbo Navajo	Elham Valley Aviation Ltd (G-SKKB)	
G-BBDT	Cessna 150H	Delta Tango Group	
G-BBDV	SIPA S.903	W. McAndrew	
G-BBEA	Luton LA-4 Minor	M. Horner	
G-BBEB	PA-28R Cherokee Arrow 200-II	R. D. Rippingale/Thruxton	
G-BBEC	PA-28 Cherokee 180	A. A. Gardner	
G-BBED	M.S.894A Rallye Minerva 220	Vista Products	
G-BBEF	PA-28 Cherokee 140	Liberty Group Assets Ltd	
G-BBEN	Bellanca 7GCBC Citabria	C. A. G. Schofield	
G-BBEX	Cessna 185A	V. M. McCarthy	
G-BBEY	PA-23 Aztec 250E	D. Nicholls	
G-BBFD	PA-28R Cherokee Arrow 200-II	C. H. Rose & A. R. Annable	
G-BBFL	GY-201 Minicab	R. Smith	
G-BBFV	PA-32 Cherokee Six 260	G-BBFV Syndicate	
G-BBGC	M.S.893E Rallye 180GT	P. M. Nolan	
G-BBGI	Fuji FA.200-160	Tandycel Co Ltd	
G-BBGL	Baby Great Lakes	F. Ball	
G-BBGR	Cameron O-65 balloon	M. L. & L. P. Willoughby	
G-BBGX	Cessna 182P Skylane	GX Group	
G-BBGZ	Cambridge Hot Air CHABA ★	British Balloon Museum Balloon Association	
G-BBHE	Enstrom F-28A	Clarke Aviation Ltd	
G-BBHF	PA-23 Aztec 250E	G. J. Williams/Sherburn	
G-BBHI	Cessna 177RG	T. G. W. Bunce	
G-BBHJ	Piper J-3C-65 Cub	R. V. Miller & J. Stanbridge	
G-BBHK	AT-16 Harvard IIB (212540/RF-40)	Bob Warner Aviation/Exeter	
G-BBHL	Sikorsky S-61N Mk II	Bristow Helicopters Ltd Glamis	
G-BBHY	PA-28 Cherokee 180	Air Operations Ltd/Guernsey	
G-BBIA	PA-28R Cherokee Arrow 200-II	G. H. Kilby/Stapleford	
G-BBIF	PA-23 Aztec 250E	D. M. Davies	
G-BBIH	Enstrom F-28A-UK	Stephenson Marine Co Ltd	
G-BBII	Fiat G-46-3B (4-97/MM52801)	Godshill Aviation/Sandown	
G-BBIL	PA-28 Cherokee 140	John West Consulting Ltd	
G-BBIO	Robin HR.100/210	R. A. King/Headcorn	
G-BBIX	PA-28 Cherokee 140	Sterling Aviation Ltd	
G-BBJB	Thunder Ax7-77 balloon	St Crispin Balloon Group	
G-BBJI	Isaacs Spitfire (RN218)	J. D. Bally	
G-BBJU	Robin DR.400/140	J. C. Lister	
G-BBJV	Cessna F.177RG	3grcomm Ltd	
G-BBJX	Cessna F.150L	L. W. Scattergood	
G-BBJY	Cessna F.172M	R. Windley	
G-BBJZ	Cessna F.172M	Burks, Green & ptnrs	
G-BBKA	Cessna F.150L	W. M. Wilson & R. L. Campbell	
G-BBKB	Cessna F.150L	Justgold Ltd/Blackpool	
G-BBKE	Cessna F.150L	Taylor Aviation Ltd	

Notes	Reg.	Type	Owner or Operator
	G-BBKF	Cessna FRA.150L	D. W. Mickleburgh
	G-BBKG	Cessna FR.172J	R. Wright
	G-BBKI	Cessna F.172M	C. W. & S. A. Burman
	G-BBKL	CP.301A Emeraude	Piel G-BBKL
	G-BBKR	Scheibe SF.24A Motorspatz	P. I. Morgans
	G-BBKU	Cessna FRA.150L	S. J. Windle
	G-BBKX	PA-28 Cherokee 180	DRA Flying Club Ltd/Farnborough
	G-BBKY	Cessna F.150L	Telesonic Ltd/Barton
	G-BBKZ	Cessna 172M	KZ Flying Group/Exeter
	G-BBLH	Piper J-3C-65 Cub (31145)	Shipping & Airlines Ltd/Biggin Hill
	G-BBLL	Cameron O-84 balloon ★	British Balloon Museum/Newbury
	G-BBLM	SOCATA Rallye 100S	J. R. Rodgers
	G-BBLS	AA-5 Traveler	A. D. Grant
	G-BBLU	PA-34-200 Seneca II	A. Phillips
	G-BBMB	Robin DR.400/180	Regent Flying Group
	G-BBMH	EAA. Sports Biplane Model P.1	I. S. Parker
	G-BBMJ	PA-23 Aztec 250E	Nationwide Caravan Rental Services Ltd
	G-BBMN	D.H.C.1 Chipmunk 22	R. Steiner/Rush Green
	G-BBMO	D.H.C.1 Chipmunk 22 (WK514)	D. M. Squires/Wellesbourne
	G-BBMR	D.H.C.1 Chipmunk 22 (WB763)	P. J. Wood
	G-BBMT	D.H.C.1 Chipmunk 22	J. Evans & D. Withers
	G-BBMV	D.H.C.1 Chipmunk 22 (WG348)	P. J. Morgan (Aviation) Ltd
	G-BBMW	D.H.C.1 Chipmunk 22 (WK628)	Mike Whiskey Group
	G-BBMX	D.H.C.1 Chipmunk 22	K. A. Doornbos/Netherlands
	G-BBMZ	D.H.C.1 Chipmunk 22	Wycombe Gliding School Syndicate/Booker
	G-BBNA	D.H.C.1 Chipmunk 22 (Lycoming)	Coventry Gliding Club Ltd/Husbands Bosworth
	G-BBNC	D.H.C.1 Chipmunk T.10 (WP790) ★	De Havilland Heritage Museum
	G-BBND	D.H.C.1 Chipmunk 22 (WD286)	Bernoulli Syndicate
	G-BBNG	Bell 206B JetRanger 2	Helicopter Crop Spraying Ltd
	G-BBNH	PA-34-200 Seneca II	M. G. D. Baverstock & ptnrs/Bournemouth
	G-BBNI	PA-34-200 Seneca II	Noisy Moose Ltd
	G-BBNJ	Cessna F.150L	Sherburn Aero Club Ltd
	G-BBNO	PA-23 Aztec 250E ★	(stored)/Biggin Hill
	G-BBNZ	Cessna F.172M	R. J. Nunn
	G-BBOA	Cessna F.172M	J. D & A. M. Black
	G-BBOC	Cameron O-77 balloon	J. A. B. Gray
	G-BBOD	Thunder O-45 balloon	B. R. & M. Boyle
	G-BBOH	Pitts S-1S Special	Venom Jet Promotions Ltd/Bournemouth
	G-BBOJ	PA-23 Aztec 250E ★	Instructional airframe/Cranfield
	G-BBOL	PA-18 Super Cub 150	N. Artt
	G-BBOO	Thunder Ax6-56 balloon	K. Meehan Tigerjack
	G-BBOR	Bell 206B JetRanger 2	M. J. Easey
	G-BBOX	Thunder Ax7-77 balloon	R. C. Weyda
	G-BBPN	Enstrom F-28A-UK	Smarta Systems Ltd
	G-BBPO	Enstrom F-28A-UK	Henfield Lodge Aviation Ltd
	G-BBPS	Jodel D.117	A. Appleby/Redhill
	G-BBPX	PA-34-200 Seneca II	Richel Investments Ltd/Guernsey
	G-BBPY	PA-28 Cherokee 180	Sunsaver Ltd
	G-BBRA	PA-23 Aztec 250D	R. C. Lough/Elstree
	G-BBRB	D.H.82A Tiger Moth (DF198)	R. Barham/Biggin Hill
	G-BBRC	Fuji FA.200-180	BBRC Ltd/Blackbushe
	G-BBRI	Bell 47G-5A	Alan Mann Helicopters Ltd/Fairoaks
	G-BBRN	Procter Kittiwake 1 (XW784/VL)	R. de H. Dobree-Carey/Yeovilton
	G-BBRV	D.H.C.1 Chipmunk 22 (WD347)	Yorkshire Vintage Flying Ltd
	G-BBRX	SIAI-Marchetti S.205-18F	R. C. West
	G-BBRZ	AA-5 Traveler	C. P. Osbourne
	G-BBSA	AA-5 Traveler	Usworth 84 Flying Associates Ltd
	G-BBSB	Beech C23 Sundowner	L. J. Welsh
	G-BBSM	PA-32 Cherokee Six 300E	G. C. Collings
	G-BBSS	D.H.C.1A Chipmunk 22	Coventry Gliding Club Ltd/Husbands Bosworth
	G-BBSW	Pietenpol Air Camper	J. K. S. Wills
	G-BBTB	Cessna FRA.150L	BBC Air Ltd/Compton Abbas
	G-BBTG	Cessna F.172M	R. W. & V. P. J. Simpson/Redhill
	G-BBTH	Cessna F.172M	Tayside Aviation Ltd
	G-BBTJ	PA-23 Aztec 250E	Cooper Aerial Surveys Ltd/Sandtoft
	G-BBTK	Cessna FRA.150L	Cleveland Flying School Ltd/Teesside
	G-BBTS	Beech V35B Bonanza	Eastern Air
	G-BBTU	ST-10 Diplomate	D. Hayden-Wright
	G-BBTY	Beech C23 Sundowner	A. W. Roderick & W. Price/Cardiff-Wales
	G-BBTZ	Cessna F.150L	Marnham Investments Ltd
	G-BBUE	AA-5 Traveler	G. A. Chadfield
	G-BBUF	AA-5 Traveler	S. & A. F. Williams

Reg.	Type	Owner or Operator
G-BBUG	PA-16 Clipper	J. Dolan
G-BBUJ	Cessna 421B	Coolflourish Ltd
G-BBUT	Western O-65 balloon	R. G. Turnbull
G-BBUU	Piper J-3C-65 Cub	C. Stokes
G-BBUW	SA.102.5 Cavalier ★	Aeroplane Collection Ltd
G-BBVA	Sikorsky S-61N Mk 2	Bristow Helicopters Ltd
G-BBVF	SA Twin Pioneer Srs 3 ★	Museum of Flight/E. Fortune
G-BBVG	PA-23 Aztec 250C ★	(stored)/Little Staughton
G-BBVJ	Beech B24R Sierra	E. J. Berek
G-BBVO	Isaacs Fury II (S1579)	J. Moore
G-BBWZ	AA-1B Trainer	W. C. Smeaton
G-BBXB	Cessna FRA.150L	M. G. & E. A. Sweet
G-BBXH	Cessna FR.172F	D. Ridley
G-BBXK	PA-34-200 Seneca	Poyston Aviation
G-BBXL	Cessna 310Q	Appleton Aviation Ltd
G-BBXS	Piper J-3C-65 Cub	M. J. Butler (G-ALMA)/Langham
G-BBXW	PA-28-151 Cherokee Warrior	Trek Air BV
G-BBXY	Bellanca 7GCBC Citabria	R. R. L. Windus
G-BBXZ	Evans VP-1	R. W. Burrows
G-BBYB	PA-18 Super Cub 95	Tiger Club (1990) Ltd/Headcorn
G-BBYH	Cessna 182P	Ramco (UK) Ltd
G-BBYM	H.P.137 Jetstream 200 ★	Aerospace Museum (G-AYWR)/Cosford
G-BBYP	PA-28 Cherokee 140	Eurocharter Aviation Ltd/Sheffield City
G-BBYS	Cessna 182P Skylane	I. M. Jones
G-BBYU	Cameron O-56 balloon	British Balloon Museum
G-BBZF	PA-28 Cherokee 140	East Coast Aviation
G-BBZH	PA-28R Cherokee Arrow 200-II	ZH Flying Ltd/Exeter
G-BBZJ	PA-34-200 Seneca II	Falcon Flying Services
G-BBZN	Fuji FA.200-180	D. Kynaston & ptnrs
G-BBZO	Fuji FA.200-160	M. S. Bird
G-BBZV	PA-28R Cherokee Arrow 200-II	P. B. Mellor/Kidlington
G-BCAH	D.H.C.1 Chipmunk 22 (WG316)	A. W. Eldridge
G-BCAP	Cameron O-56 balloon ★	Balloon Preservation Group/Lancing
G-BCAR	Thunder Ax7-77 balloon ★	British Balloon Museum/Newbury
G-BCAZ	PA-12 Super Cruiser	A. D. Williams
G-BCBG	PA-23 Aztec 250E	M. J. L. Batt
G-BCBH	Fairchild 24R-46A Argus III	Dreamticket Promotions Ltd
G-BCBJ	PA-25 Pawnee 235	Deeside Gliding Club (Aberdeenshire) Ltd/Aboyne
G-BCBL	Fairchild 24R-46A Argus III (HB751)	F. J. Cox
G-BCBR	AJEP/Wittman W.8 Tailwind	D. P. Jones
G-BCBX	Cessna F.150L	N. F. O'Neill
G-BCBZ	Cessna 337C	Envirovac 2000 Ltd
G-BCCC	Cessna F.150L	Triple Charlie Flying Group/Cranfield
G-BCCD	Cessna F.172M	Austin Aviation Ltd
G-BCCE	PA-23 Aztec 250E	Golf Charlie Echo Ltd/Shoreham
G-BCCF	PA-28 Cherokee 180	Topcat Aviation Ltd/Manchester
G-BCCG	Thunder Ax7-65 balloon	N. H. Ponsford
G-BCCJ	AA-5 Traveler	T. Needham/Woodford
G-BCCK	AA-5 Traveler	Prospect Air Ltd/Barton
G-BCCR	CP.301A Emeraude (modified)	J. H. & C. J. Waterman
G-BCCX	D.H.C.1 Chipmunk 22 (Lycoming)	RAFGSA/Dishforth
G-BCCY	Robin HR.200/100	Charlie Yankee Ltd/Filton
G-BCDJ	PA-28 Cherokee 140	R. J. Whyham
G-BCDK	Partenavia P.68B	Flyteam Aviation Ltd/Elstree
G-BCDL	Cameron O-42 balloon	D. P. & Mrs B. O. Turner Chums
G-BCDN	F.27 Friendship Mk 200 ★	Instructional airframe/Norwich
G-BCDY	Cessna FRA.150L	I. R. Chaplin
G-BCEA	Sikorsky S-61N Mk II	British International Ltd
G-BCEB	Sikorsky S-61N Mk II	Veritair Ltd
G-BCEE	AA-5 Traveler	W. A. L. Mitchell
G-BCEF	AA-5 Traveler	Enstone Flyers
G-BCEN	BN-2A-26 Islander	Atlantic Air Transport Ltd/Coventry
G-BCEP	AA-5 Traveler	G. Edelmann
G-BCER	GY-201 Minicab	D. Beaumont/Sherburn
G-BCEX	PA-23 Aztec 250E	DJ Aviation Ltd
G-BCEY	D.H.C.1 Chipmunk 22 (WG465)	Gopher Flying Group
G-BCEZ	Cameron O-84 balloon	Balloon Collection
G-BCFD	West balloon ★	British Balloon Museum Hellfire/Newbury
G-BCFF	Fuji FA-200-160	C. D. Burleigh
G-BCFO	PA-18 Super Cub 150	Portsmouth Naval Gliding Club/Lee-on-Solent

Notes	Reg.	Type	Owner or Operator
	G-BCFR	Cessna FRA.150L	Bulldog Aviation Ltd & Motorhoods Colchester Ltd/Earls Colne
	G-BCFU	Thunder Ax6-56 balloon ★	British Balloon Museum/Newbury
	G-BCFW	SAAB 91D Safir	D. R. Williams
	G-BCFY	Luton LA-4A Minor	G. Capes
	G-BCGB	Bensen B.8	J. W. Birkett
	G-BCGC	D.H.C.1 Chipmunk 22 (WP903)	J. C. Wright
	G-BCGH	SNCAN NC.854S	Nord Flying Group
	G-BCGI	PA-28 Cherokee 140	J. C. Dodd & ptnrs/Panshanger
	G-BCGJ	PA-28 Cherokee 140	Halegreen Associates Ltd & Demero Ltd
	G-BCGM	Jodel D.120	J. Pool
	G-BCGN	PA-28 Cherokee 140	Golf November Ltd/Kidlington
	G-BCGS	PA-28R Cherokee Arrow 200	Arrow Aviation Group
	G-BCGT	PA-28 Cherokee 140	L. Maikowski/Shoreham
	G-BCGW	Jodel D.11	G. H. & M. D. Chittenden
	G-BCHK	Cessna F.172H	D. Darby
	G-BCHL	D.H.C.1 Chipmunk 22A (WP788)	Shropshire Soaring Ltd/Sleap
	G-BCHM	SA.341G Gazelle 1	MW Helicopters Ltd
	G-BCHP	CP.1310-C3 Super Emeraude	G. Hughes & A. G. Just (G-JOSI)
	G-BCHT	Schleicher ASK.16	Dunstable K16 Group
	G-BCHV	D.H.C.1 Chipmunk 22	K.I. Sutherland
	G-BCID	PA-34-200 Seneca II	Shenley Farms (Aviation) Ltd
	G-BCIH	D.H.C.1 Chipmunk 22 (WD363)	J. M. Hosey/Stansted
	G-BCIJ	AA-5 Traveler	Arrow Association/Elstree
	G-BCIK	AA-5 Traveler	Trent Aviation Ltd
	G-BCIN	Thunder Ax7-77 balloon	R. A. Vale & ptnrs
	G-BCIR	PA-28-151 Warrior	P. J. Brennan
	G-BCJM	PA-28 Cherokee 140	Topcat Aviation Ltd/Manchester
	G-BCJN	PA-28 Cherokee 140	Topcat Aviation Ltd/Manchester
	G-BCJO	PA-28R Cherokee Arrow 200	R. Ross
	G-BCJP	PA-28 Cherokee 140	Omletair Flying Group
	G-BCKN	D.H.C.1A Chipmunk 22 (Lycoming)	RAFGSA/Cranwell
	G-BCKS	Fuji FA.200-180AO	J. F. Heath
	G-BCKT	Fuji FA.200-180	Kilo Tango Group
	G-BCKU	Cessna FRA.150L	Stapleford Flying Club Ltd
	G-BCKV	Cessna FRA.150L	Huck Air
	G-BCLC	Sikorsky S-61N	Bristow Helicopters/HM Coastguard
	G-BCLD	Sikorsky S-61N	Bristow Helicopters Ltd
	G-BCLI	AA-5 Traveler	W. D. Smith
	G-BCLL	PA-28 Cherokee 180	G-BCLL Group/Blackbushe
	G-BCLS	Cessna 170B	N. Simpson
	G-BCLT	M.S.894A Rallye Minerva 220	K. M. Bowen
	G-BCLU	Jodel D.117	S. J. Wynne
	G-BCLW	AA-1B Trainer	J. R. Faulkner
	G-BCMD	PA-18 Super Cub 95	P. Stephenson/Clacton
	G-BCMT	Isaacs Fury II	R.W. Burrows
	G-BCNC	GY-201 Minicab	J. R. Wraight
	G-BCNP	Cameron O-77 balloon	P. Spellward
	G-BCNX	Piper J-3C-65 Cub (540)	K. J. Lord
	G-BCNZ	Fuji FA.200-160	W. Dougan
	G-BCOB	Piper J-3C-65 Cub (329405)	R. W. & Mrs J. W. Marjoram
	G-BCOI	D.H.C.1 Chipmunk 22	M. J. Diggins
	G-BCOJ	Cameron O-56 balloon	T. J. Knott & M. J. Webber
	G-BCOL	Cessna F.172M	A. H. Creaser
	G-BCOM	Piper J-3C-65 Cub	Dougal Flying Group/Shoreham
	G-BCOO	D.H.C.1 Chipmunk 22	T. G. Fielding & M. S. Morton/Blackpool
	G-BCOR	SOCATA Rallye 100ST	P. R. W. Goslin & I. M. Speight
	G-BCOU	D.H.C.1 Chipmunk 22 (WK522)	P. J. Loweth
	G-BCOX	Bede BD-5A	H. J. Cox & B. L. Robinson
	G-BCOY	D.H.C.1 Chipmunk 22	Coventry Gliding Club Ltd/Husbands Bosworth
	G-BCPD	GY-201 Minicab	P. R. Cozens
	G-BCPG	PA-28R Cherokee Arrow 200-II	Roses Flying Group/Liverpool
	G-BCPH	Piper J-3C-65 Cub (329934)	M. J. Janaway
	G-BCPJ	Piper J-3C-65 Cub	Piper Cub Group
	G-BCPK	Cessna F.172M	D. C. C. Handley/Cranfield
	G-BCPN	AA-5 Traveler	G. K. Todd
	G-BCPU	D.H.C.1 Chipmunk 22	P. Waller/Booker
	G-BCRB	Cessna F.172M	D. E. Lamb
	G-BCRE	Cameron O-77 balloon ★	Balloon Preservation Group/Lancing
	G-BCRH	Alaparma Baldo B.75 ★	A. L. Scadding (stored)
	G-BCRI	Cameron O-65 balloon	V. J. Thorne
	G-BCRK	SA.102.5 Cavalier	P. G. R. Brown

Reg.	Type	Owner or Operator	Notes
G-BCRL	PA-28-151 Warrior	BCRL Ltd	
G-BCRP	PA-E23 Aztec 250E	Airlong Charter Ltd	
G-BCRR	AA-5B Tiger	Capulet Flying Group/Elstree	
G-BCRT	Cessna F.150M	Almat Flying Club Ltd	
G-BCRX	D.H.C.1 Chipmunk 22	Tuplin Ltd/Denham	
G-BCSA	D.H.C.1 Chipmunk 22 (Lycoming)	RAFGSA/Halton	
G-BCSL	D.H.C.1 Chipmunk 22	Chipmunk Flyers Ltd	
G-BCST	M.S.893A Rallye Commodore 180	D. R. Wilcox	
G-BCSX	Thunder Ax7-77 balloon	C. Wolstenholm	
G-BCSY	Taylor JT.2 Titch	I. L. Harding	
G-BCTF	PA-28-151 Warrior	The St. George Flying Club/Teesside	
G-BCTI	Schleicher ASK.16	Tango India Syndicate	
G-BCTJ	Cessna 310Q	D. Pearce	
G-BCTK	Cessna FR.172J	R. T. Love	
G-BCTT	Evans VP-1	E. R. G. Ludlow	
G-BCUB	Piper J-3C-65 Cub	A. L. Brown	
G-BCUF	Cessna F.172M	Howell Plant Hire & Construction	
G-BCUH	Cessna F.150M	M. G. Montgomerie	
G-BCUJ	Cessna F.150M	BCT Aircraft Leasing Ltd	
G-BCUL	SOCATA Rallye 100ST	C. A. Ussher & Fountain Estates Ltd	
G-BCUO	SA Bulldog Srs 120/122	Cranfield University	
G-BCUS	SA Bulldog Srs 120/122	S. J. & J. J. Oliver	
G-BCUV	SA Bulldog Srs 120/122	Dolphin Property (Management) Ltd	
G-BCUW	Cessna F.177RG	S. J. Westley	
G-BCUY	Cessna FRA.150M	J. C. Carpenter	
G-BCVB	PA-17 Vagabond	A. T. Nowak/Popham	
G-BCVC	SOCATA Rallye 100ST	W. Haddow	
G-BCVE	Evans VP-2	D. Masterson & D. B. Winstanley/Barton	
G-BCVF	Practavia Pilot Sprite	D. G. Hammersley	
G-BCVG	Cessna FRA.150L	G-BCVG Flying Group	
G-BCVH	Cessna FRA.150L	M. A. James	
G-BCVJ	Cessna F.172M	Rothland Ltd	
G-BCVY	PA-34-200T Seneca II	Oxford Aviation Services Ltd/Kidlington	
G-BCWB	Cessna 182P	M. F. Oliver & A. J. Mew	
G-BCWH	Practavia Pilot Sprite	R. Tasker/Blackpool	
G-BCWK	Alpavia Fournier RF-3	T. J. Hartwell & D. R. Wilkinson	
G-BCWO	BN-2A-21 Islander	Cormack (Aircraft Services) Ltd	
G-BCXB	SOCATA Rallye 100ST	A. Smails	
G-BCXE	Robin DR.400/2+2	Weald Air Services Ltd/Headcorn	
G-BCXJ	Piper L-4J Cub (480752)	W. Readman	
G-BCXN	D.H.C.1 Chipmunk 22 (WP800)	G. M. Turner/Halton	
G-BCYH	DAW Privateer Mk. 3	A. P. Paskins	
G-BCYK	Avro CF.100 Mk 4 Canuck (18393) ★	Imperial War Museum/Duxford	
G-BCYM	D.H.C.1 Chipmunk 22 (WK577)	G-BCYM Group	
G-BCYR	Cessna F.172M	Highland Flying School Ltd	
G-BCZH	D.H.C.1 Chipmunk 22 (WK622)	A. C. Byrne/Norwich	
G-BCZI	Thunder Ax7-77 balloon	R. G. Griffin & R. Blackwell	
G-BCZM	Cessna F.172M	Cornwall Flying Club Ltd/Bodmin	
G-BCZN	Cessna F.150M	Mona Aviation Ltd	
G-BCZO	Cameron O-77 balloon	W. O. T. Holmes Leo	
G-BDAD	Taylor JT.1 Monoplane	G-BDAD Group	
G-BDAG	Taylor JT.1 Monoplane	O. T. Elmer	
G-BDAH	Evans VP-1	G. H. J. Geurts	
G-BDAI	Cessna FRA.150M	D. F. Ranger	
G-BDAK	R. Commander 112	M. C. Wilson	
G-BDAL	R. 500S Shrike Commander	D. R. Tait	
G-BDAO	SIPA S.91	S. B. Churchill	
G-BDAP	AJEP Tailwind	J. Whiting	
G-BDAR	Evans VP-1	R. B. Valler	
G-BDAY	Thunder Ax5-42A balloon	T. M. Donnelly Meconium	
G-BDBD	Wittman W.8 Tailwind	Tailwindr Group	
G-BDBF	FRED Srs 2	G. E. & R. E. Collins	
G-BDBH	Bellanca 7GCBC Citabria	C. J. Gray	
G-BDBI	Cameron O-77 balloon	C. Jones	
G-BDBJ	Cessna 182P	H. C. Wilson	
G-BDBS	Short SD3-30 ★	Ulster Aviation Soc	
G-BDBU	Cessna F.150M	E. Veitch	
G-BDBV	Jodel D.11A	Seething Jodel Group	
G-BDBZ	W.S.55 Whirlwind Srs 2 ★	Ground instruction airframe/Kidlington	
G-BDCD	Piper J-3C-85 Cub (480133)	D. G. Pearce	
G-BDCI	CP.301A Emeraude	D. L. Sentance	

Notes	Reg.	Type	Owner or Operator
	G-BDCL	AA-5 Traveler	J. Crowe
	G-BDCO	B.121 Pup 1	R. J. Page & M. N. Simms
	G-BDDD	D.H.C.1 Chipmunk 22	DRA Aero Club Ltd
	G-BDDF	Jodel D.120	J. V. Thompson
	G-BDDG	Jodel D.112	J. Pool & D. G. Palmer/Sturgate
	G-BDDS	PA-25 Pawnee 235	Vale of Neath Gliding Club/Rhigos
	G-BDDX	Whittaker MW.2B Excalibur ★	Cornwall Aero Park/Helston
	G-BDDZ	CP.301A Emeraude	E. C. Mort
	G-BDEC	SOCATA Rallye 100ST	M. Mulhall
	G-BDEH	Jodel D.120A	EH Flying Group
	G-BDEI	Jodel D.9 Bebe	The Noddy Group/Booker
	G-BDEU	D.H.C.1 Chipmunk 22 (WP808)	T. E. Earl
	G-BDEX	Cessna FRA.150M	R. A. Powell
	G-BDEY	Piper J-3C-65 Cub	Ducksworth Flying Club
	G-BDEZ	Piper J-3C-65 Cub	R. J. M. Turnbull
	G-BDFB	Currie Wot	J. Jennings
	G-BDFH	Auster AOP.9 (XR240)	R. B. Webber
	G-BDFJ	Cessna F.150M	C. J. Hopewell
	G-BDFR	Fuji FA.200-160	Fuji Group/Blackpool
	G-BDFS	Fuji FA.200-160	B. Lawrence
	G-BDFU	Dragonfly MPA Mk 1 ★	Museum of Flight/E. Fortune
	G-BDFW	R. Commander 112	M. E. & E. G. Reynolds/Blackbushe
	G-BDFY	AA-5 Traveler	Grumman Group
	G-BDFZ	Cessna F.150M	L. W. Scattergood
	G-BDGB	GY-20 Minicab	D. G. Burden
	G-BDGH	Thunder Ax7-77 balloon	R. J. Mitchener & P. F. Smart
	G-BDGY	PA-28 Cherokee 140	S. J. Willcox
	G-BDHK	Piper J-3C-65 Cub (329417)	A. Liddiard
	G-BDIE	R. Commander 112	R. J. Adams
	G-BDIG	Cessna 182P	Air Group 6/Sturgate
	G-BDIH	Jodel D.117	N. D. H. Stokes
	G-BDIJ	Sikorsky S-61N	Bristow Helicopters Ltd
	G-BDIN	SA Bulldog Srs 120/125	British Disabled Flying Association
	G-BDIX	D.H.106 Comet 4C ★	Museum of Flight/E. Fortune
	G-BDJC	AJEP W.8 Tailwind	R. Bertrand/Germany
	G-BDJD	Jodel D.112	J. E. Preston
	G-BDJG	Luton LA-4A Minor	Very Slow Flying Club
	G-BDJP	Piper J-3C-90 Cub	S. T. Gilbert
	G-BDJR	SNCAN Nord NC.858	R. F. M. Marson
	G-BDKC	Cessna A185F	Bridge of Tilt Co Ltd
	G-BDKD	Enstrom F-28A	P. J. Price
	G-BDKH	CP.301A Emeraude	G-BDKH Group
	G-BDKJ	K. & S. SA.102.5 Cavalier	D. A. Garner
	G-BDKM	SIPA 903	S. W. Markham
	G-BDKW	R. Commander 112A	J. T. Klaschka
	G-BDLO	AA-5A Cheetah	S. & J. Dolan/Denham
	G-BDLS	AA-1B Trainer	A. W. Cattermole
	G-BDLT	R. Commander 112	D. L. Churchward
	G-BDLY	SA.102.5 Cavalier	P. R. Stevens/Southampton
	G-BDMM	Jodel D.11	P. N. Marshall
	G-BDMS	Piper J-3C-65 Cub (FR886)	A. T. H. Martin
	G-BDMW	Jodel DR.100A	R. O. F. Harper
	G-BDNC	Taylor JT.1 Monoplane	D. W. Mathie
	G-BDNG	Taylor JT.1 Monoplane	A. C. Beaumont
	G-BDNO	Taylor JT.1 Monoplane	S. D. Glover
	G-BDNP	BN-2A Islander ★	Ground parachute trainer/Headcorn
	G-BDNT	Jodel D.92 Bebe	R. F. Morton
	G-BDNU	Cessna F.172M	J. & K. G. McVicar
	G-BDNW	AA-1B Trainer	P. Mitchell
	G-BDNX	AA-1B Trainer	R. M. North
	G-BDOC	Sikorsky S-61N Mk II	Bristow Helicopters Ltd
	G-BDOD	Cessna F.150M	D. M. Moreau
	G-BDOE	Cessna FR.172J	D. Sansome
	G-BDOG	SA Bulldog Srs 200	D. C. Bonsall/Netherthorpe
	G-BDOL	Piper J-3C-65 Cub	L. R. Balthazor
	G-BDON	Thunder Ax7-77A balloon	M. J. Smith
	G-BDOT	BN-2A Mk.III-2 Trislander	Lyddair
	G-BDOW	Cessna FRA.150	Joystick Aviation Ltd
	G-BDPA	PA-28-151 Warrior	Prestwick Flight Centre Ltd
	G-BDPJ	PA-25 Pawnee 235B	RAFGSA/Bicester
	G-BDPK	Cameron O-56 balloon	Rango Balloon & Kite Co
	G-BDRD	Cessna FRA.150M	Prestwick Flight Centre Ltd

Reg.	Type	Owner or Operator	Notes
G-BDRG	Taylor JT.2 Titch	D. R. Gray	
G-BDRJ	D.H.C.1 Chipmunk 22 (WP857)	D. Curtis	
G-BDRK	Cameron O-65 balloon	D. L. Smith Smirk	
G-BDSB	PA-28-181 Archer II	Testair Ltd/Blackbushe	
G-BDSE	Cameron O-77 balloon	British Airways Concorde	
G-BDSF	Cameron O-56 balloon	J. H. Greensides	
G-BDSH	PA-28 Cherokee 140 (modified)	The Wright Brothers Flying Group	
G-BDSK	Cameron O-65 balloon	Southern Balloon Group Carousel II	
G-BDSL	Cessna F.150M	D. C. Bonsall	
G-BDSM	Slingsby T.31B Cadet III	F. C. J. Wevers/Netherlands	
G-BDTB	Evans VP-1	P. F. Moffatt	
G-BDTL	Evans VP-1	A. K. Lang	
G-BDTO	BN-2A Mk III-2 Trislander	Aurigny Air Services Ltd (G-RBSI/G-OTSB)	
G-BDTU	Omega III gas balloon	R. G. Turnbull	
G-BDTV	Mooney M.20F	S. Redfearn	
G-BDTX	Cessna F.150M	F. W. Ellis	
G-BDUI	Cameron V-56 balloon	D. C. Johnson	
G-BDUL	Evans VP-1	C. K. Brown	
G-BDUM	Cessna F.150M	P. Barnett & ptnrs/Earls Colne	
G-BDUN	PA-34-200T Seneca II	Air Medical Fleet Ltd	
G-BDUO	Cessna F.150M	D. F. Ranger	
G-BDUX	Slingsby T.31B Cadet III	J. C. Anderson/Cranfield	
G-BDUY	Robin DR.400/140B	J. G. Anderson	
G-BDUZ	Cameron V-56 balloon	Zebedee Balloon Service	
G-BDVA	PA-17 Vagabond	I. M. Callier	
G-BDVB	PA-15 (PA-17) Vagabond	B. P. Gardner	
G-BDVC	PA-17 Vagabond	A. R. Caveen	
G-BDWA	SOCATA Rallye 150ST	J. Thompson-Wilson	
G-BDWE	Flaglor Scooter	P. King	
G-BDWH	SOCATA Rallye 150ST	M. A. Jones	
G-BDWJ	SE-5A (replica) (F8010)	D. W. Linney	
G-BDWM	Mustang scale replica (FB226)	D. C. Bonsall	
G-BDWO	Howes Ax6 balloon	R. B. & C. Howes	
G-BDWX	Jodel D.120A	R. P. Rochester	
G-BDXE	Boeing 747-236B	European Aviation Ltd/Bournemouth	
G-BDXF	Boeing 747-236B	European Aviation Ltd/Bournemouth	
G-BDXG	Boeing 747-236B	European Aviation Ltd/Bournemouth	
G-BDXH	Boeing 747-236B	European Aviation Ltd/Bournemouth	
G-BDXJ	Boeing 747-236B	European Aviation Ltd/Bournemouth	
G-BDXO	Boeing 747-236B	European Aviation Ltd/Bournemouth	
G-BDXX	SNCAN NC.858S	K. M. Davis	
G-BDYD	R. Commander 114	J. R. Pybus	
G-BDYG	P.56 Provost T.1 (WV493) ★	Museum of Flight/E. Fortune	
G-BDYH	Cameron V-56 balloon	B. J. Godding	
G-BDZA	Scheibe SF.25E Super Falke	Hereward Flying Group/Crowland	
G-BDZC	Cessna F.150M	A. M. Lynn/Sibson	
G-BDZD	Cessna F.172M	Zephyr Group	
G-BDZU	Cessna 421C	Eagle Flying Group/E. Midlands	
G-BEAB	Jodel DR.1051	R. C. Hibberd	
G-BEAC	PA-28 Cherokee 140	Clipwing Flying Group/Humberside	
G-BEAD	WG.13 Lynx ★	Instructional airframe/Middle Wallop	
G-BEAG	PA-34-200T Seneca II	Oxford Aviation Services Ltd/Kidlington	
G-BEAH	J/2 Arrow	Bedwell Hey Flying Group	
G-BEBC	W.S.55 Whirlwind 3 (XP355) ★	Norwich Aviation Museum	
G-BEBE	AA-5A Cheetah	Bills Aviation Ltd	
G-BEBG	WSK-PZL SDZ-45A Ogar	The Ogar Syndicate	
G-BEBN	Cessna 177B	R. Turrell & P. Mason/Stapleford	
G-BEBS	Andreasson BA-4B	N. J. W. Reid	
G-BEBU	R. Commander 112A	Aeros Engineering Ltd	
G-BEBZ	PA-28-151 Warrior	Goodwood Terrena Ltd/Goodwood	
G-BECA	SOCATA Rallye 100ST	A. C. Stamp	
G-BECB	SOCATA Rallye 100ST	D. H. Tonkin	
G-BECC	SOCATA Rallye 150ST	G. R. E. Tapper	
G-BECF	Scheibe SF.25A Falke	North County Ltd	
G-BECN	Piper J-3C-65 Cub (480480)	G. Denney/Earls Colne	
G-BECS	Thunder Ax6-56A balloon	A. Sieger/Germany	
G-BECT	C.A.S.A.1.131E Jungmann 2000 (A-57)	Alpha 57 Group	
G-BECW	C.A.S.A.1.131E Jungmann 2000 (A-10)	R. A. Seeley	
G-BECZ	CAARP CAP.10B	Aerobatic Associates Ltd	
G-BEDB	Nord 1203 Norecrin ★	B. F. G. Lister (stored)/Chirk	
G-BEDD	Jodel D.117A	Dubious Group	

Notes	Reg.	Type	Owner or Operator
	G-BEDF	Boeing B-17G-105-VE (124485)	B-17 Preservation Ltd/Duxford
	G-BEDG	R. Commander 112	Hotels International Ltd
	G-BEDJ	Piper J-3C-65 Cub (44-80594)	R. Earl
	G-BEDP	BN-2A Mk.III-2 Trislander	Lyddair
	G-BEDV	V.668 Varsity T.1 (WJ945) ★	Duxford Aviation Soc
	G-BEEE	Thunder Ax6-56A balloon ★	British Balloon Museum Avie/Newbury
	G-BEEG	BN-2A-26 Islander	NW Parachute Centre Ltd/Cark
	G-BEEH	Cameron V-56 balloon	Sade Balloons Ltd
	G-BEER	Isaacs Fury II (K2075)	R. S. C. Andrews
	G-BEEU	PA-28 Cherokee 140F	E. & H. Merkado
	G-BEFA	PA-28-151 Warrior	Verran Freight
	G-BEFF	PA-28 Cherokee 140F	H. & E. Merkado
	G-BEGG	Scheibe SF.25E Super Falke	G-BEGG Flying Group
	G-BEHH	PA-32R Cherokee Lance 300	K. Swallow
	G-BEHU	PA-34-200T Seneca II	Pirin Aeronautical Ltd/Stapleford
	G-BEHV	Cessna F.172N	Edinburgh Air Centre Ltd
	G-BEIA	Cessna FRA.150M	Oxford Aviation Services Ltd/Kidlington
	G-BEIF	Cameron O-65 balloon	C. Vening
	G-BEIG	Cessna F.150M	T. J. Chapman
	G-BEII	PA-25 Pawnee 235D	Burn Gliding Club Ltd
	G-BEIL	SOCATA Rallye 150T	The Rallye Flying Group
	G-BEIP	PA-28-181 Archer II	S. Pope
	G-BEIS	Evans VP-1	P. J. Hunt
	G-BEJD	Avro 748 Srs 1	Emerald Airways Ltd *John Case*/Liverpool
	G-BEJK	Cameron S-31 balloon	Rango Balloon & Kite Co
	G-BEJL	Sikorsky S-61N	CHC Scotiia Ltd
	G-BEJV	PA-34-200T Seneca II	Oxford Aviation Services Ltd/Kidlington
	G-BEKL	Bede BD-4E-150	F. E.Tofield
	G-BEKM	Evans VP-1	G. J. McDill/Glenrothes
	G-BEKN	Cessna FRA.150M	A. L. Brown/Bourn
	G-BEKO	Cessna F.182Q	G. J. & F. J. Leese
	G-BELF	BN-2A-26 Islander	Cormack (Aircraft Services) Ltd
	G-BELP	PA-28-151 Warrior	Tatenhill Aviation Ltd
	G-BELT	Cessna F.150J	A. kumar (G-AWUV)
	G-BEMB	Cessna F.172M	Stocklaunch Ltd
	G-BEMM	Slingsby T.31B Motor Cadet III	B. J. Douglas
	G-BEMU	Thunder Ax5-42 balloon	M. A. Hall
	G-BEMW	PA-28-181 Archer II	Touch & Go Ltd
	G-BEMY	Cessna FRA.150M	M. J. L. Tondeur & M. M. E. Versyck/Belgium
	G-BEND	Cameron V-56 balloon	Dante Balloon Group Le Billet
	G-BENJ	R. Commander 112B	W. J. Homer
	G-BENK	Cessna F.172M	Graham Churchill Plant Ltd
	G-BENN	Cameron V-56 balloon	S. J. Hollingsworth & M. K. Bellamy
	G-BEOD	Cessna 180 ★	Avionics Research Ltd/Cranfield
	G-BEOE	Cessna FRA.150M	W. J. Henderson
	G-BEOH	PA-28R-201T Turbo Arrow III	G-BEOH Group/Blackbushe
	G-BEOI	PA-18 Super Cub 150	Southdown Gliding Club Ltd/Parham Park
	G-BEOK	Cessna F.150M	D. C. Bonsall
	G-BEOL	SC.7 Skyvan 3 variant 100	Invicta Aviation Ltd
	G-BEOX	L-414 Hudson IV (A16-199) ★	RAF Museum/Hendon
	G-BEOY	Cessna FRA.150L	R. W. Denny
	G-BEOZ	A.W.650 Argosy 101 ★	Aeropark/E. Midlands
	G-BEPC	SNCAN Stampe SV-4C	Papa Charlie's Flying Circus Ltd
	G-BEPF	SNCAN Stampe SV-4A	L. J. Rice
	G-BEPS	SC.5 Belfast	*(stored)*/ Southend
	G-BEPV	Fokker S.11-1 Instructor	S. W. & M. Isbister & C. Tyers
	G-BEPY	R. Commander 112B	G-BEPY Group/Blackbushe
	G-BERA	SOCATA Rallye 150ST	B. Dolby
	G-BERC	SOCATA Rallye 150ST	Severn Valley Aero Group/Welshpool
	G-BERD	Thunder Ax6-56A balloon	P. M. Gaines
	G-BERI	R. Commander 114	K. B. Harper/Blackbushe
	G-BERN	Saffrey S-330 balloon	B. Martin Beeze
	G-BERT	Cameron V-56 balloon	Southern Balloon Group *Bert*
	G-BERW	R. Commander 114	Romeo Whisky Ltd
	G-BERY	AA-1B Trainer	R. H. J. Levi
	G-BETD	Robin HR.200/100	P. I. Acott
	G-BETE	Rollason B.2A Beta	T. M. Jones/Tatenhill
	G-BETF	Cameron 'Champion' SS balloon ★	British Balloon Museum/Newbury
	G-BETG	Cessna 180K Skywagon	J. A. Hart
	G-BETI	Pitts S-1D Special	N. A. Scully (G-PIII)
	G-BETL	PA-25 Pawnee 235D	Cambridge University Gliding Trust Ltd/ Gransden Lodge

Reg.	Type	Owner or Operator	Notes
G-BETM	PA-25 Pawnee 235D	Yorkshire Gliding Club (Pty) Ltd/Sutton Bank	
G-BETO	MS.885 Super Rallye	G-BETO Group	
G-BETT	PA-34-200 Seneca II	D. F. J. Flashman	
G-BETW	Rand KR-2	S. C. Solley	
G-BEUA	PA-18 Super Cub 150	London Gliding Club (Pty) Ltd/Dunstable	
G-BEUD	Robin HR.100/285R	E. A. & L. M. C. Payton/Cranfield	
G-BEUI	Piper J-3C-65 Cub	G-BEUI Group	
G-BEUM	Taylor JT.1 Monoplane	J. M. Burgess	
G-BEUN	Cassutt Racer IIIM	R. McNulty	
G-BEUP	Robin DR.400/180	Samuels Aviation/Biggin Hill	
G-BEUU	PA-18 Super Cub 95	F. Sharples/Sandown	
G-BEUX	Cessna F.172N	Multiflight Ltd/Leeds-Bradford	
G-BEUY	Cameron N-31 balloon	M. L. & L. P. Willoughby	
G-BEVA	SOCATA Rallye 150ST	The Rallye Group	
G-BEVB	SOCATA Rallye 150ST	G. G. Hammond	
G-BEVC	SOCATA Rallye 150ST	G-BEVC Ltd	
G-BEVG	PA-34-200T-2 Seneca	A. G. & J. Wintle	
G-BEVO	Sportavia-Pützer RF-5	D. G. Hey & W. E. R. Jenkins	
G-BEVP	Evans VP-2	G. Moscrop & R. C. Crowley	
G-BEVS	Taylor JT.1 Monoplane	D. Hunter	
G-BEVT	BN-2A Mk III-2 Trislander	Aurigny Air Services Ltd/Guernsey	
G-BEVW	SOCATA Rallye 150ST	P. G. A. Sumner	
G-BEWM	Sikorsky S-61N Mk II	Brintel Helicopters	
G-BEWN	D.H.82A Tiger Moth	H. D. Labouchere	
G-BEWO	Zlin Z.326 Trener Master	P. A. Colman	
G-BEWR	Cessna F.172N	Cheshire Air Training Services Ltd/Liverpool	
G-BEWX	PA-28R-201 Arrow III	A. Vickers	
G-BEWY	Bell 206B JetRanger 3	Polo Aviation Ltd (G-CULL)	
G-BEXN	AA-1C Lynx	J. S. C. Goodale	
G-BEXO	PA-23 Apache 160	Aviation Advisory Services Ltd	
G-BEXZ	Cameron N-56 balloon	D. C. Eager & G. C. Clark	
G-BEYA	Enstrom 280C	Hovercam Ltd	
G-BEYB	Fairey Flycatcher (replica) (S1287) ★	F.A.A. Museum/Yeovilton	
G-BEYF	HPR-7 Herald 401 ★	Jet Heritage Museum/Bournemouth	
G-BEYL	PA-28 Cherokee 180	Yankee Lima Group	
G-BEYO	PA-28 Cherokee 140	Eurocharter Aviation Ltd	
G-BEYT	PA-28 Cherokee 140	G. P. & F. C. Coleman	
G-BEYV	Cessna T.210M	Austen Aviation/Edinburgh	
G-BEYW	Taylor JT.1 Monoplane	R. A. Abrahams/Barton	
G-BEYZ	Jodel DR.1051/M1	M. L. Balding	
G-BEZC	AA-5 Traveler	T. V. Montgomery/Elstree	
G-BEZE	Rutan Vari-Eze	S. K. Cockburn	
G-BEZF	AA-5 Traveler	RAF College Flying Club Ltd/Cranwell	
G-BEZG	AA-5 Traveler	M. D. R. Harling	
G-BEZH	AA-5 Traveler	L. & S. M. Sims	
G-BEZI	AA-5 Traveler	G-BEZI Flying Group/Elstree	
G-BEZK	Cessna F.172H	C. F. Strowger	
G-BEZL	PA-31-310 Turbo Navajo C	A. Jahanfar/Southend	
G-BEZO	Cessna F.172M	Staverton Flying Services Ltd	
G-BEZP	PA-32 Cherokee Six 300D	W. D. McNab & T. P. McCormack	
G-BEZR	Cessna F.172M	Kirmington Aviation Ltd	
G-BEZV	Cessna F.172M	Insch Flying Group	
G-BEZY	Rutan Vari-Eze	I. J. Pountney	
G-BEZZ	Jodel D.112	G-BEZZ Jodel Group	
G-BFAA	GY-80 Horizon 160	G. R. Williams	
G-BFAF	Aeronca 7BCM (7797)	D. C. W. Harper/Finmere	
G-BFAH	Phoenix Currie Wot	R. W. Clarke	
G-BFAI	R. Commander 114	Alpha Flying Group	
G-BFAK	GEMS MS.892A Rallye Commodore 150	J. M. Hedges	
G-BFAP	SIAI-Marchetti S.205-20R	A. O. Broin	
G-BFAS	Evans VP-1	A. I. Sutherland	
G-BFAW	D.H.C.1 Chipmunk 22	R. V. Bowles/Husbands Bosworth	
G-BFAX	D.H.C.1 Chipmunk 22 (WG422)	N. Rushton	
G-BFBA	Jodel DR.100A	W. H. Sherlock	
G-BFBB	PA-23 Aztec 250E	Air Training Services Ltd/Booker	
G-BFBC	Taylor JT.1 Monoplane	G. Heins	
G-BFBE	Robin HR.200/100	A. C. Pearson	
G-BFBM	Saffery S.330 balloon	B. Martin Beeze II	
G-BFBR	PA-28-161 Warrior II	Malcolm R Paul Racing	
G-BFBU	Partenavia P.68B	Premiair Charter Ltd	
G-BFBY	Piper J-3C-65 Cub	U. Schumacher	

Notes	Reg.	Type	Owner or Operator
	G-BFCT	Cessna Tu.206F	D. I. Schellingerhout
	G-BFDC	D.H.C.1 Chipmunk 22	N. F. O'Neill/Newtownards
	G-BFDE	Sopwith Tabloid (replica) (168) ★	RAF Museum Storage & Restoration Centre/ Cardington
	G-BFDF	SOCATA Rallye 235E	M. A. Wratten
	G-BFDI	PA-28-181 Archer II	Truman Aviation Ltd/Tollerton
	G-BFDK	PA-28-161 Warrior II	S. T. Gilbert
	G-BFDL	Piper J-3C-65 Cub (454537)	A. F. Nicholson
	G-BFDO	PA-28R-201T Turbo Arrow III	A. J. Gow
	G-BFDZ	Taylor JT.1 Monoplane	J. A. Hanslip
	G-BFEB	Jodel 150	Jodel Syndicate
	G-BFEF	Westland-Bell 47G-3B1	M. P. Wilkinson
	G-BFEH	Jodel D.117A	J. A. Crabb
	G-BFEK	Cessna F.152	Staverton Flying Services Ltd
	G-BFER	Bell 212	Bristow Helicopters Ltd
	G-BFEV	PA-25 Pawnee 235	Trent Valley Aerotowing Club Ltd/Kirton-in-Lindsey
	G-BFEW	PA-25 Pawnee 235	Cornish Gliding & Flying Club Ltd/Perranporth
	G-BFFB	Evans VP-2 ★	(stored)/Eaton Bray
	G-BFFC	Cessna F.152-II	Multiflight Ltd
	G-BFFE	Cessna F.152-II	A. J. Hastings/Edinburgh
	G-BFFJ	Sikorsky S-61N Mk II	Veritair Ltd/Penzance
	G-BFFP	PA-18 Super Cub 150 (modified)	East Sussex Gliding Club Ltd
	G-BFFT	Cameron V-56 balloon	R. I. M. Kerr & D. C. Boxall
	G-BFFW	Cessna F.152	Tayside Aviation Ltd/Dundee
	G-BFFY	Cessna F.150M	G. D. Rodmell
	G-BFGD	Cessna F.172N-II	J. T. Armstrong
	G-BFGG	Cessna FRA.150M	S. J. Windle/Bodmin
	G-BFGH	Cessna F.337G	T. Perkins/Sherburn
	G-BFGK	Jodel D.117	B. F. J. Hope
	G-BFGL	Cessna FA.152	Multiflight Ltd
	G-BFGO	Fuji FA.200-160	R. J. Everett
	G-BFGS	M.S.893E Rallye 180GT	Chiltern Flyers Ltd
	G-BFGW	Cessna F.150H	C. E. Stringer
	G-BFGX	Cessna FRA.150M	Prestwick Flight Centre Ltd
	G-BFGZ	Cessna FRA.150M	C. M. Barnes
	G-BFHH	D.H.82A Tiger Moth	P. Harrison & M. J. Gambrell/Redhill
	G-BFHI	Piper J-3C-65 Cub	N. Glass & A. J. Richardson
	G-BFHP	Champion 7GCAA Citabria	Poet Pilot Ltd
	G-BFHR	Jodel DR.220/2+2	J. E. Sweetman
	G-BFHT	Cessna F.152-II	Westward Airways (Lands End) Ltd
	G-BFHU	Cessna F.152-II	D. J. Cooke & Co Ltd
	G-BFHV	Cessna F.152-II	Falcon Flying Services/Biggin Hill
	G-BFHX	Evans VP-1	A. D. Bohanna & D. I. Trussler
	G-BFIB	PA-31 Turbo Navajo	Richard Hannon Ltd
	G-BFID	Taylor JT.2 Titch Mk III	R. W. Kilham
	G-BFIE	Cessna FRA.150M	J. P. A. Freeman
	G-BFIG	Cessna FR.172K XPII	Tenair Ltd
	G-BFIJ	AA-5A Cheetah	T. H. & M. G. Weetman
	G-BFIN	AA-5A Cheetah	Prestwick Flight Centre Ltd
	G-BFIP	Wallbro Monoplane 1909 (replica) ★	Norfolk & Suffolk Aviation Museum/Flixton, Suffolk
	G-BFIT	Thunder Ax6-56Z balloon	J. A. G. Tyson
	G-BFIU	Cessna FR.172K XP	B. M. Jobling
	G-BFIV	Cessna F.177RG	C. Fisher
	G-BFIX	Thunder Ax7-77A balloon	R. Owen
	G-BFIY	Cessna F.150M	R. J. Scott
	G-BFJJ	Evans VP-1	N. Clark
	G-BFJR	Cessna F.337G	Mannix Aviation Ltd/E. Midlands
	G-BFJZ	Robin DR.400/140B	Weald Air Services Ltd/Headcorn
	G-BFKB	Cessna F.172N	Shropshire Flying Group
	G-BFKF	Cessna FA.152	Aerolease Ltd/Conington
	G-BFKH	Cessna F.152	TG Aviation Ltd/Manston
	G-BFKL	Cameron N-56 balloon	Merrythought Toys Ltd Merrythought
	G-BFKY	PA-34-200 Seneca II	S.L.H. Construction Ltd/Biggin Hill
	G-BFLH	PA-34-200T Seneca II	Air Medical Ltd
	G-BFLI	PA-28R-201T Turbo Arrow III	J. K. Chudzicki
	G-BFLU	Cessna F.152	Atlantic Flight Training Ltd/Coventry
	G-BFLX	AA-5A Cheetah	G Force Two Ltd/Blackbushe
	G-BFLZ	Beech 95-A55 Baron	Caterite Food Service
	G-BFMG	PA-28-161 Warrior II	Stardial Ltd
	G-BFMH	Cessna 177B	J. C. Owens & A. C. Smith
	G-BFMK	Cessna FA.152	RAF Halton Aeroplane Club Ltd
	G-BFMR	PA-20 Pacer 125	J. Knight

Reg.	Type	Owner or Operator	Notes
G-BFMX	Cessna F.172N	A2Z Wholesale Fashion Jewellery Ltd	
G-BFMZ	Payne Ax6 balloon	E. G. Woolnough	
G-BFNG	Jodel D.112	A. W. Myers & J. MacGregor	
G-BFNI	PA-28-161 Warrior II	P. Elliott/Biggin Hill	
G-BFNJ	PA-28-161 Warrior II	Fleetlands Flying Association Ltd	
G-BFNK	PA-28-161 Warrior II	Oxford Aviation Services Ltd/Kidlington	
G-BFOD	Cessna F.182Q	G. N. Clarke	
G-BFOE	Cessna F.152	Redhill Air Services Ltd	
G-BFOF	Cessna F.152	Staverton Flying School Ltd	
G-BFOG	Cessna 150M	C. L. Day	
G-BFOJ	AA-1 Yankee	N. W. Thomas/Bournemouth	
G-BFOM	PA-31 Turbo Navajo C	Ashton Air Services Ltd	
G-BFOP	Jodel D.120	R. J. Wesley & G. D. Western/Ipswich	
G-BFOS	Thunder Ax6-56A balloon	N. T. Petty	
G-BFOU	Taylor JT.1 Monoplane	G. Bee	
G-BFOV	Cessna F.172N	D. J. Walker	
G-BFPA	Scheibe SF.25B Falke	R. Gibson & R. Hamilton	
G-BFPB	AA-5B Tiger	Papa Bravo Flying Group	
G-BFPH	Cessna F.172K	Linc-Air Flying Group	
G-BFPM	Cessna F.172M	M. P. Wimsey & J. W. Cope	
G-BFPO	R. Commander 112B	J. G. Hale Ltd	
G-BFPP	Bell 47J-2 Ranger	M. R. Masters	
G-BFPR	PA-25 Pawnee 235D	Booker Gliding Club Ltd	
G-BFPS	PA-25 Pawnee 235D	Kent Gliding Club Ltd/Challock	
G-BFRD	Bowers Fly-Baby 1A	R. A. Phillips	
G-BFRI	Sikorsky S-61N	Bristow Helicopters Ltd Braerich	
G-BFRR	Cessna FRA.150M	Romeo Romeo Flying Group/Tatenhill	
G-BFRS	Cessna F.172N	Poplar Hall PLC	
G-BFRV	Cessna FA.152	Solo Services Ltd	
G-BFRY	PA-25 Pawnee 260	Yorkshire Gliding Club (Pty) Ltd/Sutton Bank	
G-BFSA	Cessna F.182Q	Ensiform Aviation Ltd/Elstree	
G-BFSC	PA-25 Pawnee 235D	Essex Gliding Club Ltd/North Weald	
G-BFSD	PA-25 Pawnee 235D	Deeside Gliding Club (Aberdeenshire) Ltd/Aboyne	
G-BFSK	PA-23 Apache 160 ★	*Sub-aqua instructional airframe*/Croughton	
G-BFSR	Cessna F.150J	W. Ali	
G-BFSS	Cessna FR.172G	Albedale Farms Ltd	
G-BFSY	PA-28-181 Archer II	Downland Aviation	
G-BFTC	PA-28R-201T Turbo Arrow III	M. J. Milns/Sherburn	
G-BFTF	AA-5B Tiger	F. C. Burrow Ltd/Leeds	
G-BFTG	AA-5B Tiger	D. Hepburn & G. R. Montgomery	
G-BFTH	Cessna F.172N	J. Birkett	
G-BFTT	Cessna 421C	M. A. Ward	
G-BFUB	PA-32RT-300 Lance II	Jolida Holdings Ltd	
G-BFUD	Scheibe SF.25E Super Falke	Lakes Libelle Syndicate/Walney Island	
G-BFUZ	Cameron V-77 balloon	Servowarm Balloon Syndicate	
G-BFVG	PA-28-181 Archer II	G-BFVG Flying Group/Blackpool	
G-BFVH	D.H.2 (replica) (5964)	M. J. Kirk	
G-BFVS	AA-5B Tiger	S. W. Biroth & T. Chapman/Denham	
G-BFVU	Cessna 150L	S. D. Baker	
G-BFWB	PA-28-161 Warrior II	Mid-Anglia School of Flying	
G-BFWD	Currie Wot	D. Silsbury & B. Proctor	
G-BFWE	PA-23 Aztec 250E	Air Navigation & Trading Co Ltd/Blackpool	
G-BFXF	Andreasson BA.4B	A. Brown/Breighton	
G-BFXG	D.31 Turbulent	E. J. I. Musty & M. J. Whatley	
G-BFXK	PA-28 Cherokee 140	D. M. Wheeler	
G-BFXL	Albatross D.5A (D5397/17) ★	F.A.A. Museum/Yeovilton	
G-BFXR	Jodel D.112	J. M. Pearson	
G-BFXS	R. Commander 114	Unipak (UK) Ltd	
G-BFXW	AA-5B Tiger	Campsol Ltd	
G-BFXX	AA-5B Tiger	W. R. Gibson	
G-BFYA	MBB Bo 105DB	Sterling Helicopters Ltd/Norwich	
G-BFYC	PA-32RT-300 Lance II	A. A. Barnes	
G-BFYE	Robin HR.100/285 ★	*(stored)*/Sywell	
G-BFYI	Westland-Bell 47G-3B1	B. Walker & Co (Dursley) Ltd	
G-BFYK	Cameron V-77 balloon	L. E. Jones	
G-BFYL	Evans VP-2	W. C. Brown	
G-BFYO	SPAD XIII (replica) (1/4513) ★	American Air Museum/Duxford	
G-BFZB	Piper J-3C-85 Cub	N. Rawlinson	
G-BFZD	Cessna FR.182RG	R. B. Lewis & Co/Sleap	
G-BFZH	PA-28R Cherokee Arrow 200	Mason Aviation	
G-BFZM	R. Commander 112TC	J. A. Hart & R. J. Lamplough	
G-BFZN	Cessna FA.152	Falcon Flying Services/Biggin Hill	

Notes	Reg.	Type	Owner or Operator
	G-BFZO	AA-5A Cheetah	J. W. Cross & A. E. Kempson
	G-BFZR	AA-5B Tiger	P. C. Morrissey/Ireland
	G-BFZT	Cessna FA.152	Herefordshire Aero Club Ltd/Shobdon
	G-BFZU	Cessna FA.152	C. R. Tilley
	G-BFZV	Cessna F.172M	Wessex Flying Group Ltd
	G-BGAA	Cessna 152 II	PJC Leasing Ltd
	G-BGAB	Cessna F.152 II	TG Aviation Ltd/Manston
	G-BGAE	Cessna F.152 II	Aerolease Ltd/Conington
	G-BGAF	Cessna FA.152	G-BGAF Group/Southend
	G-BGAG	Cessna F.172N	Falcon Flying Services/Biggin Hill
	G-BGAJ	Cessna F.182Q II	Ground Airport Services Ltd/Guernsey
	G-BGAX	PA-28 Cherokee 140	C. D. Brack/Breighton
	G-BGAZ	Cameron V-77 balloon	C. J. Madigan & D. H. McGibbon
	G-BGBA	Robin R.2100A	D. Faulkner/Redhill
	G-BGBE	Jodel DR.1050	J. A. & B. Mawby
	G-BGBF	D.31A Turbulent	Eaglescott Turbulent Group
	G-BGBG	PA-28-181 Archer II	Harlow Printing Ltd/Newcastle
	G-BGBI	Cessna F.150L	C. P. Tapp
	G-BGBK	PA-38-112 Tomahawk	The Sherwood Flying Club Ltd
	G-BGBN	PA-38-112 Tomahawk	Bonus Aviation Ltd/Cranfield
	G-BGBR	Cessna F.172N	Falcon Flying Services/Biggin Hill
	G-BGBW	PA-38-112 Tomahawk	Truman Aviation Ltd/Tollerton
	G-BGBY	PA-38-112 Tomahawk	Ravenair/Liverpool
	G-BGBZ	R. Commander 114	R. S. Fenwick/Biggin Hill
	G-BGCG	Douglas C-47A ★	Datran Holdings Ltd (stored)
	G-BGCM	AA-5A Cheetah	G. & S. A. Jones
	G-BGCO	PA-44-180 Seminole	J. R. Henderson
	G-BGCY	Taylor JT.1 Monoplane	A. T. Lane
	G-BGEA	Cessna F.150M	C. J. Hopewell/Sibson
	G-BGED	Cessna U.206F	P. Marsden
	G-BGEH	Monnett Sonerai II	D. & V. T. Hubbard
	G-BGEI	Baby Great Lakes	M. T. Taylor
	G-BGEK	PA-38-112 Tomahawk	Ravenair/Liverpool
	G-BGEW	SNCAN NC.854S	R. H. Ashforth
	G-BGFC	Evans VP-2	S. W. C. Hollins
	G-BGFF	FRED Srs 2	I. Daniels
	G-BGFG	AA-5A Cheetah	T. D. Saveker
	G-BGFH	Cessna F.182Q	Rayviation Ltd
	G-BGFI	AA-5A Cheetah	I. J. Hay & A. Nayyar/Biggin Hill
	G-BGFJ	Jodel D.9 Bebe	M. D. Mold
	G-BGFT	PA-34-200T Seneca II	Oxford Aviation Services Ltd/Kidlington
	G-BGFX	Cessna F.152	Falcon Flying Services/Biggin Hill
	G-BGGA	Bellanca 7GCBC Citabria	L. A. King
	G-BGGB	Bellanca 7GCBC Citabria	G. H. N. Chamberlain
	G-BGGC	Bellanca 7GCBC Citabria	R. P. Ashfield & J. P. Stone
	G-BGGD	Bellanca 8GCBC Scout	Bristol & Gloucestershire Gliding Club/Nympsfield
	G-BGGE	PA-38-112 Tomahawk	Truman Aviation Ltd/Tollerton
	G-BGGG	PA-38-112 Tomahawk	Teesside Flight Centre Ltd
	G-BGGI	PA-38-112 Tomahawk	Truman Aviation Ltd/Tollerton
	G-BGGL	PA-38-112 Tomahawk	Grunwick Processing Laboratories Ltd/Elstree
	G-BGGM	PA-38-112 Tomahawk	Grunwick Processing Laboratories Ltd/Elstree
	G-BGGN	PA-38-112 Tomahawk	Bell Aviation Ltd
	G-BGGO	Cessna F.152	E. Midlands Flying School Ltd
	G-BGGP	Cessna F.152	E. Midlands Flying School Ltd
	G-BGGU	Wallis WA-116/RR	K. H. Wallis
	G-BGGW	Wallis WA-112	K. H. Wallis
	G-BGGY	AB-206B Jet Ranger ★	Instructional airframe/Cranfield
	G-BGHF	Westland WG.30 ★	IHM/Weston-s-Mare
	G-BGHI	Cessna F.152	V. R. McCready
	G-BGHJ	Cessna F.172N	Castle Aviation Ltd
	G-BGHM	Robin R.1180T	H. Price
	G-BGHP	Beech 76 Duchess	Magneta Ltd/Exeter
	G-BGHS	Cameron N-31 balloon	W. R. Teasdale
	G-BGHT	Falconar F-12	C. R. Coates
	G-BGHU	NA T-6G Texan (115042)	C. E. Bellhouse
	G-BGHV	Cameron V-77 balloon	E. Davies
	G-BGHY	Taylor JT.1 Monoplane	P. J. Burgess
	G-BGHZ	FRED Srs 2	FRED Group
	G-BGIB	Cessna 152 II	Redhill Air Services Ltd
	G-BGIG	PA-38-112 Tomahawk	Air Claire Ltd
	G-BGIO	Montgomerie-Bensen B.8MR	R. M. Savage

Reg.	Type	Owner or Operator	Notes
G-BGIU	Cessna F.172H	Fabricor Ltd	
G-BGIX	H.295 Super Courier	C. M. Lee	
G-BGIY	Cessna F.172N	Air Claire Ltd	
G-BGJB	PA-44-180 Seminole	Ostend Air College (G-ISFT)/Belgium	
G-BGJU	Cameron V-65 Balloon	J. A. Folkes	
G-BGKC	SOCATA Rallye 110ST	J. H. Cranmer & T. A. Timms	
G-BGKJ	MBB Bo 105D ★	*Instructional airframe*/Bourn	
G-BGKO	GY-20 Minicab	R. B. Webber	
G-BGKS	PA-28-161 Warrior II	Marnham Investments Ltd	
G-BGKT	Auster AOP.9 (XN441)	KT Group	
G-BGKU	PA-28R-201 Arrow III	Aerolease Ltd	
G-BGKV	PA-28R-201 Arrow III	R. Haverson & A. K. Lake/Shipdham	
G-BGKY	PA-38-112 Tomahawk	Top Cat Aviation Ltd	
G-BGKZ	J/5F Aiglet Trainer	R. B. Webber	
G-BGLA	PA-38-112 Tomahawk	Norwich School of Flying	
G-BGLB	Bede BD-5B ★	Science Museum/Wroughton	
G-BGLF	Evans VP-1 Srs 2	A. Stinson	
G-BGLG	Cessna 152	L. W. Scattergood	
G-BGLI	Cessna 152	Luton Flying Club	
G-BGLJ	Bell 212	FB Leasing Ltd	
G-BGLK	Monnett Sonerai 2L	J. Bradley	
G-BGLN	Cessna FA.152	Bflying Ltd	
G-BGLO	Cessna F.172N	Pulsar Yacht Services Ltd /Southend	
G-BGLS	Oldfield Super Baby Lakes	J. F. Dowe	
G-BGLW	PA-34-200 Seneca II	London Executive Aviation Ltd	
G-BGLZ	Stits SA-3A Playboy	P. C. Sheard	
G-BGMJ	GY-201 Minicab	G-BGMJ Group	
G-BGMN	H.S.748 Srs 2A	Emerald Airways Ltd/Liverpool	
G-BGMO	H.S.748 Srs 2A	Emerald Airways Ltd/Liverpool	
G-BGMP	Cessna F.172G	R. W. Collings	
G-BGMR	GY-201 Minicab	Mike Romeo Flying Group	
G-BGMS	Taylor JT.2 Titch	M. A. J. Spice	
G-BGMT	SOCATA Rallye 235E	C. G. Wheeler	
G-BGMV	Scheibe SF.25B Falke	C. A. Bloom & A. P. Twort/Shoreham	
G-BGND	Cessna F.172N	A. J. M. Freeman	
G-BGNT	Cessna F.152	Aerolease Ltd/Conington	
G-BGNV	GA-7 Cougar	G. J. Bissex	
G-BGOD	Colt 77A balloon	C. Allen & M. D. Steuer	
G-BGOG	PA-28-161 Warrior II	W. D. Moore	
G-BGOI	Cameron O-56 balloon	S. Ellis	
G-BGOJ	Cessna F.150L	D. J. Hockings (G-MABI)	
G-BGOL	PA-28R-201T Turbo Arrow III	R. G. Jackson	
G-BGON	GA-7 Cougar	Walsh Aviation	
G-BGOR	AT-6D Harvard III (14863)	P. Meyrick	
G-BGPA	Cessna 182Q	Tindon Ltd/Little Snoring	
G-BGPB	CCF T-6J Texan (1747)	J. Romain/Duxford	
G-BGPD	Piper J-3C-65 Cub (479744)	P. R. Whiteman	
G-BGPH	AA-5B Tiger	Shipping & Airlines Ltd/Biggin Hill	
G-BGPI	Plumb BGP-1	B. G. Plumb	
G-BGPJ	PA-28-161 Warrior II	W. Lancs Warrior Co Ltd/Woodvale	
G-BGPL	PA-28-161 Warrior II	TG Aviation Ltd/Manston	
G-BGPN	PA-18 Super Cub 150	D. McHugh	
G-BGPU	PA-28 Cherokee 140	Air Navigation & Trading Ltd/Blackpool	
G-BGRC	PA-28 Cherokee 140	Tecair Aviation Ltd & G. F. Haigh	
G-BGRE	Beech A200 Super King Air	Martin-Baker (Engineering) Ltd/Chalgrove	
G-BGRG	Beech 76 Duchess	Aviation Rentals/Bournemouth	
G-BGRH	Robin DR.400/22	C. R. Beard	
G-BGRI	Jodel DR.1051	B. J. L. P & W. J. A. L. de Saar	
G-BGRM	PA-38-112 Tomahawk	Classair/Biggin Hill	
G-BGRO	Cessna F.172M	Cammo Aviation	
G-BGRR	PA-38-112 Tomahawk	Goodair Leasing Ltd/Cardiff	
G-BGRS	Thunder Ax7-77Z balloon	P. M. Gaines	
G-BGRT	Steen Skybolt	O. Meier	
G-BGRX	PA-38-112 Tomahawk	Bonus Aviation Ltd	
G-BGSG	PA-44-180 Seminole	Shemburn Ltd	
G-BGSH	PA-38-112 Tomahawk	Thistle Aero Ltd/Carlisle	
G-BGSI	PA-38-112 Tomahawk	Ravenair/Liverpool	
G-BGSJ	Piper J-3C-65 Cub	A. J. Higgins	
G-BGSV	Cessna F.172N	Southwell Air Services Ltd	
G-BGSW	Beech F33 Debonair	C. Wood/Wellesbourne	
G-BGSY	GA-7 Cougar	Plane Talking Ltd/Elstree	
G-BGTB	SOCATA TB.10 Tobago ★	D. Pope *(stored)*	

Notes	Reg.	Type	Owner or Operator
	G-BGTC	Auster AOP.9 (XP282)	P. T. Bolton
	G-BGTF	PA-44-180 Seminole	NG Trustees & Nominees Ltd
	G-BGTG	PA-23 Aztec 250F	Keen Leasing (IOM) Ltd
	G-BGTI	Piper J-3C-65 Cub	A. P. Broad
	G-BGTJ	PA-28 Cherokee 180	Serendipity Aviation/Staverton
	G-BGTT	Cessna 310R	Capital Trading (Aviation) Ltd/Exeter
	G-BGTX	Jodel D.117	Madley Flying Group/Shobdon
	G-BGUB	PA-32 Cherokee Six 300E	A. P. Diplock
	G-BGVB	Robin DR.315	P. J. Leggo
	G-BGVE	CP.1310-C3 Super Emeraude	Victor Echo Group
	G-BGVH	Beech 76 Duchess	Velco Marketing
	G-BGVK	PA-28-161 Warrior II	R. S. Bristowe
	G-BGVN	PA-28RT-201 Arrow IV	C. Smith & S. Carrington
	G-BGVS	Cessna F.172M	Kirkwall Flying Club
	G-BGVV	AA-5A Cheetah	A. H. McVicar/Prestwick
	G-BGVY	AA-5B Tiger	R. J. C. Neal-Smith
	G-BGVZ	PA-28-181 Archer II	W. Walsh & S. R. Mitchell/Woodvale
	G-BGWC	Robin DR.400/180	P. R. Deacon
	G-BGWH	PA-18 Super Cub 150	Spectrum Leisure Ltd (G-ARSR)
	G-BGWJ	Sikorsky S-61N	Bristow Helicopters Ltd
	G-BGWK	Sikorsky S-61N	Bristow Helicopters Ltd
	G-BGWM	PA-28-181 Archer II	Thames Valley Flying Club Ltd
	G-BGWN	PA-38-112 Tomahawk	R. T. Callow
	G-BGWO	Jodel D.112	G-BGWO Group/Sandtoft
	G-BGWR	Cessna U.206A	The Parachute Centre Ltd (G-DISC)/Tilstock
	G-BGWS	Enstrom 280C Shark	Whisky Sierra Helicopters
	G-BGWU	PA-38-112 Tomahawk	J. S. & L. M. Markey
	G-BGWV	Aeronca 7AC Champion	RFC Flying Group/Popham
	G-BGWW	PA-23 Turbo Aztec 250E	Keen Leasing (IOM) Ltd
	G-BGWZ	Eclipse Super Eagle ★	F.A.A. Museum/Yeovilton
	G-BGXA	Piper J-3C-65 Cub (329471)	E. C. & P. King/Kemble
	G-BGXB	PA-38-112 Tomahawk	Signtest Ltd/Cardiff-Wales
	G-BGXC	SOCATA TB.10 Tobago	D. H. Courtley
	G-BGXD	SOCATA TB.10 Tobago	D. F. P. Finan
	G-BGXL	Bensen B.8MV	B. P. Triefus
	G-BGXO	PA-38-112 Tomahawk	Goodwood Terrena Ltd
	G-BGXR	Robin HR.200/100	J. R. Cross
	G-BGXS	PA-28-236 Dakota	G-BGXS Group
	G-BGXT	SOCATA TB.10 Tobago	I. R. Jones
	G-BGYN	PA-18 Super Cub 150	B. J. Dunford
	G-BGYT	EMB-110P1 Bandeirante	Keenair Airways Ltd/Liverpool
	G-BGZF	PA-38-112 Tomahawk	G. G. L. James
	G-BGZW	PA-38-112 Tomahawk	Ravenair/Liverpool
	G-BGZY	Jodel D.120	M. Hale
	G-BGZZ	Thunder Ax6-56 balloon	J. M. Eaton & K. A. Willmore
	G-BHAA	Cessna 152 II	Herefordshire Aero Club Ltd/Shobdon
	G-BHAC	Cessna A.152	Herefordshire Aero Club Ltd/Shobdon
	G-BHAD	Cessna A.152	Shropshire Aero Club Ltd/Sleap
	G-BHAI	Cessna F.152	James D. Peace & Co
	G-BHAJ	Robin DR.400/160	Rowantask Ltd
	G-BHAR	Westland-Bell 47G-3B1	E. A. L. Sturmer
	G-BHAV	Cessna F.152	T. M. & M. L. Jones/Egginton
	G-BHAW	Cessna F.172N	C. Wilson
	G-BHAX	Enstrom F-28C-UK-2	PVS (Barnsley) Ltd
	G-BHAY	PA-28RT-201 Arrow IV	Alpha Yankee Ltd
	G-BHBA	Campbell Cricket	S. N. McGovern
	G-BHBE	Westland-Bell 47G-3B1 (Soloy)	T. R. Smith (Agricultural Machinery) Ltd
	G-BHBF	Sikorsky S-76A	Bristow Helicopters Ltd
	G-BHBG	PA-32R Cherokee Lance 300	J. M. Thorpe
	G-BHBI	Mooney M.20J	G-BHBI Group
	G-BHBT	Marquart MA.5 Charger	R. G. & C. J. Maidment/Shoreham
	G-BHBZ	Partenavia P.68B	Philip Hamer & Co
	G-BHCC	Cessna 172M	D. Wood-Jenkins
	G-BHCE	Jodel D.112	D. M. Parsons
	G-BHCM	Cessna F.172H	J. Dominic
	G-BHCP	Cessna F.152	D. Copley
	G-BHCZ	PA-38-112 Tomahawk	J. E. Abbott
	G-BHDD	V.668 Varsity T.1 (WL626) ★	Aeropark/E. Midlands
	G-BHDE	SOCATA TB.10 Tobago	Alpha-Alpha Ltd
	G-BHDK	Boeing B-29A-BN (461748) ★	Imperial War Museum/Duxford
	G-BHDM	Cessna F.152 II	Big Red Kite Ltd

Reg.	Type	Owner or Operator	Notes
G-BHDP	Cessna F.182Q II	Zone Travel Ltd/White Waltham	
G-BHDR	Cessna F.152 II	James D. Peace & Co	
G-BHDS	Cessna F.152 II	Tayside Aviation Ltd/Dundee	
G-BHDU	Cessna F.152 II	Falcon Flying Services/Biggin Hill	
G-BHDV	Cameron V-77 balloon	P. Glydon	
G-BHDW	Cessna F.152 II	Tayside Aviation Ltd/Dundee	
G-BHDX	Cessna F.172N	GDX Ltd	
G-BHDZ	Cessna F.172N	Abbey Security Services Ltd	
G-BHEC	Cessna F.152 II	Stapleford Flying Club Ltd	
G-BHED	Cessna FA.152	TG Aviation Ltd/Manston	
G-BHEG	Jodel 150	D. M. Griffiths	
G-BHEK	CP.1315-C3 Super Emeraude	D. B. Winstanley/Barton	
G-BHEL	Jodel D.117	N. Wright & C. M. Kettlewell	
G-BHEM	Bensen B.8M	G. C. Kerr	
G-BHEN	Cessna FA.152	Leicestershire Aero Club Ltd	
G-BHEU	Thunder Ax7-65 balloon	D. G. Such	
G-BHEV	PA-28R Cherokee Arrow 200	7-Up Group	
G-BHEX	Colt 56A balloon	A. S. Dear & ptnrs Super Wasp	
G-BHEZ	Jodel 150	Air Yorkshire Group	
G-BHFC	Cessna F.152	Premier Flight Training Ltd	
G-BHFE	PA-44-180 Seminole	Grunwick Processing Laboratories Ltd	
G-BHFF	Jodel D.112	G. H. Gilmour-White	
G-BHFG	SNCAN Stampe SV-4C	A. D. R. Northeast & S. A. Cook	
G-BHFH	PA-34-200T Seneca II	G-WATS Aviation Ltd	
G-BHFI	Cessna F.152	BAe (Warton) Flying Group/Blackpool	
G-BHFJ	PA-28RT-201T Turbo Arrow IV	J. K. Beauchamp	
G-BHFK	PA-28-151 Warrior	Ilkeston Car Sales Ltd	
G-BHFR	Eiri PIK-20E-1	J. T. Morgan	
G-BHFS	Robin DR.400/180	C. J. Moss	
G-BHGC	PA-18 Super Cub 150	Vectis Gliding Club Ltd	
G-BHGF	Cameron V-56 balloon	P. Smallward	
G-BHGJ	Jodel D.120	Q. M. B. Oswell	
G-BHGO	PA-32 Cherokee Six 260	DOCS Ltd/Newcastle	
G-BHGP	SOCATA TB.10 Tobago	D. Suleyman	
G-BHGY	PA-28R Cherokee Arrow 200	R. J. Clark	
G-BHHB	Cameron V-77 balloon	R. Powell	
G-BHHE	Jodel DR.1051/M1	P. Bridges	
G-BHHG	Cessna F.152 II	TG Aviation Ltd/Manston	
G-BHHH	Thunder Ax7-65 balloon	C. A. Hendley (Essex) Ltd	
G-BHHK	Cameron N-77 balloon	I. S. Bridge	
G-BHHN	Cameron V-77 balloon	Itchen Valley Balloon Group	
G-BHHX	Jodel D.112	G-BHHX Group	
G-BHIB	Cessna F.182Q	S. N. Chater & B. Payne	
G-BHIC	Cessna F.182Q	Oxford Aviation Services Ltd	
G-BHIG	Colt 31A Arm Chair SS balloon	P. A. Lindstrand/Sweden	
G-BHII	Cameron V-77 balloon	R. V. Brown	
G-BHIJ	Eiri PIK-20E-1 (898)	I. W. Paterson/Portmoak	
G-BHIK	Adam RA-14 Loisirs	L. Lewis	
G-BHIL	PA-28-161 Warrior II	Falcon Flying Services (G-SSFT)/Biggin Hill	
G-BHIN	Cessna F.152	Cristal Air Ltd	
G-BHIR	PA-28R Cherokee Arrow 200	Factorcore Ltd/Barton	
G-BHIS	Thunder Ax7-65 balloon	Hedgehoppers Balloon Group	
G-BHIT	SOCATA TB.9 Tampico	C. J. P. Webster/Biggin Hill	
G-BHIY	Cessna F.150K	G. J. Ball	
G-BHJF	SOCATA TB.10 Tobago	Flying Fox Group/Blackbushe	
G-BHJI	Mooney M.20J	Hearing Centre Aarhus/Denmark	
G-BHJK	Maule M5-235C Lunar Rocket	JK Group	
G-BHJN	Fournier RF-4D	RF-4 Group	
G-BHJO	PA-28-161 Warrior II	Brackla Flying Group	
G-BHJS	Partenavia P.68B	J. J. Watts & D. Fletcher	
G-BHJU	Robin DR.400/2+2	J. Barlow & P. Crow	
G-BHKH	Cameron O-65 balloon	P. Donkin	
G-BHKJ	Cessna 421C	Totaljet Ltd	
G-BHKR	Colt 12A balloon ★	British Balloon Museum/Newbury	
G-BHKT	Jodel D.112	M. G. Davis	
G-BHLE	Robin DR.400/180	B. D. Greenwood	
G-BHLH	Robin DR.400/180	G-BHLH Group	
G-BHLJ	Saffery-Rigg S.200 balloon	I. A. Rigg	
G-BHLT	D.H.82A Tiger Moth	P. J. & A. J. Borsberry	
G-BHLU	Fournier RF-3	Skyview Systems Ltd	
G-BHLW	Cessna 120	L. W. Scattergood	
G-BHLX	AA-5B Tiger	M. D. McPherson	

Notes	Reg.	Type	Owner or Operator
	G-BHMA	SIPA 903	H. J. Taggart
	G-BHMG	Cessna FA.152	R. D. Smith
	G-BHMI	Cessna F.172N	GMI Aviation Ltd (G-WADE)
	G-BHMJ	Avenger T.200-2112 balloon	R. Light *Lord Anthony 1*
	G-BHMK	Avenger T.200-2112 balloon	P. Kinder *Lord Anthony 2*
	G-BHMR	Stinson 108-3	D. G. French/Sandown
	G-BHMT	Evans VP-1	P. E. J. Sturgeon
	G-BHMY	F.27 Friendship Mk.200 ★	Norwich Aviation Museum
	G-BHNA	Cessna F.152	Sheffield Aero Club Ltd/Netherthorpe
	G-BHNC	Cameron O-65 balloon	D. & C. Bareford
	G-BHND	Cameron N-65 balloon	S. M. Wellband
	G-BHNK	Jodel D.120A	G-BHNK Flying Group
	G-BHNL	Jodel D.112	HNL Group
	G-BHNO	PA-28-181 Archer II	Airfluid Hydraulics & Pneumatics (Wolverhampton) Ltd
	G-BHNP	Eiri PIK-20E-1	D. A. Sutton/Riseley
	G-BHNV	Westlan-Bell 47G-3B1	Leyline Helicopters Ltd
	G-BHNX	Jodel D.117	A. J. Chalkley
	G-BHOA	Robin DR.400/160	Goudhurst Service Station Ltd
	G-BHOG	Sikorsky S-61N Mk.II	Bristow Helicopters Ltd
	G-BHOH	Sikorsky S-61N Mk.II	Bristow Helicopters Ltd
	G-BHOJ	Colt 12A balloon	J. A. Folkes
	G-BHOL	Jodel DR.1050	J. E. Sharkey
	G-BHOM	PA-18 Super Cub 95	Oscar Mike Flying Group/Andrewsfield
	G-BHOO	Thunder Ax7-65 balloon	D. Livesey & J. M. Purves Scraps
	G-BHOR	PA-28-161 Warrior II	Oscar Romeo Flying Group/Biggin Hill
	G-BHOT	Cameron V-65 balloon	Dante Balloon Group
	G-BHOZ	SOCATA TB.9 Tampico	G-BHOZ Management Ltd/Kemble
	G-BHPK	Piper J-3C-65 Cub (238410/A-44)	L-4 Group
	G-BHPL	C.A.S.A. 1.131E Jungmann 1000 (E3B-350) ★	R. G. Gray/North Weald
	G-BHPM	PA-18 Super Cub 95	P. I. Morgans
	G-BHPN	Colt 14A balloon	Lindstrand Balloons Ltd/Sweden
	G-BHPS	Jodel D.120A	T. J. Price
	G-BHPY	Cessna 152 II	Halegreen Associates
	G-BHPZ	Cessna 172N	O'Brien Properties Ltd/Redhill
	G-BHRB	Cessna F.152 II	LAC (Enterprises) Ltd/Barton
	G-BHRC	PA-28-161 Warrior II	Sherwood Flying Club Ltd/Tollerton
	G-BHRH	Cessna FA.150K	Merlin Flying Club Ltd/Hucknall
	G-BHRM	Cessna F.152	Tatenhill Aviation Ltd
	G-BHRN	Cessna F.152	James D. Peace & Co
	G-BHRP	PA-44-180 Seminole	M. S. Farmers
	G-BHRR	CP.301A Emeraude	T. W. Offen
	G-BHRW	Jodel DR.221	Dauphin Flying Group
	G-BHRY	Colt 56A balloon	A. S. Davidson
	G-BHSB	Cessna 172N	SB Aviation Ltd
	G-BHSD	Scheibe SF.25E Super Falke	S. J. Filhol
	G-BHSE	R. Commander 114	604 Sqdn Flying Group Ltd
	G-BHSN	Cameron N-56 balloon	I. Bentley
	G-BHSP	Thunder Ax7-77Z balloon	Out-Of-The-Blue
	G-BHSS	Pitts S-1C Special	C. W. Burkett
	G-BHSY	Jodel DR.1050	T. R. Allebone
	G-BHTA	PA-28-236 Dakota	Dakota Ltd
	G-BHTC	Jodel DR.1050/M1	G. Clark
	G-BHTG	Thunder Ax6-56 balloon	F. R. & Mrs S. H. MacDonald
	G-BHUB	Douglas C-47A (315509) ★	Imperial War Museum/Duxford
	G-BHUE	Jodel DR.1050	M. J. Harris
	G-BHUG	Cessna 172N	FGT Aircraft Hire
	G-BHUI	Cessna 152	Galair International Ltd
	G-BHUJ	Cessna 172N	Uniform Juliet Group/Southend
	G-BHUM	D.H.82A Tiger Moth	S. G. Towers
	G-BHUR	Thunder Ax3 balloon	B. F. G. Ribbans
	G-BHUU	PA-25 Pawnee 235	Booker Gliding Club Ltd
	G-BHVB	PA-28-161 Warrior II	P. J. Clarke
	G-BHVF	Jodel 150A	J. D. Walton
	G-BHVP	Cessna 182Q	R. J. W. Wood
	G-BHVR	Cessna 172N	G-BHVR Group
	G-BHVV	Piper J-3C-65 Cub	C. A. Ward & C. A. Cash
	G-BHWA	Cessna F.152	Lincoln Aviation Ltd/Wickenby
	G-BHWB	Cessna F.152	Lincoln Aviation Ltd/Wickenby
	G-BHWH	Weedhopper JC-24A	G. A. Clephane
	G-BHWK	M.S.880B Rallye Club	L. L. Gayther

Reg.	Type	Owner or Operator	Notes
G-BHWY	PA-28R Cherokee Arrow 200-II	Kilo Foxtrot Flying Group/Sandown	
G-BHWZ	PA-28-181 Archer II	M. A. Abbott	
G-BHXA	SA Bulldog Srs 120/1210	Air Plan Flight Equipment Ltd/Barton	
G-BHXD	Jodel D.120	J. M. Fforde & M. Roberts	
G-BHXK	PA-28 Cherokee 140	GXK Flying Group	
G-BHXS	Jodel D.120	I. R. Willis	
G-BHXY	Piper J-3C-65 Cub (44-79609)	F. W. Rogers/Aldergrove	
G-BHYA	Cessna R.182RG II	B. Davies	
G-BHYC	Cessna 172RG II	IB Aeroplanes Ltd	
G-BHYD	Cessna R.172RK XP II	Sylmar Aviation Services Ltd	
G-BHYE	PA-34-200T Seneca II	Oxford Aviation Services Ltd/Kidlington	
G-BHYF	PA-34-200T Seneca II	Oxford Aviation Services Ltd/Kidlington	
G-BHYG	PA-34-200T Seneca II	Oxford Aviation Services Ltd/Kidlington	
G-BHYI	SNCAN Stampe SV-4A	G. L. Brown & A. M. Plato	
G-BHYO	Cameron N-77 balloon	Adventure Balloon Co Ltd	
G-BHYP	Cessna F.172M	Avior Ltd/Biggin Hill	
G-BHYR	Cessna F.172M	G-BHYR Group	
G-BHYV	Evans VP-1	L. Chiappi/Blackpool	
G-BHYX	Cessna 152 II	Stapleford Flying Club Ltd	
G-BHZE	PA-28-181 Archer II	Zegruppe Ltd	
G-BHZH	Cessna F.152	Plymouth School of Flying Ltd	
G-BHZK	AA-5B Tiger	ZK Group/Elstree	
G-BHZO	AA-5A Cheetah	Scotia Safari Ltd/Prestwick	
G-BHZR	SA Bulldog Srs 120/1210	White Knuckle Air Ltd	
G-BHZS	SA Bulldog Srs 120/1210	Air Plan Flight Equipment Ltd/Hawarden	
G-BHZT	SA Bulldog Srs 120/1210	D. M. Curties	
G-BHZU	Piper J-3C-65 Cub	J. K. Tomkinson	
G-BHZV	Jodel D.120A	K. J. Scott	
G-BHZX	Thunder Ax7-65A balloon	R. J. & H. M. Beattie	
G-BIAC	SOCATA Rallye 235E	D. R. Watson & A. J. Haigh	
G-BIAH	Jodel D.112	T. K. Duffy	
G-BIAI	WMB.2 Windtracker balloon	I. Chadwick	
G-BIAP	PA-16 Clipper	P. J. Bish/White Waltham	
G-BIAR	Rigg Skyliner II balloon	I. A. Rigg	
G-BIAU	Sopwith Pup (replica) (N6452) ★	F.A.A. Museum/Yeovilton	
G-BIAX	Taylor JT.2 Titch	J. T. Everest	
G-BIAY	AA-5 Traveler	M. D. Dupay & ptnrs	
G-BIBA	SOCATA TB.9 Tampico	TB Aviation Ltd	
G-BIBB	Mooney M.20C	Lefay Engineering Ltd	
G-BIBG	Sikorsky S-76A II	Bristow Helicopters Ltd	
G-BIBJ	Enstrom 280C-UK-2	Tindon Ltd/Little Snoring	
G-BIBN	Cessna FA.150K	B. V. Mayo	
G-BIBO	Cameron V-65 balloon	I. Harris	
G-BIBS	Cameron P-20 balloon	Cameron Balloons Ltd	
G-BIBT	AA-5B Tiger	Vizor Tempered Glass Ltd	
G-BIBW	Cessna F.172N	Drawflight Ltd	
G-BIBX	WMB.2 Windtracker balloon	I. A. Rigg	
G-BICD	Auster 5	T. R. Parsons	
G-BICE	AT-6C Harvard IIA (41-33275)	C. M. L. Edwards	
G-BICG	Cessna F.152 II	Falcon Flying Services/Biggin Hill	
G-BICJ	Monnett Sonerai II	I. Parr	
G-BICM	Colt 56A balloon	Avon Advertiser Balloon Club	
G-BICP	Robin DR.360	N. Thorne/Breighton	
G-BICR	Jodel D.120A	Beehive Flying Group/White Waltham	
G-BICS	Robin R.2100A	I. Young/Sandown	
G-BICU	Cameron V-56 balloon	S. D. Bather & D. Scott	
G-BICX	PA-28-161 Warrior II	D. Gellhorn/Blackbushe	
G-BICX	Maule M5-235C Lunar Rocket	A. T. Jeans & J. F. Clarkson/Old Sarum	
G-BICY	PA-23 Apache 160	A. M. Lynn/Sibson	
G-BIDD	Evans VP-1	J. Hodgkinson	
G-BIDF	Cessna F.172P	C. J. Chaplin & N. J. C. Howard	
G-BIDG	Jodel 150A	D. R. Gray/Barton	
G-BIDH	Cessna 152 II	Hull Aero Club Ltd (G-DONA)	
G-BIDI	PA-28R-201 Arrow III	Ambrit Ltd	
G-BIDJ	PA-18A Super Cub 150	Flight Solutions Ltd	
G-BIDK	PA-18 Super Cub 150	J. & M. A. McCullough	
G-BIDO	CP.301A Emeraude	A. R. Plumb	
G-BIDV	Colt 14A balloon ★	British Balloon Museum/Newbury	
G-BIDW	Sopwith 1½ Strutter (replica) (A8226) ★	RAF Museum/Hendon	
G-BIDX	Jodel D.112	P. Turton & H. C. Peake-Jones	
G-BIEF	Cameron V-77 balloon	D. S. Bush	

Notes	Reg.	Type	Owner or Operator
	G-BIEJ	Sikorsky S-76A	Bristow Helicopters Ltd
	G-BIEN	Jodel D.120A	C. A. J. Van Andel/Netherlands
	G-BIEO	Jodel D.112	Clipgate Flyers
	G-BIES	Maule M5-235C Lunar Rocket	William Proctor Farms
	G-BIET	Cameron O-77 balloon	G. M. Westley
	G-BIEY	PA-28-151 Warrior	Falcon Flying Services/Biggin Hill
	G-BIFA	Cessna 310R II	J. S. Lee
	G-BIFB	PA-28 Cherokee 150C	P. Coombs
	G-BIFN	Bensen B.8MR	B. Gunn
	G-BIFO	Evans VP-1	R. Broadhead
	G-BIFY	Cessna F.150L	Bonus Aviation Ltd
	G-BIGJ	Cessna F.172M	V. D. Speck/Clacton
	G-BIGK	Taylorcraft BC-12D	N. P. S. Ramsay
	G-BIGL	Cameron O-65 balloon	P. L. Mossman
	G-BIGP	Bensen B.8M	R. H. S. Cooper
	G-BIGR	Avenger T.200-2112 balloon	R. Light
	G-BIGZ	Scheibe SF.25B Falke	The Big Z Owners Group
	G-BIHD	Robin DR.400/160	A. J. Fieldman
	G-BIHF	SE-5A (replica) (F943)	S. H. O'connell/White Waltham
	G-BIHI	Cessna 172M	Fenland Flying School
	G-BIHO	D.H.C.6 Twin Otter 310	Isles of Scilly Skybus Ltd/St. Just
	G-BIHP	Van Den Bemden gas balloon	J. J. Harris
	G-BIHT	PA-17 Vagabond	Agri Air Services Ltd
	G-BIHU	Saffrey S.200 balloon	B. L. King
	G-BIHW	Aeronca A65TAC Defender	T. J. Ingrouille
	G-BIHX	Bensen B.8M	P. P. Willmott
	G-BIIA	Fournier RF-3	J. D. Webb & J. D. Bally
	G-BIIB	Cessna F.172M	Civil Service Flying Club (Biggin Hill) Ltd
	G-BIID	PA-18 Super Cub 95	D. A. Lacey
	G-BIIE	Cessna F.172P	Sterling Helicopters Ltd
	G-BIIK	M.S.883 Rallye 115	N. J. Garbett
	G-BIIL	Thunder Ax6-56 balloon	I. M. Ashpole
	G-BIIP	BN-2B-27 Islander	Airx Ltd/Bournemouth
	G-BIIT	PA-28-161 Warrior II	Tayside Aviation Ltd/Dundee
	G-BIIV	PA-28-181 Archer II	J. Thuret/France
	G-BIIZ	Great Lakes 2T-1A Sport Trainer	Circa 42 Ltd
	G-BIJB	PA-18 Super Cub 150	James Aero Ltd/Stapleford
	G-BIJD	Bo 208C Junior	Sikh Sydicate
	G-BIJE	Piper J-3C-65 Cub	R. L. Hayward & A. G. Scott
	G-BIJS	Luton LA-4A Minor	I. J. Smith
	G-BIJU	CP-301A Emeraude	Eastern Taildraggers Flying Clug (G-BHTX)
	G-BIJV	Cessna F.152 II	Falcon Flying Services/Biggin Hill
	G-BIJW	Cessna F.152 II	Falcon Flying Services/Biggin Hill
	G-BIJX	Cessna F.152 II	Falcon Flying Services/Biggin Hill
	G-BIKC	Boeng 757-236F	DHL Air Ltd
	G-BIKE	PA-28R Cherokee Arrow 200	R. V. Webb
	G-BIKF	Boeing 757-236F	DHL Air Ltd
	G-BIKG	Boeing 757-236F	DHL Air Ltd
	G-BIKI	Boeing 757-236F	DHL Air Ltd
	G-BIKJ	Boeing 757-236F	DHL Air Ltd
	G-BIKK	Boeing 757-236F	DHL Air Ltd
	G-BIKM	Boeing 757-236F	DHL Air Ltd
	G-BIKN	Boeing 757-236F	DHL Air Ltd
	G-BIKO	Boeing 757-236F	DHL Air Ltd
	G-BIKP	Boeing 757-236F	DHL Air Ltd
	G-BIKS	Boeing 757-236F	DHL Air Ltd
	G-BIKU	Boeing 757-236F	DHL Air Ltd
	G-BIKV	Boeing 757-236F	DHL Air Ltd
	G-BIKW	Boeing 757-236F	DHL Air Ltd
	G-BIKX	Boeing 757-236F	DHL Air Ltd
	G-BIKY	Boeing 757-236F	DHL Air Ltd
	G-BIKZ	Boeing 757-236F	DHL Air Ltd
	G-BILB	WMB.2 Windtracker balloon	B. L. King
	G-BILE	Scruggs BL.2B balloon	P. D. Ridout
	G-BILG	Scruggs BL.2B balloon	P. D. Ridout
	G-BILI	Piper J-3C-65 Cub (454467)	G-BILI Flying Group
	G-BILJ	Cessna FA.152	I. R. March
	G-BILL	PA-25 Pawnee 235	Pawnee Aviation
	G-BILR	Cessna 152 II	Shropshire Aero Club Ltd/Sleap
	G-BILS	Cessna 152 II	Mona Flying Club
	G-BILU	Cessna 172RG	Full Sutton Flying Centre Ltd
	G-BILZ	Taylor JT.1 Monoplane	A. Petherbridge

Reg.	Type	Owner or Operator	Notes
G-BIMK	Tiger T.200 Srs 1 balloon	M. K. Baron	
G-BIMM	PA-18 Super Cub 150	Spectrum Leisure Ltd/Clacton	
G-BIMN	Steen Skybolt	R. J. Thomas	
G-BIMO	SNCAN Stampe SV-4C	R. A. Roberts	
G-BIMT	Cessna FA.152	Staverton Flying Services Ltd	
G-BIMU	Sikorsky S-61N	Bristow Helicopters Ltd	
G-BIMX	Rutan Vari-Eze	D. G. Crow/Biggin Hill	
G-BIMZ	Beech 76 Duchess	R. P. Smith	
G-BING	Cessna F.172P	20th Air Training Group Ltd	
G-BINL	Scruggs BL.2B balloon	P. D. Ridout	
G-BINM	Scruggs BL.2B balloon	P. D. Ridout	
G-BINR	Unicorn UE.1A balloon	Unicorn Group	
G-BINS	Unicorn UE.2A balloon	Unicorn Group	
G-BINT	Unicorn UE.1A balloon	D. E. Bint	
G-BINX	Scruggs BL.2B balloon	P. D. Ridout	
G-BINY	Oriental balloon	J. L. Morton	
G-BIOA	Hughes 369D	AH Helicopter Services Ltd	
G-BIOB	Cessna F.172P	Simmons Aerofilms Ltd/Elstree	
G-BIOC	Cessna F.150L	Southside Flyers	
G-BIOI	Jodel DR.1051/M	R. Pidcock	
G-BIOJ	R. Commander 112TCA	A. T. Dalby	
G-BIOK	Cessna F.152	Tayside Aviation Ltd/Dundee	
G-BIOM	Cessna F.152	J. B. P. E. Fernandes	
G-BIOU	Jodel D.117A	Jemalk Group	
G-BIOW	Slingsby T.67A	A. B. Slinger/Sherburn	
G-BIPA	AA-5B Tiger	J. Campbell/Walney Island	
G-BIPH	Scruggs BL.2B balloon	C. M. Dewsnap	
G-BIPI	Everett gyroplane	C. A. Reeves	
G-BIPN	Fournier RF-3	J. C. R. Rogers & I. F. Fairhead	
G-BIPO	Mudry/CAARP CAP.20LS-200M.	The CAP-20 Group/White Waltham	
G-BIPS	SOCATA Rallye 100ST	McAully Flying Group/Little Snoring	
G-BIPT	Jodel D.112	C. R. Davies	
G-BIPV	AA-5B Tiger	Databridge Services Ltd	
G-BIPW	Avenger T.200-2112 balloon	B. L. King	
G-BIPY	Montgomerie-Bensen B.8MR	C. G. Ponsford	
G-BIRD	Pitts S-1D Special	P. Metcalfe	
G-BIRE	Colt 56 Bottle SS balloon	K. R. Gafney	
G-BIRH	PA-18 Super Cub 135 (R-163)	Aquila Gliding Club Ltd	
G-BIRI	C.A.S.A. 1.131E Jungmann 1000	M. G. & J. R. Jeffries	
G-BIRL	Avenger T.200-2112 balloon	R. Light	
G-BIRP	Arena Mk 17 Skyship balloon	A. S. Viel	
G-BIRT	Robin R.1180TD	W. D'A. Hall/Booker	
G-BIRW	M.S.505 Criquet (F+IS) ★	Museum of Flight/E. Fortune	
G-BIRZ	Zenair CH.250	D. Johnston & M. K. McGreavey	
G-BISG	FRED Srs 3	T. Littlefair	
G-BISH	Cameron V-65 balloon	P. J. Bish	
G-BISK	R. Commander 112B ★	P. A. Warner	
G-BISL	Scruggs BL.2B balloon	P. D. Ridout	
G-BISM	Scruggs BL.2B balloon	P. D. Ridout	
G-BISS	Scruggs BL.2C balloon	P. D. Ridout	
G-BIST	Scruggs BL.2C balloon	P. D. Ridout	
G-BISX	Colt 56A balloon	C. D. Steel	
G-BISZ	Sikorsky S-76A	Bristow Helicopters Ltd	
G-BITA	PA-18 Super Cub 150	Intrepid Aviation Co/North Weald	
G-BITE	SOCATA TB.10 Tobago	M. A. Smith/Fairoaks	
G-BITF	Cessna F.152 II	Tayside Aviation Ltd/Dundee	
G-BITK	FRED Srs 2	D. J. Wood	
G-BITM	Cessna F.172P	Dreamtrade Ltd	
G-BITO	Jodel D.112D	A. Dunbar/Barton	
G-BITS	Drayton B-56 balloon	M. J. Betts	
G-BITY	FD.31T balloon	A. J. Bell	
G-BIUM	Cessna F.152	Sheffield Aero Club Ltd/Netherthorpe	
G-BIUP	SNCAN NC.854S	J. Greenaway & T. D. Cooper	
G-BIUU	PA-23 Aztec 250D ★	G. Cormack/Glasgow	
G-BIUV	H.S.748 Srs 2A	Emerald Airways Ltd *City of Liverpool* (G-AYYH)/ Liverpool	
G-BIUW	PA-28-161 Warrior II	D. R. Staley/Sturgate	
G-BIUY	PA-28-181 Archer II	J. S. Devlin & Z. Islam	
G-BIVA	Robin R.2112	P. A. Richardson	
G-BIVB	Jodel D.112	B. L. Proctor & T. F. Lyddon	
G-BIVC	Jodel D.112	M. J. Barmby/Cardiff	
G-BIVF	CP.301C-3 Emeraude	R. J. Moore	

Notes	Reg.	Type	Owner or Operator
	G-BIVK	Bensen B.8M	M. J. Atyeo
	G-BIVV	AA-5A Cheetah	Robert Afia Consulting Engineer
	G-BIWB	Scruggs RS.5000 balloon	P. D. Ridout
	G-BIWC	Scruggs RS.5000 balloon	P. D. Ridout
	G-BIWF	Warren balloon	P. D. Ridout
	G-BIWG	Zelenski Mk 2 balloon	P. D. Ridout
	G-BIWJ	Unicorn UE.1A balloon	B. L. King
	G-BIWK	Cameron V-65 balloon	I. R. Williams & R. G. Bickerdike
	G-BIWL	PA-32-301 Saratoga	J. D. Richardson
	G-BIWN	Jodel D.112	C. R. Coates
	G-BIWR	Mooney M.20F	A. C. Brink
	G-BIWU	Cameron V-65 balloon	D. J. Groombridge
	G-BIWW	AA-5 Traveler	S. J. Perkins & D. Dobson
	G-BIWY	Westland WG.30 ★	Instructional airframe/Yeovil
	G-BIXA	SOCATA TB.9 Tampico	W. Maxwell
	G-BIXB	SOCATA TB.9 Tampico	Cinque Ports Aviation Ltd
	G-BIXH	Cessna F.152	Northern Aviation Ltd
	G-BIXI	Cessna 172RG Cutlass	J. F. P. Lewis/Sandown
	G-BIXL	P-51D Mustang (472216)	R. Lamplough/North Weald
	G-BIXN	Boeing Stearman A.75N1	V. S. E. Norman/Rendcomb
	G-BIXV	Bell 212	Bristow Helicopters Ltd
	G-BIXW	Colt 56B balloon	N. A. P. Bates
	G-BIXX	Pearson Srs 2 balloon	D. Pearson
	G-BIXZ	Grob G-109	D. L. Nind & I. Allum/Booker
	G-BIYI	Cameron V-65 balloon	Sarnia Balloon Group
	G-BIYJ	PA-18 Super Cub 95	S. Russell
	G-BIYK	Isaacs Fury II	S. M. Roberts
	G-BIYP	PA-20 Pacer 125	A. W. Hoy & S. W. M. Johnson
	G-BIYR	PA-18 Super Cub 150 (R-151)	Delta Foxtrot Flying Group
	G-BIYT	Colt 17A balloon	J. M. Francois/France
	G-BIYU	Fokker S.11.1 Instructor (E-15)	C. Briggs
	G-BIYW	Jodel D.112	Pollard/Balaam/Bye Flying Group
	G-BIYX	PA-28 Cherokee 140	W. B. Bateson/Blackpool
	G-BIYY	PA-18 Super Cub 95	A. E. & W. J. Taylor/Ingoldmells
	G-BIZE	SOCATA TB.9 Tampico	C. Fordham
	G-BIZF	Cessna F.172P	R. S. Bentley/Bourn
	G-BIZG	Cessna F.152	M. A. Judge
	G-BIZI	Robin DR.400/120	Headcorn Flying School Ltd
	G-BIZK	Nord 3202	A. I. Milne/Swanton Morley
	G-BIZM	Nord 3202	Global Aviation Ltd/Humberside
	G-BIZO	PA-28R Cherokee Arrow 200	Lemas Air
	G-BIZR	SOCATA TB.9 Tampico	Fenland Flying Group (G-BSEC)
	G-BIZU	Thunder Ax6-56Z balloon	M. J. Loades
	G-BIZV	PA-18 Super Cub 95 (18-2001)	A. W. & M. E. Jenkins
	G-BIZW	Champion 7GCBC Citabria	G. Read & Sons
	G-BIZY	Jodel D.112	Wayland Tunley & Associates/Cranfield
	G-BJAD	FRED Srs 2	Newark (Nottinghamshire & Lincolnshire) Air Museum
	G-BJAE	Lavadoux Starck AS.80	D. J. & S. A. E. Phillips/Coventry
	G-BJAF	Piper J-3C-65 Cub	P. J. Cottle
	G-BJAG	PA-28-181 Archer II	C. R. Chubb
	G-BJAJ	AA-5B Tiger	A. H. McVicar/Prestwick
	G-BJAL	C.A.S.A. 1.131E Jungmann 1000	I. C. Underwood & S. B. J. Chandler/Breighton
	G-BJAO	Bensen B.8M	A. P. Lay
	G-BJAP	D.H.82A Tiger Moth (K2587)	K. Knight
	G-BJAS	Rango NA.9 balloon	A. Lindsay
	G-BJAV	GY-80 Horizon 160	P. L. Lovegrove
	G-BJAW	Cameron V-65 balloon	G. W. McCarthy
	G-BJAY	Piper J-3C-65 Cub	D. W. Finlay
	G-BJBK	PA-18 Super Cub 95	M. S. Bird/Old Sarum
	G-BJBM	Monnett Sonerai I	Sonerai G-BJBM Group
	G-BJBO	Jodel DR.250/160	Wiltshire Flying Group
	G-BJBW	PA-28-161 Warrior II	152 Group
	G-BJBX	PA-28-161 Warrior II	Haimoss Ltd
	G-BJCA	PA-28-161 Warrior II	Plane Sailing (Southwest) Ltd
	G-BJCF	CP.1310-C3 Super Emeraude	K. M. Hodson & C. G. H. Gurney
	G-BJCI	PA-18 Super Cub 150 (modified)	The Borders (Milfield) Gliding Club Ltd
	G-BJCW	PA-32R-301 Saratoga SP	Golf Charlie Whisky Ltd
	G-BJDE	Cessna F.172M	Cranfield Aircraft Partnership
	G-BJDF	M.S.880B Rallye 100T	G-BJDF Group
	G-BJDJ	H.S.125 Srs 700B	TAG Farnborough Engineering Ltd (G-RCDI)

Reg.	Type	Owner or Operator	Notes
G-BJDK	European E.14 balloon	Aeroprint Tours	
G-BJDO	AA-5A Cheetah	Flying Services	
G-BJDW	Cessna F.172M	J. Rae	
G-BJEE	BN-2T Turbine Islander	Cormack (Aircraft Services) Ltd	
G-BJEF	BN-2T Turbine Islander	Cormack (Aircraft Services) Ltd	
G-BJEI	PA-18 Super Cub 95	H. J. Cox	
G-BJEJ	BN-2T Turbine Islander	Cormack (Aircraft Services) Ltd	
G-BJEL	SNCAN NC.854	N. F. & S. G. Hunter	
G-BJEV	Aeronca 11AC Chief (897)	R. F. Willcox	
G-BJEX	Bo 208C Junior	G. D. H. Crawford/Thruxton	
G-BJFC	European E.8 balloon	P. D. Ridout	
G-BJFE	PA-18 Super Cub 95	P. H. Wilmot-Allistone	
G-BJFL	Sikorsky S-76A	Bristow Helicopters Ltd	
G-BJFM	Jodel D.120	J. V. George & P. A. Smith/Popham	
G-BJGK	Cameron V-77 balloon	M. E. Orchard	
G-BJGM	Unicorn UE.1A balloon	D. Eaves & P. D. Ridout	
G-BJGX	Sikorsky S-76A	Bristow Helicopters Ltd	
G-BJGY	Cessna F.172P	K. & S. Martin	
G-BJHB	Mooney M.20J	Zitair Flying Club Ltd/Redhill	
G-BJHK	EAA Acro Sport	M. R. Holden	
G-BJHV	Voisin Replica ★	Brooklands Museum of Aviation/Weybridge	
G-BJIA	Allport balloon	D. J. Allport	
G-BJIC	Dodo 1A balloon	P. D. Ridout	
G-BJID	Osprey 1B balloon	P. D. Ridout	
G-BJIG	Slingsby T.67A	G-BJIG Slingsby Syndicate	
G-BJIR	Cessna 550 Citation II	Gator Aviation Ltd	
G-BJIV	PA-18 Super Cub 180	Yorkshire Gliding Club (Pty) Ltd/Sutton Bank	
G-BJKF	SOCATA TB.9 Tampico	P. M. A. Croton	
G-BJKW	Wills Aera II	J. K. S. Wills	
G-BJKY	Cessna F.152	Air Charter & Travel Ltd/Ronaldsway	
G-BJLB	SNCAN NC.854S	M. J. Barnaby	
G-BJLC	Monnett Sonerai IIL	A. R. Ansell	
G-BJLX	Cremer balloon	P. W. May	
G-BJLY	Cremer balloon	P. Cannon	
G-BJML	Cessna 120	D. F. Lawlor/Inverness	
G-BJMO	Taylor JT.1 Monoplane	R. C. Mark	
G-BJMR	Cessna 310R	J. McL. Robinson/Sherburn	
G-BJMW	Thunder Ax8-105 balloon	G. M. Westley	
G-BJMX	Jarre JR.3 balloon	P. D. Ridout	
G-BJMZ	European EA.8A balloon	P. D. Ridout	
G-BJNA	Arena Mk 117P balloon	P. D. Ridout	
G-BJND	Osprey Mk 1E balloon	A. Billington & D. Whitmore	
G-BJNF	Cessna F.152	D. M. & B. Cloke	
G-BJNG	Slingsby T.67AM	D. F. Hodgkinson	
G-BJNN	PA-38-112 Tomahawk	Thistle Aero Ltd/Carlisle	
G-BJNY	Aeronca 11CC Super Chief	P. I. & D. M. Morgans	
G-BJNZ	PA-23 Aztec 250F	Bonus Aviation Ltd (G-FANZ)/Cranfield	
G-BJOA	PA-28-181 Archer II	Tatenhill Aviation Ltd	
G-BJOB	Jodel D.140C	T. W. M. Beck & M. J. Smith	
G-BJOE	Jodel D.120A	Forth Flying Group	
G-BJOP	BN-2B-26 Islander	Loganair Ltd/BA Express	
G-BJOT	Jodel D.117	R. H. Ryle & ptnrs	
G-BJOV	Cessna F.150K	J. A. Boyd	
G-BJPI	Bede BD-5G	M. D. McQueen	
G-BJRA	Osprey Mk 4B balloon	E. Osborn	
G-BJRG	Osprey Mk 4B balloon	A. E. de Gruchy	
G-BJRH	Rango NA.36 balloon	N. H. Ponsford	
G-BJRP	Cremer balloon	M. D. Williams	
G-BJRR	Cremer balloon	M. D. Williams	
G-BJRV	Cremer balloon	M. D. Williams	
G-BJSS	Allport balloon	D. J. Allport	
G-BJST	CCF T-6J Harvard IV	Tuplin Holdings Ltd	
G-BJSV	PA-28-161 Warrior II	Airways Flight Training (Exeter) Ltd	
G-BJSW	Thunder Ax7-65 balloon	Sandicliffe Garage Ltd	
G-BJSZ	Piper J-3C-65 Cub	H. Gilbert	
G-BJTB	Cessna A.150M	V. D. Speck/Clacton	
G-BJTO	Piper J-3C-65 Cub	K. R. Nunn	
G-BJTP	PA-18 Super Cub 95 (115302)	J. T. Parkins	
G-BJTY	Osprey Mk 4B balloon	A. E. de Gruchy	
G-BJUB	BVS Special 01 balloon	P. G. Wild	
G-BJUC	Robinson R-22HP	JD Gallagher Estate Agents Ltd	
G-BJUD	Robin DR.400/180R	Lasham Gliding Soc Ltd	

Notes	Reg.	Type	Owner or Operator
	G-BJUR	PA-38-112 Tomahawk	Truman Aviation Ltd/Tollerton
	G-BJUS	PA-38-112 Tomahawk	Panshanger School of Flying
	G-BJUV	Cameron V-20 balloon	P. Spellward
	G-BJUY	Colt Ax7-77 Golf Ball SS balloon	Balloon Sports HB/Sweden
	G-BJVC	Evans VP-2	C. J. Morris
	G-BJVH	Cessna F.182Q	R. J. de Courcy Cuming/Wellesbourne
	G-BJVJ	Cessna F.152	Henlow Flying Club
	G-BJVK	Grob G-109	B. Kimberley/Enstone
	G-BJVM	Cessna 172N	I. C. MacLennan
	G-BJVS	CP.1310-C3 Super Emeraude	BJVS Group
	G-BJVT	Cessna F.152	Northern Aviation Ltd
	G-BJVU	Thunder Ax6-56 Bolt SS balloon	G. V. Beckwith
	G-BJVV	Robin R.1180	Medway Flying Group Ltd/Rochester
	G-BJWH	Cessna F.152 II	Linkcrest Ltd/Elstree
	G-BJWI	Cessna F.172P	Bflying Ltd
	G-BJWJ	Cameron V-65 balloon	R. G. Turnbull & S. G. Forse
	G-BJWO	BN-2A-26 Islander	Falcons Parachute Centre (G-BAXC)
	G-BJWT	Wittman W.10 Tailwind	Tailwind Group
	G-BJWV	Colt 17A balloon	D. T. Meyes
	G-BJWW	Cessna F.172N	Air Charter & Travel Ltd/Blackpool
	G-BJWX	PA-18 Super Cub 95	G-BJWX Syndicate
	G-BJWY	S-55 Whirlwind HAR.21(WV198) ★	Solway Aviation Museum/Carlisle
	G-BJWZ	PA-18 Super Cub 95	G-BJWZ Syndicate/Redhill
	G-BJXA	Slingsby T.67A	Liberty Group Assets Ltd
	G-BJXB	Slingsby T.67A	X-Ray Bravo Ltd/Barton
	G-BJXK	Fournier RF-5	G-BJXK Syndicate
	G-BJXP	Colt 56B balloon	H. J. Anderson
	G-BJXR	Auster AOP.9	I. Churm & J. Hanson
	G-BJXX	PA-23 Aztec 250E	V. Bojovic
	G-BJXZ	Cessna 172N	T. M. Jones/Egginton
	G-BJYD	Cessna F.152 II	N. J. James
	G-BJYF	Colt 56A balloon	H. Dos Santos
	G-BJYG	PA-28-161 Warrior II	P. Lodge
	G-BJYK	Jodel D.120A	T. Fox & D. A. Thorpe
	G-BJYN	PA-38-112 Tomahawk	Panshanger School of Flying Ltd (G-BJTE)
	G-BJZA	Cameron N-65 balloon	A. D. Pinner
	G-BJZB	Evans VP-2	I. P. Manley & J. Pearce
	G-BJZF	D.H.82A Tiger Moth	M. I. Lodge
	G-BJZN	Slingsby T.67A	A. R. T. Marsland
	G-BJZR	Colt 42A balloon	Selfish Balloon Group
	G-BKAE	Jodel D.120	S. J. Harris
	G-BKAF	FRED Srs 2	J. Mc. D. Robinson
	G-BKAM	Slingsby T.67M Firefly160	R. C. B. Brookhouse
	G-BKAO	Jodel D.112	R. Broadhead
	G-BKAS	PA-38-112 Tomahawk	St. George Flying Club/Teesside
	G-BKAY	R. Commander 114	The Rockwell Group
	G-BKAZ	Cessna 152	L. W. Scattergood
	G-BKBB	Hawker Fury Mk I (replica)	Brandish Holdings Ltd/O. Warden
	G-BKBD	Thunder Ax3 balloon	M. J. Casson
	G-BKBF	M.S.894A Rallye Minerva 220	K. A. Hale & L. C. Clark
	G-BKBN	SOCATA TB.10 Tobago	Dateworld Ltd
	G-BKBO	Colt 17A balloon	J. Armstrong & ptnrs
	G-BKBP	Bellanca 7GCBC Scout	M. G. & J. R. Jefferies
	G-BKBV	SOCATA TB.10 Tobago	The Studio People Ltd
	G-BKBW	SOCATA TB.10 Tobago	Merlin Aviation
	G-BKCC	PA-28 Cherokee 180	DR Flying Club Ltd
	G-BKCE	Cessna F.172P II	M. O. Loxton
	G-BKCI	Brügger MB.2 Colibri	M. R. Walters
	G-BKCJ	Oldfield Baby Great Lakes	S. V. Roberts/Sleap
	G-BKCL	PA-30 Twin Comanche 160C	R. P. Hodson
	G-BKCN	Currie Wot	N. A. A. Podmore
	G-BKCR	SOCATA TB.9 Tampico	P. A. Little
	G-BKCV	EAA Acro Sport II	M. J. Clark
	G-BKCW	Jodel D.120	Dundee Flying Group (G-BMYF)
	G-BKCX	Mudry/CAARP CAP.10B	R. Ingleton
	G-BKCY	PA-38-112 Tomahawk II ★	(stored)/Welshpool
	G-BKCZ	Jodel D.120A	J. V. George
	G-BKDC	Monnett Sonerai II	K. J. Towell
	G-BKDH	Robin DR.400/120	Dauphin Flying Group Ltd
	G-BKDI	Robin DR.400/120	Mistral Aviation Ltd
	G-BKDJ	Robin DR.400/120	Air Yorkshire Flying Group

Reg.	Type	Owner or Operator	Notes
G-BKDK	Thunder Ax7-77Z balloon	A. J. Byrne	
G-BKDP	FRED Srs 3	M. Whittaker	
G-BKDR	Pitts S-1S Special	Skyview Systems Ltd	
G-BKDT	SE-5A (replica) (F943) ★	Yorkshire Air Museum/Elvington	
G-BKDX	Jodel DR.1050	G. J. Slater	
G-BKEK	PA-32 Cherokee Six 300	S. W. Turley	
G-BKEP	Cessna F.172M	J. M. Thorpe	
G-BKER	SE-5A (replica) (F5447)	N. K. Geddes	
G-BKET	PA-18 Super Cub 95	H. M. MacKenzie	
G-BKEU	Taylor JT.1 Monoplane	R. J. Whybrow & J. M. Springham	
G-BKEV	Cessna F.172M	Echo Victor Group	
G-BKEW	Bell 206B JetRanger 3	N. R. Foster	
G-BKEY	FRED Srs 3	G. S. Taylor	
G-BKFC	Cessna F.152 II	Sulby Aerial	
G-BKFI	Evans VP-1	P. L. Naylor	
G-BKFK	Isaacs Fury II	G. G. C. Jones	
G-BKFL	Aerosport Scamp	J. Sherwood	
G-BKFM	QAC Quickie 1	G. E. Meakin	
G-BKFN	Bell 214ST	Bristow Helicopters Ltd	
G-BKFR	CP.301C Emeraude	Devonshire Flying Group	
G-BKFW	P.56 Provost T.1 (XF597)	Sylmar Aviation & Services Ltd	
G-BKFY	Beech C90 King Air	Blackbrook Aviation LLP	
G-BKFZ	PA-28R Cherokee Arrow 200	Shacklewell Flying Group	
G-BKGA	M.S.892E Rallye 150GT	BJJ Aviation	
G-BKGB	Jodel D.120	B. A. Ridgway	
G-BKGC	Maule M.6-235	B. F. Walker	
G-BKGD	Westland WG.30 Srs.100 ★	IHM/Weston-s-Mare	
G-BKGL	Beech D.18S (1164)	A. T. J. Darrah/Duxford	
G-BKGM	Beech D.18S (HB275)	Skyblue Aviation Ltd	
G-BKGR	Cameron O-65 balloon	K. Kremer & L. E. More	
G-BKGT	SOCATA Rallye 110ST	Long Marston Flying Group	
G-BKGW	Cessna F.152-II	Leicestershire Aero Club Ltd	
G-BKHA	W.S.55 Whirlwind HAR.10 (XJ763) ★	C. J. Evans	
G-BKHG	Piper J-3C-65 Cub (479766)	K. G. Wakefield	
G-BKHJ	Cessna 182P	Augur Films Ltd	
G-BKHR	Luton LA-4 Minor	C. B. Buscombe & R. Goldsworthy	
G-BKHW	Stoddard-Hamilton Glasair IIRG	D. Calabritto/Stapleford	
G-BKHY	Taylor JT.1 Monoplane	B. C. J. O'Neill	
G-BKHZ	Cessna F.172P	L. R. Leader	
G-BKIB	SOCATA TB.9 Tampico	G. A. Vickers	
G-BKIC	Cameron V-77 balloon	C. A. Butler	
G-BKIF	Fournier RF-6B	D. J. Taylor & J. T. Flint	
G-BKII	Cessna F.172M	Sealand Ap Ltd	
G-BKIJ	Cessna F.172M	V. Speck	
G-BKIK	Cameron DG-19 airship ★	Balloon Preservation Group/Lancing	
G-BKIR	Jodel D.117	R. Shaw & D. M. Hardaker/Sherburn	
G-BKIS	SOCATA TB.10 Tobago	Wessex Flyers Group	
G-BKIT	SOCATA TB.9 Tampico	K. Dowling	
G-BKIY	Thunder Ax3 balloon ★	Balloon Preservation Group/Lancing	
G-BKIZ	Cameron V-31 balloon	A. P. S. Cox	
G-BKJB	PA-18 Super Cub 135	Haimoss Ltd/O. Sarum	
G-BKJF	M.S.880B Rallye 100T	Journeyman Aviation Ltd	
G-BKJR	Hughes 269C	March Helicopters Ltd/Sywell	
G-BKJS	Jodel D.120A	Clipgate Flying Group	
G-BKJW	PA-23 Aztec 250E	Alan Williams Entertainments Ltd	
G-BKKN	Cessna 182R	R A. Marven/Elstree	
G-BKKO	Cessna 182R	E. L. King & D. S. Lightbown	
G-BKKZ	Pitts S-1D Special	J. A. Coutts	
G-BKLJ	Westland Scout AH.1 ★	N. R. Windley	
G-BKLO	Cessna F.172M	Stapleford Flying Club Ltd	
G-BKMA	Mooney M.20J Srs 201	Foxtrot Whisky Aviation	
G-BKMB	Mooney M.20J Srs 201	W. A. Cook & ptnrs	
G-BKMG	Handley Page O/400 (replica)	Paralyser Group	
G-BKMI	V.S.359 Spitfire HF.VIIIc (MT928)	Aerial Museum (North Weald) Ltd/Filton	
G-BKMT	PA-32R-301 Saratoga SP	Severn Valley Aviation Group	
G-BKMX	Short SD3-60 Variant 100	BAC Leasing Ltd	
G-BKNB	Cameron V-42 balloon	D. N. Close	
G-BKNI	GY-80 Horizon 160D	A. Hartigan & ptnrs/Fenland	
G-BKNO	Monnett Sonerai IIL	S. Hardy	
G-BKNP	Cameron V-77 balloon	E. K. K. & C. E. Odman	
G-BKNZ	CP.301A Emeraude	C. J. Bellworthy	
G-BKOA	SOCATA MS893E Rallye 180GT	P. Howick	

BRITISH CIVIL REGISTRATIONS

Notes	Reg.	Type	Owner or Operator
	G-BKOB	Z.326 Trener Master	A. L. Rae
	G-BKOT	Wassmer WA.81 Piranha	B. N. Rolfe
	G-BKOU	P.84 Jet Provost T.3 (XN637)	P. M. Grimshaw/North Weald
	G-BKPA	Hoffmann H-36 Dimona	C. I. Roberts & C. D. King
	G-BKPB	Aerosport Scamp	B. R. Thompson
	G-BKPC	Cessna A.185F	Black Knights Parachute Centre
	G-BKPD	Viking Dragonfly	E. P. Browne & G. J. Sargent
	G-BKPE	Jodel DR.250/160	J. S. & J. D. Lewer
	G-BKPN	Cameron N-77 balloon	R. H. Sanderson
	G-BKPS	AA-5B Tiger	A. E. T. Clarke
	G-BKPX	Jodel D.120A	D. M. Garrett & C. A. Jones
	G-BKPY	SAAB 91B/2 Safir (56321) ★	Newark Air Museum
	G-BKPZ	Pitts S-1T Special	M. A. Frost
	G-BKRA	NA T-6G Texan (51-15227)	First Air Ltd
	G-BKRB	Cessna 172N	Saunders Caravans Ltd
	G-BKRF	PA-18 Super Cub 95	K. M. Bishop
	G-BKRH	Brügger MB.2 Colibri	M. R. Benwell
	G-BKRK	SNCAN Stampe SV-4C	Strathgadie Stampe Group
	G-BKRL	Chichester-Miles Leopard ★	(stored)/Cranfield
	G-BKRN	Beechcraft D.18S	A. A. Marshall & P. L. Turland
	G-BKRS	Cameron V-56 balloon	D. N. & L. J. Close
	G-BKRZ	Dragon G-77 balloon	J. R. Barber
	G-BKSB	Cessna T.310Q II	G. H. Smith & Son
	G-BKSC	Saro Skeeter AOP.12 (XN351) ★	R. A. L. Falconer
	G-BKSD	Colt 56A balloon	M. J. Casson
	G-BKSE	QAC Quickie Q.1	M. D. Burns
	G-BKSP	Schleicher ASK.14	J. H. Bryson
	G-BKST	Rutan Vari-Eze	R. Towle
	G-BKSX	SNCAN Stampe SV-4C	C. A. Bailey & J. A. Carr
	G-BKTA	PA-18 Super Cub 95	M. J. Dyson & M. T. Clark
	G-BKTH	CCF Hawker Sea Hurricane IB (Z7015)	The Shuttleworth Collection/Duxford
	G-BKTM	PZL SZD-45A Ogar	J. T. Pajdak
	G-BKTR	Cameron V-77 balloon	C. Wilson
	G-BKTV	Cessna F.152	A. Jahanfar/Southend
	G-BKTZ	Slingsby T.67M Firefly	P. R. Elvidge (G-SFTV)
	G-BKUE	SOCATA TB.9 Tampico	Flying Web Ltd
	G-BKUR	CP.301A Emeraude	R. Wells
	G-BKUU	Thunder Ax7-77-1 balloon	M. A. Mould
	G-BKVA	SOCATA Rallye 180T	Buckminster Gliding Club Ltd
	G-BKVB	SOCATA Rallye 110ST	A. & K. Bishop
	G-BKVC	SOCATA TB.9 Tampico	H. P. Aubin-Parvu
	G-BKVF	FRED Srs 3	G. E. & R. E. Collins
	G-BKVG	Scheibe SF.25E Super Falke	G-BKVG Ltd
	G-BKVK	Auster AOP.9 (WZ662)	Victor Kilo Group
	G-BKVL	Robin DR.400/160	Tatenhill Aviation Ltd
	G-BKVM	PA-18 Super Cub 150 (115684)	D. G. Caffrey
	G-BKVO	Pietenpol Air Camper	M. C. Hayes
	G-BKVP	Pitts S-1D Special	S. W. Doyle
	G-BKVS	Campbell Cricket (modified)	K. Hughes
	G-BKVT	PA-23 Aztec 250E	BKS Surveys Ltd (G-HARV)
	G-BKVW	Airtour 56 balloon	L. D. & H. Vaughan
	G-BKVX	Airtour 56 balloon	P. Aldridge
	G-BKVY	Airtour 31 balloon	M. Davies
	G-BKWD	Taylor JT.2 Titch	J. F. Sully
	G-BKWR	Cameron V-65 balloon	K. J. Foster
	G-BKWW	Cameron O-77 balloon	A. M. Marten
	G-BKWY	Cessna F.152	Northern Aviation Ltd
	G-BKXA	Robin R.2100	M. Wilson
	G-BKXD	SA.365N Dauphin 2	CHC Scotia Ltd
	G-BKXF	PA-28R Cherokee Arrow 200	P. L. Brunton/Caernarfon
	G-BKXM	Colt 17A balloon	R. G. Turnbull
	G-BKXN	ICA-Brasov IS-28M2A	D. C. Wellard
	G-BKXO	Rutan LongEz	M. G. Parsons
	G-BKXP	Auster AOP.6	B. J. Ellis
	G-BKXR	D.31A Turbulent	M. B. Hill
	G-BKZB	Cameron V-77 balloon	K. B. Chapple
	G-BKZE	AS.332L Super Puma	CHC Scotia Ltd
	G-BKZF	Cameron V-56 balloon	A. D. Brice
	G-BKZG	AS.332L Super Puma	CHC Scotia Ltd
	G-BKZH	AS.332L Super Puma	CHC Scotia Ltd
	G-BKZI	Bell 206B JetRanger 2	Dolphin Property (Management) Ltd
	G-BKZT	FRED Srs 2	U. Chakravorty

Reg.	Type	Owner or Operator	Notes
G-BKZV	Bede BD-4A	G. I. J. Thomson	
G-BLAC	Cessna FA.152	D. C. C. Handley	
G-BLAF	Stolp SA.900 V-Star	P. R. Skeels	
G-BLAG	Pitts S-1D Special	G. R. J. Caunter	
G-BLAH	Thunder Ax7-77-1 balloon	T. M. Donnelly	
G-BLAI	Monnett Sonerai IIL	T. Simpon	
G-BLAM	Jodel DR.360	D. J. Durell	
G-BLAT	Jodel 150	G-BLAT Flying Group	
G-BLAX	Cessna FA.152	N. C. & M. L. Scanlan	
G-BLAY	Robin HR.100/200B	B. A. Mills	
G-BLCC	Thunder Ax7-77Z balloon	J. M. Percival	
G-BLCG	SOCATA TB.10 Tobago	Charlie Golf Flying Group (G-BHES)/Shoreham	
G-BLCH	Colt 65D balloon	Balloon Flights Club Ltd	
G-BLCI	EAA Acro Sport	M. R. Holden	
G-BLCM	SOCATA TB.9 Tampico	Charlie Mike Group	
G-BLCT	Jodel DR.220 2+2	Christopher Robin Flying Group	
G-BLCU	Scheibe SF.25B Falke	C. F. Sellers	
G-BLCV	Hoffmann H-36 Dimona	R. & M. Weaver	
G-BLCW	Evans VP-1	M. Flint	
G-BLCY	Thunder Ax7-65Z balloon	C. M. George	
G-BLDB	Taylor JT.1 Monoplane	C. J. Bush	
G-BLDD	WAG-Aero CUBy AcroTrainer	A. F. Stafford	
G-BLDG	PA-25 Pawnee 260C	Ouse Gliding Club Ltd/Rufforth	
G-BLDK	Robinson R-22	Helicentre Ltd	
G-BLDN	Rand-Robinson KR-2	S. C. Solley	
G-BLDV	BN-2B-26 Islander	Loganair Ltd	
G-BLEB	Colt 69A balloon	I. R. M. Jacobs	
G-BLEJ	PA-28-161 Warrior II	Eglinton Flying Club Ltd	
G-BLEP	Cameron V-65 balloon	D. Chapman	
G-BLES	Stolp SA.750 Acroduster Too	C. J. Kingswood	
G-BLET	Thunder Ax7-77-1 balloon	Servatruc Ltd	
G-BLEW	Cessna F.182Q	Seager Publishing Ltd	
G-BLEZ	SA.365N Dauphin 2	CHC Scotia Ltd/Aberdeen	
G-BLFI	PA-28-181 Archer II	Bonus Aviation Ltd	
G-BLFW	AA-5 Traveler	Grumman Club	
G-BLFY	Cameron V-77 balloon	A. N. F. Pertwee	
G-BLFZ	PA-31-310 Turbo Navajo C	London Executive Aviation Ltd	
G-BLGH	Robin DR.300/180R	Booker Gliding Club Ltd	
G-BLGS	SOCATA Rallye 180T	A. Waters	
G-BLGT	PA-18 Super Cub 95	Meridian Aviation Ltd/Bournemouth	
G-BLGV	Bell 206B JetRanger 3	Heliflight (UK) Ltd	
G-BLHH	Jodel DR.315	Central Certification Service Ltd	
G-BLHI	Colt 17A balloon	J. A. Folkes	
G-BLHJ	Cessna F.172P	James D. Peace & Co	
G-BLHK	Colt 105A balloon	Hale Hot-Air Balloon Club	
G-BLHM	PA-18 Super Cub 95	A. G. Edwards	
G-BLHN	Robin HR.100/285	N. P. Finch	
G-BLHR	GA-7 Cougar	T. E. Westley	
G-BLHS	Bellanca 7ECA Citabria	Hotel Sierra Group	
G-BLHW	Varga 2150A Kachina	Kachina Hotel Whiskey Group	
G-BLID	D.H.112 Venom FB.50 (J-1605) ★	P. G. Vallance Ltd/Charlwood	
G-BLIH	PA-18 Super Cub 135	I. R. F. Hammond	
G-BLIK	Wallis WA-116/F/S	K. H. Wallis	
G-BLIT	Thorp T-18 CW	A. P. Tyrwhitt-Drake	
G-BLIW	P.56 Provost T.51 (177)	Provost Flying Group/Shoreham	
G-BLIX	Saro Skeeter Mk 12 (XL809)	K. M. Scholes	
G-BLIY	M.S.892A Rallye Commodore	A. J. Brasher	
G-BLJD	Glaser-Dirks DG.400	M. I. Gee	
G-BLJF	Cameron O-65 balloon	M. D. & C. E. C. Hammond	
G-BLJH	Cameron N-77 balloon ★	Balloon Preservation Group/Lancing	
G-BLJM	Beech 95-B55 Baron	R. A. Perrot	
G-BLJO	Cessna F.152	Redhill School of Flying Ltd	
G-BLKA	D.H.112 Venom FB.54 (WR410) ★	De Havilland Heritage Museum/Swansea	
G-BLKK	Evans VP-1	N. Wright	
G-BLKL	D.31 Turbulent	D. L. Ripley	
G-BLKM	Jodel DR.1051	Kilo Mike Group	
G-BLKP	BAe Jetstream 3102	Global Aviation Ltd/Humberside	
G-BLKY	Beech 95-58 Baron	R. A. Perrot/Guernsey	
G-BLKZ	Pilatus P2-05	R. W. Hinton	
G-BLLA	Bensen B.8M	K. T. Donaghey	
G-BLLB	Bensen B.8M	D. H. Moss	

Notes	Reg.	Type	Owner or Operator
	G-BLLD	Cameron O-77 balloon	G. Birchall
	G-BLLH	Jodel DR.220A 2+2	M. D. Hughes
	G-BLLN	PA-18 Super Cub 95	P. L. Pilch & C. G. Fisher
	G-BLLR	Slingsby T.67B	R. L. Brinklow/Biggin Hill
	G-BLLS	Slingsby T.67B	Western Air (Thruxton) Ltd
	G-BLLW	Colt 56B balloon	G. Fordyce & ptnrs
	G-BLLZ	Rutan LongEz	R. S. Stoddart-Stones
	G-BLMA	Zlin 326 Trener Master	G. P. Northcott/Redhill
	G-BLMC	Avro 698 Vulcan B.2A	Aeropark/E. Midlands
	G-BLME	Robinson R-22HP	Heli Air Ltd/Wellesbourne
	G-BLMG	Grob G.109B	Mike Golf Syndicate
	G-BLMI	PA-18 Super Cub 95	B. J. Borsberry
	G-BLMN	Rutan LongEz	G-BLMN Flying Group
	G-BLMP	PA-17 Vagabond	M. Austin/Popham
	G-BLMR	PA-18 Super Cub 150	Southern Flight Centre Ltd/Shoreham
	G-BLMT	PA-18 Super Cub 135	I. S. Runnalls
	G-BLMW	T.66 Nipper 3	S. L. Millar
	G-BLMZ	Colt 105A balloon	M. D. Dickinson
	G-BLNJ	BN-2B-26 Islander	Loganair Ltd
	G-BLNO	FRED Srs 3	L. W. Smith
	G-BLOB	Colt 31A balloon	Jacques W. Soukup Enterprises Ltd/USA
	G-BLOR	PA-30 Twin Comanche 160	R. L. C. Appleton
	G-BLOS	Cessna 185A (also flown with floats)	E. Brun
	G-BLOT	Colt Ax6-56B balloon	H. J. Anderson
	G-BLOV	Thunder Ax5-42 Srs 1 balloon	A. G. R. Calder
	G-BLPA	Piper J-3C-65 Cub	A. C. Frost
	G-BLPB	Turner TSW Hot Two Wot	I. R. Hannah
	G-BLPE	PA-18 Super Cub 95	A. A. Haig-Thomas
	G-BLPF	Cessna FR.172G	P. Kohl
	G-BLPG	J/1N Alpha (16693)	D. Taylor (G-AZIH)
	G-BLPH	Cessna FRA.150L	New Aerobat Group/Shoreham
	G-BLPI	Slingsby T.67B	RAF Wyton Flying Group Ltd
	G-BLPM	AS.332L Super Puma	Bristow Helicopters Ltd
	G-BLPP	Cameron V-77 balloon	L. P. Purfield
	G-BLRA	BAe 146-100	BAE Systems (Operations) Ltd
	G-BLRC	PA-18 Super Cub 135	Supercub Group
	G-BLRD	MBB Bo 209 Monsun 150FV	T. A. Crone
	G-BLRF	Slingsby T.67C	R. C. Nicholls
	G-BLRG	Slingsby T.67B	R. L. Brinklow
	G-BLRL	CP.301C-1 Emeraude	J. A. & I. M. Macleod
	G-BLRM	Glaser-Dirks DG.400	J. A. & W. S. Y. Stephen
	G-BLRN	D.H.104 Dove 8 (WB531) ★	J. F. M. Bleeker/Netherlands
	G-BLRY	AS.332L Super Puma	Bristow Helicopters Ltd
	G-BLSD	D.H.112 Venom FB.54 (J-1758)★	R. Lamplough/North Weald
	G-BLSF	AA-5A Cheetah	Plane Talking Ltd (G-BGCK)/Elstree
	G-BLSM	H.S.125 Srs 700B	Dravidian Air Services Ltd/Heathrow
	G-BLST	Cessna 421C	Cecil Aviation Ltd/Cambridge
	G-BLSX	Cameron O-105 balloon	B. J. Petteford
	G-BLTA	Thunder Ax7-77A	K. A. Schlussler
	G-BLTC	D.31A Turbulent	S. J. Butler
	G-BLTF	Robinson R22 Alpha	Brian Seedle Helicopters Ltd
	G-BLTK	R. Commander 112TC	B. Rogalewski/Denham
	G-BLTM	Robin HR.200/100	Barton Robin Group
	G-BLTN	Thunder Ax7-65 balloon	J. A. Liddle
	G-BLTR	Scheibe SF.25B Falke	V. Mallon/Germany
	G-BLTS	Rutan LongEz	R. W. Cutler
	G-BLTU	Slingsby T.67B	RAF Wyton Flying Club Ltd
	G-BLTV	Slingsby T.67B	R. L. Brinklow
	G-BLTW	Slingsby T.67B	Cheshire Air Training Services Ltd/Liverpool
	G-BLTY	Westland WG.30 Srs 160	D. Drem-Wilson
	G-BLTZ	SOCATA TB.10 Tobago	Martin Ltd/Biggin Hill
	G-BLUI	Thunder Ax7-65 balloon	S. Johnson
	G-BLUL	CEA Jodel DR.1051/M1	J. Owen
	G-BLUM	SA.365N Dauphin 2	CHC Scotia Ltd
	G-BLUN	SA.365N Dauphin 2	CHC Scotia Ltd
	G-BLUV	Grob G.109B	The 109 Flying Group/North Weald
	G-BLUX	Slingsby T.67M Firefly 200	R. L. Brinklow
	G-BLUZ	D.H.82 Queen Bee (LF858)	The Bee Keepers Group
	G-BLVA	Airtour AH-56 balloon	A. Van Wyk
	G-BLVB	Airtour AH-56 balloon	R. W. Guild
	G-BLVI	Slingsby T.67M Firefly Mk II	Northern Aviation Ltd

Reg.	Type	Owner or Operator	Notes
G-BLVK	CAARP CAP-10B	E. K. Coventry/Earls Colne	
G-BLVL	PA-28-161 Warrior II	TG Aviation Ltd/Manston	
G-BLVS	Cessna 150M	R. Collier	
G-BLVW	Cessna F.172H	R. & D. Holloway Ltd	
G-BLWD	PA-34-200T Seneca 2	Bencray Ltd	
G-BLWF	Robin HR.100/210	Starguide Ltd	
G-BLWH	Fournier RF-6B-100	I. R. March	
G-BLWM	Bristol M.1C (replica) (C4994) ★	RAF Museum/Hendon	
G-BLWP	PA-38-112 Tomahawk	J. C. Dodd & ptnrs/Panshanger	
G-BLWT	Evans VP-1	G. Malpass	
G-BLWV	Cessna F.152	Redhill Flying Club	
G-BLWY	Robin R.2161D	K. D. Boardman	
G-BLXA	SOCATA TB.20 Trinidad	Trinidad Flyers Ltd/Blackbushe	
G-BLXG	Colt 21A balloon	A. Walker	
G-BLXH	Fournier RF-3	D. M. Boxell	
G-BLXI	CP.1310-C3 Super Emeraude	R. Howard	
G-BLXO	Jodel 150	P. R. Powell	
G-BLXP	PA-28R Cherokee Arrow 200	M. B. Hamlett	
G-BLXR	AS.332L Super Puma	Bristow Helicopters Ltd	
G-BLYD	SOCATA TB.20 Trinidad	Yankee Delta Corporation Ltd	
G-BLYE	SOCATA TB.10 Tobago	G. Hatton	
G-BLYK	PA-34-220T Seneca III	Oxford Aviation Services Ltd/Kidlington	
G-BLYP	Robin 3000/120	Weald Air Services/Headcorn	
G-BLYT	Airtour AH-77 balloon	I. J. Taylor & R. C. Kincaid	
G-BLZA	Scheibe SF.25B Falke	Chiltern Gliding Club	
G-BLZE	Cessna F.152 II	Flairhire Ltd (G-CSSC)/Redhill	
G-BLZF	Thunder Ax7-77 balloon	H. M. Savage	
G-BLZH	Cessna F.152 II	BCT Aircraft Leasing Ltd	
G-BLZJ	AS.332L Super Puma	Bristow Helicopters Ltd (G-PUMJ)	
G-BLZN	Bell 206B JetRanger	Biggin Hill Helicopters	
G-BLZP	Cessna F.152	E. Midlands Flying School Ltd	
G-BLZS	Cameron O-77 balloon	C. D. Steel	
G-BMAD	Cameron V-77 balloon	M. A. Stelling	
G-BMAL	Sikorsky S-76A	CHC Scotia Ltd	
G-BMAO	Taylor JT.1 Monoplane	S. J. Alston	
G-BMAV	AS.350B Ecureuil	PLM Dollar Group Ltd	
G-BMAX	FRED Srs 2	D. A. Arkley	
G-BMAY	PA-18 Super Cub 135	R. W. Davies	
G-BMBB	Cessna F.150L	M. Bonsall	
G-BMBJ	Schempp-Hirth Janus CM	BJ Flying Group	
G-BMBS	Colt 105A balloon	H. G. Davies	
G-BMBW	Bensen B.8MR	M. E. Vahdat	
G-BMBZ	Scheibe SF.25E Super Falke	Cornish Gliding & Flying Club Ltd/Perranporth	
G-BMCC	Thunder Ax7-77 balloon	A. K. & C. M. Russell	
G-BMCD	Cameron V-65 balloon	M. C. Drye	
G-BMCG	Grob G.109B	Lagerholm Finnimport Ltd/Booker	
G-BMCI	Cessna F.172H	A. B. Davis/Edinburgh	
G-BMCN	Cessna F.152	Eastern Air Centre Ltd	
G-BMCS	PA-22 Tri-Pacer 135	P. R. Deacon	
G-BMCV	Cessna F.152	Leicestershire Aero Club Ltd	
G-BMCW	AS.332L Super Puma	Bristow Helicopters Ltd	
G-BMCX	AS.332L Super Puma	Bristow Helicopters Ltd	
G-BMDB	SE-5A (replica) (F235)	D. Biggs	
G-BMDC	PA-32-301 Saratoga	MacLaren Aviation/Newcastle	
G-BMDE	Pietenpol Air Camper	P. B. Childs	
G-BMDJ	Price Ax7-77S balloon	R. A. Benham	
G-BMDK	PA-34-220T Seneca III	Air Medical Fleet Ltd	
G-BMDP	Partenavia P.64B Oscar 200	S. T. G. Lloyd	
G-BMDS	Jodel D.120	R. T. Mosforth	
G-BMEA	PA-18 Super Cub 95	M. J. Butler	
G-BMEE	Cameron O-105 balloon	A. G. R. Calder/Los Angeles	
G-BMEG	SOCATA TB.10 Tobago	P. Farmer	
G-BMEH	Jodel 150 Special Super Mascaret	R. J. & C. J. Lewis	
G-BMET	Taylor JT.1 Monoplane	M. K. A. Blyth	
G-BMEU	Isaacs Fury II	I. G. Harrison	
G-BMEX	Cessna A.150K	N. A. M. & R. A. Brain	
G-BMFD	PA-23 Aztec 250F	Gold Air International Ltd (G-BGYY)	
G-BMFG	Dornier Do.27A-4	R. F. Warner	
G-BMFI	PZL SZD-45A Ogar	S. L. Morrey/Andreas, IoM	
G-BMFN	QAC Quickie Tri-Q.200	A. H. Hartog	
G-BMFP	PA-28-161 Warrior II	Bravo-Mike-Fox-Papa Group/Blackbushe	

Notes	Reg.	Type	Owner or Operator
	G-BMFU	Cameron N-90 balloon	J. J. Rudoni
	G-BMFY	Grob G.109B	P. J. Shearer
	G-BMFZ	Cessna F.152 II	Cornwall Flying Club Ltd/Bodmin
	G-BMGB	PA-28R Cherokee Arrow 200	Malmesbury Specialist Cars
	G-BMGC	Fairey Swordfish Mk II (W5856)	F.A.A. Museum/Yeovilton
	G-BMGG	Cessna 152 II	Falcon Flying Services/Biggin Hill
	G-BMGR	Grob G.109B	G-BMGR Group
	G-BMHA	Rutan LongEz	S. F. Elvins
	G-BMHC	Cessna U.206F	P. Marsden
	G-BMHJ	Thunder Ax7-65 balloon	M. G. Robinson
	G-BMHL	Wittman W.8 Tailwind	C. R. Nash
	G-BMHS	Cessna F.172M	Tango X-Ray Flying Group
	G-BMID	Jodel D.120	G-BMID Flying Group
	G-BMIG	Cessna 172N	BMIG Group
	G-BMIM	Rutan LongEz	R. M. Smith
	G-BMIO	Stoddard-Hamilton Glasair RG	J. M. Ayres & S. C. Ellerton
	G-BMIP	Jodel D.112	M. T. Kinch
	G-BMIR	Westland Wasp HAS.1 (XT788)★	Park Aviation Supply/Charlwood
	G-BMIS	Monnett Sonerai II	S. R. Edwards
	G-BMIV	PA-28R-201T Turbo Arrow III	Firmbeam Ltd
	G-BMIW	PA-28-181 Archer II	Oldbus Ltd
	G-BMIY	Oldfield Baby Great Lakes	J. B. Scott (G-NOME)
	G-BMJA	PA-32R-301 Saratoga SP	H. Merkado/Panshanger
	G-BMJC	Cessna 152 II	Northern Aviation Ltd
	G-BMJD	Cessna 152 II	Donair Flying Club Ltd/E. Midlands
	G-BMJL	R. Commander 114	D. J. & S. M. Hawkins
	G-BMJM	Evans VP-1	S. E. Clarke
	G-BMJN	Cameron O-65 balloon	P. M. Traviss
	G-BMJO	PA-34-220T Seneca III	Deep Cleavage Ltd
	G-BMJR	Cessna T.337H	John Roberts Services Ltd (G-NOVA)
	G-BMJT	Beech 76 Duchess	Mike Osborne Properties Ltd
	G-BMJX	Wallis WA-116X	K. H. Wallis
	G-BMJY	Yakovlev C18M (07)	R. J. Lamplough/North Weald
	G-BMJZ	Cameron N-90 balloon	Bristol University Hot Air Ballooning Soc
	G-BMKB	PA-18 Super Cub 135	Cubair Flight Training Ltd/Redhill
	G-BMKC	Piper J-3C-65 Cub (329854)	J. W. Salter
	G-BMKD	Beech C90A King Air	A. E. Bristow
	G-BMKF	Jodel DR.221	S. T. & L. A. Gilbert
	G-BMKG	PA-38-112 Tomahawk II	APB Leasing Ltd/Welshpool
	G-BMKI	Colt 21A balloon	A. C. Booth
	G-BMKJ	Cameron V-77 balloon	R. C. Thursby
	G-BMKP	Cameron V-77 balloon	R. Bayly
	G-BMKR	PA-28-161 Warrior II	Field Flying Group (G-BGKR)/Goodwood
	G-BMKW	Cameron V-77 balloon	M. H. Redman
	G-BMKY	Cameron O-65 balloon	A. R. Rich
	G-BMLB	Jodel D.120A	C. A. Croucher
	G-BMLC	Short SD3-60 Variant 100	Emerald Airways Ltd/Liverpool
	G-BMLJ	Cameron N-77 balloon	C. J. Dunkley
	G-BMLK	Grob G.109B	Brams Syndicate/Rufforth
	G-BMLL	Grob G.109B	G-BMLL Flying Group/Denham
	G-BMLM	Beech 95-58 Baron	Atlantic Bridge Aviation Ltd/Lydd
	G-BMLS	PA-28R-201 Arrow III	R. M. Shorter
	G-BMLT	Pietenpol Air Camper	W. E. R. Jenkins
	G-BMLW	Cameron O-77 balloon	M. L. & L. P. Willoughby
	G-BMLX	Cessna F.150L	J. P. A. Freeman/Headcorn
	G-BMMC	Cessna T310Q	I. T. Cooper I
	G-BMMF	FRED Srs 2	E. C. King
	G-BMMI	Pazmany PL.4A	P. I. Morgans
	G-BMMK	Cessna 182P	G. G. Weston
	G-BMML	PA-38-112 Tomahawk	J. C. & C. H. Strong
	G-BMMM	Cessna 152 II	Luton Flight Training Ltd
	G-BMMP	Grob G.109B	E. W. Reynolds
	G-BMMV	ICA-Brasov IS-28M2A	J. F. Miles
	G-BMMW	Thunder Ax7-77 balloon	P. A. George
	G-BMMY	Thunder Ax7-77 balloon	S. W. Wade & S. E. Hadley
	G-BMNL	PA-28R Cherokee Arrow 200	Arrow Flying Group
	G-BMNV	SNCAN Stampe SV-4D	Wessex Aviation & Transport Ltd
	G-BMOE	PA-28R Cherokee Arrow 200	E. P. C. Robson
	G-BMOF	Cessna U206G	Wild Geese Skydiving Centre
	G-BMOG	Thunder Ax7-77 balloon	R. M. Boswell
	G-BMOH	Cameron N-77 balloon	P. J. Marshall & M. A. Clarke
	G-BMOI	Partenavia P.68B	Simmette Ltd

Reg.	Type	Owner or Operator	Notes
G-BMOK	ARV Super 2	R. E. Griffiths	
G-BMOL	PA-23 Aztec 250D	LDL Enterprises (G-BBSR)/Elstree	
G-BMOM	ICA-Brasov IS-28M2A	R. M. Cust	
G-BMOT	Bensen B.8M	Austin Trueman Ltd	
G-BMOV	Cameron O-105 balloon	C. Gillott	
G-BMPC	PA-28-181 Archer II	C. J. & R. J. Barnes	
G-BMPD	Cameron V-65 balloon	D. Triggs	
G-BMPL	Optica Industries OA.7 Optica	Aces High Ltd/North Weald	
G-BMPP	Cameron N-77 balloon	The Sarnia Balloon Group	
G-BMPR	PA-28R-201 Arrow III	B. Edwards	
G-BMPS	Strojnik S-2A	G. J. Green	
G-BMPY	D.H.82A Tiger Moth	S. M. F. Eisenstein	
G-BMRA	Boeing 757-236F	DHL Air Ltd	
G-BMRB	Boeing 757-236F	DHL Air Ltd	
G-BMRC	Boeing 757-236F	DHL Air Ltd	
G-BMRD	Boeing 757-236F	DHL Air Ltd	
G-BMRE	Boeing 757-236F	DHL Air Ltd	
G-BMRF	Boeing 757-236F	DHL Air Ltd	
G-BMRH	Boeing 757-236F	DHL Air Ltd	
G-BMRJ	Boeing 757-236F	DHL Air Ltd	
G-BMSA	Stinson HW.75 Voyager	M. A. Thomas (G-BCUM)/Barton	
G-BMSB	V.S.509 Spitfire IX (MJ627)	M. S. Bayliss (G-ASOZ)/Coventry	
G-BMSC	Evans VP-2	M. S. Barron	
G-BMSD	PA-28-181 Archer II	H. Merkado/Panshanger	
G-BMSE	Valentin Taifun 17E	A. J. Nurse	
G-BMSF	PA-38-112 Tomahawk	B. Catlow	
G-BMSG	SAAB 32A Lansen ★	J. E. Wilkie/Cranfield	
G-BMSL	FRED Srs 3	A. C. Coombe	
G-BMSU	Cessna 152 II	G-BMSU Group	
G-BMTA	Cessna 152 II	Alarmond Ltd	
G-BMTB	Cessna 152 II	Sky Leisure Aviation (Charters) Ltd	
G-BMTJ	Cessna 152 II	The Pilot Centre Ltd/Denham	
G-BMTN	Cameron O-77 balloon	Industrial Services (MH) Ltd	
G-BMTO	PA-38-112 Tomahawk	Falcon Flying Services/Biggin Hill	
G-BMTR	PA-28-161 Warrior II	Aeros Leasing Ltd	
G-BMTS	Cessna 172N	European Flyers/Blackbushe	
G-BMTU	Pitts S-1E Special	B. Brown & E. Evans	
G-BMTX	Cameron V-77 balloon	J. A. Langley	
G-BMUD	Cessna 182P	M. E. Taylor	
G-BMUG	Rutan LongEz	A. G. Sayers	
G-BMUJ	Colt Drachenfisch balloon	Virgin Airship & Balloon Co Ltd	
G-BMUO	Cessna A.152	Sky Leisure Aviation (Charters) Ltd	
G-BMUT	PA-34-200T Seneca II	G-DAD Air Ltd	
G-BMUU	Thunder Ax7-77 balloon	A. R. Hill	
G-BMUZ	PA-28-161 Warrior II	Newcastle-upon-Tyne Aero Club Ltd	
G-BMVA	Scheibe SF.25B Falke	M. L. Jackson	
G-BMVB	Cessna F.152	N. D. Plumb	
G-BMVG	QAC Quickie Q.1	P. M. Wright	
G-BMVL	PA-38-112 Tomahawk	Airways Aero Associations Ltd/Booker	
G-BMVM	PA-38-112 Tomahawk	Airways Aero Associations Ltd/Booker	
G-BMVS	Cameron 70 Benihana SS balloon	Benihana (UK) Ltd	
G-BMVT	Thunder Ax7-77A balloon	M. L. & L. P. Willoughby	
G-BMVU	Monnett Moni	R. M. Edworthy	
G-BMVW	Cameron O-65 balloon	S. P. Richards	
G-BMWA	Hughes 269C	R. J. H. Strong	
G-BMWE	ARV Super 2	R. J. N. Noble	
G-BMWF	ARV Super 2	N. R. Beale	
G-BMWM	ARV Super 2	P..F. Lorriman	
G-BMWR	R. Commander 112	M. & J. Edwards/Blackbushe	
G-BMWU	Cameron N-42 balloon ★	Balloon Preservation Group/Lancing	
G-BMWV	Putzer Elster B	E. A. J. Hibbard	
G-BMXA	Cessna 152 II	I. R. Chaplin/Andrewsfield	
G-BMXB	Cessna 152 II	Rybec Ltd	
G-BMXC	Cessna 152 II	Devon School of Flying Ltd	
G-BMXD	F.27 Friendship Mk 500	BAC Express Airlines Ltd	
G-BMXJ	Cessna F.150L	Arrow Aircraft Group	
G-BMXL	PA-38-112 Tomahawk	Airways Aero Associations Ltd/Booker	
G-BMXM	Colt 180A balloon	D. A. Michaud	
G-BMXX	Cessna 152 II	Evensport Ltd	
G-BMYC	SOCATA TB.10 Tobago	E. A. Grady	
G-BMYD	Beech A36 Bonanza	Seabeam Partners Ltd	
G-BMYF	Bensen B.8M	G. Callaghan	

Notes	Reg.	Type	Owner or Operator
	G-BMYG	Cessna FA.152	Greer Aviation Ltd/Prestwick
	G-BMYI	AA-5 Traveler	W. C. & S. C. Westran
	G-BMYJ	Cameron V-65 balloon	S. P. Harrowing
	G-BMYP	Fairey Gannet AEW.3 (XL502) ★	D. Copley/Sandtoft
	G-BMYS	Thunder Ax7-77Z balloon	J. E. Weidema/Netherlands
	G-BMYU	Jodel D.120	N. P. Chitty
	G-BMZB	Cameron N-77 balloon	D. C. Eager
	G-BMZE	SOCATA TB.9 Tampico	H. J. Samples
	G-BMZF	WSK-Mielec LiM-2 (MiG-15bis) (01420) ★	F.A.A. Museum/Yeovilton
	G-BMZN	Everett gyroplane	K. Ashford
	G-BMZP	Everett gyroplane	D. H. Kirton
	G-BMZS	Everett gyroplane	L. W. Cload
	G-BMZW	Bensen B.8MR	P. D. Widdicombe
	G-BMZX	Wolf W-11 Boredom Fighter	J. Nugent
	G-BNAD	Rand-Robinson KR-2	M. C. Davies
	G-BNAG	Colt 105A balloon	R. W. Batchelor
	G-BNAI	Wolf W-II Boredom Fighter (146-11083)	C. M. Bunn
	G-BNAJ	Cessna 152 II	Galair Ltd/Biggin Hill
	G-BNAN	Cameron V-65 balloon	A. M. Lindsay
	G-BNAU	Cameron V-65 balloon	4-Flight Group
	G-BNAW	Cameron V-65 balloon	A. Walker
	G-BNBL	Thunder Ax7-77 balloon	G. J. Bell
	G-BNBU	Bensen B.8MV	B. A. Lyford
	G-BNBV	Thunder Ax7-77 balloon	J. M. Robinson
	G-BNBW	Thunder Ax7-77 balloon	I. S. & S. W. Watthews
	G-BNBY	Beech 95-B55A Baron	J. Butler (G-AXXR)/France
	G-BNBZ	LET L-200D Morava	C. A. Suckling
	G-BNCB	Cameron V-77 balloon	C. W. Brown
	G-BNCC	Thunder Ax7-77 balloon	C. J. Burnhope
	G-BNCE	G.159 Gulfstream 1 ★	(stored)/Aberdeen
	G-BNCJ	Cameron V-77 balloon	D. Scott
	G-BNCL	WG.13 Lynx HAS.2 (XX469) ★	Lancashire Fire Brigade HQ/Lancaster
	G-BNCM	Cameron N-77 balloon	C. A. Stone
	G-BNCO	PA-38-112 Tomahawk	D. K. Walker
	G-BNCR	PA-28-161 Warrior II	Airways Aero Associations Ltd/Booker
	G-BNCS	Cessna 180	C. Elwell Transport Ltd
	G-BNCU	Thunder Ax7-77 balloon	W. De Bock
	G-BNCX	Hawker Hunter T.7 (XL621) ★	Brooklands Museum of Aviation/Weybridge
	G-BNCZ	Rutan LongEz	M. C. Davies
	G-BNDE	PA-38-112 Tomahawk	A. R. Willis
	G-BNDG	Wallis WA-201/R Srs1	K. H. Wallis
	G-BNDN	Cameron V-77 balloon	J. A. Smith
	G-BNDO	Cessna 152 II	Simair Ltd
	G-BNDP	Brügger MB.2 Colibri	J. P. Kynaston
	G-BNDR	SOCATA TB.10 Tobago	Delta Fire Ltd
	G-BNDT	Brügger MB.2 Colibri	Colibri Flying Group/Waddington
	G-BNDV	Cameron N-77 balloon	R. E. Jones
	G-BNDW	D.H.82A Tiger Moth	N. D. Welch
	G-BNDY	Cessna 425-1	Standard Aviation Ltd/Newcastle
	G-BNED	PA-22 Tri-Pacer 135	P. Storey
	G-BNEE	PA-28R-201 Arrow III	Britannic Management Aviation
	G-BNEJ	PA-38-112 Tomahawk II	V. C. & S. G. Swindell
	G-BNEK	PA-38-112 Tomahawk II	APB Leasing Ltd/Welshpool
	G-BNEL	PA-28-161 Warrior II	S. C. Westran
	G-BNEN	PA-34-200T Seneca II	Air Taxis Ltd
	G-BNEO	Cameron V-77 balloon	J. G. O'Connell
	G-BNES	Cameron V-77 balloon	G. Wells
	G-BNET	Cameron O-84 balloon	C. & A. I. Gibson
	G-BNEV	Viking Dragonfly	N. W. Eyre
	G-BNEX	Cameron O-120 balloon	The Balloon Club Ltd
	G-BNFG	Cameron O-77 balloon	Capital Balloon Club Ltd
	G-BNFI	Cessna 150J	A. Waters
	G-BNFM	Colt 21A balloon	M. E. Dworski/France
	G-BNFN	Cameron N-105 balloon	P. Glydon
	G-BNFO	Cameron V-77 balloon	Fox Group
	G-BNFP	Cameron O-84 balloon	B. F. G. Ribbans
	G-BNFR	Cessna 152 II	Eastern Executive Air Charter Ltd/Southend
	G-BNFS	Cessna 152 II	P. J. Clarke
	G-BNFV	Robin DR.400/120	J. P. A. Freeman
	G-BNGE	Auster AOP.6 (TW536)	M. Pocock
	G-BNGJ	Cameron N-77 balloon	Lathams Ltd

Reg.	Type	Owner or Operator	Notes
G-BNGN	Cameron N-77 balloon	C. B. Leeder	
G-BNGO	Thunder Ax7-77 balloon	J. S. Finlan	
G-BNGR	PA-38-112 Tomahawk	Teesside Flight Centre Ltd	
G-BNGS	PA-38-112 Tomahawk	Teesside Flight Centre Ltd	
G-BNGT	PA-28-181 Archer II	Edinburgh Flying Club Ltd	
G-BNGY	ARV Super 2	S. C. Smith (G-BMWL)	
G-BNHB	ARV Super 2	C. J. Challener	
G-BNHG	PA-38-112 Tomahawk II	D. A. Whitmore	
G-BNHI	Cameron V-77 balloon	C. J. Nicholls	
G-BNHJ	Cessna 152 II	The Pilot Centre Ltd/Denham	
G-BNHK	Cessna 152 II	J. P. Slack	
G-BNHN	Colt Ariel Bottle SS balloon ★	British Balloon Museum/Newbury	
G-BNHT	Fournier RF-3	G-BNHT Group	
G-BNID	Cessna 152 II	Mercia Aircraft Leasing & Sales Ltd/Coventry	
G-BNII	Cameron N-90 balloon	Topless Balloon Group	
G-BNIK	Robin HR.200/120	N. J. Wakeling	
G-BNIM	PA-38-112 Tomahawk	Aurs Aviation Ltd	
G-BNIN	Cameron V-77 balloon	Cloud Nine Balloon Group	
G-BNIO	Luscombe 8A Silvaire	W. H. Bliss	
G-BNIP	Luscombe 8A Silvaire	S. Maric	
G-BNIU	Cameron O-77 balloon	MC VH SA/Belgium	
G-BNIV	Cessna 152 II	Aerohire Ltd/Wolverhampton	
G-BNIW	Boeing Stearman PT-17	R. C. Goold	
G-BNIZ	F.27 Friendship Mk.600	Channel Express (Air Services) Ltd/Bournemouth	
G-BNJB	Cessna 152 II	Aerolease Ltd/Conington	
G-BNJC	Cessna 152 II	Stapleford Flying Club Ltd	
G-BNJD	Cessna 152 II	S. Tiernan	
G-BNJL	Bensen B.8MR	S. Ram	
G-BNGV	ARV Super 2	N. A. Onions	
G-BNGW	ARV Super 2	Southern Gas Turbines Ltd	
G-BNJG	Cameron O-77 balloon	A. M. Figiel	
G-BNJH	Cessna 152 II	J. McAuley	
G-BNJM	PA-28-161 Warrior II	Teesside Flight Centre Ltd	
G-BNJO	QAC Quickie Q.2	J. D. McKay	
G-BNJR	PA-28RT-201T Turbo Arrow IV	D. Croker	
G-BNJT	PA-28-161 Warrior II	Hawarden Flying Group	
G-BNJX	Cameron N-90 balloon	Mars UK Ltd	
G-BNJZ	Cassutt Racer IIIM	J. Cull	
G-BNKC	Cessna 152 II	Herefordshire Aero Club Ltd/Shobdon	
G-BNKD	Cessna 172N	Barnes Olsen Aero Leasing Ltd	
G-BNKE	Cessna 172N	Kilo Echo Flying Group	
G-BNKH	PA-38-112 Tomahawk	Goodwood Terrena Ltd	
G-BNKI	Cessna 152 II	RAF Halton Aeroplane Club Ltd	
G-BNKP	Cessna 152 II	Spectrum Leisure Ltd/Clacton	
G-BNKR	Cessna 152 II	Keen Leasing (IOM) Ltd	
G-BNKS	Cessna 152 II	Shropshire Aero Club Ltd/Sleap	
G-BNKT	Cameron O-77 balloon	British Airways PLC	
G-BNKV	Cessna 152 II	S. C. Westran/Shoreham	
G-BNLA	Boeing 747-436	British Airways	
G-BNLB	Boeing 747-436	British Airways	
G-BNLC	Boeing 747-436	British Airways	
G-BNLD	Boeing 747-436	British Airways	
G-BNLE	Boeing 747-436	British Airways	
G-BNLF	Boeing 747-436	British Airways	
G-BNLG	Boeing 747-436	British Airways	
G-BNLH	Boeing 747-436	British Airways	
G-BNLI	Boeing 747-436	British Asia Airways	
G-BNLJ	Boeing 747-436	British Airways	
G-BNLK	Boeing 747-436	British Airways	
G-BNLL	Boeing 747-436	British Airways	
G-BNLM	Boeing 747-436	British Airways	
G-BNLN	Boeing 747-436	British Airways	
G-BNLO	Boeing 747-436	British Airways	
G-BNLP	Boeing 747-436	British Airways	
G-BNLR	Boeing 747-436	British Airways	
G-BNLS	Boeing 747-436	British Airways	
G-BNLT	Boeing 747-436	British Airways	
G-BNLU	Boeing 747-436	British Airways	
G-BNLV	Boeing 747-436	British Airways	
G-BNLW	Boeing 747-436	British Airways	
G-BNLX	Boeing 747-436	British Airways	
G-BNLY	Boeing 747-436	British Airways	

Notes	Reg.	Type	Owner or Operator
	G-BNLZ	Boeing 747-436	British Airways
	G-BNMA	Cameron O-77 balloon	N. Woodham
	G-BNMB	PA-28-151 Warrior	Britannia Airways Ltd/Luton
	G-BNMC	Cessna 152 II	M. L. Jones/Egginton
	G-BNMD	Cessna 152 II	T. M. Jones/Egginton
	G-BNME	Cessna 152 II	Northamptonshire School of Flying Ltd/Sywell
	G-BNMF	Cessna 152 II	Central Aircraft Leasing Ltd
	G-BNMG	Cameron O-77 balloon	J. H. Turner
	G-BNMH	Pietenpol Air Camper	N. M. Hitchman
	G-BNMI	Colt Flying Fantasy SS balloon	Air 2 Air Ltd
	G-BNMK	Dornier Do.27A-1	G. Mackie
	G-BNML	Rand-Robinson KR-2	R. F. Cresswell
	G-BNMO	Cessna TR.182RG	Kenrye Developments Ltd
	G-BNMU	Short SD3-60 Variant 100	BAC Express Airlines Ltd
	G-BNMX	Thunder Ax7-77 balloon	S. A. D. Beard
	G-BNNA	Stolp SA.300 Starduster Too	Banana Group
	G-BNNE	Cameron N-77 balloon	Balloon Flights International Ltd
	G-BNNG	Cessna T.337D	Somet Ltd (G-COLD)
	G-BNNO	PA-28-161 Warrior II	Tindon Ltd/Little Snoring
	G-BNNR	Cessna 152	Sussex Flying Club Ltd/Shoreham
	G-BNNS	PA-28-161 Warrior II	S. J. French
	G-BNNT	PA-28-151 Warrior	S. T. Gilbert & D. J. Kirkwood
	G-BNNU	PA-38-112 Tomahawk	Edinburgh Flying Club Ltd
	G-BNNX	PA-28R-201T Turbo Arrow III	J. G. Freeden
	G-BNNY	PA-28-161 Warrior II	Falcon Flying Services/Biggin Hill
	G-BNNZ	PA-28-161 Warrior II	Falcon Flying Services/Biggin Hill
	G-BNOB	Wittman W.8 Tailwind	M. Robson-Robinson
	G-BNOE	PA-28-161 Warrior II	Sherburn Aero Club Ltd
	G-BNOF	PA-28-161 Warrior II	Tayside Aviation Ltd/Dundee
	G-BNOG	PA-28-161 Warrior II	Flight Training Europe SL/Spain
	G-BNOH	PA-28-161 Warrior II	Sherburn Aero Club Ltd
	G-BNOI	PA-28-161 Warrior II	BAE Systems Flight Training Ltd
	G-BNOJ	PA-28-161 Warrior II	BAE Systems (Warton) Flying Club Ltd
	G-BNOK	PA-28-161 Warrior II	Flight Training Europe SL/Spain
	G-BNOM	PA-28-161 Warrior II	Air Navigation and Trading Co Ltd
	G-BNON	PA-28-161 Warrior II	Tayside Aviation Ltd/Dundee
	G-BNOP	PA-28-161 Warrior II	BAE Systems (Warton) Flying Club Ltd
	G-BNOT	PA-28-161 Warrior II	Flight Training Europe SL/Spain
	G-BNOU	PA-28-161 Warrior II	BAE Systems Flight Training Ltd
	G-BNOZ	Cessna 152 II	CSW Flying Hire Ltd
	G-BNPE	Cameron N-77 balloon	Zebedee Balloon Service Ltd
	G-BNPF	Slingsby T.31M	S. Luck & ptnrs
	G-BNPH	P.66 Pembroke C.1 (WV740)	A. G. & G. A. G. Dixon
	G-BNPL	PA-38-112 Tomahawk	Cardiff-Wales Flying Club
	G-BNPM	PA-38-112 Tomahawk	Papa Mike Aviation
	G-BNPO	PA-28-181 Archer II	Bonus Aviation Ltd
	G-BNPU	P.66 Pembroke (XL929) ★	D-Day Museum/Shoreham
	G-BNPV	Bowers Fly-Baby 1B	J. G. Day & R. Gauld-Galliers
	G-BNPY	Cessna 152 II	Eastern Air Centre Ltd/Gamston
	G-BNPZ	Cessna 152 II	Tatenhill Aviation Ltd
	G-BNRA	SOCATA TB.10 Tobago	Double D Airgroup
	G-BNRG	PA-28-161 Warrior II	RAF Brize Norton Flying Club Ltd
	G-BNRK	Cessna 152 II	Redhill Flying Club
	G-BNRL	Cessna 152 II	Modi Aviation Ltd
	G-BNRP	PA-28-181 Archer II	Bonua Aviation Ltd/Cranfield
	G-BNRR	Cessna 172P	PHA Aviation Ltd/Elstree
	G-BNRX	PA-34-200T Seneca II	Truman Aviation Ltd/Tollerton
	G-BNRY	Cessna 182Q	Reefly Ltd
	G-BNSG	PA-28R-201 Arrow III	Arrow Flying Group
	G-BNSI	Cessna 152 II	Sky Leisure Aviation (Charters) Ltd
	G-BNSL	PA-38-112 Tomahawk II	APB Leasing Ltd/Welshpool
	G-BNSM	Cessna 152 II	Cornwall Flying Club Ltd/Bodmin
	G-BNSN	Cessna 152 II	The Pilot Centre Ltd/Denham
	G-BNSO	Slingsby T.67M Firefly Mk II	A. C. Lees
	G-BNSP	Slingsby T.67M Firefly Mk II	Slingsby Group
	G-BNSR	Slingsby T.67M Firefly Mk II	R. Harris & N. R. Thorburn
	G-BNST	Cessna 172N	CSG Bodyshop
	G-BNSU	Cessna 152 II	Channel Aviation Ltd/Bourn
	G-BNSV	Cessna 152 II	Channel Aviation Ltd/Bourn
	G-BNSY	PA-28-161 Warrior II	Carill Aviation Ltd/Southampton
	G-BNSZ	PA-28-161 Warrior II	Carill Aviation Ltd/Southampton
	G-BNTC	PA-28RT-201 Turbo Arrow IV	Halfpenny Green Flight Centre Ltd

Reg.	Type	Owner or Operator	Notes
G-BNTD	PA-28-161 Warrior II	A. M. & F. Alam/Elstree	
G-BNTP	Cessna 172N	Westnet Ltd	
G-BNTS	PA-28RT-201T Turbo Arrow IV	Nasaire Ltd/Liverpool	
G-BNTT	Beech 76 Duchess	Aviation Rentals/Bournemouth	
G-BNTW	Cameron V-77 balloon	P. Goss	
G-BNTZ	Cameron N-77 balloon	Balloon Team	
G-BNUC	Cameron O-77 balloon	T. J. Bucknall	
G-BNUI	Rutan Vari-Eze	I. T. Kennedy & K. H. McConnell	
G-BNUL	Cessna 152 II	Big Red Kite Ltd	
G-BNUN	Beech 95-58PA Baron	British Midland Airways Ltd/E. Midlands	
G-BNUO	Beech 76 Duchess	G. A. F. Tilley	
G-BNUS	Cessna 152 II	Stapleford Flying Club Ltd	
G-BNUT	Cessna 152 Turbo	Stapleford Flying Club Ltd	
G-BNUV	PA-23 Aztec 250F	L. J. Martin	
G-BNUX	Hoffmann H-36 Dimona	Buckminster Dimona Syndicate/Saltby	
G-BNUY	PA-38-112 Tomahawk II	D. J. Whitcombe	
G-BNVB	AA-5A Cheetah	Grumman Group	
G-BNVD	PA-38-112 Tomahawk	D. A. Whitmore	
G-BNVE	PA-28-181 Archer II	Solent Flight Ltd	
G-BNVT	PA-28R-201T Turbo Arrow III	Victor Tango Group	
G-BNVZ	Beech 95-B55 Baron	W. J. Forrest /White Waltham	
G-BNWA	Boeing 767-336ER	British Airways	
G-BNWC	Boeing 767-336ER	British Airways	
G-BNWD	Boeing 767-336ER	British Airways	
G-BNWH	Boeing 767-336ER	British Airways	
G-BNWI	Boeing 767-336ER	British Airways	
G-BNWM	Boeing 767-336ER	British Airways	
G-BNWN	Boeing 767-336ER	British Airways	
G-BNWO	Boeing 767-336ER	British Airways	
G-BNWR	Boeing 767-336ER	British Airways	
G-BNWS	Boeing 767-336ER	British Airways	
G-BNWT	Boeing 767-336ER	British Airways	
G-BNWU	Boeing 767-336ER	British Airways	
G-BNWV	Boeing 767-336ER	British Airways	
G-BNWX	Boeing 767-336ER	British Airways	
G-BNWZ	Boeing 767-336ER	British Airways	
G-BNXC	Cessna 152 II	Sir W. G. Armstrong Whitworth Flying Group/ Coventry	
G-BNXD	Cessna 172N	I. Chaplin	
G-BNXE	PA-28-161 Warrior II	Rugby Autobody Repairs/Coventry	
G-BNXI	Robin DR.400/180R	London Gliding Club (Pty) Ltd/Dunstable	
G-BNXK	Nott-Cameron ULD-3 balloon	J. R. P. Nott (G-BLJN)	
G-BNXL	Glaser-Dirks DG.400	G-BNXL Group	
G-BNXM	PA-18 Super Cub 95	G-BNXM Group	
G-BNXR	Cameron O-84 balloon	J. A. B. Gray	
G-BNXT	PA-28-161 Warrior II	Falcon Flying Services/Manston	
G-BNXU	PA-28-161 Warrior II	Friendly Warrior Group	
G-BNXV	PA-38-112 Tomahawk	W. B. Bateson/Blackpool	
G-BNXX	SOCATA TB.20 Trinidad	D. M. Carr	
G-BNXZ	Thunder Ax7-77 balloon	Hale Hot Air Balloon Group	
G-BNYB	PA-28-201T Turbo Dakota	G-BNYB Ltd/Blackbushe	
G-BNYD	Bell 206B JetRanger 3	Sterling Helicopters Ltd/Norwich	
G-BNYK	PA-38-112 Tomahawk	APB Leasing Ltd/Welshpool	
G-BNYL	Cessna 152 II	V. J. Freeman/Headcorn	
G-BNYM	Cessna 172N	Kestrel Syndicate	
G-BNYN	Cessna 152 II	Redhill Flying Club	
G-BNYO	Beech 76 Duchess	Harding Wragg	
G-BNYP	PA-28-181 Archer II	R. D. Cooper/Cranfield	
G-BNYS	Boeing 767-204ER	Air Atlanta Europe Ltd	
G-BNYV	PA-38-112 Tomahawk	Goodair Leasing Ltd	
G-BNYX	Denney Kitfox Mk 1	W. J. Husband	
G-BNYZ	SNCAN Stampe SV-4E	M. J. Heudebourck & D, A, Starkey	
G-BNZB	PA-28-161 Warrior II	Falcon Flying Services Ltd/Biggin Hill	
G-BNZC	D.H.C.1 Chipmunk 22 (18671)	The Shuttleworth Collection/O.Warden	
G-BNZK	Thunder Ax7-77 balloon	T. D. Marsden	
G-BNZL	Rotorway Scorpion 133	J. R. Wraight	
G-BNZM	Cessna T.210N	A. J. M. Freeman	
G-BNZN	Cameron N-56 balloon	Balloon Sports HB/Sweden	
G-BNZO	Rotorway Executive	D. Collins & R. Ayres	
G-BNZR	FRED Srs 2	R. M. Waugh/Newtownards	
G-BNZV	PA-25 Pawnee 235	D. B. Almey	
G-BNZZ	PA-28-161 Warrior II	Providence Aviation Ltd	

Notes	Reg.	Type	Owner or Operator
	G-BOAA	Concorde 102 ★	Museum Of Flight East Fortune (G-N94AA)
	G-BOAB	Concorde 102 ★	Preserved at Heathrow (G-N94AB)
	G-BOAC	Concorde 102 ★	Displayed in Viewing area Manchester International (G-N81AC)
	G-BOAF	Concorde 102 ★	Bristol Aero Collection (N94AF)/Filton
	G-BOAH	PA-28-161 Warrior II	Prestwick Flight Centre Ltd
	G-BOAI	Cessna 152 II	Galair Ltd/Biggin Hill
	G-BOAL	Cameron V-65 balloon	A. M. Lindsay
	G-BOAM	Robinson R-22B	Plane Talking Ltd/Elstree
	G-BOAO	Thunder Ax7-77 balloon	D. V. Fowler
	G-BOAS	Air Command 503 Commander	R. Robinson
	G-BOAU	Cameron V-77 balloon	G. T. Barstow
	G-BOBA	PA-28R-201 Arrow III	Atlantic Flight Training Ltd
	G-BOBB	Cameron O-120 balloon	Over The Rainbow Balloon Flights Ltd
	G-BOBH	Airtour AH-77 balloon	J. & K. Francis
	G-BOBL	PA-38-112 Tomahawk	Cardiff-Wales Aviation Services Ltd
	G-BOBR	Cameron N-77 balloon	P. A. Davies & M. Morris
	G-BOBT	Stolp SA.300 Starduster Too	G-BOBT Group
	G-BOBV	Cessna F.150M	Sheffield Aero Club Ltd/Netherthorpe
	G-BOBY	Monnett Sonerai II	R. G. Hallam (stored)/Sleap
	G-BOBZ	PA-28-181 Archer II	Trustcomms International Ltd
	G-BOCG	PA-34-200T Seneca II	Oxford Aviation Services Ltd/Kidlington
	G-BOCI	Cessna 140A	J. B. Bonnell
	G-BOCK	Sopwith Triplane (replica) (N6290)	The Shuttleworth Collection/O. Warden
	G-BOCL	Slingsby T.67C	Richard Brinklow Aviation Ltd
	G-BOCM	Slingsby T.67C	Richard Brinklow Aviation Ltd
	G-BOCN	Robinson R-22B	Auto-Rotation Ltd
	G-BOCS	PA-34-220T Seneca III	Flight Training Europe SL/Spain
	G-BODB	PA-28-161 Warrior II	Sherburn Aero Club Ltd
	G-BODC	PA-28-161 Warrior II	Sherburn Aero Club Ltd
	G-BODD	PA-28-161 Warrior II	L. W. Scattergood
	G-BODE	PA-28-161 Warrior II	Sherburn Aero Club Ltd
	G-BODH	Slingsby T.31 Motor Cadet III	M. M. Bain
	G-BODI	Stoddard-Hamilton SH-3R Glasair III	H. Arlt/Germany
	G-BODM	PA-28 Cherokee 180	R. Emery
	G-BODO	Cessna 152	A. R. Sarson
	G-BODP	PA-38-112 Tomahawk	CMA Associates Ltd
	G-BODR	PA-28-161 Warrior II	Airways Aero Associations Ltd/Booker
	G-BODS	PA-38-112 Tomahawk	Coulson Flying Services Ltd/Cranfield
	G-BODT	Jodel D.18	L. D. McPhillips
	G-BODU	Scheibe SF.25C Falke	Faulkes Flying Foundation Ltd
	G-BODX	Beech 76 Duchess	Aviation Rentals/Bournemouth
	G-BODY	Cessna 310R	Atlantic Air Transport Ltd/Coventry
	G-BODZ	Robinson R-22B	Langley Holdings PLC
	G-BOEE	PA-28-181 Archer II	T. B. Parmenter
	G-BOEG	Short SD3-60 Variant 100	BAC Express Airlines Ltd
	G-BOEH	Jodel DR.340	Piper Flyers Group
	G-BOEI	Short SD3-60 Variant 100	Aerocentury Corpn
	G-BOEK	Cameron V-77 balloon	R. I. M. Kerr & ptnrs
	G-BOEM	Aerotek-Pitts S-2A	M. Murphy
	G-BOEN	Cessna 172M	C. Barlow
	G-BOER	PA-28-161 Warrior II	M. & W. Fraser-Urquhart
	G-BOET	PA-28RT-201 Arrow IV	B. C. Chambers (G-IBEC)
	G-BOEW	Robinson R-22B	Plane Talking Ltd/Elstree
	G-BOEZ	Robinson R-22B	Plane Talking Ltd/Elstree
	G-BOFC	Beech 76 Duchess	Magenta Ltd/Kidlington
	G-BOFD	Cessna U.206G	D. M. Penny
	G-BOFE	PA-34-200T Seneca II	Alstons Upholstery Ltd
	G-BOFF	Cameron N-77 balloon	R. C. Corcoran
	G-BOFL	Cessna 152 II	GEM Rewinds Ltd/Coventry
	G-BOFM	Cessna 152 II	GEM Rewinds Ltd/Coventry
	G-BOFW	Cessna A.150M	D. F. Donovan
	G-BOFX	Cessna A.150M	N. F. O'Neill
	G-BOFY	PA-28 Cherokee 140	BCT Aircraft Leasing Ltd
	G-BOFZ	PA-28-161 Warrior II	R. W. Harris
	G-BOGC	Cessna 152 II	Mona Flying Club
	G-BOGI	Robin DR.400/180	A. L. M. Shepherd
	G-BOGK	ARV Super 2	H. N. Stone
	G-BOGM	PA-28RT-201T Turbo Arrow IV	RJP Aviation
	G-BOGO	PA-32R-301T Saratoga SP	A. S. Doman/Biggin Hill
	G-BOGP	Cameron V-77 balloon	Wealden Balloon Group
	G-BOGV	Air Command 532 Elite	G. M. Hobman

Reg.	Type	Owner or Operator	Notes
G-BOGY	Cameron V-77 balloon	R. A. Preston	
G-BOHA	PA-28-161 Warrior II	C. H. Bough & S. J. Gardiner	
G-BOHD	Colt 77A balloon	D. B. Court	
G-BOHF	Thunder Ax8-84 balloon	J. A. Harris	
G-BOHG	Air Command 532 Elite	T. E. McDonald	
G-BOHH	Cessna 172N	T. Scott	
G-BOHI	Cessna 152 II	V. D. Speck/Clacton	
G-BOHJ	Cessna 152 II	Airlaunch/Old Buckenham	
G-BOHL	Cameron A-120 balloon	J. M. Holmes	
G-BOHM	PA-28 Cherokee 180	B. F. Keogh & R. A. Scott	
G-BOHO	PA-28-161 Warrior II	Egressus Flying Group	
G-BOHR	PA-28-151 Warrior	K. L. Rivers	
G-BOHS	PA-38-112 Tomahawk	Falcon Flying Services/Biggin Hill	
G-BOHT	PA-38-112 Tomahawk	St George Flying Club/Teesside	
G-BOHU	PA-38-112 Tomahawk	D. A. Whitmore	
G-BOHV	Wittman W.8 Tailwind	R. A. Povall	
G-BOHW	Van's RV-4	P. J. Robins	
G-BOHX	PA-44-180 Seminole	Airpart Supply Ltd/Booker	
G-BOIA	Cessna 180K	R. E. Styles & ptnrs	
G-BOIB	Wittman W.10 Tailwind	R. F. Bradshaw	
G-BOIC	PA-28R-201T Turbo Arrow III	M. J. Pearson	
G-BOID	Bellanca 7ECA Citabria	D. Mallinson	
G-BOIG	PA-28-161 Warrior II	D. Vallence-Pell/Jersey	
G-BOIJ	Thunder Ax7-77 balloon	K. Dodman	
G-BOIK	Air Command 503 Commander	F. G. Shepherd	
G-BOIL	Cessna 172N	Upperstack Ltd	
G-BOIO	Cessna 152	AV Aviation Ltd	
G-BOIP	Cessna 152	Stapleford Flying Club Ltd	
G-BOIR	Cessna 152	Shropshire Aero Club Ltd/Sleap	
G-BOIT	SOCATA TB.10 Tobago	G-BOIT Flying Group	
G-BOIU	SOCATA TB.10 Tobago	R & B Aviation Ltd	
G-BOIV	Cessna 150M	M. J. Page	
G-BOIX	Cessna 172N	JR Flying Ltd	
G-BOIY	Cessna 172N	L. W. Scattergood	
G-BOIZ	PA-34-200T Seneca II	S. F. Tebby & Son	
G-BOJB	Cameron V-77 balloon	K. L. Heron & R. M. Trotter	
G-BOJD	Cameron V-77 balloon	R. S. McDonald	
G-BOJI	PA-28RT-201 Arrow IV	Arrow Two Group/Blackbushe	
G-BOJK	PA-34-220T Seneca III	Redhill Flying Club (G-BRUF)	
G-BOJM	PA-28-181 Archer II	R. P. Emms	
G-BOJR	Cessna 172P	Exeter Flying Club Ltd	
G-BOJS	Cessna 172P	I. S. H. Paul	
G-BOJU	Cameron N-77 balloon	M. A. Scholes	
G-BOJW	PA-28-161 Warrior II	Brewhamfield Farm Ltd	
G-BOJZ	PA-28-161 Warrior II	Falcon Flying Services/Biggin Hill	
G-BOKA	PA-28-201T Turbo Dakota	CBG Aviation Ltd/Biggin Hill	
G-BOKB	PA-28-161 Warrior II	Apollo Aviation Advisory Ltd/Shoreham	
G-BOKD	Bell 206B JetRanger 3	Sterling Helicopters Ltd (G-ISKY/G-PSCI)	
G-BOKF	Air Command 532 Elite	J. K. Padden	
G-BOKH	Whittaker MW.7	I. D. Evans	
G-BOKN	PA-28-161 Warrior II	BAE Systems Flight Training Ltd	
G-BOKU	PA-28-161 Warrior II	BAE Systems Flight Training Ltd	
G-BOKX	PA-28-161 Warrior II	Shenley Farms (Aviation) Ltd/Headcorn	
G-BOKY	Cessna 152 II	D. F. F. & J. E. Poore	
G-BOLB	Taylorcraft BC-12-65	A. D. Pearce, E. C. & P. King/Kemble	
G-BOLC	Fournier RF-6B-100	J. D. Cohen/Dunkeswell	
G-BOLD	PA-38-112 Tomahawk	G-BOLD Group/Eaglescott	
G-BOLE	PA-38-112 Tomahawk	J. & G. Stevenson	
G-BOLF	PA-38-112 Tomahawk	Teesside Flight Centre Ltd	
G-BOLG	Bellanca 7KCAB Citabria	B. R. Pearson/Eaglescott	
G-BOLI	Cessna 172P	Boli Flying Club	
G-BOLL	Lake LA-4 Skimmer	M. C. Holmes	
G-BOLN	Colt 21A balloon	G. Everett	
G-BOLO	Bell 206B JetRanger	Hargreaves Leasing Ltd	
G-BOLP	Colt 21A balloon	J. E. Rose	
G-BOLR	Colt 21A balloon	C. J. Sanger-Davies	
G-BOLS	FRED Srs 2	I. F. Vaughan	
G-BOLT	R. Commander 114	Harnett Air Services Ltd	
G-BOLU	Robin R.3000/120	Classair/Biggin Hill	
G-BOLV	Cessna 152 II	Falcon Flying Services/Biggin Hill	
G-BOLW	Cessna 152 II	JRB Aviation Ltd/Southend	
G-BOLX	Cessna 172N	R. J. Burrough/Headcorn	

Notes	Reg.	Type	Owner or Operator
	G-BOLY	Cessna 172N	Simair Ltd
	G-BOLZ	Rand-Robinson KR-2	B. Normington
	G-BOMB	Cassutt Racer IIIM	S. Adams
	G-BOMG	BN-2B-26 Islander	Loganair Ltd
	G-BOMN	Cessna 150F	Auburn Air Ltd
	G-BOMO	PA-38-112 Tomahawk II	APB Leasing Ltd/Welshpool
	G-BOMP	PA-28-181 Archer II	D. Carter
	G-BOMS	Cessna 172N	Almat Flying Club Ltd & Penchant Ltd
	G-BOMU	PA-28-181 Archer II	J. Sawyer
	G-BOMY	PA-28-161 Warrior II	Southern Care Maintenance
	G-BOMZ	PA-38-112 Tomahawk	BOMZ Aviation/Booker
	G-BONC	PA-28RT-201 Arrow IV	Finglow Ltd
	G-BONG	Enstrom F-28A-UK	J. G. Dunn
	G-BONO	Cessna 172N	J. D. McCandless
	G-BONP	CFM Streak Shadow	T. J. Palmer
	G-BONR	Cessna 172N	D. I. Craik/Biggin Hill
	G-BONS	Cessna 172N	BONS Group/Elstree
	G-BONT	Slingsby T.67M Mk II	Babcock Defence Services
	G-BONU	Slingsby T.67B	R. L. Brinklow
	G-BONW	Cessna 152 II	Lincoln Aero Club Ltd/Sturgate
	G-BONY	Denney Kitfox Mk 1	J. M. Vinall
	G-BONZ	Beech V35B Bonanza	P. M. Coulten
	G-BOOB	Cameron N-65 balloon	J. Rumming
	G-BOOC	PA-18 Super Cub 150	S. A. C. Whitcombe
	G-BOOD	Slingsby T.31M Motor Tutor	K. A. Hale
	G-BOOE	GA-7 Cougar	N. Gardner
	G-BOOF	PA-28-181 Archer II	H. Merkado/Panshanger
	G-BOOG	PA-28RT-201T Turbo Arrow IV	Simair Ltd
	G-BOOH	Jodel D.112	R. M. MacCormac
	G-BOOI	Cessna 152	Stapleford Flying Club Ltd
	G-BOOJ	Air Command 532 Elite II	Roger Savage Gyroplanes Ltd
	G-BOOL	Cessna 172N	Surrey & Kent Flying Club Ltd/Biggin Hill
	G-BOOV	AS.355F-2 Twin Squirrel	Atlantic Air Charters Ltd
	G-BOOW	Aerosport Scamp	D. A. Weldon/Ireland
	G-BOOX	Rutan LongEz	I. R. Wilde
	G-BOOZ	Cameron N-77 balloon	J. E. F. Kettlety
	G-BOPA	PA-28-181 Archer II	Flyco Ltd
	G-BOPB	Boeing 767-204ER	Air Atlanta Europe Ltd
	G-BOPC	PA-28-161 Warrior II	Aeros Ltd
	G-BOPD	Bede BD-4	S. T. Dauncey
	G-BOPH	Cessna TR.182RG	Grandsam Investments Ltd
	G-BOPO	Brooklands OA.7 Optica	Aces High Ltd/North Weald
	G-BOPR	Brooklands OA.7 Optica	Aces High Ltd/North Weald
	G-BOPT	Grob G.115	LAC (Enterprises) Ltd/Barton
	G-BOPU	Grob G.115	LAC (Enterprises) Ltd/Barton
	G-BOPX	Cessna A.152	Aerohire Ltd/Wolverhampton
	G-BORB	Cameron V-77 balloon	M. H. Wolff
	G-BORD	Thunder Ax7-77 balloon	D. D. Owen
	G-BORE	Colt 77A balloon	Little Secret Hot-Air Balloon Group
	G-BORG	Campbell Cricket	G. Davison & H. Hayes
	G-BORH	PA-34-200T Seneca II	Aerolease Ltd
	G-BORI	Cessna 152 II	S. Copeland
	G-BORJ	Cessna 152 II	Pool Aviation (NW) Ltd/Blackpool
	G-BORK	PA-28-161 Warrior II	The Warrior Group (G-IIIC)
	G-BORL	PA-28-161 Warrior II	Westair Flying School Ltd/Blackpool
	G-BORM	H.S.748 Srs 2B ★	Airport Fire Service/Exeter
	G-BORN	Cameron N-77 balloon	I. Chadwick
	G-BORO	Cessna 152 II	Tatenhill Aviation Ltd
	G-BORR	Thunder Ax8-90 balloon	W. J. Harris
	G-BORS	PA-28-181 Archer II	Modern Air (UK) Ltd/Fowlmere
	G-BORT	Colt 77A balloon	J. Triquet/France
	G-BORW	Cessna 172P	Briter Aviation Ltd/Coventry
	G-BORY	Cessna 150L	Alexander Aviation
	G-BOSB	Thunder Ax7-77 balloon	M. Gallagher
	G-BOSD	PA-34-200T Seneca II	Barnes Olson Aeroleasing Ltd/Bristol
	G-BOSE	PA-28-181 Archer II	G-BOSE Group
	G-BOSJ	Nord 3400 (124)	A. I. Milne
	G-BOSM	Jodel DR.253B	A. G. Stevens
	G-BOSN	AS.355F-1 Twin Squirrel	Helicopter Services/Booker
	G-BOSO	Cessna A.152	J. S. Develin & Z. Islam/Redhill
	G-BOSR	PA-28 Cherokee 140	Sierra Romeo Group
	G-BOSU	PA-28 Cherokee 140	R. A. Sands

Reg.	Type	Owner or Operator	Notes
G-BOTD	Cameron O-105 balloon	P. J. Beglan/France	
G-BOTF	PA-28-151 Warrior	G-BOTF Group/Southend	
G-BOTG	Cessna 152 II	Donington Aviation Ltd/E. Midlands	
G-BOTH	Cessna 182Q	G-BOTH Group	
G-BOTI	PA-28-151 Warrior	Falcon Flying Services/Biggin Hill	
G-BOTK	Cameron O-105 balloon	N. Woodham	
G-BOTM	Bell 206B JetRanger 3	Mexsky Ltd	
G-BOTN	PA-28-161 Warrior II	Apollo Aviation Advisory	
G-BOTO	Bellanca 7ECA Citabria	G-BOTO Group	
G-BOTP	Cessna 150J	R. F. Finnis & C. P. Williams	
G-BOTU	Piper J-3C-65 Cub	T. L. Giles	
G-BOTV	PA-32RT-300 Lance II	Robin Lance Aviation Association Ltd	
G-BOTW	Cameron V-77 balloon	M. R. Jeynes	
G-BOUD	PA-38-112 Tomahawk	A. J. Wiggins	
G-BOUE	Cessna 172N	Castleridge Ltd	
G-BOUF	Cessna 172N	B. P. & M. I. Sneap	
G-BOUJ	Cessna 150M	UJ Flying Group	
G-BOUK	PA-34-200T Seneca II	C. J. & R. J. Barnes	
G-BOUL	PA-34-200T Seneca II	Oxford Aviation Services Ltd/Kidlington	
G-BOUM	PA-34-200T Seneca II	Oxford Aviation Services Ltd/Kidlington	
G-BOUP	PA-28-161 Warrior II	Aeros Holdings Ltd	
G-BOUR	PA-28-161 Warrior II	Oxford Aviation Services Ltd/Kidlington	
G-BOUT	Colomban MC.12 Cri-Cri	C. K. Farley	
G-BOUV	Bensen B.8MR	G. C. Kerr	
G-BOUZ	Cessna 150G	Atlantic Bridge Aviation Ltd/Lydd	
G-BOVB	PA-15 Vagabond	Oscar Flying Group/Shoreham	
G-BOVC	Everett gyroplane	J. W. Highton	
G-BOVK	PA-28-161 Warrior II	Multiflight Ltd	
G-BOVR	Robinson R-22	J. O'Brien	
G-BOVS	Cessna 150M	K. J. Marchelak	
G-BOVT	Cessna 150M	C. J. Hopewell	
G-BOVU	Stoddard-Hamilton Glasair III	B. R. Chaplin	
G-BOVV	Cameron V-77 balloon	P. Glydon	
G-BOVW	Colt 69A balloon	V. Hyland	
G-BOVX	Hughes 269C	P. E. Tornberg	
G-BOWB	Cameron V-77 balloon	R. C. Stone	
G-BOWD	Cessna F.337G	D. H. G. Penney (G-BLSB)	
G-BOWE	PA-34-200T Seneca II	Oxford Aviation Services Ltd/Kidlington	
G-BOWL	Cameron V-77 balloon	P. G. & G. R. Hall	
G-BOWM	Cameron V-56 balloon	C. G. Caldecott & G. Pitt	
G-BOWN	PA-12 Super Cruiser	J. R. Kimberley	
G-BOWO	Cessna R.182	J. J. Feeney (G-BOTR)	
G-BOWP	Jodel D.120A	M. W. Bodger & M. H. Hoffman	
G-BOWU	Cameron O-84 balloon	St Elmos Fire Syndicate	
G-BOWV	Cameron V-65 balloon	R. A. Harris	
G-BOWY	PA-28RT-201T Turbo Arrow IV	J. S. Develin & Z. Islam	
G-BOWZ	Bensen B.80V	W. M. Day	
G-BOXA	PA-28-161 Warrior II	Channel Islands Aero Club (Jersey) Ltd	
G-BOXB	PA-28-161 Warrior II	First Class Ltd	
G-BOXC	PA-28-161 Warrior II	Channel Islands Aero Club (Jersey) Ltd	
G-BOXG	Cameron O-77 balloon	R. A. Wicks	
G-BOXH	Pitts S-1S Special	Pittsco Ltd	
G-BOXJ	Piper J-3C-65 Cub	J. D. Tseliki/Shoreham	
G-BOXR	GA-7 Cougar	Plane Talking Ltd/Elstree	
G-BOXT	Hughes 269C	Goldenfly Ltd	
G-BOXU	AA-5B Tiger	Marcher Aviation Group/Welshpool	
G-BOXV	Pitts S-1S Special	C. Waddington	
G-BOXW	Cassutt Racer Srs IIIM	D. I. Johnson	
G-BOYB	Cessna A.152	Northamptonshire School of Flying Ltd/Sywell	
G-BOYC	Robinson R-22B	Yorkshire Helicopters/Leeds	
G-BOYF	Sikorsky S-76B	Darley Stud Management Co Ltd/Blackbushe	
G-BOYH	PA-28-151 Warrior	Superpause Ltd/Booker	
G-BOYI	PA-28-161 Warrior II	G-BOYI Group/Welshpool	
G-BOYL	Cessna 152 II	Aerohire Ltd/Wolverhampton	
G-BOYM	Cameron O-84 balloon	M. P. Ryan	
G-BOYO	Cameron V-20 balloon	J. M. Willard	
G-BOYP	Cessna 172N	I. D. & D. Brierley	
G-BOYR	Cessna F.337G	Tri-Star Farms Ltd	
G-BOYS	Cameron N-77 balloon	Wye Valley Aviation Ltd	
G-BOYU	Cessna A.150L	Upperstack Ltd/Liverpool	
G-BOYV	PA-28R-201T Turbo Arrow III	Arrow Air Ltd	
G-BOYX	Robinson R-22B	R. Towle	

Notes	Reg.	Type	Owner or Operator
	G-BOZI	PA-28-161 Warrior II	Aerolease Ltd/Conington
	G-BOZN	Cameron N-77 balloon	Calarel Developments Ltd
	G-BOZO	AA-5B Tiger	Caslon Ltd
	G-BOZR	Cessna 152 II	GEM Rewinds Ltd/Coventry
	G-BOZS	Pitts S-1C Special	T. A. S. Rayner
	G-BOZU	Sparrow Hawk Mk II	R. V. Phillimore
	G-BOZV	CEA DR.340 Major	C. J. Turner & S. D. Kent
	G-BOZW	Bensen B.8M	M. E. Wills
	G-BOZY	Cameron RTW-120 balloon	Magical Adventures Ltd
	G-BOZZ	AA-5B Tiger	Solent Tiger Group/Southampton
	G-BPAA	Acro Advanced	Acro Engines & Airframes Ltd
	G-BPAB	Cessna 150M	M. J. Diggins/Earls Colne
	G-BPAF	PA-28-161 Warrior II	RAF Brize Norton Flying Club Ltd
	G-BPAI	Bell 47G-3B-1 (modified)	LRC Leisure Ltd
	G-BPAJ	D.H.82A Tiger Moth	P. A. Jackson (G-AOIX)
	G-BPAL	D.H.C.1 Chipmunk 22 (WG350)	K. F. & P. Tomsett (G-BCYE)
	G-BPAS	SOCATA TB.20 Trinidad	Syndicate Clerical Services Ltd
	G-BPAV	FRED Srs 2	P. A. Valentine
	G-BPAW	Cessna 150M	P. D. Sims
	G-BPAX	Cessna 150M	The Dirty Dozen
	G-BPAY	PA-28-181 Archer II	Leicestershire Aero Club Ltd
	G-BPBB	Evans VP-2 (mod)	A. Bleese
	G-BPBG	Cessna 152 II	APB Leasing Ltd/Welshpool
	G-BPBI	Cessna 152 II	B. W. Wells & Burbage Farms Ltd
	G-BPBJ	Cessna 152 II	W. Shaw & P. G. Haines
	G-BPBK	Cessna 152 II	Atlantic Flight Training Ltd
	G-BPBM	PA-28-161 Warrior II	Halfpenny Green Flight Centre Ltd
	G-BPBO	PA-28RT-201T Turbo Arrow IV	Tile Holdings Ltd
	G-BPBP	Brügger MB.2 Colibri	D. A. Preston
	G-BPBV	Cameron V-77 balloon	S. J. Farrant
	G-BPBW	Cameron O-105 balloon	R. J. Mansfield
	G-BPBY	Cameron V-77 balloon	L. Hutley (G-BPCS)
	G-BPCA	BN-2B-26 Islander	Loganair Ltd (G-BLNX)
	G-BPCF	Piper J-3C-65 Cub	T. I. Williams
	G-BPCG	Colt AS-80 airship	N. Charbonnier/Italy
	G-BPCI	Cessna R.172K	A. J. Flisher
	G-BPCK	PA-28-161 Warrior II	Compton Abbas Airfield Ltd
	G-BPCL	SA Bulldog Srs 120/128	Isohigh Ltd/Denham
	G-BPCM	Rotorway Executive	Aircare Group
	G-BPCR	Mooney M.20K	T. & R. Harris
	G-BPCV	Montgomerie-Bensen B.8MR	M. A. Hayward
	G-BPCX	PA-28-236 Dakota	D. J. Mountain
	G-BPDE	Colt 56A balloon	J. E. Weidema/Netherlands
	G-BPDF	Cameron V-77 balloon	The Ballooning Business Ltd
	G-BPDG	Cameron V-77 balloon	F. R. Battersby
	G-BPDJ	Christena Mini Coupe	J. J. Morrissey/Popham
	G-BPDK	Sorrell SNS-7 Hyperbipe ★	A. J. Cable (stored)/Barton
	G-BPDM	C.A.S.A. 1.131E Jungmann 2000	(781-32)J. D. Haslam
	G-BPDT	PA-28-161 Warrior II	Channel Islands Aero Club (Jersey) Ltd
	G-BPDV	Pitts S-1S Special	J. Vize/Sywell
	G-BPEC	Boeing 757-236	British Airways
	G-BPED	Boeing 757-236	British Airways
	G-BPEE	Boeing 757-236	British Airways
	G-BPEI	Boeing 757-236	British Airways (G-BMRK)
	G-BPEJ	Boeing 757-236	British Airways (G-BMRL)
	G-BPEK	Boeing 757-236	British Airways (G-BMRM)
	G-BPEL	PA-28-151 Warrior	R. W. Harris & A. J. Jahanfar
	G-BPEM	Cessna 150K	R. G. Lindsey & R. Strong
	G-BPEO	Cessna 152 II	Direct Helicopters/Southend
	G-BPES	PA-38-112 Tomahawk II	Sherwood Flying Club Ltd/Tollerton
	G-BPEZ	Colt 77A balloon	J. E. F. Kettley & W. J. Honey
	G-BPFB	Colt 77A balloon	S. Ingram
	G-BPFC	Mooney M.20C	D. P. Wring
	G-BPFD	Jodel D.112	M. R. Sallows
	G-BPFF	Cameron DP-70 airship	John Aimo Balloons SAS/Italy
	G-BPFH	PA-28-161 Warrior II	M. H. Kleiser
	G-BPFI	PA-28-181 Archer II	F. Teagle
	G-BPFL	Davis DA-2	B. W. Griffiths
	G-BPFM	Aeronca 7AC Champion	T. J. Roberts
	G-BPFN	Short SD3-60 Variant 100	Aurigny Air Services Ltd
	G-BPFZ	Cessna 152 II	Devon School of Flying Ltd/Dunkeswell

Reg.	Type	Owner or Operator	Notes
G-BPGC	Air Command 532 Elite	G. A. Speich	
G-BPGD	Cameron V-65 balloon	Gone With The Wind Ltd	
G-BPGE	Cessna U.206C	Scottish Parachute Club/Strathallan	
G-BPGF	Thunder Ax7-77 balloon	M. Schiavo	
G-BPGH	EAA Acro Sport II	G. M. Bradley	
G-BPGK	Aeronca 7AC Champion	D. A. Crompton	
G-BPGT	Colt AS-80 Mk II airship	P. Porati/Italy	
G-BPGU	PA-28-181 Archer II	G. Underwood/Tollerton	
G-BPGV	Robinson R-22B	G. Gazza/Monaco	
G-BPGX	SOCATA TB.9 Tampico	D. A. Lee	
G-BPGZ	Cessna 150G	J. B. Scott	
G-BPHB	PA-28-161 Warrior II	M. J. Wade	
G-BPHD	Cameron N-42 balloon	P. J. Marshall & M. A. Clarke	
G-BPHG	Robin DR.400/180	A. Hildreth	
G-BPHH	Cameron V-77 balloon	C. D. Aindow	
G-BPHI	PA-38-112 Tomahawk	J. S. Devlin & Z. Islam/Redhill	
G-BPHJ	Cameron V-77 balloon	C. W. Brown	
G-BPHK	Whittaker MW.7	J. S. Shufflebottom	
G-BPHL	PA-28-161 Warrior II	Teesside Flight Centre Ltd	
G-BPHO	Taylorcraft BC-12	A. A. Alderdice	
G-BPHP	Taylorcraft BC-12-65	J. S. Jackson	
G-BPHR	D.H.82A Tiger Moth (A17-48)	N. Parry	
G-BPHT	Cessna 152	Evensport Ltd	
G-BPHU	Thunder Ax7-77 balloon	R. P. Waite	
G-BPHW	Cessna 140	L. J. A. Bell	
G-BPHX	Cessna 140	M. McChesney	
G-BPHZ	M.S.505 Criquet (TA+RC)	Aero Vintage Ltd	
G-BPID	PA-28-161 Warrior II	J. T. Nuttall	
G-BPIF	Bensen-Parsons 2-place gyroplane	B. J. L. P. de Saar	
G-BPII	Denney Kitfox	G-BPII Group	
G-BPIJ	Brantly B.2B	Seething Brantty Group	
G-BPIK	PA-38-112 Tomahawk	M. J. White & L. Chadwick	
G-BPIL	Cessna 310B	A. L. Brown	
G-BPIN	Glaser-Dirks DG.400	J. N. Stevenson	
G-BPIO	Cessna F.152 II	I. D. McClelland	
G-BPIP	Slingsby T.31 Motor Cadet III	J. H. Beard	
G-BPIR	Scheibe SF.25E Super Falke	K. E. Ballington	
G-BPIT	Robinson R-22B	NA Air Ltd	
G-BPIU	PA-28-161 Warrior II	Golf India Uniform Group	
G-BPIV	B.149 Bolingbroke Mk IVT (R3821)	Blenheim (Duxford) Ltd	
G-BPIZ	AA-5B Tiger	D. A. Horsley	
G-BPJB	Schweizer 269C	Elborne Holdings Ltd/Portugal	
G-BPJD	SOCATA Rallye 110ST	G-BPJD Rallye Group	
G-BPJE	Cameron A-105 balloon	J. S. Eckersley	
G-BPJG	PA-18 Super Cub 150	M. W. Stein	
G-BPJH	PA-18 Super Cub 95	P. J. Heron	
G-BPJK	Colt 77A balloon	Saran UK Ltd	
G-BPJL	Cessna 152 II	K. O'Connor	
G-BPJO	PA-28-161 Cadet	Walsh Aviation	
G-BPJP	PA-28-161 Cadet	Aviation Rentals/Bournemouth	
G-BPJR	PA-28-161 Cadet	Walsh Aviation	
G-BPJU	PA-28-161 Cadet	Aviation Rentals/Bournemouth	
G-BPJV	Taylorcraft F-21	TC Flying Group	
G-BPJW	Cessna A.150K	Heald Ltd	
G-BPJZ	Cameron O-160 balloon	M. L. Gabb	
G-BPKF	Grob G.115	Dorset Aviation Group	
G-BPKK	Denney Kitfox Mk 1	D. Moffat	
G-BPKM	PA-28-161 Warrior II	R. Cass	
G-BPKO	Cessna 140	M. J. Patrick	
G-BPKR	PA-28-151 Warrior	Aeros Leasing Ltd	
G-BPKZ	Short SD3-60 Variant 100	BAC Express Airlines Ltd	
G-BPLH	Jodel DR.1051	D. W. Tovey	
G-BPLM	AIA Stampe SV-4C	C. J. Jesson/Redhill	
G-BPLR	BN-2B-26 Islander	Hebridean Air Services Ltd	
G-BPLV	Cameron V-77 balloon	MC VH SA/Belgium	
G-BPLY	Christen Pitts S-2B Special	J. L. Dixon	
G-BPLZ	Hughes 369HS	R. Kibble	
G-BPMB	Maule M5-235C Lunar Rocket	Earth Products Ltd/Crosland Moor	
G-BPME	Cessna 152 II	Seawing Flying Club/Southend	
G-BPMF	PA-28-151 Warrior	Mike Foxtrot Group	
G-BPML	Cessna 172M	N. A. Bilton	
G-BPMM	Champion 7ECA Citabria	J. Murray	

Notes	Reg.	Type	Owner or Operator
	G-BPMR	PA-28-161 Warrior II	B. McIntyre
	G-BPMU	Nord 3202B	A. I. Milne (G-BIZJ)
	G-BPMW	QAC Quickie Q.2	P. M. Wright (G-OICI/G-OGKN)
	G-BPMX	ARV Super 2	C. R. James
	G-BPNA	Cessna 150L	Wolds Flyers Syndicate
	G-BPNI	Robinson R-22B	Heliflight (UK) Ltd
	G-BPNJ	H.S.748 Srs 2A	Clewer Aviation Ltd
	G-BPNL	QAC Quickie Q.2	J. R. Jensen
	G-BPNN	Montgomerie-Bensen B.8MR	M. E. Vahdat
	G-BPNO	Zlin Z.326 Trener Master	J. A. S. Bailey & S. T. Logan
	G-BPNT	BAe 146-300	Flightline Ltd
	G-BPNU	Thunder Ax7-77 balloon	M. J. Barnes
	G-BPOA	Gloster Meteor T.7 (WF877) ★	39 Restoration Group/North Weald
	G-BPOB	Sopwith Camel F.1 (replica) (B2458)	Bianchi Aviation Film Services Ltd/Booker
	G-BPOL	Pietenpol Air Camper	G. W. Postance
	G-BPOM	PA-28-161 Warrior II	POM Flying Group
	G-BPON	PA-34-200T Seneca II	Aeroshare Ltd/Staverton
	G-BPOO	Montgomerie-Bensen B.8MR	M. E. Vahdat
	G-BPOS	Cessna 150M	Brooke Park Ltd
	G-BPOT	PA-28-181 Archer II	P. Fraser
	G-BPOU	Luscombe 8A Silvaire	Luscombe Trio
	G-BPPA	Cameron O-65 balloon	Rix Petroleum Ltd
	G-BPPD	PA-38-112 Tomahawk	Belting Products/Cardiff
	G-BPPE	PA-38-112 Tomahawk	First Air Ltd
	G-BPPF	PA-38-112 Tomahawk	Bristol Strut Flying Group
	G-BPPJ	Cameron A-180 balloon	D. J. Farrar
	G-BPPK	PA-28-151 Warrior	B & I Ltd
	G-BPPM	Beech B200 Super King Air	Gama Aviation Ltd/Fairoaks
	G-BPPO	Luscombe 8A Silvaire	M. G. Rummey
	G-BPPP	Cameron V-77 balloon	Samia Balloon Group
	G-BPPS	Mudry CAARP CAP.21	L. Van Vuuren
	G-BPPU	Air Command 532 Elite	J. Hough
	G-BPPY	Hughes 269B	D. G. Lewendon
	G-BPPZ	Taylorcraft BC-12D	Zulu Warriors Flying Group
	G-BPRA	Aeronca 11AC Chief	P. L. Clements
	G-BPRC	Cameron 77 Elephant SS balloon	A. Schneider/Germany
	G-BPRD	Pitts S-1C Special	Parrot Aerobatic Group
	G-BPRI	AS.355F-1 Twin Squirrel	MW Helicopters Ltd (G-TVPA)/Stapleford
	G-BPRJ	AS.355F-1 Twin Squirrel	PLM Dollar Group Ltd/Inverness
	G-BPRL	AS.355F-1 Twin Squirrel	Gas & Air Ltd
	G-BPRM	Cessna F.172L	BJ Aviation Ltd (G-AZKG)
	G-BPRN	PA-28-161 Warrior II	Air Navigation & Trading Co Ltd/Blackpool
	G-BPRR	Rand-Robinson KR-2	P. E. Taylor
	G-BPRX	Aeronca 11AC Chief	D. J. Dumolo & C. R. Barnes/Breighton
	G-BPRY	PA-28-161 Warrior II	White Wings Aviation
	G-BPSA	Luscombe 8A Silvaire	K. P. Gorman/Staverton
	G-BPSH	Cameron V-77 balloon	P. G. Hossack
	G-BPSJ	Thunder Ax6-56 balloon	V. Hyland
	G-BPSK	Montgomerie-Bensen B.8M	P. T. Ambrozik
	G-BPSL	Cessna 177	G-BPSL Group
	G-BPSO	Cameron N-90 balloon	J. Oberprieler/Germany
	G-BPSR	Cameron V-77 balloon	K. J. A. Maxwell
	G-BPSS	Cameron A-120 balloon	Anglian Countryside Balloons Ltd
	G-BPTA	Stinson 108-2	M. L. Ryan
	G-BPTD	Cameron V-77 balloon	J. Lippett
	G-BPTE	PA-28-181 Archer II	J. S. Develin & Z. Islam
	G-BPTG	R. Commander 112TC	L. G. Watteau
	G-BPTI	SOCATA TB.20 Trinidad	N. Davis
	G-BPTL	Cessna 172N	Cleveland Flying School Ltd/Teesside
	G-BPTS	C.A.S.A. 1.131E Jungmann 1000 (E3B-153)	Aerobatic Displays Ltd/Duxford
	G-BPTU	Cessna 152	A. M. Alam/Panshanger
	G-BPTV	Bensen B.8	C. Munro
	G-BPTX	Cameron O-120 balloon	Skybus Ballooning
	G-BPTZ	Robinson R-22B	Aero Maintenance Ltd
	G-BPUA	EAA Sport Biplane	Just Plane Trading Ltd
	G-BPUB	Cameron V-31 balloon	M. T. Evans
	G-BPUE	Air Command 532 Elite	A. H. Brent
	G-BPUF	Thunder Ax6-56Z balloon	R. C. & M. A. Trimble (G-BHRL)
	G-BPUG	Air Command 532 Elite	T. A. Holmes
	G-BPUJ	Cameron N-90 balloon	D. Grimshaw
	G-BPUL	PA-18 Super Cub 150	C. D. Duthy-James

Reg.	Type	Owner or Operator	Notes
G-BPUM	Cessna R.182RG	R. C. Chapman	
G-BPUP	Whittaker MW-7	J. H. Beard	
G-BPUR	Piper J-3L-65 Cub★	H. A. D. Monro	
G-BPUU	Cessna 140	Sherburn Aero Club Ltd	
G-BPUW	Colt 90A balloon	Gefa-Flug GmbH/Germany	
G-BPVA	Cessna 172F	S. Lancashire Flyers Group	
G-BPVC	Cameron V-77 balloon	B. D. Pettitt	
G-BPVE	Bleriot IX (replica) (1) ★	Bianchi Aviation Film Services Ltd/Booker	
G-BPVH	Cub Aircraft J-3C-65 Prospector	D. E. Cooper-Maguire	
G-BPVI	PA-32R-301 Saratoga SP	M. T. Coppen/Booker	
G-BPVK	Varga 2150A Kachina	H. W. Hall	
G-BPVM	Cameron V-77 balloon	Royal Engineers Balloon Club	
G-BPVN	PA-32R-301T Turbo Saratoga SP	Y. Leysen	
G-BPVO	Cassutt Racer IIIM	A. J. Brown	
G-BPVU	Thunder Ax7-77 balloon	B. J. Hammond	
G-BPVW	C.A.S.A. 1.131E Jungmann 2000	C. & J-W. Labeij/Netherlands	
G-BPVX	Cassutt Racer IIIM	D. D. Milne	
G-BPVY	Cessna 172D	S. J. Davies	
G-BPVZ	Luscombe 8E Silvaire	W. E. Gillham & P. Ryman	
G-BPWB	Sikorsky S-61N	Bristow Helicopters Ltd/HM Coastguard	
G-BPWC	Cameron V-77 balloon	H. B. Roberts	
G-BPWD	Cessna 120	Peregrine Flying Group	
G-BPWE	PA-28-161 Warrior II	RPR Associates Ltd/Swansea	
G-BPWF	PA-28 Cherokee 140 ★	(Static display)/1244 Sqdn ATC/Swindon	
G-BPWG	Cessna 150M	G-B Pilots Wilsford Group	
G-BPWI	Bell 206B JetRanger 3	Warren Aviation	
G-BPWK	Sportavia Fournier RF-5B	S. L. Reed/Usk	
G-BPWL	PA-25 Pawnee 235	Tecair Aviation Ltd/Shipdham	
G-BPWM	Cessna 150L	M. E. Creasey	
G-BPWN	Cessna 150L	T. P. Hadley	
G-BPWP	Rutan LongEz (modified)	D. A. Field	
G-BPWR	Cessna R.172K	Messrs Rees	
G-BPWS	Cessna 172P	Chartstone Ltd	
G-BPXA	PA-28-181 Archer II	Cherokee Flying Group/Netherthorpe	
G-BPXB	Glaser-Dirks DG.400	G. C. Westgate & Ptnrs/Parham Park	
G-BPXE	Enstrom 280C Shark	A. Healy	
G-BPXF	Cameron V-65 balloon	D. Pascall	
G-BPXH	Colt 17A balloon	Sport Promotion SRL/Italy	
G-BPXJ	PA-28RT-201T Turbo Arrow IV	K. M. Hollamby/Biggin Hill	
G-BPXX	PA-34-200T Seneca II	Yorkshire Aviation Ltd	
G-BPXY	Aeronca 11AC Chief	P. L. Turner	
G-BPYI	Cameron O-77 balloon	N. J. Logue	
G-BPYJ	Wittman W.8 Tailwind	J. Dixon	
G-BPYK	Thunder Ax7-77 balloon	A. R. Swinnerton	
G-BPYL	Hughes 369D	Morcorp (BVI) Ltd	
G-BPYN	Piper J-3C-65 Cub	The Aquila Group/White Waltham	
G-BPYO	PA-28-181 Archer II	Sherburn Aero Club Ltd	
G-BPYR	PA-31-310 Turbo Navajo	West Wales Airport Ltd (G-ECMA)/Shobdon	
G-BPYS	Cameron O-77 balloon	D. J. Goldsmith	
G-BPYT	Cameron V-77 balloon	M. H. Redman	
G-BPYV	Cameron V-77 balloon	R. J. Shortall	
G-BPYY	Cameron A-180 balloon	G. D. Fitzpatrick	
G-BPYZ	Thunder Ax7-77 balloon	J. E. Astall	
G-BPZA	Luscombe 8A Silvaire	M. J. Wright	
G-BPZB	Cessna 120	Cessna 120 Group	
G-BPZC	Luscombe 8A Silvaire	C. C. & J. M. Lovell	
G-BPZD	SNCAN NC.858S	S. J. Gaveston & ptnrs/Headcorn	
G-BPZE	Luscombe 8E Silvaire	WFG Luscombe Associates	
G-BPZK	Cameron O-120 balloon	D. L. Smith	
G-BPZM	PA-28RT-201 Arrow IV	Magenta Ltd (G-ROYW/G-CRTI)	
G-BPZP	Robin DR.400/180R	Lasham Gliding Soc. Ltd	
G-BPZU	Scheibe SF.25C Falke	Southdown Gliding Club Ltd	
G-BPZY	Pitts S-1C Special	J. S. Mitchell	
G-BPZZ	Thunder Ax8-105 balloon	Capricorn Balloons Ltd	
G-BRAF	Supermarine 394 Spitfire FR.XVIIIe	Wizard Investments Ltd	
G-BRAK	Cessna 172N	The Burnett Group Ltd/Kemble	
G-BRAM	Mikoyan MiG-21PF (503) ★	FAST Museum/Farnborough	
G-BRAR	Aeronca 7AC Champion	C. D. Ward	
G-BRAX	Payne Knight Twister 85B	R. Earl	
G-BRBA	PA-28-161 Warrior II	S. H. Pearce	
G-BRBB	PA-28-161 Warrior II	Aeros Leasing Ltd	

Notes	Reg.	Type	Owner or Operator
	G-BRBC	NA T-6G Texan	A. P. Murphy
	G-BRBD	PA-28-151 Warrior	Compton Abbas Airfield Ltd
	G-BRBE	PA-28-161 Warrior II	Solo Services Ltd/Shoreham
	G-BRBG	PA-28 Cherokee 180	Ken MacDonald & Co
	G-BRBH	Cessna 150H	J. Maffia & H. Merkado/Panshanger
	G-BRBI	Cessna 172N	G-BRBI Flying Group
	G-BRBJ	Cessna 172M	L. C. MacKnight
	G-BRBK	Robin DR.400/180	R. Kemp
	G-BRBL	Robin DR.400/180	C. A. Merren
	G-BRBM	Robin DR.400/180	R. W. Davies/Headcorn
	G-BRBN	Pitts S-1S Special	D. R. Evans
	G-BRBO	Cameron V-77 balloon	M. B. Murby
	G-BRBP	Cessna 152	Staverton Flying Services Ltd
	G-BRBS	Bensen B.8M	K. T. MacFarlane
	G-BRBT	Trotter Ax3-20 balloon	R. M. Trotter
	G-BRBV	Piper J-4A Cub Coupe	B. Schonburg
	G-BRBW	PA-28 Cherokee 140	Air Navigation and Trading Co Ltd
	G-BRBX	PA-28-181 Archer II	M. J. Ireland
	G-BRBY	Robinson R-22B	Abraxas Aviation Ltd
	G-BRCA	Jodel D.112	R. C. Jordan
	G-BRCD	Cessna A.152	D. E. Simmons/Shoreham
	G-BRCE	Pitts S-1C Special	R. O. Rogers
	G-BRCF	Montgomerie-Bensen B.8MR	J. S. Walton
	G-BRCG	Grob G.109	M. P. Flanagan
	G-BRCI	Pitts S-1C Special	G. L. A. Vandormael/Belgium
	G-BRCM	Cessna 172L	S. G. E. Plessis & D. C. C. Handley
	G-BRCT	Denney Kitfox Mk 2	M. L. Roberts
	G-BRCV	Aeronca 7AC Champion	M. A. N. Newall
	G-BRCW	Aeronca 11AC Chief	R. B. McComish
	G-BRDB	Zenair CH.701 STOL	D. L. Bowtell
	G-BRDD	Avions Mudry CAP.10B	R. D. Dickson/Gamston
	G-BRDE	Thunder Ax7-77 balloon	C. C. Brash
	G-BRDF	PA-28-161 Warrior II	White Waltham Airfield Ltd
	G-BRDG	PA-28-161 Warrior II	White Waltham Airfield Ltd
	G-BRDJ	Luscombe 8A Silvaire	P. G. Stewart
	G-BRDM	PA-28-161 Warrior II	White Waltham Airfield Ltd
	G-BRDN	M.S.880B Rallye Club	A. J. Gomes
	G-BRDO	Cessna 177B	Cardinal Aviation
	G-BRDT	Cameron DP-70 airship	Tim Balloon Promotion Airships Ltd
	G-BRDV	Viking Wood Products Spitfire Prototype replica (K5054) ★	Southampton Hall of Aviation
	G-BRDW	PA-24 Comanche 180	I. P. Gibson/Southampton
	G-BREA	Bensen B.8MR	P. Robichaud
	G-BREB	Piper J-3C-65 Cub	J. R. Wraight
	G-BREE	Whittaker MW.7	P. J. Fell
	G-BREH	Cameron V-65 balloon	S. E. & V. D. Hurst
	G-BREL	Cameron O-77 balloon	R. A. Patey
	G-BRER	Aeronca 7AC Champion	Rabbit Flight
	G-BREU	Montgomerie-Bensen B.8MR	J. S. Firth
	G-BREX	Cameron O-84 balloon	Ovolo Ltd
	G-BREY	Taylorcraft BC-12D	BREY Group
	G-BRFB	Rutan LongEz	R. Young
	G-BRFE	Cameron V-77 balloon	Esmerelda Balloon Syndicate
	G-BRFF	Colt 90A balloon	Amber Valley Aviation
	G-BRFI	Aeronca 7DC Champion	A. C. Lines
	G-BRFJ	Aeronca 11AC Chief	J. M. Mooney
	G-BRFL	PA-38-112 Tomahawk	Teesside Flight Centre Ltd
	G-BRFM	PA-28-161 Warrior II	Atlantic Flight Training Ltd
	G-BRFO	Cameron V-77 balloon	Hedge Hoppers Balloon Group
	G-BRFW	Montgomerie-Bensen B.8 2-seat	A. J. Barker
	G-BRFX	Pazmany PL.4A	D. E. Hills
	G-BRGD	Cameron O-84 balloon	D. J. Phillips
	G-BRGE	Cameron N-90 balloon	Oakfield Farm Products Ltd
	G-BRGF	Luscombe 8E Silvaire	Luscombe Flying Group
	G-BRGG	Luscombe 8A Silvaire	M. A. Lamprell
	G-BRGI	PA-28 Cherokee 180	Redhill Air Services Ltd
	G-BRGN	BAe Jetstream 3102	Cranfield University (G-BLHC)
	G-BRGO	Air Command 532 Elite	A. McCredie
	G-BRGT	PA-32 Cherokee Six 260	P. Cowley
	G-BRGW	GY-201 Minicab	R. G. White
	G-BRGX	Rotorway Executive	D. W. J. Lee
	G-BRHA	PA-32RT-300 Lance II	Lance G-BRHA Group

Reg.	Type	Owner or Operator	Notes
G-BRHB	Boeing Stearman B.75N1	D. Calabritto	
G-BRHG	Colt 90A balloon	Bath University Students' Union	
G-BRHL	Montgomerie-Bensen B.8MR	R. M. Savage & T. M. Jones	
G-BRHO	PA-34-200 Seneca	D. A. Lewis/Luton	
G-BRHP	Aeronca O-58B Grasshopper (31923)	C. J. Willis/Italy	
G-BRHR	PA-38-112 Tomahawk	The Royal Artillery Aero Club Ltd	
G-BRHT	PA-38-112 Tomahawk	Romeo Hotel Tango Group	
G-BRHW	D.H.82A Tiger Moth	P. J. & A. J. Borsberry	
G-BRHX	Luscombe 8E Silvaire	J. Lakin	
G-BRHY	Luscombe 8E Silvaire	A. R. W. Taylor/Sleap	
G-BRIA	Cessna 310L	B. J. Tucker/Kemble	
G-BRIE	Cameron N-77 balloon	S. F. Redman	
G-BRIF	Boeing 767-204ER	Thomsonfly *Horatio Nelson*	
G-BRIG	Boeing 767-204ER	Thomsonfly *Eglantyne Jebb*	
G-BRIH	Taylorcraft BC-12D	IH Flying Group	
G-BRII	Zenair CH.600 Zodiac	A. C. Bowdrey	
G-BRIJ	Taylorcraft F-19	M. W, Olliver	
G-BRIK	T.66 Nipper 3	P. R. Bentley	
G-BRIL	Piper J-5A Cub Cruiser	P. L. Jobes & D. J. Bone	
G-BRIO	Turner Super T-40A	S. Bidwell	
G-BRIR	Cameron V-56 balloon	H. G. Davies & C. Dowd	
G-BRIV	SOCATA TB.9 Tampico Club	S. J. Taft	
G-BRIY	Taylorcraft DF-65 (42-58678)	S. R. Potts	
G-BRJA	Luscombe 8A Silvaire	A. D. Keen	
G-BRJB	Zenair CH.600 Zodiac	D. Collinson	
G-BRJK	Luscombe 8A Silvaire	C. J. L. Peat & M. Richardson	
G-BRJL	PA-15 Vagabond	A. R. Williams	
G-BRJN	Pitts S-1C Special	W. Chapel	
G-BRJR	PA-38-112 Tomahawk	M. McGovern	
G-BRJT	Cessna 150H	P. K. Jenkins	
G-BRJV	PA-28-161 Cadet	Newcastle-upon-Tyne Aero Club Ltd	
G-BRJW	Bellanca 7GCBC Citabria	Juliet Whiskey Flying Club	
G-BRJX	Rand-Robinson KR-2	J. R. Bell	
G-BRJY	Rand-Robinson KR-2	R. E. Taylor	
G-BRKC	J/1 Autocrat	J. W. Conlon	
G-BRKH	PA-28-236 Dakota	A. P. H. Hay & C. C. Bennett	
G-BRKR	Cessna 182R	A. R. D. Brooker	
G-BRKW	Cameron V-77 balloon	T. J. Parker	
G-BRKX	Air Command 532 Elite	K. Davis	
G-BRKY	Viking Dragonfly Mk II	G. D. Price	
G-BRLB	Air Command 532 Elite	F. G. Shepherd	
G-BRLF	Campbell Cricket (replica)	P. G. Rawson	
G-BRLG	PA-28RT-201T Turbo Arrow IV	C. G. Westwood/Welshpool	
G-BRLI	Piper J-5A Cub Cruiser	Little Bear Ltd	
G-BRLJ	Evans VP-2	R. L. Jones	
G-BRLL	Cameron A-105 balloon	Aerosaurus Balloons Ltd	
G-BRLO	PA-38-112 Tomahawk	St. George Flight Training/Teesside	
G-BRLP	PA-38-112 Tomahawk	P. D. Brooks	
G-BRLR	Cessna 150G	D. Carr & M. R. Muter	
G-BRLS	Thunder Ax7-77 balloon	E. C. Meek	
G-BRLT	Colt 77A balloon	D. Bareford	
G-BRLV	CCF Harvard IV (93542)	Extraviation Ltd/North Weald	
G-BRMA	W.S.51 Dragonfly HR.5 (WG719) ★	IHM/Weston-s-Mare	
G-BRMB	PA-28-RT-201 Arrow IV	IHM/Weston-s-Mare	
G-BRME	PA-28-181 Archer II	Keen Leasing Ltd	
G-BRMI	Cameron V-65 balloon	M. Davies	
G-BRML	PA-38-112 Tomahawk	P. H. Rogers/Coventry	
G-BRMS	PA-28RT-201 Arrow IV	Fleetbridge Ltd	
G-BRMT	Cameron V-31 balloon	T. C. Hinton	
G-BRMU	Cameron V-77 balloon	K. J. & G. R. Ibbotson	
G-BRMV	Cameron O-77 balloon	P. D. Griffiths	
G-BRMW	Whittaker MW.7	G. S. Parsons	
G-BRNC	Cessna 150M	D. C. Bonsall	
G-BRND	Cessna 152 II	T. M. & M. L. Jones/Egginton	
G-BRNE	Cessna 152 II	Redhill Air Services Ltd	
G-BRNJ	PA-38-112 Tomahawk	Cardiff-Wales Aviation Services Ltd	
G-BRNK	Cessna 152 II	Sheffield Aero Club Ltd/Netherthorpe	
G-BRNM	Chichester-Miles Leopard	Chichester-Miles Consultants Ltd	
G-BRNN	Cessna 152 II	Sheffield Aero Club Ltd/Netherthorpe	
G-BRNT	Robin DR.400/180	M. J. Cowham	
G-BRNU	Robin DR.400/180	November Uniform Travel Syndicate Ltd/Booker	
G-BRNV	PA-28-181 Archer II	B. S. Hobbs	

Notes	Reg.	Type	Owner or Operator
	G-BRNW	Cameron V-77 balloon	N. Robertson & G. Smith
	G-BRNX	PA-22 Tri-Pacer 150	S. N. Askey
	G-BRNZ	PA-32 Cherokee Six 300B	Longfellow Flying Group
	G-BROB	Cameron V-77 balloon	J. W. Tomlinson
	G-BROE	Cameron N-65 balloon	A. I. Attwood
	G-BROG	Cameron V-65 balloon	R. Kunert
	G-BROH	Cameron O-90 balloon	P. A. Derbyshire
	G-BROI	CFM Streak Shadow Srs SA	G. J. Forshaw
	G-BROJ	Colt 31A balloon	N. J. Langley
	G-BROO	Luscombe 8E Silvaire	P. R. Bush
	G-BROP	Vans RV-4	K. E. Armstrong
	G-BROR	Piper J-3C-65 Cub	White Hart Flying Group
	G-BROX	Robinson R-22B	TLC Handling Ltd
	G-BROY	Cameron V-77 balloon	T. G. S. Dixon
	G-BROZ	PA-18 Super Cub 150	P. G. Kynsey
	G-BRPE	Cessna 120	W. B. Bateson/Blackpool
	G-BRPF	Cessna 120	A. L. Hall-Carpenter
	G-BRPG	Cessna 120	I. C. Lomax
	G-BRPH	Cessna 120	J. A. Cook
	G-BRPJ	Cameron N-90 balloon	Cloud Nine Balloon Co
	G-BRPK	PA-28 Cherokee 140	G-BRPK Group
	G-BRPM	T.66 Nipper 3	T. C. Horner
	G-BRPO	Enstrom 280C Shark	D. Jones
	G-BRPP	Brookland Hornet (modified)	B. J. L. P. & W. J. A. L. de Saar
	G-BRPR	Aeronca O-58B Grasshopper (31952)	C. S. Tolchard
	G-BRPS	Cessna 177B	R. C. Tebbett
	G-BRPT	Rans S.10 Sakota	A. R. Hawes
	G-BRPU	Beech 76 Duchess	Plane Talking Ltd/Elstree
	G-BRPV	Cessna 152	GEM Rewinds Ltd/Coventry
	G-BRPX	Taylorcraft BC-12D	The BRPX Group
	G-BRPY	PA-15 Vagabond	J. & V. Hobday
	G-BRPZ	Luscombe 8A Silvaire	S. L. & J. P. Waring
	G-BRRB	Luscombe 8E Silvaire	J. Nicholls
	G-BRRD	Scheibe SF.25B Falke	The G-BRRD Syndicate
	G-BRRF	Cameron O-77 balloon	K. P. & G. J. Storey
	G-BRRG	Glaser-Dirks DG.500M	Glider Syndicate/Sutton Bank
	G-BRRJ	PA-28RT-201T Turbo Arrow IV	M. Stower
	G-BRRK	Cessna 182Q	Werewolf Aviation Ltd
	G-BRRL	PA-18 Super Cub 95	Acebell G-BRRL Syndicate/Redhill
	G-BRRN	PA-28-161 Warrior II	G. Whitlow & I. C. Barlow
	G-BRRO	Cameron N-77 balloon	B. Birch
	G-BRRR	Cameron V-77 balloon	K. P. & G. J. Storey
	G-BRRS	Pitts S-1C Special	R. C. Atkinson
	G-BRRU	Colt 90A balloon	Reach For The Sky Ltd
	G-BRRY	Robinson R-22B	Findon Air Services
	G-BRSA	Cameron N-56 balloon	C. Wilkinson
	G-BRSC	Rans S.10 Sakota	P. Wilkinson
	G-BRSD	Cameron V-77 balloon	T. J. & J.E. Porter
	G-BRSE	PA-28-161 Warrior II	Meridian Aviation Ltd
	G-BRSF	V.S.361 Spitfire HF.9c	M. B, Phillips
	G-BRSH	C.A.S.A. 1.131E Jungmann 2000 (781-25)	L. Ness/Norway
	G-BRSJ	PA-38-112 Tomahawk II	APB Leasing Ltd/Welshpool
	G-BRSK	Boeing Stearman N2S-3 (1180)	Wymondham Engineering
	G-BRSL	Cameron N-56 balloon	S. Budd
	G-BRSN	Rand-Robinson KR-2	K. W. Darby
	G-BRSO	CFM Streak Shadow Srs SA	B. C. Norris
	G-BRSP	Air Command 532 Elite	G. M. Hobman
	G-BRSW	Luscombe 8A Silvaire	Bloody Mary Aviation/Fenland
	G-BRSX	PA-15 Vagabond	P. M. Newman
	G-BRSY	Hatz CB-1	C. Knight
	G-BRTD	Cessna 152 II	152 Group/Booker
	G-BRTH	Cameron A-180 balloon	The Ballooning Business Ltd
	G-BRTJ	Cessna 150F	Avon Aviation Ltd
	G-BRTK	Boeing Stearman E.75 (217786)	Eastern Stearman Ltd/Swanton Morley
	G-BRTL	Hughes 369E	Crewhall Ltd
	G-BRTM	PA-28-161 Warrior II	Oxford Aviation Services Ltd/Kidlington
	G-BRTP	Cessna 152 II	K. R. Emery
	G-BRTT	Schweizer 269C	Fairthorpe Ltd/Denham
	G-BRTV	Cameron O-77 balloon	M. C. Gibbons
	G-BRTW	Glaser-Dirks DG.400	I. J. Carruthers
	G-BRTX	PA-28-151 Warrior	Spectrum Flying Group

Reg.	Type	Owner or Operator	Notes
G-BRTZ	Slingsby T.31 Motor Cadet III	R. R. Walters	
G-BRUA	Cessna 152 II	BBC Air Ltd/Compton Abbas	
G-BRUB	PA-28-161 Warrior II	Flytrek Ltd/Bournemouth	
G-BRUD	PA-28-181 Archer II	Wilkins & Wilkins Special Auctions Ltd	
G-BRUG	Luscombe 8E Silvaire	K. Reeve & N. W. Barratt	
G-BRUH	Colt 105A balloon	D. C. Chipping/Portugal	
G-BRUI	PA-44-180 Seminole	M. Griffiths	
G-BRUJ	Boeing Stearman A.75N1 (6136)	M. Walker/Liverpool	
G-BRUM	Cessna A.152	Central Aircraft Leasing Ltd	
G-BRUN	Cessna 120	O. C. Brun (G-BRDH)	
G-BRUO	Taylor JT.1 Monoplane	R. Hatton	
G-BRUU	EAA Biplane Model P.1	E. C. Murgatroyd	
G-BRUV	Cameron V-77 balloon	T. W. & R. F. Benbrook	
G-BRUX	PA-44-180 Seminole	C. J. Thomas	
G-BRVB	Stolp SA.300 Starduster Too	M. N. Petchey & S. Turner	
G-BRVE	Beech D.17S	P. A. Teichman/North Weald	
G-BRVF	Colt 77A balloon	The Ballooning Business Ltd	
G-BRVG	NA SNJ-7 Texan (27)	D. Gilmour/Intrepid Aviation Co/Goodwood	
G-BRVI	Robinson R-22B	York Helicopters	
G-BRVJ	Slingsby T.31 Motor Cadet III	B. Outhwaite	
G-BRVL	Pitts S-1C Special	M. F. Pocock	
G-BRVN	Thunder Ax7-77 balloon	D. L. Beckwith	
G-BRVO	AS.350B Ecureuil	Ferns Surfacing Ltd	
G-BRVR	Barnett J4B-2 Rotorcraft	Ilkeston Contractors	
G-BRVS	Barnett J4B-2 Rotorcraft	Ilkeston Contractors	
G-BRVT	Pitts S-2B Special	R. Woollard & D. Tilley	
G-BRVU	Colt 77A balloon	J. K. Woods	
G-BRVY	Thunder Ax8-90 balloon	G. E. Morris	
G-BRVZ	Jodel D.117	L. Holland	
G-BRWA	Aeronca 7AC Champion	D. D. Smith & J. R. Edwards	
G-BRWB	NA T-6G Texan (526)	R. Clifford/Duxford	
G-BRWD	Robinson R-22B	Rotorways Helicopters	
G-BRWO	PA-28 Cherokee 140	G-BRWO Ltd	
G-BRWP	CFM Streak Shadow	R. Biffin	
G-BRWR	Aeronca 11AC Chief	A. W. Crutcher	
G-BRWT	Scheibe SF.25C Falke	Booker Gliding Club Ltd	
G-BRWU	Luton LA-4A Minor	R. B. Webber	
G-BRWV	Brügger MB.2 Colibri	M. P. Wakem	
G-BRWX	Cessna 172P	BCT Aircraft Leasing Ltd	
G-BRWZ	Cameron 90 Macaw SS balloon	Forbes Global Inc	
G-BRXA	Cameron O-120 balloon	R. J. Mansfield	
G-BRXB	Thunder Ax7-77 balloon	H. Peel	
G-BRXC	PA-28-161 Warrior II	Oxford Aviation Services Ltd/Kidlington	
G-BRXD	PA-28-181 Archer II	D. D. Stone	
G-BRXE	Taylorcraft BC-12D	W. J. Durrad	
G-BRXF	Aeronca 11AC Chief	Aeronca Flying Group	
G-BRXG	Aeronca 7AC Champion	X-Ray Golf Flying Group	
G-BRXH	Cessna 120	BRXH Group	
G-BRXL	Aeronca 11AC Chief (42-78044)	T. Smith	
G-BRXN	Montgomerie-Bensen B.8MR	C. M. Frerk	
G-BRXO	PA-34-200T Seneca II	Aviation Services Ltd	
G-BRXP	SNCAN Stampe SV-4C (modified)	T. Brown	
G-BRXS	Howard Special T Minus	A. Shuttleworth	
G-BRXV	Robinson R-22B	Heliflight (UK) Ltd	
G-BRXW	PA-24 Comanche 260	Oak Group	
G-BRXY	Pietenpol Air Camper	P. S. Ganczakowski	
G-BRYI	D.H.C.8-311 Dash Eight	British Airways CitiExpress	
G-BRYJ	D.H.C.8-311 Dash Eight	British Airways CitiExpress	
G-BRYU	D.H.C.8-311 Dash Eight	British Airways CitiExpress	
G-BRYV	D.H.C.8-311 Dash Eight	British Airways CitiExpress	
G-BRYW	D.H.C.8-311 Dash Eight	British Airways CitiExpress	
G-BRYX	D.H.C.8-311 Dash Eight	British Airways CitiExpress	
G-BRYY	D.H.C.8-311 Dash Eight	British Airways CitiExpress	
G-BRYZ	D.H.C.8-311 Dash Eight	British Airways CitiExpress	
G-BRZA	Cameron O-77 balloon	L. & R. J. Mold	
G-BRZB	Cameron A-105 balloon	Headland Services Ltd	
G-BRZD	Hapi Cygnet SF-2A	C. I. Coghill	
G-BRZE	Thunder Ax7-77 balloon	G. V. Beckwith & F. Schoeder/Germany	
G-BRZG	Enstrom F-28A	G. E. Challinor	
G-BRZI	Cameron N-180 balloon	Eastern Balloon Rides	
G-BRZK	Stinson 108-2	Voyager G-BRZK Syndicate	
G-BRZL	Pitts S-1D Special	T. R. G. Barnby	

Notes	Reg.	Type	Owner or Operator
	G-BRZO	Jodel D.18	J. D. Anson
	G-BRZS	Cessna 172P	YP Flying Group/Blackpool
	G-BRZT	Cameron V-77 balloon	B. Drawbridge
	G-BRZV	Colt Flying Apple SS balloon	Obst Vom Bodensee Marketing Gbr/Germany
	G-BRZW	Rans S.10 Sakota	D. L. Davies
	G-BRZX	Pitts S-1S Special	J. S. Dawson
	G-BRZZ	CFM Streak Shadow	Shetland Flying Group
	G-BSAI	Stoddard-Hamilton Glasair III	K. J. & P. J. Whitehead
	G-BSAJ	C.A.S.A. 1.131E Jungmann 2000	P. G. Kynsey/Redhill
	G-BSAK	Colt 21A balloon	Northern Flights
	G-BSAS	Cameron V-65 balloon	J. R. Barber
	G-BSAV	Thunder Ax7-77 balloon	I. G. & C. A. Lloyd
	G-BSAW	PA-28-161 Warrior II	Carill Aviation Ltd/Southampton
	G-BSAZ	Denney Kitfox Mk 2	A. J. Lloyd & ptnrs
	G-BSBA	PA-28-161 Warrior II	London Transport Flying Club Ltd
	G-BSBG	CCF Harvard IV (20310)	A. P. St John/Liverpool
	G-BSBH	Short SD3-30 ★	Ulster Aviation Soc Museum *(stored)*
	G-BSBI	Cameron O-77 balloon	D. M. Billing
	G-BSBN	Thunder Ax7-77 balloon	B. Pawson
	G-BSBR	Cameron V-77 balloon	R. P. Wade
	G-BSBT	Piper J-3C-65 Cub	R. W. H. Watson
	G-BSBV	Rans S.10 Sakota	D. R. G. Whitelaw
	G-BSBW	Bell 206B JetRanger 3	D. T. Sharpe
	G-BSBX	Montgomerie-Bensen B.8MR	R. J. Roan & W. Toulmin
	G-BSBZ	Cessna 150M	DTG Aviation
	G-BSCA	Cameron N-90 balloon	J. Steiner
	G-BSCB	Air Command 532 Elite	P. H. Smith
	G-BSCC	Colt 105A balloon	Capricorn Balloons Ltd
	G-BSCE	Robinson R-22B	H. Sugden
	G-BSCF	Thunder Ax7-77 balloon	V. P. Gardiner
	G-BSCG	Denney Kitfox Mk 2	S. Burrow
	G-BSCH	Denney Kitfox Mk 2	R. B. Wilson
	G-BSCI	Colt 77A balloon	J. L. & S. Wrigglesworth
	G-BSCK	Cameron H-24 balloon	J. D. Shapland
	G-BSCM	Denney Kitfox Mk 2	S. A. Hewitt
	G-BSCN	SOCATA TB.20 Trinidad	B. W. Dye
	G-BSCO	Thunder Ax7-77 balloon	F. J. Whalley
	G-BSCP	Cessna 152 II	Moray Flying Club (1990) Ltd/Kinloss
	G-BSCS	PA-28-181 Archer II	Wingtask 1995 Ltd
	G-BSCV	PA-28-161 Warrior II	Southwood Flying Group/Earls Colne
	G-BSCW	Taylorcraft BC-65	S. Leach
	G-BSCX	Thunder Ax8-105 balloon	Balloon Flights Club Ltd
	G-BSCY	PA-28-151 Warrior	Falcon Flying Services/Biggin Hill
	G-BSCZ	Cessna 152 II	The RAF Halton Aeroplane Club Ltd
	G-BSDA	Taylorcraft BC-12D	D. G. Edwards
	G-BSDB	Pitts S-1C Special	J. T. Mielech/Germany
	G-BSDD	Denney Kitfox Mk 2	D. C. Crawley
	G-BSDH	Robin DR.400/180	R. L. Brucciani
	G-BSDK	Piper J-5A Cub Cruiser	J. E. Mead
	G-BSDL	SOCATA TB.10 Tobago	Delta Lima Group/Sherburn
	G-BSDN	PA-34-200T Seneca II	McCormick Consulting Ltd
	G-BSDO	Cessna 152 II	J. Vickers//Humberside
	G-BSDP	Cessna 152 II	I. S. H. Paul
	G-BSDS	Boeing Stearman E.75 (118)	A. Basso/Switzerland
	G-BSDV	Colt 31A balloon	C. D. Monk
	G-BSDW	Cessna 182P	Parker Diving Ltd
	G-BSDZ	Enstrom 280FX	Avalon Group Ltd (G-ODSC)
	G-BSED	PA-22 Tri-Pacer 160 (modified)	Tayflite Ltd
	G-BSEE	Rans S.9	P. M. Semler
	G-BSEF	PA-28 Cherokee 180	I. D. Wakeling
	G-BSEG	Ken Brock KB-2 gyroplane	S. J. M. Ledingham
	G-BSEJ	Cessna 150M	I. Shackleton/Wolverhampton
	G-BSEK	Robinson R-22	S. J. Strange
	G-BSEL	Slingsby T.61G Super Falke	Bannerdown Glding Club/Keevil
	G-BSEP	Cessna 172	EPAviation
	G-BSER	PA-28 Cherokee 160	Yorkair Ltd
	G-BSEU	PA-28-181 Archer II	Euro Aviation 91 Ltd
	G-BSEV	Cameron O-77 balloon	P. B. Kenington
	G-BSEX	Cameron A-180 balloon	Heart of England Balloons
	G-BSEY	Beech A36 Bonanza	K. Phillips Ltd
	G-BSFA	Aero Designs Pulsar	S. Eddison & R. Minett

Reg.	Type	Owner or Operator	Notes
G-BSFB	C.A.S.A. 1.131E Jungmann 2000 (S5-B06)	C. D. Beal	
G-BSFD	Piper J-3C-65 Cub	AJD Engineering Ltd	
G-BSFE	PA-38-112 Tomahawk II	D. J. Campbell	
G-BSFF	Robin DR.400/180R	Lasham Gliding Soc Ltd	
G-BSFK	PA-28-161 Warrior II	Oxford Aviation Services Ltd/Kidlington	
G-BSFP	Cessna 152T	Walkbury Aviation Ltd/Sibson	
G-BSFR	Cessna 152 II	Galair Ltd/Biggin Hill	
G-BSFV	Woods Woody Pusher	M. J. Wells	
G-BSFW	PA-15 Vagabond	J. R. Kimberley	
G-BSFX	Denney Kitfox Mk 2	H. Hedley-Lewis	
G-BSFY	Denney Kitfox Mk 2	C. I. Bates	
G-BSGB	Gaertner Ax4 Skyranger balloon	B. Gaertner	
G-BSGD	PA-28 Cherokee 180	R. J. Cleverley	
G-BSGF	Robinson R-22B	Direct Helicopters	
G-BSGG	Denney Kitfox Mk 2	C. G. Richardson	
G-BSGH	Airtour AH-56B balloon	A. R. Hardwick	
G-BSGJ	Monnett Sonerai II	G. A. Brady	
G-BSGK	PA-34-200T Seneca II	Aeros Holdings Ltd	
G-BSGL	PA-28-161 Warrior II	Keywest Air Charter Ltd/Liverpool	
G-BSGP	Cameron N-65 balloon	T. D. Gibbs	
G-BSGS	Rans S.10 Sakota	M. R. Parr	
G-BSGT	Cessna T.210N	E. A. T. Brenninkmeyer	
G-BSHA	PA-34-200T Seneca II	Justgold Ltd/Blackpool	
G-BSHC	Colt 69A balloon	Magical Adventures Ltd	
G-BSHD	Colt 69A balloon	D. B. Court	
G-BSHH	Luscombe 8E Silvaire	S. L. Lewis	
G-BSHI	Luscombe 8F Silvaire	Calcott Garage Ltd	
G-BSHK	Denney Kitfox Mk 2	D. Doyle & C. Aherne	
G-BSHO	Cameron V-77 balloon	D. J. Duckworth & J. C. Stewart	
G-BSHP	PA-28-161 Warrior II	Plane Talking Ltd	
G-BSHR	Cessna F.172N	Deep Cleavage Ltd (G-BFGE)	
G-BSHT	Cameron V-77 balloon	ECM Construction Ltd	
G-BSHV	PA-18 Super Cub 135	G. T. Fisher	
G-BSHW	Hawker Tempest II (MW800)	-/France	
G-BSHX	Enstrom F-28A	Stephenson Aviation Ltd/Goodwood	
G-BSHY	EAA Acro Sport I	R. J. Hodder	
G-BSHZ	Enstrom F-28F	K. G. Ward	
G-BSIC	Cameron V-77 balloon	J. M. & A. Cornwall	
G-BSIF	Denney Kitfox Mk 2	P. Annable	
G-BSIG	Colt 21A balloon	S. J. Humphreys	
G-BSIH	Rutan LongEz	W. S. Allen	
G-BSII	PA-34-200T Seneca II	N. H. N. Gardner	
G-BSIJ	Cameron V-77 balloon	A. S. Jones	
G-BSIK	Denney Kitfox Mk 1	S. P. Collins	
G-BSIM	PA-28-181 Archer II	Halfpenny Green Flight Centre Ltd	
G-BSIO	Cameron 80 Shed SS balloon	R. E. Jones	
G-BSIU	Colt 90A balloon	S. Travaglia/Italy	
G-BSIY	Schleicher ASK.14	Winwick Flying Group	
G-BSIZ	PA-28-181 Archer II	A. M. L. Maxwell	
G-BSJB	Bensen B.8	J. W. Limbrick	
G-BSJU	Cessna 150M	A. C. Williamson	
G-BSJW	Everett Srs 2 gyroplane	R. Sarwan	
G-BSJX	PA-28-161 Warrior II	D. A. Shields & L. C. Brekkeflat	
G-BSJZ	Cessna 150J	BCT Aircraft Leasing Ltd	
G-BSKA	Cessna 150M	R. J. Cushing	
G-BSKD	Cameron V-77 balloon	M. J. Gunston	
G-BSKE	Cameron O-84 balloon	S. F. Redman	
G-BSKG	Maule MX-7-180	J. R. Surbey	
G-BSKI	Thunder Ax8-90 balloon	K. P. Barnes & L. A. Pibworth	
G-BSKK	PA-38-112 Tomahawk	Falcon Flying Services/Biggin Hill	
G-BSKL	PA-38-112 Tomahawk	Falcon Flying Services/Biggin Hill	
G-BSKO	Maule MXT-7-180	M. A. Ashmole	
G-BSKP	V.S.379 Spitfire F.XIV (SG-3)	Aircraft Restoration/Duxford	
G-BSKU	Cameron O-84 balloon	Alfred Bagnall & Sons (West) Ltd	
G-BSKW	PA-28-181 Archer II	Shropshire Aero Club Ltd/Sleap	
G-BSLA	Robin DR.400/180	A. B. McCoig/Biggin Hill	
G-BSLE	PA-28-161 Warrior II	Oxford Aviation Services Ltd/Kidlington	
G-BSLH	C.A.S.A. 1.131E Jungmann 2000	P. Warden/France	
G-BSLI	Cameron V-77 balloon	R. C. Corcoran	
G-BSLK	PA-28-161 Warrior II	R. A. Rose	
G-BSLM	PA-28 Cherokee 160	R. Fulton	

Notes	Reg.	Type	Owner or Operator
	G-BSLT	PA-28-161 Warrior II	L. W. Scattergood
	G-BSLU	PA-28 Cherokee 140	D. J. Budden Ltd/Shobdon
	G-BSLV	Enstrom 280FX	Keswick Outdoor Clothing Co Ltd
	G-BSLW	Bellanca 7ECA Citabria	Shoreham Citabria Group
	G-BSLX	WAR Focke-Wulf Fw.190 (replica) (4+)	Fw.190 Gruppe
	G-BSMB	Cessna U.206E	C. J. Francis
	G-BSMD	Nord 1101 Noralpha (+114)	J. W. Hardie
	G-BSME	Bo 208C1 Junior	D. J. Hampson
	G-BSMG	Montgomerie-Bensen B.8M	A. C. Timperley
	G-BSMK	Cameron O-84 balloon	G-BSMK Shareholders
	G-BSML	Schweizer 269C	K. P. Foster & B. I. Winsor
	G-BSMM	Colt 31A balloon	D. V. Fowler
	G-BSMN	CFM Streak Shadow	P. J. Porter
	G-BSMO	Denney Kitfox Mk 2	M. W. Sayers & G. P. Bridgwater
	G-BSMS	Cameron V-77 balloon	Sade Balloons Ltd
	G-BSMT	Rans S.10 Sakota	P. J. Barker
	G-BSMU	Rans S.6 Coyote II	S. Yelland (G-MWJE)
	G-BSMV	PA-17 Vagabond (modified)	A. Cheriton
	G-BSMX	Bensen B.8MR	J. S. E. R. McGregor
	G-BSND	Air Command 532 Elite	B. Gunn & W. B. Lumb
	G-BSNE	Luscombe 8E Silvaire	N. Reynolds & C. Watts
	G-BSNF	Piper J-3C-65 Cub	D. A. Hammant
	G-BSNG	Cessna 172N	A. J. & P. C. MacDonald/Edinburgh
	G-BSNJ	Cameron N-90 balloon	D. P. H. Smith/France
	G-BSNL	Bensen B.8MR	A. C. Breane
	G-BSNN	Rans S.10 Sakota	O. & S. D. Barnard
	G-BSNP	PA-28-201T Turbo Arrow III	D. F. K. Singleton/Germany
	G-BSNR	BAe 146-300A	Trident Aviation Leasing Ltd
	G-BSNT	Luscombe 8A Silvaire	Luscombe Quartet
	G-BSNU	Colt 105A balloon	Gone Ballooning
	G-BSNX	PA-28-181 Archer II	Halfpenny Green Flight Centre Ltd
	G-BSNY	Bensen B.8M	H. McCartney
	G-BSNZ	Cameron O-105 balloon	J. Francis
	G-BSOE	Luscombe 8A Silvaire	S. B. Marsden
	G-BSOF	Colt 25A balloon	L. P. Hooper
	G-BSOG	Cessna 172M	B. Chapman & A. R. Budden/Goodwood
	G-BSOI	AS.332L Super Puma	CHC Scotia Ltd/Aberdeen
	G-BSOJ	Thunder Ax7-77 balloon	R. J. S. Jones
	G-BSOK	PA-28-161 Warrior II	Aeros Leasing Ltd/Gloucestershire
	G-BSOM	Glaser-Dirks DG.400	G-BSOM Group/Tibenham
	G-BSON	Green S.25 balloon	J. J. Green
	G-BSOO	Cessna 172F	Double Oscar Flying Group
	G-BSOR	CFM Streak Shadow Srs SA	A. Parr
	G-BSOT	PA-38-112 Tomahawk II	APB Leasing Ltd/Welshpool
	G-BSOU	PA-38-112 Tomahawk II	D. J. Campbell
	G-BSOX	Luscombe 8AE Silvaire	P. S Lanary
	G-BSOZ	PA-28-161 Warrior II	The Moray Flying Club 1990)/Kinloss
	G-BSPA	QAC Quickie Q.2	G. V. McKirdy & B. K. Glover
	G-BSPB	Thunder Ax8-84 balloon	A. N. F. Pertwee
	G-BSPE	Cessna F.172P	T. W. Williamson
	G-BSPG	PA-34-200T Seneca II	Airtime Aviation Ltd
	G-BSPI	PA-28-161 Warrior II	Halegreen Associates Ltd
	G-BSPJ	Bensen B.8	D. Ross
	G-BSPK	Cessna 195A	A. G. & D. L. Bompas
	G-BSPL	CFM Streak Shadow Srs SA	G. L. Turner
	G-BSPM	PA-28-161 Warrior II	White Waltham Airfield Ltd
	G-BSPN	PA-28R-201T Turbo Arrow III	V. E. H. Taylor
	G-BSPW	Avid Speed Wing	M. J. Sewell
	G-BSRD	Cameron N-105 balloon	A. Ockelmann
	G-BSRH	Pitts S-1C Special	Glen-Davis Gorman & C. D. Swift
	G-BSRI	Lancair 235	G. Lewis/Liverpool
	G-BSRK	ARV Super 2	D. M. Blair
	G-BSRL	Campbell Cricket Mk.4 gyroplane	I. Rosewall
	G-BSRP	Rotorway Executive	R. J. Baker
	G-BSRR	Cessna 182Q	C. M. Moore
	G-BSRT	Denney Kitfox Mk 2	A. J. Lloyd
	G-BSRX	CFM Streak Shadow	P. Williams
	G-BSRZ	Air Command 532 Elite 2-seat	A. S. G. Crabb
	G-BSSA	Luscombe 8E Silvaire	Luscombe Flying Group/White Waltham
	G-BSSB	Cessna 150L	D. T. A. Rees
	G-BSSC	PA-28-161 Warrior II	Faber Developments Ltd
	G-BSSE	PA-28 Cherokee 140	A. E. Whittle & R. Murgatroyd

Reg.	Type	Owner or Operator	Notes
G-BSSF	Denney Kitfox Mk 2	A. M. Hemmings	
G-BSSI	Rans S.6 Coyote II	J. Currell (G-MWJA)	
G-BSSK	QAC Quickie Q.2	D. G. Greatrex	
G-BSSO	Cameron O-90 balloon	R. R. & J. E. Hatton	
G-BSSP	Robin DR.400/180R	Soaring (Oxford) Ltd	
G-BSST	Concorde 002 ★	F.A.A. Museum/Yeovilton	
G-BSSV	CFM Streak Shadow	R. W. Payne	
G-BSSW	PA-28-161 Warrior II	R. L. Hayward	
G-BSSX	PA-28-161 Warrior II	Airways Aero Associations Ltd/Booker	
G-BSTC	Aeronca 11AC Chief	J. Armstrong & D. Lamb	
G-BSTE	AS.355F-2 Twin Squirrel	Oscar Mayer Ltd	
G-BSTH	PA-25 Pawnee 235	Scottish Gliding Union Ltd/Portmoak	
G-BSTI	Piper J-3C-65 Cub	G. L. Nunn & J. D. Barwick	
G-BSTK	Thunder Ax8-90 balloon	M. Williams	
G-BSTL	Rand-Robinson KR-2	C. S. Hales & N. Brauns	
G-BSTM	Cessna 172L	G-BSTM Group/Cambridge	
G-BSTO	Cessna 152 II	Plymouth School of Flying Ltd	
G-BSTP	Cessna 152 II	FR Aviation Ltd/Bournemouth	
G-BSTR	AA-5 Traveler	James Allan (Aviation & Engineering) Ltd/Dundee	
G-BSTT	Rans S.6 Coyote II	D. G. Palmer	
G-BSTV	PA-32 Cherokee Six 300	B. C. Hudson	
G-BSTX	Luscombe 8A Silvaire	G. R. Nicholson	
G-BSTY	Thunder Ax8-90 balloon	Shere Ballooning Group	
G-BSTZ	PA-28 Cherokee 140	Air Navigation & Trading Co Ltd/Blackpool	
G-BSUA	Rans S.6 Coyote II	A. J. Todd	
G-BSUB	Colt 77A balloon	J. M. Foster & M. P. Hill	
G-BSUD	Luscombe 8A Silvaire	I. G. Harrison/Egginton	
G-BSUE	Cessna U.206G II	J. Dyer & I. C. Austin	
G-BSUF	PA-32RT-300 Lance II	S. T. Laffin	
G-BSUJ	Brügger MB.2 Colibri	M. A. Farrelly	
G-BSUK	Colt 77A balloon	A. J. Moore	
G-BSUO	Scheibe SF.25C Falke	Portmoak Falke Syndicate	
G-BSUT	Rans S.6-ESA Coyote II	J. Bell	
G-BSUU	Colt 180A balloon	British School of Ballooning	
G-BSUV	Cameron O-77 balloon	R. Moss	
G-BSUW	PA-34-200T Seneca II	NPD Direct Ltd	
G-BSUX	Carlson Sparrow II	J. Stephenson	
G-BSUZ	Denney Kitfox Mk 3	P. C. Avery	
G-BSVB	PA-28-181 Archer II	K. A. Boost	
G-BSVE	Binder CP.301S Smaragd	Smaragd Flying Group	
G-BSVF	PA-28-161 Warrior II	Airways Aero Associations Ltd/Booker	
G-BSVG	PA-28-161 Warrior II	Airways Aero Associations Ltd/Booker	
G-BSVH	Piper J-3C-65 Cub	C. R. & K. A. Maher	
G-BSVI	PA-16 Clipper	I. R. Blakemore	
G-BSVJ	Piper J-3C-65 Cub	V. S. E. Norman/Rendcomb	
G-BSVK	Denney Kitfox Mk 2	C. M. Looney	
G-BSVM	PA-28-161 Warrior II	EFG Flying Services/Biggin Hill	
G-BSVN	Thorp T-18	J. H. Kirkham	
G-BSVR	Schweizer 269C	Martinair Ltd	
G-BSVS	Robin DR.400/100	D. McK. Chalmers	
G-BSVW	PA-38-112 Tomahawk	Cardiff-Wales Aviation Services Ltd	
G-BSVZ	Pietenpol Air Camper	M. J. Kirk	
G-BSWB	Rans S.10 Sakota	F. A. Hewitt	
G-BSWC	Boeing Stearman E.75 (112)	Richard Thwaites Aviation Ltd	
G-BSWF	PA-16 Clipper	T. M. Storey	
G-BSWG	PA-17 Vagabond	P. E. J. Sturgeon	
G-BSWH	Cessna 152 II	Airspeed Aviation Ltd	
G-BSWI	Rans S.10 Sakota	J. M. Mooney	
G-BSWL	Slingsby T.61F Venture T.2	K. Richards	
G-BSWM	Slingsby T.61F Venture T.2	Venture Gliding Group/Bellarena	
G-BSWR	BN-2T-26 Turbine Islander	Police Authority for Northern Ireland	
G-BSWV	Cameron N-77 balloon	S. Charlish	
G-BSWX	Cameron V-90 balloon	B. J. Burrows	
G-BSWY	Cameron N-77 balloon	Nottingham Hot Air Balloon Club	
G-BSWZ	Cameron A-180 balloon	G. C. Ludlow	
G-BSXA	PA-28-161 Warrior II	Falcon Flying Services/Biggin Hill	
G-BSXB	PA-28-161 Warrior II	Aeroshow Ltd	
G-BSXC	PA-28-161 Warrior II	L. T. Halpin/Booker	
G-BSXD	Soko P-2 Kraguj (30146)	L. C. MacKnight	
G-BSXI	Mooney M.20E	A. N. Pain	
G-BSXM	Cameron V-77 balloon	C. A. Oxby	
G-BSXS	PA-28-181 Archer II	Jaxx Landing Ltd	

Notes	Reg.	Type	Owner or Operator
	G-BSXT	Piper J-5A Cub Cruiser	M. J. Kirk
	G-BSXX	Whittaker MW.7	H. J. Stanley
	G-BSXZ	BAe 146-300	Flightline Ltd (G-NJIB)/Southend
	G-BSYA	Jodel D.18	K. Wright/Isle of Man
	G-BSYB	Cameron N-120 balloon	M. Buono/Italy
	G-BSYC	PA-32R-300 Lance	Olympia Homes Ltd
	G-BSYD	Cameron A-180 balloon	A. A. Brown
	G-BSYF	Luscombe 8A Silvaire	Atlantic Aviation
	G-BSYG	PA-12 Super Cruiser	Fat Cub Group
	G-BSYH	Luscombe 8A Silvaire	N. R. Osborne
	G-BSYI	AS.355F-1 Twin Squirrel	Premiair Aviation Services Ltd
	G-BSYJ	Cameron N-77 balloon	Chubb Fire Ltd
	G-BSYO	Piper J-3C-90 Cub	C. R. Reynolds & J. D. Fuller (G-BSMJ/G-BRHE)
	G-BSYU	Robin DR.400/180	P. D. Smoothy
	G-BSYV	Cessna 150M	Fenland Flying School
	G-BSYW	Cessna 150M	B. D. Deubelbeiss
	G-BSYZ	PA-28-161 Warrior II	Yankee Zulu Group
	G-BSZB	Stolp SA.300 Starduster Too	D. T. Gethin/Swansea
	G-BSZC	Beech C-45H (51-11701A)	A. P. H. Walsh
	G-BSZD	Robin DR.400/180	R. J. Hitchman & M. Rowland
	G-BSZF	Jodel DR.250/160	J. B. Randle
	G-BSZG	Stolp SA.100 Starduster	D. F. Chapman
	G-BSZH	Thunder Ax7-77 balloon	P. K. Morris
	G-BSZI	Cessna 152 II	Eglinton Flying Club Ltd
	G-BSZJ	PA-28-181 Archer II	M. L. A. Pudney & R. D. Fuller
	G-BSZM	Montgomerie-Bensen B.8MR	A. McCredie
	G-BSZN	Bücker Bü133D-1 Jungmeister	J. Hufner
	G-BSZO	Cessna 152	Direct Helicopters/Southend
	G-BSZT	PA-28-161 Warrior II	Golf Charlie Echo Ltd
	G-BSZU	Cessna 150F	J. E. Jones
	G-BSZV	Cessna 150F	Kirmington Aviation Ltd
	G-BSZW	Cessna 152	Haimoss Ltd
	G-BTAB	BAe 125 Srs 800B	Aravco Ltd (G-BOOA)
	G-BTAG	Cameron O-77 balloon	R. A. Shapland
	G-BTAH	Bensen B.8M	C. J. Toner
	G-BTAK	EAA Acro Sport II	P. G. Harrison
	G-BTAL	Cessna F.152 II	Thanet Flying Club/Manston
	G-BTAM	PA-28-181 Archer II	Tri-Star Farms Ltd
	G-BTAN	Thunder Ax7-65Z balloon	A. S. Newham
	G-BTAS	PA-38-112 Tomahawk	M. Lowe
	G-BTAT	Denney Kitfox Mk 2	M. Lawton
	G-BTAW	PA-28-161 Warrior II	A. J. Wiggins
	G-BTAZ	Evans VP-2	G. S. Poulter
	G-BTBA	Robinson R-22B	Heliflight (UK) Ltd/Wolverhampton
	G-BTBB	Thunder Ax8-105 S2 balloon	W. J. Brogan/Austria
	G-BTBC	PA-28-161 Warrior II	Wellesbourne Flyers Ltd
	G-BTBF	Super Koala	E. A. Taylor (G-MWOZ)
	G-BTBG	Denney Kitfox Mk 2	P. D. Brookes
	G-BTBH	Ryan ST3KR (854)	P. R. Holloway
	G-BTBI	WAR P-47 Thunderbolt (replica) (85)	R. D. Myles
	G-BTBJ	Cessna 195B	P. Camus
	G-BTBL	Montgomerie-Bensen B.8MR	AES Radionic Surveillance Systems
	G-BTBN	Denney Kitfox Mk 2	R. C. Bowley
	G-BTBP	Cameron N-90 balloon	M. Catalani/Italy
	G-BTBU	PA-18 Super Cub 150	G-BTBU Syndicate
	G-BTBW	Cessna 120	M. J. Willies
	G-BTBX	Piper J-3C-65 Cub	Henlow Taildraggers
	G-BTBY	PA-17 Vagabond	G. J. Smith
	G-BTCA	PA-32R-300 Lance	Lance Group
	G-BTCB	Air Command 582 Sport	G. Scurrah
	G-BTCC	Grumman F6F-3 Hellcat (40467)	Patina Ltd/Duxford
	G-BTCD	P-51D-25-NA Mustang (44-13704/87-H)	Pelham Ltd/Duxford
	G-BTCE	Cessna 152	S. T. Gilbert
	G-BTCH	Luscombe 8E Silvaire	G-BTCH Flying Group/Popham
	G-BTCI	PA-17 Vagabond	T. R. Whittome
	G-BTCJ	Luscombe 8AE Silvaire	J. M. Lovell
	G-BTCM	Cameron N-90 balloon	G. Everett (G-BMPW)
	G-BTCR	Rans S.10 Sakota	B. J. Hewitt
	G-BTCS	Colt 90A balloon	R. C. Stone
	G-BTCZ	Cameron 84 Chateau SS balloon	Forbes Global Inc.
	G-BTDA	Slingsby T.61G Falke	Anglia Gliding ClubWattisham

Reg.	Type	Owner or Operator	Notes
G-BTDC	Denney Kitfox Mk 2	O. Smith	
G-BTDD	CFM Streak Shadow	N. D. Ewer	
G-BTDE	Cessna C-165 Airmaster	G. S. Moss	
G-BTDF	Luscombe 8A Silvaire	Delta Foxtrot Group	
G-BTDN	Denney Kitfox Mk 2	Foxy Flyers Group	
G-BTDP	TBM-3R Avenger (53319)	A. Haig-Thomas/North Weald	
G-BTDR	Aero Designs Pulsar	R. A. Blackwell	
G-BTDS	Colt 77A balloon	C. P. Witter Ltd	
G-BTDT	C.A.S.A. 1.131E Jungmann 2000	T. A. Reed	
G-BTDV	PA-28-161 Warrior II	Leeds Flying School Ltd	
G-BTDW	Cessna 152 II	J. A. Blenkharn/Carlisle	
G-BTDZ	C.A.S.A. 1.131E Jungmann 2000	R. J. Pickin & I. M. White	
G-BTEA	Cameron N-105 balloon	M. W. A. Shemilt	
G-BTEE	Cameron O-120 balloon	W. H. & J. P. Morgan	
G-BTEF	Pitts S-1 Special	C. Davidson	
G-BTEK	SOCATA TB.20 Trinidad	M. Northwood	
G-BTEL	CFM Streak Shadow	J. E. Eatwell	
G-BTES	Cessna 150H	R. A. Forward	
G-BTET	Piper J-3C-65 Cub	R. M. Jones/Blackpool	
G-BTEU	SA.365N-2 Dauphin	CHC Scotia Ltd	
G-BTEW	Cessna 120	S. D. Pryke	
G-BTEX	PA-28 Cherokee 140	McAully Flying Group Ltd/Little Snoring	
G-BTFA	Denney Kitfox Mk 2	K. R. Peek	
G-BTFC	Cessna F.152 II	Tayside Aviation Ltd/Dundee	
G-BTFE	Bensen-Parsons 2-seat gyroplane	J. R. Goldspink	
G-BTFF	Cessna T.310R II	United Sales Equipment Dealers Ltd	
G-BTFG	Boeing Stearman A.75N1 (441)	TG Aviation Ltd/Manston	
G-BTFJ	PA-15 Vagabond	C. W. Thirtle	
G-BTFK	Taylorcraft BC-12D	A. O'Rourke	
G-BTFL	Aeronca 11AC Chief	BTFL Group	
G-BTFM	Cameron O-105 balloon	Edinburgh University Hot Air Balloon Club	
G-BTFO	PA-28-161 Warrior II	Flyfar Ltd	
G-BTFP	PA-38-112 Tomahawk	Teesside Flight Centre Ltd	
G-BTFS	Cessna A.150M	P. A. James	
G-BTFT	Beech 58 Baron	Fastwing Air Charter Ltd	
G-BTFU	C?ameron N-90 balloon	J. J. Rudoni & A. C. K. Rawson	
G-BTFV	Whittaker MW.7	S. J. Luck	
G-BTFW	Montgomerie-Bensen B.8MR	J. R. J. Read	
G-BTFX	Bell 206B JetRanger 2	A. C. Watson	
G-BTGD	Rand-Robinson KR-2 (modified)	P. A. Spurr	
G-BTGG	Rans S.10 Sakota	C. A. James	
G-BTGI	Rearwin 175 Skyranger	N. A. Evans	
G-BTGJ	Smith DSA-1 Miniplane	G. J. Knowles	
G-BTGL	Light Aero Avid Flyer	I. Kazi	
G-BTGM	Aeronca 7AC Champion	G. P. Gregg/France	
G-BTGO	PA-28 Cherokee 140	Halegreen Associates Ltd & Demero Ltd	
G-BTGP	Cessna 150M	Billins Air Service Ltd	
G-BTGR	Cessna 152 II	A. J. Gomes/Shoreham	
G-BTGS	Stolp SA.300 Starduster Too	G. N. Elliott & ptnrs (G-AYMA)	
G-BTGT	CFM Streak Shadow	G. D. Bailey (G-MWPY)	
G-BTGU	PA-34-220T Seneca III	Carill Aviation Ltd	
G-BTGV	PA-34-200T Seneca II	Westflight Aviation Ltd	
G-BTGW	Cessna 152 II	Stapleford Flying Club Ltd	
G-BTGX	Cessna 152 II	Stapleford Flying Club Ltd	
G-BTGY	PA-28-161 Warrior II	Stapleford Flying Club Ltd	
G-BTGZ	PA-28-181 Archer II	Allzones Ltd/Biggin Hill	
G-BTHA	Cessna 182P	Hotel Alpha Flying Group/Liverpool	
G-BTHD	Yakovlev Yak-3U	Patina Ltd/Duxford	
G-BTHE	Cessna 150L	Humberside Police Flying Club	
G-BTHF	Cameron V-90 balloon	N. J. & S. J. Langley	
G-BTHH	Jodel DR.100A	H. R. Leefe	
G-BTHI	Robinson R-22B	Yorkshire Helicopters	
G-BTHJ	Evans VP-2	C. J. Moseley	
G-BTHK	Thunder Ax7-77 balloon	M. S.Trend	
G-BTHM	Thunder Ax8-105 balloon	J. K. Woods	
G-BTHN	Murphy Renegade 912	F. A. Purvis	
G-BTHP	Thorp T.211	M. Gardner	
G-BTHU	Light Aero Avid Flyer	R. C. Bowley	
G-BTHV	MBB Bo 105DBS/4	Bond Air Services/Aberdeen	
G-BTHW	Beech F33C Bonanza	Robin Lance Aviation Associates Ltd	
G-BTHX	Colt 105A balloon	Elmer Balloon Group	
G-BTHY	Bell 206B JetRanger 3	Sterling Helicopters Ltd	

Notes	Reg.	Type	Owner or Operator
	G-BTHZ	Cameron V-56 balloon	C. N. Marshall/Kenya
	G-BTID	PA-28-161 Warrior II	Plymouth School of Flying Ltd
	G-BTIE	SOCATA TB.10 Tobago	Aviation Spirit Ltd
	G-BTIF	Denney Kitfox Mk 3	D. A. Murchie
	G-BTII	AA-5B Tiger	BTII Group
	G-BTIJ	Luscombe 8E Silvaire	S. J. Hornsby
	G-BTIK	Cessna 152 II	I. R. Chaplin/Andrewsfield
	G-BTIL	PA-38-112 Tomahawk	B. J. Pearson/Eaglescott
	G-BTIM	PA-28-161 Cadet	Aviation Rentals
	G-BTIO	SNCAN Stampe SV-4C	M. D. & C. F. Garratt
	G-BTIR	Denney Kitfox Mk 2	R. B. Wilson
	G-BTIS	AS.355F-1 Twin Squirrel	Walsh Aviation (G-TALI)/Elstree
	G-BTIU	SOCATA MS.882A Rallye Commodore 150	Cole Aviation Ltd
	G-BTIV	PA-28-161 Warrior II	Warrior Group/Eaglescott
	G-BTIW	Jodel DR.1050/M1 ★	(stored)/Crosland Moor
	G-BTIZ	Cameron A-105 balloon	W. A. Board
	G-BTJA	Luscombe 8E Silvaire	M. W. Rudkin
	G-BTJB	Luscombe 8E Silvaire	M. Loxton
	G-BTJC	Luscombe 8F Silvaire	A. M. Noble
	G-BTJD	Thunder Ax8-90 S2 balloon	R. E. Vinten
	G-BTJF	Thunder Ax10-180 balloon	Airborne Adventures Ltd
	G-BTJH	Cameron O-77 balloon	H. Stringer
	G-BTJK	PA-38-112 Tomahawk	Western Air (Thruxton) Ltd
	G-BTJL	PA-38-112 Tomahawk	J. S. Devlin & Z. Islam/Redhill
	G-BTJO	Thunder Ax9-140 balloon	G. P. Lane
	G-BTJS	Montgomerie-Bensen B.8MR	B. F. Pearson
	G-BTJU	Cameron V-90 balloon	C. W. Jones (Floorings) Ltd
	G-BTJX	Rans S.10 Sakota	T. Scarborough
	G-BTKA	Piper J-5A Cub Cruiser	J. M. Lister
	G-BTKB	Renegade Spirit 912	P. J. Calvert
	G-BTKD	Denney Kitfox Mk 4	N. Sansom & R. A. Hills
	G-BTKG	Light Aero Avid Flyer	I. Holt
	G-BTKL	MBB Bo 105DB-4	Veritair Ltd
	G-BTKN	Cameron O-120 balloon	R. H. Etherington
	G-BTKP	CFM Streak Shadow	G. D. Martin
	G-BTKT	PA-28-161 Warrior II	Biggin Hill Flying Club Ltd
	G-BTKV	PA-22 Tri-Pacer 160	R. A. Moore
	G-BTKW	Cameron O-105 balloon	P. Spellward
	G-BTKX	PA-28-181 Archer II	D. J. Perkins
	G-BTKZ	Cameron V-77 balloon	S. P. Richards
	G-BTLB	Wassmer WA.52 Europa	Popham Flying Group G-BTLB
	G-BTLG	PA-28R Cherokee Arrow 200	W. B. Bateson/Blackpool
	G-BTLL	Pilatus P3-03 (A-806) ★	(stored)/Headcorn
	G-BTLM	PA-22 Tri-Pacer 160	F & H (Aircraft)
	G-BTLP	AA-1C Lynx	Partlease Ltd
	G-BTMA	Cessna 172N	East of England Flying Group Ltd
	G-BTMH	Colt 90A balloon	European Balloon Corporation
	G-BTMK	Cessna R.172K XPII	A. C. Barker
	G-BTMN	Thunder Ax9-120 S2 balloon	M. E. White
	G-BTMO	Colt 69A balloon	Thunder & Colt
	G-BTMP	Campbell Cricket	P. W. McLaughlin
	G-BTMR	Cessna 172M	Linley Aviation Ltd
	G-BTMS	Light Aero Avid Speedwing	F. Sayyah
	G-BTMT	Denney Kitfox Mk 1	L. G. Horne
	G-BTMV	Everett Srs 2 gyroplane	L. Armes
	G-BTMW	Zenair CH.701 STOL	L. Lewis
	G-BTMX	Denney Kitfox Mk 3	V. T. Betts
	G-BTMY	Cameron Train-80 SS balloon	Balloon Sports HB/Sweden
	G-BTNA	Robinson R-22B	The Air Group Ltd
	G-BTNC	AS.365N-2 Dauphin 2	CHC Scotia Ltd
	G-BTND	PA-38-112 Tomahawk	R. B. Turner
	G-BTNE	PA-28-161 Warrior II	D. Rowe/Wellesbourne
	G-BTNH	PA-28-161 Warrior II	Falcon Flying Services Ltd (G-DENH)/Biggin Hill
	G-BTNN	Colt 21A balloon	J. Mason
	G-BTNO	Aeronca 7AC Champion	B. J. & B. G. Robe
	G-BTNR	Denney Kitfox Mk 3	D. E. Steade
	G-BTNT	PA-28-151 Warrior	Britannia Airways Ltd/Luton
	G-BTNU	BAe 146-300	Trident Jet (Jersey) Ltd
	G-BTNV	PA-28-161 Warrior II	G. M. Bauer & A. W. Davies
	G-BTNW	Rans S.6-ESA Coyote II	R. H. Hughes
	G-BTOC	Robinson R-22B	N. Parkhouse

Reg.	Type	Owner or Operator	Notes
G-BTOG	D.H.82A Tiger Moth	P. T. Szluha	
G-BTOL	Denney Kitfox Mk 3	P. J. Gibbs	
G-BTON	PA-28 Cherokee 140	S. Quinn	
G-BTOO	Pitts S-1C Special	G. H. Matthews	
G-BTOP	Cameron V-77 balloon	J. J. Winter	
G-BTOS	Cessna 140	J. L. Kaiser/France	
G-BTOT	PA-15 Vagabond	Vagabond Flying Group	
G-BTOU	Cameron O-120 balloon	J. J. Daly	
G-BTOW	SOCATA Rallye 180GT	Cambridge University Gliding Trust Ltd/ Gransden Lodge	
G-BTOZ	Thunder Ax9-120 S2 balloon	H. G. Davies	
G-BTPA	BAe ATP	Capital Bank Leasing 12 Ltd	
G-BTPB	Cameron N-105 balloon	D. J. Farrar	
G-BTPC	BAe ATP	Capital Bank Leasing 1 Ltd	
G-BTPD	BAe ATP	Seaforth Maritime Ltd	
G-BTPE	BAe ATP	Capital Bank Leasing PLC	
G-BTPF	BAe ATP	Capital Bank Leasing 5 Ltd	
G-BTPG	BAe ATP	Capital Bank Leasing 5 Ltd	
G-BTPH	BAe ATP	Capital Bank Leasing 6 Ltd	
G-BTPJ	BAe ATP	Capital Bank Leasing 7 Ltd	
G-BTPT	Cameron N-77 balloon	Derbyshire Building Soc	
G-BTPV	Colt 90A balloon	Balloon Preservation Group	
G-BTPX	Thunder Ax8-90 balloon	E. Cordall	
G-BTPZ	Isaacs Fury II	M. A. Farrelly	
G-BTRB	Thunder Colt Mickey Mouse SS balloon	Benedikt Haggeney GmbH/Germany	
G-BTRC	Light Aero Avid Speedwing	Grangecote Ltd	
G-BTRE	Cessna F.172H	S. Clark	
G-BTRF	Aero Designs Pulsar	C. Smith	
G-BTRG	Aeronca 65C Super Chief	A. Welburn	
G-BTRI	Aeronca 11CC Super Chief	P. A. Wensak	
G-BTRK	PA-28-161 Warrior II	Stapleford Flying Club Ltd	
G-BTRL	Cameron N-105 balloon	J. Lippett	
G-BTRN	Thunder Ax9-120 S2 balloon	P. B. D. Bird	
G-BTRO	Thunder Ax8-90 balloon	Capital Balloon Club Ltd	
G-BTRP	Hughes 369E	P. C. Shann	
G-BTRR	Thunder Ax7-77 balloon	P. J. Wentworth	
G-BTRS	PA-28-161 Warrior II	Airwise Flying Group	
G-BTRT	PA-28R Cherokee Arrow 200-II	Romeo Tango Group	
G-BTRU	Robin DR.400/180	R. H. Mackay	
G-BTRW	Slingsby T.61F Venture T.2	The Falke Syndicate	
G-BTRY	PA-28-161 Warrior II	Oxford Aviation Services Ltd/Kidlington	
G-BTRZ	Jodel D.18	A. P. Aspinall	
G-BTSB	Corben Baby Ace D	J. A. MacLeod & J. Horovitz	
G-BTSJ	PA-28-161 Warrior II	Plymouth School of Flying Ltd	
G-BTSL	Cameron 70 Glass SS balloon	M. R. Humphrey & J. R .Clifton	
G-BTSM	Cessna 180A	C. Couston	
G-BTSN	Cessna 150G	C. P. Whitwell	
G-BTSP	Piper J-3C-65 Cub	J. A. Walshe & A. Corcoran	
G-BTSR	Aeronca 11AC Chief	R. D. & E. G. N. Morris	
G-BTSV	Denney Kitfox Mk 3	M. G. Dovey	
G-BTSW	Colt AS-80 Mk II airship	Gefa-Flug GmbH/Germany	
G-BTSX	Thunder Ax7-77 balloon	C. Moris-Gallimore/Portugal	
G-BTSY	EE Lightning F.6 (XR724)	Lightning Association	
G-BTSZ	Cessna 177A	W. J. Peachment	
G-BTTB	Cameron V-90 balloon	Royal Engineers Balloon Club	
G-BTTD	Montgomerie-Bensen B.8MR	A. J. P. Herculson	
G-BTTE	Cessna 150L	C. Wilson & W. B. Murray	
G-BTTL	Cameron V-90 balloon	A. J. Baird	
G-BTTO	BAe ATP	BAR Systems Inc	
G-BTTP	BAe 146-300	Trident Aviation Leasing Services (Jersey) Ltd	
G-BTTR	Aerotek Pitts S-2A Special	Extreme Air Sports Ltd	
G-BTTS	Colt 77A balloon	Rutland Balloon Club	
G-BTTW	Thunder Ax7-77 balloon	J. Kenny	
G-BTTY	Denney Kitfox Mk 2	K. J. Fleming	
G-BTTZ	Slingsby T.61F Venture T.2	M. W. Olliver	
G-BTUA	Slingsby T.61F Venture T.2	Shenington Gliding Club	
G-BTUB	Yakovlev C.11	M. G. & J. R. Jefferies	
G-BTUC	EMB-312 Tucano ★	Ulster Aviation Heritage/Langford Lodge	
G-BTUG	SOCATA Rallye 180T	Herefordshire Gliding Club Ltd/Shobdon	
G-BTUH	Cameron N-65 balloon	B. J. Godding	
G-BTUJ	Thunder Ax9-120 balloon	ECM Construction Ltd	
G-BTUK	Aerotek Pitts S-2A Special	S. H. Elkington/Wickenby	

Notes	Reg.	Type	Owner or Operator
	G-BTUL	Aerotek Pitts S-2A Special	J. M. Adams
	G-BTUM	Piper J-3C-65 Cub	G-BTUM Syndicate
	G-BTUR	PA-18 Super Cub 95 (modified)	A. P. Meredith
	G-BTUS	Whittaker MW.7	C. T. Bailey
	G-BTUU	Cameron O-120 balloon	P. Dubois-Dauphin/France
	G-BTUV	Aeronca A65TAC Defender	M. B. Hamlett & R. E. Coates/France
	G-BTUW	PA-28-151 Warrior	T. S. Kemp
	G-BTUZ	American General AG-5B Tiger	Grocontinental Ltd/Tilstock
	G-BTVA	Thunder Ax7-77 balloon	A. H. Symonds
	G-BTVB	Everett Srs 3 gyroplane	J. Pumford
	G-BTVC	Denney Kitfox Mk 2	P. Mitchell
	G-BTVE	Hawker Demon I (K8203)	Demon Displays Ltd
	G-BTVR	PA-28 Cherokee 140	Full Sutton Flying Centre Ltd
	G-BTVU	Robinson R-22B	B. Enzo/Italy
	G-BTVV	Cessna FA.337G	C. Keane
	G-BTVW	Cessna 152 II	Halegreen Associates Ltd
	G-BTVX	Cessna 152 II	Eastern Air Centre Ltd/Gamston
	G-BTWB	Denney Kitfox Mk 3	J. E. Tootell
	G-BTWC	Slingsby T.61F Venture T.2	RAFGSA/Upavon
	G-BTWD	Slingsby T.61F Venture T.2	York Gliding Centre/Rufforth
	G-BTWE	Slingsby T.61F Venture T.2	RAFGSA/Upavon
	G-BTWF	D.H.C.1 Chipmunk 22 (WK549)	J. A. & V. G. Sims
	G-BTWI	EAA Acro Sport I	S. Alexander & W. M. Coffee
	G-BTWJ	Cameron V-77 balloon	S. J. & J. A. Bellaby
	G-BTWL	WAG-Aero Acro Sport Trainer	I. M. Ashpole
	G-BTWM	Cameron V-77 balloon	R. C. Franklin
	G-BTWV	Cameron O-90 balloon	S. F. Hancke
	G-BTWX	SOCATA TB.9 Tampico	Cinque Ports Aviation Ltd
	G-BTWY	Aero Designs Pulsar	M. Stevenson
	G-BTWZ	Rans S.10 Sakota	R. V. Barber
	G-BTXB	Colt 77A balloon	A. Derbyshire
	G-BTXD	Rans S.6-ESA Coyote II	A. I. Sutherland
	G-BTXF	Cameron V-90 balloon	G. Thompson
	G-BTXG	BAe Jetstream 3102	Highland Airways Ltd/Inverness
	G-BTXH	Colt AS-56 airship	L. Kiefer/Germany
	G-BTXI	Noorduyn AT-16 Harvard IIB (FE695)	Patina Ltd/Duxford
	G-BTXK	Thunder Ax7-65 balloon	A. F. Selby
	G-BTXN	BAe 146-300	Flightline Ltd (G-JEAL)/Southend
	G-BTXS	Cameron O-120 balloon	Southern Balloon Group
	G-BTXT	Maule MXT-7-180 Star Rocket	D. Carr & Muter
	G-BTXV	Cameron A-210 balloon	The Ballooning Business Ltd
	G-BTXW	Cameron V-77 balloon	P. C. Waterhouse
	G-BTXX	Bellanca 8KCAB Decathlon	Tatenhill Aviation Ltd
	G-BTXZ	Zenair CH.250	I. Parris & P. W. J. Bull
	G-BTYC	Cessna 150L	Polestar Aviation Ltd
	G-BTYE	Cameron A-180 balloon	K. J. A. Maxwell & D. S. Messmer
	G-BTYF	Thunder Ax10-180 S2 balloon	P. Glydon
	G-BTYH	Pottier P.80S	R. Pickett
	G-BTYI	PA-28-181 Archer II	C. E. Wright
	G-BTYK	Cessna 310R	Revere Aviation Ltd
	G-BTYT	Cessna 152 II	Cristal Air Ltd
	G-BTYW	Cessna 120	G-BTYW Group
	G-BTYY	Curtiss Robin C-2	R. R. L. Windus
	G-BTYZ	Colt 210A balloon	T. M. Donnelly
	G-BTZA	Beech F33A Bonanza	G-BTZA Group/Edinburgh
	G-BTZB	Yakovlev Yak-50 (10)	D. H. Boardman
	G-BTZD	Yakovlev Yak-1	Historic Aircraft Collection Ltd
	G-BTZE	LET Yakovlev C.11	Bianchi Aviation Film Services Ltd/Booker
	G-BTZG	BAe ATP	Trident Aviation Leasing Services
	G-BTZK	BAe ATP	Trident Aviation Leasing Services
	G-BTZL	Oldfield Baby Lakes	M. R. Winter
	G-BTZN	BAe 146-300	Trident Aviation Leasing Ltd
	G-BTZO	SOCATA TB.20 Trinidad	P. R. Draper
	G-BTZP	SOCATA TB.9 Tampico	M. W. Orr
	G-BTZR	Colt 77B balloon	P. J. Fell
	G-BTZS	Colt 77A balloon	P. T. R. Ollivere
	G-BTZV	Cameron V-77 balloon	D. J. & H. M. Brown
	G-BTZX	Piper J-3C-65 Cub	ZX Cub Group
	G-BTZY	Colt 56A balloon	S. J. Wardle
	G-BTZZ	CFM Streak Shadow	D. R. Stennett
	G-BUAA	Corben Baby Ace D	M. W. Chamberlain

Reg.	Type	Owner or Operator
G-BUAB	Aeronca 11AC Chief	J. Reed
G-BUAC	Slingsby T.31 Motor Cadet III	D. A. Wilson & C. R. Partington
G-BUAF	Cameron N-77 balloon	Zebedee Balloon Service Ltd
G-BUAG	Jodel D.18	A. L. Silcox
G-BUAI	Everett Srs 3 gyroplane	I. D. Bateson
G-BUAJ	Cameron N-90 balloon	J. R. & S. J. Huggins
G-BUAM	Cameron V-77 balloon	N. Florence
G-BUAO	Luscombe 8A Silvaire	S. L. Lewis
G-BUAT	Thunder Ax9-120 balloon	J. Fenton
G-BUAV	Cameron O-105 balloon	C. D. Monk
G-BUAX	Rans S.10 Sakota	N. de Badgecoe
G-BUBL	Thunder Ax8-105 balloon ★	British Balloon Museum/Newbury
G-BUBN	BN-2B-26 Islander	Isles of Scilly Skybus Ltd/St Just
G-BUBS	Lindstrand LBL-77B balloon	B. J. Bower
G-BUBT	Stoddard Hamilton Glasair IIS RG	M. D. Evans
G-BUBU	PA-34-220T Seneca III	Brinor (Holdings) Ltd
G-BUBW	Robinson R-22B	Forth Helicopter Services Ltd/Edinburgh
G-BUBY	Thunder Ax8-105 S2 balloon	T. M. Donnelly
G-BUCA	Cessna A.150K	BUCA Group
G-BUCB	Cameron H-34 balloon	A. S. Jones
G-BUCC	C.A.S.A. 1.131E Jungmann 2000 (BU+CC)	P. L. Gaze (G-BUEM)
G-BUCG	Schleicher ASW.20L (modified)	W. B. Andrews
G-BUCH	Stinson V-77 Reliant	Gullwing Trading Ltd
G-BUCI	Auster AOP.9 (XP242)	Historic Aircraft Flight Reserve Collection/ Middle Wallop
G-BUCJ	D.H.C.2 Beaver 1 (XP772)	British Aerial Museum/Duxford
G-BUCK	C.A.S.A. 1.131E Jungmann 1000 (BU+CK)	Jungmann Flying Group/White Waltham
G-BUCM	Hawker Sea Fury FB.11	Patina Ltd/Duxford
G-BUCO	Pietenpol Air Camper	A. James
G-BUCS	Cessna 150F	Atlantic Bridge Aviation Ltd/Lydd
G-BUCT	Cessna 150L	Atlantic Bridge Aviation Ltd/Lydd
G-BUDA	Slingsby T.61F Venture T.2	Cranwell Gliding Club
G-BUDB	Slingsby T.61F Venture T.2	RAFGSA/Bicester
G-BUDC	Slingsby T.61F Venture T.2	T.61 Group
G-BUDE	PA-22 Tri-Pacer 135 (tailwheel)	B. A. Bower/Thruxton
G-BUDF	Rand-Robinson KR-2	M. Stott
G-BUDI	Aero Designs Pulsar	R. W. L. Oliver
G-BUDK	Thunder Ax7-77 balloon	W. Evans
G-BUDL	Auster 3 (NX534)	M. Pocock
G-BUDN	Cameron 90 Shoe SS balloon	Magical Adventures Ltd
G-BUDO	PZL-110 Koliber 150	A. S. Vine/Goodwood
G-BUDR	Denney Kitfox Mk 3	N. J. P. Mayled
G-BUDS	Rand-Robinson KR-2	D. W. Munday
G-BUDT	Slingsby T.61F Venture T.2	G-BUDT Group
G-BUDU	Cameron V-77 balloon	T. M. G. Amery
G-BUDW	Brügger MB.2 Colibri	D.R. Mickleburgh (G-GODS)
G-BUEC	Van's RV-6	A. H. Harper
G-BUED	Slingsby T.61F Venture T.2	617 VGS Groupd
G-BUEF	Cessna 152 II	Channel Aviation
G-BUEG	Cessna 152 II	Plymouth School of Flying Ltd
G-BUEK	Slingsby T.61F Venture T.2	Norfolk Gliding Club Ltd/Tibenham
G-BUEN	VPM M.14 Scout	F. G. Shepherd
G-BUEP	Maule MX-7-180	N. J. B. Bennett
G-BUEV	Cameron O-77 balloon	R. R. McCormack
G-BUEW	Rans S-6 Coyote II	J. D. Clabon (G-MWYE)
G-BUEX	Schweizer 269C	Tout-Saints Hotels Ltd (G-HFLR)
G-BUFA	Cameron R-77 gas balloon	Noble Adventures Ltd
G-BUFC	Cameron R-77 gas balloon	Noble Adventures Ltd
G-BUFE	Cameron R-77 gas balloon	Noble Adventures Ltd
G-BUFG	Slingsby T.61F Venture T.2	Halegreen Associates Ltd
G-BUFH	PA-28-161 Warrior II	The Tiger Leisure Group
G-BUFJ	Cameron V-90 balloon	S. P. Richards
G-BUFN	Slingsby T.61F Venture T.2	BUFN Group
G-BUFR	Slingsby T.61F Venture T.2	East Sussex Gliding Club Ltd
G-BUFT	Cameron O-120 balloon	D. Bron
G-BUFV	Light Aero Avid Speedwing Mk.4	M. & B. Gribbin
G-BUFW	AS.355F-1 Twin Squirrel	RCR Aviation Ltd
G-BUFX	Cameron N-90 balloon	Kerridge Computer Co Ltd
G-BUFY	PA-28-161 Warrior II	Bickertons Aerodromes Ltd/Denham
G-BUGB	Stolp SA.750 Acroduster Too	R. M. Chaplin

Notes	Reg.	Type	Owner or Operator
	G-BUGD	Cameron V-77 balloon	Cameron Balloons Ltd
	G-BUGE	Bellanca 7GCAA Citabria	V. Vaughan & N. O'Brien
	G-BUGG	Cessna 150F	C. P. J. Taylor & D. M. Forshaw/Panshanger
	G-BUGI	Evans VP-2	J. A. Rees
	G-BUGJ	Robin DR.400/180	W. M. Patterson
	G-BUGL	Slingsby T.61F Venture T.2	VMG Group
	G-BUGM	CFM Streak Shadow	The Shadow Group
	G-BUGO	Colt 56B balloon	Escuela de Aerostacion Mica/Spain
	G-BUGP	Cameron V-77 balloon	R. Churcher
	G-BUGS	Cameron V-77 balloon	S. J. Dymond
	G-BUGT	Slingsby T.61F Venture T.2	R. W. Hornsey/Rufforth
	G-BUGV	Slingsby T.61F Venture T.2	Oxfordshire Sportflying Ltd/Enstone
	G-BUGW	Slingsby T.61F Venture T.2	Halegreen Associates Ltd
	G-BUGY	Cameron V-90 balloon	Dante Balloon Group
	G-BUGZ	Slingsby T.61F Venture T.2	Dishforth Flying Group
	G-BUHA	Slingsby T.61F Venture T.2 (ZA634)	K. E. Ballington
	G-BUHC	BAe 146-300	Flightline Ltd (G-BTMI)
	G-BUHM	Cameron V-77 balloon	L. A. Watts
	G-BUHO	Cessna 140	W. B. Bateson/Blackpool
	G-BUHR	Slingsby T.61F Venture T.2	Connel Motor Glider Group
	G-BUHS	Stoddard-Hamilton Glasair SH-TD-1	E. J. Spalding
	G-BUHU	Cameron N-105 balloon	Unipart Balloon Club
	G-BUHY	Cameron A-210 balloon	D. J. Farrar
	G-BUHZ	Cessna 120	Cessna 140 Group
	G-BUIC	Denney Kitfox Mk 2	C. R. Northrop & B. M. Chilvers
	G-BUIE	Cameron N-90 balloon	B. Conway
	G-BUIF	PA-28-161 Warrior II	Newcastle-upon-Tyne Aero Club Ltd
	G-BUIG	Campbell Cricket (replica)	J. A. English
	G-BUIH	Slingsby T.61F Venture T.2	Yorkshire Gliding Club (Pty) Ltd/Sutton Bank
	G-BUIJ	PA-28-161 Warrior II	Tradecliff Ltd/Blackbushe
	G-BUIK	PA-28-161 Warrior II	Falcon Flying Services/Biggin Hill
	G-BUIL	CFM Streak Shadow	P. N. Bevan & L. M. Poor
	G-BUIN	Thunder Ax7-77 balloon	P. C. Johnson
	G-BUIP	Denney Kitfox Mk 2	Avcomm Developments Ltd
	G-BUIR	Light Aero Avid Speedwing Mk 4	M. C. Myers
	G-BUIU	Cameron V-90 balloon	H. Micketeit/Germany
	G-BUIZ	Cameron N-90 balloon	Balloon Preservation Flying Group
	G-BUJA	Slingsby T.61F Venture T.2	Wrekin Gliding Club/Cosford
	G-BUJB	Slingsby T.61F Venture T.2	Falke Syndicate/Shobdon
	G-BUJE	Cessna 177B	FG93 Group
	G-BUJH	Colt 77B balloon	R. P. Cross & R. Stanley
	G-BUJI	Slingsby T.61F Venture T.2	Solent Venture Syndicate Ltd
	G-BUJJ	Avid Speedwing	R. A. Dawson
	G-BUJK	Montgomerie-Bensen B.8MR	K. J. Robinson
	G-BUJL	Aero Designs Pulsar	J. J. Lynch
	G-BUJM	Cessna 120	Cessna 120 Flying Group/Yeovilton
	G-BUJN	Cessna 172N	Central Aircraft Leasing Ltd
	G-BUJO	PA-28-161 Warrior II	Falcon Flying Services/Biggin Hill
	G-BUJP	PA-28-161 Warrior II	J. M. C. Manson
	G-BUJR	Cameron A-180 balloon	Dragon Balloon Co. Ltd
	G-BUJV	Light Aero Avid Speedwing Mk 4	C. Thomas
	G-BUJW	Thunder Ax8-90 S2 balloon	R. T. Fagan
	G-BUJX	Slingsby T.61F Venture T.2	J. R. Chichester-Constable
	G-BUJY	D.H.82A Tiger Moth	P. Winters
	G-BUJZ	Rotorway Executive 90 (modified)	M. P. Swoboda
	G-BUKA	Fairchild SA227AC Metro III	Atlantic Air Transport Ltd/Coventry
	G-BUKB	Rans S.10 Sakota	M. K. Blatch
	G-BUKF	Denney Kitfox Mk 4	Kilo Foxtrot Group
	G-BUKH	D.31 Turbulent	R. B. Armitage
	G-BUKI	Thunder Ax7-77 balloon	Virgin Balloon Flights
	G-BUKK	Bücker Bü133C Jungmeister (U-80)	E. J. F. McEntee/White Waltham
	G-BUKN	PA-15 Vagabond	B. F. & M. A. Goddard
	G-BUKO	Cessna 120	S. Warrener
	G-BUKP	Denney Kitfox Mk 2	K. N. Cobb
	G-BUKR	M.S.880B Rallye Club 100T	G-BUKR Flying Group
	G-BUKS	Colt 77B balloon	R. & M. Bairstow
	G-BUKT	Luscombe 8A Silvaire	R. J. F. Swain
	G-BUKU	Luscombe 8E Silvaire	Silvaire Flying Group
	G-BUKV	Colt AS-105 Mk II Airship	A. Ockelmann/Germany
	G-BUKX	PA-28-161 Warrior II	LNP Ltd
	G-BUKZ	Evans VP-2	P. R. Farnell
	G-BULB	Thunder Ax7-77 balloon	Richard Nash Cars Ltd

Reg.	Type	Owner or Operator
G-BULC	Light Aero Avid Flyer Mk 4	C. Nice
G-BULD	Cameron N-105 balloon	C. L. Jenkins
G-BULF	Colt 77A balloon	P. Goss & T. C. Davies
G-BULG	Van's RV-4	M. J. Aldridge/Tibenham
G-BULJ	CFM Steak Shadow	C. C. Brown
G-BULK	Thunder Ax9-120 S2 balloon	Skybus Ballooning
G-BULL	SA Bulldog Srs 120/128	Solo Leisure Ltd
G-BULM	Aero Designs Pulsar	J. Lloyd
G-BULN	Colt 210A balloon	H. G. Davies
G-BULO	Luscombe 8A Silvaire	A. F. S. Caldecourt
G-BULT	Campbell Cricket	A. T. Pocklington
G-BULY	Light Aero Avid Flyer	D. R. Piercy
G-BULZ	Denney Kitfox Mk 2	T. G. F. Trenchard
G-BUMP	PA-28-181 Archer II	Marnham Investments Ltd
G-BUNB	Slingsby T.61F Venture T.2	RAFGSA/Lee-on-Solent
G-BUNC	PZL-104 Wilga 35	R. F. Goodman
G-BUND	PA-28RT-201T Turbo Arrow IV	Datalake Ltd
G-BUNG	Cameron N-77 balloon	The Bungle Balloon Group
G-BUNH	PA-28RT-201T Turbo Arrow IV	Border Air Training Ltd
G-BUNJ	Squarecraft SA.102-5 Cavalier	J. A. Smith
G-BUNM	Denney Kitfox Mk 3	P. N. Akass
G-BUNO	Lancair 320	J. Softley
G-BUNV	Thunder Ax7-77 balloon	R. M. Garnett
G-BUNZ	Thunder Ax10-180 S2 balloon	M. A. Scholes
G-BUOA	Whittaker MW.6-S Fatboy Flyer	R. Blackburn
G-BUOB	CFM Streak Shadow	N. K. & S. J. Forest
G-BUOC	Cameron A-210 balloon	Aerosaurus Balloons LLP
G-BUOD	SE-5A (replica) (B595)	M. D. Waldron/Belgium
G-BUOE	Cameron V-90 balloon	Dusters & Co
G-BUOF	D.62B Condor	R. P. Loxton
G-BUOI	PA-20 Pacer	Foley Farm Flying Group
G-BUOJ	Cessna 172N	Falcon Flying Services Ltd/Biggin Hill
G-BUOK	Rans S.6-ESA Coyote II	M. Morris
G-BUOL	Denney Kitfox Mk 3	E. C. King
G-BUON	Light Aero Avid Aerobat	S. R. Winder
G-BUOR	C.A.S.A. 1.131E Jungmann 2000	M. I. M. S. Voest/Netherlands
G-BUOS	V.S.394 Spitfire FR.XVIII (SM845/GZ-J)	Historic Flying Ltd
G-BUOW	Aero Designs Pulsar XP	T. J. Hartwell
G-BUOX	Cameron V-77 balloon	I. G. H. Woodmansey
G-BUOZ	Thunder Ax10-180 balloon	Zebedee Balloon Service Ltd
G-BUPA	Rutan LongEz	N. G. Henry
G-BUPB	Stolp SA.300 Starduster Too	Starduster PB Group
G-BUPC	Rollason Beta B.2	C. A. Rolph
G-BUPF	Bensen B.8R	P. W. Hewitt-Dean
G-BUPG	Cessna 180K	T. P. A. Norman/Rendcomb
G-BUPH	Colt 25A balloon	BAB-Ballonwerbung GmbH/Germany
G-BUPI	Cameron V-77 balloon	S. A. Masey (G-BOUC)
G-BUPJ	Fournier RF-4D	M. R. Shelton
G-BUPM	VPM M.16 Tandem Trainer	Roger Savage (Gyroplanes) Ltd
G-BUPP	Cameron V-42 balloon	L. J. Schoeman
G-BUPR	Jodel D.18	R. W. Burrows
G-BUPT	Cameron O-105 balloon	Chiltern Balloons
G-BUPU	Thunder Ax7-77 balloon	R. C. Barkworth & D. G. Maguire/USA
G-BUPV	Great Lakes 2T-1A	R. J. Fray
G-BUPW	Denney Kitfox Mk 3	Kitfox Group
G-BURD	Cessna F.172N	Tayside Aviation Ltd
G-BURE	Jodel D.9	L. J. Kingsford
G-BURG	Colt 77A balloon	S. T. Humphreys
G-BURH	Cessna 150E	C. A. Davis
G-BURI	Enstrom F-28C	India Helicopters Group
G-BURL	Colt 105A balloon	J. E. Rose
G-BURN	Cameron O-120 balloon	Innovation Ballooning Ltd
G-BURP	Rotorway Executive 90	N. K. Newman
G-BURS	Sikorsky S-76A	Premiair Aviation Services Ltd (G-OHTL)
G-BURT	PA-28-161 Warrior II	B. A. Paul
G-BURU	BAe Jetstream 3202	Trident Aviation Leasing Services
G-BURZ	Hawker Nimrod II (K3661)	Historic Aircraft Collection Ltd
G-BUSB	Airbus A.320-111	British Airways
G-BUSC	Airbus A.320-111	British Airways
G-BUSD	Airbus A.320-111	British Airways
G-BUSE	Airbus A.320-111	British Airways
G-BUSF	Airbus A.320-111	British Airways

Notes	Reg.	Type	Owner or Operator
	G-BUSG	Airbus A.320-211	British Airways
	G-BUSH	Airbus A.320-211	British Airways
	G-BUSI	Airbus A.320-211	British Airways
	G-BUSJ	Airbus A.320-211	British Airways
	G-BUSK	Airbus A.320-211	British Airways
	G-BUSN	Rotorway Executive 90	J. A. McGinley
	G-BUSR	Aero Designs Pulsar	S. S. Bateman & R. A. Watts
	G-BUSS	Cameron 90 Bus SS balloon	Magical Adventures Ltd
	G-BUSV	Colt 105A balloon	M. N. J. Kirby
	G-BUSW	R. Commander 114	J. M. J. Palmer
	G-BUSY	Thunder Ax6-56A balloon	M. E. Hooker
	G-BUTA	C.A.S.A. 1.131E Jungmann 2000	A. P. Pearson
	G-BUTB	CFM Streak Shadow	S. Vestuti
	G-BUTD	Van's RV-6	N. W. Beadle
	G-BUTE	Anderson EA-1 Kingfisher	T. Crawford (G-BRCK)
	G-BUTF	Aeronca 11AC Chief	Fox Flying Group
	G-BUTG	Zenair CH.601HD	I. J. McNally
	G-BUTH	CEA DR.220 2+2	T. V. Thorp
	G-BUTJ	Cameron O-77 balloon	A. J. A. Bubb
	G-BUTK	Murphy Rebel	G. S. Claybourn
	G-BUTM	Rans S.6-116 Coyote II	G-BUTM Group
	G-BUTW	BAe Jetstream 3202	Trident Aviation Leasing Services
	G-BUTX	C.A.S.A. 1.133C Jungmeister	A. J. E. Smith/Breighton
	G-BUTY	Brügger MB.2 Colibri	R. M. Lawday
	G-BUTZ	PA-28 Cherokee 180C	M. H. Canning (G-DARL)
	G-BUUA	Slingsby T.67M Firefly Mk II	Babcock Defence Services
	G-BUUB	Slingsby T.67M Firefly Mk II	Babcock Defence Services
	G-BUUC	Slingsby T.67M Firefly Mk II	Babcock Defence Services
	G-BUUD	Slingsby T.67M Firefly Mk II	J. T. Matthews
	G-BUUE	Slingsby T.67M Firefly Mk II	J. R. Bratty
	G-BUUF	Slingsby T.67M Firefly Mk II	Northamptonshire School of Flying Ltd
	G-BUUG	Slingsby T.67M Firefly Mk II	Flight Training Europe SL/Spain
	G-BUUI	Slingsby T.67M Firefly Mk II	P. M. Harrison
	G-BUUJ	Slingsby T.67M Firefly Mk II	R. Manning
	G-BUUK	Slingsby T.67M Firefly Mk II	Babcock Defence Services
	G-BUUL	Slingsby T.67M Firefly Mk II	Witham (Specialist Vehicles) Ltd
	G-BUUM	PA-28RT-201 Arrow IV	Bluebird Flying Group
	G-BUUO	Cameron N-90 balloon	Gone Ballooning Group
	G-BUUP	BAe ATP	Trident Aviation Leasing Ltd (G-MANU)
	G-BUUT	Interavia 70TA balloon	Aero Vintage Ltd
	G-BUUX	PA-28 Cherokee 180D	Aero Group 78/Netherthorpe
	G-BUUZ	BAe Jetstream 3202	Trident Aviation Leasing Services
	G-BUVA	PA-22 Tri-Pacer 135	Oaksey VA Group
	G-BUVC	BAe Jetstream 3206	Eastern Airways Ltd
	G-BUVD	BAe Jetstream 3206	Eastern Airways Ltd
	G-BUVE	Colt 77B balloon	G. D. Philpot
	G-BUVG	Cameron N-56 balloon	G. J. Bell
	G-BUVL	Fisher Super Koala	A. D. Malcolm
	G-BUVM	CEA DR.250/160	G-BUVM Group
	G-BUVN	C.A.S.A. 1.131E Jungmann 2000 (BI-005)	W. Van Egmond/Netherlands
	G-BUVO	Cessna F.182P	Romeo Mike Flying Group (G-WTFA)
	G-BUVR	Christen A.1 Husky	A. E. Poulson
	G-BUVS	Colt 77A balloon	S. J. Chatfield
	G-BUVT	Colt 77A balloon	N. A. Carr
	G-BUVW	Cameron N-90 balloon	Bristol Balloon Fiestas Ltd
	G-BUVX	CFM Streak Shadow	R. Barcis
	G-BUVZ	Thunder Ax10-180 S2 balloon	A. Van Wyk
	G-BUWE	SE-5A (replica) (C9533)	Airpark Flight Centre Ltd/Coventry
	G-BUWF	Cameron N-105 balloon	R. E. Jones
	G-BUWH	Parsons 2-seat gyroplane	R. V. Brunskill
	G-BUWI	Lindstrand LBL-77A balloon	Capital Balloon Club Ltd
	G-BUWJ	Pitts S-1C Special	G-BUWJ Flying Group/Portugal
	G-BUWK	Rans S.6-116 Coyote II	R. Warriner
	G-BUWL	Piper J-4A	M. L. Ryan
	G-BUWR	CFM Streak Shadow	T. Harvey
	G-BUWS	Denney Kitfox Mk 2	J. E. Brewis
	G-BUWT	Rand-Robinson KR-2	C. M. Coombe
	G-BUWU	Cameron V-77 balloon	T. R. Dews
	G-BUWZ	Robin HR.200/120B	A. N. Kaschevski
	G-BUXA	Colt 210A balloon	Balloon School International Ltd
	G-BUXC	CFM Streak Shadow	J. P. Mimnagh
	G-BUXD	Maule MXT-7-160	S. Baigent

Reg.	Type	Owner or Operator	Notes
G-BUXI	Steen Skybolt	D. Tucker	
G-BUXJ	Slingsby T.61F Venture T.2	Venture Motor Glider Club/Halton	
G-BUXK	Pietenpol Air Camper	B. P. Hogan	
G-BUXL	Taylor JT.1 Monoplane	M. Fields	
G-BUXN	Beech C23 Sundowner	Private Pilots Syndicate	
G-BUXO	Pober P-9 Pixie	P-9 Flying Group	
G-BUXR	Cameron A-250 balloon	Celebration Balloon Flights	
G-BUXS	MBB Bo 105DBS/4	Bond Air Services (G-PASA/G-BGWP)	
G-BUXV	PA-22 Tri-Pacer 160 (tailwheel)	Romeo Delta Juliet Group	
G-BUXW	Thunder Ax8-90 S2 balloon	Nottingham Hot Air Balloon Club	
G-BUXX	PA-17 Vagabond	R. H. Hunt/Old Sarum	
G-BUXY	PA-25 Pawnee 235	Bath, Wilts & North Dorset Gliding Club Ltd/ Kingston Deverill	
G-BUYB	Aero Designs Pulsar	A. P. Fenn/Shobdon	
G-BUYC	Cameron 80 Concept balloon	R. P. Cross	
G-BUYD	Thunder Ax8-90 balloon	S. & P. McGuigan	
G-BUYF	Falcon XP	G-BUYF Syndicate	
G-BUYJ	Lindstrand LBL-105A balloon	D. K. Fish & G. Fordyce	
G-BUYK	Denney Kitfox Mk 4	M. S. Shelton	
G-BUYL	RAF 2000GT gyroplane	Newtonair Gyroplanes Ltd	
G-BUYN	Cameron O-84 balloon	Reach For The Sky Ltd	
G-BUYO	Colt 77A balloon	S. F. Burden/Netherlands	
G-BUYR	Mooney M.20C	C. R. Weldon/Eire	
G-BUYS	Robin DR.400/180	G-BUYS Flying Group/Nuthampstead	
G-BUYT	Ken Brock KB-2 gyroplane	J. L. G. McLane	
G-BUYU	Bowers Fly-Baby 1A	R. Metcalfe	
G-BUYY	PA-28 Cherokee 180	G-BUYY Group	
G-BUZA	Denney Kitfox Mk 3	J. Thomas	
G-BUZB	Aero Designs Pulsar XP	S. M. Lancashire	
G-BUZC	Everett Srs 3A gyroplane	M. P. L'Hermette	
G-BUZD	AS.332L Super Puma	CHC Scotia Ltd	
G-BUZE	Light Aero Avid Speedwing	H. R. Bell	
G-BUZF	Colt 77B balloon	A. E. Austin	
G-BUZG	Zenair CH.601HD	N. C. White	
G-BUZH	Aero Designs Star-Lite SL-1	C. A. McDowall	
G-BUZJ	Lindstrand LBL-105A balloon	Eastgate Mazda	
G-BUZK	Cameron V-77 balloon	J. T. Wilkinson	
G-BUZL	VPM M.16 Tandem Trainer	Roger Savage (Photography)/Carlisle	
G-BUZM	Light Aero Avid Flyer Mk 3	R. McLuckie & O. G. Jones	
G-BUZN	Cessna 172H	H. Jones/Barton	
G-BUZO	Pietenpol Air Camper	D. A. Jones	
G-BUZR	Lindstrand LBL-77A balloon	Lindstrand Balloons Ltd	
G-BUZS	Colt Flying Pig SS balloon	Banco Bilbao Vizcaya/Spain	
G-BUZT	Kölb Twinstar Mk 3	M. D. Burns & ptnrs	
G-BUZV	Ken Brock KB-2 gyroplane	K. Hughes	
G-BUZY	Cameron A-250 balloon	P. J. D. Kerr	
G-BUZZ	AB-206B JetRanger 2	European Skytime Ltd	
G-BVAA	Light Aero Avid Aerobat Mk 4	D. T. Searchfield	
G-BVAB	Zenair CH.601HDS	N. J. Keeling & ptnrs	
G-BVAC	Zenair CH.601HD	A. G. Cozens	
G-BVAF	Piper J-3C-65 Cub	N. M. Hitchman/France	
G-BVAG	Lindstrand LBL-90A balloon	The Whitchurch Aviation Training Syndicate	
G-BVAH	Denney Kitfox Mk.3	S. Allinson	
G-BVAI	PZL-110 Koliber 150	N. J. & R. F. Morgan	
G-BVAM	Evans VP-1	R. F. Selby	
G-BVAN	M.S.892E Rallye 150	A. J. A. Weal	
G-BVAO	Colt 25A balloon	J. M. Frazer	
G-BVAW	Staaken Z-1 Flitzer (D-692)	L. R. Williams	
G-BVAX	Colt 77A balloon	P. H. Porter	
G-BVAY	Rutan Vari-Eze	D. A. Young	
G-BVAZ	Montgomerie-Bensen B.8MR	N. Steele	
G-BVBD	Sikorsky S-52-3	J. Windmill	
G-BVBF	PA-28-151 Warrior	R. K. Spence/Cardiff	
G-BVBG	PA-32R Cherokee Lance 300	R. K. Spence/Cardiff	
G-BVBP	Avro 683 Lancaster X (KB994)★	D. Copley/Sandtoft	
G-BVBR	Light Aero Avid Speedwing	P. D. Thomas	
G-BVBS	Cameron N-77 balloon	Marley Building Materials Ltd	
G-BVBU	Cameron V-77 balloon	J. Manclark	
G-BVBV	Light Aero Avid Flyer	L. W. M. Summers	
G-BVCA	Cameron N-105 balloon	Unipart Balloon Club	
G-BVCC	Monnett Sonerai 2LT	J. Eggleston	

Notes	Reg.	Type	Owner or Operator
	G-BVCG	Van's RV-6	D. Cook
	G-BVCL	Rans S.6-116 Coyote II	R. J. Powell
	G-BVCM	Cessna 525 CitationJet	Kenmore Aviation Ltd & BLP 2003-19 Ltd
	G-BVCO	FRED Srs 2	I. W. Bremner
	G-BVCP	Piper CP.1 Metisse	C. W. R. Piper
	G-BVCS	Aeronca 7BCM Champion	A. C. Lines
	G-BVCT	Denney Kitfox Mk 4	A. F. Reid
	G-BVCY	Cameron H-24 balloon	A. C. K. Rawson & J. J. Rudoni
	G-BVDB	Thunder Ax7-77 balloon	M. J. Smith & J. Towler (G-ORDY)
	G-BVDC	Van's RV-3	J. A. A. Schofield
	G-BVDD	Colt 69A balloon	R. M. Cambridge
	G-BVDH	PA-28RT-201 Arrow IV	H. M. John
	G-BVDI	Van's RV-4	J. Glen-Davis Gorman
	G-BVDJ	Campbell Cricket (replica)	S. Jennings
	G-BVDM	Cameron 60 Concept balloon	M. P. Young
	G-BVDN	PA-34-220T Seneca III	Convergence Aviation Ltd (G-IGHA/G-IPUT)
	G-BVDO	Lindstrand LBL-105A balloon	A. E. Still
	G-BVDP	Sequoia F.8L Falco	T. G. Painter
	G-BVDR	Cameron O-77 balloon	N. J. Logue
	G-BVDS	Lindstrand LBL-69A balloon	Lindstrand Hot-Air Balloons Ltd
	G-BVDT	CFM Streak Shadow	H. J. Bennet
	G-BVDW	Thunder Ax8-90 balloon	S. C. Vora
	G-BVDX	Cameron V-90 balloon	R. K. Scott
	G-BVDY	Cameron 60 Concept balloon	J. L. Bond
	G-BVDZ	Taylorcraft BC-12D	P. N. W. England
	G-BVEA	Mosler Motors N.3 Pup	N. Lynch (G-MWEA)/Breighton
	G-BVEH	Jodel D.112	M. L. Copland
	G-BVEK	Cameron 80 Concept balloon	A. D. Malcolm
	G-BVEN	Cameron 80 Concept balloon	Hildon Associates
	G-BVEP	Luscombe 8A Master	B. H. Austen
	G-BVER	D.H.C.2 Beaver 1 (XV268)	Seaflite Ltd (G-BTDM)/Duxford
	G-BVES	Cessna 340A	K. P. Gibbin & I. M. Worthington
	G-BVEU	Cameron O-105 balloon	H. C. Wright
	G-BVEV	PA-34-200 Seneca	R. W. Harris & ptnrs
	G-BVEW	Lindstrand LBL-150A balloon	A. Van Wyk
	G-BVEY	Denney Kitfox Mk 4-1200	J. H. H. Turner
	G-BVEZ	P.84 Jet Provost T.3A (XM479)	Newcastle Jet Provost Co Ltd
	G-BVFA	Rans S.10 Sakota	D. S. Wilkinson
	G-BVFB	Cameron N-31 balloon	R. Kunert
	G-BVFF	Cameron V-77 balloon	R. J. Kerr & G. P. Allen
	G-BVFM	Rans S.6-116 Coyote II	G-BVFM Flying Syndicate Ltd
	G-BVFO	Light Aero Avid Speedwing	P. Chisman
	G-BVFP	Cameron V-90 balloon	D. E. & J. M. Hartland
	G-BVFR	CFM Streak Shadow	R. W. Chatterton
	G-BVFT	Maule M5-235C	Newnham Joint Flying Syndicate
	G-BVFU	Cameron 105 Sphere SS balloon	Stichting Phoenix/Netherlands
	G-BVFY	Colt 210A balloon	Cheshire Balloon Flights
	G-BVFZ	Maule M5-180C Lunar Rocket	J. W. Macleod
	G-BVGA	Bell 206B JetRanger 3	Findon Air Services/Shoreham
	G-BVGB	Thunder Ax8-105 S2 balloon	M. E. Dunstan-Sewell
	G-BVGE	W.S.55 Whirlwind HAR.10 (XJ729)	J. F. Kelly
	G-BVGF	Shaw Europa	A. Graham & G. G. Beal
	G-BVGG	Lindstrand LBL-69A balloon	Lindstrand Hot-Air Balloons Ltd
	G-BVGH	Hawker Hunter T.7 (XL573)	Global Aviation Services Ltd/Humberside
	G-BVGI	Pereira Osprey II	A. A. Knight
	G-BVGJ	Cameron C-80 balloon	J. M. J. & V. F. Roberts
	G-BVGO	Denney Kitfox Mk 4-1200	T. Marriott
	G-BVGP	Bucker Bu.133 Jungmeister	M. V. Rijkse
	G-BVGR	RAF BE-2e (A1325)	Aero Vintage Ltd
	G-BVGS	Robinson R-22B	Hields Aviation
	G-BVGT	Auster J/1 (modified)	P. N. Birch
	G-BVGW	Luscombe 8A Silvaire	J. C. Holland
	G-BVGX	Thunder Ax8-90 S2 balloon	G-BVGX Group/New Zealand
	G-BVGY	Luscombe 8E Silvaire	M. C. Burlock
	G-BVGZ	Fokker Dr.1 (replica) (450/17) ★	R. A. Fleming
	G-BVHC	Grob G.115D-2 Heron	VT Aerospace Ltd/Plymouth
	G-BVHD	Grob G.115D-2 Heron	VT Aerospace Ltd/Plymouth
	G-BVHE	Grob G.115D-2 Heron	VT Aerospace Ltd/Plymouth
	G-BVHF	Grob G.115D-2 Heron	VT Aerospace Ltd/Plymouth
	G-BVHG	Grob G.115D-2 Heron	VT Aerospace Ltd/Plymouth
	G-BVHI	Rans S.10 Sakota	P. D. Rowley
	G-BVHJ	Cameron A-180 balloon	S. J. Boxall

Reg.	Type	Owner or Operator	Notes
G-BVHK	Cameron V-77 balloon	A. R. Rich	
G-BVHL	Nicollier HN.700 Menestrel II	W. Goldsmith	
G-BVHM	PA-38-112 Tomahawk	A. J. Gomes (G-DCAN)	
G-BVHO	Cameron V-90 balloon	N. W. B. Bews	
G-BVHP	Colt 42A balloon	Huntair Ltd	
G-BVHR	Cameron V-90 balloon	G. P. Walton	
G-BVHS	Murphy Rebel	J. R. Malpass	
G-BVHT	Light Aero Avid Speedwing Mk 4	R. S. Holt	
G-BVHV	Cameron N-105 balloon	Wye Valley Aviation Ltd	
G-BVIA	Rand-Robinson KR-2	K. Atkinson	
G-BVIC	EE Canberra B.6 (XH568)	Stored/Bruntingthorpe	
G-BVIE	PA-18 Super Cub 95 (modified)	J. C. Best (G-CLIK/G-BLMB)	
G-BVIF	Montgomerie-Bensen B.8MR	R. M. & D. Mann	
G-BVIK	Maule MXT-7-180 Star Rocket	Graveley Flying Group	
G-BVIL	Maule MXT-7-180 Star Rocket	K. & S. C. Knight	
G-BVIN	Rans S.6-ESA Coyote II	T. J. Wilkinson	
G-BVIR	Lindstrand LBL-69A balloon	Aerial Promotions Ltd	
G-BVIS	Brügger MB.2 Colibri	B. H. Shaw	
G-BVIT	Campbell Cricket	D. R. Owen	
G-BVIV	Light Aero Avid Speedwing	M. Burton	
G-BVIW	PA-18-Super Cub 150	M. J. Medland	
G-BVIX	Lindstrand LBL-180A balloon	European Balloon Display Co Ltd	
G-BVIZ	Shaw Europa	T. J. Punter & P. G. Jeffers	
G-BVJC	Fokker 100	bmi regional	
G-BVJD	Fokker 100	bmi regional	
G-BVJE	AS.350B-1 Ecureuil	PLM Dollar Group Ltd	
G-BVJF	Montgomerie-Bensen B.8MR	D. M. F. Harvey	
G-BVJG	Cyclone AX3/K	T. D. Reid (G-MYOP)	
G-BVJH	Aero Designs Pulsar	J. Stringer	
G-BVJK	Glaser-Dirks DG.800A	B. A. Eastwell & J. S. Forster	
G-BVJN	Shaw Europa	JN Europa Group	
G-BVJT	Cessna F.406	Nor Leasing	
G-BVJU	Evans VP-1	BVJU Flying Club & Associates	
G-BVJX	Marquart MA.5 Charger	E. Newsham	
G-BVJZ	PA-28-161 Warrior II	A. R. Fowkes/Denham	
G-BVKB	Boeing 737-59D	bmi Baby	
G-BVKD	Boeing 737-59D	bmi Baby	
G-BVKF	Shaw Europa	T. R. Sinclair	
G-BVKG	Colt Flying Hot Dog SS balloon	Longbreak Ltd/USA	
G-BVKH	Thunder Ax8-90 balloon	L. Ashill	
G-BVKJ	Bensen B.8	A. G. Foster	
G-BVKK	Slingsby T.61F Venture T.2	Buckminster Gliding Club Ltd	
G-BVKL	Cameron A-180 balloon	Dragon Balloon Co. Ltd	
G-BVKM	Rutan Vari-Eze	J. P. G. Lindquist/Switzerland	
G-BVKR	Sikorsky S-76A	Bristow Helicopters Ltd	
G-BVKU	Slingsby T.61F Venture T.2	G-BVKU Syndicate	
G-BVKX	Colt 14A balloon	H. C. J. Williams	
G-BVKZ	Thunder Ax9-120 balloon	D. J. Head	
G-BVLD	Campbell Cricket (replica)	C. Berry	
G-BVLE	McCandless M.4 gyroplane	H. Walls	
G-BVLF	CFM Starstreak Shadow SS-D	Skyview Systems Ltd	
G-BVLG	AS.355F-1 Twin Squirrel	PLM Dollar Group PLC	
G-BVLH	Shaw Europa	D. Barraclough	
G-BVLI	Cameron V-77 balloon	J. Lewis-Richardson/New Zealand	
G-BVLL	Lindstrand LBL-210A balloon	A. G. E. Faulkner	
G-BVLP	PA-38-112 Tomahawk	Turweston Aero Club	
G-BVLR	Van's RV-4	RV4 Group	
G-BVLT	Bellanca 7GCBC Citabria	M. D. Hinge	
G-BVLU	D.31 Turbulent	C. D. Bancroft	
G-BVLV	Shaw Europa	Euro 39 Group	
G-BVLW	Light Aero Avid Flyer Mk 4	D. M. Johnstone/Shobdon	
G-BVLX	Slingsby T.61F Venture T.2	RAFGSA/Easterton	
G-BVLZ	Lindstrand LBL-120A balloon	Balloon Flights Club Ltd	
G-BVMA	Beech 200 Super King Air	Dragonfly Aviation Services LLP (G-VPLC)	
G-BVMC	Robinson R-44 Astro	Bell Commercials	
G-BVMD	Luscombe 8E Silvaire	P. J. Kirkpatrickt	
G-BVMF	Cameron V-77 balloon	P. A. Meecham	
G-BVMH	WAG-Aero Sport Trainer (39624)	J. Mathews	
G-BVMI	PA-18 Super Cub 150	S. Sampson	
G-BVMJ	Cameron 95 Eagle SS balloon	R. D. Sargeant	
G-BVML	Lindstrand LBL-210A balloon	Ballooning Adventures Ltd	
G-BVMM	Robin HR.200/100	R. H. Ashforth	

Notes	Reg.	Type	Owner or Operator
	G-BVMN	Ken Brock KB-2 gyroplane	S. A. Scally
	G-BVMP	Bae 146-200	Trident Jet (Jersey) Ltd
	G-BVMR	Cameron V-90 balloon	I. R. Comley
	G-BVMS	Bae 146-200	Trident Jet (Jersey) Ltd
	G-BVMU	Yakovlev Yak-52 (09)	A. L. Hall-Carpenter
	G-BVNG	D.H.60G-III Moth Major	P. & G. Groves
	G-BVNI	Taylor JT-2 Titch	T. V. Adamson/Rufforth
	G-BVNL	R. Commander 114	A. W. Scragg
	G-BVNR	Cameron N-105 balloon	Liquigas SPA/Italy
	G-BVNS	PA-28-181 Archer II	Scottish Airways Flyers (Prestwick) Ltd
	G-BVNU	FLS Aerospace Sprint Club	Aces High Ltd/North Weald
	G-BVNY	Rans S.7 Courier	D. M. Byers-Jones
	G-BVOA	PA-28-181 Archer II	Millen Aviation Services
	G-BVOB	F.27 Friendship Mk 500	BAC Express Airlines Ltd
	G-BVOC	Cameron V-90 balloon	S. A. Masey
	G-BVOH	Campbell Cricket (replica)	A. Kitson
	G-BVOI	Rans S.6-116 Coyote II	D. J. Chater
	G-BVOK	Yakovlev Yak-52 (55)	Transair Aviation/Shoreham
	G-BVON	Lindstrand LBL-105A balloon	N. D. Hicks
	G-BVOO	Lindstrand LBL-105A balloon	T. G. Church
	G-BVOP	Cameron N-90 balloon	October Gold Ballooning Ltd
	G-BVOR	CFM Streak Shadow	K. Fowler
	G-BVOS	Shaw Europa	Durham Europa Group
	G-BVOU	H.S.748 Srs 2A	Emerald Airways Ltd/Liverpool
	G-BVOV	H.S.748 Srs 2A	Emerald Airways Ltd/Liverpool
	G-BVOW	Shaw Europa	N. M. Robbins
	G-BVOX	Taylorcraft F-22	Jones Samuel Ltd
	G-BVOY	Rotorway Executive 90	Southern Helicopters Ltd
	G-BVOZ	Colt 56A balloon	British School of Ballooning
	G-BVPA	Thunder Ax8-105 S2 balloon	Firefly Balloon Promotions
	G-BVPD	C.A.S.A. 1.131E Jungmann 2000	D. Bruton
	G-BVPK	Cameron O-90 balloon	D. V. Fowler
	G-BVPL	Zenair CH.601HD	J. G. Munro
	G-BVPM	Evans VP-2 Coupé	P. Marigold
	G-BVPN	Piper J-3C-65 Cub	K. I. Munro (G-TAFY)
	G-BVPP	Folland Gnat T.1 (XR993)	Red Gnat Ltd/North Weald
	G-BVPR	Robinson R-22B	Helicentre Blackpool Ltd (G-KNIT)
	G-BVPS	Jodel D.112	P. J. Sharp
	G-BVPV	Lindstrand LBL-77B balloon	A. R. Greensides
	G-BVPW	Rans S.6-116 Coyote II	J. G. Beesley
	G-BVPX	Bensen B.8 (modified) Tyro Gyro	A. W. Harvey
	G-BVPY	CFM Streak Shadow	R. J. Mitchell
	G-BVRA	Shaw Europa	N. E. Stokes
	G-BVRH	Taylorcraft BL.65	M. J. Smith
	G-BVRI	Thunder Ax6-56 balloon	P. C. Bailey
	G-BVRK	Rans S.6-ESA Coyote II	J. Secular (G-MYPK)
	G-BVRL	Lindstrand LBL-21A balloon	Exclusive Ballooning
	G-BVRR	Lindstrand LBL-77A balloon	M. Icam/France
	G-BVRU	Lindstrand LBL-105A balloon	R. P. Nash
	G-BVRV	Van's RV-4	A. Troughton
	G-BVRZ	PA-18 Super Cub 95	R. W. Davison
	G-BVSB	Team Minimax	D. G. Palmer
	G-BVSD	SE.3130 Alouette II (V-54)	M. J. Cuttell
	G-BVSF	Aero Designs Pulsar	S. N. & R. J. Freestone
	G-BVSN	Light Aero Avid Speedwing	S. J. Pemberton
	G-BVSO	Cameron A-120 balloon	Khos Ballooning
	G-BVSP	P.84 Jet Provost T.3A (XM370)	H. G. Hodges & Son Ltd
	G-BVSS	Jodel D.150	A. P. Burns
	G-BVST	Jodel D.150	A. Shipp/Breighton
	G-BVSX	Team Minimax 91	J. A. Clark
	G-BVSY	Thunder Ax9-120 balloon	G. R. Elson/Spain
	G-BVSZ	Pitts S-1E (S) Special	R. C. F. Bailey
	G-BVTA	Tri-R Kis	P. J. Webb
	G-BVTC	P.84 Jet Provost T.5A (XW333)	Global Aviation Ltd/Binbrook
	G-BVTD	CFM Streak Shadow	M. Walton
	G-BVTL	Colt 31A balloon	A. Lindsay
	G-BVTM	Cessna F.152 II	RAF Halton Aeroplane Club (G-WACS)
	G-BVTN	Cameron N-90 balloon	P. Zulehner/Austria
	G-BVTO	PA-28-151 Warrior	Falcon Flying Services (G-SEWL)/Biggin Hill
	G-BVTV	Rotorway Executive 90	M. J. Wasley
	G-BVTW	Aero Designs Pulsar	J. D. Webb
	G-BVTX	D.H.C.1 Chipmunk 22A (WP809)	TX Flying Group

Reg.	Type	Owner or Operator	Notes
G-BVUA	Cameron O-105 balloon	D. C. Eager	
G-BVUC	Colt 56A balloon	Thunder & Colt	
G-BVUG	Betts TB.1 (Stampe SV-4C)	William Tomkins Ltd (G-BEUS)	
G-BVUH	Thunder Ax6-65B balloon	Zebedee Balloon Service Ltd	
G-BVUI	Lindstrand LBL-25A balloon	J. W. Hole	
G-BVUJ	Ken Brock KB-2 gyroplane	R. J. Hutchinson	
G-BVUK	Cameron V-77 balloon	H. G. Griffiths & W. A. Steel	
G-BVUM	Rans S.6-116 Coyote II	M. A. Abbott	
G-BVUN	Van's RV-4	A. E. Kay	
G-BVUO	Cameron R-150 balloon	M. Sevrin/Belgium	
G-BVUU	Cameron C-80 balloon	T. M. C. McCoy	
G-BVUV	Shaw Europa	R. J. Mills	
G-BVUZ	Cessna 120	N. O. Anderson	
G-BVVA	Yakovlev Yak-52	T. W. Freeman/Little Gransden	
G-BVVB	Carlson Sparrow Mk II	L. M. McCullen	
G-BVVE	Jodel D.112	G. D. Gunby	
G-BVVG	Nanchang CJ-6A	Peeking Duck Group	
G-BVVH	Shaw Europa	T. G. Hoult	
G-BVVI	Hawker Audax I (K5600)	Aero Vintage Ltd	
G-BVVK	D.H.C.6 Twin Otter 310	Loganair Ltd/BA Express	
G-BVVL	EAA Acro Sport II	G. A. Breen/Portugal	
G-BVVM	Zenair CH.601HD	D. Macdonald	
G-BVVN	Brügger MB.2 Colibri	T. C. Darters	
G-BVVP	Shaw Europa	I. Mansfield	
G-BVVR	Stits SA-3A Playboy	A. D. Pearce	
G-BVVS	Van's RV-4	E. G. & N. S. C. English	
G-BVVU	Lindstrand LBL Four SS balloon	Magical Adventures Ltd	
G-BVVW	Yakovlev Yak-52	J. E. Blackman	
G-BVVZ	Corby CJ-1 Starlet	P. V. Flack	
G-BVWB	Thunder Ax8-90 S2 balloon	M. A. Stelling & K. C. Tanner	
G-BVWC	EE Canberra B.6 (WK163)	Classic Aviation Projects Ltd	
G-BVWI	Cameron 65 Light Bulb SS balloon	Balloon Preservation Flying Group	
G-BVWM	Shaw Europa	Europa Syndicate	
G-BVWP	D.H.C.1 Chipmunk 22 (WP856)	T. W. M. Beck	
G-BVWW	Lindstrand LBL-90A balloon	Drawflight Ltd	
G-BVWX	VPM M.16 Tandem Trainer	M. L. Smith	
G-BVWY	Porterfield CP.65	B. Morris	
G-BVWZ	PA-32-301 Saratoga	The Saratoga (WZ) Group	
G-BVXA	Cameron N-105 balloon	R. E. Jones	
G-BVXB	Cameron V-77 balloon	J. A. Lawton	
G-BVXC	EE Canberra B.6 (WT333)	Classic Aviation Projects Ltd/Bruntingthorpe	
G-BVXD	Cameron O-84 balloon	N. J. Langley	
G-BVXE	Steen Skybolt	J. Buglass (G-LISA)	
G-BVXF	Cameron O-120 balloon	Off The Ground Balloon Co Ltd	
G-BVXJ	C.A.S.A. 1.133 Jungmeister	A. C. Mercer	
G-BVXK	Yakovlev Yak-52 (26)	E. Gavazzi	
G-BVXM	AS.350B Ecureuil	The Berkeley Leisure Group Ltd	
G-BVXR	D.H.104 Devon C.2 (XA880)	M. Whale & M. W. A. Lunn	
G-BVXS	Taylorcraft BC-12D	D. Riley	
G-BVXW	SC.7 Skyvan Srs 3A Variant 100	Babcock Defence Services	
G-BVYF	PA-31-350 Navajo Chieftain	J. A. Rees & ptnrs (G-SAVE)	
G-BVYG	CEA DR.300/180	Ulster Gliding Club Ltd	
G-BVYK	Team Minimax	M. W. Hanley	
G-BVYM	CEA DR. 300/180	London Gliding Club (Pty) Ltd/Dunstable	
G-BVYO	Robin R.2160	G. Johnson	
G-BVYP	PA-25 Pawnee 235B	Bidford Airfield Ltd	
G-BVYR	Cameron A-250 balloon	A. Van Wyk	
G-BVYU	Cameron A-140 balloon	Balloon Flights Club Ltd	
G-BVYX	Light Aero Avid Speedwing Mk 4	M. E. Lloyd	
G-BVYY	Pietenpol Air Camper	Pietenpol G-BVYY Group	
G-BVYZ	Stemme S.10V	L. Gubbay	
G-BVZD	Tri-R Kis Cruiser	D. R. Morgan	
G-BVZE	Boeing 737-59D	bmi Baby	
G-BVZG	Boeing 737-5Q8	bmi Baby	
G-BVZH	Boeing 737-5Q8	bmi Baby	
G-BVZI	Boeing 737-5Q8	bmi Baby	
G-BVZJ	Rand-Robinson KR-2	J. P. McConnell-Wood	
G-BVZN	Cameron C-80 balloon	Sky Fly Balloons	
G-BVZO	Rans S.6-116 Coyote II	P. Atkinson	
G-BVZR	Zenair CH.601HD	J. D. White/Tollerton	
G-BVZT	Lindstrand LBL-90A balloon	Pork Farms Bowyers	
G-BVZV	Rans S.6-116 Coyote II	A. R. White	

Notes	Reg.	Type	Owner or Operator
	G-BVZX	Cameron H-34 balloon	Chianti Balloon Club/Italy
	G-BVZZ	D.H.C.1 Chipmunk 22 (WP795)	Portsmouth Naval Gliding Club/Lee-on-Solent
	G-BWAA	Cameron N-133 balloon	Bailey Balloons
	G-BWAB	Jodel D.14	W. A. Braim
	G-BWAC	Waco YKS-7	D. N. Peters
	G-BWAD	RAF 2000GT gyroplane	Newtonair Gyroplanes Ltd
	G-BWAF	Hawker Hunter F.6A (XG160) ★	Bournemouth Aviation Museum/Bournemouth
	G-BWAG	Cameron O-120 balloon	P. M. Skinner
	G-BWAH	Montgomerie-Bensen B.8MR	J. B. Allan
	G-BWAI	CFM Streak Shadow	C. M. James
	G-BWAJ	Cameron V-77 balloon	R. S. & S. H. Ham
	G-BWAK	Robinson R-22B	Caudwell Communications Ltd
	G-BWAN	Cameron N-77 balloon	Balloon Preservation Flying Group
	G-BWAO	Cameron C-80 balloon	Virgin Balloon Flights
	G-BWAP	FRED Srs 3	G. A. Shepherd
	G-BWAR	Denney Kitfox Mk 3	I. Wightman
	G-BWAT	Pietenpol Air Camper	P. W. Aitchison
	G-BWAU	Cameron V-90 balloon	K. M. & A. M. F. Hall
	G-BWAV	Schweizer 269C	Helihire
	G-BWAW	Lindstrand LBL-77A balloon	D. Bareford
	G-BWBA	Cameron V-65 balloon	Dante Balloon Group
	G-BWBB	Lindstrand LBL-14A balloon	Oxford Promotions (UK) Ltd
	G-BWBC	Cameron N-90AS balloon	Wetterauer Montgolfieren/Germany
	G-BWBE	Colt Flying Ice Cream Cone SS balloon	Benedikt Haggeney GmbH/Germany
	G-BWBF	Colt Flying Ice Cream Cone SS balloon	Benedikt Haggeney GmbH/Germany
	G-BWBG	Cvjetkovic CA-65 Skyfly	T. White & M. C. Fawkes
	G-BWBI	Taylorcraft F-22A	P. J. Wallace
	G-BWBJ	Colt 21A balloon	U. Schneider/Germany
	G-BWBT	Lindstrand LBL-90A balloon	British Telecommunications PLC
	G-BWBY	Schleicher ASH.26E	J. S. Ward
	G-BWBZ	ARV K1 Super 2	P. I. Lewis (G-BMWG)
	G-BWCA	CFM Streak Shadow	I. C. Pearson
	G-BWCC	Van Den Bemden Gas balloon	Piccard Balloon Group
	G-BWCG	Lindstrand LBL-42A balloon	Oxford Promotions (UK) Ltd
	G-BWCK	Everett Srs 2 gyroplane	A. C. S. M. Hart
	G-BWCO	Dornier Do.28D-2	Wingglider Ltd/Hibaldstow
	G-BWCS	P.84 Jet Provost T.5 (XW293)	R. E. Todd/Sandtoft
	G-BWCT	Tipsy T.66 Nipper 1	J. S. Hemmings & C. R. Steer
	G-BWCV	Shaw Europa	G. C. McKirdy
	G-BWCW	Barnett J4B rotorcraft	S. H. Kirkby
	G-BWCY	Murphy Rebel	M. Stow
	G-BWDA	Aerospatiale ATR-72-202	Aurigny Air Services Ltd
	G-BWDB	Aerospatiale ATR-72-202	Aurigny Air Services Ltd
	G-BWDE	PA-31P Pressurised Navajo	Tomkat Aviation Ltd (G-HWKN)
	G-BWDF	PZL-104 Wilga 35A	P. G. Marks/Luxembourg
	G-BWDH	Cameron N-105 balloon	Bridges Van Hire Ltd
	G-BWDM	Lindstrand LBL-120A balloon	G. D. & L. Fitzpatrick
	G-BWDO	Sikorsky S-76B	Haughey Air Ltd
	G-BWDP	Shaw Europa	W. Hueltz/Germany
	G-BWDR	P.84 Jet Provost T.3A (XM376)	W. O. Bayazid/North Weald
	G-BWDS	P.84 Jet Provost T.3A (XM424)	Area 51Aviation Services Ltd/North Weald
	G-BWDT	PA-34-220T Seneca II	H. R. Chambers (G-BKHS)/Biggin Hill
	G-BWDU	Cameron V-90 balloon	D. M. Roberts
	G-BWDV	Schweizer 269C	Helicopter One Ltd
	G-BWDX	Shaw Europa	J. B. Crane
	G-BWDZ	Sky 105-24 balloon	Skyride Balloons Ltd
	G-BWEA	Lindstrand LBL-120A balloon	S. R. Seager
	G-BWEB	P.84 Jet Provost T.5A (XW422)	S. Patrick//North Weald
	G-BWEC	Cassutt-Colson Variant	N. R. Thomason & M. P. J. Hill
	G-BWEE	Cameron V-42 balloon	J. A. Hibberd
	G-BWEF	SNCAN Stampe SV-4C	Acebell BWEF Syndicate (G-BOVL)
	G-BWEG	Shaw Europa	Wessex Europa Group
	G-BWEH	HOAC Katana DV.20	Lowlog Ltd/Elstree
	G-BWEM	V.S.358 Seafire L.IIIC (RX168)	C. J. Warrilow & S. W. Atkins
	G-BWEN	Macair Merlin GT	D. A. Hill
	G-BWEO	Lindstrand AM400 balloon	Lindstrand Balloons Ltd
	G-BWER	Lindstrand AM400 balloon	Lindstrand Balloons Ltd
	G-BWEU	Cessna F.152 II	Affair Aircraft Leasing LLP
	G-BWEV	Cessna 152 II	Haimoss Ltd
	G-BWEW	Cameron N-105 balloon	Unipart Balloon Club
	G-BWEY	Bensen B.8	F. G. Shepherd

Reg.	Type	Owner or Operator	Notes
G-BWEZ	Piper J-3C-65 Cub (436021)	PJ L-4 Group	
G-BWFG	Robin HR.200/120B	Atlantic Air Transport Ltd/Coventry	
G-BWFH	Shaw Europa	B. L. Wratten	
G-BWFI	HOAC Katana DV.20	Air Aqua Ltd	
G-BWFJ	Evans VP-1	P. A. West	
G-BWFK	Lindstrand LBL-77A balloon	Balloon Preservation Flying Group	
G-BWFM	Yakovlev Yak-50	Classic Aviation Ltd/Duxford	
G-BWFN	Hapi Cygnet SF-2A	G-BWFN Group	
G-BWFO	Colomban MC.15 Cri-Cri	O. G. Jones	
G-BWFP	Yakovlev Yak-52	M. C. Lee/Manchester	
G-BWFR	Hawker Hunter F.58 (J-4031)	The Old Flying Machine Co. Ltd/Scampton	
G-BWFT	Hawker Hunter T.8M (XL602)	T8M Group	
G-BWFV	HOAC Katana DV.20	Walsh Aviation/Cranfield	
G-BWFX	Shaw Europa	A. D. Stewart	
G-BWFY	AS.350B-1 Ecureuil	PLM Dollar Group Ltd/Inverness	
G-BWFZ	Murphy Rebel	S. Beresford (G-SAVS)	
G-BWGA	Lindstrand LBL-105A balloon	I. Chadwick	
G-BWGF	P.84 Jet Provost T.5A (XW325)	Specialscope Jet Provost Group/ Woodford	
G-BWGG	MH.1521C-1 Broussard	M. J. Burnett & R. B. Maalouf/France	
G-BWGJ	Chilton DW.1A	T. J. Harrison	
G-BWGK	Hawker Hunter GA.11 (XE689)	B. R. Pearson & GA11 Group/Exeter	
G-BWGL	Hawker Hunter T.8C (XJ615)	Elvington Events Ltd	
G-BWGM	Hawker Hunter T.8C (XE665)	The Admirals Barge/Exeter	
G-BWGN	Hawker Hunter T.8C (WT722)	T8C Group	
G-BWGO	Slingsby T.67M Firefly 200	R. Gray	
G-BWGP	Cameron C-80 balloon	D. J. Groombridge	
G-BWGR	NA TB-25N Mitchell (151632) ★	D. Copley/Sandtoft	
G-BWGS	BAC.145 Jet Provost T.5A	G-BWGS Ltd/North Weald_ (XW310)	
G-BWGT	P.84 Jet Provost T.4	R. E. Todd/Sandtoft	
G-BWGU	Cessna 150F	Goodair Leasing Ltd	
G-BWGX	Cameron N-42 balloon	Newbury Building Soc.	
G-BWGY	HOAC Katana DV.20	Stars Fly Ltd/Elstree	
G-BWGZ	HOAC Katana DV.20	Plane Talking Ltd/Cranfield	
G-BWHA	Hawker Hurricane IIB (Z5252)	Historic Flying Ltd/Duxford	
G-BWHB	Cameron O-65 balloon	G. Aimo/Italy	
G-BWHC	Cameron N-77 balloon	Travelsphere Ltd	
G-BWHD	Lindstrand LBL-31A balloon	Army Air Corps Balloon Club	
G-BWHF	PA-31-325 Navajo	Awyr Cymru Cyf/Welshpool	
G-BWHG	Cameron N-65 balloon	M. Stefanini & F. B. Alaou	
G-BWHI	D.H.C.1 Chipmunk 22A (WK624)	N. E. M. Clare	
G-BWHK	Rans S.6-116 Coyote II	D. A. Buttress	
G-BWHM	Sky 140-24 balloon	C. J. S. Limon	
G-BWHP	C.A.S.A. 1.131E Jungmann (S4+A07)	J. F. Hopkins	
G-BWHR	Tipsy Nipper T.66 Srs 1	L. R. Marnef	
G-BWHS	RAF 2000 gyroplane	J. M. Cox	
G-BWHU	Westland Scout AH.1 (XR595)	N. J. F. Boston	
G-BWHV	Denney Kitfox Mk 2	A. C. Dove	
G-BWHW	Cameron A-180 balloon	Societe Bombard SARL/France	
G-BWHY	Robinson R-22	S. Warren	
G-BWIA	Rans S.10 Sakota	J. B. Lawton	
G-BWIB	SA Bulldog Srs 120/122	B. I. Robertson	
G-BWID	D.31 Turbulent	A. M. Turney	
G-BWII	Cessna 150G	J. D. G. Hicks (G-BSKB)	
G-BWIJ	Shaw Europa	R. Lloyd	
G-BWIK	D.H.82A Tiger Moth (NL985)	B. J. Ellis	
G-BWIL	Rans S.10 Sakota	K. P. Rusling (G-WIEN)	
G-BWIP	Cameron N-90 balloon	S. H. Fell	
G-BWIR	Dornier Do.328-100	ScotAirways Ltd	
G-BWIV	Shaw Europa	T. G. Ledbury	
G-BWIW	Sky 180-24 balloon	J. A. Cooper	
G-BWIX	Sky 120-24 balloon	J. M. Percival	
G-BWJB	Thunder Ax8-105 balloon	Justerini & Brooks Ltd	
G-BWJG	Mooney M.20J	Samic Ltd	
G-BWJH	Shaw Europa	T. P. Cripps	
G-BWJI	Cameron V-90 balloon	Calarel Developments Ltd	
G-BWJM	Bristol M.1C (replica) (C4918)	The Shuttleworth Collection/O. Warden	
G-BWJN	Montgomerie-Bensen B.8	M. Johnston	
G-BWJP	Cessna 172C	T. W. R. Case	
G-BWJR	Sky 120-24 balloon	W. J. Brogan	
G-BWJT	Yakovlev Yak-50 (01385)	R. J. Luke	
G-BWJW	Westland Scout AH.Mk.1 (XV130)	S. Dadak & G. Sobell	
G-BWJY	D.H.C.1 Chipmunk 22 (WG469)	K. J. Thompson	

Notes	Reg.	Type	Owner or Operator
	G-BWKD	Cameron O-120 balloon	L. J. & M. Schoeman
	G-BWKE	Cameron AS-105GD airship	W. Arnold/Germany
	G-BWKF	Cameron N-105 balloon	R. M. M. Botti/Italy
	G-BWKG	Shaw Europa	E. H. Keppert/Austria
	G-BWKK	Auster A.O.P.9 (XP279)	C. A. Davis & D. R. White
	G-BWKR	Sky 90-24 balloon	B. Drawbridge
	G-BWKT	Stephens Akro Laser	P. D. Begley
	G-BWKU	Cameron A-250 balloon	British School of Ballooning
	G-BWKV	Cameron V-77 balloon	Poppies (UK) Ltd
	G-BWKW	Thunder Ax8-90 balloon	Venice Simplon Orient Express Ltd
	G-BWKX	Cameron A-250 balloon	Hot Airlines/Thailand
	G-BWKZ	Lindstrand LBL-77A balloon	J. H. Dobson
	G-BWLA	Lindstrand LBL-69A balloon	Balloon Preservation Flying Group
	G-BWLD	Cameron O-120 balloon	D. Pedri & ptnrs/Italy
	G-BWLF	Cessna 404	Nor Leasing (G-BNXS)
	G-BWLJ	Taylorcraft DCO-65	C. Evans
	G-BWLL	Murphy Rebel	F. W. Parker
	G-BWLM	Sky 65-24 balloon	W. J. Brogan
	G-BWLN	Cameron O-84 balloon	Reggiana Riduttori SRL/Italy
	G-BWLP	HOAC Katana DV.20	Diamond Aircraft Ltd
	G-BWLR	MH.1521M Broussard (185)	Chicory Crops Ltd
	G-BWLS	HOAC Katana DV.20-100	Shadow Aviation/Elstree
	G-BWLT	HOAC Katana DV.20	Diamond Aircraft Ltd
	G-BWLW	Light Aero Avid Speedwing Mk4	P. C. & S. A. Creswick
	G-BWLY	Rotorway Executive 90	P. W. & I. P. Bewley
	G-BWLZ	Wombat gyroplane	M. R. Harrisson
	G-BWMA	Colt 105A balloon	L. Lacroix/France
	G-BWMB	Jodel D.119	C. Hughes
	G-BWMC	Cessna 182P	Eggesford Eagles Flying Group
	G-BWMF	Gloster Meteor T.7 (WA591)	Meteor Flight (Yatesbury)
	G-BWMH	Lindstrand LBL-77B balloon	J. W. Hole
	G-BWMI	PA-28RT-201T Turbo Arrow IV	O. Cowley
	G-BWMJ	Nieuport 17/2B (replica) (B3459)	R. Gauld-Galliers & L. J. Day
	G-BWMK	D.H.82A Tiger Moth (T8191)	APB Leasing Ltd/Welshpool
	G-BWML	Cameron A-275 balloon	A. J. Street
	G-BWMN	Rans S.7 Courier	G. J. Knee
	G-BWMO	Oldfield Baby Lakes	A. G. Fowles (G-CIII)
	G-BWMS	D.H.82A Tiger Moth	Foundation Early Birds/Netherlands
	G-BWMU	Cameron 105 Monster Truck SS balloon	Magical Adventures Ltd/Canada
	G-BWMV	Colt AS-105 Mk II airship	D. Stuber/Germany
	G-BWMX	D.H.C.1 Chipmunk 22 (WG407)	407th Flying Group
	G-BWMY	Cameron Bradford & Bingley SS balloon	Magical Adventures Ltd/USA
	G-BWNB	Cessna 152 II	Galair International Ltd
	G-BWNC	Cessna 152 II	Galair International Ltd
	G-BWND	Cessna 152 II	Galair International Ltd
	G-BWNI	PA-24 Comanche 180	W. A. Stewart
	G-BWNJ	Hughes 269C	L. R. Fenwick
	G-BWNK	D,H,C,1 Chipmunk 22 (WD390)	B. Whitworth
	G-BWNM	PA-28R Cherokee Arrow 180	M. & R. C. Ramnial.
	G-BWNO	Cameron O-90 balloon	T. Knight
	G-BWNP	Cameron 90 Club SS balloon	C. J. Davies & P. Spellward
	G-BWNS	Cameron O-90 balloon	Smithair Ltd
	G-BWNT	D.H.C.1 Chipmunk 22 (WP901)	Three Point Aviation
	G-BWNU	PA-38-112 Tomahawk	Kemble Aero Club Ltd
	G-BWNY	Aeromot AMT-200 Super Ximango	A. E. Mayhew
	G-BWNZ	Agusta A.109C	Anglo Beef Processors Ltd
	G-BWOA	Sky 105-24 balloon	Akhter Group PLC
	G-BWOB	Luscombe 8F Silvaire	P. J. Tanulak & H. T. Law
	G-BWOD	Yakovlev Yak-52 (139)	Insurefast Ltd/Sywell
	G-BWOE	Yakovlev Yak-3U	Classic Aviation/Duxford (G-BUXZ)
	G-BWOF	P.84 Jet Provost T.5	Techair London Ltd
	G-BWOI	PA-28-161 Cadet	Plane Talking Ltd
	G-BWOJ	PA-28-161 Cadet	Plane Talking Ltd
	G-BWOK	Lindstrand LBL 105G balloon	Lindstrand Balloons Ltd
	G-BWON	Shaw Europa	H. J. Fish
	G-BWOR	PA-18 Super Cub 135	C. D. Baird
	G-BWOT	P.84 Jet Provost T.3A (XN459)	Red Pelicans Ltd/North Weald
	G-BWOU	Hawker Hunter F.58A (XF303)	Old Flying Machine Co Ltd/Scampton
	G-BWOV	Enstrom F-28A	P. A. Goss
	G-BWOW	Cameron N-105 balloon	Skybus Ballooning
	G-BWOX	D.H.C.1 Chipmunk 22 (WP844)	J. St. Clair-Quentin
	G-BWOY	Sky 31-24 balloon	C. Wolstenholme

Reg.	Type	Owner or Operator	Notes
G-BWOZ	CFM Streak Shadow SA	N. P. Harding	
G-BWPB	Cameron V-77 balloon	Fair Weather Friends Ballooning Co	
G-BWPC	Cameron V-77 balloon	H. Vaughan	
G-BWPE	Murphy Renegade Spirit UK	J. Hatswell/France	
G-BWPF	Sky 120-24 balloon	Zebedee Balloon Service Ltd	
G-BWPH	PA-28-181 Archer II	E. & H. Merkado	
G-BWPJ	Steen Skybolt	W. R. Penaluna	
G-BWPP	Sky 105-24 balloon	The Sarnia Balloon Group	
G-BWPS	CFM Streak Shadow SA	P. M. E. D. McNair-Wilson	
G-BWPT	Cameron N-90 balloon	G.Everett	
G-BWPV	BN-2T Defender 4000	BN Group Ltd/Bembridge	
G-BWPX	BN-2T-4S Defender 4000	Britten-Norman Aircraft Ltd	
G-BWPY	HOAC Katana DV.20	SAS Flight Services	
G-BWPZ	Cameron N-105 balloon	D. M. Moffat	
G-BWRA	Sopwith LC-1T Triplane (replica) (N500)	S. M. Truscott & J. M. Hoblyn (G-PENY)	
G-BWRC	Light Aero Avid Speedwing	B. Williams	
G-BWRM	Colt 105A balloon	N. Charbonnier/Italy	
G-BWRO	Shaw Europa	J. G. M. McDiarmid	
G-BWRP	Beech 58 Baron	Astra Aviation Ltd	
G-BWRR	Cessna 182Q	D. O. Halle	
G-BWRS	SNCAN Stampe SV-4C	G. P. J. M. Valvekens/Belgium	
G-BWRT	Cameron 60 Concept balloon	W. R. Teasdale	
G-BWRV	Lindstrand LBL-90A balloon	Flying Pictures Ltd Audi	
G-BWRY	Cameron N-105 balloon	G. Aimo/Italy	
G-BWRZ	Lindstrand LBL-105A balloon	D. J. Palmer	
G-BWSB	Lindstrand LBL-105A balloon	R. Calvert-Fisher	
G-BWSC	PA-38-112 Tomahawk II	APB Leasing Ltd/Welshpool	
G-BWSD	Campbell Cricket	R. F. G. Moyle	
G-BWSG	P.84 Jet Provost T.5 (XW324/K)	R. M. Kay	
G-BWSH	P.84 Jet Provost T.3A (XN498)	Global Aviation Ltd/Binbrook	
G-BWSI	K & S SA.102.5 Cavalier	B. W. Shaw	
G-BWSJ	Denney Kitfox Mk 3	J. M. Miller	
G-BWSL	Sky 77-24 balloon	D. Baggley	
G-BWSN	Denney Kitfox Mk 3	W. J. Forrest	
G-BWSO	Cameron 90 Apple SS balloon	Flying Pictures Ltd	
G-BWSP	Cameron 80 Carrots SS balloon	Flying Pictures Ltd	
G-BWST	Sky 200-24 balloon	S. A. Townley	
G-BWSU	Cameron N-105 balloon	A. M. Marten	
G-BWSV	Yakovlev Yak-52	M. W. Fitch	
G-BWSZ	Montgomerie-Bensen B.8MR	D. Cawkwell	
G-BWTA	HOAC Katana DV.20	Lowlog Ltd/Cranfield	
G-BWTB	Lindstrand LBL-105A balloon	Servatruc Ltd	
G-BWTC	Zlin Z.242L	Oxford Aviation Services Ltd/Kidlington	
G-BWTD	Zlin Z.242L	Oxford Aviation Services Ltd/Kidlington	
G-BWTE	Cameron O-140 balloon	R. J. & A. J. Mansfield	
G-BWTF	Lindstrand LBL Bear SS balloon	Free Enterprise Balloons Ltd/USA	
G-BWTG	D.H.C.1 Chipmunk 22 (WB671)	Chipmunk 4 Ever/Netherlands	
G-BWTH	Robinson R-22B	Helicopter Services	
G-BWTJ	Cameron V-77 balloon	A. J. Montgomery	
G-BWTK	RAF 2000 GTX-SE gyroplane	M. Love	
G-BWTN	Lindstrand LBL-90A balloon	Clarks Drainage Ltd	
G-BWTO	D.H.C.1 Chipmunk 22 (WP984)	Skycraft Services Ltd	
G-BWTR	Slingsby T.61F Venture T.2	P. R. Williams	
G-BWTW	Mooney M.20C	R. C. Volkers	
G-BWUA	Campbell Cricket	N. J. Orchard	
G-BWUB	PA-18S Super Cub 135	Caledonian Seaplanes Ltd/Cumbernauld	
G-BWUE	Hispano HA.1112M1L	R. A. Fleming(G-AWHK)/Breighton	
G-BWUF	WSK-Mielec LiM-5 (1211)	De Havilland Aviation/Bournemouth	
G-BWUH	PA-28-181 Archer III	G-BWUH Flying Group/Cambridge	
G-BWUJ	Rotorway Executive 162F	Southern Helicopters Ltd	
G-BWUK	Sky 160-24 balloon	Cameron Flights Southern Ltd	
G-BWUL	Noorduyn AT-16 Harvard IIB	Aereo Servizi Bresciana SRL/Italy	
G-BWUM	Sky 105-24 balloon	P. Stern & F. Kirchberger/Germany	
G-BWUN	D.H.C.1 Chipmunk 22 (WD310)	T. Henderson	
G-BWUP	Shaw Europa	G. A. Haines	
G-BWUR	Thunder Ax10-210 S2 balloon	T. J. Bucknall	
G-BWUS	Sky 65-24 balloon	N. A. P. Bates	
G-BWUT	D.H.C.1 Chipmunk 22 (WZ879)	Aero Vintage Ltd	
G-BWUU	Cameron N-90 balloon	South Western Electricity PLC	
G-BWUV	D.H.C.1 Chipmunk 22A (WK640)	P. Ray	
G-BWUW	P.84 Jet Provost T.5A (XW423)	Deeside College	
G-BWUZ	Campbell Cricket (replica)	K. A. Touhey	

Notes	Reg.	Type	Owner or Operator
	G-BWVB	Pietenpol Air Camper	D. Platt
	G-BWVC	Jodel D.18	R. W. J. Cripps
	G-BWVH	Robinson R-44 Astro	Derg Developments
	G-BWVI	Stern ST.80	I. M. Godfrey-Davies
	G-BWVL	Cessna 150M	A. H. Shaw
	G-BWVM	Colt AA-1050 balloon	B. B. Baxter Ltd
	G-BWVN	Whittaker MW.7	R. S. Willcox
	G-BWVR	Yakovlev Yak-52 (52)	P. Shaw
	G-BWVS	Shaw Europa	D. R. Bishop
	G-BWVT	D.H.A.82A Tiger Moth	R. Jewitt
	G-BWVU	Cameron O-90 balloon	J. Atkinson
	G-BWVV	Jodel D.18	D. S. Howarth
	G-BWVY	D.H.C.1 Chipmunk 22 (WP896)	P. W. Portelli
	G-BWVZ	D.H.C.1 Chipmunk 22A (WK590)	D. Campion/Belgium
	G-BWWA	Ultravia Pelican Club GS	Pelican Group
	G-BWWB	Shaw Europa	S. M. O'Reilly
	G-BWWC	D.H.104 Dove 7 (XM223)	Air Atlantique Ltd/Coventry
	G-BWWE	Lindstrand LBL-90A balloon	B. J. Newman
	G-BWWF	Cessna 185A	S. M. C. Harvey
	G-BWWG	SOCATA Rallye 235E	J. J. Frew
	G-BWWH	Yakovlev Yak-50	R. S. Partridge-Hicks & I. C. Austin
	G-BWWI	AS.332L Super Puma	Bristow Helicopters Ltd
	G-BWWK	Hawker Nimrod I (S1581)	Patina Ltd/Duxford
	G-BWWL	Colt Flying Egg SS balloon	Magical Adventures Ltd/USA
	G-BWWN	Isaacs Fury II (K8303)	F. J. Ball
	G-BWWP	Rans S.6-116 Coyote II	P. Lewis
	G-BWWS	RAF 2000 GTX-SE gyroplane	A. Gibbon & C. Cruickshank
	G-BWWT	Dornier Do.328-100	ScotAirways Ltd
	G-BWWU	PA-22 Tri-Pacer 150	J. D. Bally
	G-BWWW	BAe Jetstream 3102	British Aerospace PLC/Warton
	G-BWWX	Yakovlev Yak-50	J. L. Pfundt/Netherlands
	G-BWWY	Lindstrand LBL-105A balloon	M. J. Smith
	G-BWWZ	Denney Kitfox Mk 3	A. I. Eskander
	G-BWXA	Slingsby T.67M Firefly 260	Hunting Aviation Ltd/Barkston Heath
	G-BWXB	Slingsby T.67M Firefly 260	Hunting Aviation Ltd/Barkston Heath
	G-BWXC	Slingsby T.67M Firefly 260	Babcock Defence Services
	G-BWXD	Slingsby T.67M Firefly 260	Babcock Defence Services
	G-BWXE	Slingsby T.67M Firefly 260	Hunting Aviation Ltd/Barkston Heath
	G-BWXF	Slingsby T.67M Firefly 260	Babcock Defence Services
	G-BWXG	Slingsby T.67M Firefly 260	Babcock Defence Services
	G-BWXH	Slingsby T.67M Firefly 260	Hunting Aviation Ltd/Barkston Heath
	G-BWXI	Slingsby T.67M Firefly 260	Babcock Defence Services
	G-BWXJ	Slingsby T.67M Firefly 260	Babcock Defence Services
	G-BWXK	Slingsby T.67M Firefly 260	Hunting Aviation Ltd/Newton
	G-BWXL	Slingsby T.67M Firefly 260	Hunting Aviation Ltd/Barkston Heath
	G-BWXM	Slingsby T.67M Firefly 260	Hunting Aviation Ltd/Barkston Heath
	G-BWXN	Slingsby T.67M Firefly 260	Hunting Aviation Ltd/Barkston Heath
	G-BWXO	Slingsby T.67M Firefly 260	Babcock Defence Services
	G-BWXP	Slingsby T.67M Firefly 260	Hunting Aviation Ltd/Barkston Heath
	G-BWXR	Slingsby T.67M Firefly 260	Babcock Defence Services
	G-BWXS	Slingsby T.67M Firefly 260	Babcock Defence Services
	G-BWXT	Slingsby T.67M Firefly 260	Babcock Defence Services
	G-BWXU	Slingsby T.67M Firefly 260	Hunting Aviation Ltd/Barkston Heath
	G-BWXV	Slingsby T.67M Firefly 260	Hunting Aviation Ltd/Barkston Heath
	G-BWXW	Slingsby T.67M Firefly 260	Hunting Aviation Ltd/Barkston Heath
	G-BWXX	Slingsby T.67M Firefly 260	Hunting Aviation Ltd/Barkston Heath
	G-BWXY	Slingsby T.67M Firefly 260	Babcock Defence Services
	G-BWXZ	Slingsby T.67M Firefly 260	Hunting Aviation Ltd/Barkston Heath
	G-BWYB	PA-28 Cherokee 160	I. M. Latiff
	G-BWYD	Shaw Europa	F. H. Mycroft
	G-BWYE	Cessna 310R II	Air Charter Scotland Ltd/Edinburgh
	G-BWYG	Cessna 310R II	Kissair Aviation
	G-BWYH	Cessna 310R II	Air Charter Scotland Ltd/Edinburgh
	G-BWYI	Denney Kitfox Mk3	J. Adamson
	G-BWYK	Yakovlev Yak-50	Foley Farm Flying Group
	G-BWYM	HOAC Katana DV.20	Plane Talking Ltd/Elstree
	G-BWYN	Cameron O-77 balloon	W. H. Morgan (G-ODER)
	G-BWYO	Sequoia F.8L Falco	N. G. Abbott & J. Copeland
	G-BWYP	Sky 56-24 balloon	Sky High Leisure
	G-BWYR	Rans S.6-116 Coyote II	T. J. Bax
	G-BWYS	Cameron O-120 balloon	Aire Valley Balloons
	G-BWYU	Sky 120-24 balloon	D. J. Tofton

Reg.	Type	Owner or Operator	Notes
G-BWZA	Shaw Europa	T. G. Cowlishaw	
G-BWZG	Robin R.2160	Sherburn Aero Club Ltd	
G-BWZI	Agusta A. 109A-II	Pendley Farm	
G-BWZJ	Cameron A-250 balloon	Balloon Club of Great Britain	
G-BWZK	Cameron A-210 balloon	Balloon Club of Great Britain	
G-BWZT	Shaw Europa	G-BWZT Group	
G-BWZU	Lindstrand LBL-90B balloon	K. D. Pierce	
G-BWZX	AS.332L Super Puma	Bristow Helicopters Ltd	
G-BWZY	Hughes 269A	Reeve Newfields Ltd (G-FSDT)	
G-BXAB	PA-28-161 Warrior II	TG Aviation Ltd (G-BTGK)	
G-BXAC	RAF 2000 GTX-SE gyroplane	D. C. Fairbrass	
G-BXAD	Thunder Ax11-225 S2 balloon	M. W. White	
G-BXAF	Pitts S-1D Special	N. J. Watson	
G-BXAH	CP.301A Emeraude	A. P. Goodwin	
G-BXAI	Colt 120A balloon	E. F. & R. F. Casswell	
G-BXAJ	Lindstrand LBL-14A balloon	Oscair Project AB/Sweden	
G-BXAK	Yakovlev Yak-52	T. N. Jinks	
G-BXAL	Cameron 90 Bertie Bassett SS balloon	Trebor Bassett Ltd	
G-BXAM	Cameron N-90 balloon	Trebor Bassett Ltd	
G-BXAN	Scheibe SF-25C Falke	C. Falke Syndicate/Winthorpe	
G-BXAO	Avtech Jabiru SK	P. J. Thompson	
G-BXAR	Avro RJ100	British Airways CitiExpress	
G-BXAS	Avro RJ100	British Airways CitiExpress	
G-BXAU	Pitts S-1 Special	P. Tomlinson	
G-BXAV	Yakovlev Yak-52 (72)	Skytrace (UK) Ltd	
G-BXAX	Cameron N-77 balloon ★	Ballon Preservation Group	
G-BXAY	Bell 206B JetRanger 3	Viewdart Ltd	
G-BXBA	Cameron A-210 balloon	Reach For The Sky Ltd	
G-BXBB	PA-20 Pacer 150	M. E. R. Coghlan	
G-BXBC	EA.1 Kingfisher amphibian	S. Bichan	
G-BXBD	C.A.S.A. 1.131E Jungmann	P. B. Childs & B. L. Robinson	
G-BXBG	Cameron A-275 balloon	M. L. Gabb	
G-BXBI	P.84 Jet Provost T.3A	Global Aviation Ltd/Binbrook	
G-BXBK	Avions Mudry CAP-10B	S. Skipworth	
G-BXBL	Lindstrand LBL-240A balloon	Firefly Balloon Promotions	
G-BXBM	Cameron O-105 balloon	Bristol University Hot Air Ballooning Soc	
G-BXBN	Rans S.6-116 Coyote II	A. G. Foster	
G-BXBP	Denney Kitfox	G. S. Adams	
G-BXBR	Cameron A-120 balloon	M. G. Barlow	
G-BXBT	AS.355F-1 Twin Squirrel	Premiair Aviation Services Ltd (G-TMMC/G-JLCO)	
G-BXBU	Avions Mudry CAP.10B	J. F. Cosgrave & H. R. Pearson	
G-BXBY	Cameron A-105 balloon	S. P. Watkins	
G-BXBZ	PZL-104 Wilga 80	P. G. Marks/Luxembourg	
G-BXCA	Hapi Cygnet SF-2A	G. E. Collard	
G-BXCC	PA-28-201T Turbo Dakota	Greer Aviation Ltd	
G-BXCD	Team Minimax 91A	R. Davies	
G-BXCG	Jodel DR.250/160	CG Group	
G-BXCH	Shaw Europa	D. M. Stevens	
G-BXCJ	Campbell Cricket (replica)	F. Knowles	
G-BXCK	Cameron 110 Douglas SS balloon	Flying Pictures Ltd	
G-BXCL	Montgomerie-Bensen B.8MR	M. L. L. Temple	
G-BXCM	Lindstrand LBL-150A balloon	Aerosaurus Balloons Ltd	
G-BXCN	Sky 105-24 balloon	Nottingham Hot-Air Balloon Club	
G-BXCO	Colt 120A balloon	T. G. Church	
G-BXCP	D.H.C.1 Chipmunk 22 (WP859)	Echo Flying Group	
G-BXCS	Cameron N-90 balloon	Flying Pictures (Balloons) Ltd	
G-BXCT	D.H.C.1 Chipmunk 22 (WB697)	Wickenby Aviation Ltd	
G-BXCU	Rans S.6-116 Coyote II	R. S. Gent	
G-BXCV	D.H.C.1 Chipmunk 22 (WP929)	Ocean Flight Holdings Ltd/Hong Kong	
G-BXCW	Denney Kitfox Mk 3	M. J. Blanchard	
G-BXDA	D.H.C.1 Chipmunk 22 (WP860)	S. R. Cleary	
G-BXDB	Cessna U.206F	D. A. Howard (G-BMNZ)	
G-BXDD	RAF 2000GTX-SE gyroplane	A. Wane	
G-BXDE	RAF 2000GTX-SE gyroplane	A. McRedie	
G-BXDF	Beech 95-B55 Baron	Chesh-Air Ltd	
G-BXDG	D.H.C.1 Chipmunk 22 (WK630)	Felthorpe Flying Group	
G-BXDH	D.H.C.1 Chipmunk 22 (WD331)	Victory Workware Ltd	
G-BXDI	D.H.C.1 Chipmunk 22 (WD373)	Propshop Ltd/Duxford	
G-BXDL	P.84 Jet Provost T.3A (XM478)	de Havilland Aviation Ltd/Bournemouth	
G-BXDM	D.H.C.1 Chipmunk 22 (WP840)	RAF Halton Aeroplane Club Ltd	
G-BXDN	D.H.C.1 Chipmunk 22 (WK609)	W. D. Lowe & L. A. Edwards	

Notes	Reg.	Type	Owner or Operator
	G-BXDO	Rutan Cozy	R. James
	G-BXDP	D.H.C.1 Chupmunk 22 (WK642)	T. A. McBennett & J. Kelly
	G-BXDR	Lindstrand LBL-77A balloon	British Telecommunications PLC
	G-BXDS	Bell 206B JetRanger 3	Sterling Helicopters Ltd (G-OVBJ)
	G-BXDT	Robin HR.200/120B	Multiflight Ltd/Leeds-Bradford
	G-BXDU	Aero Designs Pulsar	M. P. Board
	G-BXDV	Sky 105-24 balloon	Loughborough Students Union Hot Air Balloon Club
	G-BXDY	Shaw Europa	D. G. & S. Watts
	G-BXDZ	Lindstrand LBL-105A balloon	D. J. & A. D. Sutcliffe
	G-BXEA	RAF 2000 GTX-SE gyroplane	R. Firth
	G-BXEB	RAF 2000 GTX-SE gyroplane	Penny Hydraulics Ltd
	G-BXEC	D.H.C.1 Chipmunk 22 (WK633)	K. P. & D. S. Hunt/Redhill
	G-BXEE	Enstrom 280C	R. E. Harvey
	G-BXEF	Shaw Europa	C. & W. P. Busuttil-Reynaud
	G-BXEJ	VPM M.16 Tandem Trainer	AES Radionic Surveillance Systems
	G-BXEL	MDH MD.500N	Ford Helicopters Ltd/Sywell
	G-BXEN	Cameron N-105 balloon	G. Aimo/Italy
	G-BXEP	Lindstrand LBL-14M balloon	Lindstrand Balloons Ltd
	G-BXER	PA-46-350P Malibu Mirage	Glasdon Group Ltd
	G-BXES	P.66 Pembroke C.1 (XL954)	Atlantic Air Transport Ltd/Coventry
	G-BXET	PA-38-112 Tomahawk II	APB Leasing Ltd/Welshpool
	G-BXEX	PA-28-181 Archer II	R. Mayle
	G-BXEY	Colt AS-105GD airship	D. Mayer/Germany
	G-BXEZ	Cessna 182P	Forhawk Ltd
	G-BXFB	Pitts S-1 Special	D. Dobson
	G-BXFC	Jodel D.18	B. S. Godbold
	G-BXFD	Enstrom 280C	Buckland Newton Hire Ltd
	G-BXFE	Avions Mudry CAP.10B	Avion Aerobatic Ltd
	G-BXFG	Shaw Europa	A. Rawicz-Szczerbo
	G-BXFI	Hawker Hunter T.7 (WV372)	Fox-One Ltd/Bournemouth
	G-BXFK	CFM Streak Shadow	S. J. M. French & T. I. Gorrell
	G-BXFN	Colt 77A balloon	Cameron Balloons Ltd
	G-BXFU	BAC.167 Strikemaster 83	Global Aviation Ltd/Binbrook
	G-BXFV	BAC.167 Strikemaster 83	Global Aviation Ltd/Binbrook
	G-BXFY	Cameron 90 Bierkrug SS balloon	Ballooning Bavaria
	G-BXGA	AS.350B-2 Ecureuil	PLM Dollar Group Ltd/Inverness
	G-BXGC	Cameron N-105 balloon	Cliveden Ltd
	G-BXGD	Sky 90-24 balloon	Servo & Electronic Sales Ltd
	G-BXGG	Shaw Europa	C. J. H. & P. A. J. Richardson
	G-BXGH	Diamond Katana DA.20-A1	Cumbernauld Flying School Ltd
	G-BXGK	Lindstrand LBL-203M balloon	Lindstrand Balloons Ltd
	G-BXGL	D.H.C.1 Chipmunk 22	Airways Aero Associations Ltd/Booker
	G-BXGM	D.H.C.1 Chipmunk 22 (WP928)	Chipmunk Golf Mike Group
	G-BXGO	D.H.C.1 Chipmunk 22 (WB654)	Trees Group/Booker
	G-BXGP	D.H.C.1 Chipmunk 22 (WZ882)	Eaglescott Chipmunk Group
	G-BXGS	RAF 2000 gyroplane	C. R. Gordon
	G-BXGT	I.I.I. Sky Arrow 650T	Sky Arrow (Kits) UK Ltd/Old Sarum
	G-BXGV	Cessna 172R	Skyhawk Group
	G-BXGW	Robin HR.200/120B	Multiflight Ltd/Leeds-Bradford
	G-BXGX	D.H.C.1 Chipmunk 22 (WK586)	Interflight (Air Charter) Ltd
	G-BXGY	Cameron V-65 balloon	Dante Balloon Group
	G-BXGZ	Stemme S.10V	D. Tucker & K. Lloyd
	G-BXHA	D.H.C.1 Chipmunk 22 (WP925)	F. A. de Munck & C. S. Huijers/Netherlands
	G-BXHD	Beech 76 Duchess	Aviation Rentals/Bournemouth
	G-BXHE	Lindstrand LBL-105A balloon	L. H. Ellis
	G-BXHF	D.H.C.1 Chipmunk 22 (WP930)	Hotel Fox Syndicate/Redhill
	G-BXHH	AA-5A Cheetah	Oaklands Flying/Biggin Hill
	G-BXHJ	Hapi Cygnet SF-2A	I. J. Smith
	G-BXHL	Sky 77-24 balloon	R. K. Gyselynck
	G-BXHO	Lindstrand Telewest Sphere SS balloon	Magical Adventures Ltd
	G-BXHR	Stemme S.10V	J. H. Rutherford
	G-BXHT	Bushby-Long Midget Mustang	P. P. Chapman
	G-BXHU	Campbell Cricket Mk 6	P. J. Began
	G-BXHY	Shaw Europa	Jupiter Flying Group
	G-BXIA	D.H.C.1 Chipmunk 22 (WB615)	Dales Aviation/Blackpool
	G-BXIC	Cameron A-275 balloon	Aerosaurus Balloons LLP
	G-BXID	Yakovlev Yak-52 (74)	E. S. Ewen
	G-BXIE	Colt 77B balloon	R. M. Horn
	G-BXIF	PA-28-161 Warrior II	Piper Flight Ltd
	G-BXIG	Zenair CH.701 STOL	A. J. Perry
	G-BXIH	Sky 200-24 balloon	Kent Ballooning

Reg.	Type	Owner or Operator	Notes
G-BXII	Shaw Europa	D. A. McFadyean	
G-BXIJ	Shaw Europa	D. G. Bligh	
G-BXIM	D.H.C.1 Chipmunk 22 (WK512)	A. B. Ashcroft & P. R. Joshua	
G-BXIO	Jodel DR.1050M	D. N. K. & M. A. Symon	
G-BXIT	Zebedee V-31 balloon	Zebedee Balloon Service Ltd	
G-BXIW	Sky 105-24 balloon	L. A. Watts	
G-BXIX	VPM M.16 Tandem Trainer	D. Beevers	
G-BXIY	Blake Bluetit (BAPC37)	J. Bryant	
G-BXIZ	Lindstrand LBL-31A balloon	Balloon Preservation Flying Group	
G-BXJA	Cessna 402B	Air Charter Scotland Ltd/Edinburgh	
G-BXJB	Yakovlev Yak-52 (15)	A. M. Playford & ptnrs	
G-BXJC	Cameron A-210 balloon	British School of Ballooning	
G-BXJD	PA-28 Cherokee 180C	BCT Aircraft Leasing Ltd/Filton	
G-BXJG	Lindstrand LBL-105B balloon	C. E. Wood	
G-BXJH	Cameron N-42 balloon	B. Conway	
G-BXJI	Tri-R Kis	R. M. Wakeford	
G-BXJJ	PA-28-161 Cadet	Aviation Rentals (G-GFCC)/Bournmouth	
G-BXJM	Cessna 152	I. R. Chaplin	
G-BXJO	Cameron O-90 balloon	Dragon Balloon Co. Ltd	
G-BXJP	Cameron C-80 balloon	AR. Cobaleno Pasta Fresca SRL/Italy	
G-BXJS	Schempp-Hirth Janus CM	Janus Syndicate	
G-BXJT	Sky 90-24 balloon	J. G. O'Connell	
G-BXJV	Diamond Katana DA.20-A1	Tayside Aviation Ltd/Dundee	
G-BXJW	Diamond Katana DA.20-A1	Tayside Aviation Ltd/Dundee	
G-BXJY	Van's RV-6	D. J. Sharland	
G-BXJZ	Cameron C-60 balloon	R. S. Mohr	
G-BXKA	Airbus A.320-214	Thomas Cook Airlines Ltd	
G-BXKB	Airbus A.320-214	Thomas Cook Airlines Ltd	
G-BXKC	Airbus A.320-214	Thomas Cook Airlines Ltd	
G-BXKD	Airbus A.320-214	Thomas Cook Airlines Ltd	
G-BXKF	Hawker Hunter T.7(XL577/V)	R. F. Harvey	
G-BXKH	Colt 90 Sparkasse Box SS balloon	Westfalisch-Lippischer Sparkasse UND/Germany	
G-BXKJ	Cameron A-275 balloon	Ballooning Network Ltd	
G-BXKL	Bell 206B JetRanger 3	Swattons Aviation Ltd	
G-BXKM	RAF 2000 GTX-SE gyroplane	J. R. Huggins	
G-BXKO	Sky 65-24 balloon	J-M. Reck/France	
G-BXKU	Colt AS-120 Mk II airship	D. C. Chipping/Portugal	
G-BXKW	Slingsby T.67M Firefly 200	N. A. Watling	
G-BXKX	Auster V	A. L. Jubb	
G-BXLC	Sky 120-24 balloon	Dragon Balloon Co Ltd	
G-BXLF	Lindstrand LBL-90A balloon	W. Rousell & J. Tyrrell	
G-BXLG	Cameron C-80 balloon	D. & L. S. Litchfield	
G-BXLI	Bell 206B JetRanger 3	N. K. Sherborne & R. H. Stevens (G-JODY)	
G-BXLK	Shaw Europa	R. G. Fairall	
G-BXLN	Fournier RF-4D	E. A. Wiseman & N. Thorne	
G-BXLO	P.84 Jet Provost T.4 (XR673/L)	R. F. Harvey & S. J. Davies	
G-BXLP	Sky 90-24 balloon	G. B. Lescott	
G-BXLR	PZL-110 Koliber 160A	Oakmans Systems Ltd	
G-BXLS	PZL-110 Koliber 160A	P. A. Rickells	
G-BXLT	SOCATA TB.200 Tobago XL	R. M. Shears/Blackbushe	
G-BXLW	Enstrom F-28F	M. & P. Food Products Ltd	
G-BXLY	PA-28-151 Warrior	Multiflight Ltd (G-WATZ)	
G-BXMF	Cassutt Racer IIIM	J. F. Bakewell	
G-BXMG	RAF 2000 GTX gyroplane	J. S. Wright	
G-BXMH	Beech 76 Duchess	Plane Talking Ltd/Elstree	
G-BXML	Mooney M.20A	G. Kay	
G-BXMM	Cameron A-180 balloon	B. Conway	
G-BXMV	Scheibe SF.25C Falke 1700	Falcon Flying Group/Shrivenham	
G-BXMX	Currie Wot	M. J. Hayman	
G-BXMY	Hughes 269C	Oxford Aviation Services Ltd/Kidlington	
G-BXMZ	Diamond Katana DA.20-A1	Tayside Aviation Ltd/Dundee	
G-BXNA	Light Aero Avid Flyer	G. Haynes	
G-BXNC	Shaw Europa	J. K. Cantwell	
G-BXNG	Beech 58 Baron	Bonanza Flying Club Ltd	
G-BXNH	PA-28-161 Warrior II	CC Management Associates Ltd/Redhill	
G-BXNN	D.H.C.1 Chipmunk 22 (WP983)	J. N. Robinson	
G-BXNS	Bell 206B JetRanger 3	Sterling Helicopters Ltd/Norwich	
G-BXNT	Bell 206B JetRanger 3	Sterling Helicopters Ltd/Norwich	
G-BXNV	Colt AS-105GD airship	The Sleeping Soc./Belgium	
G-BXNX	Lindstrand LBL-210A balloon	Balloon School (International) Ltd	
G-BXOA	Robinson R-22B	MG Group Ltd	
G-BXOB	Shaw Europa	S. J. Willett	

Notes	Reg.	Type	Owner or Operator
	G-BXOC	Evans VP-2	H. J. & E. M. Cox
	G-BXOF	Diamond Katana DA.20-A1	Cumbernauld Flying School Ltd
	G-BXOI	Cessna 172R	E. J. Watts
	G-BXOJ	PA-28-161 Warrior III	P. Foster
	G-BXOM	Isaacs Spitfire	J. H. Betton
	G-BXON	Auster AOP.9	C. J. & D. J. Baker
	G-BXOR	Robin HR.200/120B	L. Burrow
	G-BXOS	Cameron A-200 balloon	Airborne Balloon Management
	G-BXOT	Cameron C-70 balloon	Dante Balloon Group
	G-BXOU	CEA DR.360	S. H. & J. A. Williams/Blackpool
	G-BXOW	Colt 105A balloon	The Aerial Display Co Ltd
	G-BXOX	AA-5A Cheetah	R. L. Carter & P. J. Large
	G-BXOY	QAC Quickie Q.235	C. C. Clapham
	G-BXOZ	PA-28-181 Archer II	Spritetone Ltd
	G-BXPC	Diamond Katana DA.20-A1	Cubair Flight Training Ltd/Redhill
	G-BXPD	Diamond Katana DA.20-A1	Cubair Flight Training Ltd/Redhill
	G-BXPE	Diamond Katana DA.20-A1	Tayside Aviation Ltd/Dundee
	G-BXPF	Thorp T.211	AD Aviation Ltd/Barton
	G-BXPI	Van's RV-4	Cavendish Aviation Ltd
	G-BXPK	Cameron A-250 balloon	Richard Nash Cars Ltd
	G-BXPL	PA-28 Cherokee 140	C. R. Guggenheim
	G-BXPM	Beech 58 Baron	Foyle Flyers Ltd
	G-BXPO	Thorp T.211	AD Aviation Ltd/Barton
	G-BXPP	Sky 90-24 balloon	S. J. Farrant
	G-BXPR	Colt 110 Can SS balloon	FRB Fleishwarenfabrik Rostock-Bramow/Germany
	G-BXPS	PA-23 Aztec 250C	W. A. Moore (G-AYLI)
	G-BXPT	Ultramagic H-77 balloon	G. D. O. Bartram/Andorra
	G-BXPV	PA-34-220T Seneca IV	Oxford Aviation Services Ltd/Kidlington
	G-BXPW	PA-34-220T Seneca IV	Oxford Aviation Services Ltd/Kidlington
	G-BXPY	Robinson R-44	O. Desmet & B. Mornie/Belgium
	G-BXRA	Avions Mudry CAP.10B	B. H. D. H. Frere
	G-BXRB	Avions Mudry CAP.10B	T. T. Duhig
	G-BXRC	Avions Mudry CAP.10B	Group Alpha/Sibson
	G-BXRD	Enstrom 280FX	R. P. Bateman
	G-BXRF	CP.1310-C3 Super Emeraude	D. T. Gethin
	G-BXRG	PA-28-181 Archer II	Alderney Flying Training Ltd
	G-BXRH	Cessna 185A	R. E. M. Holmes
	G-BXRM	Cameron A-210 balloon	Dragon Balloon Co Ltd
	G-BXRO	Cessna U.206G	M. Penny
	G-BXRP	Schweizer 269C	AH Helicopters Services Ltd
	G-BXRR	Westland Scout AH.1	T. K. Phillips
	G-BXRS	Westland Scout AH.1	B-N Group Ltd/Bembridge
	G-BXRT	Robin DR.400-180	R. A. Ford
	G-BXRV	Van's RV-4	Cleeve Flying Grouip
	G-BXRY	Bell 206B JetRanger	S. Lee
	G-BXRZ	Rans S.6-116 Coyote II	G. F. M. Garner
	G-BXSC	Cameron C-80 balloon	S. J. Coates
	G-BXSD	Cessna 172R	K. K. Freeman
	G-BXSE	Cessna 172R	MK Aero Support Ltd/Andrewsfield
	G-BXSG	Robinson R-22B-2	R. M. Goodenough
	G-BXSH	Glaser-Dirks DG.800B	R. O'Conor
	G-BXSI	Avtech Jabiru SK	M. H. Molyneux
	G-BXSJ	Cameron C-80 balloon	British School of Ballooning
	G-BXSM	Cessna 172R	East Midlands Flying School Ltd
	G-BXSP	Grob G.109B	Deeside Grob Group
	G-BXSR	Cessna F172N	S. A. Parkes
	G-BXST	PA-25 Pawnee 235C	The Northumbria Gliding Club Ltd
	G-BXSU	Team Minimax 91A	M. R. Overall (G-MYGL)
	G-BXSV	SNCAN Stampe SV-4C	B. A. Bower
	G-BXSX	Cameron V-77 balloon	D. R. Medcalf
	G-BXSY	Robinson R-22B-2	N. M. G. Pearson
	G-BXTB	Cessna 152	Hairnoss Ltd
	G-BXTC	Taylor JT.1 Monoplane	R. Holden-Rushworth
	G-BXTD	Shaw Europa	P. R. Anderson
	G-BXTE	Cameron A-275 balloon	Global Ballooning Ltd
	G-BXTF	Cameron N-105 balloon	Flying Pictures Ltd *Salisbury's Strawberry*
	G-BXTG	Cameron N-42 balloon	Flying Pictures Ltd
	G-BXTH	Westland Gazelle HT.1 (XW866)	Flightline Ltd/Southend
	G-BXTI	Pitts S-1S Special	A. B. Treherne-Pollock
	G-BXTJ	Cameron N-77 balloon	Chubb Fire Ltd *Chubb*
	G-BXTK	Dornier Do.28D-2	R. Ebke/Germany
	G-BXTL	Schweizer 269C-1	Oxford Aviation Services Ltd/Kidlington

Reg.	Type	Owner or Operator	Notes
G-BXTN	Aerospatiale ATR-72-202	Aurigny Air Services Ltd	
G-BXTO	Hindustan HAL-6 Pushpak	Pushpak Flying Group	
G-BXTP	Diamond Katana DA.20-A1	Diamond Aircraft UK Ltd/Gamston	
G-BXTS	Diamond Katana DA.20-A1	I. M. Armitage	
G-BXTT	AA-5B Tiger	G-BXTT Group	
G-BXTV	Bug	B. R. Cope	
G-BXTW	PA-28-181 Archer III	Davison Plant Hire	
G-BXTY	PA-28-161 Cadet	Bflying Ltd	
G-BXTZ	PA-28-161 Cadet	Bflying Ltd	
G-BXUA	Campbell Cricket Mk.5	R. N. Bodley	
G-BXUB	Lindstrand Syrup Bottle SS balloon	Free Enterprise Balloons Ltd	
G-BXUC	Robinson R-22B	EBG (Helicopters) Ltd	
G-BXUE	Sky 240-24 balloon	Scotair Balloons	
G-BXUF	AB-206B JetRanger 3	SJ Contracting Services Ltd	
G-BXUG	Lindstrand Baby Bel SS balloon	Balloon Preservation Flying Group	
G-BXUH	Lindstrand LBL-31A balloon	Balloon Preservation Flying Groupt	
G-BXUI	Glaser-Dirks DG.800B	J. Le Coyte	
G-BXUK	Robinson R-44	Staske Construction Ltd	
G-BXUL	Goodyear FG-1D Corsair (92844/8)	J. H. Slade	
G-BXUM	Shaw Europa	D. Bosomworth	
G-BXUO	Lindstrand LBL-105A balloon	Lindstrand Balloons Ltd	
G-BXUS	Sky 65-24 balloon	PSH Skypower Ltd	
G-BXUU	Cameron V-65 balloon	M. D. Freeston & S. Mitchell	
G-BXUW	Cameron Colt 90A balloon	Zycomm Electronics Ltd	
G-BXUX	Fountain Cherry BX-2	M. F. Fountain	
G-BXUY	Cessna 310Q	Massair Ltd	
G-BXVA	SOCATA TB.200 Tobago XL	H. R. Palser/Cardiff-Wales	
G-BXVB	Cessna 152 II	PJC (Leasing) Ltd	
G-BXVD	CFM Streak Shadow SA	CFM Aircraft Ltd	
G-BXVE	Lindstrand LBL-330A balloon	Adventure Balloon Co Ltd	
G-BXVF	Thunder Ax11-250 S2 balloon	M. E. White	
G-BXVG	Sky 77-24 balloon	M. Wolf	
G-BXVH	Sky 25-16 balloon	Flying Pictures Ltd	
G-BXVJ	Cameron O-120 balloon	Aerosaurus Balloons Ltd (G-IMAX)	
G-BXVK	Robin HR.200/120B	Northamptonshire School of Flying Ltd/Sywell	
G-BXVL	Sky 180-24 balloon	Purple Balloons	
G-BXVM	Van's RV-6A	J. G. Small	
G-BXVO	Van's RV-6A	P. J. Hynes & M. E. Holden	
G-BXVP	Sky 31-24 balloon	L. Greaves	
G-BXVR	Sky 90-24 balloon	P. Hegarty	
G-BXVS	Brügger MB.2 Colibri	G. T. Snoddon	
G-BXVT	Cameron O-77 balloon	R. P. Wade	
G-BXVU	PA-28-161 Warrior II	London Ashford Airport Ltd	
G-BXVV	Cameron V-90 balloon	Floating Sensations Ltd	
G-BXVW	Colt Piggy Bank SS balloon	G. Binder/Germany	
G-BXVX	Rutan Cozy	G. E. Murray	
G-BXVY	Cessna 152	Stapleford Flying Club Ltd	
G-BXVZ	WSK-PZL Mielec TS-11 Iskra	J.Ziubrzynski	
G-BXWA	Beech 76 Duchess	Plymouth School of Flying Ltd	
G-BXWB	Robin HR.100/200B	W. A. Brunwin	
G-BXWC	Cessna 152	PJC (Leasing) Ltd/Stapleford	
G-BXWE	Fokker 100	bmi regional	
G-BXWF	Fokker 100	bmi regional	
G-BXWG	Sky 120-24 balloon	K. B. Chapple	
G-BXWH	Denney Kitfox Mk.4-1200	B. J. Finch	
G-BXWI	Cameron N-120 balloon	Flying Pictures Ltd	
G-BXWK	Rans S.6-ESA Coyote II	R. J. Teal	
G-BXWL	Sky 90-24 balloon	The Shropshire Hills Balloon Co	
G-BXWO	PA-28-181 Archer II	J. S. Develin & Z. Islam	
G-BXWP	PA-32 Cherokee Six 300	Alliance Aviation	
G-BXWR	CFM Streak Shadow	M. A. Hayward (G-MZMI)	
G-BXWT	Van's RV-6	R. C. Owen	
G-BXWU	FLS Aerospace Sprint 160	Aces High Ltd/North Weald	
G-BXWV	FLS Aerospace Sprint 160	Aces High Ltd/North Weald	
G-BXWX	Sky 25-16 balloon	Zebedee Balloon Service Ltd	
G-BXXC	Scheibe SF.25C Falke 1700	K. E. Ballington	
G-BXXD	Cessna 172R	Oxford Aviation Services Ltd/Kidlington	
G-BXXE	Rand-Robinson KR-2	N. Rawlinson	
G-BXXG	Cameron N-105 balloon	Allen Owen Ltd	
G-BXXH	Hatz CB-1	R. D. Shingler	
G-BXXI	Grob G.109B	M. N. Martin	
G-BXXJ	Colt Flying Yacht SS balloon	Magical Adventures Ltd/USA	

Notes	Reg.	Type	Owner or Operator
	G-BXXK	Cessna FR.172N	I. R. Chaplin/Andrewsfield
	G-BXXL	Cameron N-105 balloon	Flying Pictures Ltd
	G-BXXN	Robinson R-22B	Helicopter Services/Booker
	G-BXXO	Lindstrand LBL-90B balloon	K. Temple
	G-BXXP	Sky 77-24 balloon	C. J. James
	G-BXXR	Lovegrove AV-8 gyroplane	P. C. Lovegrove
	G-BXXS	Sky 105-24 balloon	Flying Pictures Ltd
	G-BXXT	Beech 76 Duchess	Pridenote Ltd
	G-BXXU	Colt 31A balloon	Sade Balloons Ltd
	G-BXXW	Enstrom F-28F	Falcon Helicopters Ltd (G-SCOX)
	G-BXYC	Schweizer 269C	Sycamore Aviation
	G-BXYD	Eurocopter EC.120B	Aero Maintenance Ltd
	G-BXYE	CP.301-C1 Emeraude	D. T. Gethin
	G-BXYF	Colt AS-105 GD airship	LN Flying Ltd
	G-BXYG	Cessna 310D	Equitus SARL/France
	G-BXYJ	Jodel DR.1050	G-BXYJ Group
	G-BXYK	Robinson R-22B	D. N. Whittlestone
	G-BXYM	PA-28 Cherokee 235	Ashurst Aviation Ltd/Shoreham
	G-BXYN	Van's RV-6	J. A. Tooley & R. M. Austin
	G-BXYO	PA-28RT-201 Arrow IV	Airways Flight Training (Exeter) Ltd
	G-BXYP	PA-28RT-201 Arrow IV	Westflight Aviation Ltd
	G-BXYR	PA-28RT-201 Arrow IV	A. Dayani
	G-BXYT	PA-28RT-201 Arrow IV	Checkflight Ltd
	G-BXYX	Van's RV-6A	A. G. Palmer
	G-BXZA	PA-38-112 Tomahawk	P. D. Brooks/Inverness
	G-BXZB	Nanchang CJ-6A (2632019)	Wingglider Ltd/Hibaldstow
	G-BXZD	Westland Gazelle HT.2 (XW895)	Gazelle Flying Group
	G-BXZF	Lindstrand LBL-90A balloon	R. G. Carrell
	G-BXZG	Cameron A-210 balloon	Société Bombard SARL/France
	G-BXZH	Cameron A-210 balloon	Société Bombard SARL/France
	G-BXZI	Lindstrand LBL-90A balloon	I. Little & J. A. Viner
	G-BXZK	MDH MD-900 Explorer	Dorset Police Air Support Unit
	G-BXZM	Cessna 182S	AB Integro Aviation Ltd
	G-BXZN	CH1 ATI	Intora-Firebird PLC
	G-BXZO	Pietenpol Air Camper	P. J. Cooke
	G-BXZS	Sikorsky S-76A (modified)	Bristow Helicopters Ltd
	G-BXZT	M.S.880B Rallye Club	Limerick Flying Club (Coonagh) Ltd
	G-BXZU	Micro Aviation Bantam B.22-S	M. E. Whapham & R. W. Hollamby
	G-BXZY	CFM Streak Shadow Srs DD	Cloudbase Aviation Services Ltd
	G-BXZZ	Sky 160-24 balloon	Skybus Ballooning
	G-BYAA	Boeing 767-204ER	Thomsonfly *Sir Matt Busby CBE*
	G-BYAB	Boeing 767-204ER	Thomsonfly *Brian Johnston CBE MC*
	G-BYAD	Boeing 757-204	Thomsonfly
	G-BYAE	Boeing 757-204	Thomsonfly
	G-BYAF	Boeing 757-204	Thomsonfly
	G-BYAH	Boeing 757-204	Thomsonfly
	G-BYAI	Boeing 757-204	Thomsonfly
	G-BYAJ	Boeing 757-204	Thomsonfly
	G-BYAK	Boeing 757-28A	Thomsonfly
	G-BYAL	Boeing 757-28A	Thomsonfly
	G-BYAN	Boeing 757-204	Thomsonfly
	G-BYAO	Boeing 757-204	Thomsonfly
	G-BYAP	Boeing 757-204	Thomsonfly
	G-BYAR	Boeing 757-204	Thomsonfly
	G-BYAS	Boeing 757-204	Thomsonfly
	G-BYAT	Boeing 757-204	Thomsonfly
	G-BYAU	Boeing 757-204	Thomsonfly
	G-BYAV	Taylor JT.1 Monoplane	J. S. Marten-Hale
	G-BYAW	Boeing 757-204	Thomsonfly *Eric Morecambe OBE*
	G-BYAX	Boeing 757-204	Thomsonfly
	G-BYAY	Boeing 757-204	Thomsonfly
	G-BYAZ	CFM Streak Shadow	A. G. Wright
	G-BYBA	AB-206B JetRanger 3	R. Forests Ltd (G-BHXV/G-OWJM)
	G-BYBC	AB-206B JetRanger 2	Sky Charter UK Ltd (G-BTWW)
	G-BYBD	Cessna F.172H	D. G. Bell & S. J. Green (G-OBHX/G-AWMU)
	G-BYBE	Jodel D.120A	R. J. Page
	G-BYBF	Robin R.2160i	D. J. R. Lloyd-Evans
	G-BYBH	PA-34-200T Seneca II	Goldspear (UK) Ltd
	G-BYBI	Bell 206B JetRanger 3	Winkburn Air Ltd
	G-BYBJ	Medway Hybred 44XLR	M. Gardner
	G-BYBK	Murphy Rebel	R. K. Hyatt

Reg.	Type	Owner or Operator	Notes
G-BYBL	GY-80 Horizon 160D	R. H. W. Beath	
G-BYBM	Avtech Jabiru SK	P. J. Hatton	
G-BYBN	Cameron N-77 balloon	M. G. & R. D. Howard	
G-BYBO	Medway Hybred 44XLR Eclipser	D. R. Purslow	
G-BYBP	Cessna A.185F	G. M. S. Scott	
G-BYBR	Rans S.6-116 Coyote II	J. B. Robinson/Blackpool	
G-BYBS	Sky 80-16 balloon	G. W. Mortimore	
G-BYBU	Renegade Spirit UK	L. C. Cook	
G-BYBV	Mainair Rapier	M. W. Robson	
G-BYBW	Team Minimax	R. M. Laver	
G-BYBX	Slingsby T.67M Firefly 260	Slingsby Aviation Ltd	
G-BYBY	Thorp T.18C Tiger	L. J. Joyce	
G-BYBZ	Jabiru SK	R. P. Lewis	
G-BYCA	PA-28 Cherokee 140D	A. Reay	
G-BYCB	Sky 21-16 balloon	S. J. Colin	
G-BYCD	Cessna 140 (modified)	G. P. James	
G-BYCE	Robinson R-44	Jim Davies Civil Engineering Ltd	
G-BYCF	Robinson R-22B-2	Teleology Ltd	
G-BYCJ	CFM Shadow Srs DD	S. R. Winter	
G-BYCL	Raj Hamsa X'Air Jabiru (1)	D. O'Keefe & ptnrs	
G-BYCM	Rans S.6-ES Coyote II	E. W. McMullan	
G-BYCN	Rans S.6-ES Coyote II	T. J. Croskery	
G-BYCP	Beech B200 Super King Air	LondonExecutive Aviation Ltd	
G-BYCS	Jodel DR.1051	Fire Defence PLC	
G-BYCT	Aero L-29 Delfin	Propeller BVBA/Belgium	
G-BYCU	Robinson R-22B	Tiger Helicopters Ltd (G-OCGJ)	
G-BYCV	Meridian Maverick 430	P. Shackleton	
G-BYCW	Mainair Blade	P. C. Watson	
G-BYCX	Westland Wasp HAS.1	M. P. Blokland	
G-BYCY	I.I.I. Sky Arrow 650T	K. A. Daniels	
G-BYCZ	Avtech Jabiru SK	R. Scroby	
G-BYDB	Grob G.115B	J. B. Baker	
G-BYDE	V.S.361 Spitfire LF. IX (PT879)	A. H. Soper	
G-BYDF	Sikorsky S-76A	Brecqhou Development Ltd	
G-BYDG	Beech C24R Sierra	Professional Flight Simulation Ltd/Bournemouth	
G-BYDI	Cameron A-210 balloon	First Flight	
G-BYDJ	Colt 120A balloon	D. K. Hempleman-Adams	
G-BYDK	SNCAN Stampe SV-4C	Bianchi Aviation Film Services Ltd/Booker	
G-BYDL	Hawker Hurricane IIB (Z5207)	R. A. Roberts	
G-BYDM	Pegasus Quantum 15-912	P. Roberts	
G-BYDT	Cameron N-90 balloon	N. J. Langley	
G-BYDV	Van's RV-6	B. F. Hill	
G-BYDW	RAF 2000 GTX-SE gyroplane	R. I. Young	
G-BYDX	American General AG-5B Tiger	Bibit Group	
G-BYDY	Beech 58 Baron	J. F. Britten	
G-BYDZ	Pegasus Quantum 15-912	G. Shaw	
G-BYEA	Cessna 172P	Plane Talking Ltd/Elstree	
G-BYEB	Cessna 172P	Plane Talking Ltd/Elstree	
G-BYEC	Glaser-Dirks DG.800B	P. R. Redshaw	
G-BYEE	Mooney M.20K	Double Echo Flying Group	
G-BYEH	CEA Jodel DR.250	E. J. Horsfall/Blackpool	
G-BYEI	Cameron 90 Chick SS balloon	Bic UK Ltd	
G-BYEJ	Scheibe SF-28A Tandem Falke	D. Shrimpton	
G-BYEK	Stoddard Hamilton Glastar	G. M. New	
G-BYEL	Van's RV-6	D. Millar	
G-BYEM	Cessna R.182 RG	Wycombe Air Centre Ltd/Booker	
G-BYEO	Zenair CH.601HDS	Cloudbase Flying Group	
G-BYEP	Lindstrand LBL-90B balloon	R. C. Barkworth & D. G. Maguire	
G-BYER	Cameron C-80 balloon	Cameron Balloons Ltd	
G-BYES	Cessna 172P	Redhill Air Services Ltd	
G-BYET	Cessna 172P	Redhill Air Services Ltd	
G-BYEW	Pegasus Quantum 15-912	R. J. Murphy	
G-BYEX	Sky 120-24 balloon	Ballongflyg Upp & Ner AB/Sweden	
G-BYEY	Lindstrand LBL-21 Silver Dream balloon	Oscair Project Ltd/Sweden	
G-BYEZ	Dyn' Aero MCR-01	J. P. Davies	
G-BYFA	Cessna F.152 II	A. J. Gomes (G-WACA)	
G-BYFC	Avtech Jabiru SK	A. C. N. Freeman	
G-BYFD	Grob G.115A	Kane Group (G-BSGE)/Ireland	
G-BYFE	Pegasus Quantum 15-912	G-BYFE Syndicate	
G-BYFF	Pegasus Quantum 15-912	Kemble Flying Club	
G-BYFG	Shaw Europa XS	P. R. Brodie	
G-BYFI	CFM Starstreak Shadow SA	D. G. Cook	

Notes	Reg.	Type	Owner or Operator
	G-BYFJ	Cameron N-105 balloon	R. J. Mercer
	G-BYFL	Diamond HK.36 TTS	Seahawk Gliding Club/Culdrose
	G-BYFM	Jodel DR.1050M-1 (replica)	A. J. Roxburgh
	G-BYFP	PA-28-181 Archer II	P. R. Nott
	G-BYFR	PA-32R-301 Saratoga II HP	Buckleton Ltd
	G-BYFT	Pietenpol Air Camper	M. W. Elliott
	G-BYFU	Lindstrand LBL-105B balloon	Balloons Lindstrand France
	G-BYFV	Team Minimax 91	W. E. Gillham
	G-BYFX	Colt 77A balloon	Flying Pictures Ltd
	G-BYFY	Avions Mudry CAP.10B	Cole Aviation
	G-BYGA	Boeing 747-436	British Airways
	G-BYGB	Boeing 747-436	British Airways
	G-BYGC	Boeing 747-436	British Airways
	G-BYGD	Boeing 747-436	British Airways
	G-BYGE	Boeing 747-436	British Airways
	G-BYGF	Boeing 747-436	British Airways
	G-BYGG	Boeing 747-436	British Airways
	G-BYHC	Cameron Z-90 balloon	S. M. Sherwin
	G-BYHE	Robinson R-22B	Kent Aviation Ltd
	G-BYHG	Dornier Do.328-100	ScotAirways Ltd
	G-BYHH	PA-28-161 Warrior III	Stapleford Flying Club Ltd
	G-BYHI	PA-28-161 Warrior II	Haimoss Ltd
	G-BYHJ	PA-28R-201 Arrow	Bflying Ltd/Bournemouth
	G-BYHK	PA-28-181 Archer III	T-Air Services
	G-BYHL	D.H.C.1 Chipmunk 22 (WG308)	M. R. & I. D. Higgins
	G-BYHM	BAe 125 Srs 800B	Bookajet Operations Ltd
	G-BYHN	Mainair Blade 912	R. Stone
	G-BYHO	Mainair Blade 912	P. J. Morton
	G-BYHP	CEA DR.253B	RB Aero Club Ltd
	G-BYHR	Pegasus Quantum 15-912	I. D. Chantler
	G-BYHS	Mainair Blade 912	C. I. Poole
	G-BYHT	Robin DR.400/180R	Deeside Robin Group
	G-BYHU	Cameron N-105 balloon	Freeup Ltd
	G-BYHV	Raj Hamsa X'Air 582	J. S. Mason
	G-BYHX	Cameron A-250 balloon	Global Ballooning
	G-BYHY	Cameron V-77 balloon	P. Spellward
	G-BYHZ	Sky 160-24 balloon	Breckland Balloons Ltd
	G-BYIA	Avtech Jabiru SK	G. M. Geary
	G-BYIB	Rans S.6-ES Coyote II	E. A. Pearson
	G-BYIC	Cessna U.206G	Wild Geese Parachute Club
	G-BYID	Rans S.6-ES Coyote II	J. A. E. Bowen
	G-BYIE	Robinson R-22B	Jepar Rotorcraft
	G-BYII	Team Minimax	J. S. R. Moodie
	G-BYIJ	C.A.S.A. 1.131E Jungmann 2000	P. R. Teager & R. N. Crosland
	G-BYIK	Shaw Europa	P. M. Davis
	G-BYIL	Cameron N-105 balloon	Oakfield Farm Products Ltd
	G-BYIM	Avtech Jabiru UL	S. D. Miller
	G-BYIN	RAF 2000 gyroplane	J. R. Legge
	G-BYIO	Colt 105A balloon	N. Charbonnier/Italy
	G-BYIP	Aerotek Pitts S-2A Special	D. P. Heather-Hayes
	G-BYIR	Aerotek Pitts S-1S Special	Hampshire Aeroplane Co Ltd/St Just
	G-BYIS	Pegasus Quantum 15-912	L. M. Tidman
	G-BYIT	Robin DR.400/500	P. R. Liddle
	G-BYIU	Cameron V-90 balloon	H. Micketeit/Germany
	G-BYIV	Cameron PM-80 balloon	A. Schneider/Germany
	G-BYIW	Cameron PM-80 balloon	A. Schneider/Germany
	G-BYIX	Cameron PM-80 balloon	A. Schneider/Germany
	G-BYIY	Lindstrand LBL-105B balloon	J. H. Dobson
	G-BYIZ	Pegasus Quantum 15-912	J. D. Gray
	G-BYJA	RAF 2000 GTX-SE gyroplane	B. Errington-Weddle
	G-BYJB	Mainair Blade 912	J. H. Bradbury
	G-BYJC	Cameron N-90 balloon	J. M. Percival
	G-BYJD	Avtech Jabiru UL	M. W. Knights
	G-BYJE	TEAM Mini-MAX 91	M. F. Cottam
	G-BYJF	Thorpe T.211	AD Aviation Ltd/Barton
	G-BYJG	Lindstrand LBL-77A balloon	Lindstrand Hot-Air Balloons Ltd
	G-BYJH	Grob G.109B	A. J. Buchanan
	G-BYJI	Shaw Europa	P. S. Jones (G-ODTI)
	G-BYJJ	Cameron C-80 balloon	Proximm Franchising SRL/Italy
	G-BYJK	Pegasus Quantum 15-912	B. S. Smy
	G-BYJL	Aero Designs Pulsar	F. A. H. Ashmead
	G-BYJM	Cyclone AX2000	Caunton AX2000 Syndicate

Reg.	Type	Owner or Operator	Notes
G-BYJN	Lindstrand LBL-105A balloon	B. Meeson	
G-BYJO	Rans S.6-ES Coyote II	G. Ferguson	
G-BYJP	Aerotek Pitts S-1S Special	Eaglescott Pitts Group	
G-BYJR	Lindstrand LBL-77B balloon	C. D. Duthy-James	
G-BYJS	SOCATA TB-20 Trinidad	J. K. Sharkey	
G-BYJT	Zenair CH.601HD	J. D. T. Tannock	
G-BYJU	Raj Hamsa X'Air 582	G. P. Morling	
G-BYJV	Cameron A-210 balloon	Societe Bombard SRL/France	
G-BYJW	Cameron 105 Sphere SS balloon	Forbes Global Inc.	
G-BYJX	Cameron C-70 balloon	B. Perona	
G-BYJZ	Lindstrand LBL-105A balloon	M. A. Webb	
G-BYKA	Lindstrand LBL-69A balloon	Aerial Promotions Ltd	
G-BYKB	R. Commander 114	A. Walton	
G-BYKC	Mainair Blade 912	P. J. Burridge	
G-BYKD	Mainair Blade 912	D. C. Boyle	
G-BYKF	Enstrom F-28F	Battle Helicopters Ltd	
G-BYKG	Pietenpol Air Camper	K. B. Hodge	
G-BYKI	Cameron N-105 balloon	J. A. Leahy/Ireland	
G-BYKJ	Westland Scout AH.1	Austen Associates	
G-BYKK	Robinson R-44	Banner Helicopters Ltd	
G-BYKL	PA-28-181 Archer II	MPFC Ltd	
G-BYKN	PA-28-161 Warrior II	Oxford Aviation Services Ltd/Kidlington	
G-BYKO	PA-28-161 Warrior II	Oxford Aviation Services Ltd/Kidlington	
G-BYKP	PA-28R-201T Turbo Arrow IV	Westflight Aviation Ltd	
G-BYKR	PA-28-161 Warrior II	Oxford Aviation Services Ltd/Kidlington	
G-BYKS	Leopoldoff L-6 Colibri	I. M. Callier	
G-BYKT	Pegasus Quantum 15-912	D. A. Bannister & N. J. Howarth	
G-BYKU	BFC Challenger II	K. W. Seedhouse	
G-BYKW	Lindstrand LBL-77B balloon	P-J. Fuseau/France	
G-BYKX	Cameron N-90 balloon	G. Davis	
G-BYKZ	Sky 140-24 balloon	D. J. Head	
G-BYLA	FRED Srs 3	R. Holden-Rushworth	
G-BYLC	Pegasus Quantum 15-912	M. Hurn	
G-BYLD	Pietenpol Air Camper	S. Bryan	
G-BYLE	PA-38-112 Tomahawk	Surrey & Kent Flying Club Ltd/Biggin Hill	
G-BYLF	Zenair CH.601HDS	G. Waters	
G-BYLG	Robin HR.200/120B	Leinster Aero Club Ltd/Ireland	
G-BYLH	Robin HR.200/120B	Multiflight Ltd	
G-BYLI	Nova Vertex 22 hang glider	M. Hay	
G-BYLJ	Letov LK-2M Sluka	W. J. McCarroll	
G-BYLL	Sequoia F.8L Falco	N. J. Langrick/Breighton	
G-BYLN	Raj Hamsa X'Air 582	R. Gillespie & S. P. McGirr	
G-BYLO	T.66 Nipper Srs 1	M. J. A. Trudgill	
G-BYLP	Rand-Robinson KR-2	C. S. Hales	
G-BYLR	Cessna 404	Air Charter Scotland Ltd/Edinburgh	
G-BYLS	Bede BD-4	G. H. Bayliss/Shobdon	
G-BYLT	Raj Hamsa X'Air 582	T. W. Phipps & B. G. Simons	
G-BYLV	Thunder Ax8-105 S2 balloon	Wind Line SRL/Italy	
G-BYLW	Lindstrand LBL-77A balloon	Associazione Gran Premio Italiano	
G-BYLX	Lindstrand LBL-105A balloon	Italiana Aeronavi/Italy	
G-BYLY	Cameron V-77 balloon (1)	R. Bayly (G-ULIA)/Italy	
G-BYLZ	Rutan Cozy	E. R. Allen	
G-BYMB	Diamond Katana DA.20-C1	Enstone Flying Club	
G-BYMC	PA-38-112 Tomahawk II	Central Aircraft Leasing Ltd	
G-BYMD	PA-38-112 Tomahawk II	Surrey & Kent Flying Club Ltd/Biggin Hill	
G-BYME	GY-80 Horizon 180	Air Venturas Ltd	
G-BYMF	Pegasus Quantum 15-912	G. R. Stockdale	
G-BYMG	Cameron A-210 balloon	Cloud Nine Balloon Co	
G-BYMH	Cessna 152	PJC (Leasing) Ltd/Stapleford	
G-BYMI	Pegasus Quantum 15	N. C. Grayson	
G-BYMJ	Cessna 152	PJC (Leasing) Ltd/Stapleford	
G-BYMK	Dornier Do.328-100	ScotAirways Ltd	
G-BYML	Dornier Do.328-100	ScotAirways Ltd	
G-BYMM	Raj Hamsa X'Air 582 (1)	R. W. F. Boarder	
G-BYMN	Rans S.6-ESA Coyote II	R. L. Barker	
G-BYMO	Campbell Cricket	D. G. Hill	
G-BYMP	Campbell Cricket Mk 1	J. J. Fitzgerald	
G-BYMR	Raj Hamsa X'Air 582	W. M. McMinn	
G-BYMT	Pegasus Quantum 15-912	C. M. Mackinnon	
G-BYMU	Rans S.6-ES Coyote II	I. R. Russell & B. Frogley	
G-BYMV	Rans S.6-ES Coyote II	G. A. Squires	
G-BYMW	Boland 52-12 balloon	C. Jones	

Notes	Reg.	Type	Owner or Operator
	G-BYMX	Cameron A-105 balloon	H. Reis/Germany
	G-BYMY	Cameron N-90 balloon	Cameron Balloons Ltd
	G-BYNA	Cessna F.172H	Heliview Ltd (G-AWTH)/Blackbushe
	G-BYND	Pegasus Quantum 15	D. G. Baker
	G-BYNE	Pilatus PC-6/B2-H4 Turbo Porter	D. M. Penny
	G-BYNF	NA-64 Yale I	R. S. Van Dijk (stored)/Duxford
	G-BYNH	Rotorway Executive 162F	R. C. MacKenzie
	G-BYNI	Rotorway Exec 90	M. Bunn
	G-BYNJ	Cameron N-77 balloon	A. Giovanni/Italy
	G-BYNK	Robin HR.200/160	Penguin Flight Group
	G-BYNL	Avtech Jabiru SK	G-BYNL Ltd
	G-BYNM	Mainair Blade 912	M. W. Holmes
	G-BYNN	Cameron V-90 balloon	M. K. Grigson
	G-BYNO	Pegasus Quantum 15-912	R. J. Newsham & G. J. Slater
	G-BYNP	Rans S.6-ES Coyote II	R. J. Lines
	G-BYNR	Avtech Jabiru UL	M. P. Maughan
	G-BYNS	Avtech Jabiru SK	D. K. Lawry
	G-BYNT	Raj Hamsa X'Air 582 (1)	G. R. Wallis
	G-BYNU	Cameron Thunder Ax7-77 balloon	P. M. Gaines
	G-BYNV	Sky 105-24 balloon	Par Rovelli Construzioni SRL/Italy
	G-BYNW	Cameron H-34 balloon	I. M. Ashpole
	G-BYNX	Cameron RX-105 balloon	Cameron Balloons Ltd
	G-BYNY	Beech 76 Duchess	Magenta Ltd/Exeter
	G-BYOB	Slingsby T.67M Firefly 260	Stapleford Flying Club Ltd
	G-BYOD	Slingsby T.67C	TDR Aviation Ltd
	G-BYOF	Robin R.2160I	Lifeskills Ltd
	G-BYOG	Pegasus Quantum 15-912	M. D. Hinge
	G-BYOH	Raj Hamsa X'Air 582 (1)	P. H. J. Kent
	G-BYOI	Sky 80-16 balloon	I. S. & S. W. Watthews
	G-BYOJ	Raj Hamsa X'Air 582 (1)	H. M. Owen
	G-BYOK	Cameron V-90 balloon	D. S. Wilson
	G-BYOM	Sikorsky S-76C (modified)	Starspeed Ltd (G-IJCB)/Blackbushe
	G-BYON	Mainair Blade	S. Mills & G. M. Hobman
	G-BYOO	CFM Streak Shadow	G. R. Eastwood
	G-BYOR	Raj Hamsa X'Air 582 (1)	K. J. Draper
	G-BYOS	Mainair Blade 912	J. L. Guy
	G-BYOT	Rans S.6-ES Coyote II	H. F. Blakeman
	G-BYOU	Rans S.6-ES Coyote II	J. R. Bramley
	G-BYOV	Pegasus Quantum 15-912	Microlight Hire Ltd
	G-BYOW	Mainair Blade	M. Forsyth
	G-BYOX	Cameron Z-90 balloon	Airship & Balloon Co. Ltd
	G-BYOY	Canadair T-33AN	K. K. Gerstorfer/North Weald
	G-BYOZ	Mainair Rapier	M. Morgan
	G-BYPA	AS.355F-2 Twin Squirrel	P. & J. Carter (G-NWPI)
	G-BYPB	Pegasus Quantum 15-912	S. Graham
	G-BYPC	Lindstrand LBL-AS2 balloon	Lindstrand Balloons Ltd
	G-BYPD	Cameron A-105 balloon	Headland Hotel Ltd
	G-BYPE	GY-80 Horizon 160D	H. I. Smith & P. R. Hendry-Smith
	G-BYPF	Thruster T.600N	Canary Syndicate
	G-BYPG	Thruster T.600N	G-BYPG Syndicate
	G-BYPH	Thruster T.600N	D. M. Canham
	G-BYPJ	Pegasus Quantum 15-912	C. K. Stow
	G-BYPL	Pegasus Quantum 15-912	I. T. Carlse
	G-BYPM	Shaw Europa XS	P. Mileham
	G-BYPN	M.S.880B Rallye Club	R. & T. C. Edwards
	G-BYPO	Raj Hamsa X'Air 582 (1)	D. W. Willis
	G-BYPP	Medway Rebel SS	J. L. Gowens
	G-BYPT	Rans S.6-ES Coyote II	B. S. Keene
	G-BYPU	PA-32R-301 Saratoga SP	AM Blatch Electrical Contractors Ltd
	G-BYPW	Raj Hamsa X'Air 583 (3)	A. D. Worrall & B. J. Ellis
	G-BYPY	Ryan ST3KR	T. Curtis-Taylor
	G-BYPZ	Rans S.6-116 Super 6	R. A. Blackbourne
	G-BYRA	BAe Jetstream 3201	Eastern Airways Ltd
	G-BYRC	WS.58 Wessex HC.2 (XT671)	D. Brem-Wilson
	G-BYRG	Rans S.6-ES Coyote II	W. H. Mills
	G-BYRH	Medway Hybred 44XLR	A. Scotney
	G-BYRJ	Pegasus Quantum 15-912	Light Flight Ltd
	G-BYRK	Cameron V-42 balloon	R. Kunert
	G-BYRM	BAe Jetstream 3201	Eastern Airways Ltd
	G-BYRO	Mainair Blade	P. W. F. Coleman
	G-BYRP	Mainair Blade 912	M. P. Middleton
	G-BYRR	Mainair Blade 912	G. R. Sharples

Reg.	Type	Owner or Operator	Notes
G-BYRS	Rans S.6-ES Coyote II	J. P. Harris	
G-BYRU	Pegasus Quantum 15-912	Sarum QTM 912 Group	
G-BYRV	Raj Hamsa X'Air 582 (1)	D. R. Darby	
G-BYRX	Westland Scout AH.1 (XT634)	Historic Helicopters Ltd	
G-BYRY	Slingsby T.67M Firefly 200	T. R. Pearson	
G-BYRZ	Lindstrand LBL-77M balloon	Challenge Transatlantique/France	
G-BYSA	Shaw Europa XS	B. Allsop	
G-BYSE	AB-206B JetRanger 2	Alspath Properties Ltd (G-BFND)	
G-BYSF	Avtech Jabiru UL	S. J. Marshall	
G-BYSG	Robin HR.200/120B	Anglian Flight Centres Ltd	
G-BYSI	WSK-PZL Koliber 160A	J. & D. F. Evans	
G-BYSJ	D.H.C.1 Chipmunk 22 (WB569)	Silver Victory BVBA/Belgium	
G-BYSK	Cameron A-275 balloon	Balloon School (International) Ltd	
G-BYSM	Cameron A-210 balloon	Balloon School (International) Ltd	
G-BYSN	Rans S.6-ES Coyote II	A. L. & A. R. Roberts	
G-BYSP	PA-28-181 Archer II	Central Aircraft Leasing Ltd	
G-BYSR	Pegasus Quantum 15-912	A. C. Stuart/Canada	
G-BYSS	Medway Rebel SS	D. W. Allen	
G-BYSV	Cameron N-120 balloon	S. Simmington	
G-BYSW	Enstrom 280FX	D. A. Marks	
G-BYSX	Pegasus Quantum 15-912	RAF Microlight Flying Association	
G-BYSY	Raj Hamsa X'Air 582 (1)	J. M. Davidson	
G-BYTA	Kölb Twinstar Mk 3 (modified)	R. E. Gray	
G-BYTB	SOCATA TB-20 Trinidad	Ottoman Empire Ltd	
G-BYTC	Pegasus Quantum 15-912	R. J. Marriott	
G-BYTD	Robinson R-22B-2	J. M. D. Moloney/Ireland	
G-BYTE	Robinson R-22B	Patriot Aviation Ltd/Cranfield	
G-BYTG	Glaser-Dirks DG.400	P. R. Williams	
G-BYTI	PA-24 Comanche 250	G-BYTI Syndicate	
G-BYTJ	Cameron C-80 balloon	M. White	
G-BYTK	Avtech Jabiru UL	S. R. Pike	
G-BYTL	Mainair Blade 912	M. E. Keefe	
G-BYTM	Dyn' Aero MCR-01	I. Lang	
G-BYTN	D.H.82A Tiger Moth	B. D. Hughes	
G-BYTR	Raj Hamsa X'Air 582 (1)	R. Dunn	
G-BYTS	Montgomerie-Bensen B.8MR gyroplane	M. G. Mee	
G-BYTT	Raj Hamsa X'Air 582 (1)	J. L. Pearson	
G-BYTU	Mainair Blade 912	K. Roberts	
G-BYTV	Avtech Jabiru UK	A. D. Tomlins	
G-BYTW	Cameron O-90 balloon	Sade Balloons Ltd	
G-BYTY	Dornier Do.328-100	ScotAirways Ltd	
G-BYTZ	Raj Hamsa X'Air 582 (1)	K. C. Millar	
G-BYUA	Grob G.115E Tutor	VT Aerospace Ltd/Wyton	
G-BYUB	Grob G.115E Tutor	VT Aerospace Ltd/Cranwell	
G-BYUC	Grob G.115E Tutor	VT Aerospace Ltd/Cranwell	
G-BYUD	Grob G.115E Tutor	VT Aerospace Ltd/Cranwell	
G-BYUE	Grob G.115E Tutor	VT Aerospace Ltd/Wyton	
G-BYUF	Grob G.115E Tutor	VT Aerospace Ltd/Cranwell	
G-BYUG	Grob G.115E Tutor	VT Aerospace Ltd/Cranwell	
G-BYUH	Grob G.115E Tutor	VT Aerospace Ltd/Cranwell	
G-BYUI	Grob G.115E Tutor	VT Aerospace Ltd/Wyton	
G-BYUJ	Grob G.115E Tutor	VT Aerospace Ltd/Wyton	
G-BYUK	Grob G.115E Tutor	VT Aerospace Ltd/Cranwell	
G-BYUL	Grob G.115E Tutor	VT Aersoapce Ltd/Wyton	
G-BYUM	Grob G.115E Tutor	VT Aerospace Ltd/Wyton	
G-BYUN	Grob G.115E Tutor	VT Aerospace Ltd/Wyton	
G-BYUO	Grob G.115E Tutor	VT Aerospace Ltd/Wyton	
G-BYUP	Grob G.115E Tutor	VT Aerospace Ltd/Leuchars	
G-BYUR	Grob G.115E Tutor	VT Aerospace Ltd/Glasgow	
G-BYUS	Grob G.115E Tutor	VT Aerospace Ltd/Benson	
G-BYUT	Grob G.115E Tutor	VT Aerospace Ltd/Benson	
G-BYUU	Grob G.115E Tutor	VT Aerospace Ltd/Glasgow	
G-BYUV	Grob G.115E Tutor	VT Aerospace Ltd	
G-BYUW	Grob G.115E Tutor	VT Aerospace Ltd	
G-BYUX	Grob G.115E Tutor	VT Aerospace Ltd	
G-BYUY	Grob G.115E Tutor	VT Aerospace Ltd	
G-BYUZ	Grob G.115E Tutor	VT Aerospace Ltd	
G-BYVA	Grob G.115E Tutor	VT Aerospace Ltd	
G-BYVB	Grob G.115E Tutor	VT Aerospace Ltd	
G-BYVC	Grob G.115E Tutor	VT Aerospace Ltd	
G-BYVD	Grob G.115E Tutor	VT Aerospace Ltd	
G-BYVE	Grob G.115E Tutor	VT Aerospace Ltd	

Notes	Reg.	Type	Owner or Operator
	G-BYVF	Grob G.115E Tutor	VT Aerospace Ltd
	G-BYVG	Grob G.115E Tutor	VT Aerospace Ltd
	G-BYVH	Grob G.115E Tutor	VT Aerospace Ltd
	G-BYVI	Grob G.115E Tutor	VT Aerospace Ltd
	G-BYVJ	Grob G.115E Tutor	VT Aerospace Ltd
	G-BYVK	Grob G.115E Tutor	VT Aerospace Ltd
	G-BYVL	Grob G.115E Tutor	VT Aerospace Ltd
	G-BYVM	Grob G.115E Tutor	VT Aerospace Ltd
	G-BYVN	Grob G.115E Tutor	VT Aerospace Ltd
	G-BYVO	Grob G.115E Tutor	VT Aerospace Ltd
	G-BYVP	Grob G.115E Tutor	VT Aerospace Ltd
	G-BYVR	Grob G.115E Tutor	VT Aerospace Ltd
	G-BYVS	Grob G.115E Tutor	VT Aerospace Ltd
	G-BYVT	Grob G.115E Tutor	VT Aerospace Ltd
	G-BYVU	Grob G.115E Tutor	VT Aerospace Ltd
	G-BYVV	Grob G.115E Tutor	VT Aerospace Ltd
	G-BYVW	Grob G.115E Tutor	VT Aerospace Ltd
	G-BYVX	Grob G.115E Tutor	VT Aerospace Ltd
	G-BYVY	Grob G.115E Tutor	VT Aerospace Ltd
	G-BYVZ	Grob G.115E Tutor	VT Aerospace Ltd
	G-BYWA	Grob G.115E Tutor	VT Aerospace Ltd
	G-BYWB	Grob G.115E Tutor	VT Aerospace Ltd
	G-BYWC	Grob G.115E Tutor	VT Aerospace Ltd
	G-BYWD	Grob G.115E Tutor	VT Aerospace Ltd
	G-BYWE	Grob G.115E Tutor	VT Aerospace Ltd
	G-BYWF	Grob G.115E Tutor	VT Aerospace Ltd
	G-BYWG	Grob G.115E Tutor	VT Aerospace Ltd
	G-BYWH	Grob G.115E Tutor	VT Aerospace Ltd
	G-BYWI	Grob G.115E Tutor	VT Aerospace Ltd
	G-BYWJ	Grob G.115E Tutor	VT Aerospace Ltd
	G-BYWK	Grob G.115E Tutor	VT Aerospace Ltd
	G-BYWL	Grob G.115E Tutor	VT Aerospace Ltd
	G-BYWM	Grob G.115E Tutor	VT Aerospace Ltd
	G-BYWN	Grob G.115E Tutor	VT Aerospace Ltd
	G-BYWO	Grob G.115E Tutor	VT Aerospace Ltd
	G-BYWP	Grob G.115E Tutor	VT Aerospace Ltd
	G-BYWR	Grob G.115E Tutor	VT Aerospace Ltd
	G-BYWS	Grob G.115E Tutor	VT Aerospace Ltd
	G-BYWT	Grob G-115E Tutor	VT Aerospace Ltd
	G-BYWU	Grob G.115E Tutor	VT Aerospace Ltd
	G-BYWV	Grob G.115E Tutor	VT Aerospace Ltd
	G-BYWW	Grob G.115E Tutor	VT Aerospace Ltd
	G-BYWX	Grob G.115E Tutor	VT Aerospace Ltd
	G-BYWY	Grob G.115E Tutor	VT Aerospace Ltd
	G-BYWZ	Grob G.115E Tutor	VT Aerospace Ltd
	G-BYXA	Grob G.115E Tutor	VT Aerospace Ltd
	G-BYXB	Grob G.115E Tutor	VT Aerospace Ltd
	G-BYXC	Grob G.115E Tutor	VT Aerospace Ltd
	G-BYXD	Grob G.115E Tutor	VT Aerospace Ltd
	G-BYXE	Grob G.115E Tutor	VT Aerospace Ltd
	G-BYXF	Grob G.115E Tutor	VT Aerospace Ltd
	G-BYXG	Grob G.115E Tutor	VT Aerospace Ltd
	G-BYXH	Grob G.115E Tutor	VT Aerospace Ltd
	G-BYXI	Grob G.115E Tutor	VT Aerospace Ltd
	G-BYXJ	Grob G.115E Tutor	VT Aerospace Ltd/Cranwell
	G-BYXK	Grob G.115E Tutor	VT Aerospace Ltd/Cranwell
	G-BYXL	Grob G.115E Tutor	VT Aerospace Ltd/Cranwell
	G-BYXM	Grob G.115E Tutor	VT Aerospace Ltd/Cranwell
	G-BYXN	Grob G.115E Tutor	VT Aerospace Ltd
	G-BYXO	Grob G.115E Tutor	VT Aerospace Ltd
	G-BYXP	Grob G.115E Tutor	VT Aerospace Ltd
	G-BYXR	Grob G.115E Tutor	VT Aerospace Ltd
	G-BYXS	Grob G.115E Tutor	VT Aerospace Ltd
	G-BYXT	Grob G.115E Tutor	VT Aerospace Ltd
	G-BYXU	PA-28-161 Warrior II	F. P. McGovern (G-BNUP)
	G-BYXV	Medway Eclipser	K. A. Christie
	G-BYXW	Medway Eclipser	G. A. Hazell
	G-BYXX	Grob G.115E Tutor	VT Aerospace Ltd
	G-BYXY	Grob G.115E Tutor	VT Aerospace Ltd
	G-BYXZ	Grob G.115E Tutor	VT Aerospace Ltd
	G-BYYA	Grob G.115E Tutor	VT Aerospace Ltd
	G-BYYB	Grob G.115E Tutor	VT Aerospace Ltd

Reg.	Type	Owner or Operator	Notes
G-BYYC	Hapi Cygnet SF-2A	C. D. Hughes	
G-BYYD	Cameron A-250 balloon	C. & J. M. Bailey	
G-BYYE	Lindstrand LBL-77A balloon	D. J. Cook	
G-BYYG	Slingsby T.67C	S. E. Marples & B. Dixon	
G-BYYI	BAe Jetstream 3107	Jetstream Executive Travel Ltd	
G-BYYK	Boeing 737-229C	European Aviation Air Charter Ltd	
G-BYYL	Avtech Jabiru UL 450	K. C. Lye	
G-BYYM	Raj Hamsa X'Air 582 (1)	B. Pilling & D. J. McCall	
G-BYYN	Pegasus Quantum 15-912	S. E. Robinson & A. Dixon	
G-BYYO	PA-28R -201 Arrow III	Stapleford Flying Club Ltd	
G-BYYP	Pegasus Quantum 15	D. A. Linsey-Bloom	
G-BYYR	Raj Hamsa X'Air 582 (4)	T. D. Bawden	
G-BYYT	Avtech Jabiru UL 450	S. C. E. Twiss	
G-BYYX	Team Minimax 91	P. L. Turner G. M. Culley	
G-BYYY	Pegasus Quantum 15-912	Redlands Airfield	
G-BYYZ	Staaken Z-21A Flitzer	P. K. Jenkins	
G-BYZA	AS.355F-2 Twin Squirrel	MMAir Ltd	
G-BYZD	Kis Cruiser	R. T. Clegg	
G-BYZE	AS.350B-2 Ecureuil	T. Clark	
G-BYZF	Raj Hamsa X'Air 582 (1)	R. P. Davies	
G-BYZG	Cameron A-275 balloon	Horizon Ballooning Ltd	
G-BYZJ	Boeing 737-3Q8	bmi baby (G-COLE)	
G-BYZL	Cameron GP-65 balloon	P. Thibo	
G-BYZM	PA-28-161 Warrior II	Goodair Leasing Ltd	
G-BYZO	Rans S.6-ES Coyote II	D. G. Matthews	
G-BYZP	Robinson R-22B	Wallis & Son	
G-BYZR	I.I.I. Sky Arrow 650TC	G-BYZR Flying Group	
G-BYZS	Avtech Jabiru UL-450	N. Fielding	
G-BYZT	Nova Vertex 26	M. Hay	
G-BYZU	Pegasus Quantum 15	N. I. Clifton	
G-BYZV	Sky 90-24 balloon	P. Farmer	
G-BYZW	Raj Hamsa X'Air 582 (2)	H. C. Lowther	
G-BYZX	Cameron R-90 balloon	D. K. Hempleman-Adams	
G-BYZZ	Robinson R-22B-2	Astra Helicopters Ltd	
G-BZAA	Mainair Blade	J. A. Kentzer	
G-BZAB	Mainair Rapier	B. J. Mould	
G-BZAD	Cessna 152	Cristal Air Ltd	
G-BZAE	Cessna 152	Jaxx Landing Ltd	
G-BZAF	Raj Hamsa X'Air 582 (1)	Y. A. Evans	
G-BZAG	Lindstrand LBL-105A balloon	A. M. Figiel	
G-BZAI	Pegasus Quantum 15	D. Paget	
G-BZAJ	WSK-PZL Koliber 160A	PZL International Marketing & Sales PLC/ North Weald	
G-BZAK	Raj Hamsa X'Air 582 (1)	R. J. Ripley	
G-BZAL	Mainair Blade 912	K. Worthington	
G-BZAO	Rans S.12XL	M. L. Robinson	
G-BZAP	Avtech Jabiru UL-450	S. Derwin	
G-BZAR	Denney Kitfox 4-1200 Speedster	C. E. Brookes (G-LEZJ)	
G-BZAS	Isaacs Fury II	H. A. Brunt & H. Frick	
G-BZAT	Avro RJ100	British Airways CitiExpress	
G-BZAU	Avro RJ100	British Airways CitiExpress	
G-BZAV	Avro RJ100	British Airways CitiExpress	
G-BZAW	Avro RJ100	British Airways CitiExpress	
G-BZAX	Avro RJ100	British Airways CitiExpress	
G-BZAY	Avro RJ100	British Airways CitiExpress	
G-BZAZ	Avro RJ100	British Airways CitiExpress	
G-BZBC	Rans S.6-ES Coyote II	A. J. Baldwin	
G-BZBE	Cameron A-210 balloon	Dragon Balloon Co. Ltd.	
G-BZBF	Cessna 172M	L. W. Scattergood	
G-BZBH	Thunder Ax6-65 balloon	R. B. & G. Clarke	
G-BZBI	Cameron V-77 balloon	C. & A. I. Gibson	
G-BZBJ	Lindstrand LBL-77A balloon	P. T. R. Ollivere	
G-BZBL	Lindstrand LBL-120A balloon	C. J. Dunkley	
G-BZBM	Cameron A-315 balloon	Listers of Coventry (Motors) Ltd	
G-BZBN	Thunder Ax9-120 balloon	K. Willie	
G-BZBO	Stoddard-Hamilton Glasair III	M. B. Hamlett/France	
G-BZBP	Raj Hamsa X'Air 582 (1)	R. J. Howlett	
G-BZBR	Pegasus Quantum 15-912	E. Lewis	
G-BZBS	PA-28-161 Warrior III	Aviation Rentals	
G-BZBT	Cameron H-34 Hopper balloon	British Telecommunications PLC	
G-BZBU	Robinson R-22	Helicentre Blackpool Ltd	

Notes	Reg.	Type	Owner or Operator
	G-BZBW	Rotorway Executive 162F	Southern Helicopters Ltd
	G-BZBX	Rans S.6-ES Coyote II	R. Johnstone
	G-BZBZ	Jodel D.9	S. Marom
	G-BZDA	PA-28-161 Warrior III	Aviation Rentals
	G-BZDB	Thruster T.600T	M. R. Jones
	G-BZDC	Mainair Blade	E. J. Wells & P. J. Smith
	G-BZDD	Mainair Blade 912	Barton Blade Group
	G-BZDE	Lindstrand LBL-210A balloon	Toucan Travel Ltd
	G-BZDF	CFM Streak Shadow SA	J. W. Beckett
	G-BZDH	PA-28R Cherokee Arrow 200-II	I. R. Chaplin
	G-BZDI	Aero L-39C Albatros	M. Gainza & E. Gavazzi/North Weald
	G-BZDJ	Cameron Z-105 balloon	BWS Security Systems Ltd
	G-BZDK	Raj Hamsa X'Air 582 (4)	B. Park
	G-BZDL	Pegasus Quantum 15-912	D. M. Holman
	G-BZDM	Stoddard-Hamilton Glastar	F. G. Miskelly
	G-BZDN	Cameron N-105 balloon	J. D. & K. Griffiths
	G-BZDP	SA Bulldog Srs 120/121	D. J. Rae
	G-BZDR	Tri-R Kis	T. J. Johnson
	G-BZDS	Pegasus Quantum 15-912	P. K. Dale
	G-BZDT	Maule MXT-7-180	Strongcrew Ltd
	G-BZDU	D.H.C.1 Chipmunk 22	M. R. Clark
	G-BZDV	Westland Gazelle HT.2	MW Helicopters Ltd/Stapleford
	G-BZDX	Cameron Colt 90 Sugarbox SS balloon	Stratos Ballooning GmbH & Co KG/Germany
	G-BZDY	Cameron Colt 90 Sugarbox SS balloon	Stratos Ballooning GmbH & Co KG/Germany
	G-BZDZ	Avtech Jabiru SP	R. M. Whiteside
	G-BZEA	Cessna A.152	Sky Leisure Aviation (Charters) Ltd
	G-BZEB	Cessna 152	Sky Leisure Aviation (Charters) Ltd
	G-BZEC	Cessna 152	Sky Leisure Aviation (Charters) Ltd
	G-BZED	Pegasus Quantum 15-912	M. P. Wimsey
	G-BZEE	AB-206B JetRanger 2	Yateley Helicopters Ltd (G-OJCB)
	G-BZEG	Mainair Blade	R. P. Cookson
	G-BZEH	PA-28 Cherokee 235B	A. D. Wood
	G-BZEJ	Raj Hamsa X'Air 582 (7)	X'Air Flying Group
	G-BZEK	Cameron C-70 balloon	Ballooning 50 Degrees Nord/Luxembourg
	G-BZEL	Mainair Blade	M. W. Bush
	G-BZEM	Glaser-Dirks DG.800G	I. M. Stromberg
	G-BZEN	Avtech Jabiru UL-450	B. W. Stockil
	G-BZEP	SA Bulldog Srs 120/121	A. J. Amato/Biggin Hill
	G-BZER	Raj Hamsa X'Air R100 (1)	N. P. Lloyd & H. Lloyd-Jones
	G-BZES	Rotorway Executive 90	Southern Helicopters Ltd (G-LUFF)
	G-BZET	Robin HR.200/120B	Anglian Flight Centres Ltd
	G-BZEU	Raj Hamsa X'Air 582 (2)	M. Bundy
	G-BZEV	Semicopter 1	M. E. Vahdat
	G-BZEW	Rans S.6-ES Coyote II	J. E. Gattrell & A. R. Trace
	G-BZEX	Raj Hamsa X'Air R.200 (2)	J. M. McCullough & R. T. Henry
	G-BZEY	Cameron N-90 balloon	Northants Auto Parts and Service Ltd
	G-BZEZ	CFM Streak Shadow	G. J. Pearce
	G-BZFB	Robin R.2112A	M. R. Brown
	G-BZFC	Pegasus Quantum 15-912	G. Addison
	G-BZFD	Cameron N-90 balloon	David Hataway Holdings Ltd
	G-BZFF	Raj Hamsa X'Air 582 (2)	D. L. Brown
	G-BZFG	Sky 105 balloon	Virgin Airship & Balloon Co. Ltd
	G-BZFH	Pegasus Quantum 15-912	Kent Scout Microlights
	G-BZFI	Avtech Jabiru UL	Group Family
	G-BZFJ	Westland Gazelle HT.2	European Marine Ltd
	G-BZFK	Team Minimax	R. C. Osler
	G-BZFN	SA Bulldog Srs 120/121	Thomas Aviation Ltd
	G-BZFO	Mainair Blade	I. Bell & M. Chambers
	G-BZFP	D.H.C.6 Twin Otter 310	Loganair Ltd
	G-BZFR	Extra EA.300/L	Powerhunt Ltd/Biggin Hill
	G-BZFS	Mainair Blade 912	S. P. Stone & F. A. Stephens
	G-BZFT	Murphy Rebel	N. A. Evans
	G-BZFU	Lindstrand LBL HS-110 airship	PNB Entreprenad AB/Sweden
	G-BZFV	Zenair CH.601UL	T. R. Sinclair & T. Clyde
	G-BZGA	D.H.C.1 Chipmunk 22 (WK585)	The Real Flying Co. Ltd/Shoreham
	G-BZGB	D.H.C.1 Chipmunk 22	Chipmunk Aviation Ltd
	G-BZGC	AS.355F-1 Twin Squirrel	McAlpine Helicopters Ltd (G-CCAO/G-SETA/ G-NEAS/G-CMMM/G-BNBJ)/Kidlington
	G-BZGD	PA-18 Super Cub 150	Proline Aviation
	G-BZGE	Medway Eclipser	G. Evans
	G-BZGF	Rans S.6-ES Coyote II	D. F. Castle
	G-BZGG	SE-3130B Alouette II	J. T. Meall (G-POSE)

Reg.	Type	Owner or Operator	Notes
G-BZGH	Cessna F.172N	Golf Hotel Group	
G-BZGI	Ultramagic M-145 balloon	European Balloon Co Ltd	
G-BZGJ	Thunder Ax10-180 S2 balloon	Merlin Balloons	
G-BZGK	NA OV-10B Bronco	Aircraft Restoration Co Ltd/Duxford	
G-BZGL	NA OV-10B Bronco	Aircraft Restoration Co Ltd/Duxford	
G-BZGM	Mainair Blade 912	D. Young	
G-BZGN	Raj Hamsa X'Air 582 (2)	C. S. Warr & P. A. Pilkington	
G-BZGO	Robinson R-44	P. Durkin	
G-BZGP	Thruster T.600N 460	M. L. Smith/Popham	
G-BZGR	Rans S.6-ES Coyote II	J. M. Benton	
G-BZGS	Mainair Blade 912	S. C. Reeve	
G-BZGT	Avtech Jabiru UL-450	P. H. Ronfell	
G-BZGU	Raj Hamsa X'Air 582 (4)	C. Kiernan	
G-BZGV	Lindstrand LBL-77A balloon	J. H. Dryden	
G-BZGW	Mainair Blade	C. S. M. Hallam	
G-BZGX	Raj Hamsa X'Air 582 (6)	A. Crowe	
G-BZGY	Dyn'Aero CR.100	D. Hayes	
G-BZGZ	Pegasus Quantum 15-912	W. H. J. Knowles	
G-BZHA	Boeing 767-336ER	British Airways	
G-BZHB	Boeing 767-336ER	British Airways	
G-BZHC	Boeing 767-336ER	British Airways	
G-BZHE	Cessna 152	Two Seven Aviation Ltd	
G-BZHF	Cessna 152	Modi Aviation Ltd	
G-BZHG	Tecnam P.92 Echo	M. & J. Turner	
G-BZHI	Enstrom F-28A-UK	Tindon Ltd (G-BPOZ)	
G-BZHJ	Raj Hamsa X'Air 582 (7)	T. Harrison-Smith	
G-BZHK	PA-28-181 Archer III	Melform Metals Ltd	
G-BZHL	Noorduyn AT-16 Harvard IIB	R. H. Cooper & S. Swallow	
G-BZHN	Pegasus Quantum 15-912	Eaglescott Microlights	
G-BZHO	Pegasus Quantum 15	R. L. Williams	
G-BZHP	Quad City Challenger II	F. Payne	
G-BZHR	Avtech Jabiru UL-450	G. W. Rowbotham	
G-BZHS	Shaw Europa	P. Waugh	
G-BZHT	PA-18A Super Cub 150	Lakes Gliding Club/Walney	
G-BZHU	Wag-Aero CUBy Sport Trainer	D. M. Lewington	
G-BZHV	PA-28-181 Archer III	M. Basson	
G-BZHW	PA-28-181 Archer III	Delta Kilo Services LLP	
G-BZHX	Thunder Ax11-250 S2 balloon	T. H. Wilson	
G-BZHY	Mainair Blade 912	M. Morris	
G-BZIA	Raj Hamsa X'Air 700 (1)	G. A. J. Salter	
G-BZIC	Lindstrand LBL Sun SS balloon	Ballongaventyr 1 Sakne AB/Sweden	
G-BZID	Montgomerie-Bensen B.8MR	A. Gault	
G-BZIG	Thruster T.600N	Ultra Air Ltd	
G-BZII	Extra EA.300/1	J. A. Carr	
G-BZIJ	Robin DR.400/500	Rob Airways Ltd	
G-BZIK	Cameron A-250 balloon	Breckland Balloons Ltd	
G-BZIL	Colt 120A balloon	Champagne Flights	
G-BZIM	Pegasus Quantum 15-912	A. & S. Cuthbertson	
G-BZIN	Robinson R-44	Helicentre Ltd	
G-BZIO	PA-28-161 Warrior III	Aviation Rentals/Bournemouth	
G-BZIP	Montgomerie-Bensen B.8MR	S. J. Boxall	
G-BZIS	Raj Hamsa X'Air 582 (2)	M. J. Badham	
G-BZIT	Beech 95-B55 Baron	Propellorhead Aviation Ltd	
G-BZIV	Avtech Jabiru UL	V. R. Leggott	
G-BZIW	Pegasus Quantum 15-912	J. M. Hodgson	
G-BZIX	Cameron N-90 balloon	Sport Promotion SRL/Italy	
G-BZIY	Raj Hamsa X'Air 582 (2)	I. K. Hogg	
G-BZIZ	Ultramagic H-31 balloon	G. D. O. Bartram	
G-BZJA	Cameron 90 Fire SS balloon	Chubb Fire Ltd	
G-BZJB	Aerostar Yakovlev Yak-52	D. Watt	
G-BZJC	Thruster T.600N	P. Johns	
G-BZJD	Thruster T.600T	P. G. Valentine/France	
G-BZJF	Pegasus Quantum 15	R. S. McMaster	
G-BZJG	Cameron A-400 balloon	Cameron Balloons Ltd	
G-BZJH	Cameron Z-90 balloon	Cameron Balloons Ltd	
G-BZJI	Nova X-Large 37 paraplane	M. Hay	
G-BZJJ	Robinson R-22B	Seiontair Ltd	
G-BZJL	Mainair Blade 912S	D. N. Powell	
G-BZJM	VPM M.16 Tandem Trainer	J. Musil	
G-BZJN	Mainair Blade 912	I. A. Forrest	
G-BZJO	Pegasus Quantum 15	J. D. Doran	
G-BZJP	Zenair CH.701UL	J. A. Ware	

Notes	Reg.	Type	Owner or Operator
	G-BZJR	Montgomerie-Bensen B.8MR	AES Radionic Surveillance Systems
	G-BZJS	Taylor JT.2 Titch	R. W. Clarke
	G-BZJU	Cameron A-200 balloon	Leeds Castle Enterprises Ltd
	G-BZJV	C.A.S.A. 1-131E Jungmann 1000	J. A. Sykes
	G-BZJW	Cessna 150F	R. J. Scott
	G-BZJX	Ultramagic N-250 balloon	Hot Air Balloons Ltd
	G-BZJZ	Pegasus Quantum 15	S. Baker
	G-BZKC	Raj Hamsa X'Air 582 (2)	P. J. Cheney
	G-BZKD	Stolp Starduster Too	P. & C. Edmunds
	G-BZKE	Lindstrand LBL-77B balloon	R. M. Cambridge
	G-BZKF	Rans S.6-ES Coyote II	A. W. Hodder
	G-BZKG	Extreme/Silex	R. M. Hardy
	G-BZKH	Flylight Airsports Doodle Bug/Target	B. Tempest
	G-BZKI	Flylight Airsports Doodle Bug/Target	S. Bond
	G-BZKJ	Flylight Airsports Doodle Bug/Target	Flylight Airsports Ltd/Sywell
	G-BZKK	Cameron V-56 balloon	P. J. Green & C. Bosley Gemini II
	G-BZKL	PA-28R-201 Arrow III	I. R. Chaplin/Andrewsfield
	G-BZKN	Campbell Cricket Mk 4	C. G. Ponsford
	G-BZKO	Rans S.6-ES Coyote II	J. A. R. Hartley
	G-BZKR	Cameron 90 Sugarbox SS balloon	Stratos Ballooning GmbH & Co KG
	G-BZKS	Ercoupe 415CD	A. H. Harper
	G-BZKT	Pegasus Quantum 15	Rochester Microlights Ltd
	G-BZKU	Cameron Z-105 balloon	N. A. Fishlock
	G-BZKV	Cameron Sky 90-24 balloon	Omega Selction Services Ltd
	G-BZKW	Ultramagic M-27 balloon	T. G. Church
	G-BZKX	Cameron V-90 balloon	Cameron Balloons Ltd
	G-BZKY	Focke Wulf Fw.189A-1	M. T. Pearce-Ware
	G-BZKZ	Lindstrand LBL-25A balloon	Lindstrand Hot-Air Balloons Ltd
	G-BZLA	SA.341G Gazelle 1	P. J. Brown
	G-BZLB	SA Bulldog Srs 120/121	L. Bax
	G-BZLC	WSK-PZL Koliber 160A	G. F. Smith
	G-BZLE	Rans S.6-ES Coyote II	R. T. P. Harris
	G-BZLF	CFM Shadow Srs CD	D. W. Stacey
	G-BZLG	Robin HR.200/120B	G. S. McNaughton
	G-BZLH	PA-28-161 Warrior II	Aviation Rentals
	G-BZLI	SOCATA TB-21 Trinidad TC	K. B. Hallam
	G-BZLK	Slingsby T.31M Motor Tutor	I. P. Manley
	G-BZLL	Pegasus Quantum 15-912	Caunton Graphites Syndicate
	G-BZLM	Mainair Blade	R. R. Celentano
	G-BZLO	Denney Kitfox Mk 2	M. W. Hanley
	G-BZLP	Robinson R-44	Regentweb Ltd
	G-BZLS	Cameron Sky 77-24 balloon	D. W. Young
	G-BZLT	Raj Hamsa X'Air 582 (1)	G. Millar
	G-BZLU	Lindstrand LBL-90A balloon	A. E. Lusty
	G-BZLV	Avtech Jabiru UL-450	G. Dalton
	G-BZLX	Pegasus Quantum 15-912	G. D. Ritchie
	G-BZLY	Grob G.109B	M. Tolson
	G-BZLZ	Pegasus Quantum 15-912	T. Bale & A. Martin
	G-BZMB	PA-28R-201 Arrow III	D. S. Seex
	G-BZMC	Avtech Jabiru UL	J. R. Banks
	G-BZMD	SA Bulldog Srs 120/121	D. M. Squires/Wellesbourne
	G-BZME	SA Bulldog Srs 120/121	B. Whitworth/Breighton
	G-BZMF	Rutan LongEz	R. A. Gardiner & A. McLaughlin
	G-BZMG	Robinson R-44	Ramsgill Aviation Ltd
	G-BZMH	SA Bulldog Srs 120/121	M. E. J. Hingley & Co Ltd
	G-BZMI	Pegasus Quantum 15-912	I. M. Bassett
	G-BZMJ	Rans S.6-ES Coyote II	Heskin Flying Group
	G-BZML	SA Bulldog Srs 120/121	I. D. Anderson
	G-BZMM	Robin DR.400/180R	N. A. C. Norman
	G-BZMO	Robinson R-22B	Sloane Helicopters Ltd/Sywell
	G-BZMR	Raj Hamsa X'Air 582 (2)	M. Grime
	G-BZMS	Mainair Blade	Beccles Buzzards
	G-BZMT	PA-28-161 Warrior III	Aviation Rentals
	G-BZMV	Cameron 80 Concept balloon	Latteria Soresinese Soc. Coop SRL/Italy
	G-BZMW	Pegasus Quantum 15-912	J. I. Greenshields
	G-BZMX	Cameron Z-90 balloon	Cameron Balloons Ltd
	G-BZMY	SPP Yakovlev Yak C-11	G. G. L. James
	G-BZMZ	CFM Streak Shadow	J. F. F. Fouche
	G-BZNB	Pegasus Quantum 15	R. C. Whittall
	G-BZNC	Pegasus Quantum 15-912	D. E. Wall
	G-BZND	Sopwith Pup (replica)	B. F. Goddard
	G-BZNE	Beech B300 Super King Air	G. Davies

Reg.	Type	Owner or Operator	Notes
G-BZNF	Colt 120A balloon	N. Charbonnier/Italy	
G-BZNG	Raj Hamsa X'Air 700 (1)	G. L. Craig	
G-BZNH	Rans S.6-ES Coyote II	R. B. Skinner	
G-BZNI	Bell 206B JetRanger 2	Trimax Ltd (G-ODIG/G-NEEP)	
G-BZNJ	Rans S.6-ES Coyote II	S. P. Read & M. H. Wise	
G-BZNK	Morane Saulnier M.S.315-D2	R. H. Cooper & S. Swallow	
G-BZNM	Pegasus Quantum 15	M. Tomlinson	
G-BZNN	Beech 76 Duchess	Aviation Rentals/Bournemouth	
G-BZNO	Ercoupe 415C	D. K. Tregilgas	
G-BZNP	Thruster T.600N	J. D. Gibbons	
G-BZNR	BAe 125 Srs 800B	RMC Group Services Ltd (G-XRMC)	
G-BZNS	Mainair Blade	M. K. B. Molyneux	
G-BZNT	Aero L-29 Delfin	H. P. Odone	
G-BZNU	Cameron A-300 balloon	Balloon School (International) Ltd	
G-BZNV	Lindstrand LBL-31A balloon	G. R. Down	
G-BZNW	Isaacs Fury II (K2048)	J. E. D. Rogerson	
G-BZNX	SOCATA M.S.880B Rallye Club	R. K. Stewart	
G-BZNY	Shaw Europa XS	W. J. Harrison	
G-BZNZ	Lindstrand LBL Cake SS balloon	Oxford Promotions (UK) Ltd	
G-BZOB	Slepcev Storch	J. E. & A. Ashby	
G-BZOD	Pegasus Quantum 15-912	S. M. Wilson	
G-BZOE	Pegasus Quantum 15	B. N. Thresher	
G-BZOF	Montgomerie-Bensen B.8MR gyroplane	S. J. M. Ledingham	
G-BZOG	Dornier Do.328-100	ScotAirways Ltd	
G-BZOI	Nicollier HN.700 Menestrel II	S. J. McCollum	
G-BZOL	Robin R.3000/140	F. Swetenham	
G-BZOM	Rotorway Executive 162F	J. A. Jackson	
G-BZON	SA Bulldog Srs 120/121	Roger Savage Gyroplanes Ltd	
G-BZOO	Pegasus Quantum 15-912	K. Brown	
G-BZOP	Robinson R-44	20:20 Logistics Ltd	
G-BZOR	Team Minimax 91	A. Watt	
G-BZOU	Pegasus Quantum 15-912	B. J. Holloway & D. R. Owen	
G-BZOV	Pegasus Quantum 15-912	D. Turner	
G-BZOW	Whittaker MW-7	G. W. Peacock	
G-BZOX	Cameron Colt 90B balloon	D. J. Head	
G-BZOY	Beech 76 Duchess	Aviation Rentals	
G-BZOZ	Van's RV-6	V. Edmondson	
G-BZPA	Mainair Blade 912S	J. McGoldrick	
G-BZPB	Hawker Hunter GA.11 (WB188 duckegg green)	B. R. Pearson	
G-BZPC	Hawker Hunter GA.11 (WB188 red)	B. R. Pearson	
G-BZPD	Cameron V-65 balloon	P. Spellward	
G-BZPE	Lindstrand LBL-310 balloon	Aerosaurus Balloons Ltd	
G-BZPF	Scheibe SF.24B Motorspatz 1	J. S. Gorrett	
G-BZPG	Beech C24R Sierra 200	Aviation Rentals	
G-BZPH	Van's RV-4	G-BZPH RV-4 Group	
G-BZPI	SOCATA TB.20 Trinidad	K. M. Brennan	
G-BZPJ	Beech 76 Duchess	Aviation Rentals	
G-BZPK	Cameron C-80 balloon	Horizon Ballooning Ltd	
G-BZPL	Robinson R-44	M. K. Shaw	
G-BZPM	Cessna 172S	Pooler-LMT Ltd	
G-BZPN	Mainair Blade 912S	S. Wing	
G-BZPP	Westland Wasp HAS.1	S. L. Negus	
G-BZPR	Ultramagic N-210 balloon	European Balloon Display Co. Ltd	
G-BZPS	SA Bulldog Srs 120/121	D. M. Squires	
G-BZPT	Ultramagic N-210 balloon	European Balloon Display Co Ltd	
G-BZPU	Cameron V-77 balloon	J. Vonka	
G-BZPV	Lindstrand LBL-90B balloon	D. P. Hopkins	
G-BZPW	Cameron V-77 balloon	J. Vonka	
G-BZPX	Ultramagic S-105 balloon	Scotair Balloons	
G-BZPY	Ultramagic H-31 balloon	Scotair Balloons	
G-BZPZ	Mainair Blade	M. C. W. Robertson	
G-BZRA	Rans S.6-ES Coyote II	A. W. Fish	
G-BZRC	D.H.115 Vampire T.11	D. Copley/Sandtoft	
G-BZRD	D.H.115 Vampire T.11	D. Copley/Sandtoft	
G-BZRE	P.56 Provost T.1	D. Copley/Sandtoft	
G-BZRF	P.56 Provost T.1	D. Copley/Sandtoft	
G-BZRG	Hunt Wing	W. G. Reynolds	
G-BZRJ	Pegasus Quantum 15-912	G-BZRJ Group	
G-BZRN	Robinson R-44	Toriamos Ltd	
G-BZRO	PA-30 Twin Comanche C	Comanche Hire Ltd	
G-BZRP	Pegasus Quantum 15-912	RAF Microlight Flying Association	

Notes	Reg.	Type	Owner or Operator
	G-BZRR	Pegasus Quantum 15-912	F. G. Green & T. Hudson
	G-BZRS	Eurocopter EC.135T-2	Bond Air Services Ltd
	G-BZRT	Beech 76 Duchess	Aviation Rentals/Bournemouth
	G-BZRU	Cameron V-90 balloon	Dragon Balloon Group
	G-BZRV	Van's RV-6	N. M. Hitchman
	G-BZRW	Mainair Blade 912S	N. D. Kube
	G-BZRX	Ultramagic M-105 balloon	Specialist Recruitment Group PLC
	G-BZRY	Rans S.6-ES Coyote II	C. J. Powell
	G-BZRZ	Thunder Ax11-250 S2 balloon	A. C. K. Rawson & J. J. Rudoni
	G-BZSA	Pegasus Quantum 15	D. W. Melville
	G-BZSB	Pitts S-1S Special	A. D. Ingold
	G-BZSC	Sopwith Camel F.1 (replica)	The Shuttleworth Collection/O.Warden
	G-BZSD	PA-46-350P Malibu Mirage	Meridian Aviation Ltd/Bournemouth
	G-BZSE	Hawker Hunter T.8B	Towerdrive Ltd
	G-BZSF	Hawker Hunter T.8B	Towerdrive Ltd
	G-BZSG	Pegasus Quantum 15-912	K. J. Gay
	G-BZSH	Ultramagic H-77 balloon	J. L. Hutsby
	G-BZSI	Pegasus Quantum 15	M. O. O'Brien
	G-BZSL	Sky 25-16 balloon	A. E. Austin
	G-BZSM	Pegasus Quantum 15	J. Walker & A. J. Johnson
	G-BZSO	Ultramagic M-77C balloon	C. C. Duppa-Miller
	G-BZSP	Stemme S.10	A. Flewelling & L. Bleaken
	G-BZSS	Pegasus Quantum 15-912	T. R. Marsh
	G-BZST	Avtech Jabiru UL	G. Hammond
	G-BZSU	Cameron A-315 balloon	Ballooning Network Ltd
	G-BZSV	Barracuda	M. J. Aherne
	G-BZSX	Pegasus Quantum 15-912	J. B. Greenwood
	G-BZSY	SNCAN Stampe SV-4A	G. J. N. Valvekens/Belgium
	G-BZSZ	Avtech Jabiru UL-450	M. P. Gurr & D. R. Burridge
	G-BZTA	Robinson R-44	Thurston Helicopters Ltd/Headcorn
	G-BZTC	Team Minimax 91	G. G. Clayton
	G-BZTD	Thruster T.600T 450 JAB	B. O. & B. C. McCartan
	G-BZTE	Cameron A-275 balloon	Richard Nash Cars Ltd
	G-BZTF	Yakovlev Yak-52	KY Flying Group
	G-BZTG	PA-34-220T Seneca V	L. R. Chiswell
	G-BZTH	Shaw Europa	T. J. Houlihan
	G-BZTI	Shaw Europa XS	W. Hoolachan
	G-BZTJ	C.A.S.A. Bu.133C Jungmeister	R. A. Seeley
	G-BZTK	Cameron V-90 balloon	E. Appollodorus
	G-BZTL	Cameron Colt Flying Ice Cream Cone SS balloon	Stratos Ballooning GmbH & Co. KG/Germany
	G-BZTM	Mainair Blade	H. N. Barrott
	G-BZTN	Shaw Europa XS	S. A. Smith
	G-BZTR	Mainair Blade	A. Raithby & ptnrs
	G-BZTS	Cameron 90 Bertie Bassett SS balloon	Trebor Bassett Ltd
	G-BZTT	Cameron A-275 balloon	Cameron Flights Southern Ltd
	G-BZTV	Mainair Blade 912S	D. J. Cook
	G-BZTW	Hunt Wing Avon 582 (1)	T. S. Walker
	G-BZTX	Mainair Blade 912	K. A. Ingham
	G-BZTY	Avtech Jabiru UL	R. P. Lewis
	G-BZTZ	MDH MD.600N	Amberley Aviation Ltd
	G-BZUB	Mainair Blade	S. E. Kearney
	G-BZUC	Pegasus Quantum 15-912	G. Breen/Portugal
	G-BZUD	Lindstrand LBL-105A balloon	Alton Aviation Ltd
	G-BZUE	Pegasus Quantum 15	W. J. Byrd
	G-BZUF	Mainair Rapier	S. J. Perry
	G-BZUG	RL.7A XP Sherwood Ranger	S. P. Sharp
	G-BZUH	Rans S.6-ES Coyote II	J. D. Sinclair-Day
	G-BZUI	Pegasus Quantum 15-912	A. P. Slade
	G-BZUK	Lindstrand LBL-31A balloon	G. R. J. Luckett/USA
	G-BZUL	Avtech Jabiru UL	P. Hawkins
	G-BZUM	Mainair Blade 912	R. B. Milton
	G-BZUN	Mainair Blade 912	E. Paxton & A. Jones
	G-BZUO	Cameron A-340HL balloon	Anglian Countryside Balloons Ltd
	G-BZUP	Raj Hamsa X'Air 582 (5)	A. A. A. Lappin
	G-BZUU	Cameron C-90 balloon	D. C. Ball & C. F. Pooley
	G-BZUV	Cameron H-24 balloon	J. N. Race
	G-BZUX	Pegasus Quantum 15	K. M. MacRae & ptnrs
	G-BZUY	Van's RV-6	R. D. Masters
	G-BZUZ	Hunt Avon-Blade R.100 (1)	J. A. Hunt
	G-BZVA	Zenair CH.701UL	M. W. Taylor
	G-BZVB	Cessna FR.172H	R. & E. M. Brereton (G-BLMX)

Reg.	Type	Owner or Operator	Notes
G-BZVC	Mickleburgh L.107	D. R. Mickleburgh	
G-BZVD	Cameron Colt 105 Forklift SS balloon	Stratos Ballooning GmbH & Co KG/Germany	
G-BZVE	Cameron N-133 balloon	I. M. Ashpole	
G-BZVF	Cessna 182T	Denston Hall Estate	
G-BZVG	Eurocopter AS.350B-3 Ecureuil	Windrush Aviation Ltd	
G-BZVH	Raj Hamsa X'Air 582 (1)	W. Bracken/Ireland	
G-BZVI	Nova Vertex 24 hang glider	M. Hay	
G-BZVJ	Pegasus Quantum 15	T. Mc.Mahon	
G-BZVK	Raj Hamsa X'Air 582 (2)	A. P. & J. M. Cadd	
G-BZVM	Rans S.6-ES Coyote II	N. N. Ducker	
G-BZVN	Van's RV-6	J. A. Booth	
G-BZVO	Cessna TR.182 RG	Swiftair Ltd	
G-BZVR	Raj Hamsa X'Air 582 (8)	Hummingbird Club	
G-BZVS	C.A.S.A. 1-131E Jungmann 2000	W. R. M. Beesley	
G-BZVT	I.I.I. Sky Arrow 650T	D. J. Goldsmith	
G-BZVU	Cameron Z-105 balloon	The Mall Balloon Team Ltd	
G-BZVV	Pegasus Quantum 15-912	D. R. Morton & R. Bain	
G-BZVW	Ilyushin IL-2 Stormovik	S. Swallow & R. H. Cooper/Sandtoft	
G-BZVX	Ilyushin IL-2 Stormovik	S. Swallow & R. H. Cooper/Sandtoft	
G-BZVZ	Eurocopter AS.355N Twin Squirrel	Iiona Ltd	
G-BZWB	Mainair Blade 912	L. Cottle	
G-BZWC	Raj Hamsa X'Air Falcon 912 (1)	C. McAfee	
G-BZWF	Colt AS-120 airship	MA Flying Ltd	
G-BZWG	PA-28 Cherokee 140	E. & H. Merkado	
G-BZWH	Cessna 152	J. & H. Aviation Services Ltd	
G-BZWI	Medway Eclipser	R. A. Keene	
G-BZWJ	CFM Streak Shadow SA	T. A. Morgan	
G-BZWK	Avtech Jabiru SK	M. J. Mines	
G-BZWM	Pegasus XL-Q	D. T. Evans	
G-BZWN	Van's RV-8	A. J. Symms & R. D. Harper	
G-BZWS	Pegasus Quantum 15-912	G-BZWS Syndicate	
G-BZWT	Technam P.92-EM Echo	R. F. Cooper	
G-BZWU	Pegasus Quantum 15-912	P. C. Hogg	
G-BZWV	Steen Skybolt	P. D. & K. Begley	
G-BZWX	Whittaker MW.5D Sorcerer	G. E. Richardson	
G-BZWY	CFM Streak Shadow SA	B. Cartwright	
G-BZWZ	Van's RV-6	J. Shanley	
G-BZXA	Raj Hamsa X'Air V2 (1)	D. W. Mullin	
G-BZXB	Van's RV-6	B. J. King-Smith & D. J. Akerman	
G-BZXC	SA Bulldog Srs 120/121	A. R. Oliver	
G-BZXD	Rotorway Executive 162F	P. G. King	
G-BZXE	D.H.C.1 Chipmunk 22	K. Moore	
G-BZXF	Cameron A-210 balloon	Off The Ground Balloon Co Ltd	
G-BZXG	Dyn' Aero MCR-01	J. Rankin	
G-BZXI	Nova Philou 26 hang glider	M. Hay	
G-BZXJ	Schweizer 269-1	Helicentre Ltd/Blackpool	
G-BZXK	Robin HR.200/120B	Aviation Rentals	
G-BZXL	Whittaker MW.5D Sorcerer	R. Hatton	
G-BZXM	Mainair Blade 912	P. Harper	
G-BZXN	Avtech Jabiru UL-450	J. Armstrong	
G-BZXO	Cameron Z-105 balloon	Airship & Balloon Co. Ltd	
G-BZXP	Kiss 400-582 (1)	E. D. Deed	
G-BZXR	Cameron N-90 balloon	Derbyshire Building Soc.	
G-BZXS	SA Bulldog Srs 120/121	K. J. Thompson	
G-BZXT	Mainair Blade 912	Barton 912 Flyers	
G-BZXV	Pegasus Quantum 15-912	S. Laws	
G-BZXW	VPM M.16 Tandem Trainer	S. J. Tyler (G-NANA)	
G-BZXX	Pegasus Quantum 15-912	G-BZXX Group	
G-BZXY	Robinson R-44	Extraviation Ltd/North Weald	
G-BZXZ	SA Bulldog Srs 120/121	C. R. Tilley	
G-BZYA	Rans S.6-ES Coyote II	M. R. Osbourn	
G-BZYB	Westland Gazelle HT.3	Gazelle Flying Group	
G-BZYD	Westland Gazelle AH.1	Aerocars Ltd	
G-BZYE	Robinson R-22B	Plane Talking Ltd/Elstree	
G-BZYG	Glaser-Dirks DG.500MB	R. C. Bromwich	
G-BZYI	Nova Phocus 123 hang glider	M. Hay	
G-BZYK	Avtech Jabiru UL	Cloudbase Aviation G-BZYK/Redhill	
G-BZYL	Rans S.6-ES Coyote II	C. B. Heslop	
G-BZYM	Raj Hamsa X'Air 700 (1A)	G. Fleck	
G-BZYN	Pegasus Quantum 15-912	G. B. Shaw	
G-BZYO	Colt 210A balloon	P. M. Forster	
G-BZYP	BAe Jetstream 3200	Trident Aviation Leasing (Ireland) Ltd *(stored)*	

Notes	Reg.	Type	Owner or Operator
	G-BZYR	Cameron N-31 balloon	N. J. Langley
	G-BZYS	Micro Aviation Bantam B.22-S	D. L. Howell
	G-BZYT	Interavia 80TA	J. King
	G-BZYU	Whittaker MW.6 Merlin	K. J. Cole
	G-BZYV	Snowbird Mk.V 582 (1)	S. Jones
	G-BZYW	Cameron N-90 balloon	Bailey Balloons
	G-BZYX	Raj Hamsa X'Air 700 (1A)	A. G. Marsh
	G-BZYY	Cameron N-90 balloon	Mason Zimbler Ltd
	G-BZZD	Cessna F.172M	R. H. M. Richardson-Bunbury (G-BDPF)
	G-CAHA	PA-34-200T Seneca II	H. & R. Marshall
	G-CAIN	CFM Shadow Srs CD	S. K. Starling (G-MTKU)
	G-CALL	PA-23 Aztec 250F	J. D. Moon
	G-CAMB	AS.355F-2 Twin Squirrel	Cambridgeshire & Essex Air Support
	G-CAMM	Hawker Cygnet (replica)	D. M. Cashmore
	G-CAMP	Cameron N-105 balloon	Hong Kong Balloon & Airship Club
	G-CAMR	BFC Challenger II	P. R. A. Walker
	G-CAPI	Mudry/CAARP CAP.10B	PI Group (G-BEXR)
	G-CAPX	Avions Mudry CAP.10B	H. J. Pessall
	G-CARS†	Pitts S-2A Special (replica)★	Toyota Ltd
	G-CBAB	SA Bulldog Srs 120/121	Propshop Ltd/Duxford
	G-CBAD	Mainair Blade 912	P. Lister
	G-CBAE	BAe 146-200	BAE Systems (Operations) Ltd
	G-CBAF	Lancair 320	R. W. Fairless
	G-CBAH	Raj Hamsa X'Air 582 (5)	D. N. B. Hearn
	G-CBAI	Flight Design CT.2K	K. D. Taylor
	G-CBAJ	D.H.C.1 Chipmunk 22	J. Lamb
	G-CBAK	Robinson R-44	J. Robinson
	G-CBAL	PA-28-161 Warrior II	Britannia Airways Ltd
	G-CBAN	SA Bulldog Srs 120/121	C. Hilliker
	G-CBAP	Zenair CH.601UL	L. J. Lowry
	G-CBAR	Stoddard-Hamilton Glastar	C. M. Barnes
	G-CBAS	Rans S.6-ES Coyote II	S. R. Green
	G-CBAT	Cameron Z-90 balloon	British Telecommunications PLC
	G-CBAU	Rand-Robinson KR-2	B. Normington
	G-CBAV	Raj Hamsa X'Air V.2 (1)	D. W. Stamp & G. J. Lampitt
	G-CBAW	Cameron A-300 balloon	D. K. Hempleman-Adams
	G-CBAX	Tecnam P.92-EM Exho	R. P. Reeves
	G-CBAZ	Rans S.6-ES Coyote II	G. V. Willder
	G-CBBA	Robin DR.400/180	H. P. K. Ferdinand/North Weald
	G-CBBB	Pegasus Quantum 15-912	Charlie Bravo Group
	G-CBBC	SA Bulldog Srs 120/121	Bulldog Support Ltd
	G-CBBF	Beech 76 Duchess	Liddell Aircraft Ltd
	G-CBBG	Mainair Blade	P. J. Donoghue
	G-CBBH	Raj Hamsa X'Air V2 (2)	R. D. Parkinson
	G-CBBL	SA Bulldog Srs 120/121	I. R. Bates
	G-CBBM	MXP-740 Savannah J (1)	C. E. Passmore
	G-CBBN	Pegasus Quantum 15-912	G-CBBN Group
	G-CBBO	Whittaker MW.5D Sorcerer	P. J. Gripton
	G-CBBP	Pegasus Quantum 15-912	P. F. Warren
	G-CBBR	SA Bulldog Srs 120/121	Elite Consultancy Corporation Ltd
	G-CBBS	SA Bulldog Srs 120/121	European Light Aviation Ltd
	G-CBBT	SA Bulldog Srs 120/121	Newcastle Bulldog Group Ltd
	G-CBBU	SA Bulldog Srs 120/121	Elite Consultancy Corporation Ltd
	G-CBBV	Westland Gazelle HT.3	C3 Consulting
	G-CBBW	SA Bulldog Srs 120/121	S. E. Robottom-Scott
	G-CBBX	Lindstrand LBL-69A balloon	J. L. F. Garcia
	G-CBBZ	Pegasus Quantum 15-912	A. J. Irving
	G-CBCA	PA-32R-301T Saratoga IITC	Thistle Aircraft Leasing Ltd
	G-CBCB	SA Bulldog Srs 120/121	The General Aviation Trading Co. Ltd
	G-CBCD	Pegasus Quantum 15	I. A. Lumley
	G-CBCE	C.A.S.A. 1-131E Jungmann (replica)	E. B. Toulson/Breighton
	G-CBCF	Pegasus Quantum 15-912	G-CBCF Group
	G-CBCH	Zenair CH.701UL	L. G. Millen
	G-CBCI	Raj Hamsa X'Air 582 (1)	A. J. Varga
	G-CBCJ	RAF 2000 GTX-SE gyroplane	J. P. Comerford
	G-CBCK	Tipsy T.66 Nipper Srs 3	N. M. Bloom (G-TEDZ)
	G-CBCL	Stoddard-Hamilton Glastar	A. H. Harper
	G-CBCM	Raj Hamsa X'Air 700 (1A)	G. Firth
	G-CBCN	Schweizer 269C-1	Helicentre Liverpool Ltd
	G-CBCP	Van's RV-6A	G-CBCP Group

Reg.	Type	Owner or Operator	Notes
G-CBCR	SA Bulldog Srs 120/121	S. C. Smith	
G-CBCS	BAe Jetstream 3202	Eastern Airways Ltd	
G-CBCT	SA Bulldog Srs 120/121	T. Brun/France	
G-CBCU	H.S. Harrier GR.3	Y. Dumortier/Belgium	
G-CBCV	SA Bulldog Srs 120/121	C. A. Patter	
G-CBCX	Pegasus Quantum 15	D. V. Lawrence	
G-CBCY	Beech C24R Sierra Super	Plane Talking Ltd	
G-CBCZ	CFM Streak Shadow SLA	J. A. Hambleton	
G-CBDA	BAe Jetstream 3217	Eastern Airways Ltd	
G-CBDC	Thruster T.600N 450-JAB	David Clarke Microlight Aircraft	
G-CBDD	Mainair Blade	R. D. Smith	
G-CBDG	Zenair CH.601HD	R. E. Lasnier	
G-CBDH	Flight Design CT.2K	K. Tuck	
G-CBDI	Denney Kitfox Mk.2	J. G. D. Barbour	
G-CBDJ	Flight Design CT.2K	P. J. Walker	
G-CBDK	SA Bulldog Srs 120/121	J. N. Randle	
G-CBDL	Mainair Blade	D. Lightwood	
G-CBDM	Tecnam P.92-EM Echo	J. J. Cozens	
G-CBDN	Mainair Blade	A. & R. W. Osborne	
G-CBDO	Raj Hamsa X'Air 582 (1)	A. Campbell	
G-CBDP	Mainair Blade 912	D. S. Parker	
G-CBDR	Agusta A.109A-II	Castle Air Charters Ltd	
G-CBDS	SA Bulldog Srs 120/121	H. R. M. Tyrrell	
G-CBDT	Zenair CH.601HD	D. G. Watt	
G-CBDU	Quad City Challenger II	Hiscox Cases Ltd	
G-CBDV	Raj Hamsa X'Air 582	R. J. Brown	
G-CBDW	Raj Hamsa X'Air Jabiru (1)	A. W. G. Ambler	
G-CBDX	Pegasus Quantum 15	C. C. Beck	
G-CBDY	Raj Hamsa X'Air V.2 (2)	P. M. Stoney	
G-CBDZ	Pegasus Quantum 15-912	C. I. D. H. Garrison	
G-CBEB	Kiss 400-582 (1)	P. R. J. & A. R. R. Williams	
G-CBEC	Cameron Z-105 balloon	A. L. Ballarino/Italy	
G-CBED	Cameron Z-90 balloon	John Aimo Balloons SAS/Italy	
G-CBEE	PA-28R Cherokee Arrow 200	IHV Aviation Ltd	
G-CBEF	SA Bulldog Srs 120/121	M. A. Wilkinson	
G-CBEG	Robinson R-44	Abel Developments Ltd	
G-CBEH	SA Bulldog Srs 120/121	R. E. Dagless	
G-CBEI	PA-22 Colt 108	M. E. Grogan	
G-CBEJ	Colt 120A balloon	J. A. Gray	
G-CBEK	SA Bulldog Srs 120/121	S. Landregan	
G-CBEL	Hawker Fury FB.11	J. A. D. Bradshaw	
G-CBEM	Mainair Blade	M. Earp	
G-CBEN	Pegasus Quantum 15-912	N. J. P. West	
G-CPEO	BAe Jetstream 3206	Trident Aviation Leasing Services Ltd	
G-CBEP	BAe Jetstream 3206	Trident Aviation Leasing Services Ltd *(stored)*	
G-CBER	BAe Jetstream 3206	Trident Aviation Leasing Services Ltd *(stored)*	
G-CBES	Shaw Europa XS	M. R. Hexley	
G-CBET	Mainair Blade 912S	D. F. Kenny	
G-CBEU	Pegasus Quantum 15-912	C. Lee	
G-CBEV	Pegasus Quantum 15-912	T. Lee	
G-CBEW	Flight Design CT.2K	Shy Talk Group	
G-CBEX	Flight Design CT.2K	B. W. T. Rood	
G-CBEY	Cameron C-80 balloon	D. V. Fowler	
G-CBEZ	Robin DR.400/180	K. V. Field	
G-CBFA	Diamond DA.40 Star	Lyrastar Ltd	
G-CBFB	Diamond DA.40 Star	Phantom Air Ltd	
G-CBFC	Diamond DA.40 Star	Diamond Aircraft (UK) Ltd/Gamston	
G-CBFE	Raj Hamsa X'Air V.2 (1)	M. L. Powell	
G-CBFF	Cameron O-120 balloon	T. M. C. McCoy	
G-CBFH	Thunder Ax8-105 S2 balloon	D. V. Fowler & A. N. F. Pertwee	
G-CBFJ	Robinson R-44	Scotia Helicopters Ltd	
G-CBFK	Murphy Rebel	P. J. Gibbs	
G-CBFM	SOCATA TB.21 Trinidad	Exec Flight Ltd	
G-CBFN	Robin DR.100/200B	Foxtrot November Group	
G-CBFO	Cessna 172S	Oxford Aviation Services Ltd/Kidlington	
G-CBFP	SA Bulldog Srs 120/121	Taylor Aviation Ltd	
G-CBFT	Raj Hamsa X'Air 582 (5)	L. Reilly	
G-CBFU	SA Bulldog Srs 120/121	J. R. & S. J. Huggins	
G-CBFV	Ikarus C.42	P. A. D. Chubb	
G-CBFW	Bensen B.8	B. F. Pearson	
G-CBFX	Rans S.6-ES Coyote II	G. P. Jones	
G-CBFY	Cameron Z-250 balloon	M. L. Gabb	

Notes	Reg.	Type	Owner or Operator
	G-CBFZ	Avtech Jabiru SPL-450	A. H. King
	G-CBGA	PZL-110 Koliber 160A	Rapidangel Ltd
	G-CBGB	Zenair CH.601UL	J. F. Woodham
	G-CBGC	SOCATA TB.10 Tobago	Tobago Aviation Ltd
	G-CBGD	Zenair CH.701UL	I. S. Walsh
	G-CBGE	Tecnam P.92-EM Echo	T. C. Robson
	G-CBGF	PA-31-310 Navajo B	Datalake Ltd
	G-CBGG	Pegasus Quantum 15	P. C. Bishop
	G-CBGH	Teverson Bisport	R. C. Teverson
	G-CBGI	CFM Streak Shadow	M. W. W. Clotworthy
	G-CBGJ	Aeroprakt A.22 Foxbat	M. J. Whiteman-Haywood
	G-CBGK	H.S. Harrier GR.3	Y- Dumortier/Belgium
	G-CBGL	MH.1521M Broussard	A. I. Milne
	G-CBGM	Mainair Blade 912	J. R. Pearce
	G-CBGN	Van's RV-4	G. A. Nash
	G-CBGO	Murphy Maverick 430	C. R. Ellis & E. A. Wrathall
	G-CBGP	Ikarus C.42 FB UK	C. F. Welby
	G-CBGR	Avtech Jabiru UL-450	K. R. Emery
	G-CBGS	Cyclone AX2000	P. J. Rooke
	G-CBGT	Mainair Blade 912	J. A. Cresswell
	G-CBGU	Thruster T.600N 450-JAB	K. Ford & M. Gill
	G-CBGV	Thruster T.600N 450-JAB	Red Arrow Syndicate
	G-CBGW	Thruster T.600N 450-JAB	N. J. S. Pitman
	G-CBGX	SA Bulldog Srs 120/121 (XX622)	G. B. Pearce
	G-CBGY	Mainair Blade 912	M. Talbot
	G-CBGZ	Westland Gazelle HT.2	D. Weatherhead Ltd
	G-CBHA	SOCATA TB.10 Tobago	Oscar Romeo Aviation Ltd
	G-CBHB	Raj Hamsa X'Air 582 (5)	R. A. J. Graham
	G-CBHC	RAF 2000 GTX-SE gyroplane	A. J. Thomas
	G-CBHD	Cameron Z-160 balloon	Ballooning 50 Degrees Nord/Luxembourg
	G-CBHE	Slingsby T.67M Firefly 260	R. Swann
	G-CBHG	Mainair Blade 912S	B. S. Hope
	G-CBHI	Shaw Europa XS	B. Price
	G-CBHJ	Mainair Blade 912	B. C. Jones
	G-CBHK	Pegasus Quantum 15 (HKS)	B. Dossett
	G-CBHL	AS.350B-2 Ecureuil	Helimac Ltd
	G-CBHM	Mainair Blade 912	W. T. Milburn
	G-CBHN	Pegasus Quantum 15-912	G. G. Cook
	G-CBHO	Gloster Gladiator II	Retro Track & Air (UK) Ltd
	G-CBHP	Corby CJ-1 Starlet	D. H. Barker
	G-CBHR	Stephens Akro Z	Acro Laser Co Ltd
	G-CBHT	Dassault Falcon 900EX	TAG Aviation (UK) Ltd (G-GPWH)
	G-CBHU	RL.5A Sherwood Ranger	M. J. Gooch
	G-CBHV	Raj Hamsa X'Air 582 (5)	M. Furniss
	G-CBHW	Cameron Z-105 balloon	Bristol Chamber of Commerce, Industry & Shipping
	G-CBHX	Cameron V-77 balloon	N. A. Apsey
	G-CBHY	Pegasus Quantum 15-912	M. W. Abbott
	G-CBHZ	RAF 2000 GTX-SE gyroplane	M. P. Donnelly
	G-CBIB	Flight Design CT.2K	J. A. Moss
	G-CBIC	Raj Hamsa X'Air V2 (2)	J. T. Blackburn & D. R. Sutton
	G-CBID	SA Bulldog Srs 120/121	D. A. Steven
	G-CBIE	Flight Design CT.2K	S. J. Page
	G-CBIF	Avtech Jabiru UL-450	J. A. Iszard
	G-CBIG	Mainair Blade 912	T. R. Clark
	G-CBIH	Cameron Z-31 balloon	Cameron Balloons Ltd
	G-CBII	Raj Hamsa X'Air 582 (2)	A. Worthington
	G-CBIJ	Ikarus C.42 FB UK Cyclone	G. S. Hill & J. D. Arthurs
	G-CBIK	Rotorway Executive 162F	J. Hodson
	G-CBIL	Cessna 182K	E. Bannister & J. R. C. Spooner (G-BFZZ)/ E. Midlands
	G-CBIM	Lindstrand LBL-90A balloon	R. K. Parsons
	G-CBIN	Team Minimax 91	K. J. Walton
	G-CBIO	Thruster T.600N 450-JAB	G. J. Slater
	G-CBIP	Thruster T.600N 450-JAB	K. D. Mitchell
	G-CBIR	Thruster T.600N 450-JAB	M. L. Smith
	G-CBIS	Raj Hamsa X'Air 582 (2)	P. T. W. T. Derges
	G-CBIT	RAF 2000 GTX-SE gyroplane	Terrafirma Services Ltd
	G-CBIU	Cameron 95 Flame SS balloon	PSH Skypower Ltd
	G-CBIV	Skyranger 912 (1)	P. M. Dewhurst & S. N. Bond
	G-CBIW	Lindstrand LBL-310A balloon	C. E. Wood
	G-CBIX	Zenair CH.601UL	R. A. & B. M. Roberts

Reg.	Type	Owner or Operator
G-CBIY	Aerotechnik EV-97 Eurostar	E. M. Middleton
G-CBIZ	Pegasus Quantum 15-912	A. P. Lambert
G-CBJA	Kiss 400-582 (1)	M. S. R. Burak
G-CBJD	Stoddard-Hamilton Glastar	K. F. Farey
G-CBJE	RAF 2000 GTX-SE gyroplane	K. F. Farey
G-CBJG	D.H.C.1 Chipmunk 20	C. J. Rees
G-CBJH	Aeroprakt A.22 Foxbat	H. Smith
G-CBJI	Cameron N-90 balloon	C. D. H. Oakland
G-CBJJ	SA Bulldog Srs 120/121 (XX525)	Elite Consultancy Corporation Ltd
G-CBJK	SA Bulldog Srs 120/121 (XX713/2)	Elite Consultancy Corporation Ltd
G-CBJL	Kiss 400-582 (1)	R. E. Morris
G-CBJM	Avtech Jabiru SP-470	A. T. Moyce
G-CBJN	RAF 2000 GTX-SE gyroplane	R. Hall
G-CBJO	Pegasus Quantum 15-912	J. H. Bradbury
G-CBJP	Zenair CH.601UL	R. E. Peirse
G-CBJR	Aerotechnik EV-97 Eurostar	B. J. Crocket
G-CBJS	Cameron C-60 balloon	J. M. Stables
G-CBJT	Mainair Blade	T. K. I. Dearden
G-CBJU	Van's RV-7A	T. W. Waltham
G-CBJV	Rotorway Executive 162F	Sacker Potatoes Ltd
G-CBJW	Ikarus C.42 FB UK	T. Collins & M. J. Male
G-CBJX	Raj Hamsa X'Air Falcon J22	M. R. Coreth
G-CBJZ	Westland Gazelle HT.3	K. G. Theurer
G-CBKA	Westland Gazelle HT.3	MW Helicopters Ltd/Stapleford
G-CBKB	Bücker Bu.181C Bestmann	W. R. & G. D. Snadden
G-CBKC	Westland Gazelle HT.3	T. E. Westley
G-CBKD	Westland Gazelle HT.2	Flying Scout Ltd
G-CBKE	Kiss 400-582 (1)	R. J. Howell
G-CBKF	Easy Raider J2.2 (1)	R. R. Armstrong
G-CBKG	Thruster T.600N 450 JAB	G. E. Hillyer-Jones
G-CBKI	Cameron Z-90 balloon	Wheatfields Park Ltd
G-CBKJ	Cameron Z-90 balloon	Invista (UK) Holdings Ltd
G-CBKK	Ultramagic S-130 balloon	Airbourne Adventures Ltd
G-CBKL	Raj Hamsa X'Air 582 (1)	Caithness X-Air Group
G-CBKM	Mainair Blade 912	N. Purdy
G-CBKN	Mainair Blade 912	S. G. Ward
G-CBKO	Mainair Blade 912S	I. Steele
G-CBKR	PA-28-161 Warrior III	Devon School of Flying Ltd
G-CBKS	Kiss 400-582 (1)	S. Kilpin
G-CBKU	Ikarus C.42 FB UK	C. Blackburn
G-CBKV	Cameron Z-77 balloon	J. F. Till
G-CBKW	Pegasus Quantum 15-912	I. W. Trench
G-CBKY	Avtech Jabiru SP-470	P. R. Sistern
G-CBLA	Aero Designs Pulsar XP	J. P. Kynaston
G-CBLB	Technam P..92-EM Echo	M. A, Lomas
G-CBLD	Mainair Blade 912S	N. E. King
G-CBLE	Robin R.2120U	Aviation Now Ltd
G-CBLF	Raj Hamsa X'Air 582 (11)	E. G. Bishop
G-CBLH	Raj Hamsa X'Air 582 (11)	S. Rance
G-CBLI	IDA Bacau Yakovlev Yak-52	R. H. Reeves
G-CBLJ	Aerostar Yakovlev Yak-52	G. H. Wilson
G-CBLK	Hawker Hind	Aero Vintage Ltd
G-CBLL	Pegasus Quantum 15-912	P. R. Jones
G-CBLM	Mainair Blade 912	G-CBLM Flying Group
G-CBLN	Cameron Z-31 balloon	N. J. Langley
G-CBLO	Lindstrand LBL-42A balloon	Balloon & Airship Co Ltd
G-CBLP	Raj Hamsa X'Air Falcon	T. W. Oakley
G-CBLT	Mainair Blade 912	G. Edwards
G-CBLU	Cameron C-90 balloon	A. G. Martin
G-CBLV	Flight Design CT.2K	A. R. Pickering
G-CBLW	Raj Hamsa X'Air Falcon V2 (1)	A. R. Cundill
G-CBLX	Kiss 400-582 (1)	J. H. Hayday
G-CBLY	Grob G.109B	G-CBLY Syndicate
G-CBLZ	Rutan LongEz	R. P. H. Hancock
G-CBMA	Raj Hamsa X'Air 582 (10)	K. Angel
G-CBMB	Cyclone Ax2000	York Microlight Centre Ltd/Rufforth
G-CBMC	Cameron Z-105 balloon	First Flight
G-CBMD	IDA Bacau Yakovlev Yak-52	R. J. Hunter
G-CBME	Cessna F.172M	Skytrax Aviation Ltd
G-CBMI	Yakovlev Yak-52	A. Burani
G-CBMJ	RAF 2000 GTX-SE gyroplane	C. D. Upsall
G-CBMK	Cameron Z-120 balloon	Flying Pictures Ltd

Notes	Reg.	Type	Owner or Operator
	G-CBML	D.H.C.6 Twin Otter 310	Isles of Scilly Skybus Ltd
	G-CBMM	Mainair Blade 912	M. R. Mosley
	G-CBMO	PA-28 Cherokee 180	C. Woodliffe
	G-CBMP	Cessna R.182	Orman (Carrolls Farm) Ltd
	G-CBMR	Medway Eclipser	A. Bradfield
	G-CBMS	Medway Eclipser	R. R. Bagge
	G-CBMT	Robin DR.400/180	A. C. Williamson
	G-CBMU	Whittaker MW.6-S Fat Boy Flyer	F. J. Brown
	G-CBMV	Pegasus Quantum 15	B. Hamilton
	G-CBMW	Zenair CH.701 UL	C. Long
	G-CBMX	Kiss 400-582 (1)	D. L. Turner
	G-CBMZ	Aerotechnik EV-97 Eurostar	P. Grenet & J. C. O'Donnell
	G-CBNA	Flight Design CT.2K	D. M. Wood
	G-CBNB	Eurocopter EC.120B	Arenberg Consultadoria E Servicos LDA/Madeira
	G-CBNC	Mainair Blade 912	A. C. Rowlands
	G-CBNF	Rans S.7 Courier	M. Rockliff
	G-CBNG	Robin R.2112	Solway Flyers Ltd/Carlisle
	G-CBNI	Lindstrand LBL-180A balloon	Cancer Research UK
	G-CBNJ	Raj Hamsa X'Air 912 (1)	912 X'Air Group
	G-CBNK	Aerotechnik EV-97 Eurostar	M. R. M. Welch
	G-CBNL	Dyn'Aero MCR-01 Club	D. H. Wilson
	G-CBNM	NA P-51D Mustang	Patina Ltd/Duxford
	G-CBNO	CFM Streak Shadow	D. J. Goldsmith
	G-CBNR	BD.700-1A10 Global Express	Marshall of Cambridge Aerospace Ltd
	G-CBNT	Pegasus Quantum 15-912	R. K. Watson
	G-CBNU	V.S.361 Spitfire LF.IX	M. Aldridge
	G-CBNV	Rans S.6-ES Coyote II	C. W. J. Davis
	G-CBNW	Cameron N-105 balloon	Bailey Balloons
	G-CBNX	Mongomerie-Bensen B.8MR	K. Ashford
	G-CBNY	Kiss 400-582 (1)	R. Redman
	G-CBNZ	Team Minimax 1700R	J. J. Penney
	G-CBOA	Auster B.8 Agricola Srs 1	C. J. Baker
	G-CBOC	Raj Hamsa X'Air 582 (5)	A. J. McAleer
	G-CBOD	Ikarus C.42 FB UK	B. Hunter
	G-CBOE	Hawker Hurricane IIB	Classic Aero Engineering Ltd
	G-CBOF	Shaw Europa XS	I. W. Ligertwood
	G-CBOG	Mainair Blade 912S	J. S. Little
	G-CBOK	Rans S.6-ES Coyote II	S. A. Clarehughr
	G-CBOL	Mainair Blade	B. Jackson
	G-CBOM	Mainair Blade 912	G. Suckling
	G-CBOO	Mainair Blade 912S	A. H. Walker
	G-CBOP	Avtech Jabiru UL-450	D. W. Batchelor
	G-CBOR	Cessna F.172N	P. Seville
	G-CBOS	Rans S.6-XS Coyote II	R. Skene
	G-CBOT	Robinson R-44	Aviation Rentals
	G-CBOU	Bensen-Parsons 2 place gyroplane	R. Collin & M.S.Sparkes
	G-CBOV	Mainair Blade	M. J. Devane
	G-CBOW	Cameron Z-120 balloon	Associated Technologies Ltd
	G-CBOY	Pegasus Quantum 15-912	T. G. Jackson
	G-CBOZ	IDA Bacau Yakovlev Yak-52	T. M. Knight
	G-CBPC	Sportavia-Putzer RF-5B Sperber	Lee RF-5B Group
	G-CBPD	Ikarus C.42 FB UK	Waxwing Group
	G-CBPE	SOCATA TB.10 Tobago	A. F. Welch
	G-CBPG	Balloon Works Firefly 7 balloon	Balloon Preservation Flying Group
	G-CBPH	Lindstrand LBL-105A balloon	I. Vastano/Italy
	G-CBPI	PA-28R-201 Arrow III	Benair Aviation Ltd
	G-CBPK	Rand-Robinson KR-2	R. J. McGoldrick
	G-CBPL	Team Minimax 93	K. M. Moores
	G-CBPM	Yakovlev Yak-50	P. W. Ansell
	G-CBPN	Thruster T.600N 450 Jabiru	J. S. Webb
	G-CBPP	Avtech Jabiru UL-450	C. J. Cullen
	G-CBPR	Avtech Jabiru UL-450	P. L. Riley & F. B. Hall
	G-CBPT	Robinson R-22B	Plane Talking Ltd/Elstree
	G-CBPU	Raj Hamsa X'Air R100 (2)	M. S. McCrudden
	G-CBPV	Zenair CH.601UL	R. D. Barnard
	G-CBPW	Lindstrand LBL-105A balloon	Flying Pictures Ltd
	G-CBPX	IDA Bacau Yakovlev Yak-52	M. Richardson
	G-CBPY	Yakovlev Yak-52	Lyttondale Associate
	G-CBPZ	Ultramagic N-300 balloon	Kent Ballooning
	G-CBRB	Ultramagic S-105 balloon	I. S. Bridge
	G-CBRC	Jodel D.18	B. W. Shaw

Reg.	Type	Owner or Operator
G-CBRD	Jodel D.18	J. D. Haslam
G-CBRE	Mainair Blade 912	R. J. Davey
G-CBRF	Ikarus C.42 FB UK Cyclone	T. W. Gale
G-CBRG	Cessna 560XL Citation Excel	Stadium City Ltd
G-CBRH	IDA Yakovlev Yak-52	B. M. Gwynnett
G-CBRJ	Mainair Blade 912S	R. W. Janion
G-CBRK	Ultramagic M-77 balloon	R. T. Revel
G-CBRL	IDA Bacau Yakovlev Yak-52	P. S. Mirams
G-CBRM	Mainair Blade	M. H. Levy
G-CBRO	Robinson R-44	R. D. Jordan
G-CBRP	IDA Bacau Yakovlev Yak-52	R. J. Pinnock
G-CBRR	Aerotechnik EV-97 Eurostar	G-CBRR Group
G-CBRT	Murphy Elite	T. W. Baylie
G-CBRU	IDA Bacau Yakovlev Yak-52	Romeo Alpha 42 Group
G-CBRV	Cameron C-90 balloon	C. J. Teall
G-CBRW	Aerostar Yakovlev Yak-52	M. A. Gainza
G-CBRX	Zenair CH.601UL Zodiac	J. B. Marshall
G-CBRY	Pegasus Quik	Cyclone Airsports
G-CBRZ	Air Creation Kiss 400/Buggy 582	B. Chantry
G-CBSB	Westland Gazelle HT.2	Flying Machinery Ltd
G-CBSC	Westland Gazelle HT.2	Flying Machinery Ltd
G-CBSD	Westland Gazelle HT.2	London Helicopter Centres Ltd/Redhill
G-CBSE	Westland Gazelle HT.2	Flying Machinery Ltd
G-CBSF	Westland Gazelle HT.2	Falcon Aviation Ltd
G-CBSH	Westland Gazelle HT.3	Alltask Ltd
G-CBSI	Westland Gazelle HT.3	P. S. Unwin
G-CBSK	Westland Gazelle HT.3	Knoland Aviation Ltd/Southend
G-CBSL	IDA Bacau Yakovlev Yak-52	N. & A. D. Barton
G-CBSM	Mainair Blade 912	Mainair Sports Ltd
G-CBSN	Aerostar Yakovlev Yak-52	P. K. Murtagh
G-CBSO	PA-28-181 Archer II	Archer One Ltd
G-CBSP	Pegasus Quantum 15-912	D. S. Carstairs
G-CBSR	Yakovlev Yak-52	L. Olivier
G-CBSS	IDA Bacau Yakovlev Yak-52	M. Chitty/Belgium
G-CBSU	Avtech Jabiru UL	P. K. Sutton
G-CBSV	Montgomerie-Bensen B.8MR	J. A. McGill
G-CBSX	Air Creation Kiss 400/Buggy 582	N- Hartley
G-CBSZ	Mainair Blade 912S	D. M. Newton
G-CBTB	I.I.I. Sky Arrow 650TS	D. A. & J. A. S. T. Hood
G-CBTD	Pegasus Quantum 15-912	D. Baillie
G-CBTE	Mainair Blade 912IS	K. J. Miles
G-CBTG	Ikarus C.42 FB UK Cyclone	Ikarus Group
G-CBTH	Flying Pictures Elson Apoly1 44000 balloon	Flying Pictures Ltd
G-CBTI	Flying Pictures Elson Apoly1 44000 balloon	Flying Pictures Ltd
G-CBTK	Raj Hamsa X'Air 582 (5)	C. D. Wood
G-CBTL	Monnett Moni	G. Dawes
G-CBTM	Mainair Blade	M. J. Ryder
G-CBTN	PA-31 Navajo C	Durban Aviation Services
G-CBTO	Rans S.6-ES Coyote II	B. J. Mould
G-CBTR	Lindstrand LBL-120A balloon	R. H. Etherington
G-CBTS	Gloster Gamecock (replica)	Retro Track & Air (UK) Ltd
G-CBTT	PA-28-181 Archer II	Citicourt Aviation Ltd (G-BFMM)
G-CBTU	Cessna 550 Citation II	Thames Aviation Ltd (G-OCDB/G-ELOT)
G-CBTV	Tri R-Kis	T. V. Thorp
G-CBTW	Mainair Blade 912	D. Hyatt
G-CBTX	Denney Kitfox Mk.2	G. I. Doake
G-CBTY	Raj Hamsa X'Air V2 (2)	K. Quigley
G-CBTZ	Pegasus Quantum 15-912	P. M. Connelly
G-CBUA	Extra EA.230	C. Butler
G-CBUC	Raj Hamsa X'Air 582 (5)	A. P. Fenn & D. R. Lewis
G-CBUD	Pegasus Quantum 15-912	A. W. Rhodes
G-CBUE	Ultramagic N-250 balloon	Elinore French Ltd
G-CBUF	Flight Design CT.2K	Cyclone Airsports Ltd
G-CBUG	Technam P.92-EM Echo	R. C. Mincik
G-CBUH	Westland Scout AH.1	C. J. Marsden
G-CBUI	Westland Wasp HAS.1	Military Helicopters Ltd/Thruxton
G-CBUJ	Raj Hamsa X'Air 582 (10)	G-CBUJ Flying Group
G-CBUK	Van's RV-6A	P. G. Greenslade
G-CBUN	Barker Charade	P. E. Barker
G-CBUO	Cameron O-90 balloon	W. J. Treacy & P. M. Smith

Notes	Reg.	Type	Owner or Operator
	G-CBUP	VPM M.16 Tandem Trainer	J. S. Firth
	G-CBUR	Zenair CH.601UL	N. A. Jack
	G-CBUS	Pegasus Quantum 15	J. Liddiard
	G-CBUU	Pegasus Quantum 15-912	K. B. Woods
	G-CBUW	Cameron Z-133 balloon	Balloon School (International) Ltd
	G-CBUX	Cyclone AX2000	J. Madhvani
	G-CBUY	Rans S.6-ES Coyote II	S. C. Jackson & J. S. Coster
	G-CBUZ	Pegasus Quantum 15	S. T. Allen
	G-CBVA	Thruster T.600N 450	G. St. C. Moseley
	G-CBVB	Robin R.2120U	Aviation Now Ltd
	G-CBVC	Raj Hamsa X'Air 582 (5)	M. J. Male
	G-CBVD	Cameron C-60 balloon	Phoenix Balloons Ltd
	G-CBVE	Raj Hamsa X'Air Falcon 912 (1)	P. K. Bennett
	G-CBVF	Murphy Maverick	J. Hopkinson
	G-CBVG	Mainair Blade 912S	A. M. Buchanan
	G-CBVH	Lindstrand LBL-120A balloon	Line Packaging & Display Ltd
	G-CBVI	Robinson R-44	N. R. Gatt
	G-CBVK	Schroeder Fire Balloons G balloon	S. Travaglia
	G-CBVL	Robinson R-22B	Heli Air Ltd/Wellesbourne
	G-CBVM	Aerotechnik EV-97 Eurostar	J. Cunliffe & A. Costello
	G-CBVN	Pegasus Quik	N. F. Mackenzie
	G-CBVO	Raj Hamsa X'Air 582 (5)	M. D. Vearncombe
	G-CBVR	Skyranger 912 (2)	S. H. Lunney
	G-CBVS	Skyranger 912 (2)	S. C. Cornock
	G-CBVT	IDA Yakovlev Yak-52	Lancair Espana SL/Spain
	G-CBVU	PA-28R Cherokee Arrow 200-II	E. W. Guess (Holdings) Ltd
	G-CBVV	Cameron N-120 balloon	John Aimo Balloons SAS/Italy
	G-CBVX	Cessna 182P	R. Martin
	G-CBVY	Ikarus C.42 FB UK	M. J. Hendra & Gossage
	G-CBVZ	Flight Design CT.2K	Mainair Sports Ltd
	G-CBWA	Flight Design CT.2K	Charlie Tango Group
	G-CBWB	PA-34-200T Seneca II	Fairoaks Airport Ltd
	G-CBWD	PA-28-161 Warrior III	Plane Talking Ltd/Elstree
	G-CBWE	Aerotechnik EV-97 Eurostar	E. Clarke
	G-CBWF	Shaw Europa XS T-G	Celebrations Ltd
	G-CBWG	Aerotechnik EV-97 Eurostar	M. Rhodes
	G-CBWI	Thruster T. 600N 450	T. Lee
	G-CBWJ	Thruster T. 600N 450	T. Harrison-Smith
	G-CBWK	Ultramagic H-77 balloon	H. C. Peel
	G-CBWM	Mainair Blade 912	C. Middleton
	G-CBWN	Campbell Cricket Mk.6	G. J. Layzell
	G-CBWO	Rotorway Executive 162F	Handyvalue Ltd
	G-CBWP	Shaw Europa	T. W. Greaves
	G-CBWR	Thunder Ax7-77 balloon	A. Lutz
	G-CBWS	Whittaker MW.6 Merlin	D. W. McCormack
	G-CBWU	Rotorway Executive 162F	Usk Valley Trout Farm
	G-CBWV	Falconar F-12A Cruiser	A. Ackland
	G-CBWW	Skyranger 912 (2)	R. L. & S. H. Tosswill
	G-CBWX	Slingsby T.67M-260	Slingsby Aviation Ltd/Kirkbymoorside
	G-CBWY	Raj Hamsa X'Air 582 (5)	G. C. Linley
	G-CBWZ	Robinson R-22B	Plane Talking Ltd/Elstree
	G-CBXA	Raj Hamsa X'Air 582 (5)	N. Stevenson-Guy
	G-CBXB	Lindstrand LBL-150A balloon	M. A. Webb
	G-CBXC	Ikarus C.42 FB UK	A. R. Lloyd
	G-CBXD	Bell 206L-3 LongRanger 3	Whirlybird Charters Ltd & Automotive and General Supply Co Ltd
	G-CBXE	Easy Raider J2.2 (2)	A. Appleby
	G-CBXF	Easy Raider J2.2	F. Colman
	G-CBXG	Thruster T.600N 450	P. J. Fahie
	G-CBXH	Thruster T.600N 450	M. L. Smith
	G-CBXJ	Cessna 172S	Caernarfon Airworld Ltd
	G-CBXK	Robinson R-22 Mariner	County Garage (Cheltenham) Ltd
	G-CBXO	Robinson R-22B	Plane Talking Ltd/Elstree
	G-CBXM	Mainair Blade	B. A. Coombe
	G-CBXN	Robinson R-22B	N. M. G. Pearson
	G-CBXR	Raj Hamsa X-Air Falcon 582 (1)	A. R. Rhodes
	G-CBXS	Skyranger J2.2 (1)	C. J. Erith
	G-CBXT	Westland Gazelle HT.3	Flying Machinery Ltd
	G-CBXU	Team Minimax 91A	C. D. Hatcher
	G-CBXV	Mainair Blade	C. Turner
	G-CBXW	Shaw Europa XS	R. G. Fairall
	G-CBXX	Robinson R-44	M. J. Magowan & C. Lilburn

Reg.	Type	Owner or Operator	Notes
G-CBXZ	Rans S.6-ES Coyote II	D. Tole	
G-CBYB	Rotorway Executive 162F	Clark Contracting	
G-CBYC	Cameron Z-275 balloon	First Flight	
G-CBYD	Rans S.6-ESA Coyote II	R. Burland	
G-CBYE	Pegasus Quik	A. D. Griffin	
G-CBYF	Mainair Blade	C. P. Lemon	
G-CBYH	Aeroprakt A.22 Foxbat	G. C. Moore	
G-CBYI	Pegasus Quantum 15-503	J. M. Hardy & M. C. Watson	
G-CBYJ	Steen Skybolt	F. G. Morris	
G-CBYM	Mainair Blade	A. Clarke	
G-CBYN	Shaw Europa XS	A. B. Milne	
G-CBYO	Pegasus Quik	L. Hogan	
G-CBYP	Whittaker MW.6-S Fat Boy Flyer	R. J. Grainger	
G-CBYS	Lindstrand LBL-21 balloon France	J. J. C. Bernardin	
G-CBYT	Thruster T.600N 450	B. E. Smith	
G-CBYU	PA-28-161 Warrior II	Stapleford Flying Club Ltd	
G-CBYV	Pegasus Quantum 15-912	N. B. Sanghrajka	
G-CBYW	Hatz CB-1	T. A. Hinton	
G-CBYX	Bell 206B JetRanger 3	RCR Aviation Ltd	
G-CBYY	Robinson R-44	Helicopter Training & Hire Ltd	
G-CBYZ	Tecnam P.92-EM Echo-Super	B. Weaver	
G-CBZA	Mainair Blade	S. Bedford	
G-CBZB	Mainair Blade	A. Bennion	
G-CBZD	Mainair Blade	J. Shaw	
G-CBZE	Robinson R-44	Blue Aviation Ltd	
G-CBZF	Robinson R-22B	Wields Aviation	
G-CBZG	Rans S.6-ES Coyote II	D. Patrick	
G-CBZH	Pegasus Quik	M. Bond	
G-CBZI	Rotorway Executive 162F	T. D. Stock	
G-CBZJ	Lindstrand LBL-25A balloon	Pegasus Ballooning	
G-CBZK	Robin DR.400/180	R. A. Fleming	
G-CBZL	Westland Gazelle HT.3	Armstrong Aviation Ltd	
G-CBZM	Avtech Jabiru SPL-450	M. E. Ledward	
G-CBZN	Rans S.6-ES Coyote II	A. James	
G-CBZP	Hawker Fury 1	Historic Aircraft Collection	
G-CBZR	PA-28R-201 Arrow III	S. J. Skilton	
G-CBZS	Aurora	J. Lynden	
G-CBZT	Pegasus Quik	M. Brown	
G-CBZU	Lindstrand LBL-180A balloon	Great Escape Ballooning Ltd	
G-CBZV	Ultramagic S-130 balloon	J. D. Griffiths	
G-CBZW	Zenair CH.701UL	T. M. Siles	
G-CBZX	Dyn' Aero MCR-01 ULC	S. L. Morris	
G-CBZY	Flight Airsports Doodle Bug	A. I. Calderhead-Lea	
G-CBZZ	Cameron Z-275 balloon	A. C. K. Rawson & J. J. Rudoni	
G-CCAB	Mainair Blade	R. W. Street	
G-CCAC	Aerotech EV-97 Eurostar	J. S. Holden	
G-CCAD	Mainair Pegasus Quik	J. M. Hardstaff	
G-CCAE	Avtech Jabiru UL-450	C. E. Daniels	
G-CCAF	Skyranger 912 (1)	D. W. & M. L. Squire	
G-CCAG	Mainair Blade 912	West Lancashire Microlight School	
G-CCAJ	Team Himax 1700R	A. P. S. John	
G-CCAK	Zenair CN.601HD	A. Kimmond	
G-CCAL	Technam P.92-EM Echo	D. Cassidy	
G-CCAM	Mainair Blade	M. D. Peacock	
G-CCAN	Cessna 182P	D. J. Hunter	
G-CCAP	Robinson R-22	HJS Helicopters Ltd	
G-CCAR	Cameron N-77 balloon	D. P. Turner	
G-CCAS	Pegasus Quik	Quik Alpha Sierra Group	
G-CCAT	AA-55A Cheetah	Plane Talking Ltd (G-OAJH/G-KILT/G-BJFA)/ Elstree	
G-CCAU	Eurocopter EC.135T-1	West Mercia Constabulary	
G-CCAV	PA-28-181 Archer II	Archer II Ltd	
G-CCAW	Mainair Blade 912	C. A. Woodhouse	
G-CCAX	Raj Hamsa X'Air 582 (1)	N. Farrell	
G-CCAY	Cameron Z-42 balloon	P. Stern	
G-CCAZ	Mainair Pegasus Quik	P. A. Bass	
G-CCBA	Skyranger R.100	Fourstrokes Group	
G-CCBB	Cameron N-90 balloon	S. C. A. & L. D. Craze	
G-CCBC	Thruster T.600N 450	M. L. Smith	
G-CCBF	Maule M.5-235C	R. Windley (G-NHVH)	
G-CCBG	Skyranger V.2 + (1)	G. R. Wallis	

Notes	Reg.	Type	Owner or Operator
	G-CCBH	PA-28 Cherokee 236	J. R. Hunt & S. M. Packer
	G-CCBI	Raj Hamsa X'Air R.100 (2)	H. Adams
	G-CCBJ	Skyranger 912 (2)	M. A. Russell
	G-CCBK	Aerotechnik EV-97 Eurostar	G. R. & J. A. Pritchard
	G-CCBL	AB-206B JetRanger 3	Salon Gold Publishing Ltd/Biggin Hill
	G-CCBM	Aerotechnik EV-97 Eurostar	W. Graves
	G-CCBN	Scale Replica SE-5A	T. H. Bishop
	G-CCBP	Lindstrand LBL-60X balloon	Lindstrand Hot-Air Balloons Ltd
	G-CCBR	Jodel D.120	R. R. Walters
	G-CCBT	Cameron Z-90 balloon	British Telecom PLC
	G-CCBU	Raj Hamsa X'Air 582 (9)	J. S. Rakker
	G-CCBV	Cameron Z-225 balloon	Compagnie Aeronautique du Grand-Duche de Luxembourg
	G-CCBW	Sherwood Ranger	P. H. Wiltshire
	G-CCBX	Raj Hamsa X'Air 582	A. D'Amico
	G-CCBY	Avtech Jabiru UL-450	D. M. Goodman
	G-CCBZ	Aero Designs Pulsar	J. M. Keane
	G-CCCA	V.S.509 Spitfire Tr.IX	Historic Flying Ltd (G-BHRH/G-TRIX)/Duxford
	G-CCCB	Thruster T.600N 450	D. G. Stanley
	G-CCCC	Cessna 172H	Springbank Aviation Ltd
	G-CCCD	Mainair Pegasus Quantum 15	R. N. Gamble
	G-CCCE	Aeroprakt A.22 Foxbat	C. V. Ellingworth
	G-CCCF	Thruster T.600N 450	Thruster Group 2004
	G-CCCG	Mainair Pegasus Quik	C, J, Gordon
	G-CCCH	Thruster T.600N 450	G-CCCH Group
	G-CCCI	Medway Eclipse R	N. H. Morley
	G-CCCJ	Nicollier HN.700 Menestrel II	R. Y. Kendal
	G-CCCK	Skyranger 912 (2)	P. L. Braniff
	G-CCCM	Skyranger 912 (2)	J. R. Moore
	G-CCCN	Robin R.3000/160	R. W. Denny
	G-CCCO	Aerotechnik EV-97 Eurostar	Connel Flying Club Eurostar Group
	G-CCCP	IDA Yakovlev Yak-52	A. H. Soper
	G-CCCR	Skyranger 912 (2)	D. M. Robbins
	G-CCCT	Ikarus C.42 FB UK Cyclone	G. A. Pentelow
	G-CCCU	Thruster T.600N 450	Medway Microlights
	G-CCCV	Raj Hamsa X'Air Falcon 133 (1)	A. J. Fraley
	G-CCCW	Pereira Osprey 2	D. J. Southward
	G-CCCX	Cameron GP-70 balloon	Cameron Balloons Ltd
	G-CCCY	Skyranger 912 (2)	D. M. Cottingham & A. Grimsley
	G-CCCZ	Raj Hamsa X'Air 582 (5)	R. A. Merrigan
	G-CCDA	PA-44-180	Saxon Logistics Ltd
	G-CCDB	Mainair Pegasus Quik	C. J. Van Dyke
	G-CCDC	Rans S.6-ES Coyote II	G. N. Smith
	G-CCDD	Mainair Pegasus Quik	M. P. Hadden & M. H. Rollins
	G-CCDE	Robinson R-22B	Wrightson Aviation and Engineering
	G-CCDF	Mainair Pegasus Quik	G-CCDF Flying Group
	G-CCDG	Skyranger 912 (1)	T. H. Filmer
	G-CCDH	Skyranger 912 (2)	P. & V. C. Reynolds
	G-CCDJ	Raj Hamsa X'Air Falcon 582 (2)	J. M. Spitz
	G-CCDK	Pegasus Quantum 15-912	S. Brock
	G-CCDL	Raj Hamsa X'Air Falcon 582 (2)	H. Burroughs
	G-CCDM	Mainair Blade	M. K. Ashmore
	G-CCDN	PA-28-181 Archer III	J. D. Scott
	G-CCDO	Pegasus Quik	A. G. Quinn
	G-CCDP	Raj Hamsa X'Air R.100 (3)	J. A. McKie
	G-CCDR	Raj Hamsa X'Air Falcon Jabiru	P. D. Sibbons
	G-CCDS	Nicollier HN.700 Menestrel II	B. W. Gowland
	G-CCDT	R. Commander 114	Simply Fly Ltd
	G-CCDU	Tecnam P.92-EM Echo	M. J. Barrett
	G-CCDV	Thruster T.600N 450	D. J. Whysall
	G-CCDW	Skyranger 582 (1)	Debts R Us Family Group
	G-CCDX	Aerotechnik EV-97 Eurostar	R. A. Morris & H. F. Breakwell
	G-CCDY	Skyranger 912 (2)	A. V. Dunne & G. S. Gee Carter
	G-CCDZ	Pegasus Quantum 15-912	K. D. Baldwin
	G-CCEA	Pegasus Quik	J. D. Ash
	G-CCEB	Thruster T.600N 450	Thruster Air Services Ltd
	G-CCEC	Evans VP-1	C. R. Harrison (G-ROSE)
	G-CCED	Zenair CH.601UL	R. P. Reynolds
	G-CCEE	PA-15 Vagabond	I. M. Callier (G-VAGA)
	G-CCEF	Shaw Europa	C. P. Garner
	G-CCEG	Rans S.6-ES Coyote II	S. G. Dalziel
	G-CCEH	Skyranger 912 (2)	A. Eastham

Reg.	Type	Owner or Operator	Notes
G-CCEI	Evans VP-2	I. P. Manley & J. Pearce (G-BJZB)	
G-CCEJ	Aerotechnik EV-97 Eurostar	C. R. Ashley	
G-CCEK	Kiss 400-582 (1)	G. S. Sage	
G-CCEL	Avtech Jabiru UL	R. Ryper	
G-CCEM	Aerotechnik EV-97 Eurostar	G. K. Kenealy	
G-CCEN	Cameron Z-120 balloon	R. Hunt	
G-CCEO	Thunder Ax10-180 S2 balloon	P. Heitzeneder/Austria	
G-CCEP	Raj Hamsa X'Air Falcon Jabiru	K. Angel	
G-CCER	Mainair Gemini Flash II	A. Hillyer	
G-CCES	Raj Hamsa X'Air 2706	G. V. McCloskey	
G-CCET	Nova Vertex 28 hang glider	M. Hay	
G-CCEU	RAF 2000 GTX-SE gyroplane	N. G. Dovaston	
G-CCEW	Mainair Pegasus Quik	C. S. Mackenzie	
G-CCEX	Cameron Z-90 balloon	Cameron Balloons Ltd	
G-CCEY	Raj Hamsa X'582 (11)	P. F. F. Spedding	
G-CCEZ	Easy Raider J2.2	S. A/ Chambers	
G-CCFA	Kiss 400-582 (1)	N. Hewitt	
G-CCFB	Mainair Pegasus Quik	W. T. Davis	
G-CCFC	Robinson R-44 II	M. Entwistle	
G-CCFD	Quad City Challenger II	W. Oswald	
G-CCFE	Tipsy Nipper T.66 Srs 2	A. R. Way	
G-CCFF	Lindstrand LBL-150A balloon	Airborne Adventures Ltd	
G-CCFG	Dyn'Aero MCR-01 Club	G. J. Sargent	
G-CCFI	PA-32 Cherokee Six 260	McManus Truck & Trailer Spares Ltd	
G-CCFJ	Kolb Twinstar Mk.3	D. Travers	
G-CCFK	Shaw Europa	C. R. Knapton	
G-CCFL	Mainair Pegasus Quik	S. A. Noble	
G-CCFM	Mainair Blade 912	J. G. I. Muncey	
G-CCFN	Cameron N-105 balloon	Procter & Gamble Ltd	
G-CCFO	Pitts S-1S Special	R. J. Anderson	
G-CCFP	Diamond DA.40D Star	N. P. de Gruchy Lambert	
G-CCFR	Diamond DA.40D Star	C. R. Lear	
G-CCFS	Diamond DA.40D Star	Principle Aircraft	
G-CCFT	Mainair Pegasus Quantum 15-912	D. Tasker	
G-CCFU	Diamond DA.40D Star	Principle Aircraft	
G-CCFV	Lindstrand LBL-77A balloon	Alton Aviation Ltd	
G-CCFW	WAR Focke-Wulf Fw.190	D. B. Conway	
G-CCFX	EAA Acrosport 2	C. D. Ward	
G-CCFY	Rotorway Executive 162F	M. Hawley	
G-CCFZ	Ikarus C.42 FB UK	B. W. Drake	
G-CCGA	Medway Eclipser	Medway Microlights	
G-CCGB	Team Minimax	A. D, Pentland	
G-CCGC	Mainair Pegasus Quik	A. Crozier	
G-CCGE	Robinson R-22B	Heli Aitch Be Ltd	
G-CCGF	Robinson R-22B	Ground Effect Ltd	
G-CCGG	Jabiru Aircraft Jabiru J.400	K. D. Pearce	
G-CCGH	Super Marine Aircraft Spitfire 26	K. D. Pearce	
G-CCGI	Mainair Pegasus Quik	M. C. Kerr	
G-CCGK	Mainair Blade	G. Kerr	
G-CCGL	SOCATA TB.20 Trinidad	Pembroke Motor Services Ltd	
G-CCGM	Kiss 450-582 (1)	A. I. Lea	
G-CCGN	Bell 206L-1 Long Ranger 2	Biggin Hill Helicopters	
G-CCGO	Medway Microlights AV8R	Medway Microlights	
G-CCGP	Bristol Type 200	R. L. Holman	
G-CCGR	Raj Hamsa X'Air 133 (1)	J. M. Weston	
G-CCGS	Dornier Do.328-100	ScotAirways Ltd	
G-CCGT	Cameron Z-425 balloon	A. A. Brown	
G-CCGU	Van's RV-9A	B. J. Main & ptnrs	
G-CCGV	Lindstrand LBL-150A balloon	Lindstrand Hot-Air Balloons Ltd	
G-CCGW	Shaw Europa	G. C. Smith	
G-CCGY	Cameron Z-105 balloon	Cameron Balloons Ltd	
G-CCGZ	Diamond DA.40D Star	Diamond Aircraft UK Ltd	
G-CCHA	Diamond DA.40D Star	Cabair College of Air Training Ltd	
G-CCHB	Diamond DA.40D Star	Cabair College of Air Training Ltd	
G-CCHC	Diamond DA.40D Star	Cabair College of Air Training Ltd	
G-CCHD	Diamond DA.40D Star	Cabair College of Air Training Ltd	
G-CCHE	Diamond DA.40D Star	Cabair College of Air Training Ltd	
G-CCHF	Diamond DA.40D Star	Cabair College of Air Training Ltd	
G-CCHG	Diamond DA.40D Star	Plane Talking Ltd	
G-CCHI	Mainair Pegasus Quik	Light Flight Ltd	
G-CCHJ	Kiss 400-582 (1)	H. C. Jones	
G-CCHK	Diamond DA.40D Star	Plane Talking Ltd	

Notes	Reg.	Type	Owner or Operator
	G-CCHL	PA-28-181 Archer iii	Archer Three Ltd
	G-CCHM	Kiss 450	W. G. Colyer
	G-CCHN	Corby CJ.1 Starlet	D. C. Mayle
	G-CCHO	Mainair Pegasus Quik	M. Allan
	G-CCHP	Cameron Z-31 balloon	M. H. Redman
	G-CCHR	Easy Raider 583 (1)	R. B. Hawkins
	G-CCHS	Raj Hamsa X'Air 582	I. Lonsdale
	G-CCHT	Cessna 152	J. S. Devlin & Z. Islam
	G-CCHU	Flight Design CT.2K	A. N. D. Arthur
	G-CCHV	Mainair Rapier	A. Butterworth
	G-CCHW	Cameron Z-77 balloon	A. Murphy
	G-CCHX	Scheibe SF.25C Falke	Lasham Gliding Soc.Ltd
	G-CCHY	Bucker Bu.131 Jungmann (A+12)	M. V. Rijkse
	G-CCHZ	Robinson R-22B	Helicopter Training and Hire Ltd
	G-CCIA	Lindstrand LBL-105A balloon	J. J-Bernadin
	G-CCIC	Thruster T.600N 450	M. L. Smith
	G-CCID	Jabiru Aircraft Jabiru J400	J. Bailey
	G-CCIE	Colt 31A balloon	T. M. Donnelly
	G-CCIF	Mainair Blade	D. J. Kennedy
	G-CCIG	Aero Designs Pulsar	P. Maguire
	G-CCIH	Mainair Pegasus Quantum 15	R. Bennett
	G-CCII	ICP Savannah Jabiru (3)	J. R. Livett & D. Chaloner
	G-CCIJ	PA-28R Cherokee Arrow 180	I. R. Chaplin/Andrewsfield
	G-CCIK	Skyranger 912 (2)	L. E. Cowling & A. P. Chapman
	G-CCIO	Skyranger 912 (2)	B. Berry
	G-CCIR	Van's RV-8	B. F. Hill
	G-CCIS	Scheibe SF.28A Tandem Falke	P. T. Ross
	G-CCIT	Zenair CH.701UL	I. M. Sinclair
	G-CCIU	Cameron N-105 balloon	Bianchi Aviation Film Services Ltd
	G-CCIV	Mainair Pegasus Quik	D. Little
	G-CCIW	Raj Hamsa X'Air 582 (2)	G. Wilkinson
	G-CCIY	Skyranger 912 (2)	L. F. Tanner
	G-CCIZ	PZL-Koliber 160A	PZL International Marketing & Sales/North Weald
	G-CCJA	Skyranger 912 (2)	T. R. Southall
	G-CCJB	Zenair CH.701 STOL	E. G. Brown
	G-CCJD	Mainair Pegasus Quantum 15	P. Clark
	G-CCJE	Schweizer 269C-1	Dragon Helicopters Ltd/Sheffield City
	G-CCJF	Cameron C-90 balloon	Balloon School International Ltd
	G-CCJG	Cameron A-200 balloon	Aire Valley Balloons
	G-CCJH	Lindstrand LBL-90A balloon	J. R. Hoare
	G-CCJI	Van's RV-6	Cavendish Aviation Ltd/Gamston
	G-CCJJ	Medway Pirana	Medway Microlights
	G-CCJK	Aerostar Yakovlev Yak-52	R. K. Howell
	G-CCJL	Super Marine Aircraft Spitfire XXVI	P. M. Whitaker
	G-CCJM	Mainair Pegasus Quik	P. Crosby
	G-CCJN	Rans S.6ES Coyote II	M. G. A. Wood
	G-CCJO	ICP-740 Savannah Jabiru 4	R. & I. Fletcher
	G-CCJS	Flying K Enterprises Easy Raider	K. Wright
	G-CCJT	Skyranger 912 (2)	J. W. Taylor
	G-CCJU	ICP MXP-740 Savannah	K. R. Wootton & A. Colverson
	G-CCJV	Aeroprakt A.22 Foxbat	Foxbat UK015 Syndicate
	G-CCJW	Skyranger 912 (2)	J. R. Walter
	G-CCJX	Shaw Europa XS	J. S. Baranski
	G-CCJY	Cameron Z-42 balloon	Cameron Balloons Ltd
	G-CCKF	Skyranger 912 (2)	T. P. M. Turnbull
	G-CCKG	Skyranger 912 (2)	J. Hannibal
	G-CCKH	Diamond DA.40D Star	Diamond Aircraft UK Ltd
	G-CCKI	Diamond DA.40D Star	S. C. Horwood
	G-CCKJ	Raj Hamsa X'Air 582 (5)	S. Thompson
	G-CCKK	Aerotechnik EV-97 Eurostar	C. Townsend
	G-CCKL	Aerotechnik EV-97 Eurostar	J. S. Liming & ptnrs
	G-CCKM	Mainair Pegasus Quik	J. W. McCarthy
	G-CCKN	Nicollier HN.700 Menestrel II	C. R. Partington
	G-CCKO	Mainair Pegasus Quik	M. J. Mawle & C. R. Bunce
	G-CCKP	Robin DR.400/120	Duxford Flying Group
	G-CCKR	Pietenpol Air Camper	T. J. Wilson
	G-CCKS	Hughes 369E	Storm Aviation Group International Ltd
	G-CCKT	Hapi Cygnet SF-2	P. W. Abraham
	G-CCKU	Canadian Home Rotors Safari	J. C. Collingwood
	G-CCKV	Isaacs Fury II	S. T. G. Ingram
	G-CCKW	PA-18 Super Cub 135	G. T. Fisher (G-GDAM)
	G-CCKX	Lindstrand LBL-210A balloon	Alba Ballooning Ltd

Reg.	Type	Owner or Operator	Notes
G-CCKY	Lindstrand LBL-240A balloon	Dance With Balloons Ltd	
G-CCKZ	Customcraft A-25 balloon	A. Van Wyk	
G-CCLB	Diamond DA.40D	The Millen Corporation	
G-CCLC	Diamond DA.40D	Diamond Aircraft UK Ltd	
G-CCLE	Aerotechnik EV-97 Eurostar	W. S. Long	
G-CCLF	Best Off Skyranger 912 (2)	G. K. R. Linney	
G-CCLG	Lindstrand LBL-105A balloon	M. A. Derbyshire	
G-CCLH	Rans S.6-ES Coyote II	K. R. Browne	
G-CCLJ	PA-28 Cherokee Cruiser 140	A. M. George	
G-CCLL	Zenair CH.601XL Zodiac	L. Lewis	
G-CCLM	Mainair Pegasus Quik	P. M. Ryder	
G-CCLO	Ultramagic H-77 balloon-	J. P. Moore	
G-CCLP	ICP MXP-740 Savannah	M. J. Kaye	
G-CCLR	Schleicher Ash 26E	C. R. Lear	
G-CCLS	Comco Ikarus C.42 FB UK	SLS Computing Services Ltd	
G-CCLU	Best Off Skyranger 912	L. Stanton	
G-CCLV	Diamond DA.40D Star	Diamond Aircraft UK Ltd/Gamston	
G-CCLW	Diamond DA.40D Star	Plane Talking Ltd	
G-CCLX	Mainair Pegasus Quik	S. D. Pain	
G-CCLY	Bell 206B JetRanger	Mulberry Homes Ltd (G-TILT/G-BRJO)	
G-CCMA	Boeing 747-267B	European Aviation Ltd (G-VCAT)/Bournemouth	
G-CCMC	Jabiru Aircraft Jabiru UL 450	J. T. McCormack	
G-CCMD	Mainair Pegasus Quik	J. T. McCormack	
G-CCME	Mainair Pegasus Quik	R. R. Nichol	
G-CCMF	Diamond DA.40D Star	Plane Talking Ltd	
G-CCMH	M.2H Hawk Major	J. A. Pothecary	
G-CCMI	SA Bulldog Srs 120/121	Just Plane Trading Ltd (G-KKKK)	
G-CCMJ	Easy Raider J2.2 (1)	G. F. Clews	
G-CCMK	Raj Hamsa X'Air Falcon	Cambridge Design Partnership Ltd	
G-CCML	Mainair Pegasus Quik	P. Ritchie	
G-CCMM	Dyn'Aero MCR-1 ULC Banbi	J. D. Harris	
G-CCMN	Cameron C90 balloon	A.E. Austin	
G-CCMO	Aerotechnik EV-97A Eurostar	E. M. Woods	
G-CCMP	Aerotechnik EV-97A Eurostar	W. K. Wilkie	
G-CCMR	Robinson R-22B	Mash Enterprises Ltd	
G-CCMS	Mainair Pegasus Quik	M. Chittenden	
G-CCMT	Thruster T.600N 450	S. P. McCaffrey	
G-CCMU	Rotorway Executive 162F	M. Irving	
G-CCMW	CFM Shadow Srs.DD	M. Wilkinson	
G-CCMX	Skyranger 912 (2)	K. J. Cole	
G-CCMZ	Best Off Skyranger 912 (2)	D. D. Appleford	
G-CCNA	Jodel DR.100A (Replica)	W. R. Davis-Smith & R. Everitt	
G-CCNB	Rans S.6ES Coyote II	M. S. Lawrence	
G-CCNC	Cameron Z-275 balloon	D. Ling	
G-CCND	Van's RV-9A	K. S. Woodard	
G-CCNE	Mainair Pegasus Quantum 15	C. D. Waldron	
G-CCNF	Raj Hamsa X'Air 582 Falcon 133	M. F. Eddington	
G-CCNG	Flight Design CT.2K	David Goode Sculpture Ltd	
G-CCNH	Rans S.6ES Coyote II	Coyote Group	
G-CCNJ	Skyranger 912 (2)	J. D. Buchanan	
G-CCNK	Robinson R-44	Aircol	
G-CCNL	Raj Hamsa X'Air Falcon 133 (1)	G. A. J. Salter	
G-CCNM	Mainair Pegasus Quik	G. T. Snoddon	
G.CCNN	Cameron Z-90 balloon	J. H. Turner	
G-CCNP	Flight Design CT.2K	M. Clare	
G-CCNR	Skyranger 912 (2)	S. N. J. Huxtable	
G-CCNS	Skyranger 912 (2)	G. G. Rowley & M. Liptrot	
G-CCNT	Ikarus C.42 FB80	Mainair Microlight School Ltd	
G-CCNU	Skyranger J2.2 (2)	D. P. Toulson & R.L. Nyman	
G-CCNV	Cameron Z-210 balloon	J. A. Cooper	
G-CCNW	Mainair Pegasus Quantum Lite	J. Childs	
G-CCNX	CAB CAP.10B	Arc Input Ltd	
G-CCNY	Robinson R-44	C. M. Evans & J. W. Blaylock	
G-CCNZ	Raj Hamsa X'Air 133 (1)	K. J. Foxall	
G-CCOB	Aero C.104 Jungmann	W. Tomkins Ltd	
G-CCOC	Mainair Pegasus Quantum 15	F. E. J. Moore	
G-CCOE	Lindstrand LBL-35A balloon	Lindstrand Hot Air Balloons Ltd	
G-CCOF	Rans S.6-ESA Coyote II	A. J. Wright & M. Govan	
G-CCOG	Mainair Pegasus Quik	J. Hood	
G-CCOH	Raj Hamsa X'Air Falcon 133(1)	A.R. Emerson & J.C. Ambrose	
G-CCOI	Lindstrand LBL-90A balloon	M.J. Warne	
G-CCOK	Mainair Pegasus Quik	E. McCallum	

Notes	Reg.	Type	Owner or Operator
	G-CCOM	Westland Lysander IIIA	Propshop Ltd
	G-CCOO	Raj Hamsa X'Air 133	A. Hipkin
	G-CCOP	Ultramagic M-105 balloon	Firefly Balloon Team
	G-CCOR	Sequoia F.8L Falco	D. J. Thoma
	G-CCOS	Cameron Z-350 balloon	M. L. Gabb
	G-CCOT	Cameron Z-105 balloon	Airborne Adventures Ltd
	G-CCOU	Mainair Pegasus Quik	J.R. Pearce
	G-CCOV	Shaw Europa XS	G.N. Drake
	G-CCOW	Mainair Pegasus Quik	J. Prentice
	G-CCOX	Piper J-3C-65 Cub	R.P. Marks
	G-CCOY	AT-6D Harvard II	Classic Aero Services Ltd
	G-CCOZ	Monnett Sonerai II	P. R. Cozens
	G-CCPA	Kiss 400-582(1)	C.P. Astridge
	G-CCPB	Mainair Blade 912	D. G. Fortune
	G-CCPC	Mainair Pegasus Quik	P. M. Coppola
	G-CCPD	Campbell Cricket Mk.4	N.C. Smith
	G-CCPE	Steen Skybolt	C. Moore
	G-CCPF	Skyranger 912 (2)	R.K. Willcox & ptnrs
	G-CCPG	Mainair Pegasus Quik	M. A. Rhodes
	G-CCPH	EV-97 Teameurostar UK	A. H. Woolley
	G-CCPI	Extra EA.300L	A. Caramella
	G-CCPJ	Aerotechnik EV-97 Teameurostar UK	S. R. Winter
	G-CCPK	Murphy Rebel	B. A. Bridgewater & D. Webb
	G-CCPL	Skyranger 912 (2)	G-CCPL Group
	G-CCPM	Mainair Blade 912	T.D. Thompson
	G-CCPN	Dyn'Aero MCR-01 Club	P. H. Nelson
	G-CCPO	Cameron N-77 balloon	A. C. Booth & M. J. Woodcock (GMITS)
	G-CCPP	Cameron 70 Concept balloon	Sarnia Balloon Group
	G-CCPS	Ikarus C.42 FB100 VLA	H. Cullens
	G-CCPT	Cameron Z-90 balloon	Blue Sky Ballooning Ltd
	G-CCPU	Pilatus PC-12/45	Technical Flight Services Ltd/Bournemouth
	G-CCPV	Jabiru J.400	J. R. Lawrence
	G-CCPW	BAe Jetstream 3102	Euromanx
	G-CCPX	Diamond DA.400 Star	R. T. Dickinson
	G-CCPY	Hughes 369D	London Air Ltd
	G-CCPZ	Cameron Z-225 balloon	Horizon Ballooning Ltd
	G-CCRA	Glaser-Dirks DG-800B	R. Arkle
	G-CCRB	Kolb Twinstar Mk.3 (modified)	R. W. Burge
	G-CCRC	Cessna Tu.206G	D. M. Penny
	G-CCRD	Robinson R-44-II	Heli Air Ltd/Wellesbourne
	G-CCRE	Bell 206L-3 LongRanger 3	Biggin Hill Helicopters Ltd
	G-CCRF	Mainair Pegasus Quantum `5	R. D. Ballard
	G-CCRG	Ultramagic M-77 balloon	Aerial Promotions Ltd
	G-CCRH	Cameron Z-315 balloon	Ballooning Network Ltd
	G-CCRI	Raj Hamsa X'Air 582 (5)	R. A. Wright
	G-CCRJ	Shaw Europa	J. F. Cliff
	G-CCRK	Luscombe 8A Silvaire	J. R. Kimberley
	G-CCRN	Thruster T.600N 450	P. Johns
	G-CCRR	Skyranger 912 (1)	J. A. Hunt
	G-CCRS	Lindstrand LBL-210A balloon	Aerosaurus Ballooning Ltd
	G-CCRT	Mainair Pegasus Quantum 15	C. R. Whitton
	G-CCRV	Skyranger 912 (2)	M. R. Mosley
	G-CCRW	Mainair Pegasus Quik	M. P. Jackson
	G-CCRX	Jabiru UL- 450	M. Everest
	G-CCSA	Cameron Z-350 balloon	Ballooning Network Ltd
	G-CCSB	Cessna FR.172H	R. J. Scott
	G-CCSD	Mainair Pegasus Quik	S. E. Dancaster
	G-CCSF	Mainair Pegasus Quik	J. S. Walton
	G-CCSG	Cameron Z-275 balloon	M. L. Gabb
	G-CCSH	Mainair Pegasus Quik	S. D. J. Harvey
	G-CCSI	Cameron Z-42 balloon	IKEA Ltd
	G-CCSJ	Cameron A-275 balloon	Dragon Balloon Co Ltd
	G-CCSK	Zenair CH.701	Thomas & Thomas Surveyors Ltd
	G-CCSL	Mainair Pegasus Quik	A. J. Harper
	G-CCSM	Lindstrand LBL-105A balloon	M. A. Webb
	G-CCSN	Cessna U.206G	K. Brady
	G-CCSO	Raj Hamsa X'Air Falcon	P. Richardson
	G-CCSP	Cameron N-77 balloon	Balloon Sports HB/Sweden
	G-CCSR	Aerotechnik EV-97A Eurostar	M. Lang
	G-CCSS	Lindstrand LBL-90A balloon	British Telecom
	G-CCST	PA-32R-301 Saratoga	G. R. Balls
	G-CCSU	IDA Bacau Yakovlev Yak-52	S. Ullrich

Reg.	Type	Owner or Operator	Notes
G-CCSV	ICP MXP-740 Savannah Jabiru	R. D. Wood	
G-CCSW	Nott PA balloon	J. R. P.Nott	
G-CCSX	Skyranger 912	T. Jackson	
G-CCSY	Mainair Pegasus Quik	D. Sykes	
G-CCTA	Zenair CH.601UL Zodiac	R. E. Gray & G. T. Harris	
G-CCTB	Avro RJ100	Trident Jet Leasing (Ireland) Ltd	
G-CCTC	Mainair Pegasus Quik	D. G. Emery & M. R. Smith	
G-CCTD	Mainair Pegasus Quik	Mainair Sports Ltd	
G-CCTE	Dyn Aero MCR-01 Banbi	G. J. Slater	
G-CCTF	Aerotek Pitts S-2A Special	Pitts Aircraft (UK) Ltd	
G-CCTG	Van's RV-3B	M. & I. G. Glenn	
G-CCTH	Aerotechnik EV-97 Teameurostar UK	Fly UK Ltd	
G-CCTI	Aerotechnik EV-97 Teameurostar	Flylight Airsports Ltd	
G-CCTJ	SAAB 2000	Eastern Airways	
G-CCTK	Glaser-Dirks DG-800B	G. W. English	
G-CCTL	Robinson R-44-II	Auto Corporation Ltd	
G-CCTM	Mainair Blade	J. N. Hanso	
G-CCTN	Ultramagic T-180 balloon	A, Derbyshire	
G-CCTO	Aerotechnik EV-97 Eurostar	A. J. Bolton	
G-CCTP	Aerotechnik EV-97 Eurostar	G. M. Yule	
G-CCTR	Skyranger 912	A. H. Trapp	
G-CCTS	Cameron Z-120 balloon	F. R. Hart	
G-CCTT	Cessna 172S	A. Reay	
G-CCTU	Mainair Pegasus Quik	B. J. Syson	
G-CCTV	Rans S.6ESA Coyote II	RIDS Ltd	
G-CCTW	Cessna 152	R. J. Dempsey	
G-CCTX	Rans S.8ES Coyote II	J. Lynch	
G-CCTY	Cameron TR-70 balloon	Cameron Balloons Ltd	
G-CCTZ	Mainair Pegasus Quik 912S	S. Baker	
G-CCUA	Mainair Pegasus Quik	H, M, Manningr	
G-CCUB	Piper J-3C-65 Cub	Cormack (Aircraft Services) Ltd	
G-CCUC	Skyranger J.2	M. Kerrison	
G-CCUD	Skyranger J.2	J. Johnston	
G-CCUE	Ultramagic T-180 balloon	Espiritu Balloon Flights Ltd	
G-CCUF	Skyranger 912	C. D. Hogbourne & D. J. Parrish	
G-CCUG	Bell 206L-1 LongRanger	Biggin Hill Helicopters	
G-CCUH	RAF 2000 GTX-SE gyroplane	J. H. Haverhals	
G-CCUI	Dyn'Aero MCR-01 Banbi	J. T. Morgan	
G-CCUJ	Cameron C-90 balloon	Rudgleigh Inn	
G-CCUK	Agusta A.109-II	Castle \Air Charters Ltd	
G-CCUL	Shaw Europa XS	Europa 6	
G-CCUN	Hughes 369D	London Air Ltd	
G-CCUO	Hughes 369D	London Air Ltd	
G-CCUP	Wessex 60 Srs 1	D. Brem-Wilson & J. Buswell	
G-CCUR	Mainair Pegasus Quantum 15-912	P. S. & N. Bewley	
G-CCUS	Diamond DA.400 Star	The Millen Corporation	
G-CCUT	Aerotechnik EV-97 Eurostar	C. K. Jones	
G-CCUU	Shiraz gyroplane	M. E. Vahdat-Hagh	
G-CCUX	BN-2T Islander	BN Group Ltd/ Bembridge	
G-CCUY	Shaw Europa	N. Evans	
G-CCUZ	Thruster T.600N 450	Fly 365 Ltd	
G-CCVA	Aerotechnik EV-97 Eurostar	A. Jones	
G-CCVB	Mainair Pegasus Quik	L. Chesworth	
G-CCVD	Cameron Z-105 balloon	Associazione Sportiva/Italy	
G-CCVE	Raj Hamsa X'Air Jabiru	G. J. Slate	
G-CCVF	Lindstrand LBL-105 balloon	Alan Patterson Design	
G-CCVG	Schweizer 269C-1	Radcliffe Engineering Services Ltd	
G-CCVH	Curtiss H-75-1	The Fighter Collection /Duxford	
G-CCVI	Zenair CH.701 SP	C. R. Hoveman	
G-CCVJ	Raj Hamsa X'Air Falcon 133	A. Davis	
G-CCVK	Aerotechnik EV-97 TeamEurostar UK	Kent Eurostar Group	
G-CCVL	Zenair CH.601XL Zodiac	A. Y-T. Leungr & G. Constantine	
G-CCVM	Van's RV-7A	J. G. Small	
G-CCVN	Jabiru SP-470	J. C. Collingwood	
G-CCVO	Bell 206B JetRanger 3	Carecoast Ltd	
G-CCVP	Beech 58	Richard Nash Cars Ltd	
G-CCVR	Skyranger 912	M. J. Batchelor	
G-CCVS	Van's RV-6A	J. Edgeworth (G-CCVC)	
G-CCVT	Zenair CH.601UL Zodiac	D. McCormack	
G-CCVU	Robinson R-22B	M. Horrell	
G-CCVW	Nicollier HN.700 Menestrel II	B. F. Enock	
G-CCVX	Mainair Tri Flyer 330	J. A. Shufflebotham	

Notes	Reg.	Type	Owner or Operator
	G-CCVY	Robinson R-22B	S. Klinge
	G-CCVZ	Cameron O-120 balloon	T. M. C. McCoy
	G-CCWA	PA-28-181 Archer III	T. P. Gooley
	G-CCWB	Aero L-39ZA Albatross	Freespirit Charters Ltd
	G-CCWC	Skyranger 912	C. Hewer
	G-CCWD	Robinson R-44	J. Henderson
	G-CCWE	Lindstrand LBL-330A balloon	Adventure Balloons Ltd
	G-CCWF	Raj Hamsa X'Air 133	N. G. Middleton
	G-CCWG	Whittaker MW.6 Merlin	D. E. Williams
	G-CCWH	Dyn'Aero MCR-01 Bambi	M. G. Rasch
	G-CCWI	Robinson R-44-II	Heli Air Ltd/Wellesbourne
	G-CCWJ	Robinson R-44-II	Heli Air Ltd/Wellesbourne
	G-CCWK	AS.355F-2 Twin Squirrel	Helicopter Express Ltd
	G-CCWL	Mainair Blade	W. D. Joyner
	G-CCWM	Robin DR.400/180	M. R. Clark
	G-CCWN	Mainair Pegasus Quantum 15-912	S. Jeffrey
	G-CCWO	Mainair Pegasus Quantum 15-912	P. K. Dean
	G-CCWP	Aerotechnik EV-97 TeamEurostar UK	C.I.D.H. Garrison
	G-CCWR	Mainair Pegasus Quik	J. A. Robinson
	G-CCWS	Balony Kubicek BB30Z balloon	H. C. J. & S. A. G. Williams
	G-CCWT	Balony Kubicek BB20GP balloon	H. C. J. & S. A. G. Williams
	G-CCWU	Skyranger 912	D. M. Lane
	G-CCWV	Mainair Pegasus Quik	R. A. Taylor
	G-CCWW	Mainair Pegasus Quantum 15-912	Double Whisky Syndicate
	G-CCWY	Pilatus PC-12/45	Meridian Aviation Group Ltd
	G-CCWZ	Raj Hamsa X'Air Falcon 133	M. A. Evans
	G-CCXA	Boeing Stearman A.75N-1 Kaydet	T. Lyons
	G-CCXC	Avion Mudry CAP10B	Skymax (Aviation) Ltd
	G-CCXD	Lindstrand LBL-105B balloon	J. H. Dobson
	G-CCXE	Cameron Z-120 balloon	Cameron Balloons Ltd
	G-CCXF	Cameron Z-90 balloon	R. G. March & T. J. Maycock
	G-CCXG	SE-5A (replica)	C. Morris
	G-CCXH	Skyranger J2.2	Skyranger UK Ltd
	G-CCXI	Thorp T.211	J. Gilro
	G-CCXJ	Cessna 340A	Caernarfon Airworld Ltd
	G-CCXK	Pitts S-1S Special	P. G. Bond
	G-CCXL	Skyranger 912	R. G. Cameron
	G-CCXM	Skyranger 912	C. J. Finnigan
	G-CCXN	Skyranger 912	C. I. Chegwen
	G-CCXO	Corgy CJ-1 Starlet	I. W. L. Aikman
	G-CCXP	ICP Savannah Jabiru	J. H. Tope & B. J. Harper
	G-CCXS	Montgomerie-Bensen B.8MR	S. A. Sharp
	G-CCXR	Pegasus Mainair Blade	M. Fowler
	G-CCXT	Mainair Pegasus Quik	G. & P. Verity
	G-CCXU	Diamond DA.40D Star	McAlpine Aviation Training Ltd/Blackbushe
	G-CCXV	Thruster T.600N 450	Thruster Air Services Ltd
	G-CCXW	Thruster T.600N 450	J. Walsh
	G-CCXX	AG-5B Tiger	Alexson Lt
	G-CCXY	BAe 146-100	Avtrade Aircraft Leasing Ltd
	G-CCXZ	Mainair Pegasus Quik	K. J. Sene
	G-CCYA	Jabiru J.450	D. J. Royce
	G-CCYB	Escapade 912	B. E. Renehan
	G-CCYC	Robinson R-44-II	Derg Developments Ltd/Ireland
	G-CCYE	Mainair Pegasus Quik	J. Lane
	G-CCYF	Aerophile 5500 tethered gas balloon	High Point Balloons Ltd
	G-CCYG	Robinson R-44	Moorland Windows
	G-CCYH	Embraer EMB-145EP	Eastern Airways
	G-CCYI	Cameron O-105 balloon	Media Balloons Ltd
	G-CCYJ	Mainair Pegasus Quik	J. Ellis
	G-CCYK	Cessna 180	K. V. McKinnon
	G-CCYL	MainairPegasus Quantum 15	M. j. L. Morris
	G-CCYM	Skyranger 912	D. McDonagh
	G-CCYN	Cameron C-80 balloon	D. R. Firkins
	G-CCYO	Christen Eagle II	J. R. Pearce
	G-CCYP	Colt 56A balloon	L. P. Hooper
	G-CCYR	Ikarus C.42 FB80	M. L. Smith
	G-CCYS	Cessna F.182Q	D. J. Colledge
	G-CCYT	Robinson R-44-II	Bell Commercials
	G-CCYU	Ultramagic S-90 balloon	A. R. Craze
	G-CCYX	Bell 412	RCR Aviation Ltd
	G-CCYY	PA-28-161 Warrior II	Flightcontrol; Ltd
	G-CCYZ	Dornier EKW C3605	William Tomkins Ltd

Reg.	Type	Owner or Operator	Notes
G-CCZB	Mainair Pegasus Quantum 15	P. Stewart	
G-CCZD	Van's RV-7	R. T. Clegg	
G-CCZE	SA Bulldog Srs 120/125	British Disabled Flying Association	
G-CCZF	SA Bulldog Srs 120/125	British Disabled Flying Association	
G-CCZG	Robinson R-44-II	Westinbrook Ltd	
G-CCZH	Robinson R-44	Heli Air Ltd/Wellesbourne	
G-CCZJ	Raj Hamsa X' Air Falcon 582	A. B. Gridley	
G-CCZK	Zenair CH.601 UL Zodiac	R. J. Hopkins	
G-CCZL	Ikarus C.42 FB 80	W. H. J. Knowles	
G-CCZM	Skyranger 912S	D. Woodward	
G-CCZN	Rans S.6-ES Coyote II	M. Taylor	
G-CCZO	Mainair Pegasus Quik	S. B. Williams	
G-CCZP	Super Marine Aircraft Spitfire 26	J. W. E. Pearson	
G-CCZR	Medway Raven Eclipse 912	R. A. Keene	
G-CCZS	Raj Hamsa X'Air Falcon 582	P. J. Sheehy	
G-CCZT	Van's RV-9A	N. A. Henderson	
G-CCZU	Diamond DA.40D Star	Diamond UK Ltd/Gamston	
G-CCZV	PA-28-151 Warrior	H. A. Barrs	
G-CCZW	Mainair Pegasus Blade	C. J. Wright	
G-CCZX	Robin DR.400/180	J. B. Mills	
G-CCZY	Van's RV-9A	Mona RV-9 Group	
G-CCZZ	Aerotechnik EV-97 Eurostar	B. M. Starck & Bastin	
G-CDAA	Mainair Pegasus Quantum 15-912	I. A. Macadam	
G-CDAB	Stoddard-Hamilton Glasair 115RG	W. L. Hitchins	
G-GDAC	Aerotechnik EV-97 TeamEurostar	Fly CB Ltd	
G-CDAD	Lindstrand LBL-25A balloon	G. J. Madelin	
G-CDAE	Van's RV-6A	K. J. Fleming	
G-CDAF	Bell 412	RCR Aviation Ltd	
G-CDAG	Mainair Blade	M. J. Gerrish	
G-CDAI	Robin DR.400/135	Cole Aviation Ltd	
G-CDAK	Zenair CH.601 UK Zodiac	K. Kerr	
G-CDAL	Zenair CH.601UL Zodiac	D. Cassidy	
G-CDAM	Sky 77-24 balloon	M. Morris & P. A. Davies	
G-CDAO	Mainair Pegasus Quantum 15 -912	A. M. Dalgetty	
G-CDAP	Aerotechnik EV-97 TeamEurostar UK	P. Parker	
G-CDAR	Mainair Pegasus Quik	A. R. Pitcher	
G-CDAT	ICP MXP-740 Savannah Jabiru	R. Simpson	
G-CDAW	Robinson R-22B	Heli Air Ltd/Wellesbourne	
G-CDAX	Mainair Pegasus Quik	M. Winship	
G-CDAY	Skyranger 912	M. E. Furniss	
G-CDAZ	Aerotechnik EV-97 Eurostar	M. C. J. Ludlow	
G-CDBA	Skyranger 912(S)	P. J. Brennan	
G-CDBB	Mainair Pegasus Quik	J. L. Ker	
G-CDBC	Aviation Enterprises Magnum	Aviation Enterprises Ltd	
G-CDBD	Jabiru J.400	S. Derwin	
G-CDBE	Montgomerie-Bensen B.8M	P. Harwood	
G-CDBF	Robinson R-22B	Hields Aviation	
G-CDBG	Robinson R-22B	Hields Aviation	
G-CDBK	Rotorway Executive 162F	NF Auto Development	
G-CDBM	Robin DR.400/180	C. M. Simmonds	
G-CDBN	Thruster T.600N 450	Thruster Air Services Ltd	
G-CDBO	Skyranger 912	A. M. Dalgetty	
G-CDBP	Bell 206B JetRanger 2	Eastern Atlantic Helicopters Ltd	
G-CDBR	Stolp SA.300 Starduster Too	R. J. Warren	
G-CDBS	MBB Bo.105DBS-4	Bond Air Services Ltd	
G-CDBT	Bell 206B JetRanger 2	Eastern Atlantic Helicopters Ltd	
G-CDBV	Skyranger 912S	K. Hall	
G-CDBX	Shaw Europa XS	R. Marston	
G-CDBY	Dyn 'Aero MCR-01 ULC	R. Germany	
G-CDBZ	Thruster T.600N 450	Thruster Air Services Ltd	
G-CDCA	Robinson R-44-II	Heli Air Ltd/Wellesbourne	
G-CDCB	Robinson R-44-II	Heli Air Ltd/Wellesbourne	
G-CDCC	Aerotechnik EV-97A Eurostar	R. E. & & N. G. Nicholson	
G-CDCD	Van's RVF-9A	RV9ers	
G-CDCE	Avions Mudry CAP.10B	The Tiger Club 1990 Ltd	
G-CDCF	Mainair Pegasus Quik	B. L. Benson	
G-CDCG	Ikarus C.42 FB UK	N. E. Ashton & R. H. J. Jenkins	
G-CDCH	Skyranger 912	K. Laud	
G-CDCJ	BN-28-20 Islander	Britten-Norman Aircraft Ltd	
G-CDCK	Mainair Pegasus Quik	S. G. Ward	
G-CDCL	PA-26-161 Warrior III	Senate Aviation Ltd	

Notes	Reg.	Type	Owner or Operator
	G-CDCM	Ikarus C..42 FB UK	S. T. Allen
	G-CDCN	Avro RJ100	Trident Jet Leasing (Ireland) Ltd
	G-CDCO	Ikarus C.42 FB UK	G. G. Bevis
	G-CDCP	Avtech Jabiru J.400	M. W. T. Wilson
	G-CDCT	Aerotechnik EV-97 TeamEurostar UK	Cosmik Aviation Ltd
	G-CDCU	Mainair Pegasus Blade	D. K. Jones
	G-CDCV	Robinson R-44- II	Heli Air Ltd/Wellesbourne
	G-CDCW	Escapade 912 (1)	P. Nicholls
	G-CDCY	Mainair Pegasus Quantum 15	G. Stocker
	G-CDCZ	Mainair Pegasus Quantum 15-912	Light Flight Ltd
	G-CDDA	SOCATA TB.20 Trinidad	Oxford Aviation Services Ltd
	G-CDDB	Grob/Schemmp-Hirth CS-11	K. D. Barber/France
	G-CDDD	Robinson R-22B	Heli Air Ltd/Wellesbourne
	G-CDDE	WSK PZL-110 Koliber 160A	PZL International Aviation Marketing and Sales PLC/North Weald
	G-CDDF	Mainair Pegasus Quantum 15-912	B. C. Blackburn
	G-CDDG	PA-26-161 Warrior II	A. Oxenham
	G-CDDH	Raj Hamsa X'Air Falcon	B. & L. Stanbridge
	G-CDDI	Thruster T..600N 450	Thruster Air Services Ltd
	G-CDDK	Cessna 172M	M. H. & P. R. Kavern
	G-CDDO	Raj Hamsa X'Air 133	R. N. Tarrant & B. J. Reynolds
	G-CDDP	Lazer Z.230	A, Smith
	G-CDDR	Skyranger 582(1)	R. J. Milward
	G-CDDS	Zenair CH..601HD	S. Foreman
	G-CDDT	SOCATA TB.20 Trinidat	Oxford Aviaiation Services Ltd/Kidlington
	G-CDDU	Skyranger 912(2)	R. C. Reynolds
	G-CDDW	Aeroprakt A.22 Foxbat	D. A. A. Wineberg
	G-CDDX	Thruster T.600N 450	L. M. Leachma
	G-CDDY	Van's RV-8	R. W. H. Cole
	G-CDDZ	Lindstrand LBL-260A balloon	Lindstrand Hot-Air Balloons Ltd
	G-CDEB	SAAB 2000	Eastern Airways
	G-CDEF	PA-28-161 Cadet	Trek Air BV
	G-CDEH	ICP MXP-740 Savannah	S. Whittaker & P. J. Wilson
	G-CDEJ	Diamond DA.40D Star	Diamond Aircraft UK Ltd/Gamston
	G-CDEK	Diamond DA.40D Star	Diamond Aircraft UK Ltd/Gamston
	G-CDEL	Diamond DA.40D Star	Diamond Aircraft UK Ltd/Gamston
	G-CDEM	Raj Hamsa X' Air 133	R. J. Froud
	G-CDEN	Mainair Pegasus Quantum 15	912 R. E. J. Patttenden
	G-CDEP	Aerotechnik EV-97 TeamEurostar	S. R. Pikle
	G-CDES	Bell 206B JetRanger 3	Apple International Inc Ltd
	G-CDET	Culver LCA Cadet	H. B. Fox/Booker
	G-CDEV	Escapade 912 (1)	M. B. Devenport
	G-CDEX	Shaw Europa	J. M. Carter
	G-CDEY	Robinson R-44-II	Hel Air Ltd/Wellesbourne
	G-CDEB	SAAB 2000	Eastern Airways
	G-CDEF	PA-28-161 Cadet	Trek Air BV
	G-CDFA	Kolb Twinstar Mk3 Extra	M. H. Moulai
	G-CDFB	Raj Hamsa X'Air 582	P. K. Morley
	G-CDFD	Scheibe SF.25C Falke	T. M. Holloway
	G-CDFF	Aerospatiale ATR-42-300	Air Wales Ltd (G-BVEF)
	G-CDFG	Mainair Pegasus Quik	D. Gabbott
	G-CDFH	Escapade 912(1)	J. Deegan
	G-CDFI	Colt 31A balloon	A. M. Holly
	G-CDFJ	Skyranger 912	W. C. Yates
	G-CDFK	Jabiru SPL-450	H. J. Bradley
	G-CDFL	Zenair CH.601UL	Caunton Zodiac Group
	G-CDFM	Raj Hamsa X'Air 582 (5)	J. Griffiths
	G-CDFP	Skyranger 912 (2)	J. M. Gammidge
	G-CDFS	Embraer EMB-135ER	City Airline
	G-CDFT	Gefa-Flug AS-105GD airship	Cameron Balloons Ltd
	G-CDFU	Rans S.6-ES	P. W. Taylor
	G-CDFX	Lindstrand LBL-90A balloon	Lindstrand Balloons Ltd r
	G-CDFZ	Zenair CH.601UL	B. Yoxall & J. P. Harris
	G-CDGA	Taylor JT.1 Monoplane	R. M. Larimore
	G-CDGB	Rans S.6-116 Coyote	S. Penoyre
	G-CDGD	Pegasus Quik	P & M Aviation Ltd
	G-CDGE	Edge XT912-IIIB	C. T. Guest
	G-CDGG	Dyn'Aero MCR-01 Club	P. Simpson & P. A. B. Morgan
	G-CDGH	Rans S.6-ES Coyote	K. T. Vinning
	G-CDGJ	American Champion 7ECA Citabria	T. A. Mann
	G-CDGM	MBB BK-117C-2	McAlpine Helicopters Ltd/Kidlington
	G-CDGO	Pegasus Quik	Mainair Sports Ltd

Reg.	Type	Owner or Operator	Notes
G-CDIL	Pegasus Quantum 15-912	J. I. Greenshields	
G-CDIY	Aerotechnik EV-97A Eurostar	G. R. Pritchard	
G-CDJM	Zenair CH.601XL	T. J. Adams-Lewis	
G-CDOG	Lindstrand LBL-Dog SS balloon	Airship and Balloon Co Ltd	
G-CDON	PA-28-161 Warrior II	East Midlands Flying School Ltd	
G-CDPD	Mainair Pegasus Quik	P. C. Davis	
G-CDPY	Shaw Europa	A. Burrill	
G-CDRU	C.A.S.A. 1.131E Jungmann 2000	P. Cunniff/White Waltham	
G-CDUO	Boeing 757-236	Thomsonfly (G-BRJI)	
G-CDUP	Boeing 757-236	Thomsonfly (G-OOOT/G-BRJJ)	
G-CDUX	PA-32 Cherokee Six 300	D. J. Mason	
G-CEAB	Airbus A.300B2-1C	European Aviation Ltd *(stored)*/Bournemouth	
G-CEAC	Boeing 737-229	Palmair European/Bournemouth	
G-CEAD	Boeing 737-229	European Aviation Ltd *(stored)*/Bournemouth	
G-CEAE	Boeing 737-229	European Aviation Ltd *(stored)*/Bournemouth	
G-CEAF	Boeing 737-229	Palmair European (G-BYRI)/Bournemouth	
G-CEAG	Boeing 737-229	European Aviation Ltd/Bournemouth	
G-CEAH	Boeing 737-229	European Aviation Ltd/Bournemouth	
G-CEAI	Boeing 737-229	European Aviation Ltd/Bournemouth	
G-CEAJ	Boeing 737-229	European Aviation Ltd/Bournemouth	
G-CEAL	Short SD3-60 Variant 100	BAC Express Airlines Ltd (G-BPXO)	
G-CEEE	Robinson R-44	R. D. Masters	
G-CEGA	PA-34-200T Seneca II	Oxford Aviation Services Ltd/Kidlington	
G-CEGP	Beech 200 Super King Air	CEGA Aviation Ltd (G-BXMA)	
G-CEGR	Beech 200 Super King Air	Henfield Lodge Aviation Ltd	
G-CEJA	Cameron V-77 balloon	L. & C. Gray (G-BTOF)	
G-CELA	Boeing 737-377	Channel Express (Air Services) Ltd *(stored)*	
G-CELB	Boeing 737-377	Channel Express (Air Services) Ltd *(stored)*	
G-CELC	Boeing 737-377	Jet 2/Channel Express (Air Services) Ltd (G-OBMA)	
G-CELD	Boeing 737-377	Jet 2/Channel Express (Air Services) Ltd (G-OBMB)	
G-CELE	Boeing 737-377	Channel Express (Air Services) Ltd (G-MONN)	
G-CELF	Boeing 737-377	Jet 2/Channel Express (Air Services)/Leeds	
G-CELG	Boeing 737-377	Jet 2/Channel Express (Air Services)/Leeds	
G-CELH	Boeing 737-330	Channel Express (Air Services) Ltd	
G-CELII	Boeing 737-330	Channel Express (Air Services) Ltd	
G-CELJ	Boeing 737-330	Channel Express (Air Services) Ltd	
G-CELP	Boeing 737-330QC	Channel Express (Air Services) Ltd	
G-CELR	Boeing 737-330QC	Channel Express (Air Services) Ltd	
G-CELS	Boeing 737-377	Jet 2/Channel Express (Air Services) Ltd	
G-CELU	Boeing 737-377	Jet 2/Channel Express (Air Services) Ltd	
G-CELV	Boeing 737-377	Jet 2/Channel Express (Air Services) Ltd	
G-CELW	Boeing 737-377	Jet 2/Channel Express (Air Services) Ltd	
G-CELX	Boeing 737-377	Jet 2/Channel Express (Air Services) Ltd	
G-CELY	Boeing 737-377	Jet 2/Channel Express (Air Services) Ltd	
G-CELZ	Boeing 737-377	Jet 2/Channel Express (Air Services) Ltd	
G-CEPT	SOCATA TB.20 Trinidad	P. J. Caiger (G-BTEK)	
G-CERI	Shaw Europa XS	S. J. M. Shepherd	
G-CERT	Mooney M.20K	K. A. Hemming/Fowlmere	
G-CEXB	F.27 Friendship Mk 500	Channel Express (Air Services) Ltd	
G-CEXE	F.27 Friendship Mk 500	Channel Express (Air Services) Ltd *(stored)*	
G-CEXG	F.27 Friendship Mk 500	Channel Express (Air Services) Ltd (G-JEAP)	
G-CEXH	Airbus A.300B4-203F	Channel Express (Air Services) Ltd	
G-CEXI	Airbus A.300B4-203F	Channel Express (Air Services) Ltd	
G-CEXJ	Airbus A.300B4-203F	Channel Express (Air Services) Ltd	
G-CEXK	Airbus A.300B4-103	Channel Express (Air Services) Ltd	
G-CEXP	HPR-7 Herald 209 ★	-	
G-CEYE	PA-32R-300 Cherokee Lance	Fleetlands Flying Association Ltd	
G-CFAA	Avro RJ100	British Airways CitiExpress Ltd	
G-CFAB	Avro RJ100	British Airways CitiExpress	
G-CFAC	Avro RJ100	British Airways CitiExpress	
G-CFAD	Avro RJ100	British Airways CitiExpress Ltd	
G-CFAE	Avro RJ100	British Airways CitiExpress	
G-CFAF	Avro RJ100	British Airways CitiExpress	
G-CFAH	Avro RJ100	British Airways CitiExpress	
G-CFBI	Colt 56A balloon	G. A. Fisher	
G-CFME	SOCATA TB.10 Tobago	Charles Funke Associates Ltd	
G-CFRA	Cessna 560XL Citation Excel	Cirrus Aviation Holding Ltd	
G-CFWR	Skyranger 912 (2)	R. W. Clarke	

Notes	Reg.	Type	Owner or Operator
	G-CGDH	Shaw Europa XS	G. D. Harding
	G-CGHM	PA-28 Cherokee 140	A. Reay
	G-CGOD	Cameron N-77 balloon	G. P. Lane
	G-CHAD	Aeroprakt A.22 Foxbat	C. J. Rossiter
	G-CHAH	Shaw Europa	T. Higgins
	G-CHAM	Cameron 90 Pot SS balloon	High Exposure Balloons
	G-CHAP	Robinson R-44	Brierley Lifting Tackle Co Ltd
	G-CHAR	Grob G.109B	RAFGSA/Bicester
	G-CHAS	PA-28-181 Archer II	C. H. Elliott
	G-CHAV	Shaw Europa	R. P. Robinson
	G-CHCD	Sikorsky S-76A (modified)	CHC Scotia Ltd (G-CBJB)
	G-CHCF	AS.332L-2 Super Puma	CHC Scotia Ltd
	G-CHCG	AS.332L-2 Super Puma	CHC Scotia Ltd
	G-CHCH	AS.332L-2 Super Puma	CHC Scotia Ltd
	G-CHEB	Shaw Europa	T. C. Butterworth
	G-CHEL	Colt 77B balloon	Chelsea Financial Services PLC
	G-CHEM	PA-34-200T Seneca II	London Executive Aviation Ltd
	G-CHER	PA-38-112 Tomahawk II	Aerohire Ltd (G-BVBL)/Wolverhampton
	G-CHET	Shaw Europa	H. P. Chetwynd-Talbot
	G-CHEZ	BN-2B-20 Islander	Cheshire Police Authority (G-BSAG)/Liverpool
	G-CHGL	Bell 206B JetRanger II	Elmridge Ltd (G-BPNG/G-ORTC)
	G-CHIK	Cessna F.152	Stapleford Flying Club Ltd (G-BHAZ)
	G-CHIP	PA-28-181 Archer II	C. M. Hough/Fairoaks
	G-CHIS	Robinson R-22B	Bradmore Helicopter Leasing
	G-CHIX	Robin DR.400/500	P. A. & R. Stephens
	G-CHKL	Cameron 120 Kookaburra SS balloon	Eagle Ltd/Australia
	G-CHKN	Kiss 400-582 (1)	D. A. Edwards
	G-CHLL	Lindstrand LBL-90A balloon	Airship and Balloon Co Ltd
	G-CHLT	Stemme S.10	J. Abbess
	G-CHMP	Bellanca 7ACA Champ	I. J. Langley
	G-CHOK	Cameron V-77 balloon	A. J. Moore
	G-CHOP	Westland-Bell 47G-3B1	Classic Rotors Ltd
	G-CHOX	Shaw Europa XS	Chocs Away Ltd
	G-CHPY	D.H.C.1 Chipmunk 22 (WB652)	Devonair Executive Business Travel Ltd
	G-CHSL	Augusta A.109A II	Swift Copter Ltd
	G-CHSU	Eurocopter EC.135T-1	Thames Valley Police Authority Chiltern Air Support Unit/Benson
	G-CHTA	AA-5A Cheetah	Quick Spin Ltd (G-BFRC)
	G-CHTG	Rotorway Executive 90	G. Cooper (G-BVAJ)
	G-CHUG	Shaw Europa	C. M. Washington
	G-CHUK	Cameron O-77 balloon	L. C. Taylor
	G-CHUM	Robinson R-44	Vitapage Ltd
	G-CHYL	Robinson R-22B	C. M. Gough-Cooper
	G-CHZN	Robinson R-22B	Cloudbase Ltd (G-GHZM/G-FENI)
	G-CIAO	I.I.I. Sky Arrow 1450-L	G. Arscott
	G-CIAS	BN-2B-21 Islander	Channel Island Air Search Ltd (G-BKJM)
	G-CIBO	Cessna 180K	CIBO Ops Ltd
	G-CICI	Cameron R-15 balloon	Noble Adventures Ltd
	G-CIDD	Bellanca 7ECA Citabria	A. & P. West
	G-CIFR	PA-28-181 Archer II	Shropshire Aero Club Ltd/Sleap
	G-CIGY	Westland-Bell 47G-3B1	R. A. Perrot (G-BGXP)
	G-CITR	Cameron Z-105 balloon	Flying Pictures Ltd
	G-CITY	PA-31-350 Navajo Chieftain	Woodgate Aviation (IOM) Ltd
	G-CIVA	Boeing 747-436	British Asia Airways
	G-CIVB	Boeing 747-436	British Asia Airways
	G-CIVC	Boeing 747-436	British Airways
	G-CIVD	Boeing 747-436	British Airways
	G-CIVE	Boeing 747-436	British Airways
	G-CIVF	Boeing 747-436	British Airways
	G-CIVG	Boeing 747-436	British Airways
	G-CIVH	Boeing 747-436	British Airways
	G-CIVI	Boeing 747-436	British Airways
	G-CIVJ	Boeing 747-436	British Airways
	G-CIVK	Boeing 747-436	British Airways
	G-CIVL	Boeing 747-436	British Airways
	G-CIVM	Boeing 747-436	British Airways
	G-CIVN	Boeing 747-436	British Airways
	G-CIVO	Boeing 747-436	British Airways
	G-CIVP	Boeing 747-436	British Airways
	G-CIVR	Boeing 747-436	British Airways

Reg.	Type	Owner or Operator	Notes
G-CIVS	Boeing 747-436	British Airways	
G-CIVT	Boeing 747-436	British Airways	
G-CIVU	Boeing 747-436	British Airways	
G-CIVV	Boeing 747-436	British Airways	
G-CIVW	Boeing 747-436	British Airways	
G-CIVX	Boeing 747-436	British Airways	
G-CIVY	Boeing 747-436	British Airways	
G-CIVZ	Boeing 747-436	British Airways	
G-CJAA	BAe 125 Srs 800B	Bookajet Operations Ltd	
		(G-HCFR/G-SHEA/G-BUWC)	
G-CJAD	Cessna 525 CitationJet	Davis Aircraft Operations	
G-CJAE	Cessna 560 Citation V	Ourjet Ltd (G-CZAR)	
G-CJAY	Mainair Pegasus Quik	Jaye Airsports	
G-CJBC	PA-28 Cherokee 180	J. B. Cave/Halfpenny Green	
G-CJCI	Pilatus P2-06 (CC+43)	Pilatus P2 Flying Group	
G-CJUD	Denney Kitfox Mk 3	AV8 Air	
G-CKCK	Enstrom 280FX	Farmax Ltd	
G-CLAC	PA-28-161 Warrior II	M. A. Steadman/Blackbushe	
G-CLAS	Short SD3-60 Variant 100	BAC Express Ltd (G-BLED)	
G-CLAV	Shaw Europa	G. Laverty	
G-CLAX	Jurca MJ.5 Sirocco F2/39	G. D. Claxton (G-AWKB)	
G-CLAY	Bell 206B JetRanger 3	Claygate Distribution Ltd (G-DENN)	
G-CLEA	PA-28-161 Warrior II	R. J. Harrison & A. R. Carpenter	
G-CLEE	Rans S.6-ES Coyote II	R. Holt	
G-CLEM	Bo 208A2 Junior	Bolkow Group (G-ASWE)	
G-CLEO	Zenair CH.601HD	K. M. Bowen	
G-CLFC	Mainair Blade	G. N. Cliffe & G. Marshall	
G-CLHD	BAe 146-200	Flightline Ltd (G-DEBF)	
G-CLIC	Cameron A-105 balloon	R. S. Mohr	
G-CLKE	Robinson R-44	Clarke Business (G-HREH)	
G-CLOE	Sky 90-24 balloon	Zebedee Balloon Service Ltd	
G-CLOS	PA-34-200 Seneca II	S. H. Kirkby	
G-CLOW	Beech 200 Super King Air	Clowes (Estates) Ltd	
G-CLRK	Sky 77-24 balloon	William Clark & Son (Parkgate) Ltd	
G-CLUB	Cessna FRA.150N	D. C. C. Handley	
G-CLUE	PA-34-200T Seneca II	Flyuk.com Ltd	
G-CLUX	Cessna F.172N	J. & K. Aviation	
G-CMED	SOCATA TB.9 Tampico	Enstone Flying Club	
G-CMGC	PA-25 Pawnee 235	Midland Gliding Club Ltd (G-BFEX)/Long Mynd	
G-CMSN	Robinson R-22B	Kuki Helicopters Ltd (G-MGEE/G-PHEL/G-RUMP)	
G-CNAB	Avtech Jabiru UL	W. A. Brighouse	
G-COAI	Cranfield A.1	Cranfield University (G-BCIT)	
G-COCO	Cessna F.172M	P. C. Sheard & R. C. Larder	
G-CODE	Bell 206B JetRanger 3	B. Wronski	
G-COIN	Bell 206B JetRanger 2	S. Pool & ptnrs	
G-COLA	Beech F33C Bonanza	J. R. C. Spooner & P. M. Scarratt (G-BUAZ)	
G-COLH	PA-28 Cherokee 140	Full Sutton Flying Centre Ltd (G-AVRT)	
G-COLL	Enstrom 280C-UK-2 Shark	Taylor Air Services Ltd	
G-COLR	Colt 69A balloon ★	British School of Ballooning/Lancing	
G-COLS	Van's RV-7A	C. Terry	
G-COMB	PA-30 Twin Comanche 160B	J. T. Bateson (G-AVBL)/Ronaldsway	
G-COMP	Cameron N-90 balloon	Computacenter Ltd	
G-CONB	Robin DR.400/180	CCB Aviation (G-BUPX)	
G-CONC	Cameron N-90 balloon	British Airways	
G-CONL	SOCATA TB.10 Tobago	J. M. Huntington	
G-CONV	Convair CV-440-54	Atlantic Air Transport Ltd/Coventry	
G-COOK	Cameron N-77 balloon	IAZ (International) Ltd	
G-COOT	Taylor Coot A	P. M. Napp	
G-COPS	Piper J-3C-65 Cub	R. W. Sproat	
G-COPZ	Van's RV-7	R. S. Horan	
G-CORB	SOCATA TB-20 Trinidad	G. D. Corbin	
G-CORD	Slingsby T.66 Nipper 3	A. V. Lamprell (G-AVTB)	
G-CORN	Bell 206B JetRanger 3	John A.Wells Ltd (G-BHTR)	
G-CORP	BAe ATP	Trident Aviation (G-BTNK)	
G-CORT	AB-206B JetRanger 3	Helicopter Training & Hire Ltd	
G-COSY	Lindstrand LBL-56A balloon	D. D. Owen	

Notes	Reg.	Type	Owner or Operator
	G-COTT	Cameron 60 Cottage SS balloon	Dragon Balloon Co. Ltd
	G-COUP	Ercoupe 415C	S. M. Gerrard
	G-COVE	Avtech Jabiru UL	A. A. Rowson
	G-COWS	ARV Super 2	T. C. Harrold (G-BONB)
	G-COXS	Aeroprakt A.22 Foxbat	S. Cox
	G-COXY	Kiss 400-582 (1)	B. G. Cox
	G-COZI	Rutan Cozy III	D. G. Machin
	G-CPCD	CEA DR.221	P. J. Taylor
	G-CPDA	D.H.106 Comet 4C	C. Walton Ltd/Bruntingthorpe
	G-CPEL	Boeing 757-236	British Airways (G-BRJE)
	G-CPEM	Boeing 757-236	British Airways
	G-CPEN	Boeing 757-236	British Airways
	G-CPEO	Boeing 757-236	British Airways
	G-CPEP	Boeing 757-2Y0	First Choice Airways Ltd
	G-CPER	Boeing 757-236	British Airways
	G-CPES	Boeing 757-236	British Airways
	G-CPET	Boeing 757-236	British Airways
	G-CPEU	Boeing 757-236	First Choice Airways Ltd
	G-CPEV	Boeing 757-236	First Choice Airways Ltd
	G-CPFC	Cessna F.152 II	Willowair Flying Club (1996) Ltd/Southend
	G-CPMK	D.H.C.1 Chipmunk 22 (WZ847)	P. A. Walley
	G-CPMS	SOCATA TB.20 Trinidad	Charlotte Park Management Services Ltd
	G-CPOL	AS.355F-1 Twin Squirrel	MW Helicopters Ltd/Stapleford
	G-CPSF	Cameron N-90 balloon	S. A. Simington & J. D. Rigden (G-OISK)
	G-CPSH	Eurocopter EC.135T-1	Thames Valley Police Authority/Booker
	G-CPTM	PA-28-151 Warrior	T. J. Mackay & C. M. Pollett (G-BTOE)
	G-CPTS	AB-206B JetRanger 2	A. R. B. Aspinall
	G-CPXC	Avions Mudry CAP.10C	J. M. Wicks
	G-CRAB	Skyranger 912 (1)	M. W. Houghton
	G-CRAY	Robinson R-22B	F.C. Owen
	G-CRDY	Agusta-Bell 206A JetRanger	Cardy Construction Ltd (G-WHAZ)
	G-CRES	Denney Kitfox Mk 3	K. M. James
	G-CRIB	Robinson R-44	Cribarth Helicopters (G-JJWL)
	G-CRIC	Colomban MC. 15 Cri-Cri	R. S. Stoddart-Stones
	G-CRIK	Colomban MC. 15 Cri-Cri	A.R. Robinson
	G-CRIL	R. Commander 112B	Rockwell Aviation Group/Cardiff
	G-CRIS	Taylor JT.1 Monoplane	C. R. Steer
	G-CRLH	Bell 206B JetRanger 3	R. L. Hartshorn (G-RJTT)
	G-CROB	Shaw Europa XS T-G	R. G. Hallam
	G-CROL	Maule MXT-7-180	N. G. P. Evans
	G-CROW	Robinson R-44	Longmoore Ltd
	G-CROY	Shaw Europa	M. T. Austin
	G-CRPH	Airbus A.320-231	My Travel Airways
	G-CRUM	Westland Scout AH.1	G-CRUM Group
	G-CRUZ	Cessna T.303	Bank Farm Ltd
	G-CSAV	Thruster T.600N 450	Thurster Air Services Ltd
	G-CSBM	Cessna F.150M	Halegreen Associates Ltd
	G-CSCS	Cessna F.172N	C.Sullivan/Stapleford
	G-CSDJ	Avtech Jabiru UL	D. W. Johnston & ptnrs
	G-CSFC	Cessna 150L	Foxtrot Charlie Flying Group
	G-CSFD	Ultramagic M-90 balloon	L. A. Watts
	G-CSFT	PA-23 Aztec 250D★	Aces High Ltd (G-AYKU)/North Weald
	G-CSIX	PA-32 Cherokee Six 300	G. A. Ponsford
	G-CSMK	Aerotechnik EV-97 Eurostar	Cosmik Aviation Ltd
	G-CSNA	Cessna 421C	Blue Swan Aviation Ltd
	G-CSWH	PA-28R Cherokee Arrow 180	The Gould Group International Ltd
	G-CTAV	Aerotechnik EV-97 Eurostar	C. M. Theakstone
	G-CTCL	SOCATA TB.10 Tobago	Gift Aviation Ltd (G-BSIV)
	G-CTCT	Flight Design CT.2K	Cyclone Airsports Ltd
	G-CTDH	Flight Design CT.2K	D. Haygreen
	G-CTEC	Stoddard-Hamilton Glastar	B. N. C. Mogg
	G-CTEL	Cameron N-90 balloon	G-CTEL Flying Group
	G-CTFF	Cessna T.206H	Rajair Ltd
	G-CTGR	Cameron N-77 balloon	T. G. Read (G-CCDI)
	G-CTIO	SOCATA TB.20 Trinidad	Cityiq Ltd
	G-CTIX	V.S.509 Spitfire T.IX (PT462)	A. A. Hodgson
	G-CTKL	Noorduyn AT-16 Harvard IIB (54137)	M. R. Simpson
	G-CTOY	Denney Kitfox Mk 3	B. McNeilly

Reg.	Type	Owner or Operator	Notes
G-CTPW	Bell 206B JetRanger 3	Aviation Rentals	
G-CTRL	Robinson R-22B	Central Helicopters Ltd	
G-CTUG	PA-25 Pawnee 235	The Borders (Milfield) Gliding Club Ltd	
G-CTWW	PA-34-200T Seneca II	Centreline Air Charter Ltd (G-ROYZ/G-GALE)	
G-CTZO	SOCATA TB.20 Trinidad GT	Trinidad Hir Ltd	
G-CUBB	PA-18 Super Cub 180	Bidford Airfield Ltd	
G-CUBE	Skyranger 912 (2)	T.R. Villa	
G-CUBI	PA-18 Super Cub 125	G. T. Fisher	
G-CUBJ	PA-18 Super Cub 150 (56-5395)	P. B. Rice	
G-CUBP	PA-18 Super Cub 150	D. W. Berger	
G-CUBS	Piper J-3C-65 Cub	Sunbeam Aviation (G-BHPT)	
G-CUBW	WAG-Aero Acro Trainer	B. G. Plumb & ptnrs	
G-CUBY	Piper J-3C-65 Cub	C. A. Bloom (G-BTZW)	
G-CUCU	Colt 180A balloon	S. R. Seage	
G-CUIK	QAC Quickie Q.200	C. S. Rayner	
G-CUPS	IAV Bacau Yakolev YAK-52	Fenland Flying School	
G-CURR	Cessna 172R II	JS Aviation Ltd (G-BXOH)	
G-CURV	Avid Speedwing	K. S. Kelso	
G-CUTE	Dyn'Aero MCR-01	E. G. Shimmin	
G-CUTY	Shaw Europa	D. J. & M. Watson	
G-CVBF	Cameron A-210 balloon	Virgin Balloon Flights Ltd	
G-CVIP	Bell 206B JetRanger	Sloane Helicopters Ltd/Skywell	
G-CVIX	D.H.110 Sea Vixen D.3 ('Red Bull')	De Havilland Aviation Ltd	
G-CVLH	PA-32-200T Seneca II	Atlantic Aviation Ltd	
G-CVPM	VPM M.16 Tandem Trainer	C. S. Teuber/Germany	
G-CVST	Jodel D.140	A. Shipp	
G-CWAG	Sequoia F. 8L Falco	I. R. Court & W. Jones	
G-CWAL	Raj Hamsa X'Air 133	C. Walsh	
G-CWBM	Phoenix Currie Wot	B. V. Mayo (G-BTVP)	
G-CWFA	PA-38-112 Tomahawk	Cardiff-Wales Flying Club Ltd (G-BTGC)	
G-CWFB	PA-38-112 Tomahawk	Cardiff-Wales Flying Club Ltd (G-OAAL)	
G-CWFC	PA-38-112 Tomahawk★	Cardiff-Wales Flying Club Ltd (G-BRTA)	
G-CWFD	PA-38-112 Tomahawk	M. Lowe & K. Hazelwood (G-BSVY)	
G-CWFE	PA-38-112 Tomahawk	Cardiff-Wales Flying Club Ltd (G-BPBR)	
G-CWFY	Cessna 152 II	Cardiff-Wales Flying Club Ltd (G-OAMY)	
G-CWFZ	PA-28-151 Warrior	Cardiff-Wales Flying Club Ltd (G-CPCH/G-BRGJ)	
G-CWIC	Mainair Pegasus Quik	R. Rainey	
G-CWIK	Mainair Pegasus Quik	L. Kirk	
G-CWIZ	AS.350B Ecureuil	PLM Dollar Group Ltd (G-DJEM/G-ZBAC/ G-SEBI/G-BMCU)	
G-CWTD	Aeroprakt A.22 Foxbat	J. V. Harris	
G-CWOT	Currie Wot	D. Doyle & H. Duggan	
G-CWVY	Mainair Pegasus Quik	I. A. Gaetan	
G-CXCX	Cameron N-90 balloon	Cathay Pacific Airways (London) Ltd	
G-CXDZ	Cassutt Speed Two	J. A. H. Chadwick	
G-CXHK	Cameron N-77 balloon	Cathay Pacific Airways (London) Ltd	
G-CXIP	Thruster T.600N	India Papa Syndicate	
G-CYLS	Cessna T.303	Gledhill Water Storage Ltd (G-BKXI)/Blackpool	
G-CYMA	GA-7 Cougar	Cyma Petroleum (UK) Ltd (G-BKOM)/Elstree	
G-CYRA	Kolb Twinstar Mk. 3 (Modified)	S. J. Fox (G-MYRA)	
G-CZAC	Zenair CH.601XL	D. Pitt	
G-CZAG	Sky 90-24 balloon	S. McCarthy	
G-CZBE	CFM Streak Shadow	M. I. M. Smith (G-MZBE)	
G-CZCZ	Avions Mudry CAP.10B	P. R. Moorhead & M. Farmer	
G-CZMI	Best Off Skyranger 912 (2)	T. W. Thiele	
G-CZNE	BN-2B-20 Islander	B-N Group Ltd (G-BWZF)/Bembridge	
G-DAAC	Canadair CL600-2B16 (604)	1247 Ltd	
G-DAAH	PA-28RT-201T Turbo Arrow IV	R. Peplow/Halfpenny Green	
G-DAAM	Robinson R-22A	Direct Helicopters/Southend	
G-DAAT	Eurocopter EC.135T-2	Bond Air Services Ltd	
G-DAAZ	PA-28RT-201T Turbo Arrow IV	Calais Ltd	
G-DABS	Robinson R-22B-2	B. Seymour	
G-DACA	P.57 Sea Prince T.1 (WF118) ★	P. G. Vallance Ltd/Charlwood	
G-DACC	Cessna 401B	Niglon Ltd (G-AYOU)/Birmingham	
G-DACF	Cessna 152 II	T. M. & M. L. Jones (G-BURY)/Egginton	

Notes	Reg.	Type	Owner or Operator
	G-DADG	PA-18-150 Super Cub	F. J. Cox
	G-DAEX	Dassault Falcon 900EX	Triair (Bermuda) Ltd
	G-DAFY	Beech 58 Baron	P. R. Earp
	G-DAIR	Luscombe 8A Silvaire	D. F. Soul (G-BURK)
	G-DAIV	Ultramagic H-77 balloon	D. Harrison-Morris
	G-DAJB	Boeing 757-2T7	Monarch Airlines Ltd/Luton
	G-DAJC	Boeing 767-31KER	My Travel Airways
	G-DAKK	Douglas C-47A	General Technics Ltd
	G-DAKO	PA-28-236 Dakota	Methods Application Ltd
	G-DAMY	Shaw Europa	U. A. Schliessler & R. J. Kelly
	G-DANA	Jodel DR.200 (replica)	F. A. Bakir (G-DAST)
	G-DAND	SOCATA TB.10 Tobago	Portway Aviation Ltd
	G-DANT	R. Commander 114	D. P. Tierney
	G-DANY	Avtech Jabiru UL	D. A. Crosbie
	G-DANZ	AS.355N Twin Squirrel	Frewton Ltd
	G-DAPH	Cessna 180K	M. R. L. Astor
	G-DARA	PA-34-220T Seneca III	SYS (Scaffolding Contractors) Ltd
	G-DARK	CFM Shadow Srs DD	R. W. Hussey
	G-DASH	R. Commander RC112	D. & M. Nelson (G-BDAJ)
	G-DASU	Cameron V-77 balloon	D. & L. S. Litchfield
	G-DATE	Agusta A.109C	Rich and Lucky Aviation Ltd (G-RNLD)
	G-DATG	Cessna F.182P	Oxford Aeroplane Co Ltd/Kidlington
	G-DATH	Aerotechnik EV-97 Eurostar	D. N. E. D'Ath
	G-DAVD	Cessna FR.172K	D. M. Driver
	G-DAVE	Jodel D.112	D. A. Porter/Sturgate
	G-DAVG	Robinson R-44 II	D. A. Gold (G-WOWW)
	G-DAVO	AA-5B Tiger	Douglas Head Cibsulting Ltd(G-GAGA/ G-BGPG/G-BGRW)/Elstree
	G-DAVT	Schleicher ASH-26E	D. A. Triplett
	G-DAWG	SA Bulldog Srs 120/121	R. H. Goldstone
	G-DAWS	Sikorsky S-61N	Air Harrods Ltd (G-LAWS/G-BHOF)
	G-DAYI	Shaw Europa	A. F. Day
	G-DAYS	Shaw Europa	D. J. Bowie
	G-DAYZ	Pietenpol Air Camper	J. G. Cronk
	G-DAZY	PA-34-200T Seneca	Centreline Air Charter Ltd
	G-DAZZ	Van's RV-8	D. M. Hartfree-Bright
	G-DBAT	Lindstrand LBL-56A balloon	G. J. Bell
	G-DBCA	Airbus A.319-131	bmi british midland
	G-DBCB	Airbus A.319-131	bmi british midland
	G-DBCC	Airbus A.319-131	bmi british midland
	G-DBCD	Airbus A.319-131	bmi british midland
	G-DBOY	Agusta A.109C	Castle Air Charters Ltd
	G-DBUG	Robinson R-44	Pietas Ltd (G-OBHI)
	G-DBYE	Mooney M.20M	A. J. Thomas
	G-DCAV	PA-32R-301Saratoga IIHP	Airquest Aviation LLP
	G-DCDB	Bell 407	Paycourt Ltd
	G-DCEA	PA-34-200T Seneca II	Barnes Olsen Aero Leasing Ltd
	G-DCKK	Cessna F.172N	J. Maffia/Panshanger
	G-DCMI	Mainair Pegasus Quik	S. J. E. Smith
	G-DCPA	MBB BK.117C-1C	Devon & Cornwall Constabulary (G-LFBA)
	G-DCSE	Robinson R-44	Foxtrot Golf Helicopters Ltd
	G-DCXL	Jodel D.140C	X-Ray Lima Group
	G-DDAY	PA-28R-201T Turbo Arrow III	G-DDAY Group (G-BPDO)/Tatenhill
	G-DDOG	SA BulldogSrs 120/121	Deltaero Ltd
	G-DDMV	NA T-6G Texan (493209)	E. A. Morgan
	G-DEAN	Solar Wings Pegasus XL-Q	A. J. Hodson (G-MVJV)
	G-DEBE	BAe 146-200	Flightline Ltd/Southend
	G-DEBR	Shaw Europa	A. J. Calvert & C. T. Smallwood
	G-DEBT	Pioneer 300	N. J. T. Tonks
	G-DECK	Cessna T.210N	R. J. Howard
	G-DECO	Dyn'Aero MCR-01 Club	A. W. Bishop & G. Castelli
	G-DEER	Robinson R-22B-2	Westinbrook Ltd/Shoreham
	G-DEFM	BAe 146-200	IMP Group (stored) (G-DEBM)
	G-DEKA	Cameron Z-90 balloon	Sport Promotion SRL
	G-DELF	Aero L-29A Delfin	B. R. Green/Manston
	G-DELT	Robinson R-22B	G-ZLLE Ltd
	G-DEMH	Cessna F.172M (modified)	M. Hammond (G-BFLO)
	G-DEMM	AS.350B-2 Ecureuil	Abbeyflight Ltd
	G-DENA	Cessna F.150G	W. M. Wilson & R. Campbell (G-AVEO)

Reg.	Type	Owner or Operator	Notes
G-DENB	Cessna F.150G	D. C. Tissington (G-ATZZ)	
G-DENC	Cessna F.150G	M. Dovey (G-AVAP)	
G-DEND	Cessna F.150M	R. N. Tate (G-WAFC/G-BDFI)	
G-DENE	PA-28 Cherokee 140	Bristol & Wessex Aeroplane Club (G-ATOS)	
G-DENH	PA-28-161 Warrior II	Plane Talking Ltd (G-BTNH)/Elstree	
G-DENI	PA-32 Cherokee Six 300	A. Bendkowski (G-BAIA)	
G-DENR	Cessna F.172N	Dorset Flying Group (G-BGNR)	
G-DENS	Binder CP.301S Smaragd	G. E. Roe & I. S. Leader	
G-DENT	Cameron N-145 balloon	Deproco UK Ltd	
G-DENZ	PA-44-180 Seminole	Horizon Ballooning Ltd (G-INDE/G-BHNM)	
G-DERB	Robinson R-22B	S. Thompson (G-BPYH)	
G-DERI	PA-46-500TP Malibu Meridian	Intesa Leasing SPA (G-PCAR)/Italy	
G-DERK	PA-46-500TP Malibu Meridian	D. Priestley	
G-DERV	Cameron Truck SS balloon	J. M. Percival	
G-DESS	Mooney M.20J	R. M. Hitchin	
G-DEST	Mooney M.20J	Allegro Aviation Ltd	
G-DEVL	Eurocopter EC.120B	Swift Frame Ltd	
G-DEVS	PA-28 Cherokee 180	180 Group (G-BGVJ)/Blackbushe	
G-DEXP	ARV Super 2	W. G. McKinnon	
G-DEZC	H.S.125 Srs 700B	Frewton Ltd (G-BWCR)	
G-DFKI	Westland Gazelle HT.2	Foremans Aviation Ltd (G-BZOT)	
G-DFLY	PA-38-112 Tomahawk	P. M. Raggett	
G-DFVA	Cessna R.172K	R. A. Plowright	
G-DGCL	Glaser-Dirks DG.800B	C. J. Lowrie	
G-DGHD	Robinson R-44-II	Ramsgill Aviation Ltd	
G-DGHI	Dyn Aero MCR-1 Club	D. G. Hall	
G-DGIV	Glaser-Dirks DG.800B	W. R. McNair	
G-DGWW	Rand-Robinson KR-2	W. Wilson/Liverpool	
G-DHCB	D.H.C.2 Beaver 1	Seaflite Ltd (G-BTDL)	
G-DHCC	D.H.C.1 Chipmunk 22	Eureka Aviation NV/Belgium	
G-DHCI	D.H.C.1 Chipmunk 22	Felthorpe Flying Group Ltd (G-BBSE)	
G-DHDV	D.H.104 Dove 8 (VP981)	Air Atlantique Ltd/Coventry	
G-DHJH	Airbus A.321-211	My Travel Airways	
G-DHLB	Cameron N-90 balloon	B. A. Bower	
G-DHLI	Colt 90 World SS balloon	Balloon Preservation Flying Group	
G-DHPM	OGMA D.H.C.1 Chipmunk 20	P. Meyrick	
G-DHSS	D.H.112 Venom FB.50	D. J. L. Wood/Bournemouth	
G-DHTM	D.H.82A Tiger Moth (replica)	E. G. Waite-Roberts	
G-DHTT	D.H.112 Venom FB.50 (WR421)	Source Classic Jet Flight (G-BMOC)	
G-DHUU	D.H.112 Venom FB.50 (WR410)	Source Classic Jet Flight (G-BMOD)	
G-DHVM	D.H.112 Venom FB.50	Air Atlantique Ltd (G-GONE)/Coventry	
G-DHVV	D.H.115 Vampire T.55 (XE897)	Source Classic Jet Flight	
G-DHWW	D.H.115 Vampire T.55 (XG775)	Source Classic Jet Flight	
G-DHXX	D.H.100 Vampire FB.6 (VT871)	Source Classic Jet Flight	
G-DHZF	D.H.82A Tiger Moth (N9192)	C. A.Parker & ptnrs (G-BSTJ)/Sywell	
G-DHZZ	D.H.115 Vampire T.55 (WZ589)	Source Classic Jet Flight	
G-DIAL	Cameron N-90 balloon	A. J. Street	
G-DIAT	PA-28 Cherokee 140	Bristol Flying Centre Ltd (G-BCGK)	
G-DICK	Thunder Ax6-56Z balloon	R. D. Sargeant	
G-DIGI	PA-32 Cherokee Six 300	D. Stokes	
G-DIKY	Murphy Rebel	R. J. P. Herivel	
G-DIMB	Boeing 767-31KER	My Travel Airways	
G-DIME	R. Commander 114	H. B. Richardson	
G-DINA	AA-5B Tiger	Portway Aviation Ltd/Shobdon	
G-DING	Colt 77A balloon	G. J. Bell	
G-DINK	Lindstrand Bulb SS balloon	Dinkelacker-Schwaben Brau AG/Germany	
G-DINO	Pegasus Quantum 15	R. A. Watering (G-MGMT)	
G-DINT	B.156 Beaufighter IF	T. E. Moore	
G-DIPI	Cameron 80 Tub SS balloon	R. A. Preston	
G-DIPM	PA-46-350P Malibu Mirage	Intesa Leasing SPA/Italy	
G-DIRK	Glaser-Dirks DG.400	G-DIRK Syndicate	
G-DISA	SA Bulldog Srs 120/125	British Disabled Flying Association	
G-DISK	PA-24 Comanche 250	A. Johnston (G-APZG)	
G-DISO	Jodel 150	P. F. Craven	
G-DIVA	Cessna R.172K XPII	R. J. Harris	
G-DIWY	PA-32 Cherokee Six 300	IFS Chemicals Ltd	
G-DIXY	PA-28-181 Archer III	M. G. Bird/Fowlmere	
G-DIZI	Escapade 912 (1)	N. Baumber	

Notes	Reg.	Type	Owner or Operator
	G-DIZO	Jodel D.120A	D. Aldersea (G-EMKM)/Breighton
	G-DIZY	PA-28R-201T Turbo Arrow III	Calverton Flying Club Ltd/Cranfield
	G-DIZZ	Hughes 369HE	H. J. Pelham
	G-DJAE	Cessna 500 Citation	Source Group Ltd (G-JEAN)/Bournemouth
	G-DJAY	Avtech Jabiru UL-450	D. J. Pearce
	G-DJCR	Varga 2150A Kachina	D. J. C. Robertson (G-BLWG)
	G-DJEA	Cessna 421C	Bettany Aircraft Holdings Ltd
	G-DJJA	PA-28-181 Archer II	Choice Aircraft/Fowlmere
	G-DJNH	Denney Kitfox Mk 3	D. J. N. Hall
	G-DJRV	Van's RV 8	S. G. Hunt
	G-DJST	Ixess 912(1)	D. J. Stimpson
	G-DKDP	Grob G.109	D. W. & J. E. Page
	G-DKGF	Viking Dragonfly	P. C. Dobwor
	G-DLCB	Shaw Europa	K. Richards
	G-DLDL	Robinson R-22B	Aeromega Ltd/Stapleford
	G-DLEE	SOCATA TB.9 Tampico Club	D. A. Lee (G-BPGX)
	G-DLFN	Aero L-29 Delfin	T. W. Freeman & N. Gooderham/North Weald
	G-DLOM	SOCATA TB.20 Trinidad	J. N. A. Adderley/Guernsey
	G-DLTR	PA-28 Cherokee 180E	BCT Aircraft Leasing Ltd (G-AYAV)
	G-DMAC	Avtech Jabiru UL	C. J. Pratt
	G-DMAH	SOCATA TB.20 Trinidad	C. A. Ringrose
	G-DMCA	Douglas DC-10-30 ★	*Forward Fuselage/Manchester Viewing Park*
	G-DMCD	Robinson R-22B	Heli Air Ltd (G-OOLI)/Wellesbourne
	G-DMCG	Robinson R-44- II	Heli Air Ltd/Wellesbourne
	G-DMCS	PA-28R Cherokee Arrow 200-II	Arrow Associates (G-CPAC)
	G-DMCT	Flight Design CT.2K	B. Zajac
	G-DMRS	Robinson R-44-II	Heli Air Ltd/Wellesbourne
	G-DMSS	Westland Gazelle HT.3	Woods of York Ltd
	G-DMWW	CFM Shadow Srs DD	Microlight Sport Aviation Ltd
	G-DNCN	AB-206A JetRanger	Flying Services/Sandown
	G-DNCS	PA-28R-201T Turbo Arrow III	BC Arrow Ltd
	G-DNGA	Balony Kubicek BB.20	G. J. Bell
	G-DNGR	Colt 31A balloon	G. J. Bell
	G-DNLB	MBB Bo 105DBS/4	Bond Air Services (G-BCDH/G-BTBD/G-BUDP)
	G-DNOP	PA-46-350P Malibu Mirage	Campbell Aviation Ltd
	G-DNVT	G.1159C Gulfstream IV	Shell Aircraft Ltd/Heathrow
	G-DOCA	Boeing 737-436	British Airways
	G-DOCB	Boeing 737-436	Air One
	G-DOCE	Boeing 737-436	British Airways
	G-DOCF	Boeing 737-436	British Airways
	G-DOCG	Boeing 737-436	British Airways
	G-DOCH	Boeing 737-436	British Airways
	G-DOCL	Boeing 737-436	British Airways
	G-DOCN	Boeing 737-436	British Airways
	G-DOCO	Boeing 737-436	British Airways
	G-DOCP	Boeing 737-436	British Airways
	G-DOCR	Boeing 737-436	Air One
	G-DOCS	Boeing 737-436	British Airways
	G-DOCT	Boeing 737-436	British Airways
	G-DOCU	Boeing 737-436	British Airways
	G-DOCV	Boeing 737-436	British Airways
	G-DOCW	Boeing 737-436	British Airways
	G-DOCX	Boeing 737-436	British Airways
	G-DOCY	Boeing 737-436	British Airways (G-BVBY)
	G-DOCZ	Boeing 737-436	Air One (G-BVBZ)
	G-DODB	Robinson R-22B	M. Gallagher
	G-DODD	Cessna F.172P-II	K. Watts/Denham
	G-DODG	Aerotechnik EV-97A Eurostar	R. Barton
	G-DODI	PA-46-350P Malibu Mirage	CAVOK SRL/Italy
	G-DODR	Robinson R-22B	Exmoor Helicopters Ltd
	G-DOEA	AA-5A Cheetah	Plane Talking Ltd (G-RJMI)/Elstree
	G-DOFY	Bell 206B JetRanger 3	Cinnamond Ltd
	G-DOGG	SA Bulldog Srs 120/121	P. Sengupta
	G-DOGZ	Horizon 1	J. E. D. Rogerson
	G-DOIN	Skyranger 912	C. D. & L. J. Church
	G-DOIT	AS.350B-1 Ecureuil	C. C. Blakey

Reg.	Type	Owner or Operator	Notes
G-DOLY	Cessna T.303	R. M. Jones (G-BJZK)	
G-DOME	PA-28-161 Warrior III	Haimoss Ltd	
G-DOMS	Aerotechnik EV-97A Eurostar	D. J. Cross	
G-DONI	AA-5B Tiger	W. P. Moritz (G-BLLT)	
G-DONS	PA-28RT-201T Turbo Arrow IV	C. E. Griffiths/Blackbushe	
G-DON'T	Zenair CH.601XL Zodiac	N. C. & A. C. J. Butcher	
G-DOOZ	AS.355F-2 Twin Squirrel	Premiair Aviation Services Ltd(G-BNSX)	
G-DORA	Focke-Wulf Fw.190-D9	G. R. Lacey	
G-DORN	EKW C-3605	R. G. Gray	
G-DOTT	CFM Streak Shadow	R. J. Bell	
G-DOVE	Cessna 182Q	Carel Investments Ltd	
G-DOVE†	D. H. 104 Devon C.2 ★	E. Surrey College (G-KOOL)/Gatton Point, Redhill	
G-DOWN	Colt 31A balloon	M. Williams	
G-DPPF	Augusta A.109E Power	Dyfed-Powys Police Authority	
G-DPST	Phillips ST.2	Speedtwin Developments Ltd	
G-DRAC	Cameron Dracula Skull SS balloon	Shiplake Investments Ltd	
G-DRAG	Cessna 152 (tailwheel)	L. A. Maynard & M. E. Scouller (G-REME/G-BRNF)	
G-DRAM	Cessna FR.172F (floatplane)	Clyde River Rats	
G-DRAW	Colt 77A balloon	C. Wolstenholme	
G-DRAY	Taylor JT.1 Monoplane	L. J. Dray	
G-DRBG	Cessna 172M	Henlow Flying Club Ltd (G-MUIL)	
G-DRCI	Jabiru UL	D. R. Calo	
G-DREX	Cameron 110 Saturn SS balloon	LRC Products Ltd	
G-DRFC	Aerospatiale ATR-42-300	Atlantic Air Transport Ltd/Coventry	
G-DRGN	Cameron N-105 balloon	W. I. Hooker & C. Parker	
G-DRGS	Cessna 182S	Walter Scott & Partners Ltd	
G-DRID	Cessna FR.172J	D. Ridley	
G-DRIV	Robinson R-44 II	Driver Hire Group Services Ltd	
G-DRKJ	Schweizer 269C	Aviation Bureau (G-BPPW)	
G-DRMM	Shaw Europa	M. W. Mason	
G-DRNT	Sikorsky S-76A	CHC Scotia Ltd	
G-DROP	Cessna U.206C	Peterborough Parachute Centre Ltd (G-UKNO/G-BAMN)/Sibson	
G-DRSV	CEA DR.315 (modified)	R. S. Voice	
G-DRYI	Cameron N-77 balloon	C. A. Butter	
G-DRYS	Cameron N-90 balloon	C. A. Butter	
G-DRZF	CEA DR.360	P. K. Kaufeler	
G-DSFT	PA-28R Cherokee Arrow 200-II	J. Jones (G-LFSE/G-BAXT)	
G-DSGC	PA-25 Pawnee 235C	Devon & Somerset Gliding Club Ltd	
G-DSID	PA-34-220T Seneca III	D. & J. Tait	
G-DSKI	Aerotechnik EV-97 Eurostar	D. R. Skill	
G-DSLL	Pegasus Quantum 15-912	R. G. Jeffery	
G-DSPI	Robinson R-44	Safe-Sec Group Ltd (G-DPS	
G-DSPZ	Robinson R-44-II	Focal Point Communications Ltd	
G-DTCP	PA-32R Cherokee Lance 300	G-DTCP Aviation Ltd (G-TEEM)	
G-DTOY	Ikarus C.42.FB100	C. W. Laske	
G-DTUG	yWag-Aero Super Sport	D. A. Bullockr	
G-DUDE	Van's RV-8	W. M. Hodgkins	
G-DUDS	C.A.S.A. 1.131E Jungmann 2000	B. J. Cox	
G-DUDZ	Robin DR.400/180	D. H. Pattison (G-BXNK)	
G-DUGE	Ikarus C.42 FB UK	D. Stevenson	
G-DUGI	Lindstrand LBL-90A balloon	D. J. Cook	
G-DUGS	Van's RV-9A	D. M. Provost	
G-DUKK	Extra EA.300/L	R. A. & K. M. Roberts	
G-DUMP	Customcraft A-25 balloon	P. C. Bailey	
G-DUNG	Sky 65-24 balloon	G. J. Bell	
G-DURO	Shaw Europa	D. J. Sagar	
G-DURX	Thunder 77A balloon	V. Trimble	
G-DUSK	D.H.115 Vampire T.11 (XE856)	R. M. A. Robinson & R. Horsfield	
G-DUST	Stolp SA.300 Starduster Too	J. V. George	
G-DUVL	Cessna F.172N	A. J. Simpson/Denham	
G-DVBF	Lindstrand LBL-210A balloon	Virgin Balloon Flights	
G-DVON	D.H.104 Devon C.2 (VP955)	C. L. Thatcher	
G-DWEL	SIPA 903	B. L. Procter (G-ASXC)	

Notes	Reg.	Type	Owner or Operator
	G-DWIA	Chilton D.W.1A	D. Elliott
	G-DWIB	Chilton D.W.1B (replica)	J. Jennings
	G-DWMS	Avtech Jabiru UL-450	D. H. S. Williams
	G-DWPF	Technam P.92-EM Echo	P. I. Franklin & D. J. M. Williams
	G-DWPH	Ultramagic M-77	Ultramagic UK
	G-DYCE	Robinson R-44 II	G. Walters (Leasing) Ltd
	G-DYKE	Dyke JD.2 Delta	M. S. Bird
	G-DYNE	Cessna 414	Commair Aviation Ltd/Tollerton
	G-DYNG	Colt 105A balloon	M. J. Gunston (G-HSHS)
	G-EAGA	Sopwith Dove (replica)	A. Wood/O. Warden
	G-EAOU†	Vickers Vimy (replica)(NX71MY)	Greenco (UK) Ltd
	G-EASD	Avro 504L	G. M. New
	G-EASQ†	Bristol Babe (replica) (BAPC87)★	Bristol Aero Collection *(stored)*/Kemble
	G-EAVX	Sopwith Pup (B1807)	K. A. M. Baker
	G-EBED†	Vickers 60 Viking (replica) (BAPC114)	Brooklands Museum of Aviation/Weybridge
	G-EBHX	D.H.53 Humming Bird	The Shuttleworth Collection/O. Warden
	G-EBIA	RAF SE-5A (F904)	The Shuttleworth Collection/O. Warden
	G-EBIB	RAF SE-5A ★	Science Museum/S. Kensington
	G-EBIC	RAF SE-5A (F938) ★	RAF Museum
	G-EBIR	D.H.51	The Shuttleworth Collection/O. Warden
	G-EBJE	Avro 504K (E449) ★	RAF Museum
	G-EBJG	Parnall Pixie III ★	Midland Aircraft Preservation Soc
	G-EBJI	Hawker Cygnet (replica)	C. J. Essex
	G-EBJO	ANEC II ★	The Shuttleworth Collection/O. Warden
	G-EBKY	Sopwith Pup (N6181)	The Shuttleworth Collection/O. Warden
	G-EBLV	D.H.60 Cirrus Moth	British Aerospace PLC/Woodford
	G-EBMB	Hawker Cygnet I ★	RAF Museum
	G-EBNV	English Electric Wren	The Shuttleworth Collection/O. Warden
	G-EBQP	D.H.53 Humming Bird (J7326) ★	Russavia Collection
	G-EBWD	D.H.60X Hermes Moth	The Shuttleworth Collection/O. Warden
	G-EBXU	D.H.60X Moth Seaplane	G. G. Pugh
	G-EBZM	Avro 594 Avian IIIA ★	Manchester Museum of Science & Industry
	G-EBZN	D.H.60X Moth	J. Hodgkinson (G-UAAP)
	G-ECAH	F.27 Friendship Mk 500	Nordic Aviation Contractor A/S (G-JEAH)/ Denmark
	G-ECAN	D.H.84 Dragon	Norman Aircraft Trust/Chilbolton
	G-ECAS	Boeing 737-36N	bmiBaby
	G-ECBH	Cessna F.150K	ECBH Flying Group
	G-ECDX	D.H.71 Tiger Moth (replica)	M. D. Souch
	G-ECGC	Cessna F.172N-II	Euroair Flying Club Ltd/Cranfield
	G-ECGO	Bo 208C1 Junior	A Flight Aviation Ltd
	G-ECHO	Enstrom 380C-UK-2	Cast Designer (G-LONS)/Norway
	G-ECJM	PA-28R-201T Turbo Arrow III	Regishire Ltd (G-FESL/G-BNRN)
	G-ECLI	Schweizer 269C	Central Communications Group
	G-ECMM	Agusta A.109E Power	The Meade Corporation (Services) Ltd (G-SIVC)
	G-ECON	Cessna 172M	Meridian Aviation Group Ltd (G-JONE)
	G-ECOX	Grega GN.1 Air Camper	H. C. Cox
	G-ECUB	PA-18 Super Cub 150	E. Hopper (G-CBFI)
	G-ECVB	Pietenpol Air Camper	K. S. Matcham
	G-EDAV	SA Bulldog Srs 120/121	Edwalton Aviation Ltd
	G-EDEN	SOCATA TB.10 Tobago	Group Eden
	G-EDES	Robinson R-44-II	A. D. Russell
	G-EDFS	Pietenpol Air Camper	D. F. Slaughter
	G-EDGE	Jodel 150	A. D. Edge
	G-EDGI	PA-28-161 Warrior II	R. A. Forster
	G-EDMC	Pegasus Quantum 15-912	M. W. Riley
	G-EDNA	PA-38-112 Tomahawk	D. J. Clucas/Manchester
	G-EDRV	Van's RV-6A	E. A. Yates
	G-EDTO	Cessna FR.172F	N. G. Hopkinson
	G-EDVL	PA-28R Cherokee Arrow 200-II	J. S. Devlin & Z. Islam (G-BXIN)
	G-EECO	Lindstrand LBL-25A balloon	P. A. Bubb & A. J. Allen
	G-EEGL	Christen Eagle II	M. P. Swoboda
	G-EEGU	PA-28-161 Warrior II	TG Aviation Ltd/Manston
	G-EEJE	PA-31 Navajo B	Geeje Ltd
	G-EELS	Cessna 208B Caravan 1	Glass Eels Ltd
	G-EENA	PA-32R-301 Saratoga SP	Gamit Ltd
	G-EENI	Shaw Europa	M. P. Grimshaw

Reg.	Type	Owner or Operator	Notes
G-EENY	GA-7 Cougar	Jade Air PLC	
G-EERH	Ruschmeyer R.90-230RG	D. Sadler	
G-EERV	Van's RV-6	C. B. Stirling (G-NESI)	
G-EESA	Shaw Europa	S. Collins (G-HIIL)	
G-EEUP	SNCAN Stampe SV-4C	A. M. Wajih	
G-EEVA	PA-23 Aztec 250	D. C. Woods (G-ASND)	
G-EEYE	Mainair Blade 912	B. J. Egerton	
G-EEZA	Robinson R-44 II	London & Wessex Ltd	
G-EEZS	Cessna 182P	W. B. Bateson	
G-EFBP	Cessna FR.172K	R. J. Howard	
G-EFGH	Robinson R-22B	Foxtrot Golf Helicopters Ltd (G-ROGG)	
G-EFIR	PA-28-181 Archer II	Leicestershire Aero Club Ltd	
G-EFOF	Robinson R-22B	NT Burton Aviation	
G-EFPA	Airbus A.321-211	My Travel Airways Ltd	
G-EFRY	Light Aero Avid Aerobat	P. A. Boyden	
G-EFSM	Slingsby T.67M Firefly 260	Pooler-LMT Ltd (G-BPLK)	
G-EFTE	Bolkow Bo 207	L. J. & A. A. Rice	
G-EFTF	AS.350B Ecureuil	T. J. French (G-CWIZ/G-DJEM/G-ZBAC/ G-SEBI/G-BMCU)	
G-EGAL	Christen Eagle II	Eagle Partners	
G-EGEE	Cessna 310Q	P. G. Lawrence (G-AZVY)	
G-EGEG	Cessna 172R	C. D. Lever	
G-EGGI	Ikarus C.42FB UK	A. G. & G. J. Higgins	
G-EGGS	Robin DR.400/180	R. Foot	
G-EGGY	Robinson R-22B	R. S. & A. G. Higgins (G-CCGD)	
G-EGHB	Ercoupe 415D	P. G. Vallance	
G-EGHH	Hawker Hunter F.58 (J-4083)	J. H. Goodwin	
G-EGHR	SOCATA TB.20 Trinidad	B. M. Prescott	
G-EGJA	SOCATA TB.20 Trinidad	D. A. Williamson/Alderney	
G-EGLE	Christen Eagle II	Eagle Group	
G-EGLS	PA-28-181 Archer III	D. J. Cooke	
G-EGLT	Cessna 310R	Capital Trading (Aviation) Ltd (G-BHTV)	
G-EGNR	PA-38-112 Tomahawk	R. Bestek	
G-EGTB	PA-28-161 Warrior II	Airways Aero Association Ltd (G-BPWA)	
G-EGTR	PA-28-161 Cadet	Stars Fly Ltd (G-BRSI)/Elstree	
G-EGUL	Christen Eagle II	G-EGUL Flying Group (G-FRYS)	
G-EGUR	Jodel D.140B	D. A. Hood/Malta	
G-EGUY	Sky 220-24 balloon	Sky Trek Ballooning	
G-EHBJ	C.A.S.A. 1.131E Jungmann 2000	E. P. Howard	
G-EHGF	PA-28-181 Archer II	Pegasus Flying Group/Barton	
G-EHIC	Jodel D.140B	M. Tolson & D. W. Smith	
G-EHLX	PA-28-181 Archer II	ASG Leasing Ltd/Guernsey	
G-EHMF	Isaacs Fury II	M. A. Farrelly	
G-EHMJ	Beech S35 Bonanza	A. L. Burton & A. J. Daley	
G-EHMM	Robin DR.400/180R	Booker Gliding Club Ltd	
G-EHMS	McD Douglas MDH-900	Virgin HEMS (London) Ltd	
G-EHUP	SA.341G Gazelle 1	MW Helicopters Ltd	
G-EHXP	R. Commander 112A	A. L. Stewart	
G-EIBM	Robinson R-22B	XL Aviation Ltd (G-BUCL)	
G-EIII	Extra EA.300	D. Dobson (G-HIII)	
G-EIKY	Shaw Europa	J. D. Milbank	
G-EIRE	Cessna T.182T	J. Byrne	
G-EISO	SOCATA MS892A Rallye Commodore 150	EISO Group	
G-EITE	Luscombe 8A Silvaire	S. R. H. Martin	
G-EIWT	Cessna FR.182RG	P. P. D. Howard-Johnston/Edinburgh	
G-EIZO	Eurocopter EC.120B	R. M. Baileyh	
G-EJEL	Cessna 550 Citation II	A. J. & E. A. Elliott	
G-EJGO	Z.226HE Trener	Aerotation Ltd/Biggin Hill	
G-EJJB	Airbus A.319-111	EasyJet Airline Co Ltd/Luton	
G-EJMG	Cessna F.150H	Bagby Aviation	
G-EJOC	AS.350B Ecureuil	Leisure & Retail Helicopters and Kensington & Chelsea Aviation Ltd (G-GEDS/G-HMAN/ G-SKIM/G-BIVP)	
G-EJRS	PA-28-161 Cadet	Small World Aviation Ltd	
G-EKIR	PA-28-262 Cadet	Aeros Leasing Ltd	
G-EKKC	Cessna FR.172G	L. B. W. & F. Hancock	

Notes	Reg.	Type	Owner or Operator
	G-EKKL	PA-28-161 Warrior II	Apollo Aviation Advisory Ltd/Shoreham
	G-EKKO	Robinson R-44	Heli Air Ltd/Wellesbourne
	G-EKMN	Zlin Z.242L	Aeroshow Ltd
	G-EKMW	Mooney M.20J Srs 205	International Aerospace Engineering
	G-EKOS	Cessna FR.182 RG	S. Charlton
	G-EKYD	Robinson R-44 II	EK Aviation Ltd
	G-ELAM	PA-30 Twin Comanche160B	Hangar 39 Ltd (G-BAWU/G-BAWV)
	G-ELDR	PA-32 Cherokee Six 260	Elder Aviation Ltd
	G-ELEE	Cameron Z-105 balloon	D. Eliot
	G-ELEN	Robin DR.400/180	N. R. & E. Foster
	G-ELIS	PA-34-200T Seneca II	Ellis & Co Restoration & Building Ltd (G-BOPV)
	G-ELIT	Bell 206L LongRanger	Henfield Lodge Aviation Ltd
	G-ELIZ	Denney Kitfox Mk 2	A. J. Ellis
	G-ELKA	Christen Eagle II	P. J. Lawton
	G-ELKS	Avid Speedwing Mk 4	H. S. Elkins
	G-ELLA	PA-32R-301 Saratoga IIHP	C. C. W. Hart
	G-ELLE	Cameron N-90 balloon	N. D. Eliot
	G-ELLI	Bell 206B JetRanger 3	R. A. Fleming Ltd
	G-ELMH	NA AT-6D Harvard III (42-84555)	M. Hammond
	G-ELMO	Robinson R44-II	Heli Air Ltd/Wellesbourne
	G-ELNX	Canadair CL.604-2B19 Challenger	Eurolynx Corporation/Stansted
	G-ELUN	Robin DR.400/180R	Cotswold DR.400 Syndicate
	G-ELUT	PA-28R Cherokee Arrow 200-II	Green Arrow Europe Ltd
	G-ELZN	PA-28-161 Warrior II	Northamptonshire School of Flying Ltd/Sywell
	G-ELZY	PA-28-161 Warrior II	Goodwood Road Racing School Ltd
	G-EMAS	Eurocopter EC.135T-1	East Midlands Air Support Unit
	G-EMAX	PA-31-350 Navajo Chieftain	AM & T Aviation Ltd
	G-EMAZ	PA-28-181 Archer II	E. J. Stanley
	G-EMBA	Embraer RJ145EU	British Airways CitiExpress
	G-EMBC	Embraer RJ145EU	British Airways CitiExpress
	G-EMBD	Embraer RJ145EU	British Airways CitiExpress
	G-EMBE	Embraer RJ145EU	British Airways CitiExpress
	G-EMBF	Embraer RJ145EU	British Airways CitiExpress
	G-EMBG	Embraer RJ145EU	British Airways CitiExpress
	G-EMBH	Embraer RJ145EU	British Airways CitiExpress
	G-EMBI	Embraer RJ145EU	British Airways CitiExpress
	G-EMBJ	Embraer RJ145EU	British Airways CitiExpress
	G-EMBK	Embraer RJ145EU	British Airways CitiExpress
	G-EMBL	Embraer RJ145EU	British Airways CitiExpress
	G-EMBM	Embraer RJ145EU	British Airways CitiExpress
	G-EMBN	Embraer RJ145EU	British Airways CitiExpress
	G-EMBO	Embraer RJ145EU	British Airways CitiExpress
	G-EMBP	Embraer RJ145EU	British Airways CitiExpress
	G-EMBS	Embraer RJ145EU	British Airways CitiExpress
	G-EMBT	Embraer RJ145EU	British Airways CitiExpress
	G-EMBU	Embraer RJ145EU	British Airways CitiExpress
	G-EMBV	Embraer RJ145EU	British Airways CitiExpress
	G-EMBW	Embraer RJ145EU	British Airways CitiExpress
	G-EMBX	Embraer RJ145EU	British Airways CitiExpress
	G-EMBY	Embraer RJ145EU	British Airways CitiExpress
	G-EMCA	Commander Aircraft 114B	Drive Tech Ltd
	G-EMDM	Diamond DA.40-P9 Star	D. J. Munson
	G-EMER	PA-34-200 Seneca II	Haimoss Ltd/Old Sarum
	G-EMHH	AS.355F-2 Twin Squirrel	Hancocks Holdings Ltd (G-BYKH)
	G-EMIL	Messerschmitt Bf.109E-3	G. R. Lacey
	G-EMIN	Shaw Europa	S. A. Lamb
	G-EMJA	C.A.S.A. 1.131E Jungmann 2000	P. J. Brand
	G-EMLE	Aerotechnik EV-97 Eurostar	A. R. White
	G-EMLY	Pegasus Quantum 15	S. J. Reid
	G-EMMI	Robinson R-44 II	Hub Of The Wheel Ltd
	G-EMMS	PA-38-112 Tomahawk	Ravenair/Liverpool
	G-EMMY	Rutan Vari-Eze	M. J. Tooze
	G-EMSB	PA-22-160 Tri-Pacer	M. S. Bird (G-ARHU)
	G-EMSI	Shaw Europa	P. W. L. Thomas
	G-EMSL	PA-28-161 Warrior II	Environmental Maintenance Services Ltd (G-TSFT/G-BLDJ)
	G-EMSY	D.H.82A Tiger Moth	B. E. Micklewright (G-ASPZ)
	G-ENCE	Partenavia P.68B	Bicton Aviation (G-OROY/G-BFSU)
	G-ENEE	CFM Streak Shadow SA	T. Green

Reg.	Type	Owner or Operator	Notes
G-ENGO	Steen Skybolt	C. Docherty	
G-ENIE	Tipsy T.66 Nipper 3	E. J. Clarke	
G-ENII	Cessna F.172M	J. Howley	
G-ENNI	Robin R.3000/180	P. F. Taylor	
G-ENNK	Cessna 172S	AK Enterprises Ltd	
G-ENNY	Cameron V-77 balloon	B. G. Jones	
G-ENOA	Cessna F.172F	M. K. Acors (G-ASZW)	
G-ENRE	Avtech Jabiru UL	J. C. Harris	
G-ENRI	Lindstrand LBL-105A balloon	P. G. Hall	
G-ENRM	Cessna 182L	P. Murray	
G-ENRY	Cameron N-105 balloon	P. G. & G. R. Hall	
G-ENSI	Beech F33A Bonanza	G. Garnett	
G-ENTS	Van's RV-9A	L. G. Johnson	
G-ENTT	Cessna F.152 II	C. & A. R. Hyett (G-BHHI)	
G-ENTW	Cessna F.152 II	Firecrest Aviation Ltd & ptnrs (G-BFLK)	
G-ENVY	Mainair Blade 912	D. L. Pollitt & P. Millership	
G-ENYA	Robinson R-44-II	GT Investigations (International) Ltd/Ireland	
G-EOFM	Cessna F.172N	20th Air Training Group Ltd	
G-EOFS	Shaw Europa	G. T. Leedham	
G-EOFW	Pegasus Quantum 15-912	G. C. Weighell	
G-EOHL	Cessna 182L	G. B. Dale & M. C. Terris	
G-EOIN	Zenair CH.701UL	D. G. Palmer	
G-EOLD	PA-28-161 Warrior II	Goodwood Road Racing Co Ltd	
G-EOMA	Airbus A.330-243	Monarch Airlines Ltd	
G-EORD	Cessna 208B	Air Medical Fleet Ltd	
G-EORG	PA-38-112 Tomahawk	Airways Aero Association/Booker	
G-EORJ	Shaw Europa	P. E. George	
G-EPAR	Robinson R-22B-2	Jepar Rotorcraft	
G-EPDI	Cameron N-77 balloon	R. Moss	
G-EPED	PA-31-350 Navajo Chieftain	Pedley Furniture International Ltd (G-BMCJ)	
G-EPFR	Airbus A.320-231	My Travel Airways (G-BVJV)	
G-EPIC	Jabiru UL-450	T. Chadwick	
G-EPMI	Robinson R-44-II	Heli Air Ltd/Wellesbourne	
G-EPOC	Jabiru UL-450	S. Cope	
G-EPOL	AS.355F-1 Twin Squirrel	Cambridge & Essex Air Support Unit (G-SASU/G-BSSM/G-BMTC/G-BKUK)	
G-EPOX	Aero Designs Pulsar XP	K. F. Farey	
G-EPPO	Robinson R-44	R. Conterno (G-JBBS)/Italy	
G-EPTR	PA-28R Cherokee Arrow 200-II	Tayflite Ltd	
G-ERAD	Beech C90A King Air	GKL Management Services Ltd	
G-ERBL	Robinson R-22B-2	G. V. Maloney	
G-ERCO	Ercoupe 415D	A. R. & M. V. Tapp	
G-ERDA	Staaken Z-21A Flitzer	J. Cresswell	
G-ERDS	D.H.82A Tiger Moth	W. A. Gerdes	
G-ERFS	PA-28-161 Warrior II	S. Harrison	
G-ERIC	R. Commander 112TC	Atomchoice Ltd	
G-ERIK	Cameron N-77 balloon	T. M. Donnelly	
G-ERIS	Hughes 369D	R. J. Howard (G-PJMD/G-BMJV)	
G-ERIW	Staaken Z-21 Flitzer	R. I. Wasey	
G-ERJA	Embraer RJ145EP	British Airways CitiExpress	
G-ERJB	Embraer RJ145EP	British Airways CitiExpress	
G-ERJC	Embraer RJ145EP	British Airways CitiExpress	
G-ERJD	Embraer RJ145EP	British Airways CitiExpress	
G-ERJE	Embraer RJ145EP	British Airways CitiExpress	
G-ERJF	Embraer RJ145EP	British Airways CitiExpress	
G-ERJG	Embraer RJ145EP	British Airways CitiExpress	
G-ERMO	ARV Super 2	T. Pond (G-BMWK)	
G-ERMS	Thunder Ax3 balloon	B. R. & M. Boyle	
G-ERNI	PA-28-181 Archer II	H. A. Daines (G-OSSY)	
G-EROL	SA.341G Gazelle 1	The Coin Group Ltd (G-NONA/G-FDAV/ G-RIFA/G-ORGE/G-BBHU)	
G-EROM	Robinson R-22B	Aeromega Ltd/Stapleford	
G-EROS	Cameron H-34 balloon	Evening Standard Co Ltd	
G-ERRY	AA-5B Tiger	Gemini Aviation (G-BFMJ)/Shobdon	
G-ESCA	Escapade Jabiru (1)	W. R. Davis-Smith	
G-ESCC	Escapade 912	G. & S. Simons	
G-ESCP	Escapade Jabiru (1)	R. G. Hughes	
G-ESEX	Eurocopter EC.135T-2	Essex Police Authority	

Notes	Reg.	Type	Owner or Operator
	G-ESFT	PA-28-161 Warrior II	Plane Talking Ltd (G-ENNA)/Elstree
	G-ESKA	Escapade 912	T. F. Francis
	G-ESKY	PA-23 Aztec 250D	Caernarfon Airworld Ltd (G-BBNN)
	G-ESLH	Agusta A.109E Power	Euroskylink Ltd
	G-ESME	Cessna R.182 II (15211)	G. C. Cherrington (G-BNOX)
	G-ESSX	PA-28-161 Warrior II	Courtenay Enterprises (G-BHYY)
	G-ESSY	Robinson R-44	EW Guess (Holdings) Ltd
	G-ESTA	Cessna 550 Citation II	Executive Aviation Services Ltd (G-GAUL)
	G-ESTR	Van's RV-6	R. M. Johnson
	G-ESUS	Rotorway Executive 162F	J. Tickner
	G-ETBY	PA-32 Cherokee Six 260	G-ETBY Group (G-AWCY)
	G-ETCW	Stoddard-Hamilton Glastar	P. G. Hayward
	G-ETDA	PA-28-161 Warrior II	T. M. Thorpe
	G-ETDC	Cessna 172P	Osprey Air Services Ltd
	G-ETHI	IDABacau Yakovlev Yak-52	Touchwood & Associates Ltd
	G-ETHU	Eurocopter EC.135T-1	Helimand Ltd
	G-ETHY	Cessna 208	N. A. Moore
	G-ETIN	Robinson R-22B	Getin Helicopters Ltd
	G-ETIV	Robin DR.400/180	J. MacGilvray
	G-ETME	Nord 1002 Pingouin (KG+EM)	108 Flying Grou
	G-ETPS	Hawker Hunter FGA.9	Skybue Aviation Ltd
	G-EUGN	Robinson R-44	Twinlite Developments Ltd
	G-EUOA	Airbus A.319-131	British Airways
	G-EUOB	Airbus A.319-131	British Airways
	G-EUOC	Airbus A.319-131	British Airways
	G-EUOD	Airbus A.319-131	British Airways
	G-EUOE	Airbus A.319-131	British Airways
	G-EUOF	Airbus A.319-131	British Airways
	G-EUOG	Airbus A.319-131	British Airways
	G-EUOH	Airbus A.319-131	British Airways
	G-EUOI	Airbus A.319-131	British Airways
	G-EUOJ	Airbus A.319-131	British Airways
	G-EUOK	Airbus A.319-131	British Airways
	G-EUOL	Airbus A.319-131	British Airways
	G-EUPA	Airbus A.319-131	British Airways
	G-EUPB	Airbus A.319-131	British Airways
	G-EUPC	Airbus A.319-131	British Airways
	G-EUPD	Airbus A.319-131	British Airways
	G-EUPE	Airbus A.319-131	British Airways
	G-EUPF	Airbus A.319-131	British Airways
	G-EUPG	Airbus A.319-131	British Airways
	G-EUPH	Airbus A.319-131	British Airways
	G-EUPJ	Airbus A.319-131	British Airways
	G-EUPK	Airbus A.319-131	British Airways
	G-EUPL	Airbus A.319-131	British Airways
	G-EUPM	Airbus A.319-131	British Airways
	G-EUPN	Airbus A.319-131	British Airways
	G-EUPO	Airbus A.319-131	British Airways
	G-EUPP	Airbus A.319-131	British Airways
	G-EUPR	Airbus A.319-131	British Airways
	G-EUPS	Airbus A.319-131	British Airways
	G-EUPT	Airbus A.319-131	British Airways
	G-EUPU	Airbus A.319-131	British Airways
	G-EUPV	Airbus A.319-131	British Airways
	G-EUPW	Airbus A.319-131	British Airways
	G-EUPX	Airbus A.319-131	British Airways
	G-EUPY	Airbus A.319-131	British Airways
	G-EUPZ	Airbus A.319-131	British Airways
	G-EURX	Shaw Europa XS	C. C. Napier
	G-EUSO	Robin DR.400/140 Major	Weald Air Services Ltd
	G-EUUA	Airbus A.320-232	British Airways
	G-EUUB	Airbus A.320-232	British Airways
	G-EUUC	Airbus A.320-232	British Airways
	G-EUUD	Airbus A.320-232	British Airways
	G-EUUE	Airbus A.320-232	British Airways
	G-EUUF	Airbus A.320-232	British Airways
	G-EUUG	Airbus A.320-232	British Airways
	G-EUUH	Airbus A.320-232	British Airways
	G-EUUI	Airbus A.320-232	British Airways
	G-EUUJ	Airbus A.320-232	British Airways

Reg.	Type	Owner or Operator	Notes
G-EUUK	Airbus A.320-232	British Airways	
G-EUUL	Airbus A.320-232	British Airways	
G-EUUM	Airbus A.320-232	British Airways	
G-EUUN	Airbus A.320-232	British Airways	
G-EUUO	Airbus A.320-232	British Airways	
G-EUUP	Airbus A.320-232	British Airways	
G-EUUR	Airbus A.320-232	British Airways	
G-EUUS	Airbus A.320-232	British Airways	
G-EUUT	Airbus A.320-232	British Airways	
G-EUUU	Airbus A.320-232	British Airways	
G-EUXC	Airbus A.321-231	British Airways	
G-EUXD	Airbus A.321-231	British Airways	
G-EUXE	Airbus A.321-231	British Airways	
G-EUXF	Airbus A.321-231	British Airways	
G-EUXG	Airbus A.321-231	British Airways	
G-EUXH	Airbus A.321-231	British Airways	
G-EUXI	Airbus A.321-231	British Airways	
G-EUXJ	Airbus A.321-231	British Airways	
G-EUXK	Airbus A.321-231	British Airways	
G-EUXL	Airbus A.321-231	British Airways	
G-EVET	Cameron 80 Concept balloon	K. J. Foster	
G-EVEY	Thruster T.600N 450-JAB	K. J. Crompton	
G-EVIE	PA-28-181 Warrior II	L. Richardson (G-ZULU)	
G-EVLE	Rearwin 8125 Cloudster	M. C. Hiscock (G-BVLK)	
G-EVLN	Gulfstream 4	Metropix Ltd	
G-EVPI	Evans VP1 Srs 2	C. P. Martyr	
G-EVRO	Aerotechnik EV-97 Eurostar	Movie-Go	
G-EVVA	PA-28R Cherokee Arrow 200	American Flight Academy Ltd (G-BAZU)	
G-EWAN	Prostar PT-2C	C. G. Shaw	
G-EWAW	Bell 206B-3 JetRanger 3	A. J. Renham (G-DORB)	
G-EWBC	Avtec Jabiru SK	E. W. B. Comber	
G-EWES	Pioneer 300	R. Y. Kendal & D. A. Ions	
G-EWHT	Robinson R.2112	Ewan Ltd	
G-EWIZ	Pitts S-2E Special	P. A. Soper	
G-EWME	PA-28 Cherokee 235	C. J. Mewis & E. S. Ewen	
G-EWRT	Eurocopter EC.135T-2	McAlpine Helicopters Ltd/Kidlington	
G-EXEA	Extra EA.300/L	Stewart Becker Aviation	
G-EXEC	PA-34-200 Seneca	Sky Air Travel Ltd	
G-EXES	Shaw Europa XS	D. Barraclough	
G-EXEX	Cessna 404	Atlantic Air Transport Ltd/Coventry	
G-EXIT	M.S.893E Rallye 180GT	M. A. Baldwin	
G-EXLL	Zenair CH.601	B. McFadden	
G-EXON	PA-28-161 Cadet	Plane Talking Ltd (G-EGLD)/Elstree	
G-EXPD	Stemme S.10-VT	Global Gliding Expeditions	
G-EXPL	Champion 7GCBC Citabria	E. J. F. McEntee	
G-EXTR	Extra EA.260	S. J. Carver	
G-EXXO	PA-28-161 Cadet	Plane Talking Ltd (G-CBXP)/Elstree	
G-EYAS	Denney Kitfox Mk 2	K. Hamnett	
G-EYCO	Robin DR.400/180	Cherokee G-AVYL Flying Group Ltd	
G-EYES	Cessna 402C	Atlantic Air Transport Ltd (G-BLCE)/Coventry	
G-EYET	Robinson R-44	Warwickshire Flight Training Centre Ltd (G-JPAD)	
G-EYLE	Bell 206L-1 LongRanger 2	Eyles Construction Ltd (G-BWCU/G-OCRP)	
G-EYNL	MBB Bo 105DBS/5	Sterling Helicopters Ltd	
G-EYOR	Van's RV-6	S. I. Fraser	
G-EYRE	Bell 206L-1 LongRanger	Hideroute Ltd (G-STVI)	
G-EZAM	Airbus A.319-111	easyJet Airline Co Ltd (G-CCKA)/Luton	
G-EZAR	Pegasus Quik	M. W. Houghton	
G-EZCL	Airbus A.319-111	easyJet Airline Co Ltd/Luton	
G-EZDC	Airbus A.319-111	easyJet Airline Co Ltd (G-CCKB)/Luton	
G-EZEA	Airbus A.319-111	easyJet Airline Co Ltd/Luton	
G-EZEB	Airbus A.319-111	easyJet Airline Co Ltd/Luton	
G-EZEC	Airbus A.319-111	easyJet Airline Co Ltd/Luton	
G-EZED	Airbus A.319-111	easyJet Airline Co Ltd/Luton	
G-EZEF	Airbus A.319-111	easyJet Airline Co Ltd/Luton	
G-EZEG	Airbus A.319-111	easyJet Airline Co Ltd/Luton	
G-EZEJ	Airbus A.319-111	easyJet Airline Co Ltd/Luton	
G-EZEK	Airbus A.319-111	easyJet Airline Co Ltd/Luton	
G-EZEL	SA.341G Gazelle 1	W. R. Pitcher (G-BAZL)	

Notes	Reg.	Type	Owner or Operator
	G-EZEO	Airbus A.319-111	easyJet Airline Co Ltd/Luton
	G-EZEP	Airbus A.319-111	easyJet Airline Co Ltd/Luton
	G-EZER	Cameron N-34 balloon	D. D. Maltby
	G-EZES	Airbus A.319-111	easyJet Airline Co Ltd/Luton
	G-EZET	Airbus A.319-111	easyJet Airline Co Ltd/Luton
	G-EZEU	Airbus A.319-111	easyJet Airline Co Ltd/Luton
	G-EZEV	Airbus A.319-111	easyJet Airline Co Ltd/Luton
	G-EZEW	Airbus A.319-111	easyJet Airline Co Ltd/Luton
	G-EZEX	Airbus A.319-111	easyJet Airline Co Ltd/Luton
	G-EZEY	Airbus A.319-111	easyJet Airline Co Ltd/Luton
	G-EZEZ	Airbus A.319-111	easyJet Airline Co Ltd/Luton
	G-EZIA	Airbus A.319-111	easyJet Airline Co Ltd/Luton
	G-EZIB	Airbus A.319-111	easyJet Airline Co Ltd/Luton
	G-EZIC	Airbus A.319-111	easyJet Airline Co Ltd/Luton
	G-EZID	Airbus A.319-111	easyJet Airline Co Ltd/Luton
	G-EZIE	Airbus A.319-111	easyJet Airline Co Ltd/Luton
	G-EZIF	Airbus A.319-111	easyJet Airline Co Ltd/Luton
	G-EZIG	Airbus A.319-111	easyJet Airline Co Ltd/Luton
	G-EZIH	Airbus A.319-111	easyJet Airline Co Ltd/Luton
	G-EZII	Airbus A.319-111	easyJet Airline Co Ltd/Luton
	G-EZIJ	Airbus A.319-111	easyJet Airline Co Ltd/Luton
	G-EZIK	Airbus A.319-111	easyJet Airline Co Ltd/Luton
	G-EZIL	Airbus A.319-111	easyJet Airline Co Ltd/Luton
	G-EZIM	Airbus A.319-111	easyJet Airline Co Ltd/Luton
	G-EZIN	Airbus A.319-111	easyJet Airline Co Ltd/Luton
	G-EZJA	Boeing 737-73V	easyJet Airline Co Ltd/Luton
	G-EZJB	Boeing 737-73V	easyJet Airline Co Ltd/Luton
	G-EZJC	Boeing 737-73V	easyJet Airline Co Ltd/Luton
	G-EZJD	Boeing 737-73V	easyJet Airline Co Ltd/Luton
	G-EZJE	Boeing 737-73V	easyJet Airline Co Ltd/Luton
	G-EZJF	Boeing 737-73V	easyJet Airline Co Ltd/Luton
	G-EZJG	Boeing 737-73V	easyJet Airline Co Ltd/Luton
	G-EZJH	Boeing 737-73V	easyJet Airline Co Ltd/Luton
	G-EZJI	Boeing 737-73V	easyJet Airline Co Ltd/Luton
	G-EZJJ	Boeing 737-73V	easyJet Airline Co Ltd/Luton
	G-EZJK	Boeing 737-73V	easyJet Airline Co Ltd/Luton
	G-EZJL	Boeing 737-73V	easyJet Airline Co Ltd/Luton
	G-EZJM	Boeing 737-73V	easyJet Airline Co Ltd/Luton
	G-EZJN	Boeing 737-73V	easyJet Airline Co Ltd/Luton
	G-EZJO	Boeing 737-73V	easyJet Airline Co Ltd/Luton
	G-EZJP	Boeing 737-73V	easyJet Airline Co Ltd/Luton
	G-EZJR	Boeing 737-73V	easyJet Airline Co Ltd/Luton
	G-EZJS	Boeing 737-73V	easyJet Airline Co Ltd/Luton
	G-EZJT	Boeing 737-73V	easyJet Airline Co Ltd/Luton
	G-EZJU	Boeing 737-73V	easyJet Airline Co Ltd/Luton
	G-EZJV	Boeing 737-73V	easyJet Airline Co Ltd/Luton
	G-EZJW	Boeing 737-73V	easyJet Airline Co Ltd/Luton
	G-EZJX	Boeing 737-73V	easyJet Airline Co.Ltd/Luton
	G-EZJY	Boeing 737-73V	easyJet Airline Co Ltd/Luton
	G-EZJZ	Boeing 737-73V	easyJet Airline Co Ltd/Luton
	G-EZKA	Boeing 737-73V	easyJet Airline Co Ltd/Luton
	G-EZKB	Boeing 737-73V	easyJet Airline Co Ltd/Luton
	G-EZKC	Boeing 737-73V	easyJet Airline Co Ltd/Luton
	G-EZKD	Boeing 737-73V	easyJet Airline Co Ltd/Luton
	G-EZKE	Boeing 737-73V	easyJet Airline Co Ltd/Luton
	G-EZKF	Boeing 737-73V	easyJet Airline Co Ltd/Luton
	G-EZKG	Boeing 737-73V	easyJet Airline Co Ltd/Luton
	G-EZMH	Airbus A.319-111	easyJet Airline Co Ltd (G-CCKD)/Luton
	G-EZMK	Airbus A.319-111	easyJet Airline Co Ltd/Luton
	G-EZNC	Airbus A.319-111	easyJet Airline Co Ltd (G-CCKC)/Luton
	G-EZNM	Airbus A.319-111	easyJet Airlibe Co Ltd/Luton
	G-EZOS	Rutan Vari-Eze	C. Moffat
	G-EZPG	Airbus A.319-111	easyJet Airline Co Ltd/Luton
	G-EZSM	Airbus A.319-111	easyJet Airline Co Ltd (G-CCKE)
	G-EZUB	Zenair CH.601HD Zodiac	R. A. C. Stephens
	G-EZVS	Colt 77B balloon	A. J. Lovell
	G-EZXO	Colt 56A balloon	A. J. Lovell
	G-EZYF	Boeing 737-375	easyJet Airline Co Ltd/Luton
	G-EZYG	Boeing 737-33V	easyJet Airline Co Ltd/Luton
	G-EZYH	Boeing 737-33V	easyJet Airline Co Ltd/Luton

Reg.	Type	Owner or Operator	Notes
G-EZYI	Boeing 737-33V	easyJet Airline Co Ltd/Luton	
G-EZYJ	Boeing 737-33V	easyJet Airline Co Ltd/Luton	
G-EZYK	Boeing 737-33V	easyJet Airline Co Ltd/Luton	
G-EZYL	Boeing 737-33V	easyJet Airline Co Ltd/Luton	
G-EZYN	Boeing 737-33V	easyJet Airline Co Ltd/Luton	
G-EZYP	Boeing 737-33V	easyJet Airline Co Ltd/Luton	
G-EZYR	Boeing 737-33V	easyJet Airline Co Ltd/Luton	
G-EZYU	PA-34-200 Seneca II	P. A. S. Dyke (G-BCDB)/Elstree	
G-EZZA	Shaw Europa XS	J. C. R. Davey	
G-FABB	Cameron V-77 balloon	P. Trumper	
G-FABI	Robinson R-44 Astro	J. Froggatt	
G-FABM	Beech 95-B55 Baron	F. B. Miles (G-JOND/G-BMVC)	
G-FABS	Thunder Ax9-120 S2 balloon	R. C. Corrall	
G-FACE	Cessna 172S	Oxford Aviation Services Ltd/Kidlington	
G-FAIR	SOCATA TB-10 Tobago	Fairwings Ltd	
G-FALC	Aeromere F.8L Falco	P. J. Jones (G-AROT)	
G-FALO	Sequoia F.8L Falco	M. J. & S. E. Aherne	
G-FAME	Starstreak Shadow SA-II	T. J. Palmer	
G-FAMH	Zenair CH.701	G. T. Neale	
G-FANC	Fairchild 24R-46 Argus III	A. T. Fines	
G-FANL	Cessna FR.172K XP-II	J. A. Rees	
G-FARE	Robinson R-44-II	Heli Air Ltd/Wellesbourne	
G-FARL	Pitts S-1E Special	F. L. McGee	
G-FARM	SOCATA Rallye 235GT	Bristol Cars Ltd	
G-FARO	Aero Designs Star-Lite SL.1	M. K. Faro	
G-FARR	Jodel 150	G. H. Farr	
G-FARY	QAC Quickie Tri-Q	F. Sayyah	
G-FATB	R. Commander 114B	James D. Pearce & Co	
G-FAUX	Cessna 182S	R. S. Faux	
G-FAVC	D.H.80A Puss Moth	R. A. Seeley	
G-FAYE	Cessna F.150M	Cheshire Air Training Services Ltd/Liverpool	
G-FBAT	Aeroprakt A.22 Foxbat	The Small Aeroplane Co Ltd	
G-FBII	Ikarus C.42 FB100	F. Beeson	
G-FBIX	D.H.100 Vampire FB.9 (WL505)	D. G. Jones	
G-FBMW	Cameron N-90 balloon	K-J. Schwer/Germany	
G-FBPI	ANEC IV Missel Thrush	R. Trickett	
G-FBRN	PA-28-181 Archer II	Herefordshire Aero Club Ltd/Shobdon	
G-FBWH	PA-28R Cherokee Arrow 180	F. A. Short	
G-FCDB	Cessna 550 Citation Bravo	Eurojet Aviation Ltd	
G-FCED	PA-31T2 Cheyenne IIXL	Air Medical Fleet Ltd/Kidlington	
G-FCLA	Boeing 757-28A	Thomas Cook Airlines	
G-FCLB	Boeing 757-28A	Thomas Cook Airlines	
G-FCLC	Boeing 757-38A	Thomas Cook Airlines	
G-FCLD	Boeing 757-25FET	Thomas Cook Airlines	
G-FCLE	Boeing 757-28A	Thomas Cook Airlines	
G-FCLF	Boeing 757-28A	Thomas Cook Airlines	
G-FCLG	Boeing 757-28A	Thomas Coor Airlines	
G-FCLH	Boeing 757-28A	Thomas Cook Airlines	
G-FCLI	Boeing 757-28A	Thomas Cook Airlines	
G-FCLJ	Boeing 757-2Y0	Thomas Cook Airlines	
G-FCLK	Boeing 757-2Y0	Thomas Cook Airlines	
G-FCSP	Robin DR.400/180	FCS Photochemicals	
G-FCUK	Pitts S-1C Special	M. O'Hearne	
G-FEAB	PA-28-181 Archer III	Feabrex Ltd	
G-FEBE	Cessna 340A	Just Plane Trading Ltd	
G-FEBY	Robinson R-22B	Astra Helicopters Ltd	
G-FEDA	Eurocopter EC.120B	Federal Aviation Ltd	
G-FEES	Eurocopter EC. 135T-2	Cairnsilver Ltd	
G-FELL	Shaw Europa	R. A. Blackwell	
G-FELT	Cameron N-77 balloon	Allan Industries Ltd	
G-FERN	Mainair Blade 912	M. H. Moulai	
G-FEWG	Fuji FA.200-160	Caseright Ltd (G-BBNV)	
G-FEZZ	Bell 206B JetRanger II	R. J. Myram	
G-FFAB	Cameron N-105 balloon	B. J. Hammond	
G-FFEN	Cessna F.150M	D. Thrower	
G-FFFT	Lindstrand LBL-31A balloon	The Aerial Display Co Ltd	
G-FFOX	Hawker Hunter T.7B	Delta Engineering Aviation Ltd/Kemble	

G-FFRA – G-FLOW

Notes	Reg.	Type	Owner or Operator
	G-FFRA	Dassault Falcon 20DC	FR Aviation Ltd/Bournemouth
	G-FFRI	AS.355F-1 Twin Squirrel	ATC Trading Ltd (G-GLOW/G-PAPA/G-CNET/G-MCAH)
	G-FFTI	SOCATA TB.20 Trinidad	R. Lenk
	G-FFTT	Lindstrand LBL Newspaper SS balloon	The Aerial Display Co Ltd
	G-FFUN	Pegasus Quantum 15	P. R. Mailer (G-MYMD)
	G-FFWD	Cessna 310R	T. S. Courtman (G-TVKE/G-EURO)
	G-FGID	Vought FG-1D Corsair (KD345)	Patina Ltd/Duxford
	G-FGSK	Cameron 120 Beer Crate SS balloon	Ballon-Sport und Luftwerbung Dresden GmbH/ Germany
	G-FHAJ	Airbus A.320-231	My Travel Lite
	G-FHAS	Scheibe SF.25E Super Falke	Burn Gliding Club Ltd
	G-FIAT	PA-28 Cherokee 140	Halegreen Associates Ltd & Demero Ltd (G-BBYW)
	G-FIBS	AS.350BA Ecureuil	Pristheath Ltd
	G-FIFE	Cessna FA.152	Tayside Aviation Ltd (G-BFYN)/Dundee
	G-FIFI	SOCATA TB.20 Trinidad	F. A. Saker (G-BMWS)
	G-FIFT	Ikarus C.42 FB 100	Fly Buy Ultralights Ltd
	G-FIGA	Cessna 152	Central Aircraft Leasing Ltd
	G-FIGB	Cessna 152	Aerohire Ltd/Wolverhampton
	G-FIII	Extra EA.300/L	G. G. Ferriman (G-RGEE)
	G-FIJJ	Cessna F.177RG	D. R. Vale (G-AZFP)
	G-FIJR	L.188PF Electra	Atlantic Airlines Ltd/Coventry
	G-FIJV	L.188CF Electra	Atlantic Airlines Ltd/Coventry
	G-FILE	PA-34-200T Seneca	A. J. Warren
	G-FILL	PA-31-310 Navajo	P. V. Naylor-Leyland
	G-FINA	Cessna F.150L	D. Norris (G-BIFT)
	G-FIND	Cessna F.406	Atlantic Airlinest Ltd/Coventry
	G-FINZ	I.I.I Sky Arrow 650T	A. G. Counsell
	G-FIRM	Cessna 550 Citation Bravo	Marshall of Cambridge Aerospace Ltd
	G-FIRS	Robinson R-22B-2	M. & S. Chantler
	G-FIRZ	Murphy Renegade Spirit UK	P. J. Houtman
	G-FISH	Cessna 310R-II	Air Charter Scotland Ltd/Edinburgh
	G-FITZ	Cessna 335	J. R. Naylor & D. Hughes (G-RIND)
	G-FIZU	L.188CF Electra	Atlantic Airlines Ltd/Coventry
	G-FIZY	Shaw Europa XS	G. N. Holland (G-DDSC)
	G-FIZZ	PA-28-161 Warrior II	Tecair Aviation Ltd
	G-FJEA	Boeing 757-23AER	Flyjet Ltd (G-LCRC/G-IEAB)
	G-FJEB	Boeing 757-23A	Flyjet Ltd (G-OOOJ)
	G-FJET	Cessna 550 Citation II	London Executive Aviation Ltd (G-DCFR/G-WYLX/G-JETD)
	G-FJMS	Partenavia P.68B	J. B. Randle (G-SVHA)
	G-FJTH	Aeroprakt A.22 Foxbat	F. J. T. Hancock
	G-FKNH	PA-15 Vagabond	M. J. Mothershaw/Liverpool
	G-FLAG	Colt 77A balloon	B. A. Williams
	G-FLAK	Beech 95-E55 Baron	D. Clarke/Swanton Morley
	G-FLAP	Cessna A.152	Walkbury Aviation Ltd (G-BHJB)
	G-FLAV	PA-28-161 Warrior II	The Crew Flying Group/Tollerton
	G-FLBI	Robinson R-44-II	Heli Air Ltd
	G-FLCA	Fleet Model 80 Canuck	E. C. Taylor
	G-FLCT	Hallam Fleche	R. G. Hallam
	G-FLDG	Skyranger 912	A. J. Gay
	G-FLEA	SOCATA TB-10 Tobago	TB Group
	G-FLEW	Lindstrand LBL-90A balloon	Lindstrand Hot-Air Balloons Ltd
	G-FLEX	Mainair Pegasus Quik	A. J. Tyler
	G-FLGT	Lindstrand LBL-105A balloon	Ballongaventyr I. Skane AB/Sweden
	G-FLII	GA-7 Cougar	Plane Talking Ltd (G-GRAC)/Elstree
	G-FLIK	Pitts S-1S Special	R. P. Millinship/Leicester
	G-FLIP	Cessna FA.152	Walkbury Aviation Ltd (G-BOES)
	G-FLIT	Rotorway Executive 162F	R. S. Snell
	G-FLIZ	Staaken Z-21 Flitzer	M. A. Wood
	G-FLKE	Scheibe SF.25C Falke	Falkes Flying Foundation Ltd
	G-FLKS	Scheibe SF.25C Falke	Falkes Flying Foundation Ltd
	G-FLOA	Cameron O-120 balloon	Floating Sensations Ltd
	G-FLOR	Shaw Europa	A. F. C. van Eldik
	G-FLOW	Cessna 172N	M. P. Dolan

Reg.	Type	Owner or Operator
G-FLOX	Shaw Europa	DPT Group
G-FLPI	R. Commander RC112	H. J. Freeman
G-FLRT	Shaw Europa	K. G. Atkinson & D. B. Southworth
G-FLSH	Yakovlev Yak-52	Flash Air Ltd
G-FLSI	FLS Aerospace Sprint 160	Aces High Ltd/North Weald
G-FLTA	BAe 146-200	Flightline Ltd
G-FLTC	BAe 146-300	Flightline Ltd (G-JEBH/G-BVTO/G-NJID)
G-FLTG	Cameron A-140 balloon	Floating Sensations Ltd
G-FLTY	EMB-110P1 Bandeirante	Keen Airways Ltd (G-ZUSS/G-REGA)/Liverpool
G-FLTZ	Beech 58 Baron	Flightline Ltd (G-PSVS)/Southend
G-FLUF	Lindstrand Bunny SS balloon	Lindstrand Balloons Ltd
G-FLUX	PA-28-181 Archer III	TEC Air Hire Ltd
G-FLVU	Cessna 501 Citation	Neonopal Ltd/Liverpool
G-FLYA	Mooney M.20J	BRF Aviation Ltd
G-FLYB	Ikarus C.42 FB100	Fly Buy Ultralights Ltd
G-FLYE	Cameron A-210 balloon	Exclusive Ballooning
G-FLYG	Slingsby T.67C	G. Laden
G-FLYH	Robinson R-22B	Cyclone Helicopters (G-BXMR)
G-FLYI	PA-34-200 Seneca II	S. Papi & Willow Air Flying Club (G-BHVO)
G-FLYP	Beagle B.206 Srs 2	Key Publishing Ltd (G-AVHO)/Cranfield
G-FLYS	Robinson R-44	Newmarket Plant Hire Ltd
G-FLYT	Shaw Europa	K. F. & R. Richardson
G-FLYY	BAC.167 Strikemaster 80A	B. T. Barber
G-FLZR	Staaken Z-21 Flitzer	J. F. Govan
G-FMAH	Fokker 100	Gazelle Ltd
G-FMAM	PA-28-151 Warrior (modified)	Lima Tango Flying Group (G-BBXV)
G-FMGG	Maule M5-235C Lunar Rocket	S. Bierbaum (G-RAGG)
G-FMKA	Diamond HK.36TC Super Dimona	D. J. M. Williams
G-FMSG	Cessna FA.150K	G. Owen (G-POTS/G-AYUY)/Gamston
G-FNEY	Cessna F.177RG	F. Ney
G-FNLD	Cessna 172N	Papa Hotel Flying Group
G-FNLY	Cessna F.172M	C. F. Dukes (G-WACX/G-BAEX)
G-FNPT	PA-28-161 Warrior III	Plane Talking Ltd
G-FOFO	Robinson R-44-II	Towers Aviation
G-FOGG	Cameron N-90 balloon	J. P. E. Money-Kyrle
G-FOGI	Shaw Europa XS	B. Fogg
G-FOGY	Robinson R-22B	Dragonfly Aviation
G-FOLD	Light Aero Avid Speedwing	B. W. & G. Evans
G-FOLI	Robinson R-22B-2	K. Duckworth
G-FOLY	Aerotek Pitts S-2A Modified	S. A. Laing
G-FONZ	Best Off Skyranger 912 (12)	A. A. Pacitti
G-FOPP	Lancair 320	Airsport (UK) Ltd
G-FORC	SNCAN Stampe SV-4C	C. C. Rollings & F. J. Hodson
G-FORD	SNCAN Stampe SV-4C	P. H. Meeson
G-FORR	PA-28-181 Archer III	Buchanan Partnership
G-FORS	Slingsby T.67C	Open Skies Partnership
G-FORZ	Pitts S-1S Special	N. W. Parkinson
G-FOSY	M.S.880B Rallye Club	A. G. Foster (G-AXAK)
G-FOTO	PA-E23 Aztec 250F	Simmons Aerofilms Ltd (G-BJDH/G-BDXV)
G-FOWL	Colt 90A balloon	G-FOWL Ballooning Group
G-FOWS	Cameron N-105 balloon	F. R. Hart
G-FOXA	PA-28-161 Cadet	Leicestershire Aero Club Ltd
G-FOXB	Aeroprakt A.22 Foxbat	M. Raflewski
G-FOXC	Denney Kitfox Mk 3	G. Hawkins
G-FOXD	Denney Kitfox Mk 2	P. P. Trangmar
G-FOXE	Denney Kitfox Mk 2	K. M. Pinkar
G-FOXF	Denney Kitfox Mk 4	M. S. Goodwin
G-FOXG	Denney Kitfox Mk 2	Kitfox Group
G-FOXI	Denney Kitfox	B. Johns
G-FOXM	Bell 206B JetRanger 2	Tyringham Charter & Group Services (G-STAK/G-BNIS)
G-FOXS	Denney Kitfox Mk 2	S. P. Watkins & C. C. Rea
G-FOXX	Denney Kitfox	R. O. F. Harper
G-FOXZ	Denney Kitfox	S. C. Goozee
G-FPIG	PA-28-151 Warrior	G. F. Strain (G-BSSR)
G-FPLA	Beech 200 Super King Air	Cobham Leasing Ltd
G-FPLB	Beech 200 Super King Air	Cobham Leasing Ltd
G-FPLC	Cessna 441	Cobham Leasing Ltd (G-FRAX/G-BMTZ)
G-FPLD	Beech 200 Super King Air	FR Aviation Ltd/Bournemouth

Notes	Reg.	Type	Owner or Operator
	G-FRAF	Dassault Falcon 20E	FR Aviation Ltd/Bournemouth
	G-FRAG	PA-32 Cherokee Six 300E	T. A. Houghton
	G-FRAH	Dassault Falcon 20DC	FR Aviation Ltd/Bournemouth
	G-FRAI	Dassault Falcon 20E	FR Aviation Ltd/Bournemouth
	G-FRAJ	Dassault Falcon 20E	FR Aviation Ltd/Bournemouth
	G-FRAK	Dassault Falcon 20DC	FR Aviation Ltd/Bournemouth
	G-FRAL	Dassault Falcon 20DC	FR Aviation Ltd/Bournemouth
	G-FRAM	Dassault Falcon 20DC	FR Aviation Ltd/Bournemouth
	G-FRAN	Piper J-3C-90 Cub(480321)	Essex L-4 Group (G-BIXY)
	G-FRAO	Dassault Falcon 20DC	FR Aviation Ltd/Bournemouth
	G-FRAP	Dassault Falcon 20DC	FR Aviation Ltd/Bournemouth
	G-FRAR	Dassault Falcon 20DC	FR Aviation Ltd/Bournemouth
	G-FRAS	Dassault Falcon 20C	FR Aviation Ltd/Bournemouth
	G-FRAT	Dassault Falcon 20C	FR Aviation Ltd/Bournemouth
	G-FRAU	Dassault Falcon 20C	FR Aviation Ltd/Bournemouth
	G-FRAW	Dassault Falcon 20ECM	FR Aviation Ltd/Bournemouth
	G-FRAY	Cassutt IIIM (modified)	C. I. Fray
	G-FRBA	Dassault Falcon 20C	FR Aviation Ltd/Bournemouth
	G-FRCE	Folland Gnat T.1 (XS104)	Airborne Innovations Ltd
	G-FRGN	PA-28-236 Dakota	Fregon Aviation Ltd
	G-FRJB	Britten Sheriff SA-1 ★	Aeropark/E. Midlands
	G-FROH	AS.350B-2 Ecureuil	Specialist Helicopters Ltd
	G-FROM	Ikarus C.42 FB100	Fly Buy Ultralights Ltd
	G-FRST	PA-44-180T Turbo Seminole	D. B. Ryder & Co. Ltd
	G-FRYI	Beech 200 Super King Air	London Executive Aviation Ltd (G-OAVX/G-IBCA/G-BMCA)/London City
	G-FRYL	Beexh 390 Premier 1	Gregg Air/ Kidlington
	G-FSHA	Denney Kitfox Mk 2	P. P. Trangmar
	G-FTAX	Cessna 421C	Gold Air International Ltd (G-BFFM)
	G-FTIL	Robin DR.400/180R	RAF Wyton Flying Club Ltd
	G-FTIM	Robin DR.400/100	Madley Flying Group
	G-FTIN	Robin DR.400/100	G. D. Clark & M. J. D. Theobold/Blackpool
	G-FTSE	BN-2A Mk.III-2 Trislander	Aurigny Air Services Ltd (G-BEPI)/Guernsey
	G-FTSL	Canadair CL.600-2B16 604	Farglobe transport Services Ltd
	G-FTUO	Van's RV-4	G-FTUO Flying Group
	G-FTWO	AS.355F-2 Twin Squirrel	McAlpine Helicopters Ltd (G-OJOR/G-BMUS)/ Hayes
	G-FUEL	Robin DR.400/180	R. Darch/Compton Abbas
	G-FUKM	Westland Gazelle AH.1	Falcon Aviation Ltd
	G-FULL	PA-28R Cherokee Arrow 200-II	Stapleford Flying Club Ltd (G-HWAY/G-JULI)
	G-FUNK	Yakovlev Yak-50	Transair (UK) Ltd
	G-FUNN	Plumb BGP-1	J. D. Anson
	G-FUSE	Cameron N-105 balloon	LE Electrical Ltd
	G-FUZY	Cameron N-77 balloon	Allan Industries Ltd
	G-FUZZ	PA-18 Super Cub 95	G. W. Cline
	G-FVBF	Lindstrand LBL-210A balloon	Virgin Airship & Balloon Co Ltd
	G-FVEL	Cameron Z-90 balloon	Fort Vale Engineering Ltd
	G-FWAY	Lindstrand LBL-90A balloon	Fairway Furniture Ltd
	G-FWPW	PA-28-236 Dakota	P. A. & F. C. Winters
	G-FXBT	Aeroprakt A.22 Foxbat	R. H. Jago
	G-FXII	V.S.366 Spitfire F.XII (EN224)	P. R. Arnold
	G-FYAN	Williams	M. D. Williams
	G-FYAO	Williams	M. D. Williams
	G-FYAU	Williams MK 2	M. D. Williams
	G-FYAV	Osprey Mk 4E2	C. D. Egan & C. Stiles
	G-FYBX	Portswood Mk XVI	I. Chadwick
	G-FYCL	Osprey Mk 4G	P. J. Rogers
	G-FYCV	Osprey Mk 4D	M. Thomson
	G-FYCZ	Osprey Mk 4D2	P. Middleton
	G-FYDF	Osprey Mk 4D	K. A. Jones
	G-FYDI	Williams Westwind Two	M. D. Williams
	G-FYDN	European 8C	P. D. Ridout
	G-FYDO	Osprey Mk 4D	N. L. Scallan
	G-FYDP	Williams Westwind Three	M. D. Williams
	G-FYDS	Osprey Mk 4D	N. L. Scallan

Reg.	Type	Owner or Operator	Notes
G-FYEK	Unicorn UE.1C	D. & D. Eaves	
G-FYEO	Eagle Mk 1	M. E. Scallan	
G-FYEV	Osprey Mk 1C	M. E. Scallan	
G-FYEZ	Firefly Mk 1	M. E. & N. L. Scallan	
G-FYFI	European E.84DS	M. Stelling	
G-FYFJ	Williams Westland 2	M. D. Williams	
G-FYFN	Osprey Saturn 2	J. & M. Woods	
G-FYFW	Rango NA-55	Rango Kite & Balloon Co	
G-FYFY	Rango NA-55RC	A. M. Lindsay	
G-FYGC	Rango NA-42B	L. J. Wardle	
G-FYGI	Rango NA-55RC	Advertair Ltd	
G-FYGJ	Airspeed 300	N. Wells	
G-FYGM	Saffrey/Smith Princess	A. Smith	
G-FZZA	General Avia F.22-A	APB Leasing Ltd/Welshpool	
G-FZZI	Cameron H-34 balloon	Magical Adventures Ltd	
G-GABD	GA-7 Cougar	C. B. Stewart/Prestwick	
G-GACA	P.57 Sea Prince T.1 ★	P. G. Vallance Ltd/Charlwood	
G-GACB	Robinson R-44 II	A. C. Barker	
G-GAFA	PA-34-200T Seneca II	Oxford Aviation Services Ltd	
G-GAFT	PA-44-180 Seminole	Atlantic Flight Training Ltd	
G-GAFX	Boeing 747-245F	Air Freight Express Ltd	
G-GAII	Hawker Hunter GA.11 (XE685)	DAT Enterprises Ltd/North Weald	
G-GAJB	AA-5B Tiger	G. A. J. Bowles (G-BHZN)	
G-GAJW	Bell 407	Titan Airways Ltd	
G-GALA	PA-28 Cherokee 180E	Flyteam Aviation Ltd/Elstree (G-AYAP)	
G-GALB	PA-28-161 Warrior II	Goodair Leasing Ltd	
G-GALL	PA-38-112 Tomahawk	M. Lowe & K. Hazelwood (G-BTEV)	
G-GAME	Cessna T.303	P. Heffron	
G-GAND	AB-206B Jet Ranger	Scotia Helicopters Ltd (G-AWMK)	
G-GANE	Sequoia F.8L Falco	S. J. Gane	
G-GASC	Hughes 369HS	Aerolease Ltd (G-WELD/G-FROG)	
G-GASP	PA-28-181 Archer II	G-GASP Flying Group	
G-GASS	Thunder Ax7-77 balloon	Servowarm Balloon Syndicate	
G-GATE	Robinson R-44-II	J. W. Gate	
G-GATT	Robinson R44 Raven II	Heli Air Ltd	
G-GAWA	Cessna 140	C140 Group (G-BRSM)	
G-GAZA	SA.341G Gazelle 1	The Auster Aircraft Co Ltd (G-RALE/G-SFTG)	
G-GAZI	SA.341G Gazelle 1	Sharpness Dock Ltd (G-BKLU)	
G-GAZL	SA.341G Gazelle HT.2	R. M. Bailey (G-CBBY)	
G-GAZZ	SA 341G Gazelle 1	Stratton Motor Co (Norfolk) Ltd	
G-GBAO	Robin R.1180TD	J. Kay-Mova	
G-GBEE	Mainair Pegasus Quik	L. G. Whitet	
G-GBFF	Cessna F.172N	Meridian Aviation Group Ltd	
G-GBFR	Cessna F.177RG	Airspeed Aviation Ltd	
G-GBGA	Scheibe SF.25C Falke	British Gliding Association Ltd	
G-GBGB	Ultramagic M.105	Universal Car Services Ltd	
G-GBHI	SOCATA TB.10 Tobago	Robert Purvis Plant Hire Ltd	
G-GBJP	Mainair Pegasus Quantum 15	R. G. Mulford	
G-GBLP	Cessna F.172M	Aviate Scotland Ltd (G-GWEN)/Edinburgh	
G-GBLR	Cessna F.150L	Almat Flying Club Ltd	
G-GBRB	PA-28 Cherokee 180C	Border Air Training Ltd	
G-GBSL	Beech 76 Duchess	M. H. Cundsy (G-BGVG)	
G-GBTA	Boeing 737-436	British Airways (G-BVHA)	
G-GBTB	Boeing 737-436	British Airways (G-BVHB)	
G-GBUE	Robin DR.400/120A	G-GBUE Group (G-BPXD)	
G-GBUN	Cessna 182T	G. M. Bunn	
G-GBXS	Shaw Europa XS	Europa Management (International) Ltd	
G-GCAC	Shaw Europa XS T-G	J. L. Gunn	
G-GCAT	PA-28 Cherokee 140B	Group Cat (G-BFRH)	
G-GCCL	Beech 76 Duchess	Aerolease Ltd	
G-GCKI	Mooney M.20K	B. Barr	
G-GCYC	Cessna F.182Q	G-GCYC Ltd	
G-GDER	Robin R.1180TD	Berkshire Aviation Services Ltd	
G-GDEZ	BAe 125-1000B	Frewton Ltd	
G-GDGR	SOCATA TB-20 Trinidad	Just Plane Trading Ltd	
G-GDMW	Beech 76 Duchess	Apollo Aviation Advisory Ltd	
G-GDOG	PA-28R Cherokee Arrow 200-II	B. Alidina (G-BDXW)	
G-GDRV	Van's RV-6	J. R. S. Heaton & R. Feather	

Notes	Reg.	Type	Owner or Operator
	G-GDTU	Avions Mudry CAP.10B	Sherburn Aero Club Ltd
	G-GEDY	Dassault Falcon 2000	Victoria Aviation Ltd
	G-GEEP	Robin R.1180TD	C. J. P. Green
	G-GEES	Cameron N-77	N. A. Carr
	G-GEEZ	Cameron N-77 balloon	Charnwood Forest Turf Accountants Ltd
	G-GEHP	PA-28RT-201 Arrow IV	Aeros Leasing Ltd
	G-GEMS	Thunder Ax8-90 S2 balloon	B. Sevenich & ptnrs/Germany
	G-GENN	GA-7 Cougar	Plane Talking Ltd (G-BNAB/G-BGYP)/Elstree
	G-GEOF	Pereira Osprey 2	G. Crossley
	G-GERT	Van's RV-7	Barnstormers
	G-GERY	Stoddard-Hamilton Glastar	G. E. Collard
	G-GFAB	Cameron N-105 balloon	L. Greaves
	G-GFCA	PA-28-161 Cadet	Aeros Leasing Ltd
	G-GFCB	PA-28-161 Cadet	AM & T Aviation Ltd/Bristol
	G-GFCD	PA-34-220T Seneca III	Stonehurst Aviation Ltd (G-KIDS)
	G-GFEY	PA-34-200T Seneca II	West Wales Airport Ltd/Aberporth
	G-GFFA	Boeing 737-59D	British Airways (G-BVZF)
	G-GFFB	Boeing 737-505	British Airways
	G-GFFC	Boeing 737-505	British Airways
	G-GFFD	Boeing 737-59D	British Airways (G-OBMY)
	G-GFFE	Boeing 737-528	British Airways
	G-GFFF	Boeing 737-53A	British Airways (G-OBMZ)
	G-GFFG	Boeing 737-505	British Airways
	G-GFFH	Boeing 737-5H6	British Airways
	G-GFFI	Boeing 737-528	British Airways
	G-GFFJ	Boeing 737-5H6	British Airways
	G-GFKY	Zenair CH.250	J. J. Beal
	G-GFLY	Cessna F.150L	Leagate Ltd
	G-GFMT	Cessna 172S	Upperstack Ltd
	G-GFTA	PA-28-161 Warrior III	One Zero Three Ltd
	G-GFTB	PA-28-161 Warrior III	One Zero Three Ltd
	G-GGGG	Thunder Ax7-77A balloon	T. A. Gilmour
	G-GGLE	PA-22 Colt 108 (tailwheel)	M. &. S. Leonard
	G-GGOW	Colt 77A balloon	G. Everett
	G-GGRR	SA Bulldog Srs 120/121	F. P. Corbett (G-CBAM)
	G-GGTT	Agusta-Bell 47G-4A	Face & Fragrance Ltd
	G-GHEE	Aerotechnik EV-97 Eurostar	C. J. Ball
	G-GHIA	Cameron N-120 balloon	J. A. Marshall
	G-GHIN	Thunder Ax7-77 balloon	N. T. Parry
	G-GHKX	PA-28-161 Warrior II	Plane Talking Ltd
	G-GHOW	Cessna F.182Q	G. How
	G-GHPG	Cesna 550 Citation 2	MCP Aviation (Charter) Ltd
	G-GHRW	PA-28RT-201 Arrow IV	Bonus Aviation Ltd (G-ONAB/G-BHAK)
	G-GHSI	PA-44-180T Turbo Seminole	M. G. Roberts
	G-GHZJ	SOCATA TB.9 Tampico	M. Haller
	G-GIDY	Shaw Europa XS	Gidy Group
	G-GIGI	M.S.893A Rallye Commodore	D. J. Moore (G-AYVX)
	G-GILT	Cessna 421C	Skymaster Air Services (G-BMZC)
	G-GIRA	H.S.125 Srs 700B	EAS Aeroserviza SAS (G-LTEC/G-BHSU)/Italy
	G-GIRY	AG-5B Tiger	Romeo Yankee Flying Group
	G-GIWT	Shaw Europa XS	A. Twigg
	G-GJCD	Robinson R-22B	J. C. Lane
	G-GJKK	Mooney M.20K	Pergola Ltd
	G-GKFC	RL-5A LW Sherwood Ranger	K. F. Crumplin (G-MYZI)
	G-GKRG	Cessna 172RG	Skytrax Aviation Ltd
	G-GLAD	Gloster G.37 Gladiator II (N5903)	Patina Ltd/Duxford
	G-GLAW	Cameron N-90 balloon	R. A. Vale
	G-GLED	Cessna 150M	Firecrest Aviation Ltd/Booker
	G-GLHI	Skyranger 912	G. L. Higgins
	G-GLIB	Robinson R-44	Helisport UK Ltd
	G-GLST	Great Lakes Sport Trainer	D. A. Graham
	G-GLSU	Bucher Bu.181B-1 Bestmann	G. R. Lacey
	G-GLTT	PA-31-350 Navajo Chieftain	Airtime Aviation Ltd
	G-GLUC	Van's RV-6	Speedfreak Ltd

Reg.	Type	Owner or Operator	Notes
G-GLUE	Cameron N-65 balloon	L. J. M. Muir & G. D. Hallett	
G-GLUG	PA-31-350 Navajo Chieftain	Champagne-Air Ltd (G-BLOE/G-NITE)/Newcastle	
G-GMAA	Learjet 45	Gama Aviation Ltd	
G-GMAB	BAe 125 Srs 1000A	Gama Aviation Ltd (G-BUWX)	
G-FMAC	Gulfstream G-IVSP	Gama Aviation Ltd	
G-GMAX	SNCAN Stampe SV-4C	Glidegold Ltd (G-BXNW)	
G-GMPA	AS.355F-2 Twin Squirrel	Police Aviation Services Ltd (G-BPOI)	
G-GMPB	BN-2T-4S Defender 4000	Greater Manchester Police Authority (G-BWPU)	
G-GMPS	MDH MD-902 Explorer	Greater Manchester Police Authority	
G-GMSI	SOCATA TB.9 Tampico	M. L. Rhodes	
G-GNAA	MD.900	Police Aviation Services Ltd	
G-GNAT	H.S. Gnat T.1 (XS101)	Brutus Holdings Ltd/Cranfield	
G-GNJW	Ikarus C.42	I. R. Westrope	
G-GNMG	Cessna U.206F	P. Marsden	
G-GNTB	SAAB SF.340A	Loganair Ltd	
G-GNTG	SAAB SF.340A	Loganair Ltd/BA	
G-GNTZ	BAe 146-200	British Airways CitiExpress (G-CLHB)	
G-GOAC	PA-34-200T Seneca II	Oxford Aviation Ltd	
G-GOBT	Colt 77A balloon	British Telecom PLC	
G-GOCX	Cameron N-90 balloon	R. D. Parry/Hong Kong	
G-GOGS	PA-34-200T Seneca II	A. Semple	
G-GOGW	Cameron N-90 balloon	S. E. Carroll	
G-GOJP	PA-46-350P Malibu Mirage	Plato Management Ltd (G-CREW)	
G-GOLF	SOCATA TB.10 Tobago	Golf Golf Group	
G-GONN	AS.355N Ecureuil 2	Taggart Homes Ltd (G-BVNW)	
G-GOOD	SOCATA TB-20 Trinidad	T. M. Sloane & M. P. Bowcock	
G-GOON	MDH.600N	Cumbrian Seafoods Ltd	
G-GORE	CFM Streak Shadow	M. S. Clinton	
G-GORF	Robin HR. 200/120B	J. A. Ingram/Tollerton	
G-GOSL	Robin DR.400/180	R. M. Gosling (G-BSDG)	
G-GOSS	Jodel DR.221	Avon Flying Group	
G-GOTC	GA-7 Cougar	WakeliteLtd	
G-GOTH	PA-28-161 Warrior III	J. Gosling	
G-GOTO	PA-32R-301T Turbo Saratoga II	J. A. Varndell	
G-GOUP	Robinson R-22B	Electric Scribe 2000 Ltd (G-DIRE)	
G-GPAG	Van's RV-6	P. A. Green	
G-GPAS	Avtech Jabiru UL-450	G. D. Allen	
G-GPEG	Sky 90-24 balloon	N. T. Parry	
G-GPFI	Boeing 737-229	European Aviation Ltd/Bournemouth	
G-GPMW	PA-28RT-201T Turbo Arrow IV	Calverton Flying Group Ltd	
G-GPST	Phillips ST.1 Speedtwin	Speedtwin Developments Ltd	
G-GREY	PA-46-350P Malibu Mirage	S. A. & K. J. Williams	
G-GRGS	Cessna 560 Citation Ultra	Houston Jet Services Ltd	
G-GRID	AS.355F-1 Twin Squirrel	National Grid Co PLC	
G-GRIN	Van's RV-6	A. Phillips	
G-GRIP	Colt 110 Bibendum SS balloon	The Aerial Display Co Ltd	
G-GROE	Grob G.115A	H. & E. Merkado	
G-GROL	Maule MXT-7-180	D. C. Croll & ptnrs	
G-GROW	Cameron N-77 balloon	Derbyshire Building Soc	
G-GRPA	Ikarus C.42 FB100	G. R Page.	
G-GRRC	PA-28-161 Warrior II	Goodwood Road Racing Co Ltd (G-BXJX)	
G-GRRR	SA Bulldog Srs 120/122	Horizons Europe Ltd (G-BXGU)	
G-GRWW	Robinson R-44-II	G. R. Williams	
G-GRYZ	Beech F33A Bonanza	J. Kawadri & M. Kaveh	
G-GSCV	Ikarus C.42 FB UK	G. Sipson	
G-GSJH	Bell 206B JetRanger 3	Interheli (G-PENT/G-IIRB)	
G-GSPN	Boeing 737-31S	Flyglobespan.com	
G-GSSA	Boeing 747-47UF	Global Supply Systems Ltd	
G-GSSB	Boeing 747-47UF	Global Supply Systems Ltd	
G-GSSC	Boeing 747-47UF	Global Supply Systems Ltd	
G-GTAX	PA-31-350 Navajo Chieftain	Hadagain Investments Ltd (G-OIAS)	
G-GTHM	PA-38-112 Tomahawk	Turweston Aero Club Ltd	
G-GUAY	Enstrom 480	Heliway Aviation	
G-GUCK	Beech C23 Sundowner 180	J. T. Francis (G-BPYG)	
G-GUFO	Cameron 80 Saucer SS balloon	Magical Adventures Ltd (G-BOUB)	

Notes	Reg.	Type	Owner or Operator
	G-GULF	Lindstrand LBL-105A balloon	M. A. Webb
	G-GULP	I.I.I. Sky Arrow 650T	S. Marriott
	G-GUMS	Cessna 182P	L. W. Scattergood (G-CBMN)
	G-GUNS	Cameron V-77 balloon	Royal School of Artillery Hot Air Balloon Club
	G-GURN	PA-31 Navajo C	Neric Ltd (G-BHGA)
	G-GURU	PA-28-161 Warrior II	Leeds Flying School Ltd
	G-GUSS	PA-28-151 Warrior	M. J. Cleaver & J. M. Newman (G-BJRY)
	G-GUST	AB-206B JetRanger 2	Gatehouse Estates Ltd (G-CBHH/G-AYBE)
	G-GUYS	PA-34-200T Seneca	R. J. & J. M. Z. Keel (G-BMWT)/Sturgate
	G-GVPI	Evans VP-1	G. Martin
	G-GWIZ	Colt Clown SS balloon	Magical Adventures Ltd
	G-GWYN	Cessna F.172M	Magic Carpet Flying Co.
	G-GYAK	Yakovlev Yak-50	M. W. Levy & M. V. Rijske
	G-GYAT	GY-80 Horizon 180	Rochester GYAT Flying Group Club
	G-GYAV	Cessna 172N	Southport & Merseyside Aero Club (1979) Ltd/ Liverpool
	G-GYBO	GY-80 Horizon 160	A. L. Fogg
	G-GYMM	PA-28R Cherokee Arrow 200	L. E.Vandervort (G-AYWW)
	G-GYRO	Campbell Cricket	J. W. Pavitt
	G-GYTO	PA-28-161 Warrior III	Wellesbourne Aviation
	G-GZDO	Cessna 172N	Cambridge Hall Aviation
	G-GZLE	SA.341G Gazelle 1	R. G. Fairall (G-PYOB/G-IYOB/G-WELA/ G-SFTD/G-RIFC)
	G-HACK	PA-18 Super Cub 150	Intrepid Aviation Co/North Weald
	G-HADA	Enstrom 480	W. B. Steele
	G-HAEC	CAC-18 Mustang 23 (472218)	R. W. Davies/Duxford
	G-HAFG	Cessna 340A	Haller & Sons (Dereham) Ltd
	G-HAIB	Aviat A-1B Husky	Aviat Aircraft (UK) Ltd
	G-HAIG	Rutan LongEz	N. M. Robbins
	G-HAIR	Robin DR.400/180	Racoon International
	G-HAJJ	Glaser-Dirks DG.400	P. W. Endean
	G-HALC	PA-28R Cherokee Arrow 200	Halcyon Aviation Ltd
	G-HALE	Robinson R-44 Astro	Barhale Surveying Ltd
	G-HALJ	Cessna 140	H. A. Lloyd-Jennings
	G-HALL	PA-22 Tri-Pacer 160	F. P. Hall (G-ARAH)
	G-HALP	SOCATA TB.10 Tobago	D. H. Halpern (stored)/Elstree
	G-HALT	Mainair Pegasus Quik	J. A. Horn
	G-HAMA	Beech 200 Super King Air	Gama Aviation Ltd/Fairoaks
	G-HAMI	Fuji FA.200-180	K. G. Cameron, M. P. Antoniak and Renzacci (UK) PLC (G-OISF/G-BAPT)
	G-HAMM	Yakovlev Yak-50	A. D. Hammond/North Weald
	G-HAMP	Bellanca 7ACA Champ	K. MacDonald
	G-HANA	WS.58 Wessex HC.2	R. A. Fidler
	G-HANS	Robin DR.400 2+2	Bagby Aviation
	G-HANY	AB-206B JetRanger 3	Swift Helicopters Ltd (G-ESAL/G-BHXW/G-JEKP)
	G-HAPI	Lindstrand LBL-105A balloon	Adventure Balloons Ltd
	G-HAPR	B.171 Sycamore HR.14 (XG547) ★	IHM/Weston-s-Mare
	G-HAPY	D.H.C.1 Chipmunk 22A (WP803)	Astrojet Ltd
	G-HARE	Cameron N-77 balloon	D. H. Sheryn & C. A. Buck
	G-HARF	G.1159C Gulfstream 4	Fayair (Jersey) 1984 Ltd
	G-HARH	Sikorsky S-76B	Air Harrods Ltd/Stansted
	G-HARI	Raj Hamsa X'Air V2 (2)	A. Chidlow
	G-HARN	PA-28-181 Archer II	J. P. & T. M. Jones (G-DENK/G-BXRJ)
	G-HARR	Robinson R-22B	Tananmera Properties Ltd
	G-HART	Cessna 152 (tailwheel)	Atlantic Air Transport Ltd (G-BPBF)/Coventry
	G-HARY	Alon A-2 Aircoupe	R. E. Dagless (G-ATWP)
	G-HASI	Cessna 421B	Chester Air Maintenance Ltd (G-BTDK)
	G-HASO	Diamond DA.400 Star	Hasso Enterprises Ltd (G-CCLZ)
	G-HATF	Thorp T-18CW	A. T. Fraser
	G-HATZ	Hatz CB-1	S. P. Rollason
	G-HAUL	Westland WG.30 Srs 300 ★	IHM/Weston-super-Mare
	G-HAUS	Hughes 369HM	Pulford Aviation (G-KBOT/G-RAMM)/Sywell
	G-HAZE	Thunder Ax8-90 balloon	T. G. Church
	G-HBBC	D.H.104 Dove 8	BBC Air Ltd (G-ALFM)
	G-HBMW	Robinson R-22	Northumbria Helicopters Ltd (G-BOFA)
	G-HBOS	Scheibe SF-25C Rotax-Falke	Coventry Gliding Club Ltd

Reg.	Type	Owner or Operator	Notes
G-HBUG	Cameron N-90 balloon	R. T. & H. Revel (G-BRCN)	
G-HCSL	PA-34-220T Seneca III	Shoreham Flight Centre Ltd	
G-HDAE	D.H.C.1 Chipmunk 22	G. P. J. M. Valvekens & ptnrs/Belgium	
G-HDEW	PA-32R-301 Saratoga SP	Plantation Stud Ltd (G-BRGZ)	
G-HDTV	Agusta A.109A-II	Castle Air Charters Ltd (G-BXWD)	
G-HDIX	Enstrom 280FX	D. H. Brown	
G-HEBE	Bell 206B JetRanger 3	M and E Building and Civil Engineering Contractors Ltd	
G-HELA	SOCATA TB.10 Tobago	Group TB.10	
G-HELE	Bell 206B JetRanger 3	B. E. E. Smith (G-OJFR)	
G-HELN	Piper PA-28-95	Helen Group	
G-HELV	D.H.115 Vampire T.55 (XJ771)	Aviation Heritage Ltd/Bournemouth	
G-HENT	SOCATA Rallye 110ST	R. J. Patton	
G-HENY	Cameron V-77 balloon	R. S. D'Alton	
G-HEPY	Robinson R-44 Astro	Beoley Mill Software Ltd	
G-HERB	PA-28R-201 Arrow III	Consort Aviation Ltd	
G-HERC	Cessna 172S	Cambridge Aero Club Ltd	
G-HERD	Lindstrand LBL-77B balloon	S. W. Herd	
G-HEVN	SOCATA TB-200 Tobago XL	I. K. Maclean	
G-HEWI	Piper J-3C-90 Cub	Denham Grasshopper Group (G-BLEN)	
G-HEWS	Hughes 369D ★	Spares' use/Sywell	
G-HEXE	Colr 17A balloon	A. Dunnington	
G-HEYY	Cameron 72 Bear SS balloon	Magical Adventures Ltd	
G-HFBM	Curtiss Robin C-2	D. M. Forshaw	
G-HFCA	Cessna A.150L	T. H. Scott	
G-HFCB	Cessna F.150L	Horizon Flying Club Ltd (G-AZVR)	
G-HFCI	Cessna F.150L	A. Modi/Earls Colne	
G-HFCL	Cessna F.152	Modi Aviation Ltd (G-BGLR)	
G-HFCT	Cessna F.152	Stapleford Flying Club Ltd	
G-HFLA	Schweizer 269C	Sterling Helicopters Ltd/Norwich	
G-HFTG	PA-23 Aztec 250E	A. Mannheim (G-BSOB/G-BCJR)	
G-HGAS	Cameron N-77 balloon	N. J. Tovey	
G-HGPI	SOCATA TB.20 Trinidad	M. J. Jackson/Bournemouth	
G-HHAA	H.S. Buccaneer S.2B (XX885)	Hawker Hunter Aviation Ltd	
G-HHAB	Hawker Hunter F.58	Hawker Hunter Aviation Ltd	
G-HHAC	Hawker Hunter F.58 (XG232)	Hawker Hunter Aviation Ltd (G-BWIU)/Scampton	
G-HHAD	Hawker Hunter F.58	Hawker Hunter Aviation Ltd (G-BWFS)	
G-HHAE	Hawker Hunter F.58	Hawker Hunter Aviation Ltd (G-BXNZ)	
G-HHAF	Hawker Hunter F.58	Hawker Hunter Aviation Ltd (G-BWKB)	
G-HHAV	M.S.894A Rallye Minerva	R. E. Dagless (G-AYDG)	
G-HHOG	Robinson R-44-II	Heli Air Ltd	
G-HIBM	Cameron N-145 balloon	Alba Ballooning Ltd	
G-HIEL	Robinson R-22B	Hields Aviation/Sherburn	
G-HIJK	Cessna 421C	Oxford Aviation Services Ltd (G-OSAL)/Kidlington	
G-HIJN	Ikarus C.42 FB100	J. R. North	
G-HILO	R. Commander 114	F. H. Parkes	
G-HILS	Cessna F.172H	Lowdon Aviation Group (G-AWCH)/Blackbushe	
G-HILT	SOCATA TB.10 Tobago	Cheshire Aircraft Leasing Ltd	
G-HIND	Maule MT-7-235	R. G. Humphries	
G-HINZ	Avtec Jabiru SK	B. Faupel	
G-HIPE	Sorrell SNS-7 Hiperbipe	J. K. Cook	
G-HIPO	Robinson R-22B	Patriot Aviation (G-BTGB)	
G-HIRE	GA-7 Cougar	London Aerial Tours Ltd (G-BGSZ)/Biggin Hill	
G-HISS	Aerotek Pitts S-2A Special	L. V. Adams & J. Maffia (G-BLVU)/Panshanger	
G-HITM	Raj Hamsa X'Air 582 (1)	G-HITM Flying Group	
G-HITS	PA-46-310P Malibu	Law 2200 Ltd (G-BMBE)	
G-HIUP	Cameron A-250 balloon	Bridges Van Hire Ltd	
G-HIVA	Cessna 337A	G. J. Banfield (G-BAES)	
G-HIVE	Cessna F.150M	M. P. Lynn (G-BCXT)/Sibson	
G-HIZZ	Robinson R-22B – II	M. J. Gee (G-CNDY/G-BXEW)	
G-HJSM	Schempp-Hirth Nimbus 4DM	60 Group (G-ROAM)	
G-HJSS	AIA Stampe SV-4C (modified)	H. J. Smith (G-AZNF)	
G-HKHM	Hughes 369B	Heli Air Ltd/Wellesbourne	

Notes	Reg.	Type	Owner or Operator
	G-HLCF	Starstreak Shadow SA-II	A. B. Atkinson
	G-HMBJ	R. Commander 114B	Bravo Juliet Aviation Ltd
	G-HMED	PA-28-161 Warrior III	H. Faizal
	G-HMEI	Dassault Falcon 900	Executive Jet Group Ltd
	G-HMJB	PA-34-220T Seneca III	Cross Atlantic Ventures Ltd
	G-HMMV	Cessna 525 CitationJet	Stellar Aviation
	G-HMPF	Robinson R-44 Astro	Mightycraft Ltd
	G-HMPH	Bell 206B JetRanger 2	Sturmer Ltd (G-BBUY)
	G-HMPT	AB-206B JetRanger 2	Helicopter Express Ltd
	G-HMSS	Bell 206B JetRanger 2	Jemaal Management Services
	G-HNTR	Hawker Hunter T.7 (XL571) ★	Yorkshire Air Museum/Elvington
	G-HOBO	Denney Kitfox Mk 4	J. P. Donovan
	G-HOBZ	Hobz Westland Gazelle HT.Mk.3	SE Hobbs (UK) Ltd (G-CBSJ)
	G-HOCK	PA-28 Cherokee 180	G-HOCK Flying Club (G-AVSH)
	G-HOFC	Shaw Europa	W. R. Mills
	G-HOFM	Cameron N-56 balloon	Magical Adventures Ltd
	G-HOGS	Cameron 90 Pig SS balloon	Magical Adventures Ltd
	G-HOHO	Colt Santa Claus SS balloon	Oxford Promotions (UK) Ltd/USA
	G-HOLY	ST.10 Diplomate	M. K. Barsham
	G-HOME	Colt 77A balloon	Anglia Balloon School Tardis
	G-HONG	Slingsby T.67M Firefly 200	Jewel Aviation Ltd
	G-HONI	Robinson R-22B	Patriot Aviation Ltd/Cranfield (G-SEGO)
	G-HONK	Cameron O-105 balloon	T. F. W. Dixon & Son Ltd
	G-HONY	Lilliput Type 1 Srs A balloon	A. E. & D. E. Thomas
	G-HOOD	SOCATA TB.20 Trinidad GT	M. J. Hoodless
	G-HOOT	AS.355F-2 Twin Squirrel	Squirrel Helicopter Hire Ltd (G-SCOW/G-POON/G-MCAL)
	G-HOOV	Cameron N-56 balloon	H. R. Evans
	G-HOPA	Lindstrand LBL-35A balloon	S. F. Burden/Netherlands
	G-HOPE	Beech F33A Bonanza	Hope Aviation
	G-HOPI	Cameron N-42 balloon	Ballonwerbung Hamburg GmbH/Germany
	G-HOPR	Lindstrand LBL-25A balloon	K. C. Tanner
	G-HOPY	Van's RV-6A	R. C. Hopkinson
	G-HORN	Cameron V-77 balloon	S. Herd
	G-HORS	Cameron Horse SS balloon	Flygande Dalahasten AB/Sweden
	G-HOTI	Colt 77A balloon	R. Ollier
	G-HOTT	Cameron O-120 balloon	D. L. Smith
	G-HOTZ	Colt 77B balloon	C. J. & S. M. Davies
	G-HOUS	Colt 31A balloon	The British Balloon Museum and Library
	G-HOWE	Thunder Ax7-77 balloon	M. F. Howe
	G-HOWL	RAF 2000 GTX-SE gyroplane	C. J. Watkinson
	G-HPAD	Bell 206B JetRanger 2	Helipad Ltd (G-CITZ/G-BRTB)
	G-HPOL	MDH MD-902 Explorer	Humberside Police Authority
	G-HPSB	R. Commander 114B	Propwash Investments Ltd
	G-HPSE	R. Commander 114B	Al Nisr Ltd
	G-HPSF	R. Commander 114B	Three Foxtrot Group/Guernsey
	G-HPSL	R. Commander 114B	Hallard (Guernsey) Ltd
	G-HPUX	Hawker Hunter T.7 (XL587)	Hawker Hunter Aviation Ltd/Scampton
	G-HRAK	AS.350B. Ecureuil	R. A. Kingston
	G-HRBS	Robinson R-22B	Universal Energy Ltd
	G-HRDS	Gulfstream GV-SP(550)	Fayair (Jersey) Ltd
	G-HRHE	Robinson R-22B	Select Helicopters Ltd (G-BTWP)
	G-HRHI	B.206 Srs 1 Basset (XS770)	M. D. Lewis
	G-HRHS	Robinson R-44	Stratus Aviation Ltd/Hong Kong
	G-HRIO	Robin HR.100/120	T. W. Evans
	G-HRLI	Hawker Hurricane 1	Hawker Restorations Ltd
	G-HRLK	SAAB 91D/2 Safir	Sylmar Aviation & Services Ltd (G-BRZY)
	G-HRLM	Brügger MB.2 Colibri	M. W. Bodger
	G-HRNT	Cessna 182S	Dingle Star Ltd
	G-HROI	R. Commander RC.112	Intereuropean Aviation Ltd
	G-HRPN	Robinson R-44	Harpin Ltd
	G-HRVD	CCF Harvard IV	Anglia Flight (G-BSBC)
	G-HRZN	Colt 77A balloon	A. J. Spindler
	G-HSDW	Bell 206B JetRanger 2	Greatsearch Ltd
	G-HSLA	Robinson R-22B	D. I. Pointon (G-BRTI)
	G-HSLB	AB-206B JetRanger 2	B3 Aviation Services Ltd

Reg.	Type	Owner or Operator	Notes
G-HSOO	Hughes 369HE	Edwards Aviation (G-BFYJ)	
G-HTAX	PA-31-350 Navajo Chieftain	Hadagain Investments Ltd	
G-HTEL	Robinson R-44	Forestdale Hotels Ltd	
G-HTRL	PA-34-220T Seneca III	Air Medical Fleet Ltd (G-BXXY)	
G-HUBB	Partenavia P.68B	G-HUBB Ltd	
G-HUCH	Cameron 80 Carrots SS balloon	Magical Adventures Ltd (G-BYPS)	
G-HUEW	Shaw Europa XS	C, R. Wright	
G-HUEY	Bell UH-1H	Argonauts Holdings Ltd	
G-HUFF	Cessna 182P	A. E. G. Cousins	
G-HUGO	Colt 240A balloon	P. G. Hall	
G-HUGS	Robinson R-22B	C. G. P. Holden (G-BYHD)	
G-HUKA	MDH Hughes 369E	B. P. Stein (G-OSOO)	
G-HULL	Cessna F.150M	Hull Aero Club Ltd	
G-HUNI	Bellanca 7GCBC Scout	T. I. M. Paul	
G-HUNK	Lindstrand LBL-77A balloon	Lindstrand Balloons Ltd	
G-HUPW	Hawker Hurricane 1	Minmere Farm Partnership	
G-HURI	CCF Hawker Hurricane XIIA (Z5140/HA-C)	Historic Aircraft Collection Ltd/Duxford	
G-HURN	Robinson R-22B	Sloane Helicopters Ltd	
G-HURR	Hawker Hurricane XIIB (BE417)	R. A. Fleming/Breighton	
G-HURY	Hawker Hurricane IV (KZ321)	Patina Ltd/Duxford	
G-HUSK	Aviat A-1B	P. H. Yarrow & A. T. Duke	
G-HUTT	Denney Kitfox Mk 2	P. C. E. Roberts	
G-HVAN	RL-5A LW Sherwood Ranger	H. T. H. van Neck	
G-HVBF	Lindstrand LBL-210A balloon	Virgin Balloon Flights	
G-HVIP	Hawker Hunter T.68	Golden Europe Jet De Luxe Club Ltd/ Bournemouth	
G-HVRD	PA-31-350 Navajo Chieftain	N. Singh (G-BEZU)	
G-HXTD	Robin DR.400/180	Hayley Aviation Ltd	
G-HYAK	IDA Bacau Yakovlev Yak-52	Goodridge (UK) Ltd	
G-HYLT	PA-32R-301 Saratoga SP	G. G. L. James	
G-HYST	Enstrom 280FX	Patten Helicopter Services Ltd	
G-IAFT	Cessna 152 II	Marnham Investments Ltd	
G-IAGD	Robinson R-22B	A & M Engineering Ltd (G-DRAI)	
G-IAMS	Cessna 560XL	Amsair Ltd	
G-IANB	Glaser-Dirks DG-800B	I. S. Bullous	
G-IANC	SOCATA TB.10 Tobago	I. Corbin (G-BIAK)	
G-IANH	SOCATA TB.10 Tobago	XD Flight Management Ltd	
G-IANI	Shaw Europa XS T-G	I. F. Rickard & I. A. Watson	
G-IANJ	Cessna F.150K	Messrs Rees of Poyston West (G-AXVW)	
G-IANN	Kolb Twinstar Mk 3	I. Newman	
G-IANW	AS.350B-3 Ecureuil	Milford Aviation Services Ltd	
G-IARC	Stoddard-Hamilton Glastar	A. A. Craig	
G-IASL	Beech 60 Duke	Applied Sweepers Ltd (G-SING)	
G-IATU	Cessna 182P	Auto Corporation Ltd (G-BIRS)	
G-IAWE	Cessna F.150L	P. P. Magire	
G-IBAZ	Ikarus C.42 FB100	B. R. Underwood	
G-IBBC	Cameron 105 Sphere SS balloon	Balloon Preservation Group	
G-IBBS	Shaw Europa	R. H. Gibbs	
G-IBED	Robinson R-22A	Brian Seedle Helicopters Blackpool (G-BMHN)	
G-IBET	Cameron 70 Can SS balloon	M. R. Humphrey & J. R. Clifton	
G-IBFC	BFC Challenger II	K. N. Dickinson	
G-IBFW	PA-28R-201 Arrow III	D. A. Triplett	
G-IBHH	Hughes 269C	Biggin Hill Helicopters (G-BSCD)	
G-IBIG	Bell 206B JetRanger 3	Big Heli-Charter Ltd (G-BORV)	
G-IBLU	Cameron Z-90 balloon	John Aimo Balloons SAS/Italy	
G-IBRI	Eurocopter EC.120B	Colibri Aviation Ltd	
G-IBRO	Cessna F.152 II	Leicestershire Aero Club Ltd	
G-IBSF	Dassault Falcon 2000	Marconda Services Ltd	
G-IBZS	Cessna 182S	Patrick Eddery Ltd/Kidlington	
G-ICAB	Robinson R-44	Northumbria Helicopters Ltd	
G-ICAS	Pitts S-2B Special	J. C. Smith	
G-ICBM	Stoddard-Hamilton Glasair III Turbine	G. V. Walters & D. N. Brown	
G-ICCL	Robinson R-22B	JK Aviation Services Ltd (G-ORZZ)	

Notes	Reg.	Type	Owner or Operator
	G-ICES	Thunder Ax6-56 balloon	British Balloon Museum & Library Ltd
	G-ICEY	Lindstrand LBL-77A balloon	G. C. Elson/Spain
	G-ICKY	Lindstrand LBL-77A balloon	M. J. Green
	G-ICOI	Lindstrand LBL-105A balloon	F. Schroeder/Germany
	G-ICOM	Cessna F.172M	C. G. Elesmore (G-BFXI)
	G-ICON	Rutan LongEz	S. J. & M. A. Carradice
	G-ICRS	Ikarus C.42 FB UK Cyclone	Ikarus Flying Group Ltd
	G-ICSG	AS.355F-1 Twin Squirrel	Stratton Motor Co (Norfolk)(G-PAMI/G-BUSA)/Ltd
	G-ICWT	Pegasus Quantum 15-912	C. W. Taylor
	G-IDAB	Cessna 550 Citation Bravo	Errigal Aviation Ltd
	G-IDAY	Skyfox CA-25N Gazelle	G. G. Johnstone
	G-IDDI	Cameron N-77 balloon	PSH Skypower Ltd
	G-IDII	Dan Rihn DR.107 One Design	C. Darlow
	G-IDPH	PA-28-181 Archer III	D. Holland
	G-IDSL	Flight Design CT.2K	D. S. Luke
	G-IDUP	Enstrom 280C Shark	Antique Buildings Ltd (G-BRZF)
	G-IDWR	Hughes 369HS	Copley Electrical Contractors (G-AXEJ)
	G-IEIO	PA-34-200T Seneca II	Jade Air PLC
	G-IEJH	Jodel 150A	A. Turner & D. Worth (G-BPAM)/Crowfield
	G-IEYE	Robin DR. 400/180	G. Wood
	G-IFAB	Cessna F.182Q	Manda Construction Ltd
	G-IFBP	AS.350B-2 Ecureuil	Frank Bird Aviation
	G-IFDM	Robinson R-44	M. Paffey
	G-IFFR	PA-32 Cherokee Six 300	D. J. D. Ritchie & ptnrs (G-BWVO)
	G-IFIT	PA-31-350 Navajo Chieftain	Dart Group PLC (G-NABI/G-MARG)/Bournemouth
	G-IFLE	Aerotechnik EV-97 TeamEurostar UK	Euravia Glight
	G-IFLI	AA-5A Cheetah	I-Fly Group Ltd
	G-IFLP	PA-34-200T Seneca II	Tayflite Ltd/Perth
	G-IFTE	H.S.125 Srs 700B	Albion Aviation Management Ltd (G-BFVI)
	G-IFTS	Robinson R-44	Frank Bird (Poultry) Ltd
	G-IGEL	Cameron N-90 balloon	Computacenter Ltd
	G-IGGL	SOCATA TB-10 Tobago	G-IGGL Flying Group (G-BYDC)/White Waltham
	G-IGHH	Enstrom 480	H. Hyndman
	G-IGII	Shaw Europa	C. D. Peacock
	G-IGLE	Cameron V-90 balloon	A. A. Laing
	G-IGLS	Champion 7GCAA Citabria	Blue Yonder Aviation Ltd/North Weald
	G-IGLZ	Champion 8KCAB	Woodgate Aviation (IOM) Ltd
	G-IGOB	Boeing 737-36Q	easyJet Airline Co Ltd
	G-IGOG	Boeing 737-3Y0	easyJet Airline Co Ltd
	G-IGOH	Boeing 737-3Y0	easyJet Airline Co Ltd
	G-IGOI	Boeing 737-33A	easyJet Airline Co Ltd (G-OBMD)
	G-IGOJ	Boeing 737-36N	easyJet Airline Co Ltd
	G-IGOK	Boeing 737-36N	easyJet Airline Co Ltd
	G-IGOL	Boeing 737-36N	easyJet Airline Co Ltd
	G-IGOM	Boeing 737-36N	easyJet Airline Co Ltd
	G-IGOO	Boeing 737-36N	easyJet Airline Co Ltd (G-SMDB)
	G-IGOP	Boeing 737-36N	easyJet Airline Co Ltd
	G-IGOR	Boeing 737-36N	easyJet Airline Co Ltd
	G-IGOS	Boeing 737-3L9	easyJet Airline Co Ltd
	G-IGOT	Boeing 737-3L9	easyJet Airline Co Ltd
	G-IGOU	Boeing 737-3L9	easyJet Airline Co Ltd
	G-IGOV	Boeing 737-3M8	easyJet Airline Co Ltd
	G-IGOW	Boeing 737-3Y0	easyJet Airline Co Ltd (G-TEAB)
	G-IGOX	Boeing 737-3L9	Titan Airways Ltd (G-BOZB)
	G-IGOY	Boeing 737-36N	easyJet Airline Co Ltd
	G-IGOZ	Boeing 737-3Q8	easyJet Airline Co Ltd (G-OBWZ)
	G-IGPW	Eurocopter EC.120B	Helihopper Ltd (G-CBRI)
	G-IGTE	SIAI Marchetti F.260	D. Fletcher & J. J. Watts
	G-IHOT	Aerotechnik EV-97 Eurostar UK	A. J. Turner
	G-IHSB	Robinson R-22B	Patriot Aviation Ltd/Cranfield
	G-IIAC	Aeronca 11AC Chief	G. R. Moore (G-BTPY)
	G-IIAN	Aero Designs Pulsar	I. G. Harrison
	G-IICI	Aviat Pitts S-2C Special	D. G. Cowden
	G-IICM	Extra EA.300/L	Phonetiques Ltd
	G-IIDI	Extra EA.300/L	Power Aerobatics Ltd (G-XTRS)

Reg.	Type	Owner or Operator	Notes
G-IIDY	Aerotek Pitts S-2B Special	The S-2B Group (G-BPVP)	
G-IIFR	Robinson R-22B-2	Wrightson Aviation & Engineering	
G-IIGI	Van's RV-4	IIGI Flying Club	
G-IIID	Dan Rihn DR.107 One Design	V. Millard	
G-IIIE	Aerotek Pitts S-2B Special	D. Dobson	
G-IIIG	Boeing Stearman A.75N1	F. & S. Vormezeele (G-BSDR)/Belgium	
G-IIII	Aerotek Pitts S-2B Special	Four Eyes Aerobatics Ltd	
G-IIIL	Pitts S-1T Special	Empyreal Airways Ltd	
G-IIIO	Schempp-Hirth Ventus 2CM	M. A. V. Gatehouse	
G-IIIR	Pitts S-1S Special	R. O. Rogers	
G-IIIS	Sukhoi SU-26M2	Airtime Aerobatics	
G-IIIT	Aerotek Pitts S-2A Special	Aerobatic Displays Ltd	
G-IIIV	Pitts Super Stinker 11-260	G. G. Ferriman	
G-IIIX	Pitts S-1S Special	D. A. Kean (G-LBAT/G-UCCI/G-BIYN)	
G-IIIZ	Sukhoi SU-26M	P. M. M. Bonhommy	
G-IIMI	Extra EA.300/L	Firebird Aerobatics Ltd/Denham	
G-IIMT	Bushby-Long Midget Mustang	M. J. A. Tredgill (G-BDGA)/Henlow	
G-IINI	Van's RV-9A	S. Sampson	
G-IIPM	AS.350B Ecureuil	Kis Associates Ltd (G-GWIL)	
G-IIPT	Robinson R-22B	Patriot Aviation Ltd (G-FUSI)	
G-IIRG	Stoddard-Hamilton Glasair IIS RG	A. C. Lang	
G-IISI	Extra EA.300/200	I. A. Scott/Sweden	
G-IITI	Extra EA.300	Aerobatic Displays Ltd/Booker	
G-IIUI	Extra EA.300/S	M. G. & J. R. Jefferies & C. Scrope (G-CCBD)	
G-IIXI	Extra EA.300/L	Southern Sailplanes	
G-IIXX	Parsons 2-seat gyroplane	J. M. Montgomerie	
G-IIYK	Yakovlev Yak-50	D. A. Hammant	
G-IIZI	Extra EA.300	S. G. Jones & Power Aerobatics Ltd	
G-IJAC	Light Aero Avid Speedwing Mk 4	I. J. A. Charlton	
G-IJBB	Enstrom 480	G. Kidger (G-LIVA/G-PBTT)	
G-IJMC	VPM M.16 Tandem Trainer	D. C. Fairbrass (G-POSA/G-BVJM)	
G-IJOE	PA-28RT-201T Turbo Arrow IV	J. H. Bailey	
G-IJYS	BAe Jetstream 3102	Eastern Airways Ltd (G-BTZT)	
G-IKAP	Cessna T.303	T. M. Beresford	
G-IKAT	Diamond DA.20C1	Diamond Aircraft UK Ltd/Gamston.	
G-IKBP	PA-28-161 Warrior II	K. B. Page	
G-IKEA	Cameron 120 Ikea SS balloon	IKEA Ltd	
G-IKEV	Jabiru UL-450	K. J. Bream	
G-IKOS	Cessna 550 Citation Bravo	London Executive Aviation Ltd/Stansted	
G-IKRK	Shaw Europa	K. R. Kesterton	
G-IKRS	Ikarus C.42 FK UK Cyclone	P. G. Walton	
G-IKUS	Ikarus C.42 FB UK Cyclone	C. I. Law	
G-ILDA	V.S.361 Spitfire HF.IX	P. W. Portelli (G-BXHZ)	
G-ILEA	PA-31-310 Navajo C	I. G. Fletcher	
G-ILEE	Colt 56A balloon	G. I. Lindsay	
G-ILES	Cameron O-90 balloon	G. N. Lantos	
G-ILLE	Boeing Stearman A.75L3 (379)	M. Minkler	
G-ILLY	PA-28-181 Archer II	R. A. & G. M. Spiers	
G-ILRS	Ikarus C.42 FB UK Cyclone	Knitsley Mill Leisure Ltd	
G-ILSE	Corby CJ-1 Starlet	S. Stride	
G-ILTS	PA-32 Cherokee Six 300	Foremans Aviation Ltd (G-CVOK)	
G-ILUM	Shaw Europa XS	A. R. Haynes	
G-IMAG	Colt 77A balloon★	Balloon Preservation Group	
G-IMAN	Colt 31A balloon	Benedikt Haggeney GmbH/Germany	
G-IMBI	QAC Quickie 1	J. D. King (G-BWIT)	
G-IMBY	Pietenpol Air Camper	P. F. Bock	
G-IMCD	Van's RV-7	I. G. McDowell	
G-IMGL	Beech B200 Super King Air	IM Aviation Ltd	
G-IMIC	IDA BacauYakovlev Yak-52	C Vogelgesang & R. Hockey	
G-IMLI	Cessna 310Q	Leisure Park Management Ltd (G-AZYK)	
G-IMME	Zenair CH.701 STOL	M. Spearman	
G-IMNY	Escapade 912	D. S. Bremner	
G-IMOK	Hoffmann HK-36R Super Dimona	A. L. Garfield	
G-IMPX	R.Commander 112B	J. C. Stewart	
G-IMPY	Light Aero Avid Flyer C	T. R. C. Griffin	

BRITISH CIVIL REGISTRATIONS

Notes	Reg.	Type	Owner or Operator
	G-INAV	Aviation Composites Mercury	I. Shaw
	G-INCA	Glaser-Dirks DG.400	K. D. Hook
	G-INCE	Skyranger 912(2)	N.P. Sleigh
	G-INDC	Cessna T.303	Crusader Aviation Ltd
	G-INDX	Robinson R-44	Kinetic Computers Ltd
	G-INDY	Robinson R-44	Lincoln Aviation
	G-INGA	Thunder Ax8-84 balloon	M. L. J. Ritchie
	G-INGE	Thruster T.600N	Thruster Air Services Ltd)
	G-INIS	Robinson R-22B	Skyscraper Aviation (G-UPMW)
	G-INNI	Jodel D.112	V. E. Murphy
	G-INNY	SE-5A (replica) (F5459)	K. S. Matcham
	G-INOW	Monnett Moni	W. C. Brown
	G-INSR	Cameron N-90 balloon	The Smith and Pinching Group Ltd & P. Phillips
	G-INTL	Boeing 747-245F (SCD)	AFX Capital Ltd (stored)
	G-INVU	AB-206B JetRanger 2	Burman Aviation Ltd (G-XXII/G-GGCC/G-BEHG)
	G-IOCO	Beech 58 Baron	Arenberg Consultadoria E Servicos LDA/Madeira
	G-IOIA	I.I.I. Sky Arrow 650T	P.J. Lynch, P.G. Ward, N.J.C. Ray
	G-IONA	Aerospatiale ATR-42-300	Highland Airways Ltd
	G-IOOI	Robin DR.400/160	N. B. Mason
	G-IOOX	Learjet 45	Hundred Percent Aviation Ltd
	G-IOPT	Cessna 182P	Indy Oscar Group
	G-IORG	Robinson R-22B	G. M. Richardson (G-ZAND)
	G-IOSI	Jodel DR.1051	Sicile Flying Group
	G-IOSO	Jodel DR.1050	A. E. Jackson
	G-IOWE	Shaw Europa XS	P. A. Lowe
	G-IPAL	Cessna 550 Citation Bravo	Pacific Aviation Ltd
	G-IPAT	Jabiru SP	M. G. Thatcher
	G-IPAX	Cessna 560XL Citation Excel	Pacific Aviation Ltd
	G-IPSI	Grob G.109B	D. G. Margetts (G-BMLO)
	G-IPSY	Rutan Vari-Eze	R. A. Fairclough/Biggin Hill
	G-IPUP	B.121 Pup 2	Skyway Group/Elstree
	G-IRAF	RAF 2000 GTX-SE gyroplane	M. S. R. Allen
	G-IRAN	Cessna 152	I. R. Chaplin
	G-IRIS	AA-5B Tiger	Carlisle Flight Centre (G-BIXU)
	G-IRJX	Avro RJX-100 ★	Manchester Heritage Museum (preserved)
	G-IRKB	PA-28R-201 Arrow III	R. K. Brierley
	G-IRLY	Colt 90A balloon	C. E. R. Smart
	G-IRON	Shaw Europa XS	T. M. Clark
	G-IROW	VPM M.16 Tandem Trainer	B. Jones (G-DBDB)
	G-IRPC	Cessna 182Q	J. W. Halfpenny (G-BSKM)
	G-IRTH	Lindstrand LBL-150A balloon	A. M. Holly (G-BZTO)
	G-ISAX	PA-28-181 Archer III	Airpark Flying Centre Ltd
	G-ISCA	PA-28RT-201 Arrow IV	D. J. & P. Pay
	G-ISDB	PA-28-161 Warrior II	Action Air Services Ltd (G-BWET)
	G-ISDN	Boeing Stearman A.75N1	D. R. L. Jones
	G-ISEH	Cessna 182R	C. M. White & R. MacFarlane (G-BIWS)
	G-ISEL	Best Off Skyranger 912 (2)	P. A. Robertson
	G-ISFC	PA-31-310 Turbo Navajo B	T. M. Latiff (G-BNEF)
	G-ISHA	PA-28-161 Warrior III	Clever Clogs (Middleton) Ltd
	G-ISKA	WSK-PZL Mielec TS-11 Iskra	P. C. Harper
	G-ISLA	BN-2A-26 Islander	Cormack (Aircraft Services) Ltd (G-BNEA)
	G-ISMO	Robinson R-22B	Moy Motorsport Ltd
	G-ISSY	Eurocopter EC.120B	D. R. Williams (G-CBCG)
	G-ISTT	Thunder Ax8-84 balloon	RAF Halton Hot Air Balloon Club
	G-ITII	Aerotech Pitts S-2A Special	Aerobatic Displays Ltd/Booker
	G-ITOI	Cameron N-90 balloon	Flying Pictures Ltd
	G-ITON	Maule MX-7-235	J. R. S. Heaton
	G-ITUG	PA-28 Cherokee 180	S. I. Tugwell (G-AVNR)
	G-ITVM	Lindstrand LBL-105A balloon	N. C. Lindsay
	G-ITWB	D.H.C.1 Chipmunk 22	I. T. Whitaker-Bethel
	G-IUAN	Cessna 525 CitationJet	R. F. Celada SPA/Italy
	G-IVAC	Airtour AH-77B balloon	T. D. Gibbs
	G-IVAN	Shaw TwinEze	A. M. Aldridge
	G-IVAR	Yakovlev Yak-50	A. H. Soper
	G-IVDM	Schempp Hirth Nimbus 4DM	G. W. Lynch
	G-IVEL	Fournier RF-4D	V. S. E. Norman (G-AVNY)/Rendcomb

Reg.	Type	Owner or Operator	Notes
G-IVEN	Robinson R-44-II	OKR Group/Ireland	
G-IVER	Shaw Europa XS	I. Phillips	
G-IVET	Shaw Europa	K. J. Fraser	
G-IVII	Vqn's RV-7	M. A. N. Newall	
G-IVIV	Robinson R-44	Heletrain Ltd	
G-IVOR	Aeronca 11AC Chief	South Western Aeronca Group/Plymouth	
G-IVYS	Parsons 2-seat gyroplane	R. M. Harris	
G-IWON	Cameron V-90 balloon	D. P. P. Jenkinson (G-BTCV)	
G-IWRC	Eurocopter EC.135T-2	McAlpine Helicopters Ltd/Kidlington	
G-IXES	Air Creation Ixess 912	Flylight Airsports Ltd	
G-IXII	Christen Eagle II	Eagle Flying Group (G-BPZI)	
G-IXIX	I.I.I. Sky Arrow 650T	W. J. De Gier	
G-IXTI	Extra EA.300/1	J. C. Merry	
G-IYAK	Yakovlev C-11	G. G. L. James	
G-IYCO	Robin DR.400/500	Timgee Holdings Ltd	
G-IZIT	Rans S.6-116 Coyote II	J. R. Caylow	
G-IZOD	Avtech Jabiru UL	D. A. Izod	
G-IZZI	Cessna T.182T	T. J. & P. S. Nicholson	
G-IZZS	Cessna 172S	Walkbury Aviation Ltd	
G-IZZY	Cesna 172R	R. Parsons (G-BXSF)	
G-IZZZ	Champion 8KCAB	A.M. Read	
G-JABB	Avtech Jabiru UL	D. J. Abbott	
G-JABI	Jabiru Aircraft Jabiru J.400	R. A. Shaw Aviation Ltd	
G-JABJ	Jabiru Aircraft Jabiru J.400	Hamsard 2668 Ltd	
G-JABO	WAR Focke-Wulf Fw.190A-3	S. P. Taylor (replica)	
G-JABS	Avtech Jabiru UL-450	Jabiru Flying Group	
G-JABY	Avtech Jabiru UL	J. T. Grant	
G-JACA	PA-28-161 Warrior II	Channel Islands Aero Club (Jersey) Ltd	
G-JACB	PA-28-181 Archer III	Channel Islands Aero Club (Jersey) Ltd (G-PNNI)	
G-JACC	PA-28-181 Archer III	Channel Islands Aero Club (Jersey) Ltd (G-IMVA/G-GIFT)	
G-JACK	Cessna 421C	JCT 600 Ltd	
G-JACO	Avtech Jabiru UL	R. Hatton	
G-JACS	PA-28-181 Archer III	Vector Air Ltd	
G-JADJ	PA-28-181 Archer III	Aviation Rentals	
G-JAEE	Van's RV-6A	J. A. E. Edser	
G-JAES	Bell 206B JetRanger 3	Heli Charter Wales Ltd (G-STOX/G-BNIR)	
G-JAGS	Cessna FRA.150L	RAF Coltishall Flying Club (G-BAUY)	
G-JAHL	Bell 206B JetRanger 3	Jet Air Helicopters	
G-JAIR	Mainair Blade	J. Loughran	
G-JAJB	AA-5A Cheetah	J. Bradley	
G-JAJK	PA-31-350 Navajo Chieftain	Keen Leasing (IOM) Ltd (G-OLDB/G-DIXI)	
G-JAJP	Avtech Jabiru UL	J. W. E. Pearson & J. Anderson	
G-JAKE	D.H.C.1 Chipmunk 22	K. Ritter (G-BBMY)	
G-JAKI	Mooney M.20R	J. M. Moss & D. M. Abrahamson	
G-JAKS	PA-28 Cherokee 160	K. Harper (G-ARVS)	
G-JALC	Boeing 757-225	My Travel Airways	
G-JAMA	Schweizer 269C-1	JWL Helicopters Ltd	
G-JAMP	PA-28-151 Warrior	Lapwing Flying Group Ltd (G-BRJU)/ White Waltham	
G-JAMY	Shaw Europa XS	J. P. Sharp	
G-JANA	PA-28-181 Archer II	Croaker Aviation/Stapleford	
G-JANB	Colt Flying Bottle SS balloon	Justerini & Brooks Ltd	
G-JANN	PA-34-220T Seneca III	MBC Aviation Ltd/Headcorn	
G-JANO	PA-28RT-201 Arrow IV	Blackpool Aviators Ltd	
G-JANS	Cessna FR.172J	I. G. Aizlewood/Luton	
G-JANT	PA-28-181 Archer II	Janair Aviation Ltd	
G-JAPS	Shaw Europa XS	P. E. Tait	
G-JARA	Robinson R-22B	Northumbria Helicopters Ltd	
G-JARV	AS.355F-1 Twin Squirrel	PLM Dollar Group (G-OGHL)	
G-JASE	PA-28-161 Warrior II	Mid-Anglia School of Flying	
G-JAST	Mooney M.20J - 201	S. J. Tillotson	
G-JATD	Robinson R-22B	Sycamore Ltd (G-HUMF)	
G-JAVO	PA-28-161 Warrior II	Victor Oscar Ltd (G-BSXW)/Wellesbourne	
G-JAWC	Pegasus Quantum 15-912	M. H. Husey	
G-JAWZ	Pitts S-1S Special	A. R. Harding	
G-JAXS	Avtech Jabiru UL	J. P. Pullin	

Notes	Reg.	Type	Owner or Operator
	G-JAYI	J/1 Autocrat	Bravo Aviation Ltd/Coventry
	G-JAYS	Skyranger 912S(1)	J. Williams
	G-JAZZ	AA-5A Cheetah	A. J. Radford
	G-JBAS	Neico Lancair 200	B. A. Slater
	G-JBBZ	AS.350B-3 Ecureuil	Fredat Ltd
	G-JBDB	AB-206B JetRanger	Dicksons Van World Ltd (G-OOPS/G-BNRD)
	G-JBDH	Robin DR.400/180	W. A. Clark
	G-JBEN	Mainair Blade 912	G. J. Bentley
	G-JBHH	Bell 206B JetRanger 2	Biggin Hill Helicopters
			(G-BBFB/G-CJHI/G-CORC/G-SCOO)
	G-JBII	Robinson R-22B	Jet Black 11 Ltd (G-BXLA)
	G-JBJB	Colt 69A balloon	Justerini & Brooks Ltd
	G-JBKA	Robinson R-44	Deeside Helicopters & Promotions
	G-JBMC	SOCATA TB.10 Tobago	J. McCloskey
	G-JBRN	Cessna 182S	Parallel Flooring Accessories Ltd (G-RITZ)
	G-JBSP	Avtech Jabiru SP-470	C. R. James
	G-JCAP	Robinson R-22B	Italian Clothes Ltd
	G-JCAR	PA-46-350P Malibu Mirage	Aquarelle Investments Ltd
	G-JCAS	PA-28-181 Archer II	Charlie Alpha Ltd
	G-JCBA	Sikorsky S-76B	J. C. Bamford Excavators Ltd/E. Midlands
	G-JCBJ	Sikorsky S-76C	J. C. Bamford Excavators Ltd/E. Midlands
	G-JCBV	Gulfstream V	J. C. Bamford Excavators Ltd/E. Midlands
	G-JCBX	Dassault Falcon 900EX	J. C. Bamford Excavators Ltd/E. Midlands
	G-JCKT	Stemme S.10VT	J. C. Taylor
	G-JCMW	Rand-Robinson KR-2	M. Wildish & J. Cook
	G-JCUB	PA-18 Super Cub 135	N. Cummins & S. Bennett
	G-JDBC	PA-34-200T Seneca II	Bowdon Aviation Ltd (G-BDEF)
	G-JDEE	SOCATA TB.20 Trinidad	A. W. Eldridge & J. A. Heard (G-BKLA)
	G-JDEL	Jodel 150	K. F. & R. Richardson (G-JDLI)
	G-JDIX	Mooney M.20B	A. L. Hall-Carpenter (G-ARTB)
	G-JDJM	PA-28 Cherokee 140	Hare Flying Group (G-HSJM/G-AYIF)
	G-JEAD	F.27 Friendship Mk 500	BAC Group Ltd
	G-JEAJ	BAe 146-200	Flybe.com British European (G-OLCA)
	G-JEAK	BAe 146-200	Flybe.com British European (G-OLCB)
	G-JEAM	BAe 146-300	Flybe.com British European/Air France (G-BTJT)
	G-JEAO	BAe 146-100	Trident Aviation Leasing Services Ltd
			(G-UKPC/G-BKXZ) *(stored)*
	G-JEAS	BAe 146-200	Flybe.com British European
			(G-OLHB/G-BSRV/G-OSUN)
	G-JEAU	BAe 146-100	Flybe.com British European/Air France (G-BVUW)
	G-JEAV	BAe 146-200	Flybe.com British European
	G-JEAW	BAe 146-200	Flybe.com.British European
	G-JEAX	BAe 146-200	Flybe.com British European
	G-JEAY	BAe-146-200	Flybe.com British European
	G-JEBA	BAe 146-300	Flybe.com British European (G-BSYR)
	G-JEBB	BAe 146-300	Flybe.com British European/Air France
	G-JEBC	BAe 146-300	Flybe.com British European
	G-JEBD	BAe 146-300	Flybe.com British European
	G-JEBE	BAe 146-300	Flybe.com British European
	G-JEBF	BAe 146-300	Flybe.com British European (G-BTUY/G-NJIC)
	G-JEBG	BAe 146-300	Flybe.com British European (G-BVCE/G-NJIE)
	G-JEBH	BAe 146-300	Flybe.com British European (G-BTVO/G-NJID)
	G-JECE	D.H.C.8Q-402 Dash Eight	Flybe.com British European
	G-JECF	D.H.C.8Q-402 Dash Eight	Flybe.com British European
	G-JECG	D.H.C.8Q-402 Dash Eight	Flybe.com British European
	G-JEDE	D.H.C.8Q-311 Dash Eight	Flybe.com British European
	G-JEDF	D.H.C.8Q-311B Dash Eight	Flybe.com British European
	G-JEDG	D.H.C.8Q-402 Dash Eight	Flybe.com British European
	G-JEDH	Robin DR.400/180	J. B. Hoolahan/Biggin Hill
	G-JEDI	D.H.C.8Q-402 Dash Eight	Flybe.com British European
	G-JEDJ	D.H.C.8Q-402 Dash Eight	Flybe.com British European
	G-JEDK	D.H.C.8Q-402 Dash Eight	Flybe.com British European
	G-JEDL	D.H.C.8Q-402 Dash Eight	Flybe.com British European
	G-JEDM	D.H.C.8Q-402 Dash Eight	Flybe.com British European
	G-JEDN	D.H.C.8Q-402 Dash Eight	Flybe.com British European
	G-JEDO	D.H.C.8Q-402 Dash Eight	Flybe.com British European
	G-JEDP	D.H.C.8Q-402 Dash Eight	Flybe com British European
	G-JEDR	D.H.C.8Q-402 Dash Eight	Flybe com British European
	G-JEDS	Andreasson BA-4B	S. B. Jedburgh (G-BEBT)

Reg.	Type	Owner or Operator	Notes
G-JEDT	D.H.C.8Q-402 Dash Eight	Flybe com British European	
G-JEDU	D.H.C.8Q-402 Dash Eight	Flybe com British European	
G-JEDV	D.H.C.8Q-402 Dash Eight	Flybe.com British European	
G-JEDW	D.H.C.8Q-402 Dash Eight	Flybe.com British European	
G-JEET	Cessna FA.152	Willowair Flying Club (1996) Ltd (G-BHMF)/ Southend	
G-JEFA	Robinson R-44	Simlot Ltd	
G-JEJE	RAF 2000 GTX-SE gyroplane	J. W. Erswell	
G-JEMA	BAe ATP	Emerald Airways Ltd	
G-JEMB	BAe ATP	Emerald Airways Ltd	
G-JEMC	BAe ATP	Emerald Airways Ltd	
G-JEMD	BAe ATP	Emerald Airways Ltd	
G-JEME	BAe ATP	Emerald Airways Ltd	
G-JEMX	Short SD3-60 Variant 100	Emerald Airways Ltd (G-SSWX/G-BNDL)	
G-JEMY	Lindstrand LBL-90A balloon	J. A. Lawton	
G-JENA	Mooney M.20K	P. Leverkuehn/Biggin Hill	
G-JENI	Cessna R.182	R. A. Bentley	
G-JENN	AA-5B Tiger	Shadow Aviation Ltd/Elstree	
G-JENO	Lindstrand LBL-105A balloon	S. F. Redman	
G-JERL	Agusta A.109E Power	Perment Ltd	
G-JERO	Shaw Europa XS	B. Robshaw & P. Jenkinson	
G-JERS	Robinson R-22B	Sloane Helicopters Ltd/Sywel	
G-JESA	Southdown Raven X (modified)	Jesa UK Ltd (G-MNLB)	
G-JESI	AS.350B Ecureuil	Staske Construction Ltd (G-JOSS/G-WILX/ G-RAHM/G-UNIC/G-COLN/G-BHIV)	
G-JESS	PA-28R-201T Turbo Arrow III	R. S. Tomlinson (G-REIS)	
G-JETC	Cessna 550 Citation II	Ability Air Ltd (G-JCFR)	
G-JETG	Learjet 35A	Gama Aviation Ltd (G-JETN/G-JJSG)	
G-JETH	Hawker Sea Hawk FGA.6 ★	P. G. Vallance Ltd/Charlwood	
G-JETI	BAe 125 Srs 800B	Ford Motor Co Ltd/Stansted	
G-JETJ	Cessna 550 Citation II	G-JETJ Ltd (G-EJET/G-DJBE)	
G-JETM	Gloster Meteor T.7 (VZ638) ★	P. G. Vallance Ltd/Charlwood	
G-JETU	AS.355F-2 Twin Squirrel	Marlborough Aviation Ltd	
G-JETX	Bell 206B JetRanger 3	AA Consultants Ltd	
G-JETZ	Hughes 369E	John Matchett Ltd	
G-JEZZ	Skyranger 582	J. W. Barwick & P. J. Harris	
G-JFMK	Zenair CH.701SP	J. D. Pearson	
G-JFRV	Van's RV-7A	J. H. Fisher	
G-JFWI	Cessna F.172N	Staryear Ltd	
G-JGBI	Bell 206L-4 LongRanger	Dorbcrest Homes Ltd	
G-JGMN	C.A.S.A. 1.131E Jungmann 2000	P. D. Scandrett/Staverton	
G-JGSI	Pegasus Quantum 15-912	P. Thomson	
G-JHAC	Cessna FRA.150L	J. H. A. Clarke (G-BACM)	
G-JHEW	Robinson R-22B	Burbage Farms Ltd	
G-JHKP	Shaw Europa XS	J. D. Heykoop	
G-JHNY	Cameron A.210 balloon	Floarting Sensations Ltd	
G-JHYS	Shaw Europa	J. D. Boyce & G. E. Walker	
G-JIII	Stolp SA.300 Starduster Too	VTIO Co/Cumbernauld	
G-JILL	R. Commander 112TCA	D. Carlton	
G-JILY	Robinson R-44	N. J. Ferris	
G-JIMB	B.121 Pup 1	K. D. H. Gray & P. G. Fowler (G-AWWF)	
G-JIMM	Shaw Europa XS	J. Riley	
G-JIVE	MDH Hughes 369E	Sleekform Ltd (G-DRAR)	
G-JJAN	PA-28-181 Archer II	Redhill Flying Club	
G-JJEN	PA-28-181 Archer III	J. E. Jenkins	
G-JJMX	Dassault Falcon 900EX	J-Max Air Services	
G-JJSI	BAe 125 Srs 800B	Gama Aviation Ltd ((G-OMGG)	
G-JJWL	Robinson R-44	Willbeth Ltd	
G-JKMF	Diamond DA.40.D Star	ADR Aviation	
G-JLCA	PA-34-200T Seneca II	C. A. S. Atha (G-BOKE)	
G-JLEE	AB-206B JetRanger 3	J. S. Lee (G-JOKE/G-CSKY/G-TALY)	
G-JLHS	Beech A36 Bonanza	I. G. Meredith	
G-JLLT	Aerotechnik EV-97 Eurostar	J. Latimer	
G-JLMW	Cameron V-77 balloon	J. L. McK. Watkins	

Notes	Reg.	Type	Owner or Operator
	G-JLRW	Beech 76 Duchess	Airways Flight Training
	G-JMAA	Boeing 757-3CQ	Thomas Cook Airlines
	G-JMAB	Boeing 757-3CQ	Thomas Coor Airlines
	G-JMAC	BAe Jetstream 4100	Liverpool John Lennon (G-JAMD/G-JXLI) (preserved)
	G-JMAN	Mainair Blade 912S	J. Manuel
	G-JMAX	Hawker 800XP	J-Max Air Services
	G-JMCD	Boeing 757-25F	Thomas Cook Airlines
	G-JMCE	Boeing 757-25F	Thomas Cook Airlines
	G-JMCF	Boeing 757-28A	Thomas Cook Airlines
	G-JMCG	Boeing 757-2G5	Thomas Cook Airlines UK Ltd
	G-JMDI	Schweizer 269C	J. J. Potter (G-FLAT)
	G-JMDW	Cessna 550 Citation II	MAS Airways Ltd
	G-JMKE	Cessna 172S	115CR (146) Ltd
	G-JMTS	Robin DR.400/180	J. R. Whiting
	G-JMTT	PA-28-201T Turbo Arrow III	N. Cooper (G-BMHM)
	G-JMXA	Agusta A.109E Power	J-Max Air Services
	G-JNAS	AA-5A Cheetah	A. L. Shore & J. R. Nutter
	G-JNET	Robinson R-22B	R. L. Hartshorn
	G-JNNB	Colt 90A balloon	N. A. P. Godfrey
	G-JODI	Agusta A.109A-II	Foxdale Consulting Ltd (G-BVCJ/G-CLRL/G-EJCB)/Isle of Man
	G-JODL	Jodel D.1050/M	D. Silsbury
	G-JOEL	Bensen B.8M	G. C. Young
	G-JOEM	Airbus A.320-231	My Travel Lite (G-OUZO)
	G-JOEY	BN-2A Mk III-2 Trislander	Aurigny Air Services (G-BDGG)/Guernsey
	G-JOJO	Cameron A-210 balloon	A. C. Rawson & J. J. Rudoni
	G-JOLY	Cessna 120	B. V. Meade
	G-JONG	Rotorway Executive 162F	J. V. George
	G-JONH	Robinson R-22B	Productivity Computer Solutions Ltd
	G-JONI	Cessna FA.152	R. F. & J. S. Pooler (G-BFTU)
	G-JONO	Colt 77A balloon	The Sandcliffe Motor Group
	G-JONY	Cyclone AX2000 HKS	K. R. Matheson
	G-JONZ	Cessna 172P	Truman Aviation Ltd/Tollerton
	G-JOOL	Mainair Blade 912	J. R. Gibson
	G-JOON	Cessna 182D	Go Skydive Group
	G-JOSH	Cameron N-105 balloon	M. White
	G-JOST	Shaw Europa	J. A. Austin
	G-JOYT	PA-28-181 Archer II	John K. Cathcart Ltd (G-BOVO)
	G-JOYZ	PA-28-181 Archer III	S. W. & J. E. Taylor
	G-JPAL	AS.355N Twin Squirrel	JPM Ltd
	G-JPAT	Robin HR.200/100	L. Giraidier & A. J. McCulloch (G-BDJN)
	G-JPMA	Avtech Jabiru UL	J. P. Metcalfe
	G-JPOT	PA-32R-301 Saratoga SP	S. W. Turley (G-BIYM)
	G-JPRO	P.84 Jet Provost T.5A (XW433)	Edwalton Aviation Ltd
	G-JPSX	Dassault Falcon 900EX Easy	Sorven Aviation Ltd
	G-JPTT	Enstrom 480	Redwood Aviation Ltd (G-PPAH)
	G-JPTV	P.84 Jet Provost T.5A	S. J. Davies
	G-JPVA	P.84 Jet Provost T.5A (XW289)	T. J. Manna (G-BVXT)/North Weald
	G-JREE	Maule MX-7-180	J. M. P. Ree
	G-JRME	Jodel D.140E	J. E. & L. L. Rex
	G-JRSL	Agusta A.109E Power	Perment Ltd
	G-JSAK	Robinson R-22B-2	Tukair Aircraft Charter
	G-JSAR	AS.332L-2 Super Puma	Bristow Helicopters Ltd
	G-JSAT	BN-2T Turbine Islander	Rhine Army Parachute Association (G-BVFK)/ Germany
	G-JSJX	Airbus A.321-211	My Travel Airways
	G-JSON	Cameron N-105 balloon	Up and Away Ballooning Ltd
	G-JSPC	BN-2T Turbine Islander	Rhine Army Parachute Association (G-BUBG)/ Germany
	G-JSPL	Avtech Jabiru SPL-450	J. A. Lord
	G-JSSD	H.P.137 Jetstream 3001 ★	Museum of Flight/E. Fortune
	G-JTCA	PA-23 Aztec 250E	J. D. Tighe (G-BBCU)/Sturgate
	G-JTEM	Van's RV-7	J. C. Bacon
	G-JTNC	Cessna 500 Citation	Eurojet Aviation Ltd (G-OEJA/G-BWFL)
	G-JTPC	Aeromot AMT-200 Super Ximango	G-JTPC Falcon 3 Group

Reg.	Type	Owner or Operator	Notes
G-JTWO	Piper J-2 Cub	C. C. Silk (G-BPZR)	
G-JTYE	Aeronca 7AC Champion	G. D. Horn	
G-JUDD	Avtech Jabiru UL-450★	C. Judd	
G-JUDE	Robin DR.400/180	Bravo India Flying Group Ltd/Liverpool	
G-JUDI	AT-6D Harvard III (FX301)	A. A. Hodgson	
G-JUDY	AA-5A Cheetah	Plane Talking Ltd/Elstree	
G-JUGE	Aerotechnik EV-97 Teameurostar UK	L. J. Appleby	
G-JUIN	Cessna 303	M. J. & J. M. Newman/Denham	
G-JULL	Stemme S.10VT	J. P. C. Fuchs	
G-JULU	Cameron V-90 balloon	N. J. Appleton	
G-JULZ	Shaw Europa	M. Parkin	
G-JUNG	C.A.S.A. 1.131E Jungmann 1000	(E3B-143)K. H. Wilson/White Waltham	
G-JUPP	PA-32RT-300 Lance II	Jupp Air Ltd (G-BNJF)	
G-JURA	BAe Jetstream 3102	Highland Airways Ltd/Inverness	
G-JURE	SOCATA TB.10 Tobago	P. M. Ireland	
G-JURG	R. Commander 114A	N. A. Southern	
G-JUST	Beech F33A Bonanza	Budge It Aviation Ltd/Elstree	
G-JVBF	Lindstrand LBL-210A balloon	Virgin Balloon Flights	
G-JWBI	AB-206B JetRanger 2	J. W. Bonser (G-RODS/G-NOEL/G-BCWN)	
G-JWCM	SA Bulldog Srs 120/1210	M. L. J. Goff (G-BHXB)	
G-JWDS	Cessna F.150G	G. Sayer (G-AVNB)	
G-JWEB	Robinson R-44	Hields Aviation Ltd	
G-JWFT	Robinson R-22B	J. P. O'Brien	
G-JWIV	Jodel DR.1051	C. M. Fitton	
G-JWJW	C.A.S.A. 1-131E Srs.2000 Jungmann 2000	J. W. & J. T. Whicher	
G-JWXS	Shaw Europa XS T-G	J. Wishart	
G-JYAK	Yakovlev Yak-50	J. W. Stow/North Weald	
G-KAAT	MDH MD-902 Explorer	Police Aviation Services Ltd (G-PASS)	
G-KAEW	Fairey Gannet AEW.3 (XL500)	T. J. Manna/North Weald	
G-KAFE	Cameron N-65 balloon	J. R. Rivers-Scott	
G-KAIR	PA-28-181 Archer II	Keen Leasing (IOM) Ltd	
G-KAMM	Hawker Hurricane XIIA (BW881)	Alpine Deer Group Ltd/New Zealand	
G-KAMP	PA-18 Super Cub 135	J. R. G. Furnell	
G-KAOM	Scheibe SF.25C Falke	Falke G-KAOM Syndicate	
G-KAOS	Van's RV-7	A. E. N. Nicholas, D. F. McGarvey	
G-KAPW	P.56 Provost T.1 (XF603)	The Shuttleworth Collection/O.Warden	
G-KARA	Brügger MB.2 Colibri	C. L. Hill (G-BMUI)	
G-KARI	Fuji FA.200-160	Scottish Civil Service Flying Club (G-BBRE) Ltd	
G-KARK	Dyn'Aero MCR-01 Club	R. Bailes-Brown	
G-KART	PA-28-161 Warrior II	Newcastle-upon-Tyne Aero Club Ltd	
G-KASX	V.S.384 Seafire Mk.XVII	T. J. Manna (G-BRMG)	
G-KATE	Westland WG.30 Srs 100 ★	(stored)/Yeovil	
G-KATI	Rans S.7 Courier	S. M. & K. E. Hall	
G-KATS	PA-28 Cherokee 140	Airlaunch/Old Buckenham (G-BIRC)	
G-KATT	Cessna 152 II	Central Aircraft Leasing Ltd (G-BMTK)	
G-KATZ	Flight Design CT.2K	A. N. D. Arthur	
G-KAUR	Colt 315A balloon	Balloon Safaris	
G-KAWA	Denney Kitfox Mk 2	D. G. Burrows	
G-KAWW	Westland Wasp HAS.1 (XT781/426)	S. J. Davies	
G-KAXF	Hawker Hunter F.6A (XF515)	Kennet Aviation/North Weald	
G-KAXT	Westland Wasp HAS.1 (XT787)	Kennet Aviation/North Weald	
G-KAYH	Extra EA.300/L	Integrated Management Practices Ltd/Netherlands	
G-KBKB	Thunder Ax8-90 S2 balloon	G. Boulden	
G-KBPI	PA-28-161 Warrior II	Goodwood Aerodrome & Motor Circuit Ltd (G-BFSZ)	
G-KCIG	Sportavia RF-5B	Deeside Fournier Group	
G-KDET	PA-28-161 Cadet	Rapidspin Ltd/Biggin Hill	
G-KDEY	Scheibe SF.25E Super Falke	Falke Syndicate	
G-KDIX	Jodel D.9 Bebe	P. M. Bowden	
G-KDLN	Zlin Z.37A-2 Cmelak	J. Richards	
G-KDMA	Cessna 560 Citation V	Gamston Aviation Ltd	
G-KDOG	SA Bulldog Srs 120/121	Gamit Ltd	

Notes	Reg.	Type	Owner or Operator
	G-KEAB	Beech 65-B80 Queen Air ★	Instructional airframe (G-BSSL/G-BFEP)/Shoreham
	G-KEAM	Schleicher ASH 26E	D. T. Reilly
	G-KEEF	Commander Aircraft 112A	K. D. Pearse
	G-KEEN	Stolp SA.300 Starduster Too	Sharp Aerobatics Ltd/Netherlands
	G-KEES	PA-28 Cherokee 180	C. N. Ellerbrook
	G-KEJY	Aerotechnik EV-97 TeamEurostar UK	Kemble Eurostar 1
	G-KELL	Van's RV-6	I. R. Thomas
	G-KELS	Van's RV-7	J. D. Kelsall
	G-KEMC	Grob G.109	Norfolk Gliding Club Ltd
	G-KEMI	PA-28-181 Archer III	K. B. Kempster
	G-KEMY	Cessna 182T	Allen Aircraft Rental Ltd
	G-KENB	Air Command 503 Commander	K. Brogden
	G-KENI	Rotorway Executive	A. J. Wheatley
	G-KENM	Luscombe 8EF Silvaire	M. G. Waters
	G-KENZ	Rutan VariEze	K. M. McConnel I (G-BNUI)
	G-KEPP	Rans S.6-ES Coyote II	S. Munday
	G-KEST	Steen Skybolt	G-KEST Syndicate
	G-KETH	Agusta-Bell 206B JetRanger 2	DAC Leasing Ltd
	G-KEVB	PA-28-181 Archer III	Palmair Ltd
	G-KEVI	Jabiru J.400	K. A. Allen
	G-KEWT	Ultramagic M.90 balloon	Kew Technik Ltd
	G-KEYS	PA-23 Aztec 250F	R. E. Myson
	G-KEYY	Cameron N-77 balloon	B. N. Trowbridge (G-BORZ)
	G-KFAN	Scheibe SF.25B Falke	R. G. & J. A. Boyes
	G-KFOX	Denney Kitfox	I. R. Lawrence
	G-KFRA	PA-32 Cherokee Six 300	West India Flying Group (G-BGII)
	G-KFZI	KFZ-1 Tigerfalck	L. R. Williams
	G-KGAO	Scheibe SF.25C Falke 1700	Falke 2000 Group
	G-KGED	Campbell Cricket Mk.4	K. G. Edwards
	G-KHOM	Aeromot AMT-200 Super Ximango	W. R. Morris
	G-KHRE	M.S.893E Rallye 150SV	D. M. Gale & K. F. Crumplin
	G-KICK	Pegasus Quantum 15-912	G. van der Gaag
	G-KIMB	Robin DR.340/140	R. M. Kimbell
	G-KIMK	Partenavia P.68B	M. Konstantinovic (G-BCPO)
	G-KIMM	Shaw Europa XS	P. A. D. Clarke
	G-KIMY	Robin DR.400/140B	P. W. & K. C. Johnson
	G-KINE	AA-5A Cheetah	Walsh Aviation
	G-KIPP	Thruster T.600 450	R. C. Kelly & S. Munday
	G-KIRK	Piper J-3C-65 Cub	M. J. Kirk
	G-KISS	Rand-Robinson KR-2	E. A. Rooney
	G-KITE	PA-28-181 Archer II	Dateworld Ltd
	G-KITF	Denney Kitfox	P. Smith
	G-KITI	Pitts S-2E Special	B. R. Cornes
	G-KITS	Shaw Europa	J. R. Evernden
	G-KITT	Curtiss P-40M Kittyhawk	Patina Ltd/Duxford
	G-KITY	Denney Kitfox Mk 2	Kitfox KFM Group/Tollerton
	G-KITZ	Shaw Europa XS	Europa Aircraft Ltd/Kirkbymoorside
	G-KIZZ	Kiss 450-582	D. C. P.Cardey & P. David
	G-KKCW	Flight Design CT2K	K. C. Wigley & Co Ltd
	G-KKER	Avtech Jabiru UL-450	W. K. Evans
	G-KKES	SOCATA TB.20 Trinidad	Knightsgate Ltd (G-BTLH)
	G-KMRV	Van-s RV-9A	G. K. Mutch
	G-KNAP	PA-28-161 Warrior II	Keen Leasing (IoM) Ltd (G-BIUX)
	G-KNEE	Ultramagic M-77C balloon	M. A.Green
	G-KNEK	Grob G.109B	Syndicate 109
	G-KNIB	Robinson R-22B-2	C. G. Knibb
	G-KNNY	ATR-42-300	Air Wales Ltd (G-ORFH)
	G-KNOB	Lindstrand LBL-180A balloon	Wye Valley Aviation Ltd
	G-KNOT	P.84 Jet Provost T.3A (XN629)	R. S. Partridge-Hicks (G-BVEG)
	G-KNOW	PA-32 Cherokee Six 300	B. R. & G. E. Mullaly
	G-KNOX	Robinson R-22B	T. A. Knox Shopfitters Ltd
	G-KNYT	Robinson R-44	Aircol
	G-KODA	Cameron O-77 balloon	P. J. Garred & J. E. Nolan
	G-KOFM	Glaser-Dirks DG.600/18M	A. Mossman

Reg.	Type	Owner or Operator	Notes
G-KOHF	Schleicher ASK.14	J. Houlihan	
G-KOKL	Hoffmann H-36 Dimona	R. Smith & R. Stembrowicz	
G-KOLB	Kölb Twinstar Mk 3A	J. L. Moar	
G-KOLI	WSK PZL-110 Koliber 150	J. R. Powell	
G-KONG	Slingsby T.67M Firefly 200	R. C. Morton	
G-KOOL	D.H.104 Devon C.2 (G-DOVE)	D. S. & K. P. Hunt	
G-KORN	Cameron 70 Berentzen SS balloon	Balloon Preservation Flying Group	
G-KOTA	PA-28-236 Dakota	JF Packaging	
G-KPAO	Robinson R-44	Avonair Ltd (G-SSSS)	
G-KPTT	SOCATA TB.20 Trinidad	IAE Ltd	
G-KRES	Stoddard-Hamilton Glasair IIS RG	A. D. Murray	
G-KRII	Rand-Robinson KR-2	M. R. Cleveley	
G-KRIS	Maule M5-235C Lunar Rocket	A. C. Vermeer	
G-KRNW	Eurocopter EC.135T-2	Bond Air Services Ltd/Aberdeen	
G-KSIR	Stoddard-Hamilton Glasair IIS RG	G. T. Grimward	
G-KSKS	Cameron N-54 balloon	Exclusive Ballooning	
G-KSKY	Sky 77-24 balloon	J. R. Howard	
G-KSVB	PA-24 Comanche 260	S. Juggler (G-ENIU/G-AVJU)	
G-KTCC	Schempp-Hirth Ventus 2CM	D. Heslop	
G-KTEE	Cameron V-77 balloon	D. C. & N. P. Bull	
G-KTKT	Sky 260-24 balloon	T. M. Donnelly	
G-KTOL	Robinson R-44	JNK 2000 (G-DCOM)	
G-KUBB	SOCATA TB.20 Trinidad GT	Offshore Marine Consultants Ltd	
G-KUIK	Mainair Pegasus Quik	G. R. Hall & P. R. Brooker	
G-KUKI	Robinson R-22B	C. C. Brook (G-BTNB)	
G-KULA	Skyranger 912ULS	C.R. Mason	
G-KUTU	Quickie Q.2	R. Nash & J. Parkinson	
G-KVBF	Cameron A-340HL balloon	Virgin Balloon Flights	
G-KVIP	Beech 200 Super King Air	Capital Aviation Ltd (G-CBFS/G-PLAT)	
G-KWAK	Scheibe SF.25C	Mendip Gliding Club Ltd	
G-KWAX	Cessna 182E	D. Shaw	
G-KWIC	Mainair Pegasus Quik	T. Southall	
G-KWIK	Partenavia P.68B	ACD Cidra BV/Belgium	
G-KWKI	QAC Quickie Q.200	B. M. Jackson	
G-KWLI	Cessna 421C	Langley Holdings PLC (G-DARR/G-BNEZ)	
G-KYAK	Yakovlev Yak C-11	M. Gainza	
G-KYDD	Robinson R-44 Astro	EK Aviation Ltd	
G-LABS	Shaw Europa	C. T. H. Pattinson	
G-LACA	PA-28-161 Warrior II	LAC (Enterprises) Ltd/Barton	
G-LACB	PA-28-161 Warrior II	LAC (Enterprises) Ltd/Barton	
G-LACD	PA-28-181 Archer III	David Brown Aviation (G-BYBG)	
G-LACE	Shaw Europa	J. H. Phillingham	
G-LACR	Denney Kitfox	C. M. Rose	
G-LADD	Enstrom 480	Combi-Lift Ltd	
G-LADI	PA-30 Twin Comanche 160	S. H. Eastwood (G-ASOO)	
G-LADS	R. Commander 114	D. F. Soul	
G-LAGR	Cameron N-90 balloon	J. R. Clifton	
G-LAIN	Robinson R-22B	Patriot Aviation Ltd/Cranfield	
G-LAIR	Stoddard-Hamilton Glasair IIS	A. I.O'Broin & S. T. Raby	
G-LAKE	Lake LA-250 Renegade	P. J. McGoldrick	
G-LAKI	Jodel DR.1050	V..Panteli (G-JWBB)	
G-LAMA	SA.315B Lama	PLM Dollar Group Ltd	
G-LAMM	Shaw Europa	S. A. Lamb	
G-LAMP	Cameron 110 Lampbulb SS balloon	LE Electrical Ltd	
G-LAMS	Cessna F.152 II	Jaxx Landing Ltd	
G-LANC	Avro 683 Lancaster X (KB889) ★	Imperial War Museum/Duxford	
G-LAND	Robinson R-22B	Heli Air Ltd/Wellesbourne	
G-LANE	Cessna F.172N	G. C. Bantin	
G-LAOK	Yakovlev Yak-52	I. F. Vaughan	
G-LAOL	PA-28RT-201 Arrow IV	G. P. Aviation Ltd	
G-LAOR	Hawker 800XP	Select Plant Hire Co Ltd	
G-LAPN	Light Aero Avid Aerobat	I. A. P. Harper	
G-LARA	Robin DR.400/180	K. D. & C. A. Brackwell	

Notes	Reg.	Type	Owner or Operator
	G-LARE	PA-39 Twin Comanche 160 C/R	Glareways (Neasden) Ltd
	G-LARK	Helton Lark 95	J. Fox
	G-LARS	Dyn'Aero MCR-01 Banbj	L. A. Oyno
	G-LARY	Robinson R-44 II	L. Behan & Sons Ltd (G-CCRZ)/Ireland
	G-LASN	Skyranger J2.2	L. C. F. Lasne
	G-LASR	Stoddard-Hamilton Glasair II	G. Lewis
	G-LASS	Rutan Vari-Eze	J. Mellor
	G-LASU	Eurocopter EC.135T-2	Lancashire Constabulary Air Support Unit
	G-LAVE	Cessna 172R	M. L. Roland (G-BYEV)
	G-LAZA	Lazer Z.200	D. G. Jenkins
	G-LAZL	PA-28-161 Warrior II	Hawk Aero Leasing/Cranfield
	G-LAZR	Cameron O-77 balloon	Laser Holdings (UK) Ltd
	G-LAZY	Lindstrand Armchair SS balloon	The Air Chair Co. Ltd/USA
	G-LAZZ	Stoddard-Hamilton Glastar	A. N. Evans
	G-LBLI	Lindstrand LBL-105A balloon	Lindstrand Balloons Ltd
	G-LBMM	PA-28-161 Warrior II	Flexi-Soft Ltd
	G-LBRC	PA-28RT-201 Arrow IV	D. J. V. Morgan
	G-LBUK	Lindstrand 7BL-77A balloon	Lindstrand Balloons Ltd
	G-LCGL	CLA.7 Swift (replica)	J. M. Greenland
	G-LCOK	Colt 69A balloon	Hot-Air Balloon Co Ltd (G-BLWI)
	G-LCYA	Dassault 900EX	London City Airport Jet Centre Ltd
	G-LDAH	Skyranger 912 (2)	A. S. Haslam & L. Dickinson
	G-LDWS	Jodel D.150	D. H. Wilson Spratt (G-BKSS)
	G-LEAF	Cessna F.406	Atlantic Air Transport Ltd/Coventry
	G-LEAM	PA-28-236 Dakota	C. S. Doherty (G-BHLS)
	G-LEAP	BN-2T Turbine Islander	Army Parachute Association (G-BLND)/ Netheravon
	G-LEAS	Sky 90-24 balloon	C. I. Humphrey
	G-LEAU	Cameron N-31 balloon	P. L. Mossman
	G-LEBE	Shaw Europa	P. Atkinson
	G-LECA	AS.355F-1 Twin Squirrel	S. W. Electricity Board (G-BNBK)/Bristol
	G-LEED	Denney Kitfox Mk 2	M. J. Beding
	G-LEEE	Avtech Jabiru UL-450	J. P. Mimnagh& L. E. G. Fekete
	G-LEEN	Aero Designs Pulsar XP	R. B. Hemsworth (G-BZMP/G-DESI)
	G-LEES	Glaser-Dirks DG.400 (800)	J. Bradley
	G-LEEZ	Bell 206L-1 LongRanger 2	Pennine Helicopters Ltd (G-BPCT)
	G-LEGG	Cessna F.182Q	P. J. Clegg (G-GOOS)/Barton
	G-LEGO	Cameron O-77 balloon	P. M. Traviss
	G-LEIC	Cessna FA.152	Leicestershire Aero Club Ltd
	G-LELE	Lindstrand LBL-31A balloon	L. E. Electrical Ltd
	G-LENA	IDA Bacau Yakovlev Yak-52	Yak-52 Ltd
	G-LENF	Mainair Blade 912S	G. D. Fuller
	G-LENI	AS.355F-1 Twin Squirrel	Grid Aviation Ltd (G-ZFDB/G-BLEV)
	G-LENN	Cameron V-56 balloon	M. Jesper/Germany
	G-LENS	Thunder Ax7-77Z balloon	R. S. Breakwell
	G-LENX	Cessna 172N	M. W, Glencross (G-BMVJ)
	G-LENY	PA-34-220T Seneca III	Air Medical Fleet Ltd
	G-LEOS	Robin DR.400/120	R. J. O. Walker
	G-LESJ	Denney Kitfox Mk 3	P. Whittingham
	G-LESZ	Denney Kitfox Mk 5	L. A. James
	G-LEVI	Aeronca 7AC Champion	G-LEVI Group
	G-LEXI	Cameron N-77 balloon	T. Gilbert
	G-LEXX	Van`s RV-8	A. A. Wordsworth
	G-LEZE	Rutan LongEz	K. G. M. Loyal & ptnrs
	G-LEZZ	Stoddard-Hamilton Glastar	L. A. James (G-BYCR)
	G-LFBW	Robinson R.44	L. Ferdinand (G-ODES)
	G-LFIX	V.S.509 Spitfire T.IX (ML407)	C. S. Grace
	G-LFSA	PA-38-112 Tomahawk	Liverpool Flying School Ltd (G-BSFC)
	G-LFSB	PA-38-112 Tomahawk	Spencer Davies Engineering Ltd (G-BLYC)
	G-LFSC	PA-28 Cherokee 140	P. Dion (G-BGTR)
	G-LFSD	PA-38-112 Tomahawk II	Liverpool Flying School Ltd (G-BNPT)
	G-LFSG	PA-28 Cherokee 180E	Liverpool Flying School Ltd (G-AYAA)
	G-LFSH	PA-38-112 Tomahawk	Liverpool Flying School Ltd (G-BOZM)
	G-LFSI	PA-28 Cherokee 140	P. S. Hoyle & S. Merriman (G-AYKV)/Humberside
	G-LFSJ	PA-28-161 Warrior II	Leeds Flying School Ltd (G-BPHE)
	G-LFSK	PA-28-181 Warrior II	Leeds Flying School Ltd
	G-LFSM	PA-38-112 Tomahawk	Liverpool Flying School Ltd (G-BWNR)

Reg.	Type	Owner or Operator	Notes
G-LFVB	V.S.349 Spitfire LF.Vb (EP120)	Patina Ltd/Duxford	
G-LFVC	V.S.349 Spitfire LF.Vc (JG891)	Historic Flying Ltd	
G-LGCA	Robin DR.400/180R		
G-LGLG	Cameron Z-210 balloon	Flying Circus SRL/Spain	
G-LGNA	SAAB SF.340B	Loganair Ltd/BA	
G-LGNB	SAAB SF.340B	Loganair Ltd/BA	
G-LGNC	SAAB SF.340B	Loganair Ltd/BA	
G-LGND	SAAB SF.340B	Loganair Ltd/BA (G-GNTH)	
G-LGNE	SAAB SF.340B	Loganair Ltd/BA (G-GNTI)	
G-LGNF	SAAB SF.340B	Loganair Ltd/BA (G-GNTJ)	
G-LGNG	SAAB SF.340B	Loganair Ltd/BA	
G-LGNH	SAAB SF.340B	Loganair Ltd/BA	
G-LGTD	Boeing 737-300	British Airways	
G-LGTE	Boeing 737-3Y0	British Airways	
G-LGTF	Boeing 737-382	British Airways	
G-LGTG	Boeing 737-3Q8	British Airways	
G-LGTH	Boeing 737-3Y0	British Airways (G-BNGL)	
G-LGTI	Boeing 737-3Y0	British Airways (G-BNGM)	
G-LGTJ	Boeing 737-300	British Airways	
G-LGTK	Boeing 737-300	British Airways	
G-LGTL	Boeing 737-300	British Airways	
G-LHCA	Robinson R-22B	Rotorcraft Ltd/Redhil	
G-LHCB	Robinson R-22B	London Helicopter Centres Ltd (G-SIVX)	
G-LHEL	AS>355F-2 Twin Squirrel	Lloyd Helicopters Europe Ltd	
G-LIBB	Cameron V-77 balloon	R. J. Mercer	
G-LIBS	Hughes 369HS	R. J. H. Strong	
G-LICK	Cessna 172N II	Sky Back Ltd (G-BNTR)	
G-LIDA	Hoffmann H36 Dimona	Bidford Airfield Ltd	
G-LIDE	PA-31-350 Navajo Chieftain	Keen Leasing (IOM) Ltd	
G-LIFE	Thunder Ax6-56Z balloon	Lakeside Lodge Golf Centre	
G-LILP	Shaw Europa XS	G. L. Jennings	
G-LILY	Bell 206B JetRanger 3	T. S. Brown (G-NTBI)	
G-LIMO	Bell 206L-1 LongRanger	Heliplayer Ltd	
G-LIMP	Cameron C-80 balloon	T. & B. Chamberlain	
G-LINC	Hughes 369HS	Wavendon Social Housing Ltd	
G-LINE	AS.355N Twin Squirrel	National Grid PLC	
G-LINN	Shaw Europa XS	T. Pond	
G-LIOA	Lockheed 10A Electra ★ (NC5171N)	Science Museum/S. Kensington	
G-LION	PA-18 Super Cub 135 (R-167)	JG Jones Haulage Ltd	
G-LIOT	Cameron O-77 balloon	N. D. Eliot	
G-LIPE	Robinson R-22B	Highland Helicopter Leasing Ltd (G-BTXJ)	
G-LIPS	Cameron 90 Lips SS balloon	Reach For The Sky Ltd (G-BZBV)	
G-LISE	Robin DR.400/500	J. Marks	
G-LISO	SIAI Marchetti SM.1019	Castiglioni Daliso/Italy	
G-LITE	R. Commander 112A	TR Air	
G-LITZ	Pitts S-1E Special	P. J. Caruth	
G-LIVH	Piper J-3C-65 Cub (330238)	M. D. Cowburn/Barton	
G-LIVR	Enstrom 480	Soil Tech BV/Netherlands	
G-LIZA	Cessna 340A II	Tayflight Ltd (G-BMDM)	
G-LIZI	PA-28 Cherokee 160	G-LIZI Group (G-ARRP)	
G-LIZY	Westland Lysander III (V9673) ★	G. A. Warner/Duxford	
G-LIZZ	PA-E23 Aztec 250E	T. J. Nathan (G-BBWM)	
G-LJCC	Murphy Rebel	P. H. Hyde	
G-LJET	Learjet 35A	Gama Aviation Ltd	
G-LKET	Cameron 100 Kindernet Dog SS balloon	G. R. J. Luckett	
G-LKTB	PA-28-181 Archer III	Top Cat Aviation Ltd	
G-LLAN	Grob G.109B	J. D. Scott	
G-LLEW	Aeromot AMT-200S Super Ximango	Echo Whiskey Ximango Syndicate	
G-LMAX	Sequoia F.8L Falco	J. Maxwell	
G-LMLV	Dyn'Aero MCR-01	L. & M. La Vecchia	
G-LNAA	MDH MD-902 Explorer	Police Aviation Services Ltd	
G-LNDS	Robinson R-44	MC Air Ltd/Wellesbourne	
G-LNIC	Robinson R-22B	Linic Consultants Ltd	
G-LNTY	AS.355F-1 Twin Squirrel	LNT Aviation Ltd (G-ECOS/G-DORL/G-BPVB))	
G-LNYS	Cessna F.177RG	Heliview Ltd (G-BDCM)	

Notes	Reg.	Type	Owner or Operator
	G-LOAN	Cameron N-77 balloon	P. Lawman
	G-LOBL	Bombardier BD-700-1A10	Global Express 1427 Ltd
	G-LOBO	Cameron O-120 balloon	Solo Aerostatics
	G-LOCH	Piper J-3C-90 Cub	J. M. Greenland
	G-LOFB	L.188CF Electra	Atlantic Airlines Ltd
	G-LOFC	L.188CF Electra	Atlantic Airlines Ltd
	G-LOFD	L.188CF Electra	Atlantic Airlines Ltd
	G-LOFE	L.188CF Electra	Atlantic Airlines Ltd
	G-LOFF	L.188CF Electra	Atlantic Airlines Ltd
	G-LOFH	L.188C Electra	Air Atlantique/Coventry
	G-LOFM	Maule MX-7-180A	Atlantic Air Transport Ltd/Coventry
	G-LOFT	Cessna 500 Citation I	Atlantic Air Transport Ltd/Coventry
	G-LOGO	Hughes 369E	Eastern Atlantic Helicopters Ltd (G-BWLC)
	G-LOIS	Avtech Jabiru UL	D. W. Newman
	G-LOKM	WSK-PZL Koliber 160A	PZL International Aviation Marketing & Sales PLC (G-BYSH)/North Weald
	G-LOLA	Beech A36 Bonanza	J. H. & L. F. Strutt
	G-LOLL	Cameron V-77 balloon	Test Valley Balloon Group
	G-LONE	Bell 206L-1 LongRanger	Patriot Aviation Ltd (G-CDAJ)
	G-LOOP	Pitts S-1C Special	D. Shutter
	G-LOOT	EMB-110P1 Bandeirante	(stored) (G-BNOC)/Southend
	G-LORC	PA-28-161 Cadet	Sherburn Aero Club Ltd
	G-LORD	PA-34-200T Seneca II	Carill Aviation Ltd & R. P. Thomas
	G-LORN	Avions Mudry CAP.10B	J. D. Gailey
	G-LORR	PA-28-181 Archer III	S. J. Sylvester
	G-LORT	Light Aero Avid Speedwing 4	G. E. Laucht
	G-LORY	Thunder Ax4-31Z balloon	A. J. Moore
	G-LOSI	Cameron Z-105 balloon	Aeropubblicita Vicenza SRL/Italy
	G-LOSM	Gloster Meteor NF.11 (WM167)	Aviation Heritage Ltd/Coventry
	G-LOST	Denney Kitfox Mk 3	J. H. S. Booth
	G-LOSY	Aerotechnik EV-97 Eurostar	J. A. Shufflebotham
	G-LOTA	Robinson R-44	Rahtol Ltd
	G-LOTI	Bleriot XI (replica) ★	Brooklands Museum Trust Ltd
	G-LOUN	AS.355N Twin Squirrel	Firstearl Ltd
	G-LOVB	BAe Jetstream 3102	London Flight Centre (Stansted) Ltd (G-BLCB)
	G-LOWS	Sky 77-24 balloon	A. J. Byrne & D. J. Bellinger
	G-LOYA	Cessna FR.172J	T. R. Scorer (G-BLVT)
	G-LOYD	SA.341G Gazelle 1	I. G. Lloyd (G-SFTC)
	G-LPAD	Lindstrand LBL-105A balloon	Line Packaging & Display Ltd
	G-LRBW	Lindstrand HS-110 airship	Croymark Ltd
	G-LRGE	Lindstrand LBL-330A balloon	Adventure Balloons Ltd
	G-LPGI	Cameron A-210 balloon	A. Derbyshire
	G-LRSN	Robinson R-44	Larsen Manufacturing Ltd
	G-LSCM	Cessna 172S	G. A. Luscombe
	G-LSFI	AA-5A Cheetah	G-LSFI Group (G-BGSK)
	G-LSFT	PA-28-161 Warrior II	Biggin Hill Flying Club Ltd (G-BXTX)
	G-LSHI	Colt 77A balloon	J. H. Dobson
	G-LSMI	Cessna F.152	Falcon Flying Services/Biggin Hill
	G-LSTR	Stoddard-Hamilton Glastar	R. Y. Kendal
	G-LTFB	PA-28 Cherokee 140	C. W. J. Cunningham (G-AVLU)
	G-LTFC	PA-28 Cherokee 140B	London Transport Flying Club Ltd (G-AXTI)/Fairoaks
	G-LTRF	Sportavia Fournier RF-7	Skyview Systems Ltd (G-EHAP)
	G-LTSB	Cameron LTSB-90 balloon	Airship & Balloon Co Ltd
	G-LUBE	Cameron N-77 balloon	A. C. K. Rawson
	G-LUCK	Cessna F.150M	Taylor Aviation Ltd/Elstree
	G-LUED	Aero Designs Pulsar	J. C. Anderson
	G-LUFT	Pützer Elster C	A. & E. A. Wiseman (G-BOPY)
	G-LUKE	Rutan LongEz	R. A. Pearson
	G-LUKI	Robinson R-44	Marcella Air Ltd (G-BZLN)
	G-LUKY	Robinson R-44	English Braids Ltd
	G-LULU	Grob G.109	A. P. Bowden
	G-LUMA	Avtech Jabiru SK	B. Luyckx
	G-LUNA	PA-32RT-300T Turbo Lance II	Lance Aviation Ltd
	G-LUND	Cessna 340 II	Prospect Developments (Northern) Ltd (G-LAST/G-UNDY/G-BBNR)
	G-LUNE	Mainair Pegasus Quik	D. Muir

Reg.	Type	Owner or Operator	Notes
G-LUSC	Luscombe 8E Silvaire	M. Fowler	
G-LUSH	PA-28-151 Warrior	A. Jahanfar	
G-LUSI	Luscombe 8F Silvaire	J. P. Hunt & D. M. Robinson	
G-LUST	Luscombe 8E Silvaire	M. Griffiths	
G-LUVY	AS.355F-1 Twin Squirrel	DNH Helicopters Ltd	
G-LUXE	BAe 146-301	BAE Systems (Operations) Ltd (G-SSSH)	
G-LVBF	Lindstrand LBL-330A balloon	Virgin Balloon Flights	
G-LVES	Cessna 182S	R. W. & A. M. Glaves (G-ELIE)	
G-LVLV	Canadair CL.604 Challenger	Gama Aviation Ltd	
G-LVPL	Edge XT912 B/Streak III/B	C. D. Connor	
G-LWAY	Robinson R-44	Glenkerrin Aviation Ltd	
G-LWNG	Aero Designs Pulsar	C. Moffat (G-OMKF)	
G-LWUK	Robinson R-44	Wyberton Developments Ltd	
G-LYAK	IDABacau Yakovlev Yak-52	Lee52 Ltd (LY-AGN)	
G-LYDA	Hoffmann H-36 Dimona	G-LYDA Flying Group/Booker	
G-LYFA	IDABacau Yakovlev Yak-52	Fox Alpha Group	
G-LYNC	Robinson R-22B-2	Traffic Management Services Ltd	
G-LYND	PA-25 Pawnee 235	York Gliding Centre Ltd (G-ASFX/G-BSFZ)/Rufforth	
G-LYNK	CFM Shadow Srs DD	J. Walton	
G-LYNX	Westland WG.13 Lynx (ZB500)	IHM/Weston-s-Mare	
G-LYPG	Avtech Jabiru UL	P. G. Gale	
G-LYTE	Thunder Ax7-77 balloon	G. M. Bulmer	
G-LZZY	PA-28RT-201T Turbo Arrow IV	J. C. Lucas (G-BMHZ)	
G-MAAH	BAC One-Eleven 488GH	Gazelle Ltd (G-BWES)	
G-MAAN	Shaw Europa XS	P. S. Mann	
G-MABE	Cessna F.150L	A. C. Saunders (G-BLJP)/Shobdon	
G-MABH	Fokker 100	Gazelle Ltd	
G-MABR	BAe 146-100	British Airways CitiExpress (G-DEBN)	
G-MACH	SIAI-Marchetti SF.260	Cheyne Motors Ltd/Old Sarum	
G-MACK	PA-28R Cherokee Arrow 200-II	Haimoss Ltd	
G-MAFA	Cessna F.406	Directflight Ltd (G-DFLT)	
G-MAFB	Cessna F.406	Directflight Ltd	
G-MAFE	Dornier Do.228-202K	FR Aviation Ltd (G-OALF/G-MLDO)/Bournemouth	
G-MAFF	BN-2T Turbine Islander	FR Aviation Ltd (G-BJEO)/Bournemouth	
G-MAFI	Dornier Do.228-202K	FR Aviation Ltd/Bournemouth	
G-MAGC	Cameron Grand Illusion SS balloon	Magical Adventures Ltd	
G-MAGG	Pitts S-1SE Special	C. A. Boardman	
G-MAGL	Sky 77-24 balloon	RCM SRL/Luxembourg	
G-MAIE	PA-32RT-301T Turbo Saratoga II TC	B. R. Sennett	
G-MAIK	PA-34-220T Seneca V	TEL (IoM) Ltd	
G-MAIN	Mainair Blade 912	W. Dawson	
G-MAIR	PA-34-200T Seneca II	A. J Warren	
G-MAJA	BAe Jetstream 4102	Eastern Airways	
G-MAJB	BAe Jetstream 4102	Eastern Airways (G-BVKT)	
G-MAJC	BAe Jetstream 4102	Eastern Airways (G-LOGJ)	
G-MAJD	BAe Jetstream 4102	Eastern Airways (G-WAWR)	
G-MAJE	BAe Jetstream 4102	Eastern Airways (G-LOGK)	
G-MAJF	BAe Jetstream 4102	Eastern Airways (G-WAWL)	
G-MAJG	BAe Jetstream 4102	Eastern Airways (G-LOGL)	
G-MAJH	BAe Jetstream 4102	Eastern Airways (G-WAYR)	
G-MAJI	BAe Jetstream 4102	Eastern Airways (G-WAND)	
G-MAJJ	BAe Jetstream 4102	Eastern Airways (G-WAFT)	
G-MAJK	BAe Jetstream 4102	Eastern Airways	
G-MAJL	BAe Jetstream 4102	Eastern Airways	
G-MAJM	BAe Jetstream 4102	Eastern Airways	
G-MAJN	BAe Jetstream 4102	Eastern Airways	
G-MAJR	D.H.C.1 Chipmunk 22 (WP805)	Chipmunk Shareholders	
G-MAJS	Airbus A.300B4-605R	Monarch Airlines Ltd/Luton	
G-MALA	PA-28-181 Archer II	M. & D. Aviation (G-BIIU)	
G-MALC	AA-5 Traveler	B. P. Hogan (G-BCPM)	
G-MALS	Mooney M.20K-231	G-MALS Group/Blackbushe	
G-MALT	Colt Flying Hop SS balloon	P. J. Stapley	
G-MAMC	Rotorway Executive 90	J. R. Carmichael	
G-MAMH	Fokker 100	Gazelle Ltd	
G-MAMK	Robinson R-44	Heli Air Ltd/Wellesbourne	
G-MAMO	Cameron V-77 balloon	The Marble Mosaic Co Ltd	

Notes	Reg.	Type	Owner or Operator
	G-MANA	BAe ATP	Trident Aviation Leasing Services (Jersey) Ltd (G-LOGH)
	G-MANE	BAe ATP	British Airways CitiExpress (G-LOGB)
	G-MANF	BAe ATP	British Airways CitiExpress (G-LOGA)
	G-MANG	BAe ATP	BAe Systems (Operations) Ltd (G-LOGD/G-OLCD)
	G-MANH	BAe ATP	BAe Systems (Operations) Ltd (G-LOGC/G-OLCC)/Isle of Man
	G-MANI	Cameron V-90 balloon	M. P. G. Papworth
	G-MANJ	BAe ATP	Loganair Ltd (G-LOGE/G-BMYL)
	G-MANL	BAe ATP	Loganair Ltd (G-ERIN/G-BMYK)
	G-MANM	BAe ATP	BAE Systems (Operations) Ltd (G-OATP/G-BZWW)
	G-MANN	SA.341G Gazelle 1	N. E. R. Brunt (G-BKLW)
	G-MANO	BAe ATP	British Airways CitiExpress (G-UIET)
	G-MANP	BAe ATP	Loganair Ltd (G-PEEL)
	G-MANS	BAe 146-200	British Airways CitiExpress (G-CLHC/G-CHSR)
	G-MANW	Tri-R Kis	M. T. Manwaring
	G-MANX	FRED Srs 2	S. Styles
	G-MAPL	Robinson R-44	M. P. Lafuente (G-BZPV)/Monaco
	G-MAPP	Cessna 402B	Simmons Mapping (UK) Ltd
	G-MAPR	Beech A36 Bonanza	Moderandum Ltd
	G-MARA	Airbus A.321-231	Monarch Airlines Ltd/Luton
	G-MARE	Schweizer 269C	The Earl of Caledon
	G-MARO	Skyranger J2.2 (2)	E, Daleki
	G-MARX	Van's RV-4	M. W. Albery
	G-MARZ	Thruster T.600N 450	D.P. Tassart
	G-MASC	Jodel 150A	K. F. & R. Richardson
	G-MASF	PA-28-181 Archer II	Mid-Anglia School of Flying
	G-MASH	Westland-Bell 47G-4A	Defence Products Ltd (G-AXKU)
	G-MASS	Cessna 152 II	MK Aero Support Ltd (G-BSHN)
	G-MASX	Masquito M.80	Masquito Aircraft NV/Belgium
	G-MASY	Masquito M.80	Masquito Aircraft NV/Belgium
	G-MASZ	Masquito M.58	Masquito Aircraft NV/Belgium
	G-MATE	Moravan Zlin Z.50LX	J. H. Askew
	G-MATS	Colt GA-42 airship	P. A. Lindstrand
	G-MATT	Robin R.2160	P. White (G-BKRC)
	G-MATY	Robinson R.22B	MT Aviation
	G-MATZ	PA-28 Cherokee 140	Midland Air Training School (G-BASI)/Coventry
	G-MAUD	BAe ATP	Loganair Ltd (G-BMYM)
	G-MAUK	Colt 77A balloon	B. Meeson
	G-MAVI	Robinson R-22B	Northumbria Helicopters Ltd
	G-MAXG	Pitts S-1S Special	Jenks Air Ltd/RAF Halton
	G-MAXI	PA-34-200T Seneca II	Draycott Seneca Syndicate Ltd
	G-MAXV	Van's RV-4	R. S. Partridge-Hicks
	G-MAYB	Robinson R.44	Heli Air Ltd/Wellesbourne
	G-MAYO	PA-28-161 Warrior II	Jermyk Engineering/Fairoaks
	G-MAZY†	D.H.82A Tiger Moth ★	Newark Air Museum
	G-MBAA	Hiway Skytrike Mk 2	-
	G-MBAB	Hovey Whing-Ding II	-
	G-MBAD	Weedhopper JC-24A	-
	G-MBAF	R. J. Swift 3	-
	G-MBAN	American Aerolights Eagle	-
	G-MBAR	Skycraft Scout	-
	G-MBAS	Typhoon Tripacer 250	-
	G-MBAU	Hiway Skytrike	-
	G-MBAW	Pterodactyl Ptraveller	-
	G-MBAZ	Rotec Rally 2B	-
	G-MBBB	Skycraft Scout 2	-
	G-MBBG	Weedhopper JC-24B	-
	G-MBBM	Eipper Quicksilver MX	-
	G-MBBT	Ultrasports Tripacer 330	-
	G-MBCA	Chargus Cyclone T.250	-
	G-MBCJ	Mainair Sports Tri-Flyer	-
	G-MBCK	Eipper Quicksilver MX	-
	G-MBCL	Sky-Trike/Typhoon	P. J. Callis
	G-MBCM	Hiway Demon 175	-
	G-MBCO	Flexiform Sealander Buggy	-
	G-MBCU	American Aerolights Eagle	-
	G-MBCX	Airwave Nimrod 165	-
	G-MBCZ	Chargus Skytrike 160	-

Reg.	Type	Owner or Operator	Notes
G-MBDE	Flexiform Skytrike	-	
G-MBDF	Rotec Rally 2B	-	
G-MBDG	Eurowing Goldwing	-	
G-MBDM	Southdown Sigma Trike	-	
G-MBDZ	Eipper Quicksilver MX	-	
G-MBEA	Hornet Nimrod	-	
G-MBED	Chargus Titan 38	-	
G-MBEG	Eipper Quicksilver MX	-	
G-MBEJ	Electraflyer Eagle	-	
G-MBEN	Eipper Quicksilver MX	-	
G-MBEP	American Aerolights Eagle	-	
G-MBES	Skyhook Cutlass	-	
G-MBET	MEA Mistral Trainer	-	
G-MBEU	Hiway Demon T.250	-	
G-MBFA	Hiway Skytrike 250	-	
G-MBFE	American Aerolights Eagle	-	
G-MBFF	Southern Aerosports Scorpion	-	
G-MBFK	Hiway Demon	-	
G-MBFM	Hiway Hang Glider	-	
G-MBFU	Ultrasports Tripacer	-	
G-MBFY	Mirage II	-	
G-MBGA	Solo Sealander	-	
G-MBGF	Twamley Trike	-	
G-MBGJ	Hiway Skytrike Mk 2	-	
G-MBGK	Electra Flyer Eagle	-	
G-MBGS	Rotec Rally 2B	-	
G-MBGX	Southdown Lightning	-	
G-MBGY	Hiway Demon Skytrike	-	
G-MBHA	Trident Trike	-	
G-MBHE	American Aerolights Eagle	-	
G-MBHK	Flexiform Skytrike	-	
G-MBHP	American Aerolights Eagle II	-	
G-MBHT	Chargus T.250	-	
G-MBHZ	Pterodactyl Ptraveller	-	
G-MBIA	Flexiform Sealander Skytrike	-	
G-MBIO	American Aerolights Eagle Z Drive	-	
G-MBIT	Hiway Demon Skytrike	-	
G-MBIV	Flexiform Skytrike	-	
G-MBIW	Hiway Demon Tri-Flyer Skytrike	-	
G-MBIY	Ultrasports Tripacer	-	
G-MBIZ	Mainair Tri-Flyer	-	
G-MBJD	American Aerolights Eagle	-	
G-MBJE	Airwave Nimrod	-	
G-MBJF	Hiway Skytrike Mk II	-	
G-MBJG	Airwave Nimrod	-	
G-MBJI	Southern Aerosports Scorpion	-	
G-MBJK	American Aerolights Eagle	-	
G-MBJL	Airwave Nimrod	-	
G-MBJM	Striplin Lone Ranger	-	
G-MBJP	Hiway Skytrike	-	
G-MBJR	American Aerolights Eagle	-	
G-MBJT	Hiway Skytrike II	-	
G-MBJU	American Eagle 215B	-	
G-MBJZ	Eurowing Catto CP.16	-	
G-MBKS	Hiway Skytrike 160	-	
G-MBKT	Mitchell Wing B.10	-	
G-MBKU	Hiway Demon Skytrike	-	
G-MBKY	American Aerolight Eagle	-	
G-MBKZ	Hiway Skytrike	-	
G-MBLB	Eipper Quicksilver MX	-	
G-MBLF	Hiway Demon 195 Tri Pacer	-	
G-MBLJ	Eipper Quicksilver MX	-	
G-MBLN	MEA Pterodactyl Ptraveller 4300	-	
G-MBLU	Southdown Lightning L.195	-	
G-MBLV	Ultrasports Hybrid	-	
G-MBLY	Flexiform Sealander Trike	-	
G-MBLZ	Southern Aerosports Scorpion	-	
G-MBME	American Aerolights Eagle Z Drive	-	
G-MBMG	Rotec Rally 2B	-	
G-MBMJ	Mainair Tri-Flyer	-	
G-MBMO	Hiway Skytrike 160	-	
G-MBMR	Ultrasports Tripacer Typhoon	-	

Notes	Reg.	Type	Owner or Operator
	G-MBMS	Hornet	-
	G-MBMU	Eurowing Goldwing	-
	G-MBNA	American Aerolights Eagle	-
	G-MBNH	Southern Airsports Scorpion	-
	G-MBNJ	Eipper Quicksilver MX	-
	G-MBNK	American Aerolights Eagle	-
	G-MBNN	Southern Microlight Gazelle P.160N	-
	G-MBOA	Flexiform Hilander	-
	G-MBOE	Solar Wing Typhoon Trike	-
	G-MBOF	Pakes Jackdaw	-
	G-MBOH	Microlight Engineering Mistral	-
	G-MBOK	Dunstable Microlight	-
	G-MBOM	Hiway Hilander	-
	G-MBOR	Chotia 460B Weedhopper	-
	G-MBOT	Hiway 250 Skytrike	-
	G-MBOX	American Aerolights Eagle	-
	G-MBPA	Weedhopper Srs 2	-
	G-MBPD	American Aerolights Eagle	-
	G-MBPG	Hunt Skytrike	-
	G-MBPJ	Moto-Delta	-
	G-MBPN	American Aerolights Eagle	-
	G-MBPO	Volnik Arrow	-
	G-MBPX	Eurowing Goldwing	-
	G-MBPY	Ultrasports Tripacer 330	-
	G-MBRB	Electraflyer Eagle 1	-
	G-MBRD	American Aerolights Eagle	-
	G-MBRH	Ultraflight Mirage Mk II	-
	G-MBRM	Hiway Demon	-
	G-MBRS	American Aerolights Eagle	-
	G-MBRV	Eurowing Goldwing	-
	G-MBSD	Southdown Puma DS	-
	G-MBSN	American Aerolights Eagle	-
	G-MBSS	Ultrasports Puma 2	-
	G-MBST	Mainair Gemini Sprint	-
	G-MBSX	Ultraflight Mirage II	-
	G-MBTA	UAS Storm Buggy 5 Mk 2	-
	G-MBTF	Mainair Tri-Flyer Skytrike	-
	G-MBTH	Whittaker MW.4	-
	G-MBTI	Hovey Whing Ding	-
	G-MBTJ	Solar Wings Microlight	-
	G-MBTO	Mainair Tri-Flyer 250	-
	G-MBTW	Raven Vector 600	-
	G-MBUA	Hiway Demon	-
	G-MBUB	Horne Sigma Skytrike	-
	G-MBUC	Huntair Pathfinder	-
	G-MBUH	Hiway Skytrike	-
	G-MBUI	Wheeler Scout Mk I	-
	G-MBUO	Southern Aerosports Scorpion	-
	G-MBUP	Hiway Skytrike	-
	G-MBUZ	Wheeler Scout Mk II	-
	G-MBVA	Volmer Jensen VJ-23E	-
	G-MBVC	American Aerolights Eagle	-
	G-MBVK	Ultraflight Mirage II	-
	G-MBVL	Southern Aerosports Scorpion	-
	G-MBVW	Skyhook Cutlass TR2	-
	G-MBWA	American Aerolights Eagle	-
	G-MBWB	Hiway Skytrike	-
	G-MBWG	Huntair Pathfinder	-
	G-MBWH	Designability Duet I	-
	G-MBWL	Huntair Pathfinder	-
	G-MBWP	Ultrasports Trike	-
	G-MBWT	Huntair Pathfinder	-
	G-MBWW	Southern Aerosports Scorpion	-
	G-MBWX	Southern Aerosports Scorpion	-
	G-MBXK	Ultrasports Puma	-
	G-MBXO	Sheffield Trident	-
	G-MBXR	Hiway Skytrike 150	-
	G-MBXT	Eipper Quicksilver MX2	-
	G-MBXX	Ultraflight Mirage II	-
	G-MBYD	American Aerolights Eagle	-
	G-MBYI	Ultraflight Lazair	-
	G-MBYL	Huntair Pathfinder 330	-

Reg.	Type	Owner or Operator	Notes
G-MBYM	Eipper Quicksilver MX	-	
G-MBYY	Southern Aerosports Scorpion	-	
G-MBZB	Hiway Skytrike	-	
G-MBZF	American Aerolights Eagle	-	
G-MBZG	Twinflight Scorpion 2 seat	-	
G-MBZH	Eurowing Goldwing	-	
G-MBZJ	Southdown Puma	-	
G-MBZK	Tri-Pacer 250	-	
G-MBZM	UAS Storm Buggy	-	
G-MBZO	Tri-Pacer 330	-	
G-MBZP	Skyhook TR2	-	
G-MBZV	American Aerolights Eagle	-	
G-MBZZ	Southern Aerosports Scorpion	-	
G-MCAI	Robinson R-44-II	Heli Air Ltd/Wellesbourne	
G-MCAP	Cameron C-80 balloon	L. D. Thurgar	
G-MCCF	Thruster T.600N	C. C. F. Fuller	
G-MCCY	IDA Bacau Yakolev Yak-52	D. P. McCoy/Ireland	
G-MCEA	Boeing 757-225	My Travel Airways	
G-MCEL	Pegasus Quantum 15-912	F. Hodgson	
G-MCJL	Pegasus Quantum 15-912	M. C. J. Ludlow	
G-MCMS	Aero Designs Pulsar	B. R. Hunter	
G-MCOX	Fuji FA.200-180AO	W. Surrey Engineering (Shepperton) Ltd	
G-MCOY	Flight Design CT.2K	Pegasus Flight Training (Cotswolds)	
G-MCPI	Bell 206B JetRanger 3	D. A. C. Pipe (G-ONTB)	
G-MCXV	Colomban MC.15 Cri-Cri	H. A. Leek	
G-MDAY	Cessna 170B	M. Day	
G-MDBC	Pegasus Quantum 15-912	D. B. Caiden	
G-MDBD	Airbus A.330-243	My Travel Airways	
G-MDCA	PA-34-220 Seneca V	MDC-Aviation LLP (G-OGOG/G-TILL)	
G-MDJN	Beech 95-B55 Baron	D. J. Nock (G-SUZI/G-BAXR)	
G-MDKD	Robinson R-22B	Brian Seedle Helicopters/Blackpool	
G-MDPI	Agusta A.109A-II	Castle Air Charters Ltd (G-PERI/G-EXEK/ G-SLNE/G-EEVS/G-OTSL))	
G-MEAH	PA-28R Cherokee Arrow 200-II	Stapleford Flying Club Ltd (G-BSNM)	
G-MEDA	Airbus A.320-231	B/Med - BA	
G-MEDB	Airbus A.320-231	B/Med - BA	
G-MEDE	Airbus A.320-232	B/Med - BA	
G-MEDF	Airbus A.321-231	B/Med - BA	
G-MEDG	Airbus A.321-231	B/Med - BA	
G-MEDH	Airbus A.320-232	B/Med - BA	
G-MEDI	Airbus A.320-231	B/Med - BA	
G-MEDJ	Airbus A.321-232	B/Med - BA	
G-	Airbus A.321-232	B/Med - BA	
G-MEGA	PA-28R-201T Turbo Arrow III	A. W. Bean	
G-MEGG	Shaw Europa XS	M. E. Mavers	
G-MELT	Cessna F.172H	J. J. Haycock & J. G. Baggott (G-AWTI)	
G-MELV	SOCATA Rallye 235E	J. W. Busby (G-BIND)	
G-MEME	PA-28R-201 Arrow III	Henry J. Clare Ltd	
G-MEOW	CFM Streak Shadow	G. J. Moor	
G-MERC	Colt 56A balloon	A. F. & C. D. Selby	
G-MERE	Lindstrand LBL-77A balloon	R. D. Baker	
G-MERF	Grob G.115A	G-MERF Group	
G-MERI	PA-28-181 Archer II	A. H. McVicar	
G-MERL	PA-28RT-201 Arrow IV	M. Giles/Cardiff-Wales	
G-MESS	SNCAN Nord 1101 Noralpha	G. R. Lacey	
G-MEUP	Cameron A-120 balloon	Innovation Ballooning Ltd	
G-MFAC	Cessna F.172H	Springbank Aviation Ltd (G-AVGZ)	
G-MFEF	Cessna FR.172J	M. & E. N. Ford	
G-MFHI	Shaw Europa	Hi Fliers	
G-MFHT	Robinson R-22B-2	MFH Ltd	
G-MFLI	Cameron V-90 balloon	J. M. Percival	
G-MFLY	Mainair Rapier	J. J. Tierney	
G-MFMF	Bell 206B JetRanger 3	S.W. Electricity Board (G-BJNJ)/Bristol	
G-MFMM	Scheibe SF.25C Falke	J. E. Selman	
G-MGAA	BFC Challenger II	-	
G-MGAG	Aviasud Mistral	-	
G-MGAN	Robinson R-44	Carnbeg Golf Ltd	
G-MGCA	Jabiru UL	-	

Notes	Reg.	Type	Owner or Operator
	G-MGCB	Pegasus XL-Q	-
	G-MGCK	Whittaker MW-6 Merlin	-
	G-MGDL	Pegasus Quantum 15	-
	G-MGEC	Rans S.6-ESD Coyote IIXL	-
	G-MGEF	Pegasus Quantum 15	-
	G-MGFK	Pegasus Quantum 15	-
	G-MGGG	Pegasus Quantum 15	-
	G-MGGT	CFM Streak Shadow SAM	-
	G-MGGV	Pegasus Quantum 15-912	-
	G-MGMC	Pegasus Quantum 15	-
	G-MGMM	PA-18 Super Cub 150	M. J. Martin
	G-MGND	Rans S.6-ESD Coyote IIXL	-
	G-MGOD	Medway Raven	-
	G-MGOO	Renegade Spirit UK Ltd	-
	G-MGPD	Pegasus XL-R	A. Armsby
	G-MGPH	CFM Streak Shadow	- (G-RSPH)
	G-MGRW	Cyclone AX3/503	-
	G-MGTG	Pegasus Quantum 15	- (G-MZIO)
	G-MGTR	Hunt Wing	-
	G-MGTW	CFM Shadow Srs DD	-
	G-MGUN	Cyclone AX2000	-
	G-MGUX	Hunt Wing	-
	G-MGUY	CFM Shadow Srs BD	-
	G-MGWH	Thruster T.300	-
	G-MGWI	Robinson R-44	N. Currie (G-BZEF)
	G-MHCB	Enstrom 280C	Springbank Aviation Ltd
	G-MHCD	Enstrom 280C-UK	S. J. Ellis (G-SHGG)
	G-MHCE	Enstrom F-28A	Wyke CommercialServices (G-BBHD)
	G-MHCF	Enstrom 280C-UK	HKC Helicopter Services (G-GSML/G-BNNV)
	G-MHCG	Enstrom 280C-UK	E. Drinkwater (G-HAYN/G-BPOX)
	G-MHCI	Enstrom 280C	Charlie India Helicopters Ltd/Barton
	G-MHCJ	Enstrom F-28C-UK	Paradise Helicopters (G-CTRN)
	G-MHCK	Enstrom 280FX	Manchester Helicopter Centre Ltd (G-BXXB)
	G-MHCL	Enstrom 280C	J. A. Newton
	G-MHGS	Stoddard-Hamilton Glastar	M. Henderson
	G-MHRV	Van's RV-6A	M. R. Harris
	G-MICH	Robinson R-22B	Tiger Helicopters Ltd (G-BNKY)/Shobdon
	G-MICI	Cessna 182S	M. J. Coleman (G-WARF)
	G-MICK	Cessna F.172N	G-MICK Flying Group
	G-MICY	Everett Srs 1 gyroplane	D. M. Hughes
	G-MIDA	Airbus A.321-231	bmi british midland
	G-MIDC	Airbus A.321-231	bmi british midland
	G-MIDD	PA-28 Cherokee 140	Midland Air Training School (G-BBDD)/Coventry
	G-MIDE	Airbus A.321-231	bmi british midland
	G-MIDF	Airbus A.321-231	bmi british midland
	G-MIDG	Midget Mustang	C. E. Bellhouse
	G-MIDH	Airbus A.321-231	bmi british midland
	G-MIDI	Airbus A.321-231	bmi british midland
	G-MIDJ	Airbus A.321-231	bmi british midland
	G-MIDK	Airbus A.321-231	bmi british midland
	G-MIDL	Airbus A.321-231	bmi british midland
	G-MIDM	Airbus A.321-231	bmi british midland
	G-MIDN	Airbus A.321-231	bmi british midland
	G-MIDO	Airbus A.321-231	bmi british midland
	G-MIDP	Airbus A.320-232	bmi british midland
	G-MIDR	Airbus A.320-232	bmi british midland
	G-MIDS	Airbus A.320-232	bmi british midland
	G-MIDT	Airbus A.320-232	bmi british midland
	G-MIDU	Airbus A.320-232	bmi british midland
	G-MIDV	Airbus A.320-232	bmi british midland
	G-MIDW	Airbus A.320-232	bmi british midland
	G-MIDX	Airbus A.320-232	bmi british midland
	G-MIDY	Airbus A.320-232	bmi british midland
	G-MIDZ	Airbus A.320-232	bmi british midland
	G-MIFF	Robin DR.400/180	G. E. Snushall
	G-MIGG	WSK-Mielec LiM-5 (1211) ★	D. Miles (G-BWUF)
	G-MIII	Extra EA.300/L	R. C. Berger
	G-MIKE	Brookland Hornet	M. H. J. Goldring
	G-MIKG	Robinson R-22 Mariner	Sloane Helicopters Ltd
	G-MIKI	Rans S.6-ESA Coyote II	S. P. Slade

Reg.	Type	Owner or Operator	Notes
G-MIKS	Robinson R-44	Direct Timber Ltd	
G-MILA	Cessna F.172N	P. J. Miller	
G-MILE	Cameron N-77 balloon	Miles Air Ltd	
G-MILI	Bell 206B JetRanger 3	Shropshire Aviation Ltd	
G-MILN	Cessna 182Q	Meon Hill Farms (Stockbridge) Ltd	
G-MILY	AA-5A Cheetah	Plane Talking Ltd (G-BFXY)/Elstree	
G-MIMA	BAe 146-200	British Airways CitiExpress (G-CNMF)	
G-MIME	Shaw Europa	N. W. Charles	
G-MIND	Cessna 404	Atlantic Air Transport Ltd (G-SKKC/G-OHUB)/ Coventry	
G-MINN	Lindstrand LBL-90A balloon	S. M. & D. Johnson	
G-MINS	Nicollier HN.700 Menestrel II	R. Fenion	
G-MINT	Pitts S-1S Special	T. G. Sanderson/Tollerton	
G-MIOO	M.100 Student ★	Museum of Berkshire Aviation (G-APLK)/Woodley	
G-MISH	Cessna 182R	L. H. Robinson (G-RFAB/G-BIXT)	
G-MISS	Taylor JT.2 Titch	P. L. A. Brenen	
G-MITE	Raj Hamsa X'Air Falcon	T. Jestico	
G-MITS	Cameron N-77 balloon	Colt Car Co Ltd	
G-MITT	Avtech Jabiru SK	N. C. Mitton	
G-MITZ	Cameron N-77 balloon	Colt Car Co Ltd	
G-MIWS	Cessna 310R II	Wilcott Sport and Construction Ltd (G-ODNP)	
G-MJAB	Ultrasports Skytrike	-	
G-MJAD	Eipper Quicksilver MX	-	
G-MJAE	American Aerolights Eagle	-	
G-MJAG	Skyhook TR1	-	
G-MJAH	American Aerolights Eagle	-	
G-MJAI	American Aerolights Eagle	-	
G-MJAJ	Eurowing Goldwing	-	
G-MJAM	Eipper Quicklsilver MX	-	
G-MJAN	Hiway Skytrike	-	
G-MJAY	Eurowing Goldwing	-	
G-MJAZ	Aerodyne Vector 610	-	
G-MJBK	Swallow B	-	
G-MJBL	American Aerolights Eagle	-	
G-MJBS	Ultralight Stormbuggy	-	
G-MJBV	American Aerolights Eagle	-	
G-MJBZ	Huntair Pathfinder	-	
G-MJCB	Hornet 330	-	
G-MJCD	Sigma Tetley Skytrike	-	
G-MJCE	Ultrasports Tripacer	-	
G-MJCI	Kruchek Firefly 440	-	
G-MJCJ	Hiway Spectrum	-	
G-MJCK	Southern Aerosports Scorpion	-	
G-MJCU	Tarjani	-	
G-MJCW	Hiway Super Scorpion	-	
G-MJCZ	Southern Aerosports Scorpion 2	-	
G-MJDE	Huntair Pathfinder	-	
G-MJDG	Hornet Supertrike	-	
G-MJDH	Huntair Pathfinder	-	
G-MJDJ	Hiway Skytrike Demon	-	
G-MJDK	American Aerolights Eagle	-	
G-MJDO	Southdown Puma 440	-	
G-MJDP	Eurowing Goldwing	-	
G-MJDR	Hiway Demon Skytrike	-	
G-MJDU	Eipper Quicksilver àMX2	-	
G-MJDW	Eipper Quicksilver MX	-	
G-MJEE	Mainair Triflyer Trike	-	
G-MJEF	Gryphon 180	-	
G-MJEJ	American Aerolights Eagle	-	
G-MJEL	GMD-01 Trike	-	
G-MJEO	American Aerolights Eagle	-	
G-MJER	Flexiform Striker	-	
G-MJET	Stratos Prototype 3 Axis 1	-	
G-MJEX	Eipper Quicksilver MX	-	
G-MJFB	Flexiform Striker	-	
G-MJFD	Ultrasports Tripacer	-	
G-MJFJ	Hiway Skytrike 250	-	
G-MJFM	Huntair Pathfinder	-	
G-MJFO	Eipper Quicksilver MX	-	
G-MJFV	Ultrasports Tripacer	-	
G-MJFX	Skyhook TR-1	-	

Notes	Reg.	Type	Owner or Operator
	G-MJFZ	Hiway Sky-Trike	A. W. Lowrie
	G-MJGI	Eipper Quicksilver MX	-
	G-MJGN	Greenslade Monotrike	-
	G-MJGT	Skyhook Cutlass Trike	-
	G-MJGV	Eipper Quicksilver MX2	-
	G-MJGW	Solar Wings TrikeB	-
	G-MJHC	Ultrasports Tripacer 330	-
	G-MJHF	Skyhook Sailwing Trike	-
	G-MJHK	Hiway Demon 195	-
	G-MJHM	Ultrasports Trike	-
	G-MJHN	American Aerolights Eagle	-
	G-MJHR	Southdown Lightning	-
	G-MJHU	Eipper Quicksilver MX	-
	G-MJHV	Hiway Demon 250	-
	G-MJHW	Ultrasports Puma 1	-
	G-MJHX	Eipper Quicksilver MX	-
	G-MJIA	Flexiform Striker	-
	G-MJIC	Ultrasports Puma 330	-
	G-MJIF	Mainair Triflyer	-
	G-MJIJ	Ultrasports Tripacer 250	-
	G-MJIK	Southdown Sailwings Lightning	-
	G-MJIN	Hiway Skytrike	-
	G-MJIR	Eipper Quicksilver MX	-
	G-MJIY	Striker/Panther	-
	G-MJIZ	Southdown Lightning	-
	G-MJJA	Huntair Pathfinder	-
	G-MJJB	Eipper Quicksilver MX	-
	G-MJJF	Solar Wings Typhoon	-
	G-MJJJ	Moyes Knight	-
	G-MJJK	Eipper Quicksilver MX2	J. A. Brumpton
	G-MJJM	Birdman Cherokee Mk 1	-
	G-MJJO	Flexiform Skytrike Dual	-
	G-MJJV	Wheeler Scoutá	-
	G-MJJX	Hiway Skytrike	-
	G-MJJY	Tirith Firefly	-
	G-MJKB	Striplin Skyranger	-
	G-MJKE	Mainair Triflyer 330	-
	G-MJKF	Hiway Demon	-
	G-MJKG	John Ivor Skytrike	-
	G-MJKJ	Eipper Quicksilver MX	-
	G-MJKO	Goldmarque 250 Skytrike	-
	G-MJKS	Mainair Triflyer	-
	G-MJKV	Hornet	-
	G-MJKX	Ultralight Skyrider Phantom	-
	G-MJLB	Ultrasports Puma 2	-
	G-MJLH	American Aerolights Eagle 2	-
	G-MJLI	Hiway Demon Skytrike	-
	G-MJLL	Hiway Demon Skytrike	-
	G-MJLR	Skyhook SK-1	-
	G-MJLS	Rotec Rally 2B	-
	G-MJLT	American Aerolights Eagle	-
	G-MJME	Ultrasports Tripacer Mega II	-
	G-MJMM	Chargus Vortex	-
	G-MJMP	Eipper Quicksilver MX	-
	G-MJMR	Solar Wings Typhoon	-
	G-MJMS	Hiway Skytrike	-
	G-MJMU	Hiway Demon	-
	G-MJMW	Eipper Quicksilver MX2	-
	G-MJNB	Hiway Skytrike	-
	G-MJNE	Hornet Supreme Dual Trike	-
	G-MJNK	Hiway Skytrike	-
	G-MJNL	American Aerolights Eagle	-
	G-MJNM	American Aerolights Double Eagle	-
	G-MJNN	Ultraflight Mirage II	-
	G-MJNO	American Aerolights Double Eagle	-
	G-MJNR	Ultralight Solar Buggy	-
	G-MJNT	Hiway Skytrike	-
	G-MJNU	Skyhook Cutlass	-
	G-MJNY	Skyhook Sabre Trike	-
	G-MJOC	Huntair Pathfinder	-
	G-MJOE	Eurowing Goldwing	-
	G-MJOG	American Aerolights Eagle	-

Reg.	Type	Owner or Operator	Notes
G-MJOI	Hiway Demon	-	
G-MJOJ	Flexiform Skytrike	-	
G-MJOL	Skyhook Cutlass	-	
G-MJOM	Southdown Puma 40F	-	
G-MJOW	Eipper Quicksilver MX	-	
G-MJPA	Rotec Rally 2B	-	
G-MJPE	Hiway Demon Skytrike	-	
G-MJPG	American Aerolights Eagle 430R	-	
G-MJPI	Flexiform Striker	-	
G-MJPK	Hiway Vulcan	-	
G-MJPO	Eurowing Goldwing	-	
G-MJPT	Dragon	-	
G-MJPV	Eipper Quicksilver MX	-	
G-MJRE	Hiway Demon	-	
G-MJRI	American Aerolights Eagle	-	
G-MJRK	Flexiform Striker	-	
G-MJRL	Eurowing Goldwing	-	
G-MJRN	Flexiform Striker	-	
G-MJRO	Eurowing Goldwing	-	
G-MJRR	Striplin Skyranger Srs 1	-	
G-MJRS	Eurowing Goldwing	R. M. Newlands	
G-MJRU	MBA Tiger Cub 440	-	
G-MJRX	Ultrasports Puma II	-	
G-MJSA	Mainair 2-Seat Trike	-	
G-MJSE	Skyrider Airsports Phantom	K. H. A. Negal	
G-MJSF	Skyrider Airsports Phantom	-	
G-MJSL	Dragon 200	-	
G-MJSO	Hiway Skytrike	-	
G-MJSP	Romain Tiger Cub 440	-	
G-MJSS	American Aerolights Eagle	-	
G-MJST	Pterodactyl Ptraveller	-	
G-MJSV	MBA Tiger Cub	-	
G-MJSY	Eurowing Goldwing	-	
G-MJSZ	DH Wasp	-	
G-MJTC	Ultrasports Tri-Pacer	-	
G-MJTD	Gardner T-M Scout	-	
G-MJTE	Skyrider Phantom	-	
G-MJTF	Gryphon Wing	-	
G-MJTM	Aerostructure Pipistrelle 2B	-	
G-MJTN	Eipper Quicksilver MX	-	
G-MJTP	Flexiform Striker	-	
G-MJTR	Southdown Puma DS Mk 1	-	
G-MJTW	Eurowing Trike	-	
G-MJTX	Skyrider Airsports Phantom	-	
G-MJTZ	Skyrider Airsports Phantom	-	
G-MJUC	MBA Tiger Cub 440	P. C. Avery	
G-MJUI	Flexiform Striker	-	
G-MJUS	MBA Tiger Cub 440	-	
G-MJUT	Eurowing Goldwing	-	
G-MJUU	Eurowing Goldwing	-	
G-MJUV	Huntair Pathfinder 1	-	
G-MJUW	MBA Tiger Cub 440	-	
G-MJUX	Skyrider Airsports Phantom	-	
G-MJUZ	Dragon Srs 150	-	
G-MJVA	Skyrider Airsports Phantom	-	
G-MJVE	Hybred Skytrike	-	
G-MJVF	CFM Shadow	-	
G-MJVG	Hiway Skytrike	-	
G-MJVJ	Flexiform Striker Dual	-	
G-MJVM	Dragon 150	-	
G-MJVN	Ultrasports Puma 440	-	
G-MJVP	Eipper Quicksilver MX II	-	
G-MJVR	Flexiform Striker	-	
G-MJVU	Eipper Quicksilver MX II	-	
G-MJVW	Airwave Nimrod	-	
G-MJVX	Skyrider Phantom	-	
G-MJVY	Dragon Srs 150	-	
G-MJVZ	Hiway Demon Tripacer	-	
G-MJWB	Eurowing Goldwing	-	
G-MJWF	Tiger Cub 440	-	
G-MJWJ	MBA Tiger Cub 440	-	
G-MJWK	Huntair Pathfinder	-	

Notes	Reg.	Type	Owner or Operator
	G-MJWN	Flexiform Striker	-
	G-MJWR	MBA Tiger Cub 440	-
	G-MJWU	Maxair Hummer TX	-
	G-MJWZ	Ultrasports Panther XL	-
	G-MJXD	MBA Tiger Cub 440	-
	G-MJXF	MBA Tiger Cub 440	-
	G-MJXM	Hiway Skytrike	-
	G-MJXR	Huntair Pathfinder II	-
	G-MJXS	Huntair Pathfinder II	-
	G-MJXV	Flexiform Striker	-
	G-MJXY	Hiway Demon Skytrike	-
	G-MJYA	Huntair Pathfinder	-
	G-MJYD	MBA Tiger Cub 440	-
	G-MJYG	Skyhook Orion Canard	-
	G-MJYM	Southdown Puma Sprint	-
	G-MJYP	Mainair Triflyer 440	-
	G-MJYR	Catto CP.16	-
	G-MJYS	Southdown Puma Sprint	-
	G-MJYT	Southdown Puma Sprint	-
	G-MJYV	Mainair Triflyer 2 Seat	-
	G-MJYW	Wasp Gryphon III	-
	G-MJYX	Mainair Triflyer	-
	G-MJYY	Hiway Demon	-
	G-MJZA	MBA Tiger Cub	-
	G-MJZD	Mainair Gemini Flash	-
	G-MJZE	MBA Tiger Cub 440	-
	G-MJZJ	Hiway Cutlass Skytrike	-
	G-MJZK	Southdown Puma Sprint 440	-
	G-MJZL	Eipper Quicksilver MX II	-
	G-MJZT	Flexiform Striker	-
	G-MJZU	Flexiform Striker	-
	G-MJZW	Eipper Quicksilver MX II	-
	G-MJZX	Hummer TX	-
	G-MKAK	Colt 77A balloon	M. Kendrick
	G-MKAS	PA-28 Cherokee 140	MK Aero Support Ltd (G-BKVR)
	G-MKIA	V.S.300 Spitfire 1 (P9374)	S. J. Marsh/Italy
	G-MKSS	H.S.125 Srs 700B	Markoss Aviation Ltd
	G-MKVB	V.S.349 Spitfire LF.VB (BM597)	Historic Aircraft Collection/Duxford
	G-MKVI	D.H. Vampire FB.6 (WL505)	C. T. Topen
	G-MKXI	VX.365 Spitfire PR.XI	P. A. Teichman
	G-MLAS	Cessna 182E ★	Parachute jump trainer/St Merryn
	G-MLFF	PA-23 Aztec 250E	W. C. Cullinane (G-WEBB/G-BJBU)
	G-MLGL	Colt 21A balloon	H. C. J. Williams
	G-MLHI	Maule MX-7-180 Star Rocket	Maulehigh (G-BTMJ)
	G-MLJL	Airbus A.330-243	My Travel Airways
	G-MLSN	Hughes 369E	Molson Holdings Ltd (G-HMAC)
	G-MLTY	AS.365N-2 Dauphin 2	Multiflight Ltd/Leeds-Bradford
	G-MLWI	Thunder Ax7-77 balloon	M. L. & L. P. Willoughby
	G-MMAC	Dragon Srs 150	-
	G-MMAE	Dragon Srs 150	-
	G-MMAG	MBA Tiger Cub 440	-
	G-MMAH	Eipper Quicksilver MX II	-
	G-MMAI	Dragon Srs 2	-
	G-MMAN	Flexiform Striker	-
	G-MMAP	Hummer TX	-
	G-MMAR	Southdown Puma Sprint	-
	G-MMAT	Southdown Puma Sprint	-
	G-MMAW	Mainair 330	-
	G-MMAZ	Southdown Puma Sprint	-
	G-MMBD	Spectrum 330	-
	G-MMBE	MBA Tiger Cub 440	-
	G-MMBH	MBA Super Tiger Cub 440	-
	G-MMBL	Southdown Puma	-
	G-MMBN	Eurowing Goldwing	-
	G-MMBT	MBA Tiger Cub 440	-
	G-MMBU	Eipper Quicksilver MX II	-
	G-MMBV	Huntair Pathfinder	-
	G-MMBX	MBA Tiger Cub 440	-
	G-MMBY	Solar Wings Panther XL	-

Reg.	Type	Owner or Operator	Notes
G-MMBZ	Solar Wings Typhoon P	-	
G-MMCD	Southdown Lightning DS	-	
G-MMCE	MBA Tiger Cub 440	-	
G-MMCF	Solar Wings Panther 330	-	
G-MMCI	Southdown Puma Sprint	-	
G-MMCM	Southdown Puma Sprint	-	
G-MMCN	Hiway Skytrike	-	
G-MMCS	Southdown Puma Sprint	-	
G-MMCV	Solar Wings Typhoon III	-	
G-MMCX	MBA Super Tiger Cub 440	-	
G-MMCY	Flexiform Striker	-	
G-MMCZ	Flexiform Striker	-	
G-MMDC	Eipper Quicksilver MXII	-	
G-MMDE	Solar Wings Typhoon	-	
G-MMDF	Southdown Lightning II	-	
G-MMDK	Flexiform Striker	-	
G-MMDN	Flexiform Striker	-	
G-MMDO	Southdown Sprint	-	
G-MMDP	Southdown Sprint	-	
G-MMDR	Huntair Pathfinder II	-	
G-MMDV	Ultrasports Panther	-	
G-MMDW	Pterodactyl Pfledgling	-	
G-MMDX	Solar Wings Typhoon	-	
G-MMDY	Puma Sprint X	-	
G-MMDZ	Flexiform Dual Strike	-	
G-MMEE	American Aerolights Eagle	-	
G-MMEF	Hiway Super Scorpion	-	
G-MMEJ	Striker/TriFlyer	-	
G-MMEK	Medway Hybred 44XL	-	
G-MMEN	Solar Wings Typhoon XL2	-	
G-MMEP	MBA Tiger Cub 440	-	
G-MMET	Skyhook Sabre TR-1 Mk II	-	
G-MMEW	MBA Tiger Cub 440	-	
G-MMEY	MBA Tiger Cub 440	-	
G-MMFD	Flexiform Striker	-	
G-MMFE	Flexiform Striker	-	
G-MMFG	Flexiform Striker	-	
G-MMFI	Flexiform Striker	-	
G-MMFL	Flexiform Striker	-	
G-MMFS	MBA Tiger Cub 440	-	
G-MMFT	MBA Tiger Cub 440	-	
G-MMFV	Flexiform Striker	-	
G-MMFY	Flexiform Dual Striker	-	
G-MMGA	Bass Gosling	-	
G-MMGB	Southdown Puma Sprint	-	
G-MMGC	Southdown Puma Sprint	-	
G-MMGD	Southdown Puma Sprint	-	
G-MMGE	Hiway Super Scorpion	-	
G-MMGF	MBA Tiger Cub 440	-	
G-MMGL	MBA Tiger Cub 440	-	
G-MMGN	Southdown Puma Sprint	-	
G-MMGP	Southdown Puma Sprint	-	
G-MMGS	Solar Wings Panther XL	-	
G-MMGT	Solar Wings Typhoon	-	
G-MMGU	Flexiform Sealander	-	
G-MMGV	Whittaker MW-5 Sorcerer	-	
G-MMGX	Southdown Puma	-	
G-MMHE	Southdown Puma Sprint	-	
G-MMHK	Hiway Super Scorpion	-	
G-MMHL	Hiway Super Scorpion	-	
G-MMHM	Goldmarque Gyr	-	
G-MMHN	MBA Tiger Cub 440	-	
G-MMHP	Hiway Demon	-	
G-MMHS	SMD Viper	-	
G-MMHX	Hornet Invader 440	-	
G-MMHY	Hornet Invader 440	-	
G-MMHZ	Solar Wings Typhoon XL	-	
G-MMIC	Luscombe Vitality	-	
G-MMIE	MBA Tiger Cub 440	-	
G-MMIF	Wasp Gryphon	-	
G-MMIH	MBA Tiger Cub 440	-	
G-MMIL	Eipper Quicksilver MX II	-	

Notes	Reg.	Type	Owner or Operator
	G-MMIR	Mainair Tri-Flyer 440	-
	G-MMIW	Southdown Puma Sprint	-
	G-MMIX	MBA Tiger Cub 440	-
	G-MMIY	Eurowing Goldwing	-
	G-MMJD	Southdown Puma Sprint	-
	G-MMJE	Southdown Puma Sprint	-
	G-MMJF	Ultrasports Panther Dual 440	-
	G-MMJG	Mainair Tri-Flyer 440	-
	G-MMJJ	Solar Wings Typhoon	-
	G-MMJM	Southdown Puma Sprint	-
	G-MMJN	Eipper Quicksilver MX II	-
	G-MMJT	Southdown Puma Sprint	-
	G-MMJU	Hiway Demon	-
	G-MMJV	MBA Tiger Cub 440	-
	G-MMJX	Teman Mono-Fly	-
	G-MMJY	MBA Tiger Cub 440	-
	G-MMKA	Ultrasports Panther Dual	-
	G-MMKE	Birdman Chinook WT-11	-
	G-MMKG	Solar Wings Typhoon XL	-
	G-MMKH	Solar Wings Typhoon XL	-
	G-MMKI	Ultrasports Panther 330	-
	G-MMKL	Mainair Gemini Flash	-
	G-MMKM	Flexiform Dual Striker	-
	G-MMKP	MBA Tiger Cub 440	-
	G-MMKR	Southdown Lightning DS	-
	G-MMKU	Southdown Puma Sprint	-
	G-MMKV	Southdown Puma Sprint	-
	G-MMKW	Solar Wings Storm	-
	G-MMKX	Skyrider Phantom 330	-
	G-MMKZ	Ultrasports Puma 440	-
	G-MMLB	MBA Tiger Cub 440	-
	G-MMLE	Eurowing Goldwing SP	-
	G-MMLH	Hiway Demon	-
	G-MMLM	MBA Tiger Cub 440	-
	G-MMLO	Skyhook Pixie	-
	G-MMMB	Mainair Tri-Flyer	-
	G-MMMG	Eipper Quicksilver MXL	-
	G-MMMH	Hadland Willow	-
	G-MMMJ	Southdown Sprint	-
	G-MMMK	Hornet Invader	-
	G-MMML	Dragon 150	-
	G-MMMN	Ultrasports Panther Dual 440	-
	G-MMMP	Flexiform Dual Striker	-
	G-MMNB	Eipper Quicksilver MX	-
	G-MMNC	Eipper Quicksilver MX	-
	G-MMND	Eipper Quicksilver MX II-Q2	-
	G-MMNF	Hornet	-
	G-MMNG	Solar Wings Typhoon XL	-
	G-MMNH	Dragon 150	-
	G-MMNN	Buzzard	-
	G-MMNS	Mitchell U-2 Super Wing	-
	G-MMNT	Flexiform Striker	-
	G-MMOB	Southdown Sprint	-
	G-MMOH	Solar Wings Typhoon XL	-
	G-MMOI	MBA Tiger Cub 440	-
	G-MMOK	Solar Wings Panther XL	-
	G-MMOL	Skycraft Scout R3	-
	G-MMOW	Mainair Gemini Flash	-
	G-MMOY	Mainair Gemini Sprint	-
	G-MMPG	Southdown Puma	-
	G-MMPH	Southdown Puma Sprint	D. A. Frank
	G-MMPI	Pterodactyl Ptraveller	-
	G-MMPJ	Mainair Tri-Flyer 440	-
	G-MMPL	Flexiform Dual Striker	-
	G-MMPN	Chargus T250	-
	G-MMPO	Mainair Gemini Flash	-
	G-MMPT	SMD Gazelle	-
	G-MMPU	Ultrasports Tripacer 250	-
	G-MMPW	Airwave Nimrod	-
	G-MMPX	Ultrasports Panther Dual 440	-
	G-MMPZ	Teman Mono-Fly	-
	G-MMRA	Mainair Tri-Flyer 250	-

Reg.	Type	Owner or Operator	Notes
G-MMRF	MBA Tiger Cub 440	-	
G-MMRH	Hiway Demon	-	
G-MMRJ	Solar Wings Panther XL	-	
G-MMRK	Solar Wings Panther XL-S	-	
G-MMRL	Solar Wings Panther XL	-	
G-MMRN	Southdown Puma Sprint	-	
G-MMRP	Mainair Gemini	-	
G-MMRU	Tirith Firebird FB-2	-	
G-MMRW	Flexiform Dual Striker	-	
G-MMRY	Chargus T.250	-	
G-MMRZ	Ultrasports Panther Dual 440	-	
G-MMSA	Ultrasports Panther XL	-	
G-MMSC	Mainair Gemini	-	
G-MMSE	Eipper Quicksilver MX	-	
G-MMSG	Solar Wings Panther XL-S	-	
G-MMSH	Solar Wings Panther XL	-	
G-MMSO	Mainair Tri-Flyer 440	-	
G-MMSP	Mainair Gemini Flash	-	
G-MMSR	MBA Tiger Cub 440	-	
G-MMSS	Southdown Lightning	-	
G-MMSW	MBA Tiger Cub 440	-	
G-MMSZ	Medway Half Pint	-	
G-MMTA	Ultrasports Panther XL	-	
G-MMTC	Ultrasports Panther Dual	-	
G-MMTD	Mainair Tri-Flyer 330	-	
G-MMTG	Mainair Gemini Sprint	-	
G-MMTH	Southdown Puma Sprint	-	
G-MMTI	Southdown Puma Sprint	-	
G-MMTJ	Southdown Puma Sprint	-	
G-MMTL	Mainair Gemini	-	
G-MMTR	Ultrasports Panther	-	
G-MMTS	Solar Wings Panther XL	-	
G-MMTT	Solar Wings Panther XL-S	-	
G-MMTV	American Aerolights Eagle	-	
G-MMTX	Mainair Gemini 440	A. Worthington	
G-MMTY	Fisher FP.202U	-	
G-MMTZ	Eurowing Goldwing	-	
G-MMUA	Southdown Puma Sprint	-	
G-MMUC	Mainair Gemini 440	-	
G-MMUG	Mainair Tri-Flyer	-	
G-MMUH	Mainair Tri-Flyer	-	
G-MMUK	Mainair Tri-Flyer	-	
G-MMUL	Ward Elf E.47	-	
G-MMUM	MBA Tiger Cub 440	-	
G-MMUO	Mainair Gemini Flash	-	
G-MMUP	Airwave Nimrod 140	-	
G-MMUU	ParaPlane PM-1	-	
G-MMUV	Southdown Puma Sprint	-	
G-MMUW	Mainair Gemini Flash	-	
G-MMVA	Southdown Puma Sprint	-	
G-MMVC	Ultrasports Panther XL	-	
G-MMVH	Southdown Raven	-	
G-MMVI	Southdown Puma Sprint	-	
G-MMVL	Ultrasports Panther XL-S	-	
G-MMVM	Whiteley Orion 1	-	
G-MMVR	Hiway Skytrike 1	-	
G-MMVS	Skyhook Pixie	-	
G-MMVX	Southdown Puma Sprint	-	
G-MMVZ	Southdown Puma Sprint	-	
G-MMWA	Mainair Gemini Flash	N. Roberts	
G-MMWC	Eipper Quicksilver MXII	-	
G-MMWF	Hiway Skytrike 250	-	
G-MMWG	Greenslade Mono-Trike	-	
G-MMWI	Southdown Lightning	-	
G-MMWL	Eurowing Goldwing	-	
G-MMWN	Ultrasports Tripacer	-	
G-MMWS	Mainair Tri-Flyer	-	
G-MMWX	Southdown Puma Sprint	-	
G-MMXD	Mainair Gemini Flash	-	
G-MMXE	Mainair Gemini Flash	-	
G-MMXG	Mainair Gemini Flash	-	
G-MMXI	Horizon Prototype	-	

Notes	Reg.	Type	Owner or Operator
	G-MMXJ	Mainair Gemini Flash	-
	G-MMXL	Mainair Gemini Flash	-
	G-MMXM	Mainair Gemini Flash	-
	G-MMXN	Southdown Puma Sprint	-
	G-MMXO	Southdown Puma Sprint	-
	G-MMXU	Mainair Gemini Flash	-
	G-MMXV	Mainair Gemini Flash	K. C. Beattie
	G-MMXW	Mainair Gemini	-
	G-MMXX	Mainair Gemini	-
	G-MMYA	Solar Wings Pegasus XL	J. North
	G-MMYB	Solar Wings Pegasus XL	-
	G-MMYD	CFM Shadow Srs B	-
	G-MMYF	Southdown Puma Sprint	M. Campbell
	G-MMYI	Southdown Puma Sprint	-
	G-MMYL	Cyclone 70	-
	G-MMYN	Ultrasports Panther XL	-
	G-MMYO	Southdown Puma Sprint	-
	G-MMYR	Eipper Quicksilver MXII	-
	G-MMYS	Southdown Puma Sprint	-
	G-MMYT	Southdown Puma Sprint	-
	G-MMYU	Southdown Puma Sprint	-
	G-MMYV	Webb Trike	-
	G-MMYY	Southdown Puma Sprint	D. J. Whittle
	G-MMZA	Mainair Gemini Flash	-
	G-MMZB	Mainair Gemini Flash	-
	G-MMZE	Mainair Gemini Flas	-
	G-MMZF	Mainair Gemini Flash	J. Tait
	G-MMZG	Ultrasports Panther XL-S	-
	G-MMZI	Medway 130SX	-
	G-MMZJ	Mainair Gemini Flas	-
	G-MMZK	Mainair Gemini Flash	-
	G-MMZL	Mainair Gemini Flash	-
	G-MMZM	Mainair Gemini Flash	-
	G-MMZN	Mainair Gemini Flash	-
	G-MMZO	Microflight Spectrum	-
	G-MMZP	Ultrasports Panther XL	-
	G-MMZR	Southdown Puma Sprint	-
	G-MMZS	Eipper Quicksilver MX1	-
	G-MMZV	Mainair Gemini Flash	-
	G-MMZW	Southdown Puma Sprint	-
	G-MMZY	Ultrasports Tripacer 330	-
	G-MNAA	Striplin Sky Ranger	-
	G-MNAC	Mainair Gemini Flash	C. Bayliss
	G-MNAE	Mainair Gemini Flash	-
	G-MNAF	Solar Wings Panther XL-S	-
	G-MNAH	Solar Wings Panther XL	-
	G-MNAI	Ultrasports Panther XL-S	-
	G-MNAK	Solar Wings Panther XL-S	-
	G-MNAM	Solar Wings Panther XL-S	-
	G-MNAO	Solar Wings Panther XL-S	-
	G-MNAT	Solar Wings Pegasus XL-R	-
	G-MNAV	Southdown Puma Sprint	-
	G-MNAW	Solar Wings Pegasus XL-R	-
	G-MNAX	Solar Wings Pegasus XL-R	-
	G-MNAY	Ultrasports Panther XL-S	-
	G-MNAZ	Solar Wings Pegasus XL-R	-
	G-MNBA	Solar Wings Pegasus XL-R	-
	G-MNBB	Solar Wings Pegasus XL-R	R. Piper
	G-MNBC	Solar Wings Pegasus XL-R	-
	G-MNBD	Mainair Gemini Flash	-
	G-MNBE	Southdown Puma Sprint	-
	G-MNBF	Mainair Gemini Flash	-
	G-MNBG	Mainair Gemini Flash	-
	G-MNBI	Ultrasports Panther XL	A-M. Whelan/Ireland
	G-MNBJ	Skyhook Pixie	-
	G-MNBM	Southdown Puma Sprint	-
	G-MNBN	Mainair Gemini Flash	-
	G-MNBP	Mainair Gemini Flash	-
	G-MNBS	Mainair Gemini Flash	-
	G-MNBT	Mainair Gemini Flash	T. H. Parr
	G-MNBV	Mainair Gemini Flash	-

Reg.	Type	Owner or Operator	Notes
G-MNCA	Hiway Demon 175	M. A. Sirant	
G-MNCF	Mainair Gemini Flash	M. Atkinson	
G-MNCG	Mainair Gemini Flash	-	
G-MNCI	Southdown Puma Sprint	-	
G-MNCJ	Mainair Gemini Flash	-	
G-MNCL	Southdown Puma Sprint	-	
G-MNCM	CFM Shadow Srs B	-	
G-MNCO	Eipper Quicksilver MXII	-	
G-MNCP	Southdown Puma Sprint	-	
G-MNCR	Flexiform Striker	-	
G-MNCS	Skyrider Airsports Phantom	-	
G-MNCU	Medway Hybred 44XL	-	
G-MNCV	Medway Hybred 44XL	-	
G-MNCZ	Solar Wings Pegasus XL	-	
G-MNDA	Thruster TST	-	
G-MNDD	Mainair Scorcher Solo	L. Hurman	
G-MNDE	Medway Half Pint	-	
G-MNDF	Mainair Gemini Flash	R. Bowden	
G-MNDG	Southdown Puma Sprint	-	
G-MNDH	Hiway Skytrike	-	
G-MNDI	MBA Tiger Cub 440	-	
G-MNDO	Mainair Flash	-	
G-MNDU	Midland Sirocco 377GB	-	
G-MNDV	Midland Sirocco 377GB	-	
G-MNDY	Southdown Puma Sprint	-	
G-MNEF	Mainair Gemini Flash	-	
G-MNEG	Mainair Gemini Flash	A. Sexton/Ireland	
G-MNEH	Mainair Gemini Flash	-	
G-MNEI	Medway Hybred 440	-	
G-MNEK	Medway Half Pint	-	
G-MNEP	Aerostructure Pipstrelle P.2B	-	
G-MNER	CFM Shadow Srs B	-	
G-MNET	Mainair Gemini Flash	-	
G-MNEV	Mainair Gemini Flash	-	
G-MNEY	Mainair Gemini Flash	-	
G-MNEZ	Skyhook TR1 Mk 2	-	
G-MNFB	Southdown Puma Sprint	-	
G-MNFE	Mainair Gemini Flash	-	
G-MNFF	Mainair Gemini Flash	-	
G-MNFG	Southdown Puma Sprint	-	
G-MNFH	Mainair Gemini Flash	-	
G-MNFL	AMF Chevvron	-	
G-MNFM	Mainair Gemini Flash	-	
G-MNFN	Mainair Gemini Flash	-	
G-MNFP	Mainair Gemini Flash	-	
G-MNFW	Medway Hybred 44XL	-	
G-MNFX	Southdown Puma Sprint	-	
G-MNFY	Hornet 250	-	
G-MNGA	Aerial Arts Chaser 110SX	-	
G-MNGD	Quest Air Services	-	
G-MNGF	Solar Wings Pegasus	R. G. Smith	
G-MNGG	Solar Wings Pegasus XL-R	-	
G-MNGH	Skyhook Pixie	-	
G-MNGJ	Skyhook Zipper	-	
G-MNGK	Mainair Gemini Flash	-	
G-MNGL	Mainair Gemini Flash	-	
G-MNGM	Mainair Gemini Flash	-	
G-MNGN	Mainair Gemini Flash	-	
G-MNGO	Solar Wings Storm	-	
G-MNGS	Southdown Puma 330	-	
G-MNGT	Mainair Gemini Flash	-	
G-MNGU	Mainair Gemini Flash	-	
G-MNGW	Mainair Gemini Flash	-	
G-MNGX	Southdown Puma Sprint	-	
G-MNHB	Solar Wings Pegasus XL-R	-	
G-MNHC	Solar Wings Pegasus XL-R	-	
G-MNHD	Solar Wings Pegasus XL-R	-	
G-MNHE	Solar Wings Pegasus XL-R	-	
G-MNHF	Solar Wings Pegasus XL-R	J. E. Cox	
G-MNHH	Solar Wings Panther XL-S	-	
G-MNHI	Solar Wings Pegasus XL-R	-	
G-MNHJ	Solar Wings Pegasus XL-R	-	

Notes	Reg.	Type	Owner or Operator
	G-MNHK	Solar Wings Pegasus XL-R	-
	G-MNHL	Solar Wings Pegasus XL-R	-
	G-MNHM	Solar Wings Pegasus XL-R	A. C. Bell
	G-MNHN	Solar Wings Pegasus XL-R	-
	G-MNHR	Solar Wings Pegasus XL-R	-
	G-MNHS	Solar Wings Pegasus XL-R	-
	G-MNHT	Solar Wings Pegasus XL-R	-
	G-MNHV	Solar Wings Pegasus XL-R	-
	G-MNHX	Solar Wings Typhoon S4	-
	G-MNHZ	Mainair Gemini Flash	-
	G-MNIA	Mainair Gemini Flash	-
	G-MNIE	Mainair Gemini Flash	-
	G-MNIF	Mainair Gemini Flash	-
	G-MNIG	Mainair Gemini Flash	-
	G-MNIH	Mainair Gemini Flash	-
	G-MNII	Mainair Gemini Flash	-
	G-MNIK	Pegasus Photon	J. Grotrian
	G-MNIL	Southdown Puma Sprint	A. Bishop
	G-MNIM	Maxair Hummer	-
	G-MNIO	Mainair Gemini Flash	-
	G-MNIP	Mainair Gemini Flash	-
	G-MNIS	CFM Shadow Srs B	-
	G-MNIT	Aerial Arts 130SX	-
	G-MNIU	Solar Wings Pegasus Photon	-
	G-MNIV	Solar Wings Typhoon	-
	G-MNIW	Airwave Nimrod 165	-
	G-MNIX	Mainair Gemini Flash	-
	G-MNIY	Skyhook Pixie Zipper	-
	G-MNIZ	Mainair Gemini Flash	-
	G-MNJB	Southdown Raven	-
	G-MNJC	MBA Tiger Cub 440	-
	G-MNJD	Southdown Puma Sprint	-
	G-MNJF	Dragon 150	-
	G-MNJG	Mainair Tri-Flyer	-
	G-MNJH	SW Pegasus Flash	-
	G-MNJI	SW Pegasus Flash	-
	G-MNJJ	SW Pegasus Flash	-
	G-MNJL	SW Pegasus Flash	-
	G-MNJM	SW Pegasus Flash	-
	G-MNJN	SW Pegasus Flash	-
	G-MNJO	SW Pegasus Flash	-
	G-MNJR	SW Pegasus Flash	-
	G-MNJS	Southdown Puma Sprint	-
	G-MNJT	Southdown Raven	-
	G-MNJU	Mainair Gemini Flash	-
	G-MNJV	Medway Half Pint	-
	G-MNJX	Medway Hybred 44XL	-
	G-MNKB	SW Pegasus Photon	-
	G-MNKC	SW Pegasus Photon	C. Murphy
	G-MNKD	SW Pegasus Photon	-
	G-MNKE	SW Pegasus Photon	-
	G-MNKG	SW Pegasus Photon	-
	G-MNKK	SW Pegasus Photon	-
	G-MNKM	MBA Tiger Cub 440	A. R. Sunley
	G-MNKO	SW Pegasus Flash	-
	G-MNKP	SW Pegasus Flash	-
	G-MNKR	SW Pegasus Flash	T. R. Murfet
	G-MNKT	Solar Wings Typhoon S4	-
	G-MNKU	Southdown Puma Sprint	-
	G-MNKV	SW Pegasus Flash	G. Wakerley
	G-MNKW	SW Pegasus Flash	-
	G-MNKX	SW Pegasus Flash	-
	G-MNKZ	Southdown Raven	-
	G-MNLE	Southdown Raven X	-
	G-MNLH	Romain Cobra Biplane	-
	G-MNLI	Mainair Gemini Flash	-
	G-MNLK	Southdown Raven	-
	G-MNLM	Southdown Raven	-
	G-MNLN	Southdown Raven	-
	G-MNLP	Southdown Raven	-
	G-MNLT	Southdown Raven	-
	G-MNLU	Southdown Raven	-

Reg.	Type	Owner or Operator	Notes
G-MNLV	Southdown Raven X	-	
G-MNLY	Mainair Gemini Flash	-	
G-MNLZ	Southdown Raven	-	
G-MNMC	Mainair Gemini Sprint	-	
G-MNMD	Southdown Raven	-	
G-MNMG	Mainair Gemini Flash	-	
G-MNMI	Mainair Gemini Flash	-	
G-MNMK	Solar Wings Pegasus XL-R	-	
G-MNML	Southdown Puma Sprint	-	
G-MNMM	Aerotech MW.5 Sorcerer	S. F. N. Warnell	
G-MNMN	Medway Hybred 44XLR	-	
G-MNMR	Solar Wings Typhoon 180	-	
G-MNMT	Southdown Raven	-	
G-MNMU	Southdown Raven	-	
G-MNMV	Mainair Gemini Flash	-	
G-MNMW	Aerotech MW.6 Merlin	-	
G-MNMY	Cyclone 70	-	
G-MNNA	Southdown Raven	-	
G-MNNB	Southdown Raven	-	
G-MNNC	Southdown Raven	-	
G-MNNF	Mainair Gemini Flash	-	
G-MNNG	Solar Wings Photon	-	
G-MNNI	Mainair Gemini Flash	-	
G-MNNJ	Mainair Gemini Flash II	H. D. Lynch	
G-MNNK	Mainair Gemini Flash	-	
G-MNNL	Mainair Gemini Flash	-	
G-MNNM	Mainair Scorcher Solo	-	
G-MNNO	Southdown Raven	-	
G-MNNP	Mainair Gemini Flash	-	
G-MNNR	Mainair Gemini Flash	-	
G-MNNS	Eurowing Goldwing	-	
G-MNNV	Mainair Gemini Flash	-	
G-MNNY	SW Pegasus Flash	-	
G-MNNZ	SW Pegasus Flash	-	
G-MNPA	SW Pegasus Flash	-	
G-MNPC	Mainair Gemini Flash	-	
G-MNPF	Mainair Gemini Flash	-	
G-MNPG	Mainair Gemini Flash	-	
G-MNPH	Flexiform Dual Striker	-	
G-MNPL	Ultrasports Panther 330	-	
G-MNPV	Mainair Scorcher Solo	-	
G-MNPW	AMF Chevvron	-	
G-MNPY	Mainair Scorcher Solo	-	
G-MNPZ	Mainair Scorcher Solo	-	
G-MNRA	CFM Shadow Srs B	-	
G-MNRD	Ultraflight Lazair IIIE	-	
G-MNRE	Mainair Scorcher Solo	-	
G-MNRF	Mainair Scorcher Solo	-	
G-MNRI	Hornet Dual Trainer	R. H. Goll	
G-MNRK	Hornet Dual Trainer	-	
G-MNRM	Hornet Dual Trainer	-	
G-MNRN	Hornet Dual Trainer	-	
G-MNRP	Southdown Raven	-	
G-MNRS	Southdown Raven	-	
G-MNRT	Midland Ultralights Sirocco	-	
G-MNRW	Mainair Gemini Flash II	-	
G-MNRX	Mainair Gemini Flash II	-	
G-MNRZ	Mainair Scorcher Solo	-	
G-MNSA	Mainair Gemini Flash	-	
G-MNSB	Southdown Puma Sprint	-	
G-MNSD	Solar Wings Typhoon S4	-	
G-MNSF	Hornet Dual Trainer	-	
G-MNSH	SW Pegasus Flash II	-	
G-MNSI	Mainair Gemini Flash II	-	
G-MNSJ	Mainair Gemini Flash	-	
G-MNSL	Southdown Raven X	-	
G-MNSM	Hornet Demon	-	
G-MNSN	SW Pegasus Flash II	-	
G-MNSP	Aerial Arts 130SX	-	
G-MNSR	Mainair Gemini Flash	-	
G-MNSS	American Aerolights Eagle	-	
G-MNSV	CFM Shadow Srs B	-	

Notes	Reg.	Type	Owner or Operator
	G-MNSX	Southdown Raven X	-
	G-MNSY	Southdown Raven X	-
	G-MNTC	Southdown Raven X	-
	G-MNTD	Aerial Arts Chaser 110SX	-
	G-MNTE	Southdown Raven X	-
	G-MNTF	Southdown Raven X	-
	G-MNTH	Mainair Gemini Flash	-
	G-MNTI	Mainair Gemini Flas	-
	G-MNTK	CFM Shadow Srs B	-
	G-MNTM	Southdown Raven X	-
	G-MNTN	Southdown Raven X	-
	G-MNTP	CFM Shadow Srs B	-
	G-MNTT	Medway Half Pint	-
	G-MNTU	Mainair Gemini Flash II	-
	G-MNTV	Mainair Gemini Flash II	-
	G-MNTW	Mainair Gemini Flash II	-
	G-MNTX	Mainair Gemini Flash II	-
	G-MNTY	Southdown Raven X	-
	G-MNTZ	Mainair Gemini Flash II	-
	G-MNUA	Mainair Gemini Flash II	-
	G-MNUD	SW Pegasus Flash II	-
	G-MNUE	SW Pegasus Flash II	-
	G-MNUF	Mainair Gemini Flash II	-
	G-MNUG	Mainair Gemini Flash II	A. S. Nader
	G-MNUI	Skyhook Cutlass Dual	-
	G-MNUO	Mainair Gemini Flash II	-
	G-MNUR	Mainair Gemini Flash II	-
	G-MNUT	Southdown Raven X	-
	G-MNUU	Southdown Raven X	-
	G-MNUX	Solar Wings Pegasus XL-R	-
	G-MNVA	Solar Wings Pegasus XL-R	-
	G-MNVB	Solar Wings Pegasus XL-R	-
	G-MNVC	Solar Wings Pegasus XL-R	-
	G-MNVE	Solar Wings Pegasus XL-R	-
	G-MNVG	SW Pegasus Flash II	-
	G-MNVH	SW Pegasus Flash II	-
	G-MNVI	CFM Shadow Srs B	-
	G-MNVJ	CFM Shadow Srs CD	-
	G-MNVK	CFM Shadow Srs B	-
	G-MNVN	Southdown Raven X	-
	G-MNVO	Hovey Whing-Ding II	-
	G-MNVT	Mainair Gemini Flash II	-
	G-MNVU	Mainair Gemini Flash II	-
	G-MNVV	Mainair Gemini Flash II	-
	G-MNVW	Mainair Gemini Flash II	-
	G-MNVZ	SW Pegasus Photon	-
	G-MNWD	Mainair Gemini Flash	-
	G-MNWF	Southdown Raven X	-
	G-MNWG	Southdown Raven X	-
	G-MNWI	Mainair Gemini Flash II	-
	G-MNWJ	Mainair Gemini Flash II	-
	G-MNWK	CFM Shadow Srs B	-
	G-MNWL	Aerial Arts 130SX	-
	G-MNWN	Mainair Gemini Flash II	-
	G-MNWP	SW Pegasus Flash II	-
	G-MNWU	SW Pegasus Flash II	-
	G-MNWV	SW Pegasus Flash II	-
	G-MNWW	Solar Wings Pegasus XL-R	-
	G-MNWY	CFM Shadown Srs B	-
	G-MNWZ	Mainair Gemini Flash II	-
	G-MNXA	Southdown Raven X	-
	G-MNXB	Solar Wings Photon	-
	G-MNXD	Southdown Raven X	-
	G-MNXE	Southdown Raven X	-
	G-MNXF	Southdown Raven X	-
	G-MNXG	Southdown Raven X	-
	G-MNXI	Southdown Raven X	-
	G-MNXM	Medway Hybred 44XLR	-
	G-MNXO	Medway Hybred 44XLR	-
	G-MNXP	Pegasus Flash II	I. K. Priestley
	G-MNXS	Mainair Gemini Flash II	-
	G-MNXU	Mainair Gemini Flash II	-

Reg.	Type	Owner or Operator	Notes
G-MNXX	CFM Shadow Srs BD	P. J. Mogg	
G-MNXZ	Whittaker MW.5 Sorcerer	A. J. Glynn	
G-MNYA	SW Pegasus Flash II	-	
G-MNYB	Solar Wings Pegasus XL-R	G. Hanna/Ireland	
G-MNYC	Solar Wings Pegasus XL-R	-	
G-MNYD	Aerial Arts 110SX Chaser	-	
G-MNYE	Aerial Arts 110SX Chaser	-	
G-MNYF	Aerial Arts 110SX Chaser	-	
G-MNYG	Southdown Raven	-	
G-MNYH	Southdown Puma Sprint	-	
G-MNYI	Southdown Raven X	-	
G-MNYJ	Mainair Gemini Flash II	-	
G-MNYK	Mainair Gemini Flash II	-	
G-MNYL	Southdown Raven X	-	
G-MNYM	Southdown Raven X	R. L. Davis	
G-MNYO	Southdown Raven X	-	
G-MNYP	Southdown Raven X	-	
G-MNYS	Southdown Raven X	-	
G-MNYU	Pegasus XL-R	J-B. Weber	
G-MNYV	Solar Wings Pegasus XL-R	-	
G-MNYW	Solar Wings Pegasus XL-R	-	
G-MNYX	Solar Wings Pegasus XL-R	-	
G-MNYZ	SW Pegasus Flash	-	
G-MNZB	Mainair Gemini Flash II	-	
G-MNZC	Mainair Gemini Flash II	-	
G-MNZD	Mainair Gemini Flash II	-	
G-MNZF	Mainair Gemini Flash II	-	
G-MNZI	Solar Wings Typhoon	-	
G-MNZJ	CFM Shadow Srs BD	-	
G-MNZK	Solar Wings Pegasus XL-R	-	
G-MNZM	Solar Wings Pegasus XL-R	-	
G-MNZN	SW Pegasus Flash II	-	
G-MNZP	CFM Shadow Srs B	-	
G-MNZR	CFM Shadown Srs BD	P. J. Watson	
G-MNZS	Aerial Arts 130SX	-	
G-MNZU	Eurowing Goldwing	P. D. Coppin & P. R. Millen	
G-MNZW	Southdown Raven X	-	
G-MNZX	Southdown Raven X	-	
G-MNZZ	CFM Shadow Srs B	-	
G-MOAC	Beech F33A Bonanza	R. M. Camrass	
G-MOAN	Aeromot AMT-200S Super Ximango	A. J. Leigh	
G-MODE	Eurocopter EC.120B	McAlpine Helicopters Ltd/Kidlington	
G-MOFB	Cameron O-120 balloon	D. M. Moffat	
G-MOFF	Cameron O-77 balloon	D. M. Moffat	
G-MOFZ	Cameron O-90 balloon	D. M. Moffat	
G-MOGI	AA-5A Cheetah	MOGI Flying Group (G-BFMU)	
G-MOGY	Robinson R-22B	Northumbria Helicopters Ltd	
G-MOHS	PA-31-350 Navajo Chieftain	Sky Air Travel Ltd (G-BWOC)	
G-MOKE	Cameron V-77 balloon	D. D. Owen/Luxembourg	
G-MOLE	Taylor JT.2 Titch	K. R. H. Wingate	
G-MOLI	Cameron A-250 balloon	Wickers Air Balloon Co	
G-MOLL	PA-32-301T Turbo Saratoga	M. S. Bennett	
G-MOLY	PA-23 Apache 160	J. L. Thorogood (G-APFV)	
G-MOMA	Thruster T.600N 450	Turley Farms Ltd (G-CCIB)	
G-MOMO	Agusta A.109E Power Elite	Air Harrods Ltd/Stansted	
G-MONB	Boeing 757-2T7	Monarch Airlines Ltd/Luton	
G-MONC	Boeing 757-2T7	Monarch Airlines Ltd/Luton	
G-MOND	Boeing 757-2T7	Monarch Airlines Ltd/Luton	
G-MONE	Boeing 757-2T7	Monarch Airlines Ltd/Luton	
G-MONI	Monnett Moni	R. M. Edworthy	
G-MONJ	Boeing 757-2T7	Monarch Airlines Ltd/Luton	
G-MONK	Boeing 757-2T7	Monarch Airlines Ltd/Luton	
G-MONR	Airbus A.300-605R	Monarch Airlines Ltd/Luton	
G-MONS	Airbus A.300-605R	Monarch Airlines Ltd/Luton	
G-MONX	Airbus A.320-212	Monarch Airlines Ltd/Luton	
G-MONY	Airbus A.320-212	Monarch Airlines Ltd/Skyservice (C-GVNY)/Luton	
G-MOOR	SOCATA TB.10 Tobago	M. Watkin (G-MILK)	
G-MOOS	P.56 Provost T.1 (XF690)	T. J. Manna (G-BGKA)/Cranfield	
G-MOPB	Diamond DA.40 Star	Papa Bravo Aviation Ltd	
G-MOSS	Beech 95-D55 Baron	S. C. Tysoe (G-AWAD)	
G-MOSY	Cameron O-84 balloon	P. L. Mossman	

Notes	Reg.	Type	Owner or Operator
	G-MOTA	Bell 206B JetRanger 3	J. W. Sandle
	G-MOTH	D.H.82A Tiger Moth (K2567)	M. C. Russell
	G-MOTI	Robin DR.400/500	Tango India Flying Group
	G-MOTO	PA-24 Comanche 180	L. T. & S. Evans (G-EDHE/G-ASFH)/Sandown
	G-MOUL	Maule M6-235	M. Klinge
	G-MOUN	Beech B200 Super King Air	G. & H. L. Mountain
	G-MOUR	H.S. Gnat T.1 (XR991)	Yellowjack Group/Kemble
	G-MOUT	Cessna 182T	C. Mountain
	G-MOVE	PA-60-601P Aerostar	A. Cazaz & A1 Hydraulics Ltd
	G-MOVI	PA-32R-301 Saratoga SP	G-BOON Ltd (G-MARI)
	G-MOZZ	Avions Mudry CAP.10B	N. Skipworth & J. R. W. Luxton
	G-MPAC	Ultravia Pelican PL	M. J. Craven
	G-MPBH	Cessna FA.152	The Moray Flying Club (1996) Ltd (G-FLIC/G-BILV)
	G-MPBI	Cessna 310R	M. P. Bolshaw
	G-MPCD	Airbus A.320-212	Monarch Airlines Ltd
	G-MPRL	Cessna 210M	Myriad Public Relations Ltd
	G-MPWI	Robin HR.100/210	P. G. Clarkson & S. King
	G-MPWT	PA-34-220T Seneca III	Modern Air (UK) Ltd
	G-MRAJ	Hughes 369E	A. Jardine
	G-MRAM	Mignet HM.1000 Balerit	R. A. Marven
	G-MRED	Christavia Mk 1	E. Hewett
	G-MRJJ	Mainair Pegasus Quik	J.H. Sparks
	G-MRKI	Extra EA.200/300	Extra 200 Ltd
	G-MRKS	Robinson R-44	TJD Trade Ltd (G-RAYC)
	G-MRKT	Lindstrand LBL-90A balloon	Marketplace Public Relations (London) Ltd
	G-MRLN	Sky 240-24 balloon	Merlin Balloons
	G-MRMR	PA-31-350 Navajo Chieftain	MRMR (Flight Services) (G-WROX/G-BNZI)
	G-MROC	Pegasus Quantum 15-912	M. Convine
	G-MROY	Ikarus C.42	D. M. Jobbins & K. R. Rowland
	G-MRSN	Robinson R-22B	Yorkshire Helicopters Ltd
	G-MRST	PA-28 RT-201 Arrow IV	Calverton Flying Group Ltd
	G-MRTN	SOCATA TB.10 Tobago	Underwood Kitchens Ltd (G-BHET)
	G-MRTY	Cameron N-77 balloon	R. A. Vale & ptnrs
	G-MSAL	MS.733 Alcyon (143)	North Weald Flying Services Ltd
	G-MSFC	PA-38-112 Tomahawk	Sherwood Flying Club Ltd/Tollerton
	G-MSFT	PA-28-161 Warrior II	Western Air (Thruxton) Ltd (G-MUMS)
	G-MSIX	Glaser-Dirks DG.800B	G-MSIX Group
	G-MSKY	Ikarus C.42	G-MSKY Group
	G-MSOO	Mini-500	R. H. Ryan
	G-MSPY	Pegasus Quantum 15-912	J. Madhvani & R. K. Green
	G-MSTC	AA-5A Cheetah	Association of Manx Pilots (G-BIJT)/Isle of Man
	G-MSTG	NA P-51D Mustang (414419)	M. Hammond
	G-MSTR	Cameron 110 Monster SS	Airship & Balloon Co. Ltd (G-OJOB)
	G-MTAA	Solar Wings Pegasus XL-R	-
	G-MTAB	Mainair Gemini Flash II	-
	G-MTAE	Mainair Gemini Flash II	C. E. Hannigan
	G-MTAF	Mainair Gemini Flash II	-
	G-MTAG	Mainair Gemini Flash II	-
	G-MTAH	Mainair Gemini Flash II	-
	G-MTAI	Solar Wings Pegasus XL-R	-
	G-MTAJ	Solar Wings Pegasus XL-R	-
	G-MTAK	Solar Wings Pegasus XL-R	-
	G-MTAL	Solar Wings Photon	-
	G-MTAM	SW Pegasus Flash	-
	G-MTAO	Solar Wings Pegasus XL-R	-
	G-MTAP	Southdown Raven X	-
	G-MTAR	Mainair Gemini Flash II	-
	G-MTAS	Whittaker MW.5 Sorcerer	G. J. Jones
	G-MTAV	Solar Wings Pegasus XL-R	-
	G-MTAW	Solar Wings Pegasus XL-R	-
	G-MTAX	Solar Wings Pegasus XL-R	-
	G-MTAY	Solar Wings Pegasus XL-R	-
	G-MTAZ	Solar Wings Pegasus XL-R	-
	G-MTBA	Solar Wings Pegasus XL-R	-
	G-MTBB	Southdown Raven X	-
	G-MTBD	Mainair Gemini Flash II	-
	G-MTBE	CFM Shadow Srs BD	-

Reg.	Type	Owner or Operator	Notes
G-MTBF	Mirage Mk II	-	
G-MTBH	Mainair Gemini Flash II	-	
G-MTBI	Mainair Gemini Flash II	-	
G-MTBJ	Mainair Gemini Flash II	-	
G-MTBK	Southdown Raven X	I. Garwood	
G-MTBL	Solar Wings Pegasus XL-R	-	
G-MTBN	Southdown Raven X	-	
G-MTBO	Southdown Raven X	-	
G-MTBP	Aerotech MW.5 Sorcerer	-	
G-MTBR	Aerotech MW.5 Sorcerer	-	
G-MTBS	Aerotech MW.5 Sorcerer	-	
G-MTBV	Solar Wings Pegasus XL-R	-	
G-MTBX	Mainair Gemini Flash II	A. F. Grimwood	
G-MTBY	Mainair Gemini Flash II	D. Pearson	
G-MTBZ	Southdown Raven X	-	
G-MTCA	CFM Shadow Srs B	-	
G-MTCB	Snowbird Mk III	-	
G-MTCE	Mainair Gemini Flash II	-	
G-MTCG	Solar Wings Pegasus XL-R	-	
G-MTCK	SW Pegasus Flash	J. J. Littler	
G-MTCL	Southdown Raven X	-	
G-MTCM	Southdown Raven X	-	
G-MTCN	Solar Wings Pegasus XL-R	-	
G-MTCO	Solar Wings Pegasus XL-R	-	
G-MTCP	Aerial Arts Chaser 110SX	-	
G-MTCR	Solar Wings Pegasus XL-R	G. M. Cruise-Smith	
G-MTCT	CFM Shadow Srs BD	-	
G-MTCU	Mainair Gemini Flash II	-	
G-MTCV	Microflight Spectrum	-	
G-MTCW	Mainair Gemini Flash	S. B. Walters	
G-MTCX	Solar Wings Pegasus XL-R	-	
G-MTCZ	Ultrasports Tripacer 250	-	
G-MTDA	Hornet Dual Trainer	-	
G-MTDD	Aerial Arts Chaser 110SX	-	
G-MTDE	American Aerolights 110SX	-	
G-MTDF	Mainair Gemini Flash II	P. G. Barnes	
G-MTDG	Solar Wings Pegasus XL-R	-	
G-MTDH	Solar Wings Pegasus XL-R	F. J. McVey	
G-MTDI	Solar Wings Pegasus XL-R	D. Allan	
G-MTDK	Aerotech MW.5 Sorcerer	-	
G-MTDM	Mainair Gemini Flash II	-	
G-MTDN	Ultraflight Lazair IIIE	-	
G-MTDO	Eipper Quicksilver MXII	-	
G-MTDR	Mainair Gemini Flash II	J. W. & C. Richardson	
G-MTDT	Solar Wings Pegasus XL-R	-	
G-MTDU	CFM Shadow Srs BD	-	
G-MTDW	Mainair Gemini Flash II	-	
G-MTDX	CFM Shadow Srs BD	-	
G-MTDY	Mainair Gemini Flash II	-	
G-MTDZ	Eipper Quicksilver MXII	-	
G-MTEA	Solar Wings Pegasus XL-R	-	
G-MTEB	Solar Wings Pegasus XL-R	-	
G-MTEC	Solar Wings Pegasus XL-R	-	
G-MTEE	Solar Wings Pegasus XL-R	M. J. Moulton	
G-MTEG	Mainair Gemini Flash II	-	
G-MTEJ	Mainair Gemini Flash II	-	
G-MTEK	Mainair Gemini Flash II	-	
G-MTEN	Mainair Gemini Flash II	-	
G-MTEO	Midland Ultralight Sirocco 337	-	
G-MTER	Solar Wings Pegasus XL-R	-	
G-MTES	Solar Wings Pegasus XL-R	-	
G-MTET	Solar Wings Pegasus XL-R	-	
G-MTEU	Solar Wings Pegasus XL-R	-	
G-MTEW	Solar Wings Pegasus XL-R	-	
G-MTEX	Solar Wings Pegasus XL-R	-	
G-MTEY	Mainair Gemini Flash II	-	
G-MTFA	Pegasus XL-R	-	
G-MTFB	Solar Wings Pegasus XL-R	-	
G-MTFC	Medway Hybred 44XLR	-	
G-MTFE	Solar Wings Pegasus XL-R	-	
G-MTFF	Mainair Gemini Flash II	-	
G-MTFG	AMF Chevvron 232	-	

Notes	Reg.	Type	Owner or Operator
	G-MTFI	Mainair Gemini Flash II	-
	G-MTFL	AMF Lazair IIIE	-
	G-MTFM	Solar Wings Pegasus XL-R	-
	G-MTFN	Aerotech MW.5 Sorcerer	R. D. Davidson
	G-MTFO	Solar Wings Pegasus XL-R	-
	G-MTFP	Solar Wings Pegasus XL-R	-
	G-MTFR	Solar Wings Pegasus XL-R	-
	G-MTFS	Solar Wings Pegasus XL-R	-
	G-MTFT	Solar Wings Pegasus XL-R	-
	G-MTFX	Mainair Gemini Flash	-
	G-MTFZ	CFM Shadow Srs BD	-
	G-MTGA	Mainair Gemini Flash	-
	G-MTGB	Thruster TST Mk 1	-
	G-MTGC	Thruster TST Mk 1	-
	G-MTGD	Thruster TST Mk 1	-
	G-MTGE	Thruster TST Mk 1	-
	G-MTGF	Thruster TST Mk 1	-
	G-MTGH	Mainair Gemini Flash IIA	-
	G-MTGJ	Solar Wings Pegasus XL-R	-
	G-MTGK	Solar Wings Pegasus XL-R	-
	G-MTGL	Solar Wings Pegasus XL-R	-
	G-MTGM	Solar Wings Pegasus XL-R	-
	G-MTGO	Mainair Gemini Flash	-
	G-MTGR	Thruster TST Mk 1	-
	G-MTGS	Thruster TST Mk 1	-
	G-MTGT	Thruster TST Mk 1	-
	G-MTGU	Thruster TST Mk 1	-
	G-MTGV	CFM Shadow Srs BD	-
	G-MTGW	CFM Shadow Srs BD	-
	G-MTGX	Hornet Dual Trainer	-
	G-MTGY	Southdown Lightning	-
	G-MTHB	Aerotech MW.5B Sorcerer	-
	G-MTHC	Raven X	-
	G-MTHG	Solar Wings Pegasus XL-R	-
	G-MTHH	Solar Wings Pegasus XL-R	-
	G-MTHI	Solar Wings Pegasus XL-R	-
	G-MTHJ	Solar Wings Pegasus XL-R	-
	G-MTHK	Solar Wings Pegasus XL-R	-
	G-MTHN	Solar Wings Pegasus XL-R	-
	G-MTHT	CFM Shadow Srs BD	C. A. S. Powell
	G-MTHU	Hornet Dual Trainer	-
	G-MTHV	CFM Shadow Srs BD	-
	G-MTHW	Mainair Gemini Flash II	-
	G-MTHZ	Mainair Gemini Flash IIA	-
	G-MTIA	Mainair Gemini Flash IIA	-
	G-MTIB	Mainair Gemini Flash IIA	-
	G-MTIE	Solar Wings Pegasus XL-R	P. Wibberley
	G-MTIH	Solar Wings Pegasus XL-R	B. Chapman
	G-MTIJ	Solar Wings Pegasus XL-R	-
	G-MTIK	Southdown Raven X	-
	G-MTIL	Mainair Gemini Flash IIA	P. G. Nolan
	G-MTIM	Mainair Gemini Flash IIA	-
	G-MTIN	Mainair Gemini Flash IIA	-
	G-MTIO	Solar Wings Pegasus XL-R	J. W. Mount
	G-MTIP	Solar Wings Pegasus XL-R	-
	G-MTIR	Solar Wings Pegasus XL-R	-
	G-MTIS	Solar Wings Pegasus XL-R	-
	G-MTIT	Solar Wings Pegasus XL-R	-
	G-MTIU	Solar Wings Pegasus XL-R	D. E. Pedder
	G-MTIV	Solar Wings Pegasus XL-R	-
	G-MTIW	Solar Wings Pegasus XL-R	-
	G-MTIX	Solar Wings Pegasus XL-R	-
	G-MTIY	Solar Wings Pegasus XL-R	-
	G-MTIZ	Solar Wings Pegasus XL-R	-
	G-MTJA	Mainair Gemini Flash IIA	K. A. Brunton
	G-MTJB	Mainair Gemini Flash IIA	B. Skidmore
	G-MTJC	Mainair Gemini Flash IIA	-
	G-MTJD	Mainair Gemini Flash IIA	-
	G-MTJE	Mainair Gemini Flash IIA	-
	G-MTJG	Medway Hybred 44XLR	-
	G-MTJH	SW Pegasus Flash	-
	G-MTJK	Mainair Gemini Flash IIA	-

Reg.	Type	Owner or Operator	Notes
G-MTJL	Mainair Gemini Flash IIA	J. Murphy	
G-MTJM	Mainair Gemini Flash IIA	-	
G-MTJN	Midland Ultralights Sirocco 377GB	-	
G-MTJP	Medway Hybred 44XLR	-	
G-MTJS	Solar Wings Pegasus XL-Q	-	
G-MTJT	Mainair Gemini Flash IIA	-	
G-MTJV	Mainair Gemini Flash IIA	-	
G-MTJW	Mainair Gemini Flash IIA	-	
G-MTJX	Raven/Dual Trainer	-	
G-MTJZ	Mainair Gemini Flash IIA	-	
G-MTKA	Thruster TST Mk 1	-	
G-MTKB	Thruster TST Mk 1	-	
G-MTKD	Thruster TST Mk 1	E. Spain/Ireland	
G-MTKE	Thruster TST Mk 1	-	
G-MTKG	Solar Wings Pegasus XL-R	-	
G-MTKH	Solar Wings Pegasus XL-R	-	
G-MTKI	Solar Wings Pegasus XL-R	-	
G-MTKM	Gardner T-M Scout S.2	-	
G-MTKN	Mainair Gemini Flash IIA	-	
G-MTKR	CFM Shadow Srs BD	-	
G-MTKS	CFM Shadow Srs BD	-	
G-MTKW	Mainair Gemini Flash IIA	-	
G-MTKX	Mainair Gemini Flash IIA	-	
G-MTKZ	Mainair Gemini Flash IIA	W. J. F. McLean & I. S. McNeill	
G-MTLB	Mainair Gemini Flash IIA	-	
G-MTLC	Mainair Gemini Flash IIA	-	
G-MTLD	Mainair Gemini Flash IIA	C. Kearney	
G-MTLE	-	-	
G-MTLG	Solar Wings Pegasus XL-R	G. J. Simoni	
G-MTLI	Solar Wings Pegasus XL-R	-	
G-MTLJ	Solar Wings Pegasus XL-R	-	
G-MTLK	Raven X	-	
G-MTLL	Mainair Gemini Flash IIA	-	
G-MTLM	Thruster TST Mk 1	-	
G-MTLN	Thruster TST Mk 1	A. R. Brew	
G-MTLR	Thruster TST Mk 1	-	
G-MTLS	Solar Wings Pegasus XL-R	-	
G-MTLT	Solar Wings Pegasus XL-R	-	
G-MTLU	Solar Wings Pegasus XL-R	-	
G-MTLV	Solar Wings Pegasus XL-R	-	
G-MTLX	Medway Hybred 44XLR	-	
G-MTLY	Solar Wings Pegasus XL-R	-	
G-MTLZ	Whittaker MW.5 Sorceror	-	
G-MTMA	Mainair Gemini Flash IIA	-	
G-MTMB	Mainair Gemini Flash IIA	-	
G-MTMC	Mainair Gemini Flash IIA	-	
G-MTME	Solar Wings Pegasus XL-R	-	
G-MTMF	Solar Wings Pegasus XL-R	-	
G-MTMG	Solar Wings Pegasus XL-R	-	
G-MTMI	Solar Wings Pegasus XL-R	-	
G-MTMK	Raven X	-	
G-MTML	Mainair Gemini Flash IIA	-	
G-MTMO	Raven X	-	
G-MTMP	Hornet Dual Trainer/Raven	-	
G-MTMR	Hornet Dual Trainer/Raven	-	
G-MTMT	Mainair Gemini Flash IIA	-	
G-MTMV	Mainair Gemini Flash IIA	-	
G-MTMW	Mainair Gemini Flash IIA	-	
G-MTMX	CFM Shadow Srs BD	-	
G-MTMY	CFM Shadow Srs BD	-	
G-MTNC	Mainair Gemini Flash IIA	-	
G-MTND	Medway Hybred 44XLR	-	
G-MTNE	Medway Hybred 44XLR	-	
G-MTNF	Medway Hybred 44XLR	-	
G-MTNG	Mainair Gemini Flash IIA	-	
G-MTNH	Mainair Gemini Flash IIA	-	
G-MTNI	Mainair Gemini Flash IIA	-	
G-MTNK	Weedhopper JC-24B	S. R. Davis	
G-MTNL	Mainair Gemini Flash IIA	-	
G-MTNM	Mainair Gemini Flash IIA	-	
G-MTNO	Solar Wings Pegasus XL-Q	-	
G-MTNP	Solar Wings Pegasus XL-Q	-	

Notes	Reg.	Type	Owner or Operator
	G-MTNR	Thruster TST Mk 1	A. M. Sirant
	G-MTNS	Thruster TST Mk 1	-
	G-MTNT	Thruster TST Mk 1	M. J. Clifford
	G-MTNU	Thruster TST Mk 1	J. L. A. Campbell
	G-MTNV	Thruster TST Mk 1	-
	G-MTNX	Mainair Gemini Flash II	-
	G-MTNY	Mainair Gemini Flash IIA	-
	G-MTOA	Solar Wings Pegasus XL-R	-
	G-MTOB	Solar Wings Pegasus XL-R	-
	G-MTOD	Solar Wings Pegasus XL-R	-
	G-MTOE	Solar Wings Pegasus XL-R	-
	G-MTOG	Solar Wings Pegasus XL-R	-
	G-MTOH	Solar Wings Pegasus XL-R	-
	G-MTOI	Solar Wings Pegasus XL-R	-
	G-MTOJ	Solar Wings Pegasus XL-R	S. Jelley
	G-MTOK	Solar Wings Pegasus XL-R	-
	G-MTOM	Solar Wings Pegasus XL-R	-
	G-MTON	Solar Wings Pegasus XL-R	-
	G-MTOO	Solar Wings Pegasus XL-R	-
	G-MTOP	Solar Wings Pegasus XL-R	-
	G-MTOS	Solar Wings Pegasus XL-R	-
	G-MTOT	Solar Wings Pegasus XL-R	A. J. Lloyd
	G-MTOU	Solar Wings Pegasus XL-R	-
	G-MTOV	Solar Wings Pegasus XL-R	-
	G-MTOY	Solar Wings Pegasus XL-R	-
	G-MTOZ	Solar Wings Pegasus XL-R	M. A. Furber
	G-MTPA	Mainair Gemini Flash IIA	-
	G-MTPC	Raven X	-
	G-MTPE	Solar Wings Pegasus XL-R	-
	G-MTPF	Solar Wings Pegasus XL-R	-
	G-MTPG	Solar Wings Pegasus XL-R	-
	G-MTPH	Solar Wings Pegasus XL-R	-
	G-MTPI	Solar Wings Pegasus XL-R	-
	G-MTPJ	Solar Wings Pegasus XL-R	-
	G-MTPK	Solar Wings Pegasus XL-R	-
	G-MTPL	Solar Wings Pegasus XL-R	-
	G-MTPM	Solar Wings Pegasus XL-R	-
	G-MTPN	Solar Wings Pegasus XL-Q	H. N. Graham
	G-MTPP	Solar Wings Pegasus XL-R	-
	G-MTPR	Solar Wings Pegasus XL-R	-
	G-MTPS	Solar Wings Pegasus XL-Q	-
	G-MTPT	Thruster TST Mk 1	-
	G-MTPU	Thruster TST Mk 1	-
	G-MTPW	Thruster TST Mk 1	-
	G-MTPX	Thruster TST Mk 1	-
	G-MTPY	Thruster TST Mk 1	C. M. Bradford
	G-MTRA	Mainair Gemini Flash IIA	-
	G-MTRC	Midlands Ultralights Sirocco 377G	-
	G-MTRD	Midlands Ultralights Sirocco 377G	-
	G-MTRJ	AMF Chevvron 232	-
	G-MTRL	Hornet Dual Trainer	-
	G-MTRM	Solar Wings Pegasus XL-R	M. S. Ahmadu
	G-MTRN	Solar Wings Pegasus XL-R	-
	G-MTRO	Solar Wings Pegasus XL-R	-
	G-MTRP	Solar Wings Pegasus XL-R	-
	G-MTRS	Solar Wings Pegasus XL-R	J. J. R. Tickle
	G-MTRT	Raven X	-
	G-MTRU	Solar Wings Pegasus XL-Q	-
	G-MTRV	Solar Wings Pegasus XL-Q	R. M. Adams
	G-MTRW	Raven X	-
	G-MTRX	Whittaker MW.5 Sorceror	-
	G-MTRZ	Mainair Gemini Flash IIA	-
	G-MTSB	Mainair Gemini Flash IIA	-
	G-MTSC	Mainair Gemini Flash IIA	-
	G-MTSD	Raven X	-
	G-MTSG	CFM Shadow Srs BD	-
	G-MTSH	Thruster TST Mk 1	-
	G-MTSJ	Thruster TST Mk 1	-
	G-MTSK	Thruster TST Mk 1	-
	G-MTSM	Thruster TST Mk 1	-
	G-MTSN	Solar Wings Pegasus XL-R	-
	G-MTSP	Solar Wings Pegasus XL-R	-

Reg.	Type	Owner or Operator	Notes
G-MTSR	Solar Wings Pegasus XL-R	-	
G-MTSS	Solar Wings Pegasus XL-R	-	
G-MTST	Thruster TST Mk 1	-	
G-MTSU	Solar Wings Pegasus XL-R	-	
G-MTSV	Solar Wings Pegasus XL-R	-	
G-MTSX	Solar Wings Pegasus XL-R	-	
G-MTSY	Solar Wings Pegasus XL-R	-	
G-MTSZ	Solar Wings Pegasus XL-R	-	
G-MTTA	Solar Wings Pegasus XL-R	-	
G-MTTB	Solar Wings Pegasus XL-R	-	
G-MTTD	Pegasus XL-Q	-	
G-MTTE	Solar Wings Pegasus XL-R	-	
G-MTTF	Aerotech MW.6 Merlin	-	
G-MTTH	CFM Shadow Srs BD	-	
G-MTTI	Mainair Gemini Flash IIA	-	
G-MTTL	Hiway Sky-Trike	-	
G-MTTM	Mainair Gemini Flash IIA	-	
G-MTTN	Ultralight Flight Phantom	-	
G-MTTO	Mainair Gemini Flash IIA	-	
G-MTTP	Mainair Gemini Flash IIA	-	
G-MTTR	Mainair Gemini Flash IIA	-	
G-MTTS	Mainair Gemini Flash IIA	-	
G-MTTU	Solar Wings Pegasus XL-R	A. Friend	
G-MTTW	Mainair Gemini Flash IIA	A. F. Glover	
G-MTTX	Solar Wings Pegasus XL-	-	
G-MTTY	Solar Wings Pegasus XL-Q	-	
G-MTTZ	Solar Wings Pegasus XL-Q	-	
G-MTUA	Solar Wings Pegasus XL-R	M. D. Reardon	
G-MTUB	Thruster TST Mk 1	-	
G-MTUC	Thruster TST Mk 1	-	
G-MTUD	Thruster TST Mk 1	-	
G-MTUF	Thruster TST Mk 1	-	
G-MTUH	Solar Wings Pegasus XL-R	-	
G-MTUI	Solar Wings Pegasus XL-R	R. Green	
G-MTUJ	Solar Wings Pegasus XL-R	-	
G-MTUK	Solar Wings Pegasus XL-R	-	
G-MTUL	Solar Wings Pegasus XL-R	-	
G-MTUN	Solar Wings Pegasus XL-Q	P. Boardman	
G-MTUP	Solar Wings Pegasus XL-Q	-	
G-MTUR	Solar Wings Pegasus XL-Q	-	
G-MTUS	Solar Wings Pegasus XL-Q	-	
G-MTUT	Solar Wings Pegasus XL-Q	-	
G-MTUU	Mainair Gemini Flash IIA	-	
G-MTUV	Mainair Gemini Flash IIA	-	
G-MTUX	Medway Hybred 44XLR	-	
G-MTUY	Solar Wings Pegasus XL-Q	-	
G-MTVA	Solar Wings Pegasus XL-R	-	
G-MTVB	Solar Wings Pegasus XL-R	-	
G-MTVE	Solar Wings Pegasus XL-R	-	
G-MTVF	Solar Wings Pegasus XL-R	-	
G-MTVH	Mainair Gemini Flash IIA	-	
G-MTVI	Mainair Gemini Flash IIA	-	
G-MTVJ	Mainair Gemini Flash IIA	D. W. Buck	
G-MTVK	Solar Wings Pegasus XL-R	-	
G-MTVL	Solar Wings Pegasus XL-R	-	
G-MTVM	Solar Wings Pegasus XL-R	-	
G-MTVN	Solar Wings Pegasus XL-R	-	
G-MTVO	Solar Wings Pegasus XL-R	-	
G-MTVP	Thruster TST Mk 1	-	
G-MTVR	Thruster TST Mk 1	G. Hawes	
G-MTVS	Thruster TST Mk 1	J. G. McMinn	
G-MTVT	Thruster TST Mk 1	-	
G-MTVV	Thruster TST Mk 1	C. Jones	
G-MTVX	Solar Wings Pegasus XL-Q	D. A. Foster	
G-MTWA	Solar Wings Pegasus XL-R	-	
G-MTWB	Solar Wings Pegasus XL-R	-	
G-MTWD	Solar Wings Pegasus XL-R	-	
G-MTWF	Mainair Gemini Flash IIA	-	
G-MTWG	Mainair Gemini Flash IIA	-	
G-MTWH	CFM Shadow Srs BD	-	
G-MTWK	CFM Shadow Srs BD	Bagby Shadow Whiskey Kilo Group	
G-MTWL	CFM Shadow Srs BD	-	

Notes	Reg.	Type	Owner or Operator
	G-MTWN	CFM Shadow Srs BD	-
	G-MTWP	CFM Shadow Srs BD	-
	G-MTWR	Mainair Gemini Flash IIA	-
	G-MTWS	Mainair Gemini Flash IIA	-
	G-MTWX	Mainair Gemini Flash IIA	-
	G-MTWY	Thruster TST Mk 1	-
	G-MTWZ	Thruster TST Mk 1	-
	G-MTXA	Thruster TST Mk 1	J. Upex
	G-MTXB	Thruster TST Mk 1	-
	G-MTXC	Thruster TST Mk 1	-
	G-MTXD	Thruster TST Mk 1	-
	G-MTXE	Hornet Dual Trainer	-
	G-MTXH	Solar Wings Pegasus XL-Q	-
	G-MTXI	Solar Wings Pegasus XL-Q	-
	G-MTXJ	Solar Wings Pegasus XL-Q	-
	G-MTXK	Solar Wings Pegasus XL-Q	-
	G-MTXL	Noble Hardman Snowbird Mk IV	-
	G-MTXM	Mainair Gemini Flash IIA	-
	G-MTXO	Whittaker MW.6	-
	G-MTXP	Mainair Gemini Flash IIA	G. S. Duerden
	G-MTXR	CFM Shadow Srs BD	-
	G-MTXS	Mainair Gemini Flash IIA	J. Kennedy
	G-MTXT	MBA Tiger Cub 440	-
	G-MTXU	Snowbird Mk.IV	-
	G-MTXW	Snowbird Mk.IV	-
	G-MTXY	Hornet Dual Trainer	-
	G-MTXZ	Mainair Gemini Flash IIA	-
	G-MTYA	Solar Wings Pegasus XL-Q	-
	G-MTYC	Solar Wings Pegasus XL-Q	-
	G-MTYD	Solar Wings Pegasus XL-Q	-
	G-MTYE	Solar Wings Pegasus XL-Q	-
	G-MTYF	Solar Wings Pegasus XL-Q	-
	G-MTYG	Solar Wings Pegasus XL-Q	-
	G-MTYI	Solar Wings Pegasus XL-Q	-
	G-MTYL	Solar Wings Pegasus XL-Q	-
	G-MTYP	Solar Wings Pegasus XL-Q	-
	G-MTYR	Solar Wings Pegasus XL-Q	-
	G-MTYS	Solar Wings Pegasus XL-Q	-
	G-MTYU	Solar Wings Pegasus XL-Q	S. East
	G-MTYW	Raven X	-
	G-MTYX	Raven X	-
	G-MTYY	Solar Wings Pegasus XL-R	-
	G-MTZA	Thruster TST Mk 1	-
	G-MTZB	Thruster TST Mk 1	-
	G-MTZC	Thruster TST Mk 1	-
	G-MTZD	Thruster TST Mk 1	A. Spence
	G-MTZE	Thruster TST Mk 1	B. S. P. Finch
	G-MTZF	Thruster TST Mk 1	D. C. Marsh
	G-MTZG	Mainair Gemini Flash IIA	-
	G-MTZH	Mainair Gemini Flash IIA	-
	G-MTZJ	Solar Wings Pegasus XL-R	-
	G-MTZK	Solar Wings Pegasus XL-R	-
	G-MTZL	Mainair Gemini Flash IIA	-
	G-MTZK	Solar Wings Pegasus XL-R	G. F. Jones
	G-MTZM	Mainair Gemini Flash IIA	-
	G-MTZN	Mainair Gemini Flash IIA	-
	G-MTZO	Mainair Gemini Flash IIA	-
	G-MTZP	Solar Wings Pegasus XL-Q	-
	G-MTZR	Solar Wings Pegasus XL-Q	-
	G-MTZS	Solar Wings Pegasus XL-Q	-
	G-MTZT	Solar Wings Pegasus XL-Q	-
	G-MTZV	Mainair Gemini Flash IIA	-
	G-MTZW	Mainair Gemini Flash IIA	-
	G-MTZX	Mainair Gemini Flash IIA	-
	G-MTZY	Mainair Gemini Flash IIA	T. C. Palmer
	G-MTZZ	Mainair Gemini Flash IIA	-
	G-MUFY	Robinson R-22B	Rotormurf Ltd
	G-MUIR	Cameron V-65 balloon	L. C. M. Muir
	G-MULT	Beech 76 Duchess	Folada Aero & Technical Services Ltd
	G-MUNI	Mooney M.20J	M. W. Fane
	G-MURP	AS.350 Ecureuil	M. Murphy

Reg.	Type	Owner or Operator	Notes
G-MURR	Whittaker MW.6 Merlin	D. Murray	
G-MURY	Robinson R-44	Simlot Ltd	
G-MUSH	Robinson R-44-II	Heli Air Ltd/Wellesbourne	
G-MUSO	Rutan LongEz	C. J. Tadjeran/Sweden	
G-MUTE	Colt 31A balloon	Redmalt Ltd	
G-MUTZ	Avtech Jabiru J.400	N. C. Dean	
G-MUVG	Cessna 421C	Martin Collins Aviation Ltd	
G-MVAA	Mainair Gemini Flash IIA	-	
G-MVAB	Mainair Gemini Flash IIA	B. Hindley	
G-MVAC	CFM Shadow Srs BD	-	
G-MVAD	Mainair Gemini Flash IIA	N. D. Fox & C. J. Hemmingway	
G-MVAF	Southdown Puma Sprint	-	
G-MVAG	Thruster TST Mk 1	-	
G-MVAH	Thruster TST Mk 1	-	
G-MVAI	Thruster TST Mk 1	-	
G-MVAJ	Thruster TST Mk 1	-	
G-MVAK	Thruster TST Mk 1	-	
G-MVAL	Thruster TST Mk 1	-	
G-MVAM	CFM Shadow Srs BD	-	
G-MVAN	CFM Shadow Srs BD	-	
G-MVAO	Mainair Gemini Flash IIA	-	
G-MVAP	Mainair Gemini Flash IIA	-	
G-MVAR	Solar Wings Pegasus XL-R	A. J. Thomas	
G-MVAT	Solar Wings Pegasus XL-R	-	
G-MVAV	Solar Wings Pegasus XL-R	-	
G-MVAW	Solar Wings Pegasus XL-Q	-	
G-MVAX	Solar Wings Pegasus XL-Q	-	
G-MVAY	Solar Wings Pegasus XL-Q	-	
G-MVAZ	Solar Wings Pegasus XL-Q	-	
G-MVBA	Solar Wings Pegasus XL-Q	-	
G-MVBB	CFM Shadow Srs BD	-	
G-MVBC	Aerial Arts Tri-Flyer 130SX	-	
G-MVBD	Mainair Gemini Flash IIA	-	
G-MVBF	Mainair Gemini Flash IIA	-	
G-MVBG	Mainair Gemini Flash IIA	-	
G-MVBI	Mainair Gemini Flash IIA	-	
G-MVBK	Mainair Gemini Flash IIA	-	
G-MVBL	Mainair Gemini Flash IIA	-	
G-MVBM	Mainair Gemini Flash IIA	-	
G-MVBN	Mainair Gemini Flash IIA	-	
G-MVBO	Mainair Gemini Flash IIA	-	
G-MVBP	Thruster TST Mk 1	-	
G-MVBS	Thruster TST Mk 1	-	
G-MVBT	Thruster TST Mk 1	-	
G-MVBY	Solar Wings Pegasus XL-R	-	
G-MVBZ	Solar Wings Pegasus XL-R	-	
G-MVCA	Solar Wings Pegasus XL-R	-	
G-MVCB	Solar Wings Pegasus XL-R	-	
G-MVCC	CFM Shadow Srs BD	S. J. Payne	
G-MVCD	Medway Hybred 44XLR	-	
G-MVCE	Mainair Gemini Flash IIA	-	
G-MVCF	Mainair Gemini Flash IIA	J. S. Harris	
G-MVCH	Noble Hardman Snowbird Mk IV	-	
G-MVCI	Noble Hardman Snowbird Mk IV	-	
G-MVCJ	Noble Hardman Snowbird Mk IV	-	
G-MVCL	Solar Wings Pegasus XL-Q	-	
G-MVCM	Solar Wings Pegasus XL-Q	-	
G-MVCN	Solar Wings Pegasus XL-Q	-	
G-MVCP	Solar Wings Pegasus XL-Q	-	
G-MVCR	Solar Wings Pegasus XL-Q	-	
G-MVCS	Solar Wings Pegasus XL-Q	-	
G-MVCT	Solar Wings Pegasus XL-Q	-	
G-MVCV	Solar Wings Pegasus XL-Q	-	
G-MVCW	CFM Shadow Srs BD	-	
G-MVCY	Mainair Gemini Flash IIA	B. Hall	
G-MVCZ	Mainair Gemini Flash IIA	-	
G-MVDA	Mainair Gemini Flash IIA	-	
G-MVDD	Thruster TST Mk 1	-	
G-MVDE	Thruster TST Mk 1	-	
G-MVDF	Thruster TST Mk 1	-	
G-MVDG	Thruster TST Mk 1	-	

Notes	Reg.	Type	Owner or Operator
	G-MVDH	Thruster TST Mk 1	-
	G-MVDJ	Medway Hybred 44XLR	-
	G-MVDK	Aerial Arts Chaser S	-
	G-MVDL	Aerial Arts Chaser S	-
	G-MVDO	Aerial Arts Chaser S	-
	G-MVDP	Aerial Arts Chaser S	-
	G-MVDR	Aerial Arts Chaser S.447	A. M. Sutton
	G-MVDT	Mainair Gemini Flash IIA	-
	G-MVDU	Solar Wings Pegasus XL-R	-
	G-MVDV	Solar Wings Pegasus XL-R	I. Hutchinson
	G-MVDW	Solar Wings Pegasus XL-R	-
	G-MVDX	Solar Wings Pegasus XL-R	-
	G-MVDY	Solar Wings Pegasus XL-R	-
	G-MVDZ	Solar Wings Pegasus XL-R	-
	G-MVEC	Solar Wings Pegasus XL-R	-
	G-MVED	Solar Wings Pegasus XL-R	-
	G-MVEE	Medway Hybred 44XLR	-
	G-MVEF	Solar Wings Pegasus XL-R	-
	G-MVEG	Solar Wings Pegasus XL-R	-
	G-MVEH	Mainair Gemini Flash IIA	K. Bailey
	G-MVEI	CFM Shadow Srs BD	-
	G-MVEJ	Mainair Gemini Flash IIA	-
	G-MVEK	Mainair Gemini Flash IIA	R. M. Rea
	G-MVEL	Mainair Gemini Flash IIA	-
	G-MVEN	CFM Shadow Srs BD	N. J. Mepham
	G-MVEO	Mainair Gemini Flash IIA	-
	G-MVER	Mainair Gemini Flash IIA	-
	G-MVES	Mainair Gemini Flash IIA	-
	G-MVET	Mainair Gemini Flash IIA	-
	G-MVEV	Mainair Gemini Flash IIA	-
	G-MVEW	Mainair Gemini Flash IIA	-
	G-MVEX	Solar Wings Pegasus XL-Q	-
	G-MVEZ	Solar Wings Pegasus XL-Q	-
	G-MVFA	Solar Wings Pegasus XL-Q	-
	G-MVFB	Solar Wings Pegasus XL-Q	-
	G-MVFC	Solar Wings Pegasus XL-Q	-
	G-MVFD	Solar Wings Pegasus XL-Q	-
	G-MVFE	Solar Wings Pegasus XL-Q	-
	G-MVFF	Solar Wings Pegasus XL-Q	-
	G-MVFH	CFM Shadow Srs BD	-
	G-MVFJ	Thruster TST Mk 1	-
	G-MVFK	Thruster TST Mk 1	-
	G-MVFL	Thruster TST Mk 1	-
	G-MVFM	Thruster TST Mk 1	-
	G-MVFN	Thruster TST Mk 1	-
	G-MVFO	Thruster TST Mk 1	-
	G-MVFP	Solar Wings Pegasus XL-R	-
	G-MVFR	Solar Wings Pegasus XL-R	-
	G-MVFS	Solar Wings Pegasus XL-R	-
	G-MVFT	Solar Wings Pegasus XL-R	-
	G-MVFV	Solar Wings Pegasus XL-R	L. R. M. Grigg
	G-MVFW	Solar Wings Pegasus XL-R	-
	G-MVFX	Solar Wings Pegasus XL-R	-
	G-MVFY	Solar Wings Pegasus XL-R	-
	G-MVFZ	Solar Wings Pegasus XL-R	-
	G-MVGA	Aerial Arts Chaser S	-
	G-MVGB	Medway Hybred 44XLR	-
	G-MVGC	AMF Chevvron 2-32	W. Fletcher
	G-MVGD	AMF Chevvron 2-32	-
	G-MVGE	AMF Chevvron 2-32	-
	G-MVGF	Aerial Arts Chaser S	-
	G-MVGG	Aerial Arts Chaser S	J. J. Bowen
	G-MVGH	Aerial Arts Chaser S	-
	G-MVGM	Mainair Gemini Flash IIA	-
	G-MVGN	Solar Wings Pegasus XL-R	M. J. Smith
	G-MVGO	Solar Wings Pegasus XL-R	-
	G-MVGU	Solar Wings Pegasus XL-Q	-
	G-MVGW	Solar Wings Pegasus XL-Q	-
	G-MVGX	Solar Wings Pegasus XL-Q	-
	G-MVGY	Medway Hybred 44XL	-
	G-MVGZ	Ultraflight Lazair IIIE	-
	G-MVHA	Aerial Arts Chaser S	-

Reg.	Type	Owner or Operator	Notes
G-MVHB	Powerchute Raider	-	
G-MVHC	Powerchute Raider	-	
G-MVHD	CFM Shadow Srs BD	-	
G-MVHE	Mainair Gemini Flash IIA	-	
G-MVHF	Mainair Gemini Flash IIA	-	
G-MVHG	Mainair Gemini Flash II	-	
G-MVHH	Mainair Gemini Flash IIA	I. J. Cleland & M. D. Calder	
G-MVHI	Thruster TST Mk 1	R. A. Samulis	
G-MVHJ	Thruster TST Mk 1	-	
G-MVHK	Thruster TST Mk 1	-	
G-MVHL	Thruster TST Mk 1	-	
G-MVHN	Aerial Arts Chaser S	-	
G-MVHP	Solar Wings Pegasus XL-Q	-	
G-MVHR	Solar Wings Pegasus XL-Q	-	
G-MVHS	Solar Wings Pegasus XL-Q	-	
G-MVHV	Solar Wings Pegasus XL-Q	-	
G-MVHW	Solar Wings Pegasus XL-Q	-	
G-MVHX	Solar Wings Pegasus XL-Q	-	
G-MVHY	Solar Wings Pegasus XL-Q	-	
G-MVHZ	Hornet Dual Trainer	J. M. Addison	
G-MVIA	Solar Wings Pegasus XL-R	-	
G-MVIB	Mainair Gemini Flash IIA	LSA Systems	
G-MVID	Aerial Arts Chaser 5	-	
G-MVIE	Aerial Arts Chaser S	-	
G-MVIF	Medway Raven X	-	
G-MVIG	CFM Shadow Srs B	-	
G-MVIH	Mainair Gemini Flash IIA	-	
G-MVIL	Noble Hardman Snowbird Mk IV	-	
G-MVIM	Noble Hardman Snowbird Mk.IV	-	
G-MVIN	Noble Hardman Snowbird Mk.IV	C. P. Dawes	
G-MVIO	Noble Hardman Snowbird Mk IV	-	
G-MVIP	AMF Chevvron 232	P. C. Avery	
G-MVIR	Thruster TST Mk 1	-	
G-MVIT	Thruster TST Mk 1	-	
G-MVIU	Thruster TST Mk 1	-	
G-MVIV	Thruster TST Mk 1	-	
G-MVIW	Thruster TST Mk 1	-	
G-MVIX	Mainair Gemini Flash IIA	-	
G-MVIY	Mainair Gemini Flash IIA	-	
G-MVIZ	Mainair Gemini Flash IIA	-	
G-MVJA	Mainair Gemini Flash IIA	-	
G-MVJC	Mainair Gemini Flash IIA	-	
G-MVJD	Solar Wings Pegasus XL-R	-	
G-MVJE	Mainair Gemini Flash IIA	-	
G-MVJF	Aerial Arts Chaser S	V. S. Rudham	
G-MVJG	Aerial Arts Chaser S	-	
G-MVJH	Aerial Arts Chaser S	-	
G-MVJI	Aerial Arts Chaser S	-	
G-MVJJ	Aerial Arts Chaser S	-	
G-MVJK	Aerial Arts Chaser S	K. J. Samuels	
G-MVJL	Mainair Gemini Flash IIA	-	
G-MVJM	Microflight Spectrum	-	
G-MVJN	Solar Wings Pegasus XL-Q	R. A. Paintain	
G-MVJO	Solar Wings Pegasus XL-Q	C. M. Wilkes	
G-MVJP	Solar Wings Pegasus XL-Q	-	
G-MVJR	Solar Wings Pegasus XL-Q	B. Goldsmith	
G-MVJS	Solar Wings Pegasus XL-Q	-	
G-MVJT	Solar Wings Pegasus XL-Q	-	
G-MVJU	Solar Wings Pegasus XL-Q	-	
G-MVJW	Solar Wings Pegasus XL-Q	-	
G-MVJZ	Birdman Cherokee	-	
G-MVKB	Medway Hybred 44XLR	-	
G-MVKC	Mainair Gemini Flash IIA	-	
G-MVKF	Solar Wings Pegasus XL-R	-	
G-MVKH	Solar Wings Pegasus XL-R	-	
G-MVKJ	Solar Wings Pegasus XL-R	-	
G-MVKK	Solar Wings Pegasus XL-R	-	
G-MVKL	Solar Wings Pegasus XL-R	-	
G-MVKM	Solar Wings Pegasus XL-R	A. E. Dobson	
G-MVKN	Solar Wings Pegasus XL-R	-	
G-MVKO	Solar Wings Pegasus XL-Q	C. R. Bunce	
G-MVKP	Solar Wings Pegasus XL-Q	P. Mokryk & S. King	

Notes	Reg.	Type	Owner or Operator
	G-MVKR	Solar Wings Pegasus XL-Q	-
	G-MVKS	Solar Wings Pegasus XL-Q	-
	G-MVKT	Solar Wings Pegasus XL-Q	-
	G-MVKU	Solar Wings Pegasus XL-Q	-
	G-MVKV	Solar Wings Pegasus XL-Q	D. M. Taylor
	G-MVKW	Solar Wings Pegasus XL-Q	-
	G-MVKY	Aerial Arts Chaser S	-
	G-MVKZ	Aerial Arts Chaser S	-
	G-MVLA	Aerial Arts Chaser S	-
	G-MVLB	Aerial Arts Chaser S	R. P. Wilkinson
	G-MVLC	Aerial Arts Chaser S	-
	G-MVLD	Aerial Arts Chaser S	A. W. Leadley
	G-MVLE	Aerial Arts Chaser S	-
	G-MVLF	Aerial Arts Chaser S	I. B. Smith
	G-MVLG	Aerial Arts Chaser S	-
	G-MVLH	Aerial Arts Chaser S	-
	G-MVLJ	CFM Shadow Srs B	-
	G-MVLL	Mainair Gemini Flash IIA	M. J. A. New
	G-MVLP	CFM Shadow Srs BD	-
	G-MVLR	Mainair Gemini Flash IIA	-
	G-MVLS	Aerial Arts Chaser S	-
	G-MVLT	Aerial Arts Chaser S	-
	G-MVLW	Aerial Arts Chaser S	-
	G-MVLX	Solar Wings Pegasus XL-Q	-
	G-MVLY	Solar Wings Pegasus XL-Q	-
	G-MVMA	Solar Wings Pegasus XL-Q	-
	G-MVMC	Solar Wings Pegasus XL-Q	P. Smith & I. W. Barlow
	G-MVMD	Powerchute Raider	-
	G-MVME	Thruster TST Mk 1	P. J. Edwards
	G-MVMG	Thruster TST Mk 1	A. D. McCaldin
	G-MVMI	Thruster TST Mk 1	-
	G-MVMK	Medway Hybred 44XLR	-
	G-MVML	Aerial Arts Chaser S	-
	G-MVMM	Aerial Arts Chaser S	-
	G-MVMO	Mainair Gemini Flash IIA	-
	G-MVMR	Mainair Gemini Flash IIA	-
	G-MVMT	Mainair Gemini Flash IIA	-
	G-MVMU	Mainair Gemini Flash IIA	-
	G-MVMV	Aerotech MW.5 (K) Sorcerer	-
	G-MVMW	Mainair Gemini Flash IIA	-
	G-MVMX	Mainair Gemini Flash IIA	N. M. Corr
	G-MVMY	Mainair Gemini Flash IIA	-
	G-MVMZ	Mainair Gemini Flash IIA	-
	G-MVNA	Powerchute Raider	-
	G-MVNB	Powerchute Raider	-
	G-MVNC	Powerchute Raider	-
	G-MVNF	Powerchute Raider	-
	G-MVNI	Powerchute Raider	-
	G-MVNK	Powerchute Raider	-
	G-MVNL	Powerchute Raider?	-
	G-MVNO	Aerotech MW.5 (K) Sorcerer	-
	G-MVNP	Aerotech MW.5 (K) Sorcerer	-
	G-MVNR	Aerotech MW.5 (K) Sorcerer	E. I. Rowlands-Jones
	G-MVNS	Aerotech MW.5 (K) Sorcerer	-
	G-MVNT	Whittaker MW.5 (K) Sorcerer	-
	G-MVNU	Aerotech MW.5 Sorcerer	-
	G-MVNV	Aerotech MW.5 Sorcerer	-
	G-MVNW	Mainair Gemini Flash IIA	-
	G-MVNX	Mainair Gemini Flash IIA	-
	G-MVNY	Mainair Gemini Flash IIA	-
	G-MVNZ	Mainair Gemini Flash IIA	J. Howarth
	G-MVOA	Aerial Arts Alligator	-
	G-MVOB	Mainair Gemini Flash IIA	-
	G-MVOD	Aerial Arts Chaser 110SX	-
	G-MVOF	Mainair Gemini Flash IIA	P. J. Nolan
	G-MVOH	CFM Shadow Srs B	-
	G-MVOI	Noble Hardman Snowbird Mk IV	-
	G-MVOJ	Noble Hardman Snowbird Mk IV	-
	G-MVOK	Noble Hardman Snowbird Mk IV	-
	G-MVOL	Noble Hardman Snowbird Mk IV	-
	G-MVON	Mainair Gemini Flash IIA	-
	G-MVOO	AMF Chevvron 2-32	-

Reg.	Type	Owner or Operator	Notes
G-MVOP	Aerial Arts Chaser S	-	
G-MVOR	Mainair Gemini Flash IIA		
G-MVOT	Thruster TST Mk 1	-	
G-MVOU	Thruster TST Mk 1	-	
G-MVOV	Thruster TST Mk 1	-	
G-MVOW	Thruster TST Mk 1	-	
G-MVOX	Thruster TST Mk 1	-	
G-MVOY	Thruster TST Mk 1	-	
G-MVPA	Mainair Gemini Flash IIA		
G-MVPB	Mainair Gemini Flash IIA	O. Carter	
G-MVPC	Mainair Gemini Flash IIA	-	
G-MVPD	Mainair Gemini Flash IIA		
G-MVPE	Mainair Gemini Flash IIA	A. Croucher	
G-MVPF	Medway Hybred 44XLR	-	
G-MVPG	Medway Hybred 44XLR	-	
G-MVPH	Whittaker MW.6 Merlin	-	
G-MVPI	Mainair Gemini Flash IIA	-	
G-MVPJ	Rans S.5	-	
G-MVPK	CFM Shadow Srs B	-	
G-MVPL	Medway Hybred 44XLR	-	
G-MVPM	Whittaker MW.6 Merlin	-	
G-MVPN	Whittaker MW.6 Merlin	-	
G-MVPO	Mainair Gemini Flash IIA	-	
G-MVPR	Solar Wings Pegasus XL-Q	-	
G-MVPS	Solar Wings Pegasus XL-Q	-	
G-MVPT	Solar Wings Pegasus XL-Q	-	
G-MVPU	Solar Wings Pegasus XL-Q	-	
G-MVPW	Solar Wings Pegasus XL-R	-	
G-MVPX	Solar Wings Pegasus XL-Q	-	
G-MVPY	Solar Wings Pegasus XL-Q	-	
G-MVRA	Mainair Gemini Flash IIA	-	
G-MVRB	Mainair Gemini Flash	J. Walshe/Northern Ireland	
G-MVRC	Mainair Gemini Flash IIA	-	
G-MVRD	Mainair Gemini Flash IIA	-	
G-MVRE	CFM Shadow Srs BD	-	
G-MVRF	Rotec Rally 2B	-	
G-MVRG	Aerial Arts Chaser S	-	
G-MVRH	Solar Wings Pegasus XL-Q	-	
G-MVRI	Solar Wings Pegasus XL-Q	-	
G-MVRJ	Solar Wings Pegasus XL-Q	J. Goldsmith-Ryan	
G-MVRL	Aerial Arts Chaser S	-	
G-MVRM	Mainair Gemini Flash IIA	G. G. Wood	
G-MVRO	CFM Shadow Srs CD	-	
G-MVRP	CFM Shadow Srs BD	B. Barrass	
G-MVRR	CFM Shadow Srs BD	-	
G-MVRT	CFM Shadow Srs BD	-	
G-MVRU	Solar Wings Pegasus XL-Q	P. Copping	
G-MVRV	Powerchute Kestrel	-	
G-MVRW	Solar Wings Pegasus XL-Q	-	
G-MVRX	Solar Wings Pegasus XL-Q	-	
G-MVRY	Medway Hybred 44XLR	-	
G-MVRZ	Medway Hybred 44XLR	-	
G-MVSA	Solar Wings Pegasus XL-Q	-	
G-MVSB	Solar Wings Pegasus XL-Q	-	
G-MVSD	Solar Wings Pegasus XL-Q	-	
G-MVSE	Solar Wings Pegasus XL-Q	-	
G-MVSG	Aerial Arts Chaser S	-	
G-MVSI	Medway Hybred 44XLR	-	
G-MVSJ	Aviasud Mistral 532	-	
G-MVSK	Aerial Arts Chaser S	-	
G-MVSM	Midland Ultralights Sirocco	-	
G-MVSN	Mainair Gemini Flash IIA	D. W. Watson	
G-MVSO	Mainair Gemini Flash IIA	-	
G-MVSP	Mainair Gemini Flash IIA	-	
G-MVSR	Medway Hybred 44XLR	-	
G-MVSS	Hornet RS-ZA	-	
G-MVST	Mainair Gemini Flash IIA	-	
G-MVSV	Mainair Gemini Flash IIA	-	
G-MVSW	Solar Wings Pegasus XL-Q	-	
G-MVSX	Solar Wings Pegasus XL-Q	-	
G-MVSY	Solar Wings Pegasus XL-Q	-	
G-MVSZ	Solar Wings Pegasus XL-Q	D. J. Ackroyd	

Notes	Reg.	Type	Owner or Operator
	G-MVTA	Solar Wings Pegasus XL-Q	P. Hanby
	G-MVTC	Mainair Gemini Flash IIA	-
	G-MVTD	Whittaker MW.6 Merlin	-
	G-MVTF	Aerial Arts Chaser S	-
	G-MVTI	Solar Wings Pegasus XL-Q	-
	G-MVTJ	Solar Wings Pegasus XL-Q	-
	G-MVTK	Solar Wings Pegasus XL-Q	I. P. Sissons
	G-MVTL	Aerial Arts Chaser S	-
	G-MVTM	Aerial Arts Chaser S	-
	G-MVUA	Mainair Gemini Flash IIA	-
	G-MVUB	Thruster T.300	-
	G-MVUC	Medway Hybred 44XLR	-
	G-MVUD	Medway Hybred 44XLR	-
	G-MVUE	Solar Wings Pegasus XL-Q	-
	G-MVUF	Solar Wings Pegasus XL-Q	-
	G-MVUG	Solar Wings Pegasus XL-Q	-
	G-MVUI	Solar Wings Pegasus XL-Q	-
	G-MVUJ	Solar Wings Pegasus XL-Q	-
	G-MVUK	Solar Wings Pegasus XL-Q	-
	G-MVUL	Solar Wings Pegasus XL-Q	-
	G-MVUM	Solar Wings Pegasus XL-Q	-
	G-MVUN	Solar Wings Pegasus XL-Q	-
	G-MVUO	AMF Chevvron 2-32	-
	G-MVUP	Aviasud Mistral	-
	G-MVUR	Hornet ZA	-
	G-MVUS	Aerial Arts Chaser S	-
	G-MVUT	Aerial Arts Chaser S	-
	G-MVUU	Hornet ZA	-
	G-MVVF	Medway Hybred 44XLR	-
	G-MVVG	Medway Hybred 44XLR	-
	G-MVVH	Medway Hybred 44XLR	-
	G-MVVI	Medway Hybred 44XLR	-
	G-MVVK	Solar Wings Pegasus XL-R	-
	G-MVVM	Solar Wings Pegasus XL-R	A. F. Cunningham
	G-MVVN	Solar Wings Pegasus XL-Q	-
	G-MVVP	Solar Wings Pegasus XL-Q	-
	G-MVVT	CFM Shadow Srs BD	-
	G-MVVU	Aerial Arts Chaser S	-
	G-MVVV	AMF Chevvron 2-32	-
	G-MVVZ	Powerchute Raider	-
	G-MVWD	Powerchute Raider	-
	G-MVWE	Powerchute Raider	-
	G-MVWJ	Powerchute Raider	-
	G-MVWN	Thruster T.300	-
	G-MVWO	Thruster T.300	-
	G-MVWR	Thruster T.300	-
	G-MVWS	Thruster T.300	-
	G-MVWV	Medway Hybred 44XLR	-
	G-MVWW	Aviasud Mistral	-
	G-MVWZ	Aviasud Mistral	Chilbolton Mistral Group
	G-MVXA	Whittaker MW.6 Merlin	-
	G-MVXB	Mainair Gemini Flash IIA	-
	G-MVXC	Mainair Gemini Flash IIA	-
	G-MVXD	Medway Hybred 44XLR	-
	G-MVXE	Medway Hybred 44XLR	-
	G-MVXI	Medway Hybred 44XLR	T. de Landro
	G-MVXJ	Medway Hybred 44XLR	-
	G-MVXL	Thruster TST Mk 1	-
	G-MVXM	Medway Hybred 44XLR	-
	G-MVXN	Aviasud Mistral	-
	G-MVXR	Mainair Gemini Flash IIA	-
	G-MVXS	Mainair Gemini Flash IIA	-
	G-MVXT	Mainair Gemini Flash IIA	-
	G-MVXV	Aviasud Mistral	-
	G-MVXW	Rans S.4 Coyote	-
	G-MVXX	AMF Chevvron 232	-
	G-MVYA	Aerial Arts Chaser S	-
	G-MVYC	Solar Wings Pegasus XL-Q	-
	G-MVYD	Solar Wings Pegasus XL-Q	-
	G-MVYE	Thruster TST Mk 1	-
	G-MVYG	Hornet R-ZA	-
	G-MVYI	Hornet R-ZA	-

Reg.	Type	Owner or Operator	Notes
G-MVYK	Hornet R-ZA	-	
G-MVYL	Hornet R-ZA	-	
G-MVYN	Hornet R-ZA	-	
G-MVYP	Medway Hybred 44XLR	-	
G-MVYR	Medway Hybred 44XLR	-	
G-MVYS	Mainair Gemini Flash IIA	-	
G-MVYT	Noble Hardman Snowbird Mk IV	-	
G-MVYU	Noble Hardman Snowbird Mk IV	B. Foster & P. Meah	
G-MVYV	Noble Hardman Snowbird Mk IV	-	
G-MVYW	Noble Hardman Snowbird Mk IV	-	
G-MVYX	Noble Hardman Snowbird Mk IV	-	
G-MVYY	Aerial Arts Chaser S508	-	
G-MVYZ	CFM Shadow Srs BD	A. P. Worbey	
G-MVZA	Thruster T.300	-	
G-MVZB	Thruster T.300	-	
G-MVZC	Thruster T.300	-	
G-MVZD	Thruster T.300	-	
G-MVZE	Thruster T.300	-	
G-MVZG	Thruster T.300	-	
G-MVZI	Thruster T.300	-	
G-MVZJ	Solar Wings Pegasus XL-Q	-	
G-MVZK	Challenger II	G-MVZK Group	
G-MVZL	Solar Wings Pegasus XL-Q	-	
G-MVZM	Aerial Arts Chaser S	J. L. Parker	
G-MVZN	Aerial Arts Chaser S	-	
G-MVZO	Medway Hybred 44XLR	-	
G-MVZP	Murphy Renegade Spirit UK	-	
G-MVZR	Aviasud Mistral	-	
G-MVZS	Mainair Gemini Flash IIA	-	
G-MVZT	Solar Wings Pegasus XL-Q	-	
G-MVZU	Solar Wings Pegasus XL-Q	-	
G-MVZV	Solar Wings Pegasus XL-Q	-	
G-MVZW	Hornet R-ZA	-	
G-MVZX	Renegade Spirit UK	-	
G-MVZZ	AMF Chevvron 232	-	
G-MWAB	Mainair Gemini Flash IIA	-	
G-MWAC	Solar Wings Pegasus XL-Q	-	
G-MWAD	Solar Wings Pegasus XL-Q	-	
G-MWAE	CFM Shadow Srs BD	-	
G-MWAF	Solar Wings Pegasus XL-R	I. W. Skeldon	
G-MWAG	Solar Wings Pegasus XL-R	X. Norman	
G-MWAH	Hornet RS-ZA	-	
G-MWAJ	Renegade Spirit UK	-	
G-MWAL	Solar Wings Pegasus XL-Q	-	
G-MWAM	Thruster T.300	-	
G-MWAN	Thruster T.300	R. B. Hawkins	
G-MWAP	Thruster T.300	-	
G-MWAR	Thruster T.300	-	
G-MWAT	Solar Wings Pegasus XL-Q	-	
G-MWAU	Mainair Gemini Flash IIA	-	
G-MWAV	Solar Wings Pegasus XL-R	-	
G-MWAW	Whittaker MW.6 Merlin	-	
G-MWBH	Hornet RS-ZA	-	
G-MWBI	Medway Hybred 44XLR	-	
G-MWBJ	Medway Sprint	-	
G-MWBK	Solar Wings Pegasus XL-Q	-	
G-MWBL	Solar Wings Pegasus XL-Q	-	
G-MWBO	Rans S.4 Coyote	G-MWBO Group	
G-MWBP	Hornet RS-ZA	S. Brader	
G-MWBR	Hornet RS-ZA	-	
G-MWBS	Hornet RS-ZA	-	
G-MWBW	Hornet RS-ZA	-	
G-MWBX	Hornet RS-ZA	-	
G-MWBY	Hornet RS-ZA	-	
G-MWBZ	Hornet RS-ZA	-	
G-MWCB	Solar Wings Pegasus XL-Q	-	
G-MWCC	Solar Wings Pegasus XL-R	L. Robinson	
G-MWCE	Mainair Gemini Flash IIA	-	
G-MWCF	Solar Wings Pegasus XL-R	-	
G-MWCG	Microflight Spectrum	M. W. Shepherd	
G-MWCH	Rans S.6 Coyote	G-MWCH Group	

Notes	Reg.	Type	Owner or Operator
	G-MWCI	Powerchute Kestrel	-
	G-MWCJ	Powerchute Kestrel	-
	G-MWCK	Powerchute Kestrel	F. W. Downham
	G-MWCM	Powerchute Kestrel	-
	G-MWCN	Powerchute Kestrel	-
	G-MWCO	Powerchute Kestrel	-
	G-MWCP	Powerchute Kestrel	-
	G-MWCR	Southdown Puma Sprint	-
	G-MWCS	Powerchute Kestrel	R. S. McFadyen
	G-MWCU	Solar Wings Pegasus XL-R	T. K. Duffy
	G-MWCW	Mainair Gemini Flash IIA	-
	G-MWCX	Medway Hybred 44XLR	-
	G-MWCY	Medway Hybred 44XLR	-
	G-MWCZ	Medway Hybred 44XLR	-
	G-MWDB	CFM Shadow Srs BD	-
	G-MWDC	Solar Wings Pegasus XL-R	-
	G-MWDD	Solar Wings Pegasus XL-Q	D. J. Billham
	G-MWDE	Hornet RS-ZA	-
	G-MWDF	Hornet RS-ZA	-
	G-MWDG	Hornet RS-ZA	-
	G-MWDH	Hornet RS-ZA	-
	G-MWDI	Hornet RS-ZA	-
	G-MWDJ	Mainair Gemini Flash IIA	-
	G-MWDK	Solar Wings Pegasus XL-R	-
	G-MWDL	Solar Wings Pegasus XL-R	-
	G-MWDM	Renegade Spirit UK	-
	G-MWDN	CFM Shadow Srs BD	-
	G-MWDP	Thruster TST Mk 1	-
	G-MWDS	Thruster T.300	-
	G-MWDZ	Eipper Quicksilver MXL II	-
	G-MWEE	Solar Wings Pegasus XL-Q	-
	G-MWEF	Solar Wings Pegasus XL-Q	-
	G-MWEG	Solar Wings Pegasus XL-Q	-
	G-MWEH	Solar Wings Pegasus XL-Q	-
	G-MWEK	Whittaker MW.5 Sorcerer	-
	G-MWEL	Mainair Gemini Flash IIA	-
	G-MWEN	CFM Shadow Srs BD	-
	G-MWEO	Whittaker MW.5 Sorcerer	-
	G-MWEP	Rans S.4 Coyote	-
	G-MWER	Solar Wings Pegasus XL-Q	-
	G-MWES	Rans S.4 Coyote	-
	G-MWEU	Hornet RS-ZA	-
	G-MWEY	Hornet RS-ZA	-
	G-MWEZ	CFM Shadow Srs CD	-
	G-MWFA	Solar Wings Pegasus XL-R	-
	G-MWFB	CFM Shadow Srs BD	K. W. E. Brunnenkant
	G-MWFC	Team Minimax (G-BTXC)	-
	G-MWFD	Team Minimax	J. Flanagan
	G-MWFE	Robin 330/Lightning 195	-
	G-MWFF	Rans S.4 Coyote	J. S. Sweetingham
	G-MWFG	Powerchute Kestrel	R. I. Simpson
	G-MWFH	Powerchute Kestrel	-
	G-MWFI	Powerchute Kestrel	-
	G-MWFL	Powerchute Kestrel	-
	G-MWFP	Solar Wings Pegasus XL-R	-
	G-MWFS	Solar Wings Pegasus XL-Q	-
	G-MWFT	MBA Tiger Cub 440	-
	G-MWFU	Quad City Challenger II UK	-
	G-MWFV	Quad City Challenger II UK	P. Bowers
	G-MWFW	Rans S.4 Coyote	-
	G-MWFX	Quad City Challenger II UK	-
	G-MWFY	Quad City Challenger II UK	-
	G-MWFZ	Quad City Challenger II UK	-
	G-MWGA	Rans S.5 Coyote	-
	G-MWGC	Medway Hybred 44XLR	C. Spalding
	G-MWGF	Renegade Spirit UK	-
	G-MWGG	Mainair Gemini Flash IIA	-
	G-MWGI	Whittaker MW.5 (K) Sorcerer	-
	G-MWGJ	Whittaker MW.5 (K) Sorcerer	-
	G-MWGK	Whittaker MW.5 (K) Sorcerer	-
	G-MWGL	Solar Wings Pegasus XL-Q	G. D. Haimes & A. R. Campbell
	G-MWGM	Solar Wings Pegasus XL-Q	-

Reg.	Type	Owner or Operator	Notes
G-MWGN	Rans S.4 Coyote II	-	
G-MWGO	Aerial Arts Chaser 110SX	-	
G-MWGR	Solar Wings Pegasus XL-Q	A. Maskell	
G-MWGU	Powerchute Kestrel	-	
G-MWGW	Powerchute Kestrel	-	
G-MWGY	Powerchute Kestrel	-	
G-MWGZ	Powerchute Kestrel	-	
G-MWHC	Solar Wings Pegasus XL-Q	-	
G-MWHD	Microflight Spectrum	-	
G-MWHE	Microflight Spectrum	-	
G-MWHF	Solar Wings Pegasus XL-Q	-	
G-MWHG	Solar Wings Pegasus XL-Q	-	
G-MWHH	Team Minimax	-	
G-MWHI	Mainair Gmini Flash	P. Harwood	
G-MWHL	Solar Wings Pegasus XL-Q	-	
G-MWHM	Whittaker MW.6 Merlin	-	
G-MWHO	Mainair Gemini Flash IIA	-	
G-MWHP	Rans S.6-ESD Coyote	-	
G-MWHR	Mainair Gemini Flash IIA	-	
G-MWHT	SW Pegasus Quasar	E. H. Gatehouse	
G-MWHU	SW Pegasus Quasar	-	
G-MWHV	SW Pegasus Quasar	-	
G-MWHX	Solar Wings Pegasus XL-Q	-	
G-MWHZ	Trion J-1	-	
G-MWIA	Mainair Gemini Flash IIA	-	
G-MWIB	Aviasud Mistral	-	
G-MWIC	Whittaker MW.5 Sorcerer	A. M. Witt	
G-MWIE	Solar Wings Pegasus XL-Q	-	
G-MWIF	Rans S.6-ESD Coyote II	-	
G-MWIG	Mainair Gemini Flash IIA	A. P. Purbrick	
G-MWIH	Mainair Gemini Flash IIA	-	
G-MWIK	Medway Hybred 44XLR	-	
G-MWIL	Medway Hybred 44XLR	-	
G-MWIM	SW Pegasus Quasar	-	
G-MWIO	Rans S.4 Coyote	-	
G-MWIP	Whittaker MW.6 Merlin	-	
G-MWIR	Solar Wings Pegasus XL-Q	-	
G-MWIS	Solar Wings Pegasus XL-Q	-	
G-MWIU	Solar Wings Pegasus XL-Q	-	
G-MWIV	Mainair Gemini Flash IIA	-	
G-MWIW	SW Pegasus Quasar	-	
G-MWIX	SW Pegasus Quasar	-	
G-MWIY	SW Pegasus Quasar	R. J. Coppin	
G-MWIZ	CFM Shadow Srs BD	-	
G-MWJD	SW Pegasus Quasar	-	
G-MWJF	CFM Shadow Srs BD	-	
G-MWJG	Solar Wings Pegasus XL-R	-	
G-MWJH	SW Pegasus Quasar	-	
G-MWJI	SW Pegasus Quasar	-	
G-MWJJ	SW Pegasus Quasar	-	
G-MWJK	SW Pegasus Quasar	-	
G-MWJM	AMF Chevvron 232	-	
G-MWJN	Solar Wings Pegasus XL-Q	-	
G-MWJP	Medway Hybred 44XLR	-	
G-MWJR	Medway Hybred 44XLR	T. G. Almond	
G-MWJS	SW Pegasus Quasar	-	
G-MWJT	SW Pegasus Quasar	-	
G-MWJV	SW Pegasus Quasar	-	
G-MWJW	Whittaker MW.5 Sorcerer	-	
G-MWJX	Medway Puma Sprint	-	
G-MWJY	Mainair Gemini Flash IIA	-	
G-MWKA	Renegade Spirit UK	-	
G-MWKE	Hornet R-ZA	-	
G-MWKO	Solar Wings Pegasus XL-Q	R. M. Nutt	
G-MWKP	Solar Wings Pegasus XL-Q	-	
G-MWKW	Microflight Spectrum	P. L. Stribling	
G-MWKX	Microflight Spectrum	-	
G-MWKY	Solar Wings Pegasus XL-Q	-	
G-MWKZ	Solar Wings Pegasus XL-Q	-	
G-MWLA	Rans S.4 Coyote	D. C. Lees	
G-MWLB	Medway Hybred 44XLR	-	
G-MWLC	Medway Hybred 44XLR	-	

Notes	Reg.	Type	Owner or Operator
	G-MWLD	CFM Shadow Srs BD	-
	G-MWLE	Solar Wings Pegasus XL-R	-
	G-MWLG	Solar Wings Pegasus XL-R	A. S. Docherty
	G-MWLH	Solar Wings Pegasus XL-R	-
	G-MWLJ	SW Pegasus Quasar	-
	G-MWLK	SW Pegasus Quasar	D. J. Shippen
	G-MWLL	Solar Wings Pegasus XL-Q	-
	G-MWLM	Solar Wings Pegasus XL-Q	A. A. Judge
	G-MWLN	Whittaker MW.6-S Fatboy Flyer	-
	G-MWLO	Whittaker MW.6 Merlin	-
	G-MWLP	Mainair Gemini Flash IIA	-
	G-MWLR	Mainair Gemini Flash IIA	-
	G-MWLS	Medway Hybred 44XLR	M. A. Oliver
	G-MWLT	Mainair Gemini Flash IIA	-
	G-MWLU	Solar Wings Pegasus XL-R	-
	G-MWLW	TEAM miniMAX	-
	G-MWLX	Mainair Gemini Flash IIA	G. Good & E. J. Douglas
	G-MWLZ	Rans S.4 Coyote	-
	G-MWMA	Powerchute Kestrel	-
	G-MWMB	Powerchute Kestrel	-
	G-MWMC	Powerchute Kestrel	-
	G-MWMD	Powerchute Kestrel	-
	G-MWMF	Powerchute Kestrel	-
	G-MWMG	Powerchute Kestrel	-
	G-MWMH	Powerchute Kestrel	-
	G-MWMI	SW Pegasus Quasar	A. R. Winton
	G-MWMJ	SW Pegasus Quasar	-
	G-MWMK	SW Pegasus Quasar TC	-
	G-MWML	SW Pegasus Quasar	-
	G-MWMM	Mainair Gemini Flash IIA	-
	G-MWMN	Solar Wings Pegasus XL-Q	-
	G-MWMO	Solar Wings Pegasus XL-Q	D. S. F. McNair
	G-MWMP	Solar Wings Pegasus XL-Q	-
	G-MWMR	Solar Wings Pegasus XL-R	-
	G-MWMS	Mainair Gemini Flash	A. Gannon
	G-MWMT	Mainair Gemini Flash IIA	R. Findlay
	G-MWMU	CFM Shadow Srs C	-
	G-MWMV	Solar Wings Pegasus XL-R	M. A. Oakley
	G-MWMW	Renegade Spirit UK	-
	G-MWMX	Mainair Gemini Flash IIA	-
	G-MWMY	Mainair Gemini Flash IIA	P. J. Harrison
	G-MWMZ	Solar Wings Pegasus XL-Q	-
	G-MWNA	Solar Wings Pegasus XL-Q	-
	G-MWNB	Solar Wings Pegasus XL-Q	-
	G-MWNC	Solar Wings Pegasus XL-Q	M. E. Oakman
	G-MWND	Tiger Cub Developments RL.5A	-
	G-MWNE	Mainair Gemini Flash IIA	-
	G-MWNF	Renegade Spirit UK	-
	G-MWNG	Solar Wings Pegasus XL-Q	-
	G-MWNK	SW Pegasus Quasar	-
	G-MWNL	SW Pegasus Quasar	R. J. Humphries
	G-MWNM	SW Pegasus Quasar	-
	G-MWNN	SW Pegasus Quasar	-
	G-MWNO	AMF Chevvron 232	-
	G-MWNP	AMF Chevvron 232	-
	G-MWNR	Renegade Spirit UK	-
	G-MWNS	Mainair Gemini Flash IIA	-
	G-MWNT	Mainair Gemini Flash IIA	-
	G-MWNU	Mainair Gemini Flash IIA	-
	G-MWNV	Powerchute Kestrel	-
	G-MWNX	Powerchute Kestrel	-
	G-MWOB	Powerchute Kestrel	-
	G-MWOC	Powerchute Kestrel	R. S. McFadyen
	G-MWOD	Powerchute Kestre	-
	G-MWOE	Powerchute Kestrel	-
	G-MWOF	Microflight Spectrum	-
	G-MWOH	Solar Wings Pegasus XL-R	-
	G-MWOI	Solar Wings Pegasus XL-R	B. T. Geoghegan
	G-MWOJ	Mainair Gemini Flash IIA	-
	G-MWOK	Mainair Gemini Flash IIA	-
	G-MWOM	SW Pegasus Quasar TC	-
	G-MWON	CFM Shadow Srs CD	-

Reg.	Type	Owner or Operator	Notes
G-MWOO	Renegade Spirit UK	-	
G-MWOP	SW Pegasus Quasar	-	
G-MWOR	Solar Wings Pegasus XL-Q	-	
G-MWOS	Cosmos Chronos	-	
G-MWOV	Whittaker MW.6 Merlin	I. R. Hodgson	
G-MWOY	Solar Wings Pegasus XL-Q	-	
G-MWPA	Mainair Gemini Flash IIA	-	
G-MWPB	Mainair Gemini Flash IIA		
G-MWPC	Mainair Gemini Flash IIA	M. Johnson	
G-MWPD	Mainair Gemini Flash IIA		
G-MWPE	Solar Wings Pegasus XL-Q	-	
G-MWPF	Mainair Gemini Flash IIA	L. H. Black	
G-MWPG	Microflight Spectrum	-	
G-MWPH	Microflight Spectrum	S. Rickett & M. J. Deacon	
G-MWPJ	Solar Wings Pegasus XL-Q	-	
G-MWPK	Solar Wings Pegasus XL-Q	-	
G-MWPL	MBA Tiger Cub 440	-	
G-MWPP	CFM Streak Shadow	A. J. Thomas (G-BTEM)	
G-MWPR	Whittaker MW.6 Merlin	-	
G-MWPS	Renegade Spirit UK	-	
G-MWPT	Hunt Wing	-	
G-MWPU	SW Pegasus Quasar TC	-	
G-MWPX	Solar Wings Pegasus XL-R	-	
G-MWPZ	Renegade Spirit UK	-	
G-MWRB	Mainair Gemini Flash IIA	-	
G-MWRC	Mainair Gemini Flash IIA	-	
G-MWRD	Mainair Gemini Flash IIA	-	
G-MWRE	Mainair Gemini Flash IIA	-	
G-MWRF	Mainair Gemini Flash IIA	N. Hay	
G-MWRG	Mainair Gemini Flash IIA	F. J. Clarehugh	
G-MWRH	Mainair Gemini Flash IIA		
G-MWRI	Mainair Gemini Flash IIA	-	
G-MWRJ	Mainair Gemini Flash IIA	-	
G-MWRL	CFM Shadow Srs.CD	R. A. & C. A. Allen	
G-MWRM	Medway Hybred 44XLR	-	
G-MWRN	Solar Wings Pegasus XL-R	-	
G-MWRO	Solar Wings Pegasus XL-R	-	
G-MWRP	Solar Wings Pegasus XL-R	-	
G-MWRR	Mainair Gemini Flash IIA	-	
G-MWRS	Ultravia Super Pelican	-	
G-MWRT	Solar Wings Pegasus XL-R	-	
G-MWRU	Solar Wings Pegasus XL-R	-	
G-MWRV	Solar Wings Pegasus XL-R	-	
G-MWRW	Solar Wings Pegasus XL-Q	-	
G-MWRX	Solar Wings Pegasus XL-Q	-	
G-MWRY	CFM Shadow Srs CD	-	
G-MWRZ	AMF Chevvron 232	-	
G-MWSA	Team Minimax	-	
G-MWSB	Mainair Gemini Flash IIA	G. R. Whittemore	
G-MWSC	Rans S.6-ESD Coyote II	-	
G-MWSD	Solar Wings Pegasus XL-Q	-	
G-MWSE	Solar Wings Pegasus XL-R	-	
G-MWSF	Solar Wings Pegasus XL-R	N. A. & F. W. Milne	
G-MWSH	SW Pegasus Quasar TC	G. Jones	
G-MWSI	SW Pegasus Quasar TC	-	
G-MWSJ	Solar Wings Pegasus XL-Q	-	
G-MWSK	Solar Wings Pegasus XL-Q	-	
G-MWSL	Mainair Gemini Flash IIA	-	
G-MWSM	Mainair Gemini Flash IIA	R. M. Wall	
G-MWSN	SW Pegasus Quasar TC	-	
G-MWSO	Solar Wings Pegasus XL-R	-	
G-MWSP	Solar Wings Pegasus XL-R	-	
G-MWSR	Solar Wings Pegasus XL-R	-	
G-MWSS	Medway Hybred 44XLR	-	
G-MWST	Medway Hybred 44XLR	-	
G-MWSU	Medway Hybred 44XLR	-	
G-MWSV	SW Pegasus Quasar TC	-	
G-MWSW	Whittaker MW.6 Merlin	-	
G-MWSX	Whittaker MW.5 Sorcerer	-	
G-MWSY	Whittaker MW.5 Sorcerer	-	
G-MWSZ	CFM Shadow Srs CD	-	
G-MWTA	Solar Wings Pegasus XL-Q	-	

Notes	Reg.	Type	Owner or Operator
	G-MWTB	Solar Wings Pegasus XL-Q	-
	G-MWTC	Solar Wings Pegasus XL-Q	-
	G-MWTD	Microflight Spectrum	-
	G-MWTE	Microflight Spectrum	-
	G-MWTG	Mainair Gemini Flash IIA	-
	G-MWTH	Mainair Gemini Flash IIA	A. Strang
	G-MWTI	Solar Wings Pegasus XL-Q	-
	G-MWTJ	CFM Shadow Srs CD	-
	G-MWTK	Solar Wings Pegasus XL-R	-
	G-MWTL	Solar Wings Pegasus XL-R	-
	G-MWTM	Solar Wings Pegasus XL-R	-
	G-MWTN	CFM Shadow Srs CD	-
	G-MWTO	Mainair Gemini Flash IIA	-
	G-MWTP	CFM Shadow Srs CD	-
	G-MWTR	Mainair Gemini Flash IIA	-
	G-MWTT	Rans S.6-ESD Coyote II	-
	G-MWTU	Solar Wings Pegasus XL-R	-
	G-MWTY	Mainair Gemini Flash IIA	-
	G-MWTZ	Mainair Gemini Flash IIA	-
	G-MWUA	CFM Shadow Srs CD	-
	G-MWUB	Solar Wings Pegasus XL-R	-
	G-MWUC	Solar Wings Pegasus XL-R	M. A. Hicks
	G-MWUD	Solar Wings Pegasus XL-R	-
	G-MWUF	Solar Wings Pegasus XL-R	-
	G-MWUH	Renegade Spirit UK	-
	G-MWUI	AMF Chevvron 2-32C	-
	G-MWUK	Rans S.6-ESD Coyote II	-
	G-MWUL	Rans S.6-ESD Coyote II	-
	G-MWUO	Solar Wings Pegasus XL-Q	-
	G-MWUP	Solar Wings Pegasus XL-R	-
	G-MWUR	Solar Wings Pegasus XL-R	Nottingham Aerotow Club
	G-MWUS	Solar Wings Pegasus XL-R	-
	G-MWUU	Solar Wings Pegasus XL-R	-
	G-MWUV	Solar Wings Pegasus XL-R	C. D. Baines
	G-MWUW	Solar Wings Pegasus XL-R	-
	G-MWUX	Solar Wings Pegasus XL-Q	-
	G-MWUY	Solar Wings Pegasus XL-Q	-
	G-MWUZ	Solar Wings Pegasus XL-Q	S. R. Nanson
	G-MWVA	Solar Wings Pegasus XL-Q	-
	G-MWVB	Solar Wings Pegasus XL-R	-
	G-MWVE	Solar Wings Pegasus XL-R	-
	G-MWVF	Solar Wings Pegasus XL-R	-
	G-MWVG	CFM Shadow Srs CD	-
	G-MWVH	CFM Shadow Srs CD	-
	G-MWVK	Mainair Mercury	-
	G-MWVL	Rans S.6-ESD Coyote II	-
	G-MWVM	SW Pegasus Quasar II	-
	G-MWVN	Mainair Gemini Flash IIA	-
	G-MWVO	Mainair Gemini Flash IIA	P. Webb
	G-MWVP	Renegade Spirit UK	-
	G-MWVS	Mainair Gemini Flash IIA	M. J. A. New & A. Clift
	G-MWVT	Mainair Gemini Flash IIA	-
	G-MWVU	Medway Hybred 44XLR	-
	G-MWVW	Mainair Gemini Flash IIA	-
	G-MWVY	Mainair Gemini Flash IIA	-
	G-MWVZ	Mainair Gemini Flash IIA	-
	G-MWWB	Mainair Gemini Flash IIA	-
	G-MWWC	Mainair Gemini Flash IIA	-
	G-MWWD	Renegade Spirit	-
	G-MWWE	Team Minimax	-
	G-MWWF	Kolb Twinstar Mk 3	-
	G-MWWG	Solar Wings Pegasus XL-Q	-
	G-MWWH	Solar Wings Pegasus XL-Q	-
	G-MWWI	Mainair Gemini Flash IIA	-
	G-MWWJ	Mainair Gemini Flash IIA	-
	G-MWWK	Mainair Gemini Flash IIA	-
	G-MWWL	Rans S.6-ESD Coyote II	S. Munday
	G-MWWM	Kolb Twinstar Mk 2	-
	G-MWWN	Mainair Gemini Flash IIA	-
	G-MWWP	Rans S.4 Coyote	-
	G-MWVR	Mainair Gemini Flash IIA	-
	G-MWWR	Microflight Spectrum	-

Reg.	Type	Owner or Operator	Notes
G-MWWS	Thruster T.300	D. P. Wring	
G-MWWU	Air Creation Fun 18 GTBI	-	
G-MWWV	Solar Wings Pegasus XL-Q	-	
G-MWWW	Whittaker MW.6-S Fatboy Flyer	-	
G-MWWZ	Cyclone Chaser S	-	
G-MWXB	Mainair Gemini Flash IIA	-	
G-MWXC	Mainair Gemini Flash IIA	-	
G-MWXE	Flexiform Skytrike	-	
G-MWXF	Mainair Mercury	-	
G-MWXG	SW Pegasus Quasar IITC	-	
G-MWXH	SW Pegasus Quasar IITC	-	
G-MWXJ	Mainair Mercury	-	
G-MWXK	Mainair Mercury	-	
G-MWXL	Mainair Gemini Flash IIA	-	
G-MWXN	Mainair Gemini Flash IIA	-	
G-MWXO	Mainair Gemini Flash IIA	-	
G-MWXP	Solar Wings Pegasus XL-Q	-	
G-MWXR	Solar Wings Pegasus XL-Q	-	
G-MWXU	Mainair Gemini Flash IIA	-	
G-MWXV	Mainair Gemini Flash IIA	A. W. Lowrie	
G-MWXW	Cyclone Chaser S	-	
G-MWXX	Cyclone Chaser S 447	P. I. Frost	
G-MWXY	Cyclone Chaser S 447	-	
G-MWXZ	Cyclone Chaser S 508	-	
G-MWYA	Mainair Gemini Flash IIA	-	
G-MWYB	Solar Wings Pegasus XL-Q	-	
G-MWYC	Solar Wings Pegasus XL-Q	-	
G-MWYD	CFM Shadow Srs C	-	
G-MWYE	Rans S.6-ESD Coyote II	-	
G-MWYF	Rans S.6 Coyote II	-	
G-MWYG	Mainair Gemini Flash IIA	M. J. Burns	
G-MWYH	Mainair Gemini Flash IIA	-	
G-MWYI	SW Pegasus Quasar II	-	
G-MWYJ	SW Pegasus Quasar IITC	-	
G-MWYL	Mainair Gemini Flash IIA	-	
G-MWYM	Cyclone Chaser S 1000	-	
G-MWYN	Rans S.6-ESD Coyote	W. R. Tull	
G-MWYS	CGS Hawk 1 Arrow	-	
G-MWYT	Mainair Gemini Flash IIA	-	
G-MWYU	Solar Wings Pegasus XL-Q	-	
G-MWYV	Mainair Gemini Flash IIA	-	
G-MWYX	Mainair Gemini Flash IIA	-	
G-MWYY	Mainair Gemini Flash IIA	-	
G-MWYZ	Solar Wings Pegasus XL-Q	-	
G-MWZA	Mainair Mercury	-	
G-MWZB	AMF Chevvron 2-32C	C. M. Lewis & M. J. Wilson	
G-MWZC	Mainair Gemini Flash IIA	-	
G-MWZD	SW Pegasus Quasar IITC	-	
G-MWZE	SW Pegasus Quasar IITC	-	
G-MWZF	SW Pegasus Quasar IITC	-	
G-MWZG	Mainair Gemini Flash IIA	C. J. O'Sullivan/Ireland	
G-MWZH	Solar Wings Pegasus XL-R	-	
G-MWZI	Solar Wings Pegasus XL-R	-	
G-MWZJ	Solar Wings Pegasus XL-R	-	
G-MWZK	Solar Wings Pegasus XL-R	-	
G-MWZL	Mainair Gemini Flash IIA	-	
G-MWZM	Team Minimax 91	-	
G-MWZN	Mainair Gemini Flash IIA	-	
G-MWZO	SW Pegasus Quasar IITC	-	
G-MWZP	SW Pegasus Quasar IITC	-	
G-MWZR	SW Pegasus Quasar IITC	P. K. Appleton	
G-MWZS	SW Pegasus Quasar IITC	-	
G-MWZT	Solar Wings Pegasus XL-R	-	
G-MWZU	Solar Wings Pegasus XL-R	-	
G-MWZV	Solar Wings Pegasus XL-R	-	
G-MWZY	Solar Wings Pegasus XL-R	-	
G-MWZZ	Solar Wings Pegasus XL-R	-	
G-MXVI	V.S.361 Spitfire LF.XVIe (TE184)	De Cadenet Motor Racing Ltd	
G-MYAB	Solar Wings Pegasus XL-R	-	
G-MYAC	Solar Wings Pegasus XL-Q	-	

Notes	Reg.	Type	Owner or Operator
	G-MYAE	Solar Wings Pegasus XL-Q	-
	G-MYAF	Solar Wings Pegasus XL-Q	-
	G-MYAG	Quad City Challenger II	-
	G-MYAH	Whittaker MW.5 Sorcerer	-
	G-MYAI	Mainair Mercury	-
	G-MYAJ	Rans S.6-ESD Coyote II	-
	G-MYAK	SW Pegasus Quasar IITC	-
	G-MYAM	Renegade Spirit UK	-
	G-MYAN	Whittaker MW.5 (K) Sorcerer	-
	G-MYAO	Mainair Gemini Flash IIA	K. Fowler
	G-MYAP	Thruster T.300	-
	G-MYAR	Thruster T.300	-
	G-MYAS	Mainair Gemini Flash IIA	-
	G-MYAT	Team Minimax	-
	G-MYAU	Mainair Gemini Flash IIA	-
	G-MYAV	Mainair Mercury	-
	G-MYAY	Microflight Spectrum	P. F. Craggs
	G-MYAZ	Renegade Spirit UK	-
	G-MYBA	Rans S.6-ESD Coyote II	-
	G-MYBB	Maxair Drifter	-
	G-MYBC	CFM Shadow Srs CD	-
	G-MYBD	SW Pegasus Quasar IITC	A. Gunn
	G-MYBE	SW Pegasus Quasar IITC	D. Lumsdon
	G-MYBF	Solar Wings Pegasus XL-Q	-
	G-MYBG	Solar Wings Pegasus XL-Q	-
	G-MYBI	Rans S.6-ESD Coyote II	N. C. Tambiah
	G-MYBJ	Mainair Gemini Flash IIA	T. J. Mellor
	G-MYBK	SW Pegasus Quasar IITC	-
	G-MYBL	CFM Shadow Srs C	-
	G-MYBM	Team Minimax	-
	G-MYBN	Hiway Demon 175	-
	G-MYBO	Solar Wings Pegasus XL-R	-
	G-MYBP	Solar Wings Pegasus XL-R	-
	G-MYBR	Solar Wings Pegasus XL-Q	-
	G-MYBS	Solar Wings Pegasus XL-Q	T. Smith
	G-MYBT	SW Pegasus Quasar IITC	-
	G-MYBU	Cyclone Chaser S 447	-
	G-MYBV	Solar Wings Pegasus XL-Q	-
	G-MYBW	Solar Wings Pegasus XL-Q	-
	G-MYBY	Solar Wings Pegasus XL-Q	I. D. A. Spanton
	G-MYBZ	Solar Wings Pegasus XL-Q	A. J. Blackwell
	G-MYCA	Whittaker MW.6 Merlin	-
	G-MYCB	Cyclone Chaser S 447	-
	G-MYCE	SW Pegasus Quasar IITC	-
	G-MYCF	SW Pegasus Quasar IITC	-
	G-MYCJ	Mainair Mercury	J. Agnew
	G-MYCK	Mainair Gemini Flash IIA	-
	G-MYCL	Mainair Mercury	-
	G-MYCM	CFM Shadow Srs CD	-
	G-MYCN	Mainair Mercury	-
	G-MYCO	Renegade Spirit UK	M. J. Downes
	G-MYCP	Whittaker MW.6 Merlin	-
	G-MYCR	Mainair Gemini Flash IIA	-
	G-MYCS	Mainair Gemini Flash IIA	-
	G-MYCT	Team Minimax	D. D. Rayment
	G-MYCV	Mainair Mercury	-
	G-MYCW	Powerchute Kestrel	-
	G-MYCX	Powerchute Kestrel	-
	G-MYCY	Powerchute Kestrel	-
	G-MYDA	Powerchute Kestrel	-
	G-MYDB	Powerchute Kestrel	-
	G-MYDC	Mainair Mercury	-
	G-MYDD	CFM Shadow Srs CD	-
	G-MYDE	CFM Shadow Srs CD	-
	G-MYDF	Team Minimax	A. M. Hughes
	G-MYDG	Solar Wings Pegasus XL-R	-
	G-MYDI	Solar Wings Pegasus XL-R	-
	G-MYDJ	Solar Wings Pegasus XL-R	Cambridgeshire Aerotow Club
	G-MYDK	Rans S.6-ESD Coyote II	-
	G-MYDL	Whittaker MW.5 (K) Sorcerer	-
	G-MYDM	Whittaker MW.6-S Fatboy Flyer	-
	G-MYDN	Quad City Challenger II	-

Reg.	Type	Owner or Operator	Notes
G-MYDO	Rans S.5 Coyote	-	
G-MYDP	Kolb Twinstar Mk 3		
G-MYDR	Thruster Tn.300		
G-MYDS	Quad City Challenger II	L. R. Graham	
G-MYDU	Thruster T.300	S. Collins	
G-MYDV	Thruster T.300	-	
G-MYDW	Whittaker MW.6 Merlin	-	
G-MYDX	Rans S.6-ESD Coyote II	-	
G-MYDZ	Mignet HM.1000 Balerit	-	
G-MYEA	Solar Wings Pegasus XL-Q	-	
G-MYEC	Solar Wings Pegasus XL-Q	J. I. King	
G-MYED	Solar Wings Pegasus XL-R	-	
G-MYEG	Solar Wings Pegasus XL-R	-	
G-MYEH	Solar Wings Pegasus XL-R	-	
G-MYEI	Cyclone Chaser S447	-	
G-MYEJ	Cyclone Chaser S447	-	
G-MYEK	SW Pegasus Quasar IITC	-	
G-MYEL	SW Pegasus Quasar IITC	-	
G-MYEM	SW Pegasus Quasar IITC		
G-MYEN	SW Pegasus Quasar IITC	J. C. Higham	
G-MYEO	SW Pegasus Quasar IITC	-	
G-MYEP	CFM Shadow Srs CD	-	
G-MYER	Cyclone AX3/503	-	
G-MYES	Rans S.6-ESD Coyote II	-	
G-MYET	Whittaker MW.6 Merlin		
G-MYEU	Mainair Gemini Flash IIA	G. J. Webster	
G-MYEV	Whittaker MW.6 Merlin	-	
G-MYEW	Powerchute Kestrel	-	
G-MYEX	Powerchute Kestrel	-	
G-MYFA	Powerchute Kestrel	-	
G-MYFG	Hunt Avon Skytrike	-	
G-MYFH	Quad City Challenger II	-	
G-MYFI	Cyclone AX3/503		
G-MYFK	SW Pegasus Quasar IITC	R. L. Harris	
G-MYFL	SW Pegasus Quasar IITC	-	
G-MYFM	Renegade Spirit UK	-	
G-MYFN	Rans S.5 Coyote	-	
G-MYFO	Cyclone Chaser S	-	
G-MYFP	Mainair Gemini Flash IIA	-	
G-MYFR	Mainair Gemini Flash IIA		
G-MYFS	Pegasus XL-R	D. J. Middleton	
G-MYFT	Mainair Scorcher	-	
G-MYFU	Mainair Gemini Flash IIA	-	
G-MYFV	Cyclone AX3/503	-	
G-MYFW	Cyclone AX3/503	-	
G-MYFX	Solar Wings Pegasus XL-Q	-	
G-MYFY	Cyclone AX3/503	-	
G-MYFZ	Cyclone AX3/503	-	
G-MYGD	Cyclone AX3/503	-	
G-MYGE	Whittaker MW.6 Merlin	-	
G-MYGF	Team Minimax	-	
G-MYGG	Mainair Mercury		
G-MYGH	Rans S.6ESD Coyote II	L. J. Field	
G-MYGI	Cyclone Chaser S 447		
G-MYGJ	Mainair Mercury	J. R. Harnett	
G-MYGK	Cyclone Chaser S 508	-	
G-MYGL	Team Minimax	-	
G-MYGM	Quad City Challenger II		
G-MYGN	AMF Chevvron 2-32C	A. C. Barber	
G-MYGO	CFM Shadow Srs CD	-	
G-MYGP	Rans S.6-ESD Coyote II	-	
G-MYGR	Rans S.6-ESD Coyote II	-	
G-MYGS	Whittaker MW.5 (K) Sorcerer	-	
G-MYGT	Solar Wings Pegasus XL-R	-	
G-MYGU	Solar Wings Pegasus XL-R	-	
G-MYGV	Solar Wings Pegasus XL-R	-	
G-MYGZ	Mainair Gemini Flash IIA	-	
G-MYHF	Mainair Gemini Flash IIA	-	
G-MYHG	Cyclone AX/503	-	
G-MYHH	Cyclone AX/503	-	
G-MYHI	Rans S.6-ESD Coyote II	-	
G-MYHK	Rans S.6-ESD Coyote II	-	

Notes	Reg.	Type	Owner or Operator
	G-MYHL	Mainair Gemini Flash IIA	-
	G-MYHM	Cyclone AX3/503	-
	G-MYHN	Mainair Gemini Flash IIA	-
	G-MYHP	Rans S.6-ESD Coyote II	-
	G-MYHR	Cyclone AX3/503	-
	G-MYHS	Powerchute Kestrel	-
	G-MYHX	Mainair Gemini Flash IIA	-
	G-MYIA	Quad City Challenger II	-
	G-MYIE	Whittaker MW.6 Merlin	K. Gair & J. G. E. Lane
	G-MYIF	CFM Shadow Srs CD	-
	G-MYIH	Mainair Gemini Flash IIA	-
	G-MYII	Team Minimax	-
	G-MYIJ	Cyclone AX3/503	-
	G-MYIK	Kolb Twinstar Mk 3	-
	G-MYIL	Cyclone Chaser S 508	-
	G-MYIN	SW Pegasus Quasar IITC	-
	G-MYIO	SW Pegasus Quasar IITC	E. Foster & J. H. Peet
	G-MYIP	CFM Shadow Srs CD	-
	G-MYIR	Rans S.6-ESD Coyote II	-
	G-MYIS	Rans S.6-ESD Coyote II	-
	G-MYIT	Cyclone Chaser S 508	-
	G-MYIU	Cyclone AX3/503	-
	G-MYIV	Mainair Gemini Flash IIA	-
	G-MYIW	Mainair Mercury	-
	G-MYIX	Quad City Challenger II	-
	G-MYIY	Mainair Gemini Flash IIA	-
	G-MYIZ	Team Minimax 2	-
	G-MYJA	—	-
	G-MYJB	Mainair Gemini Flash IIA	-
	G-MYJC	Mainair Gemini Flash IIA	-
	G-MYJD	Rans S.6-ESD Coyote II	-
	G-MYJF	Thruster T.300	-
	G-MYJG	Thruster Super T.300	-
	G-MYJJ	SW Pegasus Quasar IITC	D. Murray
	G-MYJK	SW Pegasus Quasar IITC	B. Hall
	G-MYJM	Mainair Gemini Flash IIA	-
	G-MYJO	Cyclone Chaser S 508	-
	G-MYJR	Mainair Mercury	-
	G-MYJS	SW Pegasus Quasar IITC	-
	G-MYJT	SW Pegasus Quasar IITC	-
	G-MYJU	SW Pegasus Quasar IITC	-
	G-MYJW	Cyclone Chaser S 508	-
	G-MYJY	Rans S.6-ESD Coyote II	-
	G-MYJZ	Whittaker MW.5D Sorcerer	-
	G-MYKA	Cyclone AX3/503	K. Stevens
	G-MYKB	Kölb Twinstar Mk 3	-
	G-MYKC	Mainair Gemini Flash IIA	R. W. Lenthall
	G-MYKD	Cyclone Chaser S 447	-
	G-MYKE	CFM Shadow Srs BD	-
	G-MYKG	Mainair Gemini Flash IIA	-
	G-MYKH	Mainair Gemini Flash IIA	-
	G-MYKI	Mainair Mercury	-
	G-MYKJ	Team Minimax	-
	G-MYKL	Medway Raven	-
	G-MYKN	Rans S.6-ESD Coyote II	S. E. Hartles
	G-MYKO	Whittaker MW.6-S Fat Boy Flyer	K. R. Challis & C. S. Andersson
	G-MYKP	SW Pegasus Quasar IITC	R. F. Dye & G. S. B. Airth
	G-MYKR	SW Pegasus Quasar IITC	-
	G-MYKS	SW Pegasus Quasar IITC	N. Groome
	G-MYKT	Cyclone AX3/503	-
	G-MYKV	Mainair Gemini Flash IIA	P. J. Gulliver
	G-MYKW	Mainair Mercury	C. Foster
	G-MYKX	Mainair Mercury	-
	G-MYKY	Mainair Mercury	-
	G-MYKZ	Team Minimax (G-BVAV)	G. H. Hills
	G-MYLB	Team Minimax	-
	G-MYLC	SW Pegasus Quantum 15	-
	G-MYLD	Rans S.6-ESD Coyote II	-
	G-MYLE	SW Pegasus Quantum 15	-
	G-MYLF	Rans S.6-ESD Coyote II	A. J. Spencer
	G-MYLG	Mainair Gemini Flash IIA	-
	G-MYLH	SW Pegasus Quantum 15	-

Reg.	Type	Owner or Operator	Notes
G-MYLI	SW Pegasus Quantum 15	-	
G-MYLJ	Cyclone Chaser S 447	-	
G-MYLK	SW Pegasus Quantum 15	-	
G-MYLM	SW Pegasus Quasar IITC	P. A. Ashton	
G-MYLN	Kölb Twinstar Mk 3	-	
G-MYLO	Rans S.6-ESD Coyote II	-	
G-MYLP	Kölb Twinstar Mk 3 (G-BVCR)	-	
G-MYLR	Mainair Gemini Flash IIA	-	
G-MYLS	Mainair Mercury	-	
G-MYLT	Mainair Blade	-	
G-MYLU	Experience/Hunt Wing	-	
G-MYLV	CFM Shadow Srs CD	-	
G-MYLW	Rans S.6-ESD Coyote II	-	
G-MYLX	Medway Raven	-	
G-MYLY	Medway Raven	-	
G-MYLZ	SW Pegasus Quantum 15	-	
G-MYMB	SW Pegasus Quantum 15	-	
G-MYMC	SW Pegasus Quantum 15	B. J. Topham	
G-MYME	Cyclone AX3/503	-	
G-MYMH	Rans S.6-ESD Coyote II	-	
G-MYMI	Kölb Twinstar Mk 3	-	
G-MYMJ	Medway Raven	-	
G-MYMK	Mainair Gemini Flash IIA	-	
G-MYML	Mainair Mercury	-	
G-MYMM	Ultraflight Fun 18S	-	
G-MYMN	Whittaker MW.6 Merlin	-	
G-MYMO	Mainair Gemini Flash IIA	S. P. Moores	
G-MYMP	Rans S.6-ESD Coyote II	F. M. Pearce (G-CHAZ)	
G-MYMR	Rans S.6-ESD Coyote II	-	
G-MYMS	Rans S.6-ESD Coyote II	-	
G-MYMT	Mainair Mercury	-	
G-MYMV	Mainair Gemini Flash IIA	-	
G-MYMW	Cyclone AX3/503	-	
G-MYMX	SW Pegasus Quantum 15	-	
G-MYMY	Cyclone Chaser S 508	-	
G-MYMZ	Cyclone AX3/503	-	
G-MYNA	CFM Shadow Srs C	-	
G-MYNB	SW Pegasus Quantum 15	-	
G-MYNC	Mainair Mercury	-	
G-MYND	Mainair Gemini Flash IIA	R. Dowall	
G-MYNE	Rans S.6-ESD Coyote II	-	
G-MYNF	Mainair Mercury	-	
G-MYNH	Rans S.6-ESD Coyote II	-	
G-MYNI	Team Minimax	-	
G-MYNJ	Mainair Mercury	-	
G-MYNK	SW Pegasus Quantum 15	-	
G-MYNL	SW Pegasus Quantum 15	C. Hodgkiss	
G-MYNN	SW Pegasus Quantum 15	-	
G-MYNO	SW Pegasus Quantum 15	L. C. Stockman	
G-MYNP	SW Pegasus Quantum 15	-	
G-MYNR	SW Pegasus Quantum 15	A. S. Wason	
G-MYNS	SW Pegasus Quantum 15	-	
G-MYNT	SW Pegasus Quantum 15	-	
G-MYNV	SW Pegasus Quantum 15	M. D. Gregory	
G-MYNX	CFM Streak Shadow Srs S-A1	-	
G-MYNY	Kölb Twinstar Mk 3	-	
G-MYNZ	SW Pegasus Quantum 15	G. Jones	
G-MYOA	Rans S.6-ESD Coyote II	-	
G-MYOB	Mainair Mercury	-	
G-MYOF	Mainair Mercury	-	
G-MYOG	Kölb Twinstar Mk 3	-	
G-MYOH	CFM Shadow Srs CD	-	
G-MYOI	Rans S.6-ESD Coyote II	-	
G-MYOL	Air Creation Fun 18S GTBIS	-	
G-MYOM	Mainair Gemini Flash IIA	-	
G-MYON	CFM Shadow Srs CD	-	
G-MYOO	Kölb Twinstar Mk 3	-	
G-MYOR	Kölb Twinstar Mk 3	-	
G-MYOU	SW Pegasus Quantum 15	M. Botten & O. Kent	
G-MYOV	Mainair Mercury	-	
G-MYOW	Mainair Gemini Flash IIA	-	
G-MYOX	Mainair Mercury	-	

Notes	Reg.	Type	Owner or Operator
	G-MYOY	Cyclone AX3/503	-
	G-MYOZ	Quad City Challenger II UK	J. J. Littler
	G-MYPA	Rans S.6-ESD Coyote II	-
	G-MYPB	Cyclone Chaser S 447	-
	G-MYPC	Kölb Twinstar Mk 3	-
	G-MYPD	Mainair Mercury	-
	G-MYPE	Mainair Gemini Flash IIA	-
	G-MYPG	SW Pegasus XL-Q	I. A. G. Hull
	G-MYPH	SW Pegasus Quantum 15	-
	G-MYPI	SW Pegasus Quantum 15	-
	G-MYPJ	Rans S.6-ESD Coyote II	K. A. Eden
	G-MYPK	Rans S.6-ESD Coyote II	-
	G-MYPL	CFM Shadow Srs CD	-
	G-MYPM	Cyclone AX3/503	-
	G-MYPN	SW Pegasus Quantum 15	-
	G-MYPO	Hunt Wing/Experience	-
	G-MYPP	Whittaker MW.6-S Fat Boy Flyer	-
	G-MYPR	Cyclone AX3/503	-
	G-MYPS	Whittaker MW.6 Merlin	-
	G-MYPT	CFM Shadow Srs CD	-
	G-MYPV	Mainair Mercury	-
	G-MYPW	Mainair Gemini Flash IIA	-
	G-MYPX	SW Pegasus Quantum 15	-
	G-MYPY	SW Pegasus Quantum 15	-
	G-MYPZ	Quad City Challenger II	-
	G-MYRB	Whittaker MW.5 Sorcerer	-
	G-MYRC	Mainair Blade	-
	G-MYRD	Mainair Blade	-
	G-MYRE	Cyclone Chaser S	-
	G-MYRF	SW Pegasus Quantum 15	-
	G-MYRG	Team Minimax	-
	G-MYRH	Quad City Challenger II	C. M. Gray
	G-MYRI	Medway 44XLR	-
	G-MYRJ	Quad City Challenger II	-
	G-MYRK	Renegade Spirit UK	-
	G-MYRL	Team Minimax	-
	G-MYRM	SW Pegasus Quantum 15	-
	G-MYRN	SW Pegasus Quantum 15	-
	G-MYRO	Cyclone AX3/503	-
	G-MYRP	Letov LK-2M Sluka	-
	G-MYRR	Letov LK-2M Sluka	-
	G-MYRS	SW Pegasus Quantum 15	-
	G-MYRT	SW Pegasus Quantum 15	-
	G-MYRU	Cyclone AX3/503	-
	G-MYRV	Cyclone AX3/503	-
	G-MYRW	Mainair Mercury	-
	G-MYRY	SW Pegasus Quantum 15	-
	G-MYRZ	SW Pegasus Quantum 15	-
	G-MYSA	Cyclone Chaser S 508	-
	G-MYSB	SW Pegasus Quantum 15	-
	G-MYSC	SW Pegasus Quantum 15	-
	G-MYSD	BFC Challlenger II	-
	G-MYSG	Mainair Mercury	-
	G-MYSI	HM14/93	-
	G-MYSJ	Mainair Gemini Flash IIA	-
	G-MYSK	Team Minimax	-
	G-MYSL	Aviasud Mistral	-
	G-MYSM	CFM Shadow Srs CD	-
	G-MYSN	Whittaker MW.6 Merlin	-
	G-MYSO	Cyclone AX3/50	-
	G-MYSP	Rans S.6-ESD Coyote II	-
	G-MYSR	SW Pegasus Quatum 15	W. G. Craig
	G-MYSU	Rans S.6-ESD Coyote II	-
	G-MYSV	Aerial Arts Chaser	-
	G-MYSW	SW Pegasus Quantum 1	D. J. Cornelius
	G-MYSX	SW Pegasus Quantum 15	-
	G-MYSY	SW Pegasus Quantum 15	Premier Aviation (UK) Ltd
	G-MYSZ	Mainair Mercury	-
	G-MYTA	Team Minimax 91	-
	G-MYTB	Mainair Mercur	-
	G-MYTC	SW Pegasus XL-Q	-
	G-MYTD	Mainair Blade	-

Reg.	Type	Owner or Operator	Notes
G-MYTE	Rans S.6-ESD Coyote II	The Rans Flying Group	
G-MYTG	Mainair Blade	O. P. Farrell	
G-MYTH	CFM Shadow Srs CD	-	
G-MYTI	Pegasus Quantum 15	-	
G-MYTJ	SW Pegasus Quantum 15	-	
G-MYTK	Mainair Mercury	-	
G-MYTL	Mainair Blade	-	
G-MYTM	Cyclone AX3/503	-	
G-MYTN	SW Pegasus Quantum 15	-	
G-MYTO	Quad City Challenger II	-	
G-MYTP	Arrowflight Hawk II	-	
G-MYTR	Pegasus Quantum 15	-	
G-MYTT	Quad City Challenger II	-	
G-MYTU	Mainair Blade	-	
G-MYTV	Hunt Avon Skytrike	-	
G-MYTW	Mainair Blade	-	
G-MYTX	Mainair Mercury	-	
G-MYTY	CFM Streak Shadow Srs M	-	
G-MYTZ	Air Creation Fun 18S GTBIS	-	
G-MYUA	Air Creation Fun 18S GTBIS	-	
G-MYUB	Mainair Mercury	-	
G-MYUC	Mainair Blade	-	
G-MYUD	Mainair Mercury	-	
G-MYUE	Mainair Mercury	-	
G-MYUF	Renegade Spirit	-	
G-MYUH	SW Pegasus XL-Q	-	
G-MYUK	Mainair Mercury	-	
G-MYUL	Quad City Challenger II	-	
G-MYUN	Mainair Blade	-	
G-MYUO	Pegasus Quantum 15	S. M. Neil	
G-MYUP	Letov LK-2M Sluka	-	
G-MYUR	Hunt Wing	T. C. Saltmarsh	
G-MYUS	CFM Shadow Srs CD	-	
G-MYUT	Hunt Wing	-	
G-MYUU	Pegasus Quantum 15	J. A. Slocombe	
G-MYUV	Pegasus Quantum 15	D. W. Wilson	
G-MYUW	Mainair Mercury	G. C. Hobson	
G-MYUZ	Rans S.6-ESD Coyote II	-	
G-MYVA	Kolb Twinstar Mk 3	-	
G-MYVB	Mainair Blade	M. & P. L. Eardley	
G-MYVC	Pegasus Quantum 15	-	
G-MYVE	Mainair Blade	-	
G-MYVF	Pegasus Quantum 15	-	
G-MYVG	Letov LK-2M Sluka	-	
G-MYVH	Mainair Mercury	S. E. Wilks	
G-MYVI	Air Creation Fun 18S GTBIS	-	
G-MYVJ	Pegasus Quantum 15	-	
G-MYVK	Pegasus Quantum 15	-	
G-MYVM	Pegasus Quantum 15	T. P. Hunt	
G-MYVN	Cyclone AX3/503	-	
G-MYVO	Mainair Blade	-	
G-MYVP	Rans S.6-ESD Coyote II	-	
G-MYVR	Pegasus Quantum 15	-	
G-MYVS	Mainair Mercury	-	
G-MYVT	Letov LK-2M Sluka	-	
G-MYVU	Medway Raven	-	
G-MYVV	Medway Hybred 44XLR	-	
G-MYVW	Medway Raven	-	
G-MYVX	Medway Hybred 44XLR	-	
G-MYVY	Mainair Blade	-	
G-MYVZ	Mainair Blade	-	
G-MYWA	Mainair Mercury	-	
G-MYWC	Hunt Wing	M. A. Coffin	
G-MYWD	Thruster T.600	-	
G-MYWE	Thruster T.600	-	
G-MYWF	CFM Shadow Srs CD	-	
G-MYWG	Pegasus Quantum 15	-	
G-MYWH	Hunt Wing/Experience	-	
G-MYWI	Pegasus Quantum 15	-	
G-MYWJ	Pegasus Quantum 15	-	
G-MYWK	Pegasus Quantum 15	-	
G-MYWL	Pegasus Quantum 15	E. Smith	

Notes	Reg.	Type	Owner or Operator
	G-MYWM	CFM Shadow Srs CD	-
	G-MYWN	Cyclone Chaser S 508	-
	G-MYWO	Pegasus Quantum 15	-
	G-MYWP	Kolb Twinstar Mk 3	-
	G-MYWR	Pegasus Quantum 15	-
	G-MYWS	Cyclone Chaser S 447	-
	G-MYWT	Pegasus Quantum 1	-
	G-MYWU	Pegasus Quantum 15	-
	G-MYWV	Rans S.4 Coyote	D. Cassidy
	G-MYWW	Pegasus Quantum 15	-
	G-MYWX	Pegasus Quantum 15	-
	G-MYWY	Pegasus Quantum 15	A. Czajka
	G-MYWZ	Thruster TST Mk 1	-
	G-MYXA	Team Minimax 91	-
	G-MYXB	Rans S.6-ESD Coyote II	V. G. J. Davies & D. A. Hall
	G-MYXC	Quad City Challenger I	-
	G-MYXD	Pegasus Quasar IITC	-
	G-MYXE	Pegasus Quantum 15	-
	G-MYXF	Air Creation Fun GT503	-
	G-MYXH	Cyclone AX3/503	-
	G-MYXI	Aries 1	-
	G-MYXJ	Mainair Blade	-
	G-MYXK	Quad City Challenger II	-
	G-MYXL	Mignet HM.1000 Baleri	-
	G-MYXM	Mainair Blade	-
	G-MYXN	Mainair Blade	-
	G-MYXO	Letov LK-2M Sluka	-
	G-MYXP	Rans S.6-ESD Coyote II	-
	G-MYXR	Renegade Spirit UK	-
	G-MYXS	Kolb Twinstar Mk 3	-
	G-MYXT	Pegasus Quantum 15	A. Bloomfield & A. Underwood
	G-MYXU	Thruster T.300	-
	G-MYXV	Quad City Challenger II	-
	G-MYXW	Pegasus Quantum 15	-
	G-MYXX	Pegasus Quantum 15	-
	G-MYXY	CFM Shadow Srs CD	-
	G-MYXZ	Pegasus Quantum 15	-
	G-MYYA	Mainair Blade	-
	G-MYYB	Pegasus Quantum 15	-
	G-MYYC	Pegasus Quantum 15	G. A. Dennett
	G-MYYD	Cyclone Chaser S 447	K. A. Armstrong
	G-MYYF	Quad City Challenger II	-
	G-MYYG	Mainair Blade	-
	G-MYYH	Mainair Blade	-
	G-MYYI	Pegasus Quantum 15	-
	G-MYYJ	Hunt Wing	-
	G-MYYK	Pegasus Quantum 15	R. Noble
	G-MYYL	Cyclone AX3/503	-
	G-MYYM	Microchute Motor 27	-
	G-MYYN	Pegasus Quantum 15	J. Darby
	G-MYYO	Medway Raven X	-
	G-MYYP	AMF Chevvron 2-45CS	I. D. Smith
	G-MYYR	Team Minimax 91	-
	G-MYYS	Team Minimax	-
	G-MYYU	Mainair Mercury	-
	G-MYYV	Rans S.6-ESD Coyote IIXL	M. B. Buttle
	G-MYYW	Mainair Blade	A. Morris-Jones
	G-MYYX	Pegasus Quantum 15	-
	G-MYYY	Mainair Blade	-
	G-MYYZ	Medway Raven X	-
	G-MYZA	Whittaker MW.6 Merlin	-
	G-MYZB	Pegasus Quantum 15	-
	G-MYZC	Cyclone AX3/503	-
	G-MYZD	Pegasus Quantum 15	-
	G-MYZE	Team Minimax	-
	G-MYZF	Cyclone AX3/503	-
	G-MYZJ	Pegasus Quantum 1	-
	G-MYZK	Pegasus Quantum 15	-
	G-MYZL	Pegasus Quantum 15	-
	G-MYZM	Pegasus Quantum 15	-
	G-MYZN	Whittaker MW.6-S Fatboy Flyer	-
	G-MYZO	Medway Raven X	-

Reg.	Type	Owner or Operator	Notes
G-MYZP	CFM Shadow Srs DD	-	
G-MYZR	Rans S.6-ESD Coyote II	-	
G-MYZV	Rans S.6-ESD Coyote II	-	
G-MYZW	Cyclone Chaser S 508	-	
G-MYZY	Pegasus Quantum 15	-	
G-MYZZ	Pegasus Quantum 15	-	
G-MZAA	Mainair Blade	P. R. Proost	
G-MZAC	Quad City Challenger II	-	
G-MZAE	Mainair Blade	-	
G-MZAF	Mainair Blade	-	
G-MZAG	Mainair Blade	-	
G-MZAH	Rans S.6-ESD Coyote II	-	
G-MZAI	Mainair Blade 912	-	
G-MZAJ	Mainair Blade	-	
G-MZAK	Mainair Mercury	-	
G-MZAL	Mainair Blade	-	
G-MZAM	Mainair Blade	-	
G-MZAN	Pegasus Quantum 15	-	
G-MZAP	Mainair Blade	-	
G-MZAR	Mainair Blade	D. A. Valentine	
G-MZAS	Mainair Blade	-	
G-MZAT	Mainair Blade	-	
G-MZAU	Mainair Blade	-	
G-MZAV	Mainair Blade	-	
G-MZAW	Pegasus Quantum 15	K. A. O'Neill	
G-MZAY	Mainair Blade	-	
G-MZAZ	Mainair Blade	T. Porter & D. Whiteley	
G-MZBA	Mainair Blade 912	S. Stone	
G-MZBB	Pegasus Quantum 15	-	
G-MZBC	Pegasus Quantum 15	-	
G-MZBD	Rans S.6-ESD Coyote II	-	
G-MZBF	Letov LK-2M Sluka	-	
G-MZBG	Whittaker MW.6-S Fatboy Flye	-	
G-MZBH	Rans S.6-ESD Coyote II	-	
G-MZBI	Pegasus Quantum 15	-	
G-MZBK	Letov LK-2M Sluka	-	
G-MZBL	Mainair Blade	-	
G-MZBM	Pegasus Quantum 15	W. Gray	
G-MZBN	CFM Shadow Srs B	-	
G-MZBO	Pegasus Quantum 15	J. E. Davis	
G-MZBR	Southdown Raven	-	
G-MZBS	CFM Shadow Srs D	-	
G-MZBT	Pegasus Quantum 15	-	
G-MZBU	Rans S.6-ESD Coyote II	-	
G-MZBV	Rans S.6-ESD Coyote II	-	
G-MZBW	Quad City Challenger II UK	-	
G-MZBX	Whittaker MW.6-S Fatboy Flye	-	
G-MZBY	Pegasus Quantum 15	L. G. Wray	
G-MZBZ	Quad City Challenger II UK	T. R. Gregory	
G-MZCA	Rans S.6-ESD Coyote II	G. R. Inston	
G-MZCB	Cyclone Chaser S 447	-	
G-MZCC	Mainair Blade 912	-	
G-MZCD	Mainair Blade	-	
G-MZCE	Mainair Blade	C. T. Halliday	
G-MZCF	Mainair Blade	-	
G-MZCG	Mainair Blade	-	
G-MZCH	Whittaker MW.6-S Fatboy Flyer	-	
G-MZCI	Pegasus Quantum 15	-	
G-MZCJ	Pegasus Quantum 15	-	
G-MZCK	AMF Chevvron 2-32C	-	
G-MZCM	Pegasus Quantum 15	-	
G-MZCN	Mainair Blade	-	
G-MZCO	Mainair Mercury	-	
G-MZCP	SW Pegasus XL-Q	-	
G-MZCR	Pegasus Quantum 15	-	
G-MZCS	Team Minimax	-	
G-MZCT	CFM Shadow Srs CD	-	
G-MZCU	Mainair Blade	-	
G-MZCV	Pegasus Quantum 15	-	
G-MZCW	Pegasus Quantum 15	-	
G-MZCX	Hunt Wing	-	

Notes	Reg.	Type	Owner or Operator
	G-MZCY	Pegasus Quantum 15	-
	G-MZDA	Rans S.6-ESD Coyote IIXL	-
	G-MZDB	Pegasus Quantum 15	Scottish Aerotow Club
	G-MZDC	Pegasus Quantum 15	-
	G-MZDD	Pegasus Quantum 15	-
	G-MZDE	Pegasus Quantum 15	-
	G-MZDF	Mainair Blade	M. Liptrot
	G-MZDG	Rans S.6-ESD Coyote IIXL	-
	G-MZDH	Pegasus Quantum 15	B. J. Palfreyman
	G-MZDI	Whittaker MW.6-S Fatboy Flyer	-(G-BUNN)
	G-MZDJ	Medway Raven X	-
	G-MZDK	Mainair Blade	-
	G-MZDL	Whittaker MW.6-S Fatboy Flyer	-
	G-MZDM	Rans S.6-ESD Coyote II	-
	G-MZDN	Pegasus Quantum 15	-
	G-MZDO	Cyclone AX3/503	-
	G-MZDP	AMF Chevvron 2-32	-
	G-MZDR	Rans S.6-ESD Coyote IIXL	-
	G-MZDS	Cyclone AX3/503	D. S. Parker
	G-MZDT	Mainair Blade	-
	G-MZDU	Pegasus Quantum 15	-
	G-MZDV	Pegasus Quantum 15	-
	G-MZDX	Letov LK-2M Sluka	A. L. Brown
	G-MZDY	Pegasus Quantum 15	-
	G-MZDZ	Hunt Wing	-
	G-MZEA	Quad City Challenger II UK	-
	G-MZEB	Mainair Blade	-
	G-MZEC	Pegasus Quantum 15	-
	G-MZED	Mainair Blade	-
	G-MZEE	Pegasus Quantum 15	-
	G-MZEG	Mainair Blade	R. & A. Soltysik
	G-MZEH	Pegasus Quantum 15	-
	G-MZEI	Whittaker MW.5-D Sorcerer	S. A. Gill
	G-MZEJ	Mainair Blade	-
	G-MZEK	Mainair Mercury	-
	G-MZEL	Cyclone AX3/503	-
	G-MZEM	Pegasus Quantum 15	D. E. J. McVicker
	G-MZEN	Rans S.6-ESD Coyote II	-
	G-MZEO	Rans S.6-ESD Coyote IIXL	-
	G-MZEP	Mainair Rapier	-
	G-MZER	Cyclone AX2000	J. H. Keep
	G-MZES	Letov LK-2N Sluka	-
	G-MZEU	Rans S.6-ESD Coyote IIXL	-
	G-MZEV	Mainair Rapier	G-MZEV Syndicate
	G-MZEW	Mainair Blade	-
	G-MZEX	Pegasus Quantum 15	M. P. Duckett
	G-MZEY	Micro Bantam B.22	K. T. Bettington & D. Harris
	G-MZEZ	Pegasus Quantum 15	-
	G-MZFA	Cyclone AX2000	R. S. McMaster
	G-MZFB	Mainair Blade	A. J. Plant
	G-MZFC	Letov LK-2M Sluka	-
	G-MZFD	Mainair Rapier	-
	G-MZFE	Hunt Wing	-
	G-MZFF	Hunt Wing	-
	G-MZFG	Pegasus Quantum 15	-
	G-MZFH	AMF Chevvron 2-32C	D. R. Gooby
	G-MZFI	Iolaire	-
	G-MZFK	Whittaker MW.6 Merlin	-
	G-MZFL	Rans S.6-ESD Coyote IIXL	G-MZFL Flying Group
	G-MZFM	Pegasus Quantum 15	-
	G-MZFN	Rans S.6.ESD Coyote IIXL	-
	G-MZFO	Thruster T.600N	-
	G-MZFR	Thruster T.600N	-
	G-MZFS	Mainair Blade	-
	G-MZFT	Pegasus Quantum 15	C. R. Cawley
	G-MZFU	Thruster T.600N	-
	G-MZFV	Pegasus Quantum 15	-
	G-MZFX	Cyclone AX2000	-
	G-MZFY	Rans S.6-ESD Coyote IIXL	-
	G-MZFZ	Mainair Blade	-
	G-MZGA	Cyclone AX2000	-
	G-MZGB	Cyclone AX2000	-

Reg.	Type	Owner or Operator	Notes
G-MZGC	Cyclone AX2000	-	
G-MZGD	Rans S.6 Coyote II	-	
G-MZGF	Letov LK-2M Sluka	-	
G-MZGG	Pegasus Quantum 15	-	
G-MZGH	Hunt Wing/Avon 462	-	
G-MZGI	Mainair Blade 912	-	
G-MZGJ	Kolb Twinstar Mk 1	-	
G-MZGK	Pegasus Quantum 15	-	
G-MZGL	Mainair Rapier	-	
G-MZGM	Cyclone AX2000	-	
G-MZGN	Pegasus Quantum 15	G. Taylor	
G-MZGO	Pegasus Quantum 15	-	
G-MZGP	Cyclone AX2000	-	
G-MZGR	Team Minimax	-	
G-MZGS	CFM Shadow Srs BD	E. Fogarty	
G-MZGT	RH78 Tiger Light	-	
G-MZGU	Arrowflight Hawk II (UK)	J. N. Holden	
G-MZGV	Pegasus Quantum 15	-	
G-MZGW	Mainair Blade	-	
G-MZGX	Thruster T.600N	-	
G-MZGY	Thruster T.600N	-	
G-MZGZ	Thruster T.600N	B. E. J. Badger	
G-MZHA	Thruster T.600N	-	
G-MZHB	Mainair Blade	-	
G-MZHC	Thruster T.600N-	-	
G-MZHD	Thruster T.600N	-	
G-MZHE	Thruster T.600N	-	
G-MZHF	Thruster T.600N	-	
G-MZHG	Whittaker MW.6-T	-	
G-MZHI	Pegasus Quantum 15	-	
G-MZHJ	Mainair Rapier	-	
G-MZHK	Pegasus Quantum 15	O. Goodwin	
G-MZHL	Mainair Rapier	-	
G-MZHM	Team Himax 1700R	-	
G-MZHN	Pegasus Quantum 15	F. Omarie-Hamdanie	
G-MZHO	Quad City Challenger II	-	
G-MZHP	Pegasus Quantum 15	-	
G-MZHR	Cyclone AX2000	-	
G-MZHS	Thruster T.600T	-	
G-MZHT	Whittaker MW.6 Merlin	G. J. Chadwick	
G-MZHU	Thruster T.600T	-	
G-MZHV	Thruster T.600T	-	
G-MZHW	Thruster T.600N	-	
G-MZHX	Thruster T.600N	-	
G-MZHY	Thruster T.600N	-	
G-MZIA	Team Himax 1700R	-	
G-MZIB	Pegasus Quantum 15	-	
G-MZIC	Pegasus Quantum 15	-	
G-MZID	Whittaker MW.6 Merlin	-	
G-MZIE	Pegasus Quantum 15	-	
G-MZIF	Pegasus Quantum 15	-	
G-MZII	TEAM MiniMax 88	-	
G-MZIJ	Pegasus Quantum 15	-	
G-MZIK	Pegasus Quantum 15	L. A. Read	
G-MZIL	Mainair Rapier	G. S. Highley	
G-MZIM	Mainair Rapier	-	
G-MZIN	Whittaker MW-6 Merlin	-	
G-MZIP	Renegade Spirit UK	-	
G-MZIR	Mainair Blade	-	
G-MZIS	Mainair Blade	-	
G-MZIT	Mainair Blade 912	-	
G-MZIU	Pegasus Quantum 15	-	
G-MZIV	Cyclone AX2000	-	
G-MZIW	Mainair Blade	-	
G-MZIX	Mignet HM.1000 Balerit	-	
G-MZIY	Rans S.6-ESD Coyote II	-	
G-MZIZ	Renegade Spirit UK (G-MWGP)	R. J. Collins	
G-MZJA	Mainair Blade	-	
G-MZJB	Aviasud Mistral	J. M. Whitham	
G-MZJC	Micro Bantam B22-S	-	
G-MZJD	Mainair Blade	A. R. Vincent & R. W. Neal	
G-MZJE	Mainair Rapier	-	

Notes	Reg.	Type	Owner or Operator
	G-MZJF	Cyclone AX2000	P. W. Hastings
	G-MZJG	Pegasus Quantum 15	-
	G-MZJH	Pegasus Quantum 15	-
	G-MZJI	Rans S.6-ESD Coyote II	M. A. Newbould & C. Topp
	G-MZJJ	Maverick	-
	G-MZJL	Cyclone AX2000	M. H. Owen
	G-MZJM	Rans S.6-ESD Coyote IIXL	K. A. Hastie
	G-MZJN	Pegasus Quantum 15	-
	G-MZJO	Pegasus Quantum 15	-
	G-MZJP	Whittaker MW.6-S Fatboy Flyer	-
	G-MZJR	Cyclone AX2000	-
	G-MZJS	Meridian Maverick	-
	G-MZJT	Pegasus Quantum 15	-
	G-MZJV	Mainair Blade 912	-
	G-MZJW	Pegasus Quantum 15	-
	G-MZJX	Mainair Blade	A. D. Taylor
	G-MZJY	Pegasus Quantum 15	E. J. Childs/Ireland
	G-MZJZ	Mainair Blade	-
	G-MZKA	Pegasus Quantum 15	-
	G-MZKC	Cyclone AX2000	Broad Farm Flyers
	G-MZKD	Pegasus Quantum 15	-
	G-MZKE	Rans S.6-ESD Coyote IIXL	-
	G-MZKF	Pegasus Quantum 15	-
	G-MZKG	Mainair Blade	-
	G-MZKH	CFM Shadow Srs DD	S. P. H. Calvert & L. J. Praill
	G-MZKI	Mainair Blade	-
	G-MZKJ	Mainair Blade	-
	G-MZKK	Mainair Blade 912	A. S. Markey
	G-MZKL	Pegasus Quantum 15	G. Williams
	G-MZKM	Mainair Blade 912	-
	G-MZKN	Mainair Rapier	-
	G-MZKO	Mainair Blade	-
	G-MZKP	Thruster T.600N	-
	G-MZKR	Thruster T.600N	-
	G-MZKS	Thruster T.600N	-
	G-MZKT	Thruster T.600N	-
	G-MZKU	Thruster T.600N	-
	G-MZKV	Mainair Blade 912	-
	G-MZKW	Quad City Challenger II	-
	G-MZKX	Pegasus Quantum 15	D. B. Jones
	G-MZKY	Pegasus Quantum 15	-
	G-MZKZ	Mainair Blade	-
	G-MZLA	Pegasus Quantum 15	-
	G-MZLB	Hunt Wing	-
	G-MZLC	Mainair Blade 912	-
	G-MZLD	Pegasus Quantum 15	S. G. McLean
	G-MZLE	Maverick (G-BXSZ)	J. Smith
	G-MZLF	Pegasus Quantum 15	S. Seymour
	G-MZLG	Rans S.6-ESD Coyote IIXL	G-MZLG Group
	G-MZLH	Pegasus Quantum 15	-
	G-MZLI	Mignet HM.1000 Balerit	-
	G-MZLJ	Pegasus Quantum 15	-
	G-MZLL	Rans S.6-ESD Coyote II	-
	G-MZLM	Cyclone AX2000	-
	G-MZLN	Pegasus Quantum 15	-
	G-MZLP	CFM Shadow Srs O	-
	G-MZLR	SW Pegasus XL-Q	I. W. Barlow & L. M. Courtney
	G-MZLS	Cyclone AX2000	M. J. A. New
	G-MZLT	Pegasus Quantum 15	-
	G-MZLU	Cyclone AX2000	E. Pashley
	G-MZLV	Pegasus Quantum 15	-
	G-MZLW	Pegasus Quantum 15	-
	G-MZLX	Micro Bantam B.22-5	-
	G-MZLY	Letov LK-2M Sluka	-
	G-MZLZ	Mainair Blade 912	-
	G-MZMA	SW Pegasus Quasar IITC	-
	G-MZMB	Mainair Blade-	-
	G-MZMC	Pegasus Quantum 15	-
	G-MZMD	Mainair Blade 912	-
	G-MZMG	Pegasus Quantum 15	-
	G-MZMH	Pegasus Quantum 15	-
	G-MZMJ	Mainair Blade	T. F. R. Calladine

Reg.	Type	Owner or Operator	Notes
G-MZMK	Chevvron 2-32C	-	
G-MZML	Mainair Blade 912	-	
G-MZMM	Mainair Blade 912	-	
G-MZMN	Pegasus Quantum 912	-	
G-MZMO	Team Minimax 91	-	
G-MZMP	Mainair Blade	-	
G-MZMR	Rans S.6-ESA Coyote II	-	
G-MZMS	Rans S.6-ESD Coyote II	-	
G-MZMT	Pegasus Quantum 15	-	
G-MZMU	Rans S.6-ESD Coyote II	S. Bishop	
G-MZMV	Mainair Blade	-	
G-MZMW	Mignet HM.1000 Balerit	-	
G-MZMX	Cyclone AX2000	P. A. Tarplee	
G-MZMY	Mainair Blade	-	
G-MZMZ	Mainair Blade	-	
G-MZNA	Quad City Challenger II UK	-	
G-MZNB	Pegasus Quantum 15	-	
G-MZNC	Mainair Blade 912	-	
G-MZND	Mainair Rapier	-	
G-MZNE	Whittaker MW.6-S Fatboy Flyer	-	
G-MZNG	Pegasus Quantum 15	-	
G-MZNH	CFM Shadow Srs DD	-	
G-MZNI	Mainair Blade 912	-	
G-MZNJ	Mainair Blade	-	
G-MZNK	Mainair Blade 912	R.P. Taylor	
G-MZNL	Mainair Blade 912	M. A. Williams	
G-MZNM	Team Minimax	-	
G-MZNN	Team Minimax	-	
G-MZNO	Mainair Blade	-	
G-MZNP	Pegasus Quantum 15	-	
G-MZNR	Pegasus Quantum 15	-	
G-MZNS	Pegasus Quantum 15	-	
G-MZNT	Pegasus Quantum 15-912	J. Rodgers	
G-MZNU	Mainair Rapier	-	
G-MZNV	Rans S.6-ESD Coyote II	-	
G-MZNW	Thruster T.600N HKS	-	
G-MZNX	Thruster T.600N		
G-MZNY	Thruster T.600N	L. O. Partington & G. Price	
G-MZOC	Mainair Blade	-	
G-MZOD	Pegasus Quantum 15	-	
G-MZOE	Cyclone AX2000	-	
G-MZOF	Mainair Blade	R. M. Ellis	
G-MZOG	Pegasus Quantum 15-912	J. Urwin	
G-MZOH	Whittaker MW.5D Sorcerer	-	
G-MZOI	Letov LK-2M Sluka	B. S. P. Finch	
G-MZOJ	Pegasus Quantum 15	-	
G-MZOK	Whittaker MW.6 Merlin	G-MZOK Syndicate	
G-MZOM	CFM Shadow Srs DD	-	
G-MZON	Mainair Rapier	-	
G-MZOP	Mainair Blade 912	-	
G-MZOR	Mainair Blade 912	-	
G-MZOS	Pegasus Quantum 15-912	-	
G-MZOT	Letov LK-2M Sluka	-	
G-MZOV	Pegasus Quantum 15	Pegasus XL Group	
G-MZOW	Pegasus Quantum 15-912	I. A. Macadam	
G-MZOX	Letov LK-2M Sluka	-	
G-MZOZ	Rans S.6-ESD Coyote IIXL	-	
G-MZPB	Mignet HM.1000 Balerit	-	
G-MZPD	Pegasus Quantum 15	-	
G-MZPH	Mainair Blade	J. D. Hoyland	
G-MZPJ	Team Minimax	-	
G-MZPW	Pegasus Quantum 15	-	
G-MZRC	Pegasus Quantum 15	-	
G-MZRH	Pegasus Quantum 15	-	
G-MZRM	Pegasus Quantum 15	-	
G-MZRS	CFM Shadow Srs CD	-	
G-MZSC	Pegasus Quantum 15	-	
G-MZSD	Mainair Blade 912	-	
G-MZSM	Mainair Blade	-	
G-MZTA	Mignet HM.1000 Balerit	-	
G-MZTS	Aerial Arts Chaser S	-(G-MVDM)	
G-MZUB	Rans S.6-ESD Coyote IIXL	-	

Notes	Reg.	Type	Owner or Operator
	G-MZZT	Kolb Twinstar Mk 3	-
	G-MZZY	Mainair Blade 912	-
	G-NAAA	MBB Bo.105DBS/4	Bond Air Services (G-BUTN/G-AZTI)/Aberdeen
	G-NAAB	MBB Bo.105DBS/4	Bond Air Services/Aberdeen
	G-NAAS	AS.355F-1 Twin Squirrel	Police Aviation Services Ltd (G-BPRG/G-NWPA)
	G-NACA	Norman NAC.2 Freelance 180	NDN Aircraft Ltd/Sandown
	G-NACI	Norman NAC.1 Srs 100	L. J. Martin (G-AXFB)
	G-NACL	Norman NAC.6 Fieldmaster	EPA Aircraft Co Ltd (G-BNEG)
	G-NACO	Norman NAC-6 Fieldmaster	EPA Aircraft Co Ltd
	G-NACP	Norman NAC-6 Fieldmaster	EPA Aircraft Co Ltd
	G-NADS	Team Minimax 91	S. Stockill
	G-NAPO	Pegasus Quantum 15-912	A. J. Varga
	G-NAPP	Van's RV-7	R. J. Napp
	G-NARO	Cassutt Racer	D. A. Wirdnam (G-BTXR)
	G-NATT	R. Commander 114A	Northgleam Ltd
	G-NATX	Cameron O-65 balloon	A. G. E. Faulkner
	G-NATY	H.S. Gnat T.1 (XR537)★	F. C. Hackett-Jones/Bournemouth
	G-NBDD	Robin DR.400/180	J. N. Binks & I. H. Taylor
	G-NBSI	Cameron N-77 balloon	Nottingham Hot-Air Balloon Club
	G-NCFC	PA-38-112 Tomahawk II	S. J. Elvery (G-BNOA)
	G-NCFE	PA-38-112 Tomahawk	R. M. Browes (G-BKMK)
	G-NCUB	Piper J-3C-65 Cub	L. W. Usherwood (G-BGXV)/Norwich
	G-NDGC	Grob G.109	J. E. Bedford & M. Mathieson
	G-NDNI	NDN.1 Firecracker	N. W. G. Marsh
	G-NDOL	Shaw Europa	S. Longstaff
	G-NDOT	Thruster T.600N	P. C. Bailey
	G-NEAL	PA-32 Cherokee Six 260	VSD Group (G-BFPY)
	G-NEAT	Shaw Europa	M. Burton
	G-NEAU	Eurocopter EC.135T-2	McAlpine Helicopters Ltd/Kidlington
	G-NEEL	Rotorway Executive 90	I. C. Bedford
	G-NEGG	Acrosport 2	D. K. Keays & R.S. Goodwin
	G-NEGS	Thunder Ax7-77 balloon	M. Rowlands
	G-NEIL	Thunder Ax3 balloon	N. A. Robertson
	G-NELI	PA-28R Cherokee Arrow 180	European Light Aviation Ltd
	G-NEMO	Raj Hamsa X'Air Jabiru	D. G. Smith
	G-NEON	PA-32 Cherokee Six 300B	S. C. A. Lever
	G-NEPB	Cameron N-77 balloon	The Post Office
	G-NERC	PA-31-350 Navajo Chieftain	Natural Environment Research Council (G-BBXX)/ Coventry
	G-NESA	Shaw Europa XS	K. G. & V. E. Summerhill
	G-NESU	BN-2B-20 Islander	Northumbria Police Authority (G-BTVN)/Teesside
	G-NESV	Eurocopter EC.135T-1	Northumbria Police Authority
	G-NESW	PA-34-220T Seneca III	Scot Wings Ltd
	G-NESY	PA-18 Super Cub 95	V. Fisher
	G-NETA	Cessna 560XL Citation Excel	Houston Air Taxis Ltd
	G-NETY	PA-18 Super Cub 150	N. B. Mason
	G-NEUF	Bell 206L-1 LongRanger 2	Yendle Roberts Ltd (G-BVVV)
	G-NEWR	PA-31-350 Navajo Chieftain	Eastern Air Executive Ltd/Sturgate
	G-NEWS	Bell 206B JetRanger 3	Lanthwaite Aviation
	G-NEWT	Beech 35 Bonanza	J. S. Allison (G-APVW)
	G-NEWZ	Bell 206B JetRanger 3	Peter Press Ltd
	G-NFLC	H.P.137 Jetstream 1★ (G-AXUI)	Instructional airframe/Perth
	G-NFNF	Robin DR.400/180	N. French
	G-NGRM	Spezio DAL.1 Tuholer	S. H. Crook
	G-NHRH	PA-28 Cherokee 140	J. E. & I. Parkinson
	G-NHRJ	Shaw Europa XS	D. A. Lowe
	G-NICC	Aerotechnik EV-97 Team Eurostar UK	Pickup & Son Property Maintenance Ltd
	G-NIGC	Avtech Jabiru UL-450	N. Creeney
	G-NIGE	Luscombe 8E Silvaire	Garden Party Ltd (G-BSHG)
	G-NIGL	Shaw Europa	N. M. Graham
	G-NIGS	Thunder Ax7-65 balloon	A. N. F. Pertwee
	G-NIJM	PA-28R Cheroklee Arrow 180	Eagle MachineryHoldings Ltd
	G-NIKE	PA-28-181 Archer II	Key Properties Ltd/White Waltham
	G-NIKO	Airbus A.321-211	My Travel Airways
	G-NINA	PA-28-161 Warrior II	A. P. Gorrod (G-BEUC)

Reg.	Type	Owner or Operator	Notes
G-NINB	PA-28 Cherokee 180G	P. A. Layzell	
G-NINC	PA-28 Cherokee 180G	P. A. Layzell	
G-NINE	Murphy Renegade 912	R. F. Bond	
G-NIOG	Robinson R-44-II	Farm Aviation Ltd	
G-NIOS	PA-32R-301 Saratoga SP	Plant Aviation	
G-NIPA	Slingsby T.66 Nipper 3	R. J. O. Walker (G-AWDD)	
G-NIPP	Slingsby T.66 Nipper 3	T. Dale (G-AVKJ)	
G-NIPY	Hughes 369HS	Jet Aviation (Northwest) Ltd	
G-NITA	PA-28 Cherokee 180	T. Clifford (G-AVVG)	
G-NJAG	Cessna 207	G. H. Nolan Ltd	
G-NJSH	Robinson R-22B	A. J. Hawes	
G-NLEE	Cessna 182Q	G. Hall (G-TLTD)	
G-NLYB	Cameron N-105 balloon	P. H. E. Van Overwalle/Belgium	
G-NMID	Eurocopter EC.135T-2	Derbyshire Constabulary	
G-NMOS	Cameron C-80 balloon	C. J. Thomas & M. C. East	
G-NNAC	PA-18 Super Cub 135	PAW Flying Services Ltd	
G-NNON	Mainair Blade	A. Gannon	
G-NOBI	Spezio HES-1 Tuholer Sport	M. G. Parsons	
G-NOCK	Cessna FR.182RG II	F. J. Whidbourne (G-BGTK)	
G-NODE	AA-5B Tiger	Strategic Telecom Networks Ltd	
G-NODY	American General AG-5B Tiger	Curd & Green Ltd/Elstree	
G-NOIR	Bell 222	Arlington Property Developments (2003) (G-OJLC/G-OSEB/G-BNDA)	
G-NOIZ	Yakovlev Yak-55M	S. C. Cattlin	
G-NOMO	Cameron O-31 balloon	Tim Balloon Promotion Airships Ltd	
G-NONE	Dyn' Aero MCR-01 ULC	J. Fisher	
G-NONI	AA-5 Traveler	November India Flying Group (G-BBDA)	
G-NOOK	Mainair Blade 912S	P. M. Knight	
G-NOOR	Commander 114B	As-Al Ltd	
G-NORD	SNCAN NC.854	W. J. McCollum	
G-NORT	Robinson R-22	Plane Talking Ltd	
G-NOSE	Cessna 402B	Atlantic Air Transport Ltd (G-MPCU)/Coventry	
G-NOSY	Robinson R-44	Cotswold Helicopters Ltd (G-LATK/G-BVMK))	
G-NOTE	PA-28-181 Archer III	J. Beach	
G-NOTR	MDH MD.500N	Eastern Atlantic Helicopters Ltd	
G-NOTT	Nott ULD-2 balloon	J. R. P. Nott	
G-NOTY	Westland Scout AH.1	R. P. Coplestone	
G-NOVO	Colt AS-56 airship	J. R. Huggins	
G-NOWW	Mainair Blade 912	C. Bodill	
G-NPKJ	Van's RV-6	H. M. Darlington	
G-NPPL	Comco Ikarus C.42 FB.100	Papa Lima Group	
G-NROY	PA-32RT-300 Lance II	Roy West Cars (G-LYNN/G-BGNY)	
G-NRRA	SIAI-Marchetti SF.260W	G. Boot	
G-NRSC	PA-23 Aztec 250E	Infoterra Ltd (G-BSFL)	
G-NRYL	Mooney M.20R	C. D. Wood	
G-NSBB	Ikarus C.42 FB-100 VLA	B. Bayes & N.E. Sams	
G-NSEW	Robinson R-44	Captive Audience (UK) Ltd	
G-NSOF	Robin HR.200/120B	Northamptonshire School of Flying Ltd/Sywell	
G-NSTG	Cessna F.150F	Westair Flying Services Ltd (G-ATNI)/Blackpool	
G-NSUK	PA-34-220T	Genus PLC	
G-NUDE	Robinson R-44	The Last Great Journey Ltd (G-NSYT)	
G-NUKA	PA-28-181 Archer II	N. Ibrahim	
G-NULA	Flight Design CT.2K	G-NULA Flying Group	
G-NUTS	Cameron 35SS balloon	Bristol Balloons	
G-NUTY	AS.350B Ecureuil	Arena Aviation Ltd (G-BXKT)	
G-NVBF	Lindstrand LBL-210A balloon	Virgin Balloon Flights	
G-NVSA	D.H.C.8-311 Dash Eight	British Airways CitiExpress	
G-NVSB	D.H.C.8-311 Dash Eight	British Airways CitiExpress	
G-NWPR	Cameron N-77 balloon	D. B. Court	
G-NWPS	Eurocopter EC.135T-1	North-West Police Authority	
G-NYMF	PA-25 Pawnee 235D	Bristol Gliding Club Pty Ltd/Nympsfield	

Notes	Reg.	Type	Owner or Operator
	G-NYTE	Cessna F.337G	I. M. Latiff (G-BATH)
	G-NYZS	Cessna 182G	P. Ragg (G-ASRR)
	G-NZGL	Cameron O-105 balloon	R. A. Vale & ptnrs
	G-NZSS	Boeing Stearman N2S-5 (343251)	Anglian Aircraft Co Ltd
	G-OAAA	PA-28-161 Warrior II	Halfpenny Green Flight Centre Ltd
	G-OAAC	Airtour AH-77B balloon ★	Director Army Air Corp/Middle Wallop
	G-OABB	Jodel D.150	K. Manley
	G-OABC	Colt 69A balloon	P. A. C. Stuart-Kregor
	G-OABO	Enstrom F-28A	ABO Ltd (G-BAIB)
	G-OABR	AG-5B Tiger	Vulcan House Management UK Ltd
	G-OACA	PA-44-180 Seminole	Plane Talking Ltd (G-GSFT)/Elstree
	G-OACC	PA-44-180 Seminole	Plane Talking Ltd (G-FSFT)/Elstree
	G-OACE	Valentin Taifun 17E	D. R. Piercy
	G-OACF	Robin DR.400/180	A. C. Fletcher
	G-OACG	PA-34-200T Seneca II	Cega Aviation Ltd (G-BUNR)
	G-OACI	M.S.893E Rallye 180GT	A. M. Quayle (G-DOOR)
	G-OACP	OGMA D.H.C.1 Chipmunk 20	Aeroclub de Portugal
	G-OADY	Beech 76 Duchess	Multiflight Ltd
	G-OAER	Lindstrand LBL-105A balloon	T. M. Donnelly
	G-OAFT	Cessna 152 II	Evensport Ltd (G-BNKM)
	G-OAHC	Beech F33C Bonanza	V. D. Speck (G-BTTF)/Clacton
	G-OAJB	Cyclone AX2000	G. K. R. Linney (G-MZFJ)
	G-OAJC	Robinson R-44	A. J. Cain
	G-OAJL	Ikarus C.42 FB100	A.JL Driver Training
	G-OAJS	PA-39 Twin Comanche 160 C/R	Go-AJS Ltd (G-BCIO)
	G-OAKJ	BAe Jetstream 3202	Eastern Airways Ltd (G-BOTJ
	G-OAKR	Cessna 172S	A. K. Robson)
	G-OALB	Aero L-39C Albatros	Starindale Ltd
	G-OALD	SOCATA TB.20 Trinidad	Gold Aviation/Biggin Hill
	G-OALH	Tecnam P.92-EM Echo	L. Hill
	G-OAMF	Pegasus Quantum 15-912	I. B. Smith
	G-OAMG	Bell 206B JetRanger 3	Alan Mann Helicopters Ltd/Fairoaks
	G-OAMI	Bell 206B JetRanger 2	Techno Solutions Ltd (G-BAUN)
	G-OAML	Cameron AML-105 balloon	Stratton Motor Co. (Norfolk) Ltd
	G-OAMP	Cessna F.177RG	Vale Aero Group (G-AYPF)
	G-OAMT	PA-31-350 Navajo Chieftain	AM & T Solutions Ltd (G-BXKS)/Bristol;
	G-OANI	PA-28-161 Warrior II	J. F. Mitchell
	G-OANN	Zenair CH.601HDS	E. W. Chapman
	G-OAPE	Cessna T.303	C. Twiston-Davies & P. L. Drew
	G-OAPR	Brantly B.2B	Helicopter International Magazine
	G-OAPW	Glaser-Dirks DG.400	D. Bonucci
	G-OARA	PA-28R-201 Arrow III	Airsure
	G-OARC	PA-28RT-201 Arrow IV	Aviation Rentals (G-BMVE)
	G-OARG	Cameron C-80 balloon	G. & R. Madelin
	G-OARI	PA-28R-201 Arrow III	Plane Talking Ltd/Elstree
	G-OARO	PA-28R-201 Arrow III	Plane Talking Ltd/Elstree
	G-OART	PA-23 Aztec 250D	Levenmere Ltd (G-AXKD)
	G-OARU	PA-28R-201 Arrow III	Mind Power Consultancy Ltd
	G-OARV	ARV Super 2	N. R. Beale
	G-OARW	PA-32R Saratoga IIHP	A. R. Ward
	G-OASH	Robinson R-22B	J. C. Lane
	G-OASJ	Thruster T.600N 450	A. S. Johnson
	G-OASP	AS.355F-2 Twin Squirrel	Helicopter Services Ltd
	G-OATE	Mainair Pegasus Quantum 15-912	S. J. Goate
	G-OATG	Advanced Technologies AT-10	Advanced Technologies Group Ltd
	G-OATS	PA-38-112 Tomahawk	Truman Aviation Ltd/Tollerton
	G-OATV	Cameron V-77 balloon	W. G. Andrews
	G-OAVA	Robinson R-22B	J. G. M. McDiarmid
	G-OAVB	Boeing 757-2G5	Astraeus Ltd (G-OOOI)
	G-OAWS	Colt 77A balloon	E. A,. & H. A. Evans
	G-OBAK	PA-28R-201T Turbo Arrow III	DP Group Aviation
	G-OBAL	Mooney M.20J	Britannia Airways Ltd/Luton
	G-OBAM	Bell 206B JetRanger 3	Cherwell Tobacco Ltd
	G-OBAN	Jodel D.140B	S. R. Cameron (G-ATSU)/North Connel
	G-OBAX	Thruster T.600N 450-JAB	Baxby Airsports Club
	G-OBAZ	Best Off Skyranger 912	B. J. Marsh
	G-OBBC	Colt 90A balloon	R. A. & M. A. Riley
	G-OBBJ	Boeing 737-8DR	Multiflight Jet Charter Ltd
	G-OBBO	Cessna 182S	F. Friedenberg

Reg.	Type	Owner or Operator	Notes
G-OBBY	Robinson R-44	P. C. & J. A. Twigg	
G-OBDA	Diamond Katana DA.20-A1	Oscar Papa Ltd	
G-OBDM	Shaw Europa XS	B. D. McHugh	
G-OBDN	PA-28-161 Warrior II	Barn Air Ltd	
G-OBEI	SOCATA TB.200 Tobago XL	Air Touring Ltd	
G-OBEN	Cessna 152 II	Airbase Aircraft Ltd (G-NALI/G-BHVM)	
G-OBET	Sky 77-24 balloon	P. M. Watkins & S. M. Carden	
G-OBEV	Shaw Europa	M. B. Hill & N. I. Wingfield	
G-OBFC	PA-28-161 Warrior II	Bflying Ltd/Bournemouth	
G-OBFE	Sky 120-24 balloon	H. Schmidt	
G-OBFS	PA-28-161 Warrior III	Plane Talking Ltd/Elstree	
G-OBGC	SOCATA TB-20 Trinidad	Bidford Airfield Ltd	
G-OBHD	Short SD3-60 Variant 100	Emerald Airways Ltd (G-BNDK)/Liverpool	
G-OBHL	AS.355F-2 Twin Squirrel	Brands Hatch Leisure Group Ltd (G-HARO/G-DAFT/G-BNNN)	
G-OBIB	Colt 120A balloon	The Aerial Display Co Ltd	
G-OBIL	Robinson R-22B	C. A. Rosenberg	
G-OBIO	Robinson R-22B	A. E. Churchill	
G-OBJB	Lindstrand LBL-90A balloon	B. J. Bower	
G-OBJP	Pegasus Quantum 15-912	S. J. Baker	
G-OBJT	Shaw Europa	B. J. Tarmar (G-MUZO)	
G-OBLC	Beech 76 Duchess	Pridenote Ltd	
G-OBLN	D.H.115 Vampire T.11 (XE956)	De Havilland Aviation Ltd/Swansea	
G-OBLU	Cameron H-34 balloon	John Aimo Balloons SAS/Italy	
G-OBMI	Mainair Blade	P. Clark	
G-OBMP	Boeing 737-3Q8	bmi Baby	
G-OBMS	Cessna F.172N	D. Beverley & ptnrs	
G-OBMW	AA-5 Traveler	Fretcourt Ltd (G-BDFV)	
G-OBNA	PA-34-220T Seneca V	Palmair Ltd	
G-OBNW	PA-31-350 Navajo Chieftain	British North West Airlines (G-BFDA)	
G-OBRI	Medway Eclipser	B. D. Campbell	
G-OBRY	Cameron N-180 balloon	A. C. K. Rawson & J. J. Rudoni	
G-OBUN	Cameron A-250 balloon	A. C. K. Rawson & J. J. Roudoni	
G-OBUY	Colt 69A balloon	Virgin Airship & Balloon Co Ltd	
G-OBWP	BAe ATP	Trident Aviation Leasing Services Ltd (G-BTPO)	
G-OBWR	BAe ATP	Trident Aviation Leasing Services Ltd (G-BUWP)	
G-OBYB	Boeing 767-304ER	Thomsonfly	
G-OBYC	Boeing 767-304ER	Thomsonfly	
G-OBYD	Boeing 767-304ER	Thomsonfly	
G-OBYE	Boeing 767-304ER	Thomsonfly	
G-OBYF	Boeing 767-304ER	Thomsonfly	
G-OBYG	Boeing 767-3Q8ER	Thomsonfly	
G-OBYH	Boeing 767-304ER	Thomsonfly	
G-OBYI	Boeing 767-304ER	Thomsonfly	
G-OBYJ	Boeing 767-304ER	Thomsonfly	
G-OBYT	AB-206A JetRanger	R. J. Everett (G-BNRC)	
G-OCAD	Sequoia F.8L Falco	Falco Flying Group	
G-OCAM	AA-5A Cheetah	Plane Talking Ltd (G-BLHO)/Elstree	
G-OCAR	Colt 77A balloon	S. C. J. Derham	
G-OCBA	H. S. 125 Srs 3B	R. J. Everett (G-MRFB/G-AZVS)	
G-OCBI	Schweizer 269C-1	JWL Helicopters Ltd	
G-OCBS	Lindstrand LBL-210A balloon	G. Binder	
G-OCDS	Aviamilano F.8L Falco II (modified)	C. O. P. Barth (G-VEGL)	
G-OCDW	Jabiru UL	C. D. Wood	
G-OCEA	Short SD3-60 Variant 100	BAC Express Airlines Ltd (G-BRMX)	
G-OCFC	Robin R.2160	Cornwall Flying Club Ltd/Bodmin	
G-OCFD	Bell 206B JetRanger 3	Cranfield Helicopters Ltd (G-WGAL/G-OICS/	
G-OCFM	PA-34-200 Seneca	Stapleford Flying Club Ltd (G-ELBC/G-BANS)	
G-OCHM	Robinson R-44	Westleigh Developments Ltd	
G-OCIT	Cessna 208B	Fly CI Ltd/Jersey	
G-OCJK	Schweizer 269C	P. Crawley	
G-OCMI	Aerotechnik EV-97 Teameurostar	C. M. Theakstone	
G-OCMM	Agusta A.109A II	Maison Air (G-BXCB/G-ISEB/G-IADT/G-HBCA)	
G-OCOV	Robinson R-22B	Flight Training Ltd	
G-OCPC	Cessna FA.152	Westward Airways (Lands End) Ltd/St Just	
G-OCRI	Colomban MC.15 Cri-Cri	M. J. J. Dunning	
G-OCST	AB-206B JetRanger 3	Lift West Ltd (G-BMKM)	
G-OCTI	PA-32 Cherokee Six 260	D. G. Williams (G-BGZX)	
G-OCUB	Piper J-3C-90 Cub	C. A. Foss & P. A. Brook/Shoreham	
G-ODAC	Cessna F.152 II	T. M. Jones (G-BITG)/Egginton	

Notes	Reg.	Type	Owner or Operator
	G-ODAD	Colt 77A balloon	K. Meehan
	G-ODAK	PA-28-236 Dakota	Airways Aero Associations Ltd/Booker
	G-ODAT	Aero L-29 Delfin	Graniteweb Ltd
	G-ODAV	Aerotechnik EV-97 Eurostar	B. R. Davies
	G-ODAY	Cameron N-56 balloon	British Balloon Museum & Library
	G-ODBN	Lindstrand LBL Flowers SS balloon	Magical Adventures Ltd
	G-ODCS	Robinson R-22B-2	Heli Air Ltd/Wellesbourne
	G-ODDY	Lindstrand LBL-105A balloon	P. & T. Huckle
	G-ODEB	Cameron A-250 balloon	A. Derbyshire
	G-ODEE	Van's RV-6	D. Powell
	G-ODEL	Falconar F-11-3	G. F. Brummell
	G-ODEN	PA-28-161 Cadet	Atrium Aviation Ltd
	G-ODGS	Avtech Jabiru UL-450	D. G. Salt
	G-ODHG	Robinson R-44	Sloane Helicopters Ltd
	G-ODHL	Cameron N-77 balloon	DHL International (UK) Ltd
	G-ODIN	Avions Mudry CAP.10B	R. P. W. Steele
	G-ODJB	Robinson R-22B	N. T. Burton
	G-ODJD	Raj Hamsa X'Air 582 (7)	K. P. Roper
	G-ODJG	Shaw Europa	D. J. Goldsmith
	G-ODJH	Mooney M.20C	R. M. Schweitzer (G-BMLH)/Netherlands
	G-ODLY	Cessna 310J	Card Tech Ltd (G-TUBY/G-ASZZ)
	G-ODMC	AS.350B-1 Ecureuil	D. M. Coombs (G-BPVF)/Denham
	G-ODNH	Schweizer 269C-1	Oxford Aviation Services Ltd/Kidlington
	G-ODOC	Robinson R-44	Gas & Air Ltd
	G-ODOD	MDH MD-600N Explorer	HPM Investments Ltd
	G-ODOG	PA-28R Cherokee Arrow 200-II	Advanced Investments Ltd (G-BAAR)
	G-ODOT	Robinson R-22B-2	Ribbands Explosives
	G-ODPJ	VPM M.16 Tandem Trainer	A. P. Wilkinson (G-BVWX)
	G-ODSK	Boeing 737-37Q	bmi Baby
	G-ODTW	Shaw Europa	D. T. Walters
	G-ODUD	PA-28-181 Archer II	D. Rogg (G-IBBO)
	G-ODUS	Boeing 737-36Q	easyJet Airline Co Ltd
	G-ODVB	CFM Shadow Srs DD	D. V. Brunt (G-MGDB)
	G-OEAC	Mooney M.20J	S. Lovatt
	G-OEAT	Robinson R-22B	C. Y. O. Seeds Ltd (G-RACH)
	G-OECH	AA-5A Cheetah	Plane Talking Ltd (G-BKBE)/Elstree
	G-OECM	Commander 114B	ECM (Vehicle Delivery Service) Ltd
	G-OEDB	PA-38-112 Tomahawk	Euroskylink Ltd (G-BGGJ)
	G-OEDP	Cameron N-77 balloon	M. J. Betts
	G-OEGL	Christen Eagle II	The Eagle Flight Syndicate/Shoreham
	G-OELD	Pegasus Quantum 15-912	K. M. Sullivan
	G-OERR	Lindstrand LBL-60A balloon	P. C. Gooch
	G-OERX	Cameron O-65 balloon	R. Roehsler/Austria
	G-OESY	Easy Raider J2.2 (1)	G. C. Long
	G-OETI	Bell 206B JetRanger 3	Elec-Track Installations Ltd (G-RMIE/G-BPIE)
	G-OETV	PA-31-350 Navajo Chieftain	European Executive Ltd
	G-OEVA	PA-32-260 Cherokee Six	M. G. Cookson(G-FLJA/G-AVTJ)
	G-OEWA	D.H.104 Dove 8	D. C. Hunter (G-DDCD/G-ARUM)
	G-OEYE	Rans S.10 Sakota	I. M. J. Mitchell
	G-OEZI	Easy Raider J.2	M. A. Claydon
	G-OEZY	Shaw Europa	A. W. Wakefield
	G-OFAS	Robinson R-22B	Findon Air Services/Shoreham
	G-OFBJ	Thunder Ax7-77 balloon	N. D. Hicks
	G-OFBU	Ikarus C.42	Old Sarum C42 Group
	G-OFCM	Cessna F.172L	F. C. M. Aviation Ltd (G-AZUN)/Guernsey
	G-OFER	PA-18 Super Cub 150	M. S. W. Meagher/Edgehill
	G-OFFA	Pietenpol Air Camper	OFFA Group
	G-OFIL	Robinson R-44	North Helicopters
	G-OFIT	SOCATA TB.10 Tobago	GFI Aviation Group (G-BRIU)
	G-OFLG	SOCATA TB.10 Tobago	MRR Aviation Ltd (G-JMWT)
	G-OFLI	Colt 105A balloon	Virgin Airship & Balloon Co Ltd
	G-OFLT	EMB-110P1 Bandeirante★	Rescue trainer(G-MOBL/G-BGCS/Aveley, Essex
	G-OFLY	Cessna 210M	A. P. Mothew/Stapleford
	G-OFMB	Rand-Robinson KR-2	F. M. & S. I. Burden
	G-OFOA	BAe 146-100	Formula One Administration Ltd (G-BKMN/G-ODAN)
	G-OFOM	BAe 146-100	Formula One Management Ltd (G-BSLP/G-BRLM)
	G-OFOX	Denney Kitfox	P. R. Skeels
	G-OFRA	Boeing 737-36Q	easyJet Airline Co Ltd

Reg.	Type	Owner or Operator	Notes
G-OFRB	Everett gyroplane	-	
G-OFRY	Cessna 152	Devon School of Flying Ltd	
G-OFTI	PA-28 Cherokee 140	G. S. A. Spencer (G-BRKU)	
G-OGAN	Shaw Europa	G-OGAN Group	
G-OGAR	PZL SZD-45A Ogar	N. C. Grayson	
G-OGAS	Westland WG.30 Srs 100 ★	(stored) (G-BKNW)/Yeovil	
G-OGAZ	SA.341G Gazelle 1	Killochries Fold (G-OCJR/G-BRGS)	
G-OGBE	Boeing 737-3L9	bmi Baby	
G-OGCA	PA-28-161 Warrior II	Cardiff-Wales Aviation Services Ltd	
G-OGEM	PA-28-181 Archer II	GEM Rewinds Ltd	
G-OGEO	SA.341G Gazelle 1	MW Helicopters Ltd (G-BXJK)	
G-OGES	Enstrom 280FX Shark	Eastern Atlantic Helicopters Ltd (G-CBYL)	
G-OGGS	Thunder Ax8-84 balloon	G. Gamble & Sons (Quorn) Ltd	
G-OGGY	Aviat A.1B	Chris Irvine Aviation Ltd	
G-OGIL	Short SD3-30 Variant 100 ★	N.E. Aircraft Museum (G-BITV)/Usworth	
G-OGJM	Cameron C-80 balloon	G. F. Madelin	
G-OGJP	Commander 114B	Speedsport Ltd	
G-OGJS	Puffer Cozy	G. J. Stamper	
G-OGOA	AS.350B Ecureuil	Lomas Brothers Ltd (G-PLMD/G-NIAL)	
G-OGOB	Schweizer 269C	Kingfisher Helicopters Ltd (G-GLEE/G-BRUW)	
G-OGOH	Robinson R-22B-2	Lomas Helicopters (G-IPDM/G-OMSG)	
G-OGOS	Everett gyroplane	N. A. Seymour	
G-OGPN	Thompson Cassutt Special	S. Alexander (G-OMFI/G-BKCH)	
G-OGRG	Cessna 560 Citation Ultra V	Houston Jet Services Ltd (G-RIBV)	
G-OGSA	Avtech Jabiru UL 450	A. Knape	
G-OGSS	Lindstrand LBL-120A balloon	R. Klarer/Germany	
G-OGTS	Air Command 532 Elite	GTS Engineering (Coventry) Ltd	
G-OHAC	Cessna F.182Q	The RAF Halton Aeroplane Club	
G-OHAJ	Boeing 737-36Q	easyJet Airline Co Ltd	
G-OHAL	Pietenpol Air Camper	H. C. Danby	
G-OHCP	AS.355F-1 Twin Squirrel	AJJ Developments Ltd	
		(G-BTVS/ G-STVE/G-TOFF/G-BKJX)	
G-OHDC	Colt Film Cassette SS balloon★	Balloon Preservation Group	
G-OHFT	Robinson R-22B	Heliflight (UK) Ltd (G-TYPO/G-JBWI)	
G-OHGC	Scheibe SF.25C Falke	Heron Gliding Club	
G-OHHI	Bell 206L-1 LongRanger	Bradmore Helicopters (G-BWYJ)	
G-OHIG	EMB-110P1 Bandeirante ★	Valley Nurseries (G-OPPP)/Alton	
G-OHKS	Pegasus Quantum 15-912	York Microlight Centre Ltd	
G-OHMS	AS.355F-1 Twin Squirrel	S.W. Electricity PLC	
G-OHOV	Rotorway Executive 162F	M. G. Bird	
G-OHRH	Lindstrand LBL-150A balloon	Exclusive Ballooning	
G-OHSA	Cameron N-77 balloon	D. N. & L. J. Close	
G-OHSL	Robinson R-22B	Astons of Kempsey (Holdings) Ltd (G-BPNF)	
G-OHVA	Mainair Blade 912	M. C. Metatidj	
G-OHWV	Raj Hamsa X'Air 582 (5)	H. W. Vasey	
G-OHYE	Thruster T.600N 450	G-OHYE Group (G-CCRO)	
G-OIBM	R. Commander 114	E. J. Percival (G-BLVZ)	
G-OIBO	PA-28 Cherokee 180	Britannia Airways Ltd (G-AVAZ)/Luton	
G-OIDW	Cessna F.150G	T. S. Sheridan-McGinnitty	
G-OIFM	Cameron 90 Dude SS balloon	Magical Adventures	
G-OIIO	Robinson R-22B	Whizzard Helicopters (G-ULAB)	
G-OIMC	Cessna 152 II	E. Midlands Flying School Ltd	
G-OING	AA-5A Cheetah ★	Abraxas Aviation Ltd (G-BFPD)/Denham	
G-OINK	Piper J-3C-65 Cub	A. R. Harding (G-BILD/G-KERK)	
G-OINV	BAe 146-300	British Airways CitiExpress	
G-OIOZ	Thunder Ax9-120 S2 balloon	C. M. Hodges	
G-OIPB	IDA Bacau Yakalev Yak-52	Mastercraft Ltd	
G-OISO	Cessna FRA.150L	L. A. Mills & B. A. Mills (G-BBJW)	
G-OITN	AS.355F-1 Twin Squirrel	Lynton Aviation Ltd/Denham	
G-OITV	Enstrom 280C-UK-2	C. W. Brierley Jones (G-HRVY/G-DUGY/ G-BEEL)	
G-OJAB	Avtech Jabiru SK	P. A. Brigstock	
G-OJAC	Mooney M.20J	Hornet Engineering Ltd	
G-OJAE	Hughes 269C	J. A. & C. M. Wilson	
G-OJAN	Robinson R-22B	Heliflight (UK) Ltd (G-SANS/G-BUHX)	
G-OJAS	Auster J/1U Workmaster	K. P. & D. S. Hunt_ (Lycoming)	
G-OJAV	BN-2A Mk III-2 Trislander	Lyddair Ltd/Lydd (G-BDOS)	
G-OJBB	Enstrom 280FX	Pendragon (Design & Build) Ltd	
G-OJBM	Cameron N-90 balloon	P. Spinlove	
G-OJBS	Cameron N-105A balloon	Up & Away Ballooning Ltd	

Notes	Reg.	Type	Owner or Operator
	G-OJBW	Lindstrand LBL J & B Bottle SS balloon	N. A. P. Godfrey
	G-OJCW	PA-32RT-300 Lance II	CW Group
	G-OJDA	EAA Acrosport II	D. B. Almey
	G-OJDC	Thunder Ax7-77 balloon	J. Crosby
	G-OJEG	Airbus A.321-231	Monarch Airlines Ltd/Luton
	G-OJEH	PA-28-181 Archer II	P. C. & M. A. Greenaway
	G-OJEN	Cameron V-77 balloon	D. J. Geddes
	G-OJDR	Yakovlev Yak-50	J. D. Rooney
	G-OJDS	Ikarus C.42 FB 80	J. D. Smith
	G-OJGT	Maule M.5-235C	J. G. Townsend
	G-OJHB	Colt Flying Ice Cream Cone SS balloon	Benedikt Haggeney GmbH/Germany
	G-OJHL	Shaw Europa	J. H. Lace
	G-OJIL	PA-31-350 Navajo Chieftain	Redhill Aviation Ltd
	G-OJIM	PA-28R-201T Turbo Arrow III	Piper Arrow Group
	G-OJJB	Mooney M.20K	G. Italiano/Italy
	G-OJJF	D.31 Turbulent	J. J. Ferguson
	G-OJKM	Rans S.7 Courier	M. Jackson
	G-OJLH	Team Minimax 91	J. L. Hamer (G-MYAW)
	G-OJMB	Airbus A.330-243	Thomas Cook Airlines
	G-OJMC	Airbus A.330-243	Thomas Cook Airlines
	G-OJMF	Enstrom 280FX	JMF Ltd (G-DDOD)
	G-OJMR	Airbus A.300B4-605R	Monarch Airlines Ltd/Luton
	G-OJNB	Linsdstrand LBL-21A balloon	N. A. P. Godfrey
	G-OJOD	Jodel D.18	D. Hawkes & C. Poundes
	G-OJON	Taylor JT.2 Titch	A. Donald
	G-OJRH	Robinson R-44	Holgate Construction Ltd
	G-OJRM	Cessna T.182T	SPD Ltd/Old Sarum
	G-OJSH	Thruster T.600N 450 JAB	November Whiskey Flying Club
	G-OJTA	Stemme S.10V	OJT Associates
	G-OJTW	Boeing 737-36N	bmiBaby (G-JTWF)
	G-OJVA	Van's RV-6	J. A. Village
	G-OJVH	Cessna F.150H	A. W. Cairns (G-AWJZ)
	G-OJVL	Van's RV-6	S. E. Tomlinson
	G-OJWS	PA-28-161 Warrior II	P. J. Ward
	G-OKAG	PA-28R Cherokee Arrow 180	B. R. Green
	G-OKAY	Pitts S-1E Special	D. S. T. Eggleton
	G-OKBT	Colt 25A Mk II balloon	British Telecommunications PLC
	G-OKCC	Cameron N-90 balloon	D. J. Head
	G-OKED	Cessna 150L	L. J. Pluck
	G-OKEM	Pegasus Mainair Quik	Kemble Quik
	G-OKEN	PA-28R-201T Turbo Arrow III	W. B. Bateson/Blackpool
	G-OKER	Van's RV-7	R. M. Johnson
	G-OKES	Robinson R-44	Direct Helicopters/Southend
	G-OKEV	Shaw Europa	K. R. Pilcher
	G-OKEY	Robinson R-22B	Key Properties Ltd/Booker
	G-OKGB	IDA Yakovlev Yak-52	W. Hanekom
	G-OKIM	Best Off Sykyranger 912 (2)	K. P. Taylor
	G-OKIS	Tri-R Kis	M. R. Cleveley
	G-OKJN	Boeing 727-225RE	Cougar Airlines Ltd
	G-OKMA	Tri-R Kis	K. Miller
	G-OKPW	Tri-R Kis	K. P. Wordsworth
	G-OKYA	Cameron V-77 balloon	Army Balloon Club
	G-OKYM	PA-28 Cherokee 140	Hi-Fliers Aviation Ltd (G-AVLS)
	G-OLAU	Robinson R-22B	MPW Aviation Ltd
	G-OLAW	Lindstrand LBL-25A balloon	George Law Plant
	G-OLDC	Learjet 45	Gold Air International Ltd
	G-OLDD	BAe 125 Srs 800B	Gold Air International Ltd
	G-OLDF	Learjet 45	Gold Air International Ltd (G-JRJR)
	G-OLDG	Cessna T.182T	Gold Air International Ltd (G-CBTJ)
	G-OLDH	SA.341G Gazelle 1	- (G-UTZY)
	G-OLDJ	Learjet 45	Gold Air International Ltd
	G-OLDL	Learjet 45	Gold Air International Ltd
	G-OLDM	Pegasus Quantum 15-912	A. P. Watkins
	G-OLDN	Bell 206L LongRanger	Von Essen Aviation Ltd (G-TBCA/G-BFAL)
	G-OLDP	Mainair Pegasus Quik	M. J. Wilson & G. Lace
	G-OLDR	Learjet 45	Gold Air International Ltd
	G-OLDX	Cessna 182T	Gold Air International Ltd (G-IBZT)
	G-OLEE	Cessna F.152	Redhill Air Services Ltd
	G-OLEL	American Blimp Corp. A-60 airship	Keelex 195 Ltd
	G-OLEM	Jodel D.18	D. G. H. Oswald (G-BSBP)

Reg.	Type	Owner or Operator	Notes
G-OLEO	Thunder Ax10-210 S2 balloon	P. J. Waller	
G-OLEZ	Piper J-3C-65 Cub	L. Powell (G-BSAX)	
G-OLFB	Pegasus Quantum 15-912	A. J. Boyd	
G-OLFC	PA-38-112 Tomahawk	M. W. Glencross (G-BGZG)	
G-OLFO	Robinson R-44	Crinstown Aviation Ltd	
G-OLFT	R. Commander 114	D. A. Tubby (G-WJMN)	
G-OLGA	CFM Starstreak Shadow SA-II	N. F. Smith	
G-OLIZ	Robinson R-22B	D. McGarrity	
G-OLJT	Mainair Gemini Flash IIA	A. Wraith (G-MTKY)	
G-OLLI	Cameron O-31 SS balloon	N. A. Robertson	
G-OLLS	Cessna U.206H Floatplane	Loch Lomond Seaplanes Ltd	
G-OLMA	Partenavia P.68B	C. M. Evans (G-BGBT)	
G-OLOW	Robinson R-44	Hields Aviation	
G-OLRT	Robinson R-22B	First Degree Air/Tatenhill	
G-OLSF	PA-28-161 Cadet	Bflying Ltd (G-OTYJ)	
G-OMAC	Cessna FR.172E	S. G. Shilling	
G-OMAF	Dornier Do.228-200	FR Aviation Ltd/Bournemouth	
G-OMAK	Airbus A.319-132	Twinjet Aircraft Sales Ltd	
G-OMAL	Thruster T.600N 450	M. Howland	
G-OMAP	R. Commander 685	Cooper Aerial Surveys Ltd/Sandtoft	
G-OMAT	PA-28 Cherokee 140	Midland Air Training School (G-JIMY/G-AYUG)/Coventry	
G-OMAX	Brantly B.2B	P. D. Benmax (G-AVJN)	
G-OMCC (LOLY)	AS.350B Ecureuil	Michael Car Centres Ltd (G-JTCM/G-HLEN/G-	
G-OMCD	Robinson R-44 II	McDiarmid Partnership	
G-OMDB	Van's V-6A	D. A. Roseblade	
G-OMDD	Thunder Ax8-90 S2 balloon	M. D. Dickinson	
G-OMDG	Hoffmann H-36 Dimona	P. Turner/Halesland	
G-OMDH	Hughes 369E	Stilgate Ltd/Booker	
G-OMDR	AB-206B JetRanger 3	Atlas Helicopters Ltd (G-HRAY/G-VANG/G-BIZA)	
G-OMEL	Robinson R-44	Coolen-Huybregts Vof (G-BVPB)/Netherlands	
G-OMEN	Cameron Z-90 balloon	Cameron Balloons Ltd	
G-OMEX	Zenair CH.701 STOL	S. J. Perry	
G-OMEZ	Zenair CH.601HDS	C. J. Gow	
G-OMFG	Cameron A-120 balloon	M. F. Glue	
G-OMGH	Robinson R-44 Clipper II	Universal Energy Ltd	
G-OMHC	PA-28RT-201 Arrow IV	Tatenhill Aviation Ltd	
G-OMHI	Mills MH-1	J. P. Mills	
G-OMHP	Avtech Jabiru UL	M. H. Player	
G-OMIA	M.S.893A Rallye Commodore 180	P. W. Portelli	
G-OMIK	Shaw Europa	M. J. Clews	
G-OMJC	Beech 390 Premier 1	Manhattan Jet Charter/Blackbushe	
G-OMJT	Rutan LongEz	M. J. Timmons	
G-OMMG	Robinson R-22B	Preston Associates Ltd (G-BPYX)	
G-OMMM	Colt 90A balloon	V. Trimble	
G-OMMT	Robinson R-44	Morrison Motors (Turriff) (G-IBKA/G-USTE)	
G-OMNH	Beech 200 Super King Air	Newborne Ltd	
G-OMNI	PA-28R Cherokee Arrow 200D	Air Gloster Ltd (G-BAWA)	
G-OMOL	Maule MX-7-180C	Aeromarine Ltd	
G-OMRB	Cameron V-77 balloon	I. J. Jevons	
G-OMRG	Hoffmann H-36 Dimona	M. R. Grimwood (G-BLHG)	
G-OMST	PA-28-161 Warrior III	Mid-Sussex Timber Co. Ltd (G-BZUA)	
G-OMUC	Boeing 737-36Q	easyJet Airline Co Ltd	
G-OMUM	R. Commander 114	C. E. Campbell	
G-OMWE	Zenair CH.601HD	G. Cockburn (G-BVXU)	
G-OMYT	Airbus A.330-243	My Travel Airways (G-MOJO)	
G-ONAF	Naval Aircraft Factory N3N-3	R. P. W. Steel & J. E. Hutchinson	
G-ONAV	PA-31-310 Turbo Navajo C	Panther Aviation Ltd (G-IGAR)	
G-ONCB	Lindstrand LBL-31A balloon	R. J. Mole	
G-ONCL	Colt 77A balloon	D. R. Pearce	
G-ONCM	Partenavia P.68C	A. R. Arnell (G-TELE/G-DORE)	
G-ONEB	Westland Scout AH.1	E. R. Meredith & E. M. Smith (G-BXOE)	
G-ONES	Slingsby T.67M Firefly 200	L. J. Jones	
G-ONET	PA-28 Cherokee 180E	Hatfield Flying Club/Elstree (G-AYAU)	
G-ONFL	Meridian Maverick	R. Foster (G-MYUJ)	
G-ONGA	Robinson R.44- II	Mash Enterprises Ltd	
G-ONGC	Robin DR.400/180R	Norfolk Gliding Club Ltd/Tibenham	
G-ONHH	Forney F-1A Aircoupe	R. D. I. Tarry (G-ARHA)	
G-ONIG	Murphy Elite	N. S. Smith	

Notes	Reg.	Type	Owner or Operator
	G-ONIX	Cameron C-80 balloon	T. L. Gorman
	G-ONKA	Aeronca K	N. J. R. Minchin
	G-ONMT	Robinson R-22B-2	Redcourt Enterprises Ltd
	G-ONON	RAF 2000 GTX-SE gyroplane	M. P. Lhermette
	G-ONOW	Bell 206A JetRanger 2	J. Lucketti (G-AYMX)
	G-ONPA	PA-31-350 Navajo Chieftain	West Wales Airport Ltd/Shobdon
	G-ONSF	PA-28R-201 Arrow III	Northamptonshire School of Flying Ltd (G-EMAK)
	G-ONTV	AB-206B JetRanger 3	Castle Air Charters Ltd
	G-ONUN	Van's RV-6A	R. E. Nunn
	G-ONUP	Enstrom F-28C-UK	S. Brophy (G-MHCA/G-SHWW/G-SMUJ/G-BHTF)
	G-ONYX	Bell 206B JetRanger 3	N. C. Wheelwright (G-BXPN)
	G-ONZO	Cameron N-77 balloon	K. Temple
	G-OOAE	Airbus A.321-211	First Choice Airways Ltd (G-UNIF)
	G-OOAF	Airbus A.321-211	First Choice Airways Ltd (G-UNID/G-UKLO)
	G-OOAH	Airbus A.321-211	First Choice Airways Ltd
	G-OOAL	Boeing 767-38AER	First Choice Airways Ltd
	G-OOAM	Boeing 767-38AER	First Choice Airways Ltd
	G-OOAN	Boeing 767-39HER	First Choice Airways Ltd (G-UKLH)
	G-OOAP	Airbus A.320-214	First Choice Airways Ltd
	G-OOAR	Airbus A.320-214	First Choice Airways Ltd
	G-OOAT	Airbus A.320-214	First Choice Airways Ltd (C-GTDH)
	G-OOAU	Airbus A.320-214	First Choice Airways Ltd
	G-OOAV	Airbus A.321-211	First Choice Airways Ltd
	G-OOAW	Airbus A.320-214	First Choice Airways Ltd
	G-OOAX	Airbus A.320-214	First Choice Airways Ltd
	G-OOBA	Boeing 757-26N	First Choice Airways Ltd
	G-OOBB	Boeing 757-26N	First Choice Airways Ltd
	G-OOBC	Boeing 757-28A	First Choice Airways Ltd
	G-OOBD	Boeing 757-28A	First Choice Airways Ltd
	G-OOBE	Boeing 757-28A	First Choice Airways Ltd
	G-OOBF	Boeing 757-28A	First Choice Airways Ltd
	G-OOBI	Boeing 757-2B7	First Choice Airways Ltd
	G-OOBJ	Boeing 757-2B7	First Choice Airways Ltd
	G-OOBK	Boeing 767-324ER	First Choice Airways Ltd
	G-OOCS	Hughes 369E	R & S Fire and Security Ltd (G-OTDB/G-BXUR)
	G-OODE	SNCAN Stampe SV-4C	A. R. Radford (G-AZNN)
	G-OODI	Pitts S-1D Special	R. M. Buchan (G-BBBU)
	G-OODW	PA-28-181 Archer II	Goodwood Terrena Ltd
	G-OOER	Lindstrand LBL-25A balloon	Airborne Adventures Ltd
	G-OOFE	Thruster T.600N 450	Rochester Microlights Ltd
	G-OOFT	PA-28-161 Warrior III	Lyrical Computing Ltd
	G-OOGA	GA-7 Cougar	Cougar Aviation Ltd/Elstree
	G-OOGI	GA-7 Cougar	Plane Talking Ltd (G-PLAS/G-BGHL)
	G-OOGO	GA-7 Cougar	Leonard F. Jollye (Brookmans Park) Ltd/Elstree
	G-OOGS	GA-7 Cougar	Leeds Flying School Ltd (G-BGJW)
	G-OOHO	Bell 206B JetRanger 3	Into Space Ltd
			(G-OCHC/G-KLEE/G-SIZL/G-BOSW)
	G-OOIO	AS.350B-3 Ecureuil	Hovering Ltd
	G-OOJC	Bensen B.8MR	J. R. Cooper
	G-OOJP	Commander 114B	Plato Management Ltd
	G-OOLE	Cessna 172M	P. S. Eccersley (G-BOSI)
	G-OOMW	Ikarus C.42 FV UK	R. O'Malley-White
	G-OONE	Mooney M.20J	Go One Aviation Ltd
	G-OONI	Thunder Ax7-77 balloon	Fivedata Ltd
	G-OONY	PA-28-161 Warrior II	D. A. Field & P. B. Jenkins
	G-OOOB	Boeing 757-28A	Astraeus/Daallo Airlines
	G-OOOC	Boeing 757-28A	First Choice Airways Ltd
	G-OOOJ	Boeing 757-23A	First Choice Airways Ltd
	G-OOON	PA-34-220T Seneca III	Goon Aviation Ltd
	G-OOOX	Boeing 757-2Y0	First Choice Airways Ltd
	G-OOOY	Boeing 757-2Q8	First Choice Airways Ltd
	G-OOSE	Rutan Vari-Eze	B. O. Smith & J. A. Towers
	G-OOSI	Cessna 404	Cooper Aerial Surveys Ltd
	G-OOSY	D.H.82A Tiger Moth	M. Goose
	G-OOTB	SOCATA TB.20 Trinidad	Select Helicopters Ltd
	G-OOTC	PA-28R-201T Turbo Arrow III	D. G. & C. M. King (G-CLIV)
	G-OOTW	Cameron Z-275 balloon	Airborne Balloon Management Ltd
	G-OOUT	Colt Flying Shuttlecock SS balloon	Shiplake Investments Ltd
	G-OOXP	Aero Designs Pulsar XP	T. D. Baker

Reg.	Type	Owner or Operator	Notes
G-OPAG	PA-34-200 Seneca II	A. H. Lavender (G-BNGB)/Biggin Hill	
G-OPAL	Robinson R-22B	Hell Air Ltd/Wellesbourne	
G-OPAM	Cessna F.152 II (tailwheel)	PJC Leasing Ltd (G-BFZS)	
G-OPAT	Beech 76 Duchess	R. D. J. Axford (G-BHAO)	
G-OPAZ	Pazmany PL.2	K. Morris	
G-OPCG	Cessna 182T	Pye Consulting Group Ltd/Blackpool	
G-OPCS	Hughes 369E	Productivity Computer Solutions Ltd	
G-OPDS	Denney Kitfox Mk 4	P. D. Sparling	
G-OPEP	PA-28RT-201T Turbo Arrow IV	Sam Aviation/Cranfield	
G-OPET	PA-28-181 Archer II	Cambrian Flying Group Ltd	
G-OPFA	Pioneer 300	S. Eddison & R. Minett	
G-OPFW	H.S.748 Srs 2A	Emerald Airways Ltd (G-BMFT)/Liverpool	
G-OPHA	Robinson R-44	Simax Services Ltd	
G-OPHR	Diamond DA.40 Star	MC Air Ltd/Wellesbourne	
G-OPHT	Schleicher ASH-26E	P. Turner	
G-OPIC	Cessna FRA.150L	Air Survey (G-BGNZ)	
G-OPIK	Eiri PIK-20E	A. J. McWilliam/Newtownards	
G-OPIT	CFM Streak Shadow Srs SA	I. Sinnett	
G-OPJC	Cessna 152 II	PJC Leasing Ltd/Stapleford	
G-OPJD	PA-28RT-201T Turbo Arrow IV	J. M. McMillan	
G-OPJH	D.62B Condor	P. J. Hall (G-AVDW)	
G-OPJK	Shaw Europa	P. J. Kember	
G-OPJM	Bell 206B JetRanger 2	PJM Helicopters Ltd	
G-OPJS	Pietenpol Air Camper	P. J. Shenton	
G-OPKF	Cameron 90 Bowler SS balloon	D. K. Fish	
G-OPLB	Cessna 340A II	Just Plane Trading Ltd (G-FCHJ/G-BJLS)	
G-OPLC	D.H.104 Dove 8	W. G. T. Pritchard (G-BLRB)	
G-OPME	PA-23 Aztec 250D	Portway Aviation Ltd (G-ODIR/G-AZGB)	
G-OPMT	Lindstrand LBL-105A balloon	Pace Micro Technology PLC	
G-OPNH	Stoddard-Hamilton Glasair IIRG	J. L. Mangelschots (G-CINY)/Belgium	
G-OPPL	AA-5A Cheetah	Plane Talking Ltd (G-BGNN)/Elstree	
G-OPRC	Shaw Europa XS	M. J. Ashby-Arnold	
G-OPSF	PA-38-112 Tomahawk	Panshanger School of Flying (G-BGZI)	
G-OPSL	PA-32R-301 Saratoga SP	P. R. Tomkins (G-IMPW)	
G-OPSS	Cirrus SR.20	Caseright Ltd	
G-OPST	Cessna 182R	Lota Ltd/Shoreham	
G-OPTF	Robinson R-44 II	Heli Air Ltd/Wellesbourne	
G-OPUB	Slingsby T.67M Firefly 160	P. M. Barker (G-DLTA/G-SFTX)	
G-OPUP	B.121 Pup 2	A. Brinkley (G-AXEU)	
G-OPUS	Avtech Jabiru SK	H. H. R. Lagache	
G-OPWK	AA-5A Cheetah	A. H. McVicar (G-OAEL)/Prestwick	
G-OPWS	Mooney M.20K	A. R. Mills	
G-OPYE	Cessna 172S	Far North Aviation/Wick	
G-ORAC	Cameron 110 Van SS balloon	A. G. Kennedy	
G-ORAE	Van's RV-7	R. W. Eaton	
G-ORAF	CFM Streak Shadow	A. P. Hunn	
G-ORAL	H.S.748 Srs 2A	Emerald Airways Ltd (G-BPDA/G-GLAS)/Liverpool	
G-ORAR	PA-28-181 Archer III	P. N. & S. M. Thornton	
G-ORAS	Clutton FRED Srs 2	A. I. Sutherland	
G-ORAY	Cessna F.182Q II	Yorkshire Estates Ltd (G-BHDN)/Isle of Man	
G-ORBD	Van's RV-6A	O. R. B. Dixon (G-BVRE)	
G-ORBK	Robinson R-44-II	GTC (UK) Ltd (G-CCNO)	
G-ORBS	Mainair Blade	J. W. Dodson	
G-ORCA	Van's RV-4	M. R. H. Wishart	
G-ORCP	H.S.748 Srs 2A	Emerald Airways Ltd	
G-ORDB	Cessna 550 Citation Bravo	Equipe Air Ltd/Gamston	
G-ORDO	PA-30 Twin Comanche	Avcorp Ltd	
G-ORDS	Thruster T.600N 450	Thruster Air Services Ltd	
G-ORED	BN-2T Turbine Islander	Fly BN Ltd (G-BJYW)	
G-OREV	Mini -500	R. H. Everett	
G-ORFC	Jurca MJ.5 Sirocco	D. J. Phillips	
G-ORGY	Cameron Z-210 balloon	A. M. Holly	
G-ORIG	Glaser-Dirks DG.800A	I. Godfrey	
G-ORIX	ARV K1 Super 2	T. M. Lyons (G-BUXH/G-BNVK)	
G-ORJA	Beech B.200 Super King Air	Airwest Ltd	
G-ORJB	Cessna 500 Citation	Coalpower Ltd (G-OKSP)/Gamston	
G-ORJW	Laverda F.8L Falco Srs 4	W. R. M. Sutton	
G-ORMA	AS.355F-1 Twin Squirrel	MW Helicopters Ltd (G-SITE/ G-BPHC)	
G-ORMB	Robinson R-22B	Scotia Helicopters Ltd	
G-ORMG	Cessna 172R II	J. R. T. Royle	
G-OROB	Robinson R-22B	Corniche Helicopters (G-TBFC)	
G-OROD	PA-18 Super Cub 150	B. W. Faulkner	

Notes	Reg.	Type	Owner or Operator
	G-ORON	Cameron 77A balloon	Orion Hot Air Balloon Group
	G-ORPC	Shaw Europa XS	P. W. Churms
	G-ORPR	Cameron O-77 balloon	T. Strauss & A. Sheehan
	G-ORRR	Hughes 369HS	The Lower Mill Estate Ltd (G-BKTK/G-STEF)
	G-ORTH	Beech E90 King Air	Kilo Aviation Ltd (G-DEXY)
	G-ORTM	Glaser-Dirks DG.400	A. R. Garcia/Spain
	G-ORUG	Thruster T.600N 450	I. Shulver
	G-ORVB	McCulloch J-2	R. V. Bowles (G-BLGI/G-BKKL)
	G-ORVG	Van's RV-6	R. J. Fray
	G-ORVR	Partenavia P.68B	Ravenair (G-BFBD)/Liverpool
	G-OSAT	Cameron Z-105 balloon	Lotus Balloons Ltd
	G-OSAW	QAC Quickie Q.2	S. A. Wilson (G-BVYT)
	G-OSCC	PA-32 Cherokee Six 300	BG & G Airlines Ltd (G-BGFD)
	G-OSCH	Cessna 421C	Northern Aviation Ltd (G-SALI)
	G-OSDI	Beech 95-58 Baron	D. Darling (G-BHFY)
	G-OSEA	BN-2B-26 Islander	W. T. Johnson & Sons (Huddersfield) Ltd (G-BKOL)
	G-OSEE	Robinson R-22B	Aero-Charter Ltd
	G-OSEP	Mainair Blade 912	J. D. Smith
	G-OSFA	Diamond HK.36TC Super Dimona	Oxfordshire Sportflying Ltd
	G-OSFC	Cessna F.152	Stapleford Flying Club Ltd (G-BIVJ)
	G-OSFS	Cessan F.177RG	Cardinal Sin Ltd
	G-OSGB	PA-31-350 Navajo Chieftain	Aerial Support Services Ltd (G-YSKY)
	G-OSHL	Robinson R-22B	Sloane Helicopters Ltd/Sywell
	G-OSIC	Pitts S-1C Special	J. A. Dodd (G-BUAW)
	G-OSII	Cessna 172N	K. J. Abrams (G-BIVY)
	G-OSIP	Robinson R-22B-2	Heli Air Ltd
	G-OSIS	Pitts S-1S Special	C. Butler
	G-OSIT	Pitts S-1T Special	P. Shaw
	G-OSIX	PA-32 Cherokee Six 260	J. T. Le Bon (G-AZMO)
	G-OSJN	Shaw Europa XS	S. J. Nash
	G-OSKP	Enstrom 480	Churchill Stairlifts Ltd
	G-OSKR	Skyranger 912 (2)	Skyranger UK Ltd
	G-OSKY	Cessna 172M	Skyhawk Leasing Ltd/Wellesbourne
	G-OSLD	Shaw Europa XS	Opus Software Ltd
	G-OSLO	Schweizer 269C	A. H. Helicopter Services Ltd
	G-OSMD	Bell 206B JetRanger 2	Stuart Aviation Ltd (G-LTEK/G-BMIB)
	G-OSMS	Robinson R-22B	Heliflight (UK) Ltd (G-BXYW)
	G-OSND	Cessna FRA.150M	Wilkins & Wilkins Special Auctions Ltd (G-BDOU)
	G-OSNI	PA-23 Aztec 250C	Marham Investments Ltd (G-AWER)
	G-OSOE	H.S.748 Srs 2A	Emerald Airways Ltd (G-AYYG)/Liverpool
	G-OSPD	Aerotechnik EV-97 Teameurostar UK	V. C. Garwood
	G-OSPG	BAe 125 Srs 800B	Houston Jet Services Ltd (G-ETOM/G-BVFC/G-TPHK/G-FDSL)
	G-OSPS	PA-18 Super Cub 95	J. P. Orrissey
	G-OSSA	Cessna Tu.206B	Skydive St.Andrews Ltd
	G-OSSF	AA-5A Cheetah	Direct Helicopters (G-MELD/G-BHCB)
	G-OSSI	Robinson R-44-II	Goss Air Ltd
	G-OSST	Colt 77A balloon	British Airways PLC
	G-OSTC	AA-5A Cheetah	5th Generation Designs Ltd
	G-OSTU	AA-5A Cheetah	Direct Helicopters (G-BGCL)
	G-OSTY	Cessna F.150G	C. R. Guggenheim (G-AVCU)
	G-OSUP	Lindstrand LBL-90A balloon	British Airways Balloon Club
	G-OSUS	Mooney M.20K	J. B. King/Goodwood
	G-OSZB	Christen Pitts S-2B Special	P. M. Ambrose (G-OGEE)
	G-OTAL	ARV Super 2	N. R. Beale (G-BNGZ)
	G-OTAM	Cessna 172M	G. V. White
	G-OTAN	PA-18 Super Cub 135	S. D. Turner
	G-OTBA	H.S.748 Srs 2A	Emerald Airways Ltd/Liverpool
	G-OTBY	PA-32 Cherokee Six 300	M. J. Willing
	G-OTCH	CFM Streak Shadow	H. E. Gotch
	G-OTCV	Skyranger 912S (1)	T. C. Viner
	G-OTDA	Boeing 737-31S	Flyglobespan.com
	G-OTDI	Diamond DA.40D Star	Diamond Aircraft UK Ltd
	G-OTEL	Thunder Ax8-90 balloon	D. N. Belton
	G-OTFT	PA-38-112 Tomahawk	P. Tribble (G-BNKW)
	G-OTGA	PA-28R-201 Arrow III	TG Aviation Ltd
	G-OTHE	Enstrom 280C-UK Shark	National Technologies Ltd (G-OPJT/G-BKCO)
	G-OTIB	Robin DR.400/180R	Norfolk Gliding Club Ltd/Tibenham
	G-OTIG	AA-5B Tiger	D. H. Green (G-PENN)/Elstree
	G-OTIM	Bensen B.8MV	T. J. Deane

Reg.	Type	Owner or Operator	Notes
G-OTIS	Cessna 550 Citation II	The Streamline Partnership Ltd	
G-OTJB	Robinson R-44	T. J. Burke	
G-OTJH	Pegasus Quantum 15-912	T. J. Hector	
G-OTOE	Aeronca 7AC Champion	J. M. Gale (G-BRWW)	
G-OTOO	Stolp SA.300 Starduster Too	I. M. Castle	
G-OTOY	Robinson R-22B	Tickstock Ltd (G-BPEW)	
G-OTRG	Cessna TR.182RG	P. Mather	
G-OTRV	Van's RV-6	W. R. C. Williams-Wynne	
G-OTSP	AS.355F-1 Twin Squirrel	Aeromega Aviation PLC (G-XPOL/G-BPRF)	
G-OTTI	Cameron 34 Otti SS balloon	Ballonwerbung Hamburg GmbH/Germany	
G-OTTO	Cameron 82 Katalog SS balloon	Ballonwerbung Hamburg GmbH/Germany	
G-OTUG	PA-18 Super Cub 150	B. F. Walker	
G-OTUI	SOCATA TB-20 Trinidad	P. F. Rothwell (G-KKDL/G-BSHU)	
G-OTUN	Aerotechnik EV97 Eurostar	E. O. Otun	
G-OTUP	Lindstrand LBL-180A balloon	Airborne Adventures Ltd	
G-OTWO	Rutan Defiant	A. J. Baggerley	
G-OTYE	Aerotechnik EV-97 Eurostar	A. B. Godber & J. Tye	
G-OTYP	PA-28 Cherokee 180	I. R. Chaplin	
G-OUCH	Cameron N-105 balloon	Flying Pictures Ltd	
G-OUHI	Shaw Europa XS	Airplan Flight Equipment Ltd	
G-OUIK	Mainair Pegasus Quik	N. A. Harwood	
G-OUMC	Lindstrand LBL-105A balloon	Executive Ballooning	
G-OURA	BAe 125 Srs 800B	Ourjet Ltd (G-ICFR/G-BUCR)	
G-OURB	H.S.125 Srs 700B	Ourjet Ltd (G-NCFR/G-BVJY)	
G-OURO	Shaw Europa	M. Crunden	
G-OURS	Sky 120-24 balloon	M. P. A. Sevrin	
G-OUVI	Cameron O-105 balloon	Bristol University Hot Air Ballooning Soc	
G-OVAA	Colt Jumbo SS balloon	Virgin Airship & Balloon Co Ltd	
G-OVAG	Tipsy Nipper T.66 Srs 1	L. D, Johnston	
G-OVAL	Ikarus C.42 FB100	J. I. Greenshields	
G-OVAX	Colt AS-80 Mk II airship	Gefa-Flug GmbH/Germany	
G-OVBF	Cameron A-250 balloon	Virgin Balloon Flights	
G-OVBL	Lindstrand LBL-150A balloon	R. J. Henderson	
G-OVET	Cameron O-56 balloon	E. J. A. Macholc	
G-OVFM	Cessna 120	A. P. Bacon & A. Sutherland	
G-OVFR	Cessna F.172N	Western Air (Thruxton) Ltd	
G-OVIA	Lindstrand LBL-105A balloon	N. C. Lindsey	
G-OVIC	Cameron A-250 balloon	M. E. White/Ireland	
G-OVID	Light Aero Avid Flyer	G. G. Ansell	
G-OVII	Van's RV-7		
G-OVIN	Rockwell Commander 112TC	G-OVIN Aviation Ltd	
G-OVLA	Ikarus C.42 FB	Webb Plant Sales	
G-OVMC	Cessna F.152 II	S. C. Moss	
G-OVNR	Robinson R-22B	Helicopter Training & Hire Ltd	
G-OWAC	Cessna F.152	Aviation South West Ltd (G-BHEB)	
G-OWAK	Cessna F.152	Falcon Flying Services (G-BHEA)	
G-OWAL	PA-34-220T Seneca III	R. G. & W. Allison	
G-OWAR	PA-28-161 Warrior II	Bickertons Aerodromes Ltd	
G-OWAX	Beech 200 Super King Air	Context GB Ltd/Blackpool	
G-OWAZ	Pitts S-1C Special	P. E. S. Latham (G-BRPI)	
G-OWCS	Cessna 182J	P. Ragg	
G-OWDB	H.S.125 Srs 700B	Bizair Ltd (G-BYFO/G-OWEB)	
G-OWEL	Colt 105A balloon	S. R. Seager	
G-OWEN	K & S Jungster	R. C. Owen	
G-OWET	Thurston TSC-1A2 Teal	D. Nieman	
G-OWFS	Cessna A.152	MAMM Ltd (G-DESY/G-BNJE)/Blackpool	
G-OWGC	Slingsby T.61F Venture T.2	Wolds Gliding Club Ltd/Pocklington	
G-OWLC	PA-31 Turbo Navajo	Channel Airways Ltd (G-AYFZ)	
G-OWMC	Thruster T.600N	Wilts Microlight Centre	
G-OWND	Robinson R-44 Astro	W. N. Dore	
G-OWOW	Cessna 152 II	Falcon Flying Services (G-BMSZ)/Biggin Hill	
G-OWRC	Cessna F152 II	Unimat SA/France	
G-OWRT	Cessna 182G	Blackpool & Flyde Aero Club Ltd (G-ASUL)	
G-OWWW	Shaw Europa	Whisky Group	
G-OWYE	Lindstrand LBL-240A balloon	Wye Valley Aviation Ltd	
G-OWYN	Aviamilano F.14 Nibbio	R. Nash	
G-OXBC	Cameron A-140 balloon	J. E. Rose	
G-OXBY	Cameron N-90 balloon	C. A. Oxby	

Notes	Reg.	Type	Owner or Operator
	G-OXKB	Cameron 110 Sports Car SS balloon	D. M. Moffat
	G-OXOM	PA-28-161 Cadet	Plane Talking Ltd (G-BRSG)/Elstree
	G-OXTC	PA-23 Aztec 250D	Falcon Flying Services (G-AZOD)/Biggin Hill
	G-OXVI	V.S.361 Spitfire LF.XVIe (TD248)	Silver Victory BVBA/Belgium
	G-OXXL	Cameron A-300 balloon	Exclusive Ballooning Ltd
	G-OYAK	Yakovlev C-11 (27)	A. H. Soper/Earls Colne
	G-OYES	Mainair Blade 912	J. Crowe
	G-OYST	AB-206B JetRanger 2	Oyster Leasing Ltd
			(G-JIMW/G-UNIK/G-TPPH/G-BCYP)
	G-OYTE	Rans S.6ES Coyote II	I. M. Vass
	G-OZAR	Enstrom 480	Lancroft Air Ltd (G-BWFF)
	G-OZBB	Airbus A.320-212	Monarch Airlines Ltd/Luton
	G-OZBE	Airbus A.321-231	Monarch Airlines Ltd/Luton
	G-OZBF	Airbus A.321-231	Monarch Airlines Ltd/Luton
	G-OZBG	Airbus A.321-231	Monarch Airlines Ltd/Luton
	G-OZBH	Airbus A.321-231	Monarch Airlines Ltd/Luton
	G-OZBI	Airbus A.321-231	Monarch Airlines Ltd/Luton
	G-OZEE	Light Aero Avid Speedwing Mk 4	S. C. Goozee
	G-OZEF	Shaw Europa XS	Z. M. Ahmad
	G-OZOI	Cessna R.182	J. R. G. & F. L. G. Fleming (G-ROBK)
	G-OZOO	Cessna 172N	Atlantic Bridge Aviation Ltd (G-BWEI)/Lydd
	G-OZRH	BAe 146-200	Flightline Ltd
	G-OZZI	Jabiru SK	A. H. Godfrey
	G-OZZY	Robinson R-22B	G. T. Kozlowski (G-PWEL)
	G-PACE	Robin R.1180T	Millicron Instruments Ltd/Coventry
	G-PACL	Robinson R-22B	R. Wharam
	G-PACT	PA-28-181 Archer III	Burscombe Consulting Ltd
	G-PADD	AA-5A Cheetah	BPAD Ltd (G-ESTE/G-GHNC)
	G-PADE	Escapade 912	C. L. G. Innocent
	G-PADI	Cameron V-77 balloon	R. F. Penney
	G-PADS	Commander 114B	Echo Delta Ltd
	G-PAGS	SA.341G Gazelle 1	P. A. G. Seers (G-OAFY/G-SFTH/G-BLAP)
	G-PAIZ	PA-12 Super Cruiser	B. R. Pearson/Eaglescott
	G-PALS	Enstrom 280C-UK-2 Shark	G. Firbank
	G-PAPS	PA-32R-301T Turbo Saratoga SP	Nicol Aviation
	G-PARG	Pitts S-1C Special	M. Kotsageridis
	G-PARI	Cessna 172RG Cutlass	Applied Signs Ltd/Tatenhill
	G-PART	Partenavia P.68B	Springbank Aviation Ltd
	G-PASF	AS.355F-1 Twin Squirrel	Police Aviation Services Ltd (G-SCHU)/Newcastle
	G-PASG	MBB Bo 105DBS/4	Police Aviation Services Ltd (G-MHSL)/Staverton
	G-PASH	AS.355F-1 Twin Squirrel	Police Aviation Services Ltd/Staverton
	G-PASV	BN-2B-21 Islander	Police Aviation Services Ltd (G-BKJH)/Teesside
	G-PASX	MBB Bo 105DBS/4	Police Aviation Services Ltd/Shoreham
	G-PATF	Shaw Europa	E. P. Farrell
	G-PATG	Cameron O-90 balloon	Bath University Students Union
	G-PATI	Cessna F.172M	Nigel Kenny Aviation Ltd (G-WACZ/G-BCUK)
	G-PATN	SOCATA TB.10 Tobago	N. Robson (G-LUAR)
	G-PATP	Lindstrand LBL-77A balloon	P. Pruchnickyj
	G-PATS	Shaw Europa	D. J. D. Kesterton
	G-PATX	Lindstrand LBL-90A balloon	P. A. Bubb
	G-PATZ	Shaw Europa	H. P. H. Griffin
	G-PAVL	Robin R.3000/120	S. Baker
	G-PAWL	PA-28 Cherokee 140	G-PAWL Group (G-AWEU)
	G-PAWN	PA-25 Pawnee 260C	A. P. Meredith (G-BEHS)/Lasham
	G-PAWS	AA-5A Cheetah	Direct Helicopters
	G-PAXX	PA-20 Pacer 135	D. W. Grace
	G-PAYD	Robin DR.400/180	A. Head
	G-PAZY	Pazmany PL.4A	C. R. Nash (G-BLAJ)
	G-PBEE	Robinson R-44	P. Barnard
	G-PBEK	Agusta A.109A	P. Beck (G-BXIV)
	G-PBEL	CFM Shadow Srs DD	S. J. Joseph
	G-PBUS	Avtech Jabiru SK	G. R. Pybus
	G-PBYA	Consolidated PBY-5A Catalina	Catalina Aircraft Ltd
	G-PBYY	Enstrom 280FX	Hogan Holdings Ltd (G-BXKV)
	G-PCAF	Pietenpol Air Camper	C. C. & F. M. Barley
	G-PCAM	BN-2A Mk.III-2 Trislander	Aurigny Air Services Ltd (G-BEPH)
	G-PCAT	SOCATA TB-10 Tobago	D. P. Boyle (G-BHER)

Reg.	Type	Owner or Operator	Notes
G-PCCC	Pioneer 300	Pioneer Aviation UK Ltd	
G-PCDP	Zlin Z.526F Trener Master	J. Mann	
G-PDGE	Eurocopter EC.120B	Cadenza Helicopters Ltd	
G-PDGG	Aeromere F.8L Falco Srs 3	P. D. G. Grist	
G-PDGN	SA.365N Dauphin 2	PLM Dollar Group Ltd (G-TRAF/G-BLDR)	
G-PDHJ	Cessna T.182R	P. G. Vallance Ltd	
G-PDOC	PA-44-180 Seminole	Medicare (G-PVAF)/Newcastle	
G-PDOG	Cessna O-1E Bird Dog	J. D. Needham	
G-PDSI	Cessna 172N	DA Flying Group	
G-PDWI	Mini-500	P. Waterhouse	
G-PEAK	AB-206B JetRanger 2	Techanimation (G-BLJE)	
G-PECK	PA-32-300 Cherokee Six D	H. Peck (G-ETAV/G-MCAR/G-LADA/G-AYWK)	
G-PEGA	Pegasus Quantum 15-912	B. A. Showell	
G-PEGG	Colt 90A balloon	Ballon Vole Association/France	
G-PEGI	PA-34-200T Seneca II	Tayflite Ltd	
G-PEGY	Shaw Europa	M. T. Dawson	
G-PEJM	PA-28-181 Archer III	D. A. Earle	
G-PEKT	SOCATA TB.20 Trinidad	A. J. Dales	
G-PEPA	Cessna 206H	R. D. Lygo (G-MGMG)	
G-PEPL	MDH MD.600N	Blue Anchor Leisure Ltd	
G-PERC	Cameron N-90 balloon	P. A. Foot & I. R. Warrington	
G-PERE	Robinson R22 Beta	Central Helicopters Ltd	
G-PERI	Agusta A.109A-II	Castle Air Charters Ltd	
G-PERR	Cameron 60 Bottle SS balloon ★	British Balloon Museum/Newbury	
G-PERZ	Bell 206B JetRanger 3	C. P. Lockyer	
G-PEST	Hawker Tempest II (MW401)	Tempest Two Ltd	
G-PETH	PA-24 Comanche 260C	S. H. Petherbridge	
G-PETR	PA-28 Cherokee 140	A. A. Gardner (G-BCJL)	
G-PFAA	EAA Biplane Model P	R. J. Marshall	
G-PFAF	FRED Srs 2	M. S. Perkins	
G-PFAG	Evans VP-1	J. A. Hatch	
G-PFAH	Evans VP-1	J. A. Scott	
G-PFAL	FRED Srs 2	J. McD. Robinson/Bann Foot	
G-PFAO	Evans VP-1	P. W. Price	
G-PFAP	Currie Wot/SE-5A (C1904)	J. H. Seed	
G-PFAR	Isaacs Fury II (K2059)	J. W. Hale & R. Cooper	
G-PFAT	Monnett Sonerai II	H. B. Carter	
G-PFAW	Evans VP-1	R. F. Shingler	
G-PFCL	Cessna 172S	Prestwick Flight Centre	
G-PFFN	Beech 200 Super King Air	The Puffin Club Ltd	
G-PFKD	WSK Yakovlev Yak-12M	M. J. Kirk	
G-PFML	Robinson R-44	M. J. Magowan	
G-PFSL	Cessna F.152	P. A. Simon	
G-PGAC	Dyn Aero MCR-01	D. T. S. Walsh & G. A. Coatesworth	
G-PGFG	Tecnam P.92-EM Echo	P. G. Fitzgerald	
G-PGHM	Air Creation Kiss 450	P. G. H. Millbank	
G-PGSA	Thruster T.600N	A. J. A. Hitchcock	
G-PGSI	Robin R.2160	M. A. Spencer	
G-PGUY	Sky 70-16 balloon	Black Sheep Balloons (G-BXZJ)	
G-PHAA	Cessna F.150M	PHA Aviation Ltd (G-BCPE)	
G-PHIL	Brookland Hornet	A. J. Philpotts	
G-PHLB	RAF 2000GTX-SE gyroplane		
G-PHSI	Colt 90A balloon	P. H. Strickland	
G-PHTG	SOCATA TB.10 Tobago	A. J. Baggarley	
G-PHXS	Shaw Europa XS	P. Handford	
G-PHYL	Denney Kitfox Mk 4	J. Dunn	
G-PIAF	Thunder Ax7-65 balloon	L. Battersley	
G-PICT	Colt 180A balloon	J. L. Guy	
G-PIDG	Robinson R-44	P. J. Rogers	
G-PIDS	Boeing 757-225	My Travel Airways	
G-PIEL	CP.301A Emeraude	P. R. Thorne (G-BARY)	
G-PIES	Thunder Ax7-77Z balloon	Pork Farms Ltd	
G-PIET	Pietenpol Air Camper	N. D. Marshall	
G-PIGG	Lindstrand LBL Pig SS balloon	I. Heidenreich/Germany	
G-PIGS	SOCATA Rallye 150ST	Boonhill Flying Group (G-BDWB)	
G-PIGY	SC.7 Skyvan Srs 3A Variant 100	Babcock Defence Services	

Notes	Reg.	Type	Owner or Operator
	G-PIIX	Cessna P.210N	J. R. Colthurst (G-KATH)
	G-PIKE	Robinson R-22 Mariner	Sloane Helicopters Ltd/Sywell
	G-PIKK	PA-28 Cherokee 140	Coventry Aviators Flying Group (G-AVLA)
	G-PILE	Rotorway Executive 90	J. B. Russell
	G-PILL	Light Aero Avid Flyer Mk 4	D. R. Meston
	G-PINC	Cameron Z-90 balloon	M. Cowling
	G-PING	AA-5A Cheetah	Plane Talking Ltd (G-OCWC/G-WULL)
	G-PINT	Cameron 65 Barrel SS balloon	D. K. Fish
	G-PINX	Lindstrand Pink Panther SS balloon	Magical Adventures Ltd/USA
	G-PIPR	PA-18 Super Cub 95	D. S. Sweet (G-BCDC)
	G-PIPS	Van's RV-4	C. J. Marsh
	G-PIPY	Cameron 105 Pipe SS balloon	Cameron Balloons Ltd
	G-PITS	Pitts S-2AE Special	The Eitlean Group
	G-PITZ	Pitts S-2A Special	J. A. Coutts
	G-PIXE	Colt 31A balloon	N. D. Eliot
	G-PIXI	Pegasus Quantum 15-912	G. R. Craig
	G-PIXS	Cessna 336	Atlantic Bridge Aviation Ltd/Lydd
	G-PIXX	Robinson R-44-II	Flying TV Ltd
	G-PIZZ	Lindstrand LBL-105A balloon	HD Bargain SRL/Italy
	G-PJMT	Lancair 320	M. T. Holland
	G-PJNZ	Commander 114B	P. D. Jackson
	G-PJSY	Van's RV-6	P. J. York
	G-PJTM	Cessna FR.172K II	Jane Air (G-BFIF)
	G-PKPK	Schweizer 269C	C. H. Dobson
	G-PLAC	PA-31-350 Navajo Chieftain	D. B. Harper (G-OLDA/ G-BNDS)
	G-PLAH	BAe Jetstream 3102	Jetstream Executive Travel Ltd
			(G-LOVA/G-OAKA/G-BUFM/G-LAKH)
	G-PLAJ	BAe Jetstream 3102	Jetstream Executive Travel Ltd
	G-PLAN	Cessna F.150L	G-PLAN Flying Group
	G-PLAY	Robin R.2100A	D. R. Austin
	G-PLAZ	R. Commander 112	Simat Marketing Ltd (G-RDCI/G-BFWG)
	G-PLBI	Cessna 172S	Grandfort Properties Ltd/Booker
	G-PLEE	Cessna 182Q	Peterlee Parachute Centre
	G-PLIV	Pazmany PL.4A	B. P. North
	G-PLMB	AS.350B Ecureuil	PLM Dollar Group Ltd (G-BMMB)
	G-PLMH	AS.350B-2 Ecureuil	PLM Dollar Group Ltd
	G-PLMI	SA.365C-1 Dauphin	PLM Dollar Group Ltd
	G-PLOD	Tecnam P.92-EM Echo	C. M. Jupp & S. P. Pearson
	G-PLOW	Hughes 269B	Sulby Aerial Surveys (G-AVUM)
	G-PLPC	Schweizer Hughes 269C	Power Lines, Pipes & Cables Ltd (G-JMAT)
	G-PLPM	Shaw Europa XS	P. L. P. Mansfield
	G-PLSA	Aero Designs Pulsar XP	Air Ads Ltd (G-NEVS)
	G-PLXI	BAe ATP/Jetstream 61	BAe (Operations) Ltd (G-MATP)/Woodford
	G-PMAM	Cameron V-65 balloon	P. A. Meecham
	G-PMAX	PA-31-350 Navajo Chieftain	Liberty Group Assets Ltd (G-GRAM/G-BRHF)
	G-PMNF	V.S.361 Spitfire HF.IX (TA805)	P. R. Monk
	G-PNEU	Colt 110 Bibendum SS balloon	Balloon Preservation Group
	G-PNIX	Cessna FRA.150L	Dukeries Aviation (G-BBEO)
	G-POCO	Cessna 152	K. M. Watts
	G-POGO	Flight Design CT.2K	P. A. & M. W. Aston
	G-POLL	Skyranger 912 (1)	Thorne Thorne Engineering Ltd
	G-POLY	Cameron N-77 balloon	Empty Wallets Balloon Group
	G-POND	Oldfield Baby Lakes	C. Bellmer/Germany
	G-POOH	Piper J-3C-65 Cub	P. & H. Robinson
	G-POOL	ARV Super 2	P. A. Dawson (G-BNHA)
	G-POOP	Dyn Aero MCR-01	Eurodata Computer Supplies
	G-POPA	Beech A36 Bonanza	C. J. O'Sullivan
	G-POPE	Eiri PIK-20E-1	P. Rees
	G-POPI	SOCATA TB.10 Tobago	I. S. Hacon & C. J. Earle (G-BKEN)
	G-POPS	PA-34-220T Seneca III	Alpine Ltd
	G-POPW	Cessna 182S	D. L. Price
	G-PORK	AA-5B Tiger	C. M. M. Grange & D. Thomas (G-BFHS)
	G-PORT	Bell 206B JetRanger 3	J. Poole
	G-POSH	Colt 56A balloon	B. K. Rippon (G-BMPT)
	G-POTT	Robinson R-44 Astro	S. J. A. Brown
	G-POWL	Cessna 182R	Hillhouse Estates Ltd

Reg.	Type	Owner or Operator	Notes
G-POZA	Escapade Jabiru	M. R. Jones	
G-PPLL	Van's RV-7A	P. G. Leonard	
G-PPPP	Denney Kitfox Mk 3	R. Powers	
G-PPTS	Robinson R-44	Supablast Nationwide Ltd	
G-PRAG	Brügger MB.2 Colibri	Colibri Flying Group	
G-PRAH	Flight Design CT.2K	P. R. A. Hammond	
G-PRET	Robinson R-44	R. D. Masters	
G-PREY	Pereira Osprey II	D. W. Gibson (G-BEPB)	
G-PREZ	Robin DR.400/500	M. A. Wilkinson	
G-PRII	Hawker Hunter PR.11	Stick & Rudder Aviation Ltd/Belgium	
G-PRIM	PA-38-112 Tomahawk	Braddock Ltd	
G-PRIT	Cameron N-90 balloon	R. D. Stagg (G-HTVI)	
G-PRLY	Avtech Jabiru SK	N. C. Cowell (G-BYKY)	
G-PRNT	Cameron V-90 balloon	E. K. Gray	
G-PROB	AS.350B-2 Ecureuil	Irvine Aviation Ltd (G-PROD)	
G-PROF	Lindstrand LBL-90A balloon	S. J. Wardle	
G-PROM	AS.350B Ecureuil	General Cabins & Engineering Ltd (G-MAGY/G-BIYC)	
G-PROP	AA-5A Cheetah	Fortune Technology Ltd (G-BHKU)	
G-PROV	P.84 Jet Provost T.52A (T.4)	Provost Group	
G-PROW	Aerotechnik EV-97 Eurostar	G. M. Prowling	
G-PRSI	Pegasus Quantum 15-912	J. C. Kitchen	
G-PRTT	Cameron N-31 balloon	J. M. Albury	
G-PSAX	Lindstrand LBL-77B balloon	P. A. Sachs	
G-PSGC	PA-25 Pawnee 260C (modified)	Peterborough & Spalding Gliding Club Ltd (G-BDDT)	
G-PSIC	NA P-51C Mustang (2106449)	Patina Ltd/Duxford	
G-PSNI	Eurocopter EC.135T-2	McAlpine Helicopters Ltd/Kidlington	
G-PSON	Colt Cylinder One SS balloon	Balloon Preservation Flying Group	
G-PSRT	PA-28-151 Warrior	P. A. S. Dyke (G-BSGN)	
G-PSST	Hunter F.58A	Heritage Aviation Developments Ltd/Bournemouth	
G-PSUE	CFM Shadow Srs CD	D. A. Crosbie (G-MYAA)	
G-PSUK	Thruster T.600N 450	A. J. Dunlop	
G-PTAG	Shaw Europa	R. C. Harrison	
G-PTRE	SOCATA TB.20 Trinidad	Trantshore Ltd (G-BNKU)	
G-PTTS	Aerotek Pitts S-2A	D. C. Avery	
G-PTWB	Cessna T.303	F. Kratky (G-BYNG)	
G-PTWO	Pilatus P2-05 (U-110)	Bulldog Aviation Ltd/Earls Colne	
G-PTYE	Shaw Europa	Hitech International	
G-PUDL	PA-18 Super Cub 150	R. A. Roberts	
G-PUDS	Shaw Europa	I. Milner	
G-PUFF	Thunder Ax7-77A balloon	Intervarsity Balloon Club	
G-PUFN	Cessna 340A	G. R. Case	
G-PUGS	Cessna 182H	N. C. & M. F. Shaw	
G-PUKA	Jabiru Aircraft Jabiru J400	D. P. Harris	
G-PUMA	AS.332L Super Puma	CHC Scotia Ltd	
G-PUMB	AS.332L Super Puma	CHC Scotia Ltd	
G-PUMD	AS.332L Super Puma	CHC Scotia Ltd	
G-PUME	AS.332L Super Puma	CHC Scotia Ltd	
G-PUMH	AS.332L Super Puma	Bristow Helicopters Ltd	
G-PUML	AS.332L Super Puma	CHC Scotia Ltd	
G-PUMN	AS.332L Super Puma	CHC Scotia Ltd	
G-PUMO	AS.332L-2 Super Puma	CHC Scotia Ltd	
G-PUMS	AS.332L-2 Super Puma	CHC Scotia Ltd	
G-PUNK	Thunder Ax8-105 balloon	S. C. Kinsey	
G-PUPP	B.121 Pup 2	M. D. O'Brien (G-BASD)	
G-PUPY	Shaw Europa XS	P. G. Johnson	
G-PURR	AA-5A Cheetah	Nabco Retail Display (G-BJDN)	
G-PURS	Rotorway Executive	J. E. Houseman	
G-PUSH	Rutan LongEz	E. G. Peterson	
G-PUSI	Cessna T.303	Crusader Aviation Ltd/Kidlington	
G-PUSK	PA-32R-301 Saratoga IIHP	HN Consultancy (UK) Ltd	
G-PUSS	Cameron N-77 balloon	L. D. Thurgar	
G-PUSY	RL-5A LW Sherwood Ranger	S. C. Briggs(G-MZNF)	
G-PUTT	Cameron 76 Golf SS balloon	Lakeside Lodge Golf Centre	
G-PVBF	Lindstrand LBL-260S balloon	Virgin Balloon Flights	

Notes	Reg.	Type	Owner or Operator
	G-PVET	D.H.C.1 Chipmunk 22 (WB565)	Connect Properties Ltd
	G-PVIP	Cessna 421C	Passion 4 Health International Ltd (G-RLMC)
	G-PVST	Thruster T.600N 450	P. V. Stevens
	G-PWBE	D.H.82A Tiger Moth	P. W. Beales
	G-PWIT	Bell 206L-1 LongRanger	Formal Graphics Ltd (G-DWMI)
	G-PWUL	Van's RV-6	P. C. Woolley
	G-PYNE	Thruster T.600N 450	R. Dereham
	G-PYRO	Cameron N-65 balloon	A. C. Booth
	G-PZAZ	PA-31-350 Navajo Chieftain	Air Medical Fleet Ltd (G-VTAX/G-UTAX)
	G-PZIZ	PA-31-350 Navajo Chieftain	Air Medical Fleet Ltd (G-CAFZ/G-BPPT)
	G-RABA	Cessna FR.172H	
	G-RACA	P.57 Sea Prince T.1 (571/CU)	*(stored)*/Long Marston
	G-RACI	Beech C90 King Air (modified)	King Air Ltd (G-SHAM)
	G-RACO	PA-28R Cherokee Arrow 200-II	Graco Group Ltd
	G-RACY	Cessna 182S	N. J. Fuller
	G-RADA	Soko P-2 Kraguj (30140)	Flight Consultancy Services
	G-RADI	PA-28-181 Archer II	G. S. & D. V. Foster
	G-RADR	Douglas AD-4NA Skyraider (126922)	T. J. Manna (G-RAID)/North Weald
	G-RAEM	Rutan LongEz	G. F. H. Singleton
	G-RAES	Boeing 777-236	British Airways
	G-RAFA	Grob G.115	RAF College Flying Club Ltd/Cranwell
	G-RAFB	Grob G.115	RAF College Flying Club Ltd/Cranwell
	G-RAFC	Robin R.2112	RAF Charlie Group
	G-RAFE	Thunder Ax7-77 balloon	Giraffe Balloon Syndicate
	G-RAFG	Slingsby T.67C Firefly	Arrow Flying Ltd
	G-RAFH	Thruster T.600N 450	RAF Microlight Flying Association/Halton
	G-RAFI	P.84 Jet Provost T.4 (XP672)	R. J. Everett/North Weald
	G-RAFJ	Beech B.200 Super King Air	Serco Ltd/Cranwell
	G-RAFK	Beech B.200 Super King Air	Serco Ltd/Cranwell
	G-RAFL	Beech B.200 Super King Air	Serco Ltd/Cranwell
	G-RAFM	Beech B.200 Super King Air	Serco Ltd/Cranwell
	G-RAFN	Beech B.200 Super King Air	Serco Ltd/Cranwell
	G-RAFO	Beech B.200 Super King Air	Serco Ltd/Cranwell
	G-RAFP	Beech B.200 Super King Air	Serco Ltd/Cranwell
	G-RAFR	Skyranger J2.2(1)	RAF Microlight Flying Association/Halton
	G-RAFS	Thruster T.600N 450	RAF Microlight Association/Halton
	G-RAFT	Rutan LongEz	B. Wronsk
	G-RAFV	Avid Speedwing	A. F. Vizoso
	G-RAFW	Mooney M.20E	Vinola (Knitwear) Manufacturing Co. Ltd (G-ATHW)
	G-RAFZ	RAF 2000 GTX-SE	John Pavitt (Engineers) Ltd
	G-RAGS	Pietenpol Air Camper	R. F. Billington
	G-RAIL	Colt 105A balloon	Ballooning World Ltd
	G-RAIN	Maule M5-235C Lunar Rocket	D. S. McKay & J. A. Rayment/Hinton-in-the-Hedges
	G-RAIX	CCF AT-16 Harvard 4 (KF584)	M. R. Paul & P. A. Shaw (G-BIWX)
	G-RAJA	Raj Hamsa X'Air 582 (2)	M. Quarterman
	G-RALD	Robinson R-22HP	Heli Air Ltd (G-CHIL)
	G-RAME	Bell 206B JetRanger 3	R & M Management Services (G-CTEK/G-BWZW)
	G-RAMI	Bell 206B JetRanger 3	Yorkshire Helicopters/Leeds
	G-RAMP	Piper J-3C-65 Cub	R. N. Whittall
	G-RAMS	PA-32R-301 Saratoga SP	Air Tobago Ltd/Netherthorpe
	G-RAMY	Bell 206B JetRanger 2	Lincair Ltd
	G-RANI	AS.355N Twin Squirrel	Errigal Helicopters LLP (G-CCIN)
	G-RANS	Rans S.10 Sakota	J. D. Weller
	G-RANZ	Rans S-10 Sakota	O. M. Dismore
	G-RAPH	Cameron O-77 balloon	M. E. Mason
	G-RAPI	Lindstrand LBL-105A balloon	Rapido Balloons
	G-RAPP	Cameron H-34 balloon	Cameron Balloons Ltd
	G-RARB	Cessna 172N	Richlyn Aviation Ltd (G-BOII)
	G-RARE	Thunder Ax5-42 SS balloon ★	Balloon Preservation Group
	G-RASC	Evans VP-2	K. A. Stewart & G. Oldfield
	G-RASH	Grob G.109E	G-RASH Syndicate
	G-RATE	AA-5A Cheetah	Plane Talking Ltd (G-BIFF)
	G-RATH	Rotorway Executive 162F	M. S. Cole
	G-RATZ	Shaw Europa	W. Goldsmith
	G-RAVE	Southdown Raven X	M. J. Robbins (G-MNZV)
	G-RAVN	Robinson R-44	Heli Air Ltd/Wellesbourne
	G-RAWS	Rotorway Executive 162F	Raw Sports Ltd

Reg.	Type	Owner or Operator	Notes
G-RAYA	Denney Kitfox Mk 4	A. K. Ray	
G-RAYE	PA-32 Cherokee Six 260	G-RAYE Group (G-ATTY)	
G-RAYH	Zenair CH.701UL	R. Horner	
G-RAYO	Lindstrand LBL-90A balloon	R. Owen	
G-RAYS	Zenair CH.250	M. J. Malbon	
G-RAZY	PA-28-181 Archer II	R. W. Cooper (G-REXS)	
G-RAZZ	Maule MX-7-180	Airtime Aviation Ltd	
G-RBBB	Shaw Europa	T. J. Hartwell	
G-RBCI	BN-2A Mk.III-2 Trislander	Aurigny Air Services Ltd (G-BDWV)	
G-RBJW	Swah Europa XS	J. Worthington & R. J. Bull	
G-RBMV	Cameron O-31 balloon	P. D. Griffiths	
G-RBOS	Colt AS-105 airship ★	Science Museum/Wroughton	
G-RBOW	Thunder Ax-7-65 balloon	R. S. McDonald	
G-RBSN	Ikarus C.42 FB80	P. B. & M. Robinson	
G-RBSG	Dassault Falcon 900EX	Royal Bank of Scotland PLC	
G-RCED	R. Commander 114	D. G. Welch	
G-RCEJ	BAe 125 Srs 800B	Albion Aviation Management Ltd (G-GEIL)	
G-RCHY	Aerotechnik EV-97 Eurostar	N. McKenzie	
G-RCKT	Harmon Rocket II	K. E. Armstrong	
G-RCMC	Murphy Renegade 912	R. C. M. Collisson	
G-RCMF	Cameron V-77 balloon	J. M. Percival	
G-RCML	Sky 77-24 balloon	R. C. M. Sarl/Luxembourg	
G-RCMS	Agusta A.109E Power	Stolkin Helicopters Ltd (G-BZEI)	
G-RCNB	Eurocopter EC.120B	J. R. Clark Ltd	
G-RCOM	Bell 206L-3 LongRanger 3	3GRComm Ltd	
G-RDBS	Cessna 550 Citation II	Albion Aviation Management Ltd (G-JETA)	
G-RDCI	Rockwell Commander R.C.112	Simat Marketing Ltd	
G-RDCO	Avtech Jabiru J.400	RDCO (International) LLP	
G-RDEL	Robinson R-44	Jara Aviation	
G-RDHS	Shaw Europa XS	R. D. H. Spencer	
G-RDNS	Rans S.6-S Super Coyote	G. J. McDill	
G-READ	Colt 77A balloon	J. Keena	
G-REAH	PA-32R-301 Saratoga SP	M. Q. Tolbod & S. J. Rogers (G-CELL)	
G-REAL	AS.350B-2 Ecureuil	Imagine Leisure Ltd (G-DRHL)	
G-REAP	Pitts S-1S Special	R. Dixon	
G-REAR	Lindstrand LBL-69X balloon	Exclusive Ballooning Ltd	
G-REAS	Van's RV-6A	T. J. Smith	
G-REAT	GA-7 Cougar	Goodtechnique Ltd	
G-REBA	RAF 2000 GTX-SE gyroplane	D. J. Pearce	
G-REBL	Hughes 269B	Farmax Ltd	
G-RECK	PA-28 Cherokee 140B	R. J. Grantham & D. Boatswain (G-AXJW)	
G-RECO	Jurca MJ-5L Sirocco	J. D. Tseliki	
G-RECS	PA-38-112 Tomahawk	S. H. & C. L. Maynard	
G-REDB	Cessna 310Q	Red Baron Haulage Ltd (G-BBIC)	
G-REDC	Pegasus Quantum 15-912	R. F. Richardson	
G-REDD	Cessna 310R II	G. Wightman (G-BMGT)	
G-REDI	Robinson R-44	Redeye.com Ltd	
G-REDJ	Eurocopter AS.332L-2 Super Puma	International Aviation Leasing Ltd	
G-REDK	Eurocopter AS.332L-2 Super Puma	International Aviation Leasing Ltd	
G-REDL	Eurocopter AS.332L-2 Super Puma	International Aviation Leasing Ltd	
G-REDM	Eurocopter AS.332L-2 Super Puma	International Aviation Leasing Ltd	
G-REDN	Eurocopter AS.332L-2 Super Puma	International Aviation Leasing Ltd	
G-REDS	Cessna 560XL Citation Excel	Ferron Trading Ltd	
G-REDX	Experimental Aviation Berkut	G. V. Waters	
G-REDY	Robinson R-22B	Plane Talking Ltd (G-CBXO)	
G-REDZ	Thruster T.600N 450	S. L. & W. J. Smith	
G-REEC	Sequoia F.8L Falco	J. D. Tseliki	
G-REED	Mainair Blade 912S	D. Jessop	
G-REEF	Mainair Blade 912S	G. Mowll	
G-REEK	AA-5A Cheetah	J. & A. Pearson	
G-REEM	AS.355F-1 Twin Squirrel	Heliking Ltd (G-EMAN/G-WEKR/G-CHLA)	
G-REEN	Cessna 340	Just Plane Trading Ltd (G-AZYR)	
G-REES	Jodel D.140C	G-REES Flying Group	
G-REGI	Cyclone Chaser S508	G. S. Stokes (G-MYZW)	
G-REKO	Pegasus Quasar IITC	M. Sims (G-MWWA)	
G-RENE	Murphy Renegade 912	P. M. Whitaker	
G-RENO	SOCATA TB.10 Tobago	Lamond Ltd	
G-REPH	Pegasus Quantum 15-912	R. S. Partridge-Hicks	

Notes	Reg.	Type	Owner or Operator
	G-RESG	Dyn'aero MCR-01 Club	R. E. S. Greenwood
	G-REST	Beech P35 Bonanza	C. R. Taylor (G-ASFJ)
	G-RETA	C.A.S.A. 1.131 Jungmann 2000	Richard Shuttleworth Trustees (G-BGZC)
	G-REVO	Skyranger 912 (2)	R. T. Henry
	G-REYS	Canadair CL.604 Challenger	Greyscape Ltd
	G-RFDS	Agusta A.109A-II	Castle Air Charters Ltd (G-BOLA)
	G-RFIO	Aeromot AMT-200 Super Ximango	M. D. Evans
	G-RFSB	Sportavia RF-5B	J. F. Mcaulay & A. A. Jury
	G-RFUN	Robinson R-44	BDW Fuels
	G-RGEN	Cessna T.337D	Legoprint SpA (G-EDOT/G-BJIY)/Italy
	G-RGUS	Fairchild 24A-46A Argus III (44-83184)	Fenlands Ltd
	G-RHCB	Schweizer 269C-1	Helicopter One Ltd
	G-RHHT	PA-32RT-300 Lance II	G. R. Bright
	G-RHOP	BN-2A Mk III-2 Trislander	B-N Group Ltd (G-BEFP/G-WEAC)
	G-RHYM	PA-31-310 Turbo Navajo B	ATC Trading Ltd (G-BJLO)
	G-RHYS	Rotorway Executive 90	A. K. Voase & K. Matthews
	G-RIAN	AB-206A JetRanger	B. J. Green (G-SOOR/G-FMAL/G-BHSG)
	G-RIAT	Robinson R-22B-2	RMJ Helicopters
	G-RIBS	Diamond Katana DA.20-A1	D. M. Green (G-BWWM)
	G-RIBZ	Enstrom 480B	Premiair Aviation Group Ltd
	G-RICC	AS.350B-2 Ecureuil	Specialist Helicopters Ltd (G-BTXA)
	G-RICE	Robinson R-22B	Silverstar Components Ltd
	G-RICK	Beech 95-B55 Baron	James Jack Lifting Services Ltd (G-BAAG)
	G-RICO	AG-\5B Tiger	Plane Talking Ltd
	G-RICS	Shaw Europa	The Flying Property Doctor
	G-RIDD	Robinson R-22B	Essex Match Co Ltd
	G-RIDE	Stephens Akro	R. Mitchell/Coventry
	G-RIDL	Robinson R-22B	Peterborough Helicopter Hire Ltd
	G-RIET	Hoffmann H.36 Dimona	Dimona Gliding Group
	G-RIFB	Hughes 269C	J. McHugh & Son (Civil Engineering Contractors) Ltd
	G-RIFN	Avion Mudry CAP.10B	R. A. G. Spurrell
	G-RIGB	Thunder Ax7-77 balloon	N. J. Bettin
	G-RIGH	PA-32R-301 Saratoga IIHP	Right Aviation Ltd
	G-RIGS	PA-60 Aerostar 601P	G. G. Caravatti & P. G. Penati/Italy
	G-RIHN	Dan Rihn DR.107 One Design	J. P. Brown
	G-RIIN	WSK PZL-104MN Wilga 2000	E. A. M. Austin
	G-RIKI	Mainair Blade 912	R. Cook
	G-RIKS	Shaw Europa XS	R. Morris
	G-RIKY	Mainair Pegasus Quik	R. J. Cook
	G-RIMB	Lindstrand LBL-105A balloon	D. Grimshaw
	G-RIME	Lindstrand LBL-25A balloon	Poppies (UK) Ltd
	G-RIMM	Westland Wasp HAS.1 (XT435/430)	M. P. Grimshaw & T. Martin
	G-RING	SA Bulldog Srs.100/101	C. S. Beevers (G-AZMR)/North Weald
	G-RINN	Mainair Blade	J. P. Lang
	G-RINO	Thunder Ax7-77 balloon	D. J. Head
	G-RINS	Rans S.6-ESD Coyote II	D. Watt
	G-RINT	CFM Streak Shadow	D. Grint
	G-RIPS	Cameron 110 Parachutist SS balloon ★	Balloon Preservation Group
	G-RISE	Cameron V-77 balloon	D. L. Smith
	G-RIST	Cessna 310R II	F. B. Spriggs (G-DATS
	G-RIVE	Jodel D.150	P. Fines
	G-RIVR	Thruster T.600N	Thruster Air Services Ltd
	G-RIVT	Van's RV-6	N. Reddish
	G-RIXS	Shaw Europa XS	R. Iddon
	G-RIZE	Cameron O-90 balloon	S. F. Burden/Netherlands
	G-RIZI	Cameron N-90 balloon	R. Wiles
	G-RIZZ	PA-28-161 Warrior II	Northamptonshire School of Flying Ltd/Sywell
	G-RJAH	Boeing Stearman A.75N1	R. J. Horne
	G-RJAM	Sequoia F.8C Falco	R. J. Marks
	G-RJCP	R. Commander 114B	Heltor Ltd
	G-RJGR	Boeing 757-225	My Travel Airways
	G-RJMS	PA-28R-201 Arrow III	M. G. Hill
	G-RJTT	Bell 206B JetRanger 3	Air Deluxe
	G-RJWW	Maule M5-235C Lunar Rocket	PAW Flying Services Ltd (G-BRWG)
	G-RJWX	Shaw Europa XS	J. R. Jones
	G-RJXA	Embraer RJ145EP	bmi regional

Reg.	Type	Owner or Operator	Notes
G-RJXB	Embraer RJ145EP	bmi regional	
G-RJXC	Embraer RJ145EP	bmi regional	
G-RJXD	Embraer RJ145EP	bmi regional	
G-RJXE	Embraer RJ145EP	bmi regional	
G-RJXF	Embraer RJ145EP	bmi regional	
G-RJXG	Embraer RJ145EP	bmi regional	
G-RJXH	Embraer RJ145EP	bmi regional	
G-RJXI	Embraer RJ145EP	bmi regional	
G-RJXJ	Embraer RJ135LR	bmi regional	
G-RJXK	Embraer RJ135LR	bmi regional	
G-RJXL	Embraer RJ135LR	bmi regional	
G-RJXM	Embraer RJ135LR	bmi regional	
G-RKEL	AB-206B JetRanger 3	Nunkeeling Ltd	
G-RKET	Taylor JT.2 Titch	P. A. Dunkley (G-BIBK)	
G-RKJT	PA-46-500TP Malibu Meridian	Harpin Ltd	
G-RLFI	Cessna FA.152	Tayside Aviation Ltd (G-DFTS)/Aberdeen	
G-RLON	BN-2A Mk III-2 Trislander	Aurigny Air Services Ltd (G-ITEX/G-OCTA/G-BCXW)	
G-RMAC	Shaw Europa	P. J. Lawless	
G-RMAN	Aero Designs Pulsar	M. B. Redman	
G-RMAX	Cameron C-80 balloon	M. Quinn & D. Curtain	
G-RMIT	Van's RV-4	J. P. Kloos	
G-RMPY	Aerotechnik EV-97 Eurostar	N. R. Beale	
G-RMUG	Cameron 90 Mug SS balloon	Nestle UK Ltd	
G-RNAC	IDA Bacau Yakovlev Yak-52	RNAEC Ltd	
G-RNAS	D.H.104 Sea Devon C.20 (XK896) ★	Airport Fire Service/Filton	
G-RNBW	Bell 206B JetRanger 2	Rainbow Helicopters Ltd	
G-RNDD	Robin DR.400/500	Sterna Aviation Ltd	
G-RNGO	Robinson R-22B-2	B. E. Llewellyn	
G-RNIE	Cameron 70 Ball SS balloon	N. J. Bland	
G-RNLI	V.S.236 Walrus I (W2718) ★	R. E. Melton	
G-RNRM	Cessna A.185F	Skydive St. Andrews Ltd	
G-RNRS	SA Bulldog Srs.100/101	Power Aerobatics Ltd (G-AZIT)	
G-ROAR	Cessna 401	Special Scope Ltd (G-BZFL/G-AWSF)	
G-ROBD	Shaw Europa	R. D. Davies	
G-ROBN	Robin R.1180T	Bustard Flying Club Ltd	
G-ROBT	Hawker Hurricane I	R. A. Roberts	
G-ROBY	Colt 17A balloon	Virgin Airship & Balloon Co Ltd	
G-ROCH	Cessna T.303	R. S. Bentley	
G-ROCK	Thunder Ax7-77 balloon	M. A. Green	
G-ROCR	Schweizer 269C	C. J. Williams	
G-RODC	Steen Skybolt	R. G. Cameron	
G-RODD	Cessna 310R II	R. J. Herbert Engineering Ltd (G-TEDD/G-MADI)	
G-RODG	Avtech Jabiru UL	S. Jackson	
G-RODI	Isaacs Fury (K3731)	M. R. Baker/Shoreham	
G-ROGY	Cameron 60 Concept balloon	S. A. Laing	
G-ROKT	Cessna FR.172E	Sylmar Aviation & Services Ltd	
G-ROLF	PA-32R-301 Saratoga SP	P. F. Larkins	
G-ROLL	Pitts S-2A Special	Aerial & Aerobatic Services	
G-ROLO	Robinson R-22B	Plane Talking Ltd/Elstree	
G-ROLY	Cessna F.172N	Beaufort Construction Ltd (G-BHIH	
G-ROME	I.I.I. Sky Arrow 650TC	Sky Arrow (Kits) UK Ltd	
G-ROMW	Cyclone AX2000	L. P. Taylor	
G-RONA	Shaw Europa	C. M. Noakes	
G-ROND	Short SD3-60 Variant 100	Emerald Airways Ltd (G-OLAH/G-BPCO/G-RMSS/G-BKKU)/Liverpool	
G-RONG	PA-28R Cherokee Arrow 200-II	E. Tang	
G-RONI	Cameron V-77 balloon	R. E. Simpson	
G-RONN	Robinson R-44 Astro	R. Hallam & S. E. Watts	
G-RONS	Robin DR.400/180	R. & K. Baker	
G-RONW	FRED Srs 2	V. Magee	
G-ROOK	Cessna F.172P	Rolim Ltd	
G-ROOV	Shaw Europa XS	E. Sheridan & P. W. Hawkins	
G-RORI	Folland Gnat T.1 (01)	Delta Engineering Aviation Ltd/Kemble	
G-RORY	Piaggio FWP.149D	Bushfire Investments Ltd (G-TOWN)/North Weald	
G-ROSI	Thunder Ax7-77 balloon	J. E. Rose	
G-ROSS	Practavia Pilot Sprite	A. D. Janaway	

Notes	Reg.	Type	Owner or Operator
	G-ROTI	Luscombe 8A Silvaire	A. L. Chapman & R. Ludgate
	G-ROTR	Brantly B.2B	P. G. R. Brown
	G-ROTS	CFM Streak Shadow Srs SA	A. G. Vallis & C. J. Kendal
	G-ROUP	Cessna F.172M	Stapleford Flying Club Ltd (G-BDPH)
	G-ROUS	PA-34-200T Seneca II	Oxford Aviation Services Ltd/Kidlington
	G-ROUT	Robinson R-22B	Preston Associatres Ltd
	G-ROVE	PA-18 Super Cub 135	S. J. Gaveston
	G-ROVY	Robinson R-22B-2	Plane Talking Ltd
	G-ROWE	Cessna F.182P	D. Rowe/Liverpool
	G-ROWI	Shaw Europa XS	R. M. Carson
	G-ROWL	AA-5B Tiger	Airhouse Corporation
	G-ROWN	Beech 200 Super King Air	Valentia Air Ltd (G-BHLC)
	G-ROWR	Robinson R-44	R. A. Oldworth
	G-ROWS	PA-28-151 Warrior	N. J. Arney
	G-ROXY	Skystar Kitbox Mk.7	P. N. Akass
	G-ROZI	Robinson R-44	Milford Garage Ltd
	G-ROZY	Cameron R.36 balloon	Jacques W. Soukup Enterprises Ltd/USA
	G-ROZZ	Ikarus C.42 FB80	A. J. Blackwell
	G-RPBM	Cameron Z-210 balloon	First Flight
	G-RPEZ	Rutan LongEz	D. G. Foreman
	G-RPRV	Van's RV-9A	G. R. Pybus
	G-RRCU	CEA DR.221B Dauphin	Merlin Flying Club Ltd
	G-RRFC	SOCATA TB.20 Trinidad GT	C. A. Hawkins
	G-RRGN	V.S.390 Spitfire PR.XIX (PS853)	Rolls-Royce PLC (G-MXIX)/Filton
	G-RROB	Robinson R-44 II	Ranc Helicopters Ltd
	G-RROD	PA-30 Twin Comanche 160B	R. P. Coplestone (G-SHAW)
	G-RSCJ	Cessna 525 CitationJet	Pektron Group Ltd
	G-RSFT	PA-28-161 Warrior II	Deep Cleavage Ltd (G-WARI)
	G-RSKR	PA-28-161 Warrior II	Krown Group (G-BOJY)
	G-RSKY	Skyranger 912(2)	C. G. Benham
	G-RSSF	Denney Kitfox Mk 2	R. W. Somerville
	G-RSVP	Robinson R-22B-2	Plane Talking Ltd
	G-RSWO	Cessna 172R	AC Management Associates Ltd
	G-RSWW	Robinson R-22B	Woodstock Enterprises
	G-RTBI	Thunder Ax6-56 balloon	P. J. Waller
	G-RTMS	Rans S.6 ES Coyote II	C. J. Arthur
	G-RTWW	Robinson R-44 Astro	Rotorvation
	G-RUBB	AA-5B Tiger	D. E. Gee/Blackbushe
	G-RUBI	Thunder Ax7-77 balloon	Warren & Johnson
	G-RUBY	PA-28RT-201T Turbo Arrow IV	Arrow Aircraft Group (G-BROU)
	G-RUDD	Cameron V-65 balloon	N. A. Apsey
	G-RUES	Robin HR.100/210	R. H. R. Rue
	G-RUFF	Mainair Blade 912	J. C. Townsend
	G-RUFS	Avtech Jabiru UL	S. Richens
	G-RUGS	Campbell Cricket Mk 4 gyroplane	J. L. G. McLane
	G-RUIA	Cessna F.172N	Knockin Flying Club Ltd
	G-RUMI	Noble Harman Snowbird Mk.IV	G. Crossley (G-MVOI)
	G-RUMM	Grumman F8F-2P Bearcat (21714)	Patina Ltd/Duxford
	G-RUMN	AA-1A Trainer	M. T. Manwaring
	G-RUMT	Grumman F7F-3P Tigercat (80425)	Patina Ltd/Duxford
	G-RUMW	Grumman FM-2 Wildcat (JV579)	Patina Ltd/Duxford
	G-RUNG	SAAB SF.340A	Aurigny Air Services Ltd
	G-RUNT	Cassutt Racer IIIM	N. A. Scully
	G-RUSA	Pegasus Quantum 15-912	A. D. Stewart
	G-RUSL	Van's RV-6A	G. R. Russell
	G-RUSO	Robinson R-22B	R. M. Barnes-Gorell
	G-RUSS	Cessna 172N ★	Leisure Lease (stored)/Southend
	G-RUVI	Zenair CH.601UL	P. G. Depper
	G-RUVY	Van's RV-9A	R. D. Taylor
	G-RUZZ	Robinson R-44 II	Russell Harrison PLC
	G-RVAB	Van's RV-7	I. M. Belmore & A. T. Banks
	G-RVAL	Van's RV-8	R. N. York
	G-RVAN	Van's RV-6	D. Broom
	G-RVAW	Van's RV-6	High Flatts RV Group
	G-RVBA	Van's RV-8A	S. Hawksworth
	G-RVBC	Van's RV-6A	B. J. Clifford

Reg.	Type	Owner or Operator	Notes
G-RVBF	Cameron A-340 balloon	Virgin Balloon Flights	
G-RVCE	Van's RV-6A	M. D. Barnard & C. Voelger	
G-RVCG	Van's RV-6A	C. J. Griffin	
G-RVCL	Van's RV-6	C. T. Lamb	
G-RVDJ	Van's RV-6	J. D. Jewitt	
G-RVDP	Van's RV-4	D. H. Pattison	
G-RVDR	Van's RV-6A	T. M. Norman	
G-RVDS	Van's RV-4	D. F. Sargant	
G-RVEE	Van's RV-6	J. C. A. Wheeler	
G-RVET	Van's RV-6	D. R. Coleman	
G-RVGA	Van's RV-6A	D. P. Dawson	
G-RVHT	Cessna 550 Citation 2	Ravenheat Manufacturing Ltd	
G-RVIA	Van's RV-6A	K. R. Emery	
G-RVIB	Van's RV-6	K. Martin & P. Gorman	
G-RVIC	Van's RV-6A	I. T. Corse	
G-RVII	Van's RV-7	P. H. C. Hall	
G-RVIN	Van's RV-6	R. G. Jines	
G-RVIS	Van's RV-8	I. V. Sharman	
G-RVIT	Van's RV-6	P. J. Shotbolt	
G-RVIV	Van's RV-4	G. S. Scott	
G-RVIX	Van's RV-9A	R. E. Garforth	
G-RVJM	Van's RV-6	M. D. Challoner	
G-RVMC	Van's RV-7	M. R. McNeil	
G-RVMJ	Van's RV-4	M. J. de Ruiter	
G-RVMT	Van's RV-6	M. R. Tingle	
G-RVMZ	Van's RV-8	M. W. Zipfel	
G-RVPH	Van's RV-8	J. C. P. Herbert	
G-RVPL	Van's RV-8	A. P. Lawton	
G-RVPW	Van's RV-6A	P. Waldron	
G-RVRA	PA-28 Cherokee 140	Mona Flying Club (G-OWVA)	
G-RVRB	PA-34-200T Seneca II	Ravenair (G-BTAJ)/Liverpool	
G-RVRC	PA-23 Aztec 250E	Ravenair (G-BNPD)/Liverpool	
G-RVRD	PA-23 Aztec 250E	Ravenair (G-BRAV/G-BBCM)/Liverpool	
G-RVRE	Partenavia P.68B	Ravenair/Liverpool	
G-RVRF	PA-38-112 Tomahawk	Ravenair (G-BGEL)/Liverpool	
G-RVRG	PA-38-112 Tomahawk	Ravenair (G-BHAF)/Liverpool	
G-RVRJ	PA-E23 Aztec 250E	Ravenair (G-BBGB)/Liverpool	
G-RVRP	Van's RV-7	R. C. Parris	
G-RVRV	Van's RV-4	P. Jenkins	
G-RVRW	PA-23 Aztec 250E	Cheshire Flying Services Ltd (G-BAVZ)	
G-RVSA	Van's RV-6A	W. H. Knott	
G-RVSG	Van's RV-9A	S. Gerrish	
G-RVSH	Van's RV-6A	S. J. D. Hall	
G-RVSX	Van's RV-6	R. L. & V. A. West	
G-RVVI	Van's RV-6	J. E. Alsford & J. N. Parr	
G-RWAY	Rotorway Executive 162F	S. Andrews (G-URCH)	
G-RWHC	Cameron A-180 balloon	Wickers World Hot Air Balloon Co	
G-RWIN	Rearwin 175	A. B. Bourne & N. D. Battye	
G-RWLY	Shaw Europa XS	C. R. Arcle	
G-RWRW	Ultramagic M-77 balloon	Flying Pictures Ltd	
G-RWSS	Denney Kitfox Mk 2	R. W. Somerville	
G-RWWW	W.S.55 Whirlwind HCC.12 (XR486) ★	IHM/Weston-s-Mare	
G-RXUK	Lindstrand LBL-105A balloon	P. A. Hames	
G-RYAL	Avtech Jabiru UL	A. C. Ryall	
G-RYPH	Mainair Blade 912	I. A. Cunningham	
G-SAAB	R. Commander 112TC	J. B. Barbour (G-BEFS)	
G-SAAM	Cessna T.182R	M. D. Harvey & ptnrs (G-TAGL)	
G-SABA	PA-28R-201T Turbo Arrow III	C. A. Burton (G-BFEN)	
G-SABB	Eurocopter EC.135T-1	Bond Air Services Ltd	
G-SABR	NA F-86A Sabre (8178)	Golden Apple Operations Ltd/Bournemouth	
G-SACB	Cessna F.152 II	P. Wilson (G-BFRB)	
G-SACD	Cessna F.172H	Northbrook College of Design & Technology (G-AVCD)/Shoreham	
G-SACH	Stoddard-Hamilton Glastar	R. S. Holt	
G-SACI	PA-28-161 Warrior II	PJC (Leasing) Ltd	
G-SACK	Robin R.2160	Sherburn Aero Club Ltd	
G-SACO	PA-28-161 Warrior II	M. & D. C. Brooks	

G-SACR – G-SDLW

BRITISH CIVIL REGISTRATIONS

Notes	Reg.	Type	Owner or Operator
	G-SACR	PA-28-161 Cadet	Sherburn Aero Club Ltd
	G-SACS	PA-28-161 Cadet	Sherburn Aero Club Ltd
	G-SACT	PA-28-161 Cadet	Sherburn Aero Club Ltd
	G-SACZ	PA-28-161 Warrior II	J. R. Santamaria
	G-SADE	Cessna F.150L	N. E. Sams (G-AZJW)
	G-SAFE	Cameron N-77 balloon	P. J. Waller
	G-SAFI	CP.1320 Super Emeraude	C. S. Carleton-Smith
	G-SAFR	SAAB 91D Safir	Sylmar Aviation & Services Ltd
	G-SAGA	Grob G.109B	G-GROB Ltd/Booker
	G-SAGE	Luscombe 8A Silvaire	C. Howell & J. O'Brien (G-AKTL)
	G-SAHI	Trago Mills SAH-1	Aces High Ltd/North Weald
	G-SAIR	Cessna 421C	Air Support Aviation Services Ltd (G-OBCA)
	G-SAIX	Cameron N-77 balloon	C. Walther & ptnrs
	G-SALA	PA-32 Cherokee Six 300E	Stonebold Ltd
	G-SALL	Cessna F.150L (Tailwheel)	D. & P. A. Hailey
	G-SAMG	Grob G.109B	RAFGSA/Bicester
	G-SAMI	Cameron N-90 balloon	Flying Pictures Ltd (G-BWSE)
	G-SAMJ	Partenavia P.68B	G-SAMJ Group
	G-SAMM	Cessna 340A	Calverton Flying Group Ltd
	G-SAMY	Shaw Europa	K. R. Tallent
	G-SAMZ	Cessna 150D	Bonus Aviation Ltd (G-ASSO)/Cranfield
	G-SAPM	SOCATA TB.20 Trinidad (G-EWFN)	Trinidair Ltd
	G-SARA	PA-28-181 Archer II	R. P. Lewis
	G-SARH	PA-28-161 Warrior II	Sussex Flying Club Ltd/Shoreham
	G-SARK	BAC.167 Strikemaster Mk 84 (311)	Tubetime Ltd
	G-SARO	Saro Skeeter Mk 12 (XL812)	B. Chamberlain
	G-SARV	Van's RV-4	S. N. Aston
	G-SASA	Eurocopter EC.135T-1	Bond Air Services Ltd
	G-SASB	Eurocopter EC.135T-2	Bond Air Services Ltd
	G-SASK	PA-31P Pressurised Navajo	Middle East Business Club Ltd (G-BFAM)
	G-SATL	Cameron 105 Sphere SS balloon	Ballonwerbung Hamburg GmbH/Germany
	G-SATR	Boeing 767-204ER	Air Atlanta Europe Ltd (TF-ATR/G-BPFV)
	G-SAUF	Colt 90A balloon	K. H. Medau
	G-SAWI	PA-32RT-300T Turbo Lance II	S. T. Day
	G-SAXO	Cameron N-105 balloon	Flying Pictures Ltd
	G-SAYS	RAF 2000 GTX-SE gyroplane	Aziz Corporation Ltd
	G-SAZY	Avtech Jabiru J.400	JC Aviation
	G-SAZZ	CP.328 Super Emeraude	D. J. Long
	G-SBAE	Cessna F.172P	Warton Flying Club/Blackpool
	G-SBAS	Beech B200 Super King Air	Gama Aviation Ltd (G-BJJV)
	G-SBHH	Schweizer 269C	Biggin Hill Helicopters Ltd (G-XALP)
	G-SBIZ	Cameron Z-90 balloon	Snow Business International Ltd
	G-SBKR	SOCATA TB-10 Tobago	S. C. M. Bagley
	G-SBLT	Steen Skybolt	Skybolt Group
	G-SBMM	PA-28R Cherokee Arrow 180	K. S. Kalsi (G-BBEL)
	G-SBMO	Robin R.2160I	D. Henderson & ptnrs
	G-SBRA	Robinson R-44 II	Sabretooth Aviation Ltd
	G-SBUS	BN-2A-26 Islander	Isles of Scilly Skybus Ltd (G-BMMH)/St Just
	G-SBUT	Robinson R-22B-2	Steve Butler Ltd (G-BXMT)
	G-SCAN	Vinten-Wallis WA-116/100	K. H. Wallis
	G-SCAT	Cessna F.150F (tailwheel)	Westward Airways (Lands End) Ltd (G-ATRN)
	G-SCBI	SOCATA TB.20 Trinidad	Ace Services
	G-SCFO	Cameron O-77 balloon	M. K. Grigson
	G-SCHI	AS.350B-2 Ecureuil	Patriot Aviation Ltd
	G-SCIP	SOCATA TB.20 Trinidad GT	J. C. White
	G-SCLX	FLS Aerospace Sprint 160	Aces High Ltd (G-PLYM)/North Weald
	G-SCOI	Agusta A.109E Power	Trustair Ltd (G-HPWH/G-HWPH)
	G-SCPD	Escapade 912 (1)	R. Gibson
	G-SCPL	PA-28 Cherokee 140	Aeros Leasing Ltd (G-BPVL)
	G-SCRU	Cameron A-250 balloon	Societe Bombard SARL (G-BWWO)/France
	G-SCTA	Westland Scout AH.1 (XV126)	G. R. Harrison
	G-SCUB	PA-18 Super Cub 135 (542447)	C. L.. Needham
	G-SCUD	Montgomerie-Bensen B.8MR	D. Taylor
	G-SCUL	Rutan Cozy	K. R. W. Scull
	G-SCUR	Eurocopter EC.120B	JS Aviation Ltd
	G-SDCI	Bell 206B JetRanger 2	S. D. Coomes (G-GHCL/G-SHVV)
	G-SDEV	D.H. 104 Sea Devon C.20 (XK895)	Wyndeham Press Group PLC/Shoreham
	G-SDFM	Aerotechnik EV-97 Eurostar	D. F. Randall
	G-SDLW	Cameron O-105 balloon	P. J. Smart

264

Reg.	Type	Owner or Operator
G-SDOI	Aeroprakt A.22 Foxbat	S. A. Owen
G-SDOZ	Tecnam P.92-EA Echo Super	S. P. S. Dornan
G-SEAI	Cessna U.206G (amphibian)	K. O'Conner
G-SEDO	Cameron N-105 balloon	Flying Pictures Ltd
G-SEED	Piper J-3C-65 Cub	J. H. Seed
G-SEEK	Cessna T.210N	A. Hopper
G-SEFI	Robinson R-44 II	Heli Air Ltd
G-SEGA	Cameron 90 Sonic SS balloon	Balloon Preservation Flying Group
G-SEJW	PA-28-161 Warrior II	Keen Leasing Ltd
G-SELF	Shaw Europa	N. D. Crisp & ptnrs
G-SELL	Robin DR.400/180	G-SELL Regent Group
G-SELY	AB-206B JetRanger 3	CT_Rental Ltd
G-SEMI	PA-44-180 Seminole	Halfpenny Green Flight Centre Ltd (G-DENW)
G-SENA	Rutan LongEz	G. Bennett
G-SEND	Colt 90A balloon	J-P. Barre/France
G-SENE	PA-34-200T Seneca II	R. Clarke
G-SENX	PA-34-200T Seneca II	Katotech Ltd (G-DARE/G-WOTS/G-SEVL)
G-SEPA	AS.355N Twin Squirrel	Metropolitan Police (G-METD/G-BUJF)
G-SEPB	AS.355N Twin Squirrel	Metropolitan Police (G-BVSE)
G-SEPC	AS.355N Twin Squirrel	Metropolitan Police (G-BWGV)
G-SEPT	Cameron N-105 balloon	P. Gooch
G-SERA	Enstrom F-28A-UK	W. R. Pitcher (G-BAHU)
G-SERL	SOCATA TB.10 Tobago	R. J. Searle (G-LANA)/Rochester
G-SERV	Cameron N-105 balloon	PSH Skypower Ltd
G-SETI	Cameron Sky 80-16 balloon	R. P. Allan
G-SEVA	SE-5A (replica) (F141)	I. D. Gregory
G-SEVE	Cessna 172N	MK Aero Support Ltd
G-SEVN	Van's RV-7	N. Reddish
G-SEWP	AS.355F-2 Twin Squirrel	Veritair Ltd (G-OFIN/G-DANS/G-BTNM)
G-SEXE	Scheibe SF.25C Falke	Repulor Ltd
G-SEXX	PA-28-161 Warrior II	A. M. Blatch, Electrical Contractors Ltd
G-SEXY	AA-1 Yankee ★	(stored)/Liverpool (G-AYLM)
G-SFCJ	Cessna 525 CitationJet	Sureflight Aviation Ltd
G-SFHR	PA-23 Aztec 250F	Comed Aviation Ltd (G-BHSO)/Blackpool
G-SFLY	Diamond DA.40 Star	F. Pilkington & L. Turner
G-SFOX	Rotorway Executive 90	Magpie Technology Ltd (G-BUAH)
G-SFPA	Cessna F.406	Scottish Fisheries Protection Agency
G-SFPB	Cessna F.406	Scottish Fisheries Protection Agency
G-SFRY	Thunder Ax7-77 balloon	M. Rowlands
G-SFSG	Beech E90 King Air	Premiair Charter Ltd
G-SFSL	Cameron Z-105	Exclusive Ballooning
G-SFTZ	Slingsby T.67M Firefly 160	Western Air (Thruxton) Ltd
G-SGAS	Colt 77A balloon	A. Derbyshire
G-SGEC	Beech B.200 Super King Air	Bridgtown Plant Ltd
G-SGEN	Ikarus C.42 FB80	Fly Buy Ultralights Ltd
G-SGSE	PA-28-181 Archer II	D. Masson (G-BOJX)
G-SHAA	Enstrom 280-UK	ELT Radio Telephones
G-SHAH	Cessna F.152	I. R. Chaplin/Andrewsfield
G-SHAY	PA-28R-201T Turbo Arrow III	Alpha Yankee Flying Group (G-BFDG/G-JEFS)
G-SHCB	Schweizer 269C-1	Oxford Aviation Services Ltd/Kidlington
G-SHED	PA-28-181 Archer II	G-SHED Flying Group (G-BRAU)
G-SHEZ	Mainair Pegasus Quik	A. Anderson
G-SHIM	CFM Streak Shadow	K. R. Anderson
G-SHIP	PA-23 Aztec 250F ★	Midland Air Museum/Coventry
G-SHOG	Colomban MC.15 Cri-Cri	V. S. E. Norman (G-PFAB)/Rendcomb
G-SHOW	M.S.733 Alycon	Vintage Aircraft Team/Cranfield
G-SHPP	Hughes TH-55A	Helirouge Ltd
G-SHRK	Enstrom 280C-UK	Aviation Bureau (G-BGMX)
G-SHRT	Robinson R-44-II	Overby Ltd
G-SHSH	Shaw Europa	D. G. Hillam
G-SHSP	Cessna 172S	Shropshire Aero Club Ltd/Sleap
G-SHUF	Mainair Blade	R. G. Bradley
G-SHUG	PA-28R-201T Turbo Arrow III	G-SHUG Ltd
G-SHUU	Enstrom 280C-UK-2	D. Ellis (G-OMCP/G-KENY/G-BJFG)
G-SHUV	Aerosport Woody Pusher	J. R. Wraigh
G-SHWK	Cessna 172S	Cambridge Aero Club Ltd

Notes	Reg.	Type	Owner or Operator
	G-SIAI	SIAI-Marchetti SF.260W	D. Gage
	G-SIAL	Hawker Hunter F.58 (J-4090)	Old Flying Machine Club/Scapmton
	G-SIAM	Cameron V-90 balloon	D. Tuck (G-BXBS)
	G-SIAX	Cameron Z-210 balloon	Societe Bombard SRL/France
	G-SIGN	PA-39 Twin Comanche 160 C/R	D. Buttle/Blackbushe
	G-SIIA	Pitts S-2A Special	A. P. Crumpholt
	G-SIIB	Pitts S-2B Special	T. H. Bishop & J. H. Milne (G-BUVY)
	G-SIID	Sukhoi Su.26M2	Gold Star International Ltd
	G-SIIE	Christen Pitts S-2B Special	Technoforce Ltd (G-SKYD)
	G-SIII	Extra EA.300	Callmast Ltd
	G-SIIS	Pitts S-1S Special	I. H. Searson (G-RIPE)
	G-SIJJ	North American P-51D-NA Mustang	P. A. Teichman
	G-SIJW	SA Bulldog Srs 120/121 (XX630)	M. Miles
	G-SILS	Pietenpol Skyscout	D. Silsbury
	G-SILY	Pegasus Quantum 15	L. Harland
	G-SIMI	Cameron A-315 balloon	Balloon Safaris
	G-SIMM	Ikarus C.42 FB100 VLA	D. Simmons
	G-SIMN	Robinson R-22B-2	Heli Air Ltd
	G-SIMP	Avtech Jabiru SP	J. C. Simpson
	G-SIMS	Robinson R-22B	Heli-One
	G-SIMY	PA-32-300 Cherokee Six	I. Simpson (G-OCPF/G-BOCH)
	G-SIPA	SIPA 903	A. C. Leak & J. H. Dilland (G-BGBM)
	G-SIRS	Cessna 560XL Citation Excel	Amsair Ltd
	G-SITA	Pegasus Quantum 15-912	A. R. Oliver
	G-SIVJ	Westland Gazelle HT.2	C. J. Siva-Jothy (G-CBSG)
	G-SIVN	MD 500N	Mandarin Aviation Ltd
	G-SIVR	MDH MD.900	Mandarin Aviation Ltd
	G-SIVW	Lake LA-250	C. J. Siva-Jothy
	G-SIXC	Douglas DC-6A	Atlantic Air Transport Ltd/Coventry
	G-SIXD	PA-32 Cherokee Six 300D	M. B. Paine & I. Gordon
	G-SIXS	Whittaker MW-6S Fat Boy Flyer	R. H. Braithwaite
	G-SIXX	Colt 77A balloon	M. Dear & M. Taylor
	G-SIXY	Van's RV-6	C. J. Hall & C. R. P. Hamlett
	G-SJCH	BN-2T-4S Defender 4000	Hampshire Police Authority (G-BWPK)
	G-SJDI	Robinson R-44	Helicopter Support Ltd
	G-SJKR	Lindstrand LBL-90A balloon	S. J. Roake
	G-SJEN	Ikarus C.42 FB80	Charles Henry Services
	G-SJMC	Boeing 767-31KER	My Travel Airways
	G-SKAN	Cessna F.172M	Bustard Flying Club Ltd (G-BFKT)
	G-SKCI	Rutan Vari-Eze	S. K. Cockburn
	G-SKEW	Mudry CAP. 232	J. H. Askew
	G-SKIE	Steen Skybolt	S. Gray
	G-SKII	Augusta-Bell 206B JetRanger III	C & M Coldstores/Ireland
	G-SKNT	Aerotek S-2A	T. G. Lloyd (G-PEAL)
	G-SKOT	Cameron V-42 balloon	S. A. Laing
	G-SKPG	Best Off Skyranger 912 (2)	P. Gibbs
	G-SKRG	Best Off Skyranger 912 (2)	R. W. Goddin (G-AZIT)
	G-SKUL	SA.341G Gazelle	Flaming Skull Aviation Ltd
	G-SKYC	Slingsby T.67M Firefly	T. W. Cassells (G-BLDP)
	G-SKYE	Cessna TU.206G	RAF Sport Parachute Association
	G-SKYF	SOCATA TB.10 Tobago	Air Touring Ltd/Biggin Hill
	G-SKYG	I.I.I. Sky Arrow 650TC	R. Jones
	G-SKYK	Cameron A-275 balloon	Cameron Flights Southern Ltd
	G-SKYL	Cessna 182S	Skylane Aviation Ltd/Sherburn
	G-SKYM	Cessna F.337E	Bencray Ltd (G-AYHW) (stored)/Blackpool
	G-SKYN	AS.355F-1 Twin Squirrel	Arena Aviation Ltd (G-OGRK/G-BWZC/G-MODZ)
	G-SKYO	Slingsby T.67M-200	E. D. Fern
	G-SKYR	Cameron A-180 balloon	Cameron Flights Southern Ltd
	G-SKYT	I.I.I. Sky Arrow 650TC	W. M. Bell & S. J. Brooks
	G-SKYU	Cameron A-210 balloon	Cameron Flights Southern Ltd
	G-SKYV	PA-28RT-201T Turbo Arrow IV	Skyviews Pictures Ltd (G-BNZG)
	G-SKYX	Cameron A-210 balloon	Cameron Flights Southern Ltd
	G-SKYY	Cameron A-275 balloon	Cameron Flights Southern Ltd
	G-SLCE	Cameron C-80 balloon	A. M. Holly
	G-SLEA	Mudry/CAARP CAP.10B	M. J. M. Jenkins
	G-SLII	Cameron O-90 balloon	R. B. & A. M. Harris
	G-SLIP	Easy Raider	J. S. Harris
	G-SLMG	Diamond HK.36 TTC Super Dimona	B. D. James
	G-SLOW	Pietenpol Air Camper	C. Newton

Reg.	Type	Owner or Operator
G-SLTN	SOCATA TB.20 Trinidad	Oceana Air Ltd
G-SLYN	PA-28-161 Warrior II	Haimoss Ltd
G-SMAN	Airbus A.330-243	Monarch Airlines Ltd
G-SMBM	Pegasus Quantum 15-912	M. C. Watson
G-SMDH	Shaw Europa XS	S. W. Pitt
G-SMDJ	AS.350B-2 Ecureuil	Denis Ferranti Hoverknights Ltd
G-SMIG	Cameron O-65 balloon	R. D. Parry
G-SMJJ	Cessna 414A	Gull Air Ltd/Guernsey
G-SMTC	Colt Flying Hut SS balloon	Shiplake Investments Ltd/Switzerland
G-SMTH	PA-28 Cherokee 140	Aerosen/Southend (G-AYJS)
G-SMTJ	Airbus A.321-211	My Travel Airways
G-SNAK	Lindstrand LBL-105A balloon	Ballooning Adventures Ltd
G-SNAP	Cameron V-77 balloon	C. J. S. Limon
G-SNEV	CFM Streak Shadow SA	Aviation for Paraplegics & Tetraplegics Trust
G-SNOG	Kiss 400-582 (1)	B. H. Ashman
G-SNOW	Cameron V-77 balloon	M. J. Ball
G-SNOZ	Shaw Europa	M. P. Wiseman (G-DONZ)
G-SNUZ	PA-28-161 Warrior II	J. C. O. & C. A. Adams (G-PSFT/G-BPDS)
G-SOAY	Cessna T.303	Wrekin Construction Co. Ltd
G-SOBI	PA-28-181 Archer II	Northern Aviation Ltd
G-SOCK	Mainair Pegasus Quik	P. W. Lupton
G-SOCT	Yakovlev Yak-50	C. R. Turton
G-SOEI	H.S.748 Srs 2A	Emerald Airways Ltd/Liverpool
G-SOFT	Thunder Ax7-77 balloon	A. J. Bowen
G-SOHI	Agusta A.109E	Tri-Ventures Group Ltd
G-SOHO	Diamond DA.400 Star	Diamond Aircraft UK Ltd/Gamston
G-SOKO	Soko P-2 Kraguj (30149)	A. G. & G. A. G. Dixon (G-BRXK)
G-SOLA	Aero Designs Star-Lite SL.1	J. P. Roberts-Lethaby
G-SOLH	Bell 47G-5	SOL Helicopters Ltd (G-AZMB)
G-SOLO	Pitts S-2S Special	H. Staltmeir/Germany
G-SONA	SOCATA TB.10 Tobago	M. Kelly (G-BIBI)/Breighton
G-SOOC	Hughes 369HS	R.J.H. Strong (G-BRRX)
G-SOOE	Hughes 369E	R. W. Nash
G-SOOS	Colt 21A balloon	P. J. Stapley
G-SOOT	PA-28 Cherokee 180	J. A. Bridger/Exeter (G-AVNM)
G-SOPP	Enstrom 280FX	F. J. Sopp (G-OSAB)
G-SORT	Cameron N-90 balloon	A. Brown
G-SOUL	Cessna 310R	Atlantic Air Transport Ltd/Coventry
G-SPAL	Robinson R-44-II	Productive Investments Ltd
G-SPAM	Light Aero Avid Aerobat	J. Lee
G-SPAT	Aero AT-3 R100	S2T Aero Ltd
G-SPDR	D.H.115 Sea Vampire T.22 (N6-766)	M. J. Cobb/Bournemouth
G-SPEE	Robinson R-22B	Verve Systems Ltd (G-BPJC)
G-SPEL	Sky 220-24 balloon	Pendle Balloon Co
G-SPEY	AB-206B JetRanger 3	Castle Air Charters Ltd (G-BIGO)
G-SPFX	Rutan Cozy	B. D. Tutty
G-SPHU	Eurocopter EC.135T-2	Bond Air Services Ltd
G-SPIN	Pitts S-2A Special	N. M. R. Richards
G-SPIT	V.S.379 Spitfire FR.XIV (MV268)	Patina Ltd (G-BGHB)/Duxford
G-SPOG	Jodel DR.1050	A. C. Frost (G-AXVS)
G-SPUR	Cessna 550 Citation II	Amsail Ltd
G-SPYI	Bell 206B JetRanger 3	Heli-bott Ltd (G-BVRC/G-BSJC)
G-SRII	Easy Raider 503	Sierra Romeo India India Group
G-SROE	Westland Scout AH.1 (XP907)	Bolenda Engineering Ltd
G-SRVO	Cameron N-90 balloon	Servo & Electronic Sales Ltd
G-SRWN	PA-28-181 Warrior II	S. Smith (G-MAND/G-BRKT)
G-SRYY	Shaw Europa XS	S. R. Young
G-SSAS	Airbus A.320-231	My Travel Lite (G-BYES)
G-SSCL	MDH Hughes 369E	Shaun Stevens Contractors Ltd
G-SSEA	ATR 42-300	Air Wales Ltd
G-SSIX	Rans S.6-116 Coyote II	R. I. Kelly
G-SSKY	BN-2B-26 Islander	Isles of Scilly Skybus Ltd (G-BSWT)
G-SSLF	Lindstrand LBL-210A balloon	Exclusive Ballooning
G-SSPP	Sky Science Powerhawk L70/500	Sky Science Powered Parachutes Ltd
G-SSSC	Sikorsky S-76C	CHC Scotia Ltd
G-SSSD	Sikorsky S-76C	CHC Scotia Ltd

Notes	Reg.	Type	Owner or Operator
	G-SSSE	Sikorsky S-76C	CHC Scotia Ltd
	G-SSTI	Cameron N-105 balloon	British Airways
	G-SSWA	Short SD3-30 Variant 100	Emerald Airways Ltd (G-BHHU)
	G-SSWB	Short SD3-60 Variant 100	Emerald Airways Ltd (G-BMLE)
	G-SSWC	Short SD3-60 Variant 100	Emerald Airways Ltd (G-BMHX)
	G-SSWE	Short SD3-60 Variant 100	Emerald Airways Ltd
	G-SSWM	Short SD3-60 Variant 100	Emerald Airways Ltd (G-OAAS/G-BLIL)
	G-SSWO	Short SD3-60 Variant 100	Emerald Airways Ltd (G-BKMY)
	G-SSWR	Short SD3-60 Variant 100	Emerald Airways Ltd (G-BLWJ)
	G-SSWV	Sportavia Fournier RF-5B	N. Fisher & Arhey
	G-SSXX	Eurocopter EC.135 T-2	Bond Air Services Ltd(G-SSSX)
	G-STAF	Van's RV-7A	A. F. Stafford
	G-STAY	Cessna FR.172K	G. A. Owston
	G-STCH	Fiesler Fi 156A-1 Storch	G. R. Lacey
	G-STEA	PA-28R Cherokee Arrow 200	D. J. Brown
	G-STEM	Stemme S.10V	G-STEM Group
	G-STEN	Stemme S.10 (4)	G-STEN Syndicate
	G-STEP	Schweizer 269C	M. Johnson
	G-STER	Bell 206B JetRanger 3	Maintopic Ltd
	G-STEV	Jodel DR.221	S. W. Talbot/Long Marston
	G-STIG	Focke Wulf Fw-44J Steiglitz	G. R. Lacey
	G-STMP	SNCAN Stampe SV-4A	A,. C. Thorne
	G-STOK	Colt 77B balloon	A. C. Booth
	G-STOO	Stolp Starduster Too	K. F. Crumplin
	G-STOR	Fieseler Fi 156D-0 Storch	G. R. Lacey
	G-STOT	Robinson R-44-II	Howard Stott Demolition Ltd
	G-STOW	Cameron 90 Wine Box SS balloon	Flying Enterprise Partnership
	G-STPI	Cameron A-210 balloon	A. D. Pinner
	G-STRA	Boeing 737-3S3	Astraeus Ltd (G-OBWY/G-DEBZ/G-BNPB)
	G-STRB	Boeing 737-3Y0	Astraeus Ltd (G-OBWY/G-MONL)
	G-STRE	Boeing 737-36N	Astraeus Ltd (G-XBHX)
	G-STRF	Boeing 737-76N	Astraeus Ltd
	G-STRH	Boeing 737-36N	Astraeus Ltd
	G-STRK	CFM Streak Shadow Srs SA	E. J. Hadley/Switzerland
	G-STRL	AS.355N Twin Squirrel	
	G-STRM	Cameron N-90 balloon	High Profile Balloons
	G-STUA	Aerotek Pitts S-2A Special (modified)	Rollquick Group
	G-STUB	Christen Pitts S-2B Special	P. T. Borchert
	G-STUK	Junkers Ju.87/R4	G. R. Lacey
	G-STUY	Robinson R-44- II	Heli Air Ltd/Wellesbourne
	G-STWO	ARV Super 2	G. E. Morris
	G-STYL	Pitts S-1S Special	P. D. Albrow
	G-STYX	Mainair Pegasus Quik	T. A. E. Stewart
	G-SUCH	Cameron V-77 balloon	D. G. Such (G-BIGD)
	G-SUCK	Cameron Z-105 balloon	Airship & Balloon Co Ltd
	G-SUEB	PA-28-181	Saxon Logistics Ltd
	G-SUED	Thunder Ax8-90 balloon	E. C. Lubbock & S. A. Kidd (G-PINE)
	G-SUEW	Airbus A.320-214	My Travel Airways
	G-SUEY	Bell 206L-1 Long Ranger	Aerospeed Ltd
	G-SUEZ	AB-206B JetRanger 2	Aerospeed Ltd
	G-SUFF	Eurocopter EC.135T-1	Suffolk Constabulary Air Support Unit
	G-SUKI	PA-38-112 Tomahawk	Western Air (Thruxton) Ltd (G-BPNV)
	G-SUMT	Robinson R-22B	Aero Maintenance Ltd (G-BUKD)
	G-SUMX	Robinson R-22B	Frankham Bros Ltd
	G-SUMZ	Robinson R-44	Frankham Bros Ltd
	G-SUNN	Robinson R-44	Helicentre Ltd/Blackpool
	G-SUPA	PA-18 Super Cub 150	Supa Group
	G-SURG	PA-30 Twin Comanche 160B	A. R. Taylor (G-VIST/G-AVHG)/Kidlington
	G-SURY	Eurocopter EC.135T-2	Surrey Police Authority
	G-SUSE	Shaw Europa XS	P. R. Tunney
	G-SUSI	Cameron V-77 balloon	J. H. Dryden
	G-SUSX	MDH MD-902 Explorer	Sussex Police Authority
	G-SUSY	P-51D-25-NA Mustang (472773)	E. A. Morgan
	G-SUTN	I.I.I. Sky Arrow 650TC	G. C. Sutton & M. A. Coltman
	G-SUZN	PA-28-161 Warrior II	The St. George Flying Club/Teesside
	G-SUZY	Taylor JT.1 Monoplane	N. Gregson
	G-SVEA	PA-28-161 Warrior II	E-C. V. Dunning
	G-SVET	Yakovlev Yak-50	Yak 52 Ltd
	G-SVIV	SNCAN Stampe SV-4C	R. Taylor

Reg.	Type	Owner or Operator
G-SWEB	Cameron N-90 balloon	South Western Electricity PLC
G-SWEE	Beech 95-B55 Baron	Mirage Aircraft Leasing Ltd (G-AZDK)
G-SWEL	Hughes 369HS	M. A. Crook & A. E. Wright (G-RBUT)
G-SWIF	V.S.541 Swift F.7 (XF114) ★	Southampton Hall of Aviation (Solent Sky)
G-SWIS	D.H.100 Vampire FB.6 (J-1149) ★	Hunter Wing/Bournemouth
G-SWOT	Currie Wot (C3011)	P. M. Flint
G-SWPR	Cameron N-56 balloon	A. Brown
G-SWUN	Pitts S-1 Special (modified)	J. E. Rands (G-BSXH)
G-SWWM	Westland Gazelle HT.Mk.2	M. S. Beaton
G-SYCO	Shaw Europa	SYCO Syndicate
G-SYDD	PA-28-181 Archer III	Sherborne Aviation Ltd
G-SYFW	Focke-Wulf Fw.190 replica (2+1)	R. P. Cross
G-SYPA	AS.355F-2 Twin Squirrel	British International (G-BPRE)
G-SYPS	MDH MD.900 Explorer	South Yorkshire Police Authority
G-SYTN	Robinson R-44	Silverstar Components Ltd
G-TAAL	Cessna 172R	Standard Aviation Ltd
G-TABS	EMB-110P1 Bandeirante	Skydrift Ltd (G-PBAC)
G-TACK	Grob G.109B	A. P. Mayne
G-TADC	Aeroprakt A.22 Foxbat	R. J. Sharp
G-TAFF	C.A.S.A. 1.131E Jungmann 1000	A. J. E. Smith (G-BFNE)/Breighton
G-TAFI	Bücker Bu133C Jungmeister	R.P. Lamplough
G-TAGG	Eurocopter EC.135T-2	McAlpine Helicopters Ltd/Kidlington
G-TAGR	Shaw Europa XS	A. G. Rackstraw
G-TAGS	PA-28-161 Warrior II	Oxford Aviation Services Ltd/Kidlington
G-TAIL	Cessna 150J	L. I. D. Denham-Brown
G-TAIR	PA-34-200T Seneca II	Nigel Kenny Aviation Ltd
G-TAIT	Cessna 172R	Centenary Flying Group Ltd (G-DREY)
G-TAJF	Lindstrand LBL-77A balloon	T. A. J. Fowles
G-TAMR	Cessna 172S	Tamair Leasing
G-TAMS	Beech A23-24 Musketeer Super	Aerograde Ltd
G-TAMY	Cessna 421B	Charniere Ltd
G-TAND	Robinson R-44	Southwest Helicharter Ltd
G-TANI	GA-7 Cougar	S. Spier (G-VJAI/G-OCAB/G-BICF)/Elstree
G-TANK	Cameron N-90 balloon	Hoyers (UK) Ltd
G-TANS	SOCATA TB-20 Trinidad	Tettenhall Leisure
G-TANY	EAA Acrosport 2	P. J. Tanulak
G-TAPE	PA-23 Aztec 250D	D. J. Hare (G-AWVW)
G-TAPS	PA-28RT-201T Turbo Arrow IV	P. G. Doble
G-TARN	Pietenpol Air Camper	P. J. Heilbron
G-TART	PA-28-236 Dakota	Prescot Planes Ltd
G-TARV	ARV Super 2	M. F. Filer (G-OARV)
G-TASH	Cessna 172N (modified)	A. Ashpitel
G-TASK	Cessna 404	Bravo Aviation Ltd
G-TASS	Schweizer 269C	A. Tasker
G-TATT	GY-20 Minicab	Tatt's Group
G-TATY	Robinson R-44	W. R. Walker
G-TAWE	Aerospatiale ATR-42-300	Air Wales Ltd (G-BVJP)
G-TAXI	PA-23 Aztec 250E	SWL Leasing Ltd/Leeds
G-TAYI	Grob G.115	K. P. Widdowson & K. Hackshall (G-DODO)
G-TAYS	Cessna F.152 II	Tayside Aviation Ltd (G-LFCA)/Aberdeen
G-TBAE	BAe 146-200	BAe Systems (Corporate Travel Ltd) (G-HWPB/G-BSRU/G-OSKI/G-JEAR)
G-TBAG	Murphy Renegade II	M. R. Tetley
G-TBAH	Bell 206B JetRanger 2	RB Helicopters (G-OMJB)
G-TBBC	Pegasus Quantum 15-912	J. Horn
G-TBEE	Dyn'Aero MCR-01	A. D. S. Baker
G-TBGL	Agusta A.109A-II	Bulford Holdings Ltd (G-VJCB/G-BOUA)
G-TBGT	SOCATA TB.10 Tobago GT	P. G. Sherry & A. J. Simmonds/Liverpool
G-TBIC	BAe 146-200	Flightline Ltd
G-TBIO	SOCATA TB.10 Togago	Delta Bird Aviation Ltd
G-TBJP	Mainair Pegasus Quik	B. J. Partridge
G-TBMW	Murphy Renegade Spirit	S. J. & M. J. Spavins (G-MYIG)
G-TBOK	SOCATA TB.10 Tobago	TB -10 Ltd
G-TBRD	Lockheed T-33A (21261)	Golden Apple Operations Ltd (G-JETT/G-OAHB)/ Duxford
G-TBTN	SOCATA TB.10 Tobago	J. S. Chaggar (G-BKIA)
G-TBXX	SOCATA TB.20 Trinidad	D. A. Phillips
G-TBZI	SOCATA TB.21 Trinidad TC	Skypartners UK Ltd
G-TBZO	SOCATA TB.20 Tobago	R. P. Lewis & D. L. Clarke

Notes	Reg.	Type	Owner or Operator
	G-TCAN	Colt 69A balloon	H. C. J. Williams
	G-TCAP	BAe 125 Srs 800B	BAE Systems Ltd
	G-TCKE	Airbus A.320-214	Thomas Cook Airlines UK Ltd
	G-TCMM	Agusta-Bell 206B-3 Jet Ranger III	Westair Aviation Ltd (G-JMVB/G-OIML)
	G-TCNM	Tecnam P.92-EA Echo	J. Quaife
	G-TCNY	Mainair Pegasus Quik	T. Butler
	G-TCOM	PA-30 Twin Comanche 160B	C. A. C. Burrough
	G-TCTC	PA-28RT-200 Arrow IV	T. Haigh
	G-TCUB	Piper J-3C-65 Cub	C. Kirk
	G-TDOG	SA Bulldog Srs 120/121	G. S. Taylor
	G-TDVB	Dyn' Aero MCR-01ULC	D. V. Brunt
	G-TEBZ	PA-28R-201 Arrow III	S. F. Tebby & Son
	G-TECC	Aeronca 7AC Champion	N. J. Orchard-Armitage
	G-TECH	R. Commander 114	P. A. Reed (G-BEDH)/Denham
	G-TEDI	Best Off Skyranger	K. Lorenzen
	G-TECK	Cameron V-77 balloon	M. W. A. Shemilt
	G-TECM	Tecnam P.92-EM Echo	D. A. Lawrence
	G-TECS	Tecnam P.92-EA Echo Super	D. A. Lawrence
	G-TEDF	Cameron N-90 balloon	Fort Vale Engineering Ltd
	G-TEDS	SOCATA TB.10 Tobago	G-TEDS Group Aviation (G-BHCO)
	G-TEDW	Kiss 450-582 (2)	D.J. Wood
	G-TEDY	Evans VP-1	N. K. Marston (G-BHGN)
	G-TEFC	PA-28 Cherokee 140	P. M. Havard
	G-TEHL	CFM Streak Shadow Srs M	A. K. Paterson (G-MYJE)
	G-TELY	Agusta A.109A-II	Castle Air Charters Ltd
	G-TEMP	PA-28 Cherokee 180	BEV Group (G-AYBK)/Andrewsfield
	G-TEMT	Hawker Tempest II (MW763)	Tempest Two Ltd/Gamston
	G-TENG	Extra EA.300/L	10G Aerobatics Ltd
	G-TENT	J/1N Alpha	R. Callaway-Lewis (G-AKJU)
	G-TERN	Shaw Europa	J. Smith
	G-TERR	Mainair Pegasus Quik	T. R. Thomas
	G-TERY	PA-28-181 Archer II	J. R. Bratherton (G-BOXZ)
	G-TEST	PA-34-200 Seneca	Stapleford Flying Club Ltd (G-BLCD)
	G-TETI	Cameron N-90 balloon	Teti SPA/Italy
	G-TEWS	PA-28 Cherokee 140	G-TEWS Flying Group (G-KEAN/G-AWTM)
	G-TEXS	Van's RV-6	W. H. Greenwood
	G-REXT	Robinson R-44-II	Heli Air Ltd/Wellesbourne
	G-TFCI	Cessna FA-152	Tayside Aviation Ltd/Dundee
	G-TFIN	PA-32RT-300T Turbo Lance II	M. D. Parker
	G-TFIX	Pegasus Mainair Quantum 15-912	
	G-TFOX	Denney Kitfox Mk.2	F. A. Bakir
	G-TFRB	Air Command 532 Elite	F. R. Blennerhassett
	G-TFSA	Cessna F.152 II	S. J. George (G-BITH)
	G-TFUN	Valentin Taifun 17E	North West Taifun Group
	G-TFYN	PA-32RT-300 Lance II	R. C. Poolman
	G-TGAS	Cameron O-160 balloon	Zebedee Balloon Service
	G-TGER	AA-5B Tiger	Photonic Science Ltd (G-BFZP)/Biggin Hil
	G-TGGR	Eurocopter EC.120B	Blue Five Aviation Ltdl
	G-TGRA	Agusta A.109A	Tiger Helicopters Ltd
	G-TGRD	Robinson R-22B-II	Tiger Helicopters Ltd (G-OPTS)
	G-TGRE	Robinson R-22A	Tiger Helicopters Ltd (G-SOLD)
	G-TGRR	Robinson R-22B	Tiger Helicopters Ltd (G-BSZS)
	G-TGRZ	Bell 206B JetRanger 3	Tiger Helicopters Ltd (G-BXZX)/Shobdon
	G-THAI	CFM Shadow Srs D	D. L. Hendry
	G-THAT	Raj Hamsa X'Air Falcon 912 (1)	M. G. Thatcher
	G-THEL	Robinson R-44	N. Parkhouse (G-OCCB/G-STMM)
	G-THEO	Team Minimax 91	C. Fletcher
	G-THIN	Cessan FR.172E	IC. A. Ussher (G-BXYY)
	G-THLA	Robinson R-22B	Thurston Helicopters Ltd
	G-THOA	Boeing 737-5L9	Thomsonfly (G-MSKA)
	G-THOB	Boeing 737-5L9	Thomsonfly (G-MSKB
	G-THOC	Boeing 737-59D	Thomsonfly (G-BVKA)
	G-THOD	Boeing 737-59D	Thomsonfly (G-BVKC)
	G-THOE	Boeing 737-3Q8	Thomsonfly (G-BZZZH)
	G-THOF	Boeing 737-3Q8	Thomsonfly (G-BZZI)
	G-THOM	Thunder Ax-6-56 balloon	T. H. Wilson
	G-THOS	Thunder Ax7-77 balloon	C. E. A. Breton
	G-THOT	Avtech Jabiru SK	D. J. Reed

Reg.	Type	Owner or Operator
G-THRE	Cessna 182S	S. J. Mole
G-THSL	PA-28R-201 Arrow III	D. M. Markscheffe
G-THUN	Republic P-47D Thunderbolt (226671)	Patina Ltd/Duxford
G-THZL	SOCATA TB.20	Thistle Aviation Ltd
G-TIDS	Jodel 150	M. R. Parker
G-TIGA	D.H.82 Tiger Moth	D. E. Leatherland (G-AOEG)/Tollerton
G-TIGB	AS.332L Super Puma	Bristow Helicopters Ltd (G-BJXC)
G-TIGC	AS.332L Super Puma	Bristow Helicopters Ltd (G-BJYH)
G-TIGE	AS.332L Super Puma	Bristow Helicopters Ltd (G-BJYJ)
G-TIGF	AS.332L Super Puma	Bristow Helicopters Ltd
G-TIGG	AS.332L Super Puma	Bristow Helicopters Ltd
G-TIGI	AS.332L Super Puma	Bristow Helicopters Ltd
G-TIGJ	AS.332L Super Puma	Bristow Helicopters Ltd
G-TIGO	AS.332L Super Puma	Bristow Helicopters Ltd
G-TIGP	AS.332L Super Puma	Bristow Helicopters Ltd
G-TIGR	AS.332L Super Puma	Bristow Helicopters Ltd
G-TIGS	AS.332L Super Puma	Bristow Helicopters Ltd
G-TIGT	AS.332L Super Puma	Bristow Helicopters Ltd
G-TIGV	AS.332L Super Puma	Bristow Helicopters Ltd
G-TIGZ	AS.332L Super Puma	Bristow Helicopters Ltd
G-TIII	Aerotek Pitts S-2A Special	D. G. Cowden (G-BGSE)
G-TIKO	Hatz CB-1	Tiko Architecture
G-TILE	Robinson R-22B	Fenland Helicopters Ltd
G-TILI	Bell 206B JetRanger 2	CIM Helicopters
G-TIMB	Rutan Vari-Eze	T. M. Bailey (G-BKXJ)
G-TIME	Ted Smith Aerostar 601P	T & G Engineering Co. Ltd
G-TIMK	PA-28-181 Archer II	T. Baker
G-TIMM	Folland Gnat T.1 (XM693)	T. J. Manna/Cranfield
G-TIMP	Aeronca 7BCM Champion	M. G. Rummey
G-TIMS	Falconar F-12A	T. Sheridan
G-TIMY	GY080 Horizon 160	R. G. Whyte
G-TINA	SOCATA TB.10 Tobago	A. Lister
G-TING	Cameron O-120 balloon	Floating Sensations Ltd
G-TINS	Cameron N-90 balloon	J. R. Clifton
G-TINY	Z.526F Trener Master	D. Evans
G-TIPS	Tipsy T.66 Nipper Srs 5	R. F. L. Cuypers/Belgium
G-TIVS	Rans S.6ES Coyote II	S. Hoyle
G-TJAL	Jabiru SPL-430	D. W. Cross
G-TJAV	Mainair Pegasus Quik	Red Communications Ltd
G-TJAY	PA-22 Tri-Pacer 135	D. D. Saint
G-TKAY	Shaw Europa	A. M. Kay
G-TKGR	Lindstrand LBL Racing Car SS balloon	Brown & Williams Tobacco Corporation (Export) Ltd/USA
G-TKIS	Tri-R Kis	T. J. Bone
G-TKPZ	Cessna 310R	Fraggle Leasing Ltd (G-BRAH)
G-TLBC	SOCATA MS.892A Rallye 150	Film Funding Inc
G-TLDK	PA-22 Tri-Pacer 150	A. M. Thomson
G-TLEL	American Blimp Corpn A.60+ airship	Keelex 195 Ltd
G-TLET	PA-28-161 Cadet	Meridiian Aviiation Sales Ltd (G-GFCF/G-RHBH)
G-TMCB	Best Off Skyranger 912 (2)	A. H. McBreen
G-TMCC	Cameron N-90 balloon	Prudential Assurance Co. Ltd
G-TMKI	P.56 Provost T.1 (WW453)	B. L. Robinson
G-TMOL	SOCATA TB.20 Trinidad	West Wales Airport Ltd
G-TNTN	Thunder Ax6-56 balloon	H. M. Savage & J. F. Trehern
G-TOAD	Jodel D.140B	J. H. Stevens
G-TOAK	SOCATA TB.20 Trinidad	Phoenix Group
G-TOBA	SOCATA TB.10 Tobago	E. Downing
G-TOBI	Cessna F.172K	A. I. Bird (G-AYVB)
G-TODD	ICA IS-28M2A	C. I. Roberts & C. D. King/Shobdon
G-TODE	Ruschmeyer R.90-230RG	A. I. D. Rich
G-TOFT	Colt 90A balloon	C. S. Perceval
G-TOGO	Van's RV-6	G. Schwetz
G-TOHS	Cameron V-31 balloon	J. P. Moore
G-TOLL	PA-28R-201 Arrow III	Plymouth School of Flying Ltd
G-TOLY	Robinson R-22B	Helicopter Services Ltd (G-NSHR)
G-TOMC	NA AT-6D Harvard III	A. A. Marshall

Notes	Reg.	Type	Owner or Operator
	G-TOMJ	Flight Design CT.2K	P. T. Knight
	G-TOMM	Robinson R-22B	Airfleet Aircraft Leasing Ltd
	G-TOMS	PA-38-112 Tomahawk	P. Millar
	G-TOMZ	Denney Kitfox Mk.2	S. Austen
	G-TONN	Mainair Pegasus Quik	The Windmill Kennels & Cattery Ltd
	G-TONS	Slingsby T.67M-200	D. I. Stanbridge
	G-TOOL	Thunder Ax8-105 balloon	D. V. Howard
	G-TOOT	Dyn'Aero MCR-01	E. K. Griffin
	G-TOPC	AS.355F-1 Twin Squirrel	Bridge Street Nominees Ltd
	G-TOPK	Shaw Europa	P. J. Kember
	G-TOPS	AS.355F-1 Twin Squirrel	Sterling Helicopters (G-BPRH)
	G-TORC	PA-28R Cherokee Arrow 200	Haimoss Ltd
	G-TORE	P.84 Jet Provost T.3A (XM405) ★	Instructional airframe City University
	G-TORS	Robinson R-22B	IW Aviation Ltd
	G-TOSH	Robinson R-22B	Heli Air Ltd/Wellesbourne
	G-TOTN	Cessna 210M	Just Plane Trading Ltd (G-BVZM)
	G-TOTO	Cessna F.177RG	Horizon Flyers Ltd (G-OADE/G-AZKH)
	G-TOUR	Robin R.2112	Mardenair Ltd
	G-TOWS	PA-25 Pawnee 260	Lasham Gliding Soc Ltd
	G-TOYA	Boeing 737-3Q8	bmi Baby (G-BZZE)
	G-TOYB	Boeing 737-3Q8	bmi Baby (G-BZZF)
	G-TOYC	Boeing 737-3Q8	bmi Baby (G-BZZG)
	G-TOYD	Boeing 737-33V	bmi Baby (G-EZYT)
	G-TOYZ	Bell 206B JetRanger 3	A. R. Pocock (G-RGER)
	G-TPSL	Cessna 182S	A. N. Purslow/Blackbushe
	G-TRAC	Robinson R-44	C. J. Sharples
	G-TRAM	Pegasus Quantum 15-912	I. W. Barlow
	G-TRAN	Beech 76 Duchess	Multiflight Ltd (G-NIFR)
	G-TRCY	Robinson R-44	Beauville BV/Netherlands
	G-TRDM	SOCATA TB.20 Trinidad	West Wales Airport Ltd/Shobdon
	G-TREC	Cessna 421C	Sovereign Business Integration PLC (G-TLOL)
	G-TRED	Colt 110 Bibendum balloon	The Aerial Display Co. Ltd
	G-TREE	Bell 206B JetRanger 3	Bush Woodlands
	G-TREK	Jodel D.18	R. H. Mole/Leicester
	G-TRIB	Lindstrand HS-110 airship	J. Addison
	G-TRIC	D.H.C.1 Chipmunk 22A (18013)	D. M. Barnett (G-AOSZ)
	G-TRIG	Cameron Z-90 balloon	Trigger Concepts Ltd
	G-TRIM	Monnett Moni	J. E. Bennell
	G-TRIN	SOCATA TB.20 Trinidad	TL Aviation Ltd
	G-TRIO	Cessna 172M	C. M. B. Reid (G-BNXY)
	G-TRNT	Robinson R-44-II	Charles Trent Ltd
	G-TROP	Cessna 310R II	D. E. Carpenter/Shoreham
	G-TROY	NA T-28A Fennec (51-7692)	S. G. Howell & S. Tilling
	G-TRUD	Enstrom 480	Sussex Aviation Ltd
	G-TRUE	MDH Hughes 369E	Bailey Employment Services Ltd
	G-TRUK	Stoddard-Hamilton Glasair RG	M. P. Jackson
	G-TRUX	Colt 77A balloon	M. J. Forster
	G-TRYG	Robinson R-44	Nottinghamshire Helicopters (2004) Ltd
	G-TRYK	Kiss 400-582 (1)	S. Elsbury
	G-TSAM	BAe 125 Srs 800B	BAE Systems (Operations) Ltd/Warton
	G-TSGJ	PA-28-181 Archer II	Golf Juliet Flying Club
	G-TSIX	AT-6C Harvard IIA (111836)	S. J. Davies
	G-TSKD	Raj Hamsa X'Air Jabiru J.2.2.	T. Sexton & K. B. Dupuy
	G-TSKY	B.121 Pup 2	R. G. Hayes (G-AWDY)
	G-TSOB	Rans S.6ES Coyote II	S. Luck
	G-TSOL	EAA Acrosport 1	A. G. Fowles (G-BPKI)
	G-TTAC	SOCATA TB.20 Trinidad GT	AC Aviation Ltd
	G-TTDD	Zenair CH.701 STOL	D. B. Dainton & V. D. Asque
	G-TTFN	Cessna 560 Citation V	Corporate Administration Management Ltd
	G-TTHC	Robinson R-22B	Multiflight Ltd/Leeds-Bradford
	G-TTIA	Airbus A.321-231	GB Airways Ltd
	G-TTIB	Airbus A.321-231	GB Airways Ltd
	G-TTIC	Airbus A.321-231	GB Airways Ltd
	G-TTID	Airbus A.321-231	GB Airways Ltd
	G-TTIE	Airbus A.321-231	GB Airways Ltd
	G-TTMB	Bell 206B JetRanger 3	Helirentals (G-RNME/G-CBDF)/Chelmsford
	G-TTOA	Airbus A.320-232	GB Airways Ltd
	G-TTOB	Airbus A.320-232	GB Airways Ltd

Reg.	Type	Owner or Operator	Notes
G-TTOC	Airbus A.320-232	GB Airways Ltd	
G-TTOD	Airbus A.320-232	GB Airways Ltd	
G-TTOE	Airbus A.320-232	GB Airways Ltd	
G-TTOF	Airbus A.320-232	GB Airways Ltd	
G-TTOG	Airbus A.320-232	GB Airways Ltd	
G-TTOH	Airbus A.320-232	GB Airways Ltd	
G-TTOI	Airbus A.320-232	GB Airways Ltd	
G-TTOJ	Airbus A.320-232	GB Airways Ltd	
G-TTOY	CFM Streak Shadow SA	D. & B. D. C Barnard.	
G-TUBB	Avtech Jabiru UL	A. H. Bower	
G-TUCK	Van's RV-8	M. A. Tuck	
G-TUDR	Cameron V-77 balloon	Jacques W. Soukup Enterprises Ltd	
G-TUGG	PA-18 Super Cub 150	Ulster Gliding Club Ltd/Bellarena	
G-TUGY	Robin DR.400/180	Buckminster Gliding Club/Saltby	
G-TULP	Lindstrand LBL Tulips SS balloon	Oxford Promotions (UK) Ltd	
G-TUNE	Robinson R-22B	Ecurie Ecosse (Scotland) Ltd (G-OJVI)	
G-TURF	Cessna F.406	Atlantic Air Transport Ltd/Coventry	
G-TURN	Steen Skybolt	G-TURN Flying Group	
G-TUSA	Pegasus Quantum 15-912	N. J. Holt	
G-TUSK	Bell 206B JetRanger 3	Heli Aviation Ltd (G-BWZH)/Blackbushe	
G-TVAA	Agusta A.109E Power	Agusta SpA/Italy	
G-TVAM	MBB Bo105DBS-4	Bond Air Services Ltd (G-SPOL)	
G-TVBF	Lindstrand LBL-310A balloon	Virgin Balloons Flights	
G-TVII	Hawker Hunter T.7 (XX467)	G-TVII Group/Exeter	
G-TVIJ	CCF Harvard IV (T-6J) (28521)	R. W. Davies (G-BSBE)	
G-TVIP	Cessna 404	Capital Trading (Aviation) Ltd (G-KIWI/G-BHNI)	
G-TVTV	Cameron 90 TV SS balloon	J. Krebs/Germany	
G-TWEL	PA-28-181 Archer II	International Aerospace Engineering Ltd	
G-TWEY	Colt 69A balloon	N. Bland	
G-TWIG	Cessna F.406	Bravo Aviation Ltd	
G-TWIN	PA-44-180 Seminole	Bonus Aviation Ltd/Cranfield	
G-TWIZ	R. Commander 114	B. C. & P. M. Cox	
G-TWST	Silence Twister	Zulu Glasstek Ltd	
G-TWTW	Denney Kitfox Mk.2	T. Willford	
G-TXSE	RAF 2000 GTX-SE gyroplane	G. J. Layzell	
G-TYAK	IDABacau Yakovlev Yak-52	S. J. Ducker	
G-TYCN	Agusta A.109E Power	A. J. Walter (Aviation) Ltd (G-VMCO)	
G-TYER	Robin DR.400/500	Alfred Graham Ltd	
G-TYGA	AA-5B Tiger	D. H. & R. J. Carman (G-BHNZ)	
G-TYGR	Best Off Skyranger 912S (1)	M. J. Poole	
G-TYKE	Avtech Jabiru UL-450	A. Parker	
G-TYNE	SOCATA TB.20 Trinidad	D. T. Watkins	
G-TYRE	Cessna F.172M	Staverton Flying School	
G-TZEE	SOCATA TB.10 Tobago	Zytech Ltd	
G-TZII	Thorp T.211B	AD Aviation Ltd/Barton	
G-UACA	Skyranger R.100	R. G. Openshaw	
G-UAKE	NA P-51D-5-NA Mustang	Mustang Restoration Co Ltd	
G-UANT	PA-28 Cherokee 140	Air Navigation & Trading Co Ltd/Blackpool	
G-UAPA	Robin DR.400/140B	Carlos Saraive Lda/Portugal	
G-UAPO	Ruschmeyer R.90-230RG	P. Randall	
G-UAVA	PA-30 Twin Comanche	Small World Aviation Ltd	
G-UCCC	Cameron 90 Sign SS balloon	B. Conway	
G-UDAY	Robinson R-22B	D. J. Fowler	
G-UDGE	Thruster T.600N	G-UDGE Syndicate (G-BYPI)	
G-UDOG	SA Bulldog Srs 120/121	Gamit Ltd	
G-UEST	Bell 206B JetRanger 2	Leisure and Retail Helicopters (G-RYOB/G-BLWU)	
G-UESY	Robinson R-22B-2	E. W. Guess (Holdings) Ltd	
G-UFAW	Raj Hamsa X'Air 582 (5)	T. J. Butler	
G-UFCB	Cessna 172S	The Cambridge Aero Club Ltd	
G-UFCC	Cessna 172S	Oxford Aviation Services Ltd	

Notes	Reg.	Type	Owner or Operator
	G-UFCD	Cessna 172S	Oxford Aviation Services Ltd (G-OYZK)/Kidlington
	G-UFCG	Cessna 172S	Ulster Flying Club (1961) Ltd
	G-UFCH	Cessna 172S	Ulster Flying Club (1961) Ltd
	G-UFLY	Cessna F.150H	Westair Flying Services Ltd (G-AVVY)/Blackpool
	G-UGLY	SE.313B Alouette II	Helicopter Services (G-BSFN)
	G-UILD	Grob G.109B	Runnymede Consultants Ltd
	G-UILE	Lancair 320	R. J. Martin
	G-UILT	Cessna T.303	Rock Seat Ltd (G-EDRY)
	G-UINN	Stolp SA.300 Starduster Too	J. D. H. Gordon
	G-UIST	BAe Jetstream 3102	Highland Airways Ltd
	G-UJAB	Avtech Jabiru UL	C. A. Thomas
	G-UJGK	Avtech Jabiru UL	W. G. Upton & J. G. Kosak
	G-UKAG	BAe 146-300	(stored)
	G-UKAT	Aero AT-3	G-UKAT Group
	G-UKOZ	Avtech Jabiru SK	D. J. Burnett
	G-UKRB	Colt 105A balloon	Virgin Airship & Balloon Co Ltd
	G-UKRC	BAe 146-300	(stored)
	G-UKSC	BAe 146-300	Buzz Stansted Ltd
	G-UKTA	Fokker 50	(stored)
	G-UKTB	Fokker 50	(stored)
	G-UKTD	Fokker 50	(stored)
	G-UKUK	Head Ax8-105 balloon	P. A. George
	G-ULAS	D.H.C.1 Chipmunk 22 (WK517)	ULAS Flying Club Ltd/Denham
	G-ULES	AS.355 F.-2 Twin Squirrel	Select Plant Hire Company Ltd (G-OBHL)
	G-ULHI	SA Bulldog Srs.100/101	Power Aerobatics Ltd (G-OPOD/G-AZMS)
	G-ULIA	Cameron V-77 balloon	J. M. Dean
	G-ULLS	Lindstrand LBL-90A balloon	J. R. Clifton
	G-ULLY	Thruster T600N 450	P. J. Fahie (G-CCRP)
	G-ULPS	Everett Srs 1 gyroplane	C. J. Watkinson (G-BMNY)
	G-ULSY	Ikarus C.42 FB80	P. J. Fahie
	G-ULTR	Cameron A-105 balloon	P. Glydon
	G-UMBO	Thunder Ax7-77A balloon	Virgin Airship & Balloon Co Ltd
	G-UMMI	PA-31-310 Turbo Navajo	Messrs Rees of Poynston West (G-BGSO)
	G-UNDD	PA-23 Aztec 250E	G. J. & D. P. Deadman (G-BATX)
	G-UNER	Lindstrand LBL-90A balloon	Royal Artillery Display Troop
	G-UNGE	Lindstrand LBL-90A balloon	Silver Ghost Balloon Club (G-BVPJ)
	G-UNGO	Pietenpol Air Camper	A. R. Wyatt
	G-UNIV	Montgomerie-Parsons 2-seat gyroplane	University of Glasgow (G-BWTP)
	G-UNNY	BAC.167 Strikemaster 87	Transair (UK) Ltd (G-AYHR)/North Weald
	G-UNRL	Lindstrand LBL-21A balloon	Virgin Balloon & Airship Co. Ltd
	G-UNYT	Robinson R-22B	D. I. Pointon (G-BWZV/G-LIAN)
	G-UORO	Shaw Europa	D. Dufton
	G-UPFS	Waco UPS-7	D. N. Peters & N. R. Finlayson
	G-UPHL	Cameron 80 Concept SS balloon	Uphill Motor Co
	G-UPPI	BAC.167 Strikemaster Mk.80	Gower Jets Ltd (G-CBPB)
	G-UPPP	Colt 77A balloon	M. Williams
	G-UPPY	Cameron DP-80 airship	Jacques W. Soukup Enterprises Ltd/USA
	G-UPUP	Cameron V-77 balloon	S. F. Burden/Netherlands
	G-UPUZ	Lindstrand LBL-120A balloon	C.J. Sanger-Davies
	G-UROP	Beech 95-B55 Baron	Pooler International Ltd/Sleap
	G-URRR	Air Command 582 Sport	L. Armes
	G-URUH	Robinson R-44	Heli Air Ltd/Wellesbourne
	G-URUS	Maule MX7-180B Super Rocket	Broomco Ltd
	G-USAM	Cameron Uncle Sam SS balloon	Corn Palace Balloon Club Ltd
	G-USIL	Thunder Ax7-77 balloon	Window On The World Ltd
	G-USKY	Aviat A-18 Husky	ADR Aviation
	G-USMC	Cameron 90 Chestie SS balloon	Jacques W. Soukup Enterprises Ltd/USA
	G-USRV	Vans RV 6	W. H. Greenwood
	G-USSI	Stoddard-Hamilton Glasair III	Lord Rotherwick
	G-USSR	Cameron 90 Doll SS balloon	Corn Palace Balloon Club Ltd
	G-USSY	PA-28-181 Archer II	Western Air (Thruxton) Ltd
	G-USTB	Agusta A.109A	Newton Aviation Ltd

Reg.	Type	Owner or Operator	Notes
G-USTC	Agusta A.109C	MW Helicopters Ltd	
G-USTS	Agusta A.109A-II	MB Air Ltd (G-MKSF)	
G-USTY	FRED Srs 2	R. G. Hallam	
G-UTSI	Rand-Robinson KR-2	K. B. Gutridge/Thruxton	
G-UTSY	PA-28R-201 Arrow III	Arrow Aviation Ltd	
G-UTTS	Robinson R-44	Heli Hire Ltd (G-ROAP)	
G-UTZI	Robinson R-44-II	Heli Air Ltd	
G-UTZY	SA341G Gazelle 1	Gold Air International Ltd	
G-UVIP	Cessna 421C	Capital Trading Aviation (G-BSKH)/Filton	
G-UVNA	PA-24-260 Comanche	D. G. Sheppard (G-BAHG)	
G-UVNR	BAC.167 Strikemaster Mk 87	Global Aviation Services Ltd (G-BXFS)	
G-UZEL	SA.341G Gazelle 1	Fairalls of Godstone Ltd (G-BRNH)	
G-UZLE	Colt 77A balloon	Flying Pictures Ltd	
G-UZZY	Enstrom 480	Shoreham Helicopters (G-BWMD)	
G-VAEL	Airbus A.340-311	Virgin Atlantic Airways Ltd *Maiden Toulouse*	
G-VAGA	PA-15 Vagabond	I. M. Callier/White Waltham	
G-VAIR	Airbus A.340-313	Virgin Atlantic Airways Ltd *Maiden Tokyo*	
G-VALS	Pietenpol Air Camper	I. G. & V. A. Brice	
G-VALV	Robinson R-44	Valve Train Components	
G-VALY	SOCATA TB-21 Trinidad GT Turbo	Valley Flying Co. Ltd	
G-VALZ	Cameron N-120 balloon	D. Ling	
G-VANN	Van's RV-7A	D. N. & J. A. Carnegie	
G-VANS	Van's RV-4	M. Swanborough & D. Jones	
G-VANZ	Van's RV-6A	S. J. Baxter	
G-VARG	Varga 2150A Kachina	R. A. Denton	
G-VART	Rotorway Executive 90	G. Varty (G-BSUR)	
G-VAST	Boeing 747-41R	Virgin Atlantic Airways Ltd *Ladybird*	
G-VATL	Airbus A.340-642	Virgin Atlantic Airways Ltd	
G-VAUN	Cessna 340	A. J. Barham & G. S. Austin	
G-VBAC	Short SD3-60 Variant 100	BAC Leasing Ltd (G-BOEJ)	
G-VBEE	Boeing 747-219B	Air Atlanta Europe	
G-VBIG	Boeing 747-4Q8	Virgin Atlantic Airways Ltd *Tinker Belle*	
G-VBUS	Airbus A.340-311	Virgin Atlantic Airways Ltd *Lady in Red*	
G-VCED	Airbus A.320-231	My Travel Airways	
G-VCIO	EAA Acro Sport II	V. Millard	
G-VCML	Beech 58 Baron	St Angelo Aviation Ltd	
G-VDIR	Cessna T.310R	J. Driver	
G-VECD	Robin R.1180T	Mistral Aviation Ltd	
G-VECE	Robin R.2120U	Mistral Aviation Ltd	
G-VECG	Robin R.2160	Mistral Aviation Ltd	
G-VEEE	Robinson R-22B	Veee Helicopters Ltd (G-REDA)	
G-VEGA	Slingsby T.65A Vega	R. A. Rice (G-BFZN)	
G-VEIL	Airbus A.340-642	Virgin Atlantic Airways Ltd	
G-VEIT	Robinson R-44-II	Heli Air Ltd	
G-VELA	SIAI-Marchetti S.205-22R	Broadland Flyers Ltd	
G-VELD	Airbus A.340-313	Virgin Atlantic Airways Ltd *African Queen*	
G-VENI	D.H.112 Venom FB.50 (WE402)	Lindsay Wood Promotions Ltd/Bournemouth	
G-VENM	D.H.112 Venom FB.50 (WK436)	Kennet Aviation *(stored)* (G-BLIE)/Yeovilton	
G-VENT	Schempp-Hirth Ventus 2CM	D. Rance	
G-VERA	GY-201 Minicab	D. K. Shipton	
G-VERN	PA-32R-300 Cherokee Lance	P. M. Moyle (G-BVBG)	
G-VETA	Hawker Hunter T7	Gower Jets Ltd (G-BVWN)	
G-VETS	Enstrom 280C-UK Shark	A. J. Warburton (G-FSDC/G-BKTG)	
G-VEYE	Robinson R-22	RK Transport Services Ltd (G-BPTP)	
G-VEZE	Rutan Vari-Eze	S. D. Brown & ptnrs	
G-VFAB	Boeing 747-4Q8	Virgin Atlantic Airways Ltd *Lady Penelope*	
G-VFAR	Airbus A.340-313	Virgin Atlantic Airways Ltd *Diana*	
G-VFLY	Airbus A.340-311	Virgin Atlantic Airways Ltd *Dragon Lady*	
G-VFOX	Airbus A.340-642	Virgin Atlantic Airways Ltd	
G-VGAL	Boeing 747-443	Virgin Atlantic Airways Ltd	
G-VGMC	Eurocopter AS355N Twin Squirrel	Finlay (Holdings) Ltd (G-HEMH)	

Notes	Reg.	Type	Owner or Operator
	G-VGOA	Airbus A.340-642	Virgin Atlantic Airways Ltd
	G-VHOL	Airbus A.340-311	Virgin Atlantic Airways Ltd *Jetstreamer*
	G-VHOT	Boeing 747-4Q8	Virgin Atlantic Airways Ltd *Tubular Belle*
	G-VIBA	Cameron DP-80 airship	Jacques W. Soukup Enterprises Ltd/USA
	G-VIBE	Boeing 747-219B	*(stored)*
	G-VICC	PA-28-161 Warrior II	Charlie Charlie Syndicate (G-JFHL)
	G-VICE	MDH Hughes 369E	B. T. Anderson
	G-VICI	D.H.112 Venom FB.50 (J-1573)	Lindsay Wood Promotions Ltd/Bournemouth
	G-VICM	Beech F33C Bonanza	Velocity Engineering Ltd
	G-VICS	Commander 114B	Millennium Aviation Ltd
	G-VICT	PA-31-310 Turbo Navajo	Heliquick Ltd (G-BBZI)
	G-VIEW	Vinten-Wallis WA-116/100	K. H. Wallis
	G-VIIA	Boeing 777-236	British Airways
	G-VIIB	Boeing 777-236	British Airways
	G-VIIC	Boeing 777-236	British Airways
	G-VIID	Boeing 777-236	British Airways
	G-VIIE	Boeing 777-236	British Airways
	G-VIIF	Boeing 777-236	British Airways
	G-VIIG	Boeing 777-236	British Airways
	G-VIIH	Boeing 777-236	British Airways
	G-VIIJ	Boeing 777-236	British Airways
	G-VIIK	Boeing 777-236	British Airways
	G-VIIL	Boeing 777-236	British Airways
	G-VIIM	Boeing 777-236	British Airways
	G-VIIN	Boeing 777-236	British Airways
	G-VIIO	Boeing 777-236	British Airways
	G-VIIP	Boeing 777-236	British Airways
	G-VIIR	Boeing 777-236	British Airways
	G-VIIS	Boeing 777-236	British Airways
	G-VIIT	Boeing 777-236	British Airways
	G-VIIU	Boeing 777-236	British Airways
	G-VIIV	Boeing 777-236	British Airways
	G-VIIW	Boeing 777-236	British Airways
	G-VIIX	Boeing 777-236	British Airways
	G-VIIY	Boeing 777-236	British Airways
	G-VIKE	Bellanca 1730A Viking	W. G. Prout
	G-VIKY	Cameron A-120 balloon	P. J. Stanley
	G-VILA	Avtech Jabiru UL	R. W. Sage & T. R. Villa (G-BYIF)
	G-VILL	Lazer Z.200 (modified)	M. G. Jefferies (G-BOYZ)/Little Gransden
	G-VINO	Sky 90-24 balloon	Fivedata Ltd
	G-VINS	Cameron N-90 balloon	Les Montgolfieres du Sud SARL/France
	G-VIPA	Cessna 182S	Stallingborough Aviation Ltd
	G-VIPH	Agusta A.109C	Cheqair Ltd(G-BVNH/G-LAXO)
	G-VIPI	BAe 125 Srs 800B	Yeates of Leicester Ltd
	G-VIPP	PA-31-350 Navajo Chieftain	Capital Trading Aviation (G-OGRV/G-BMPX)/ Filton
	G-VIPY	PA-31-350 Navajo Chieftain	Capital Trading Aviation (G-POLO)/Filton
	G-VITE	Robin R.1180T	G-VITE Flying Group
	G-VITL	Lindstrand LBL-105A balloon	Vital Resources
	G-VITO	BAe Jetstream 3112	Aceline Air Ltd (G-BSIW/G-OEDL/G-OAKK)
	G-VIVA	Thunder Ax7-65 balloon	R. J. Mitchener
	G-VIVI	Taylor JT.2 Titch	D. G. Tucker
	G-VIVM	P.84 Jet Provost T.5	The Skys The Ltd (G-BVWF)/North Weald
	G-VIXN	D.H.110 Sea Vixen FAW.2 (XS587) ★	P. G. Vallance Ltd/Charlwood
	G-VIZZ	Sportavia RS.180 Sportsman	Exeter Fournier Group
	G-VJAB	Avtech Jabiru UL	A. S. R. Milner
	G-VJET	Avro 698 Vulcan B.2 (XL426) ★	Vulcan Restoration Trust/Southend
	G-VJIM	Colt 77 Jumbo Jim SS balloon	Magical Adventures Ltd/USA
	G-VKIT	Shaw Europa	T. H. Crow
	G-VKNG	Boeing 767-3Z9ER	Air Atlanta Ltd
	G-VLAD	Yakovlev Yak-50	M. B. Smith/Booker
	G-VLCN	Avro 698 Vulcan B.2 (XH558) ★	C. Walton Ltd/Bruntingthorpe
	G-VLIP	Boeing 747-443	Virgin Atlantic Airways Ltd
	G-VMCG	PA-38-112 Tomahawk	J. McGarry (G-BSVX)
	G-VMDE	Cessna P.210N	P. L. Goldberg
	G-VMEG	Airbus A.340-642	Virgin Atlantic Airways Ltd *Mystic Maiden*

Reg.	Type	Owner or Operator	Notes
G-VMJM	SOCATA TB.10 Tobago	S. C. Brown (G-BTOK)	
G-VMPR	D.H.115 Vampire T.11 (XE920)	de Havilland Aviation Ltd/Bournemouth	
G-VMSL	Robinson R-22A	L. L. F. Smith (G-KILY)	
G-VNOM	D.H.112 Venom FB.50 (J-1632)★	de Havilland Heritage Museum	
G-VNUS	Hughes 269C	Enable international Ltd (G-BATT)	
G-VOAR	PA-28-181 Archer III	Solent Flight Ltd	
G-VODA	Cameron N-77 balloon	Racal Telecom PLC	
G-VOGE	Airbus A.340-642	Virgin Atlantic Airways Ltd *Cover Maiden*	
G-VOID	PA-28RT-201 Arrow IV	Newbus Aviation Ltd	
G-VOLK	Bell 206L-1 LongRanger 2	Chester Air Maintenance Ltd (G-GBAY/G-CSWL/G-SIRI)	
G-VONA	Sikorsky S-76A	Von Essen Aviation Ltd (G-BUXB)	
G-VONB	Sikorsky S-76B	Von Essen Aviation Ltd (G-POAH)	
G-VONE	Eurocopter AS355N Twin Squirrel	Von Essen Aviation Ltd (G-LCON)	
G-VOND	Bell 222	Von Essen Aviation Ltd (G-OWCG/G-VERT/G-JLBZ/G-BNGB)	
G-VONS	PA-32R-301T Saratoga IITC	W. S. Stanley	
G-VOOM	Pitts S-1S Special	P. G. Roberts	
G-VOTE	Ultramagic M-77 balloon	Flying Pictures Ltd	
G-VPAT	Evans VP-1 Srs 2	A. P. Twort	
G-VPSJ	Shaw Europa	J. D. Bean	
G-VPUF	Boeing 747-219B	*(stored)*	
G-VROC	Boeing 747-41R	Virgin Atlantic Airways Ltd	
G-VROE	Avro 652A Anson T.21 (WD413)	Air Atlantique Ltd (G-BFIR)/Coventry	
G-VROM	Boeing 747-443	Virgin Atlantic Airways Ltd	
G-VROS	Boeing 747-443	Virgin Atlantic Airways Ltd	
G-VROY	Boeing 747-443	Virgin Atlantic Airways Ltd	
G-VRST	PA-46-350P Malibu Mirage	Winchfield Enterprises Ltd	
G-VRTX	Enstrom 280FX	Bladerunner Aviation Ltd (G-CBNH)	
G-VRVI	Cameron O-90 balloon	SNT Property Ltd	
G-VSBC	Beech B200 Super King Air	Vickers Shipbuilding & Engineering Ltd/ Walney Island	
G-VSEA	Airbus A.340-311	Virgin Atlantic Airways Ltd *Plane Sailing*	
G-VSGE	Cameron O-105 balloon	G. Aimo (G-BSSD)	
G-VSHY	Airbus A.340-642	Virgin Atlantic Airways Ltd	
G-VSSH	Airbus A.340-642	Virgin Atlantic Airways Ltd	
G-VSSS	Boeing 747-219B	Air Atlanta Europe Ltd/Virgin Atlantic	
G-VSUN	Airbus A.340-313	Virgin Atlantic Airways Ltd *Rainbow Lady*	
G-VTII	D.H.115 Vampire T.11 (WZ507)	de Havilland Aviation Ltd/Bournemouth	
G-VTOL	H.S. Harrier T.52 (ZA250) ★	Brooklands Museum of Aviation/Weybridge	
G-VTOP	Boeing 747-4Q8	Virgin Atlantic Airways Ltd *Virginia Plain*	
G-VUEA	Cessna 550 Citation II	AD Aviation Ltd (G-BWOM)	
G-VULC	Avro 698 Vulcan B.2A (XM655)★	Radarmoor Ltd/Wellesbourne	
G-VVBF	Colt 315A balloon	Virgin Balloon Flights	
G-VVBK	PA-34-200T Seneca II	Ravenair (G-BSBS/G-BDRI)/Liverpool	
G-VVIP	Cessna 421C	Air Deluxe (G-BMWB)	
G-VVVV	Skyranger 912 (2)	J. Thomas & J. B. Hobbs	
G-VVWW	Enstrom 280C Shark	P. J. Odendaal	
G-VWOW	Boeing 747-41R	Virgin Atlantic Airways Ltd	
G-VXLG	Boeing 747-41R	Virgin Atlantic Airways Ltd *Ruby Tuesday*	
G-VYGR	Colt 120A balloon	A. van Wyk	
G-VZZZ	Boeing 747-219B	*(stored)*	
G-WAAC	Cameron N-56 balloon	N. P. Hemsley	
G-WAAN	MBB Bo.105DB	PLM Dollar Group Ltd (G-AZOR)	
G-WAAS	MBB Bo.105DBS-4	Bond Air Services Ltd (G-ESAM/G-BUIB/G-BDYZ)	
G-WABH	Cessna 172S	P. Green	
G-WACB	Cessna F.152 II	Wycombe Air Centre Ltd	
G-WACE	Cessna F.152 II	Wycombe Air Centre Ltd	
G-WACF	Cessna 152 II	Wycombe Air Centre Ltd	
G-WACG	Cessna 152 II	Wycombe Air Centre Ltd	

Notes	Reg.	Type	Owner or Operator
	G-WACH	Cessna FA.152 II	Wycombe Air Centre Ltd
	G-WACI	Beech 76 Duchess	Wycombe Air Centre Ltd
	G-WACJ	Beech 76 Duchess	Wycombe Air Centre Ltd
	G-WACL	Cessna F.172N	A. G. Arthur (G-BHGG)
	G-WACM	Cessna 172S	Wycombe Air Centre Ltd
	G-WACO	Waco UPF-7	RGV (Aircraft Services) & Co/Staverton
	G-WACP	PA-28 Cherokee 180	Big Red Kite Ltd (G-BBPP)
	G-WACR	PA-28 Cherokee 180	Lees Avionics Ltd (G-BCZF)/Booker
	G-WACT	Cessna F.152 II	The Exeter Flying Club Ltd (G-BKFT)
	G-WACU	Cessna FA.152	Wycombe Air Centre Ltd (G-BJZU)
	G-WACW	Cessna 172P	Wycombe Air Centre Ltd
	G-WACY	Cessna F.172P	Wycombe Air Centre Ltd
	G-WADI	PA-46-350P Malibu Mirage	H. J. D. S. Baioes/Cranfield
	G-WADS	Robinson R-22B	Whizzard Helicopters (G-NICO)
	G-WAFU	Robinson R-44	Magic Aviation Ltd
	G-WAGG	Robinson R-22B-2	N. J. Wagstaff Leasing
	G-WAHL	QAC Quickie	A. A. A. Wahlberg
	G-WAIR	PA-32-301 Saratoga	Thorne Aviation
	G-WAIT	Cameron V-77 balloon	C. P. Brown
	G-WAKE	Mainair Blade 912	B. W. Webster
	G-WAKY	Cyclone AX2000	Cyclone Airsports Ltd
	G-WALY	Maule MX-7-180	A. J. West
	G-WAMS	PA-28R-201 Arrow	Amsair Ltd
	G-WARB	PA-28-161 Warrior III	Muller Aircraft Leasing Ltd
	G-WARC	PA-28-161 Warrior III	Plane Talking Ltd/Elstree
	G-WARD	Taylor JT.1 Monoplane	R. P. J. Hunter
	G-WARE	PA-28-161 Warrior II	W. B. Ware/Filton
	G-WARH	PA-28-161 Warrior III	Newcastle-Upon-Tyne Aero Club Ltd
	G-WARK	Schweizer 269C	K. Sutcliffe
	G-WARR	PA-28-161 Warrior II	T. J. & G. M. Laundy
	G-WARS	PA-28-161 Warrior III	Blaneby Ltd
	G-WARV	PA-28-161 Warrior III	Plane Talking Ltd/Elstree
	G-WARW	PA-28-161 Warrior III	C. J. Simmonds
	G-WARX	PA-28-161 Warrior III	C. M. A. Clark
	G-WARY	PA-28-161 Warrior III	Armstrong Aviation Ltd
	G-WARZ	PA-28-161 Warrior III	Aviation Rentals/Bournemouth
	G-WATR	Christen A1 Husky	S. N. Gregory
	G-WAVA	Robin HR.200/120B	Wellesbourne Aviation
	G-WAVE	Grob G.109B	M. L. Murdoch/Cranfield
	G-WAVI	Robin HR.200/120B	Wellesbourne Flyers Ltd (G-BZDG)
	G-WAVN	Robin HR.200/120B	Wellesbourne Flyers Ltd (G-VECA)
	G-WAVT	Robin R.21601	Wellesbourne Flyers Ltd (G-CBLG)
	G-WAZP	Skyranger 912	K. H. A. Negal
	G-WAZZ	Pitts S-1S Special	D. T. Knight (G-BRRP)
	G-WBAT	Wombat gyroplane	M. R. Harrisson (G-BSID)
	G-WBEV	Cameron N-77 balloon	T. J. & M. Turner (G-PVCU)
	G-WBLY	Pegasus Mainair Quik	A. J. Lindsey
	G-WBMG	Cameron N Ele-90 SS balloon	P. H. E. van Overwalle (G-BUYV)/Belgium
	G-WBTS	Falconar F-11	W. C. Brown (G-BDPL)
	G-WBVS	Diamond DA.4D Star	G. W. Beavis
	G-WCAO	Eurocopter EC.135T-2	Avon & Somerset Constabulary & Gloucestershire Constabulary
	G-WCAT	Colt Flying Mitt SS balloon	Balloon Preservation Flying Group
	G-WCEI	M.S.894E Rallye 220GT	R. A. L. Lucas (G-BAOC)
	G-WCRD	SA.341G Gazelle	Wickford Development Co. Ltd
	G-WCUB	PA-18 Super Cub 150	P. A. Walley
	G-WDEB	Thunder Ax-7-77 balloon	A. Heginbottom
	G-WDEV	SA.341G Gazelle 1	Mentorvale Construction Ltd (G-IZEL/G-BBHW)/Ireland
	G-WEEM	Shaw Europa XS	N. A. Harrison
	G-WEGO	Robinson R-44-II	Clifton Helicopter Hire
	G-WELI	Cameron N-77 balloon	M. A. Shannon
	G-WELL	Beech E90 King Air	Colt Group Ltd
	G-WELS	Cameron N-65 balloon	K. J. Vickery
	G-WENA	AS.355F-2 Twin Squirrel	Kensington & Chelsea Aviation Ltd (G-CORR/G-MUFF/G-MOBI)
	G-WEND	PA-28RT-201 Arrow IV	Tayside Aviation Ltd/Dundee
	G-WERY	SOCATA TB.20 Trinidad	Fastour Aviation Ltd

Reg.	Type	Owner or Operator	Notes
G-WEST	Agusta A.109A	Westland Helicopters Ltd/Yeovil	
G-WESX	CFM Streak Shadow	M. Catania	
G-WFFW	PA-28-161 Warrior II	N. F. Duke	
G-WFLY	Mainair Pegasus Quik	D. E. Lord	
G-WFOX	Robinson R-22B-2	Fox Air	
G-WGCS	PA-18 Super Cub 95	S. C. Thompson	
G-WGHB	Canadair T-33AN Silver Star 3	Parkhouse Aviation	
G-WGSC	Pilatus PC-6/B2-H4 Turbo Porter	D. M. Penny	
G-WHAL	QAC Quickie	A. A, M. Wahiberg	
G-WHAM	AS.350B-3 Ecureuil	Horizon Helicopter Hire Ltd/Kidlington	
G-WHAT	Colt 77A balloon	M. A. Scholes	
G-WHEE	Pegasus Quantum 15-912	Airways Airsports Ltd	
G-WHEN	Tecnam P.92-EM Echo	E. Windle	
G-WHIM	Colt 77A balloon	D. L. Morgan	
G-WHOG	CFM Streak Shadow	B. R. Cannell	
G-WHOO	Rotorway Executive 162F	C. A. Saul	
G-WHRL	Schweizer 269C	M. Gardiner	
G-WHST	AS.350B2 Ecureuil	Hawkrise Ltd (G-BWYA)	
G-WIBB	Jodel D.18	D. Dobson	
G-WIBS	C.A.S.A. 1-131E Jungmann 2000	C. Willoughby	
G-WIFE	Cessna R.182 RG II	Wife Group (G-BGVT)	
G-WIFI	Cameron Z-90 balloon	Trigger Concepts Ltd	
G-WIIZ	Augusta-Bell 206B JetRanger 2	Action Vehicles Ltd (G-DBHH/G-AWVO)	
G-WIKY	Cessna 208B Grand Caravan	Provident Partners Ltd	
G-WILD	Pitts S-1T Special	A. McClean	
G-WILG	PZL-104 Wilga 35	M. H. Bletsoe-Brown (G-AZYJ)	
G-WILS	PA-28RT-201T Turbo Arrow IV	B. Walker & Co (Dursley) Ltd	
G-WILY	Rutan LongEz	W. S. Allen	
G-WIMP	Colt 56A balloon	T. & B. Chamberlain	
G-WINA	Cessna 560XL Citation XL	Ability Air Ltd	
G-WINE	Thunder Ax7-77Z balloon ★	Balloon Preservation Group/Lancing	
G-WINI	SA Bulldog Srs.120/121	B. Robinson (G-CBCO)	
G-WINK	AA-5B Tiger	B. St J. Cooke	
G-WINS	PA-32 Cherokee Six 300	Cheyenne Ltd	
G-WIRE	AS.355F-1 Twin Squirrel	National Grid Co PLC (G-CEGB/G-BLJL)	
G-WIRL	Robinson R-22B	Rivermead Aviation Ltd/Switzerland	
G-WISH	Lindstrand LBL Cake SS balloon	Oxford Promotions (UK) Ltd/USA	
G-WIXI	Avions Mudry CAP.10B	Meridian Aviation Group Ltd	
G-WIZA	Robinson R-22B	Patriot Aviation Ltd (G-PERL)/Cranfield	
G-WIZB	Grob G.115A	R. N. R. Bellamy	
G-WIZD	Lindstrand LBL-180A balloon	Wizard Balloons Cambridge Ltd	
G-WIZI	Enstrom 280FX	C. A. V. Witheridge	
G-WIZO	PA-34-220T Seneca III	B. J. Booty	
G-WIZR	Robinson R-22B-2	Findon Air Services/Shoreham	
G-WIZS	Mainair Pegasus Quik	G. R. Barker	
G-WIZY	Robinson R-22B	B. J. North (G-BMWX)	
G-WIZZ	AB-206B JetRanger 2	Rivermead Aviation Ltd	
G-WKRD	AS.350B-2 Ecureuil	Wickford Development Co (G-BUJG/G-HEAR)	
G-WLAC	PA-18 Super Cub 150	White Waltham Airfield Ltd (G-HAHA/G-BSWE)	
G-WLGA	PZL-104 Wilga 80	Agri Air Services Ltd	
G-WLLY	Bell 206B JetRanger	G and N Aviation	
		(G-OBHH/G-RODY/G-ROGR/G-AXMM)	
G-WLMS	Mainair Blade 912	I. D. Smart & C. J. Coggins	
G-WLSH	ATR-42-300	Air Wales Ltd (G-BXEG)	
G-WMAA	MBB Bo 105DBS/4	W. Midlands Air Ambulance (G-PASB/G-BDMC)	
G-WMAN	SA.341G Gazelle 1	J. Wightman	
G-WMAS	Eurocopter EC.135T-1	Bond Air Services Ltd/Aberdeen	
G-WMBT	Robinson R-44-II	P. Winslow	
G-WMID	MDH MD-900 Explorer	W. Midlands Police Authority	
G-WMLT	Cessna 182Q	G. Wimlett (G-BOPG)	
G-WMPA	AS.355F-2 Twin Squirrel	Police Aviation Services Ltd	
G-WMTM	AA-5B Tiger	Falcon Flying Group	
G-WMWM	Robinson R-44	MM Air Ltd	

Notes	Reg.	Type	Owner or Operator
	G-WNAA	Agusta A.109E Power	Sloane Helicopters Ltd (G-TVAC)
	G-WNGS	Cameron N-105 balloon	R. M. Horn
	G-WOLF	PA-28 Cherokee 140	Aircraft Management Services Ltd
	G-WOOD	Beech 95-B55A Baron	T. D. Broadhurst (G-AYID)/Sleap
	G-WOOF	Enstrom 480	Netcopter.co.uk Ltd & Curvature Ltd
	G-WOOL	Colt 77A balloon	Whacko Balloon Group
	G-WORM	Thruster T.600N	West Lancashire Microlight School
	G-WOSY	MBB Bo.105DBS/4	Redwood Aviation Ltd (G-PASD/G-BNRS)
	G-WOTG	BN-2T Turbine Islander	RAF Sports Parachute Association (G-BJYT)
	G-WOWA	D.H.C.8-311 Dash Eight	Air Southwest Ltd (G-BRYS)
	G-WOWB	D.H.C.8-311 Dash Eight	Air Southwest Ltd (G-BRYT)
	G-WOWC	D.H.C.8-311 Dash Eight	Air Southwest Ltd (G-BRYO)
	G-WPAS	MDH MD-900 Explorer	Police Aviation Services Ltd
	G-WREN	Pitts S-2A Special	Northamptonshire School of Flying Ltd/Sywell
	G-WRFM	Enstrom 280C-UK Shark	W. F. Blake (G-CTSI/G-BKIO)
	G-WRIT	Thunder Ax7-77A balloon	G. Pusey
	G-WRLY	Robinson R-22B	Burman Aviation Ltd (G-OFJS/G-BNXJ)/Cranfield
	G-WRWR	Robinson R-22B-2	AJS Helicopters Ltd
	G-WSEC	Enstrom F-28C	AJD Engineering Ltd (G-BONF)
	G-WSKY	Enstrom 280C-UK-2 Shark	M. I. Edwards Engineers (G-BEEK)
	G-WUFF	Shaw Europa	M. A. Barker
	G-WULF	WAR Focke-Wulf Fw.190 (8+)	A. Howe
	G-WUSH	Eurocopter EC.120B	First Degree Air
	G-WVBF	Lindstrand LBL-210A balloon	Virgin Balloon Flights Ltd
	G-WVIP	Beech B.200 Super King Air	Capital Trading (Aviation) Ltd
	G-WWAL	PA-28R Cherokee Arrow 180	White Waltham Airfield Ltd (G-AZSH)
	G-WWAS	PA-34-220T Seneca III	D. Intzevidis (G-BPPB)/Greece
	G-WWAY	Piper PA-28-181 Archer II	R. A. Witchell
	G-WWBB	Airbus A.330-243	bmi british midland/South African Airways
	G-WWBC	Airbus A.330-243	bmi british midland
	G-WWBD	Airbus A.330-243	bmi british midland/South African Airways
	G-WWBM	Airbus A.330-243	bmi british midland
	G-WWIZ	Beech 95-58 Baron	Scenestage Ltd (G-GAMA/G-BBSD)/ Bournemouth
	G-WWWG	Shaw Europa	C. F. Williams-Wynne
	G-WYAT	CFM Streak Shadow Srs SA	M. G. Whyatt
	G-WYCH	Cameron 90 Witch SS balloon	Corn Palace Balloon Club Ltd
	G-WYLE	Rans S.6-ES Coyote II	A. & R. W. Osborne
	G-WYMR	Robinson R-44	Heli Air Ltd/Wellesbourne
	G-WYND	Wittman W.8 Tailwind	Forge Group
	G-WYNS	Aero Designs Pulsar XP	S. L. Bauza/Majorca
	G-WYNT	Cameron N-56 balloon	S. L. G. Williams
	G-WYPA	MBB Bo 105DBS/4	Police Aviation Services Ltd/Gloucestershire
	G-WYSP	Robinson R-44	Calderbrook Estates Ltd
	G-WZOL	RL.5B LWS Sherwood Ranger	G. W. F. Webb (G-MZOL)
	G-WZZZ	Colt AS-56 airship	Lindstrand Balloons Ltd
	G-XARV	ARV Super 2	D. J. Burton (G-OPIG/G-BMSJ)
	G-XATS	Aerotek Pitts S-2A Special	Air Training Services Ltd/Booker
	G-XAXA	BN-2A-26 Islander	Airx Ltd (G-LOTO/G-BDWG)
	G-XAYR	Raj Hamsa X'Air 582 (6)	D. L. Connolly & R. V. Barber
	G-XBCI	Bell 206B JetRanger 3	BCI Helicopter Charters Ltd
	G-XCCC	Extra EA.300/L	P. T. Fellows
	G-XCIT	Pioneer 300	A. Thomas
	G-XCUB	PA-18 Super Cub 150	M. C. Barraclough
	G-XENA	PA-28-161 Warrior II	Braddock Ltd
	G-XIII	Van's RV-7	G-XIII Group
	G-XIIX	Robinson R-22B ★	(Static exhibit)
	G-XINE	PA-28-161 Warrior II	P. Tee (G-BPAC)
	G-XIOO	Raj Hamsa X'Air R.100 (2)	R. Paton

Reg.	Type	Owner or Operator	Notes
G-XKEN	PA-34-200T Seneca III	Choicecircle Ltd	
G-XLAA	Boeing 737-8Q8	Excel Airways Ltd (G-OKDN)	
G-XLAB	Boeing 737-8Q8	Excel Airways Ltd (G-OJSW)	
G-XLAE	Boeing 737-8Q8	Excel Airways Ltd (G-OKJW)	
G-XLAF	Boeing 737-86N	Excel Airways Ltd	
G-XLAG	Boeing 737-86N	Excel Airways Ltd	
G-XLAH	Boeing 737-8Q8	Excel Airways Ltd	
G-XLIV	Robinson R-44	Rotorcraft Ltd	
G-XLMB	Cessna 560XL Citation Excel	Aviation Beauport Ltd	
G-XLNT	Zenair CH.601XL	Zenair G-XLNT Group	
G-XLTG	Cessna 182S	D. H. Morgan	
G-XLXL	Robin DR.400/160	L. R. Marchant (G-BAUD)/Biggin Hill	
G-XMEN	Eurocopter AS.350B-3 Ecureuil	Corporate Estates Ltd (G-ZWRC)	
G-XMGO	Aeromot AMT-200S Super Ximango	G. McLean & R. P. Beck	
G-XMII	Eurocopter EC.135T-1	Merseyside Police Authority/Woodvale	
G-XOIL	AS355N Twin Squirrel	Firstearl Marine and Aviation Ltd (G-LOUN)	
G-XPBI	Letov LK-2M Sluka	K. Harness	
G-XPRS	Bombardier BD-700-1A10 Global Express	Leferson Holdings Ltd	
G-XPSS	Short SD3-60 Variant 100	BAC Express Airlines Ltd (G-OBOH/G-BNDJ)	
G-XPXP	Aero Designs Pulsar XP	B. J. Edwards	
G-XRAF	Raj Hamsa X'Air 582 (2)	R. G. Kirkland	
G-XRAY	Rand-Robinson KR-2	R. S. Smith	
G-XRLD	Cameron A-250 balloon	Red Letter Days Ltd	
G-XRXR	Raj Hamsa X'Air 582 (1)	R. J. Philpotts	
G-XSAM	Van's RV-9A	D. G. Lucas	
G-XSDJ	Shaw Europa XS	D. N. Joyce	
G-XSFT	PA-23 Aztec 250F	T. L. B. Dykes (G-CPPC/G-BGBH)/Goodwood	
G-XSKY	Cameron N-77 balloon	T. D. Gibbs	
G-XTEE	Airborne XT912 Steak III	Airborne Australia UK	
G-XTEK	Robinson R-44	PLM Properties Ltd	
G-XTOR	BN-2A Mk III-2 Trislander	Aurigny Air Services Ltd (G-BAXD)	
G-XTUN	Westland-Bell 47G-3B1 (XT223)	Hields Aviation (G-BGZK)	
G-XVBF	Lindstrand LBL-330A balloon	Virgin Balloon Flights	
G-XVOM	Van's RV-6	A. Baker-Munton	
G-XXEA	Sikorsky S-76C	Director of Royal Travel/Blackbushe	
G-XXIV	AB-206B JetRanger 3	Bart Fiftyt Nine Ltd	
G-XXTR	Extra EA.300/L	Airpark Flight Centre Ltd (G-ECCC)	
G-XXVI	Sukhoi Su-26M	A. N. Onn/Headcorn	
G-XYAK	IDABacau Yakovlev Yak-52	NRG Technology Ltd	
G-XYJY	Best Off Skyranger 912 (2)	A. V. Francis	
G-YACB	Robinson R-22B	Heli Air Ltd (G-VOSL)	
G-YAKA	Yakovlev Yak-50	M. Chapman	
G-YAKB	Aerostar Yakovlev Yak-52	Kemble Air Services Ltd	
G-YAKC	Yakovlev Yak-52	T. J. Wilson	
G-YAKH	IDABacau Yakovlev Yak-52	Plus 7 minus 5 Ltd	
G-YAKI	Yakovlev Yak-52 (100)	Yak One Ltd/White Waltham	
G-YAKM	Yakovlev Yak-50 (66 Soviet AF)	Airborne Services Ltd	
G-YAKN	IDA Bacau Yakovlev Yak-52	Airborne Services Ltd	
G-YAKO	Yakovlev Yak-52	M. K. Shaw	
G-YAKR	IDA Bacau Yakovlev Yak-52	A. S. Nottage & R. A. Alexander	
G-YAKS	Yakovlev Yak-52 (2)	Two Bees Associates Ltd	
G-YAKT	Yakovlev Yak-52	G-YAKT Group	
G-YAKU	Yakovlev Yak-50	D. J. Hopkinson (G-BXND)	
G-YAKV	Aerostar Yakovlev Yak-52	P. D. Scandrett	
G-YAKW	IDA Bacau Yakovlev Yak-52	Cuddesdon Flying Group	
G-YAKX	Aerostar Yakovlev Yak-52	The X-Flyers Ltd	
G-YAKY	Aerostar Yakovlev Yak-52	W. T. Marriott	
G-YAKZ	Yakovlev Yak-50	Greenhouse New Media Ltd	
G-YANK	PA-28-181 Archer II	G-YANK Flying Group	
G-YARR	Mainair Rapier	D. Yarr	
G-YARV	ARV Super 2	P. R. Snowden (G-BMDO)	

Notes	Reg.	Type	Owner or Operator
	G-YAWW	PA-28RT-201T Turbo Arrow IV	Barton Aviation Ltd
	G-YBAA	Cessna FR.172J	A. Evans
	G-YCII	LET Yakovlev C-11	R. W. Davies
	G-YCUB	PA-18 Super Cub 150	F. W. Rogers Garage (Saltash) Ltd
	G-YEAH	Robinson R-44-II	Turboprop Leasing LLP
	G-YELL	Murphy Rebel	A. D. Keen
	G-YEOM	PA-31-350 Navajo Chieftain	Foster Yeoman Ltd/Exeter
	G-YEWS	Rotorway Executive 152	R. Turrell & P. Mason
	G-YFLY	VPM M.16 Tandem Trainer	A. J. Unwin (G-BWGI)
	G-YFUT	Yakovlev Yak-52	R. Oliver
	G-YFZT	Cessna 172S	AB Integro
	G-YIII	Cessna F.150L	Sherburn Aero Club Ltd
	G-YIIK	Robinson R-44	The Websiteshop (UK) Ltd
	G-YIPI	Cessan FR.172K	A. J. G. Davis
	G-YJET	Montgomerie-Bensen B.8MR	A. Shuttleworth (G-BMUH)
	G-YKCT	Aerostar Yakovlev Yak-52	C. R. Turton
	G-YKSO	Yakovlev Yak-50	Classic Display (Scotland) Ltd
	G-YKSZ	Yakovlev Yak-52	J. Tzarina Group
	G-YKYK	Aerostar Yakovlev Yak-52	K. J. Pilling
	G-YLYB	Cameron N-105 balloon	Airship & Balloon Co Ltd
	G-YMBO	Robinson R-22M Mariner	J. Robinson
	G-YMFC	Waco YMF	S. J. Brenchley
	G-YMMA	Boeing 777-236ER	British Airways
	G-YMMB	Boeing 777-236ER	British Airways
	G-YMMC	Boeing 777-236ER	British Airways
	G-YMMD	Boeing 777-236ER	British Airways
	G-YMME	Boeing 777-236ER	British Airways
	G-YMMF	Boeing 777-236ER	British Airways
	G-YMMG	Boeing 777-236ER	British Airways
	G-YMMH	Boeing 777-236ER	British Airways
	G-YMMI	Boeing 777-236ER	British Airways
	G-YMMJ	Boeing 777-236ER	British Airways
	G-YMMK	Boeing 777-236ER	British Airways
	G-YMML	Boeing 777-236ER	British Airways
	G-YMMM	Boeing 777-236ER	British Airways
	G-YMMN	Boeing 777-236ER	British Airways
	G-YMMO	Boeing 777-236ER	British Airways
	G-YMMP	Boeing 777-236ER	British Airways
	G-YNOT	D.62B Condor	T. Littlefair (G-AYFH)
	G-YOGI	Robin DR.400/140B	M. M. Pepper (G-BDME)
	G-YORK	Cessna F.172M	H. Waetjen
	G-YOTS	IDA Bacau Yakovlev Yak-52	J. D. F. Barke
	G-YOYO	Pitts S-1E Special	J. D. L. Richardson (G-OTSW/G-BLHE)
	G-YPOL	MDH MD-900 Explorer	West Yorkshire Police Authority
	G-YPSY	Andreasson BA-4B	R. W. Hinton
	G-YRAF	RAF 2000 GTX-SE gyroplane	C. V. King
	G-YRIL	Luscombe 8E Silvaire	C. Potter
	G-YROI	Air Command 532 Elite	W. B. Lumb
	G-YROJ	RAF 2000 GTX-SE gyroplane	J. R. Mercer
	G-YROO	RAF 2000 GTX-SE gyroplane	K. D. Rhodes & C. S. Oakes
	G-YROS	Montgomerie-Bensen B.8M	Flight Academy (Girocopters) Ltd
	G-YROW	VPM M.16 Tandem Trainer	B. Jones
	G-YROY	Montgomerie-Bensen B.8MR	S. Brennan
	G-YRUS	Jodel D.140E	W. E. Massam (G-YRNS)
	G-YSMO	Pegasus Mainair Quik	M. P. & R. A. Wells
	G-YSPY	Cessna 172Q	J. Henderson
	G-YSTT	PA-32R-301 Saratoga II HP	A. W. Kendrick
	G-YTUK	Cameron A-210 balloon	Societe Bombard SRL/France

Reg.	Type	Owner or Operator	Notes
G-YUGO	H.S.125 Srs 1B/R-522 ★	Fire Section (G-ATWH)/Dunsfold	
G-YULL	PA-28 Cherokee 180E	G. Watkinson-Yull (G-BEAJ)	
G-YUMM	Cameron N-90 balloon	Wunderbar Ltd	
G-YUPI	Cameron N-90 balloon	MCVH SA/Belgium	
G-YURO	Shaw Europa ★	Yorkshire Air Museum/Elvington	
G-YVBF	Lindstrand LBL-317S balloon	Virgin Balloon Flights	
G-YVES	Alpi Pioneer 300	M. C, Birchall	
G-YVET	Cameron V-90 balloon	K. J. Foster	
G-YVFS	D.H.82A Tiger Moth	Yorkshire Vintage Flying Ltd (G-ANDE)	
G-YYAK	Aerostar SA Yak-52	J. Armstrong & D. W. Lamb	
G-YYYY	MH.1521C-1 Broussard	Aerosuperbatics Ltd/Rendcomb	
G-YZMO	Champion 8KCAB	Blue Yonder Aviation Ltd	
G-YZYZ	Mainair Blade 912	R. Chadwick	
G-ZAAZ	Van's RV-8	P. A. Soper	
G-ZABC	Sky 90-24 balloon	P. Donnelly	
G-ZACE	Cessna 172S	M. C. Tonsbeek	
G-ZACH	Robin DR.400/100	A. P. Wellings (G-FTIO)/Sandown	
G-ZAIR	Zenair CH 601HD	J. R. Standring	
G-ZANG	PA-28 Cherokee 140	A. J. Gale	
G-ZANY	Diamond DA.40D Star	Altair Aviation Ltd	
G-ZAPH	Bell 206B JetRanger 4	Northern Flights Ltd (G-DBMW)/Stansted	
G-ZAPK	BAe 146-200QC	Titan Airways Ltd (G-BTIA/G-PRIN)/Stansted	
G-ZAPM	Boeing 737-33A	Titan Airways Ltd	
G-ZAPN	BAe 146-200QC	Titan Airways Ltd (G-BPBT)/Stansted	
G-ZAPO	BAe 146-200QC	Titan Airways Ltd (G-BWLG/G-PRCS)/Stansted	
G-ZAPR	BAe 146-200F	Titan Airways Ltd/Stansted (G-BOXE)	
G-ZAPT	Beech B.200C Super King Air	Titan Airways Ltd/Stansted	
G-ZAPU	Boeing 757-2Y0	Titan Airways Ltd/Stansted	
G-ZAPV	Boeing 737-3Y0	Titan Airways Ltd/Royal Mail (G-IGOC)/Stansted	
G-ZAPY	Robinson R-22B	Heli Air Ltd (G-INGB)/Wellesbourne	
G-ZARI	AA-5B Tiger	ZARI Aviation Ltd (G-BHVY)	
G-ZARV	ARV Super 2	P. R. Snowden	
G-ZAZA	PA-18 Super Cub 95	Airborne Taxi Services Ltd	
G-ZBED	Robinson R-22B	P. D. Spinks	
G-ZBLT	Cessna 182S	Entee Global Services Ltd	
G-ZEBO	Thunder Ax8-105 S2 balloon	S. M. Waterton	
G-ZEBY	PA-28 Cherokee 140	A. Buchanan (G-BFBF)	
G-ZEIN	Slingsby T.67M Firefly 260	R. C. P. Brookhouse	
G-ZELE	Westland Gazelle HT.Mk.2	London Helicopter Centres Ltd (G-CBSA)	
G-ZENA	Zenair CH.701UL	A. N. Aston	
G-ZEPI	Colt GA-42 gas airship	P. A. Lindstrand (G-ISPY/G-BPRB)	
G-ZERO	AA-5B Tiger	G-ZERO Syndicate	
G-ZETA	Lindstrand LBL-105A balloon	S. Travaglia/Italy	
G-ZHKF	Escapade 912(1)	C. D. Wills	
G-ZHWH	Rotorway Executive 162F	B. Alexander	
G-ZIGI	Robin DR.400/180	R. J. Dix/France	
G-ZINT	Cameron Z-77 balloon	Film Production Consultants SRL	
G-ZIPA	R. Commander 114A	A. C. Lees (G-BHRA)	
G-ZIPI	Robin DR.400/180	H. U. & D. C. Stahlberg/Headcorn	
G-ZIPY	Wittman W.8 Tailwind	M. J. Butler	
G-ZITZ	AS.355F-2 Twin Squirrel	Heli Aviation Ltd	
G-ZIZI	Cessna 525 CitationJet	Ortac Air Ltd	
G-ZLIN	Z.526 Trener Master	N. J. Arthur	
G-ZLLE	SA.341G Gazelle	G-ZLLE Ltd	
G-ZLOJ	Beech A36 Bonanza	W. D. Gray	
G-ZLYN	Z.526F Trener Master	H. Philippart	
G-ZMAM	PA-28-181 Archer II	Z. Mahmood (G-BNPN)	
G-ZODI	Zenair CH.601UL	B. McFadden	
G-ZODY	Zenair CH.601UL Zodiac	Sarum AX2000 Group	
G-ZONK	Robinson R-44	CCB Aviation Ltd (G-EDIE)	
G-ZOOL	Cessna FA.152	G. G. Hammond (G-BGXZ)	

Notes	Reg.	Type	Owner or Operator
	G-ZORO	Shaw Europa	N. T. Read
	G-ZTED	Shaw Europa	J. J. Kennedy & E. W. Gladstone
	G-ZUMI	Van`s RV-8	P. M. Wells
	G-ZVBF	Cameron A-400 balloon	Virgin Balloon Flights
	G-ZWAR	Eurocopter EC.120B	Hedgeton Trading Ltd
	G-ZXZX	Learjet 45	Gama Aviation Ltd
	G-ZZAP	Champion 8KCAB	L. Maikowski & ptnrs
	G-ZZEL	Westland Gazelle AH.1	Military Helicopters Ltd
	G-ZZLE	SA.341G Gazelle 1	G-ZZLE Ltd
	G-ZZOE	Eurocopter EC.120B	J. F. H. James
	G-ZZWW	Enstrom 280FX	DB International (UK) Ltd (G-BSIE)
	G-ZZZA	Boeing 777-236	British Airways
	G-ZZZB	Boeing 777-236	British Airways
	G-ZZZC	Boeing 777-236	British Airways

G-AGJG, D.H.89A Dragon Rapide. *M. J. F. Bowyer*

G-DBCB, Airbus A.319-132 of bmi British Midland. *AJW*

Military to Civil Cross-Reference

Serial carried	Civil identity	Serial carried	Civil identity
(DOSAAF)	G-BXAV	18393 (RCAF)	G-BCYK
1	G-BPVE	18671 (671 RCAF)	G-BNZC
2 (CIS)	G-YAKS	20310 (310 RCAF)	G-BSBG
09 (DOSAAF)	G-BVMU	21261 (RCAF)	G-TBRD
10 (DOSAAF)	G-BTZB	21714 (201-B USN)	G-RUMM
15 (DOSAAF)	G-BXJB	28521 (TA-521 USAF)	G-TVIJ
26 (US)	G-BAVO	30140 (Yugoslav Army)	G-RADA
26 (DOSAAF)	G-BVXK	30146 (Yugoslav Army)	G-BSXD
27 (CIS)	G-YAKX	30149 (Yugoslav Army)	G-SOKO
27 (USN)	G-BRVG	31145 (G-26 USAAF)	G-BBLH
27 (CIS)	G-OYAK	3-1923 (USAAC)	G-BRHP
27 (USAAC)	G-AGYY	31952 (USAAC)	G-BRPR
43 (SC USAF)	G-AZSC	39624 (D-39 USAAF)	G-BVMH
44 (K-33 USAAF)	G-BJLH	40467 (19 USN)	G-BTCC
52	G-BWVR	53319 (319/RB USN)	G-BTDP
55 (DOSAAF)	G-BVOK	56321 (U-AB RNorAF)	G-BKPY
69 (DOSAAF)	G-BTZB	80425 (WT-14 USN)	G-RUMT
71 (DOSAAF)	G-BXAV	86711 (USN)	G-RUMW
74 (DOSAAF)	G-BXID	91007 (USAF)	G-NASA
85 (USAAF)	G-BTBI	92844 (8) RNZAF)	G-BXUL
100 (DOSAAF)	G-YAKI	93542 (LTA-542 USAF)	G-BRLV
112 (USAAC)	G-BSWC	111836 (JZ-6 USN)	G-TSIX
118 (USAAC)	G-BSDS	115042 (TA-042 USAF)	G-BGHU
124 (Fr AF)	G-BOSJ	115302 (TP USAAF)	G-BJTP
139 (DOSAAF)	G-BWOD	115684 (D-C USAAF)	G-BKVM
143 (Fr AF)	G-MSAL	124485 (DF-A USAAF)	G-BEDF
152/17	G-ATJM	126922 (402/AK USN)	G-RADR
168	G-BFDE	150225 (123 USAAF)	G-AWOX
177 (Irish AC)	G-BLIW	151632 (USAAF)	G-BWGR
185 (Fr AF)	G-BWLR	18-2001 (USAAF)	G-BIZV
311 (SingaporeAF)	G-SARK	18-5395 (CDG Fr AF)	G-CUBJ
379 (USAAC)	G-ILLE	212540 (RF-40)	G-BBHK
422-15	G-AVJO	217786 (25 USAAF)	G-BRTK
423 / 427 (RNorAF)	G-AMRK	226671 (MX-X USAAF)	G-THUN
441 (USN)	G-BTFG	238410 (A-44 USAAF)	G-BHPK
450/17	G-BVGZ	2632019 (China AF)	G-BXZB
503 (Hungarian AF)	G-BRAM	314887 (USAAF)	G-AJPI
540 (USAAF)	G-BCNX	315509 (W7-S USAAF)	G-BHUB
781-25 (Span AF)	G-BRSH	329405 (A-23 USAAF)	G-BCOB
781-32 (Span AF)	G-BPDM	329417 (USAAF)	G-BDHK
854 (USAAC)	G-BTBH	329471 (F-44 USAAF)	G-BGXA
897 (E USN)	G-BJEV	329601 (D-44 USAAF)	G-AXHR
1102 (102 USN)	G-AZLE	329854 (R-44 USAAF)	G-BMKC
1164 (64 USAAC)	G-BKGL	329934 (B-72 USAAF)	G-BCPH
1180 (USN)	G-BRSK	330238 (A-24 USAAF)	G-LIVH
1211(N. Korean AF)	G-BWUF	330485 (C-44 USAAF)	G-AJES
1377 (Portuguese AF)	G-BARS	343251 (27 USAAC)	G-NZSS
1747 (Portuguese AF)	G-BGPB	414419 (LH-F USAAF)	G-MSTG
2345	G-ATVP	436021 (USAAF)	G-BWEZ
2807 (V-103 USN)	G-BHTH	44-1307 (87-H USAAF)	G-BTCD
3066	G-AETA	454467 (J-44 USAAF)	G-BILI
5964	G-BFVH	454537 (J-04 USAAF)	G-BFDL
6136 (205 USN)	G-BRUJ	461748 (Y USAF)	G-BHDK
7198/18	G-AANJ	472216 (HO M USAAF)	G-BIXL
7797 (USAAF)	G-BFAF	472218 (WZ-I USAAF)	G-HAEC
8178 (FU-178 USAF)	G-SABR	472773 (QP-M USAAF)	G-SUSY
8449M	G-ASWJ	479744 (M-49 USAAF)	G-BGPD
01385 (CIS)	G-BWJT	479766 (D-63 USAAF)	G-BKHG
01420 (Korean AF)	G-BMZF	480015 (M-44 USAAF)	G-AKIB
1/4513 (Fr AF)	G-BFYO	480133 (B-44 USAAF)	G-BDCD
14863 (USAAF)	G-BGOR	480321 (H-44 USAAF)	G-FRAN
16693 (693 RCAF)	G-BLPG	480480 (E-44 USAAF)	G-BECN
18013 (013 RCAF)	G-TRIC	480636 (A-58 USAAF)	G-AXHP

Serial carried	Civil identity	Serial carried	Civil identity
480752 (E-39 USAAF)	G-BCXJ	K2048	G-BZNW
493209 (US ANG)	G-DDMV	K2050	G-ASCM
41-33275 (CE USAAC)	G-BICE	K2059	G-PFAR
42-58678 (IY USAAC)	G-BRIY	K2060	G-BKZM
42-78044 (USAAC)	G-BRXL	K2075	G-BEER
42-84555 (EP-H)	G-ELMH	K2227	G-ABBB
43-5802 (49 USAAC)	G-KITT	K2567	G-MOTH
44-14419 (LH-F USAAF)	G-MSTG	K2572	G-AOZH
44-79609 (44-S PR USAAF)	G-BHXY	K2587	G-BJAP
44-80594 (USAAF)	G-BEDJ	K3215	G-AHSA
44-83184 (USAAF)	G-RGUS	K3661	G-BURZ
51-7692 (Fr AF)	G-TROY	K3731	G-RODI
51-11701A (AF258 USAF)	G-BSZC	K4259 (71)	G-ANMO
51-14526 (526 USAF)	G-BRWB	K5054	G-BRDV
51-15227 (10 USN)	G-BKRA	K5414 (XV)	G-AENP
54-2447 (USAF)	G-SCUB	K5600	G-BVVI
607327 (09-L USAAF)	G-ARAO	K8203	G-BTVE
A-10 (Swiss AF)	G-BECW	K8303 (D)	G-BWWN
A16-199 (SF-R RAAF)	G-BEOX	L2301	G-AIZG
A17-48 (RAAF)	G-BPHR	L6906	G-AKKY
A-57 (Swiss AF)	G-BECT	N500	G-BWRA
A-806 (Swiss AF)	G-BTLL	N1854	G-AIBE
A1325	G-BVGR	N4877 (VX-F)	G-AMDA
A8226	G-BIDW	N5182	G-APUP
B595 (W)	G-BUOD	N5195	G-ABOX
B1807	G-EAVX	N5903 (H)	G-GLAD
B2458	G-BPOB	N6-766 (R Australian N)	G-SPDR
B3459	G-BWMJ	N6181	G-EBKY
B6401	G-AWYY	N6290	G-BOCK
B7270	G-BFCZ	N6452	G-BIAU
C1904 (Z)	G-PFAP	N6532	G-ANTS
C3011 (S)	G-SWOT	N6537	G-AOHY
C4918	G-BWJM	N6797	G-ANEH
C4994	G-BLWM	N6847	G-APAL
C9533 (M)	G-BUWE	N6965 (FL-J)	G-AJTW
D-692	G-BVAW	N6985	G-AHMN
D5397/17	G-BFXL	N9191	G-ALND
D7889	G-AANM	N9192 (RCO-N)	G-DHZF
D8084	G-ACAA	P9374	G-MKIA
D8096 (D)	G-AEPH	N9389	G-ANJA
E-15 (RNethAF)	G-BIYU	P6382	G-AJRS
E3B-143 (Span AF)	G-JUNG	R-151 (RNethAF)	G-BIYR
E3B-153 (781-75 Span AF)	G-BPTS	R-163 (RNethAF)	G-BIRH
E3B-350 (05-97 Span AF)	G-BHPL	R-167 (RNethAF)	G-LION
E449	G-EBJE	R1914	G-AHUJ
F141 (G)	G-SEVA	R3281 (UX-N)	G-BPIV
F235 (B)	G-BMDB	R5250	G-AODT
F904	G-EBIA	S1287	G-BEYB
F938	G-EBIC	S1579 (571)	G-BBVO
F943	G-BIHF	S1581 (573)	G-BWWK
F943	G-BKDT	T5672	G-ALRI
F5447 (N)	G-BKER	T5854	G-ANKK
F5459 (Y)	G-INNY	T5879	G-AXBW
F8010 (Z)	G-BDWJ	T6313	G-AHVU
F8614	G-AWAU	T6562	G-ANTE
G-48-1 (Class B)	G-ALSX	T6818 (91)	G-ANKT
H5199	G-ADEV	T7230	G-AFVE
J-1149 (Swiss AF)	G-SWIS	T7281	G-ARTL
J-1573 (Swiss AF)	G-VICI	T7328	G-APPN
J-1605 (Swiss AF)	G-BLID	T7404 (04)	G-ANMV
J-1632 (Swiss AF)	G-VNOM	T7793	G-ANKV
J-1758 (Swiss AF)	G-BLSD	T7842	G-AMTF
J-4021	G-HHAC	T7909	G-ANON
J-4083 (Swiss AF)	G-EGHH	T8191	G-BWMK
J-4090 (Swiss AF)	G-SIAL	T9707	G-AKKR
J7326	G-EBQP	T9738	G-AKAT
J9941 (57)	G-ABMR	U-0247 (Class B identity)	G-AGOY
K1786	G-AFTA	U-80 (Swiss AF)	G-BUKK

Serial carried	Civil identity	Serial carried	Civil identity
U-99 (Swiss AF)	G-AXMT	MJ627 (G9-P)	G-BMSB
U-108 (Swiss AF)	G-BJAX	ML407 (OU-V)	G-LFIX
U-110 (Swiss AF)	G-PTWO	MP425	G-AITB
V-54 (Swiss AF)	G-BVSD	MS824 (Fr AF)	G-AWBU
V1075	G-AKPF	MT438	G-AREI
V3388	G-AHTW	MT928 (ZX-M)	G-BKMI
V9367 (MA-B)	G-AZWT	MV268 (JE-J)	G-SPIT
V9673 (MA-J)	G-LIZY	MW401	G-PEST
W7 (Italian AF)	G-AGFT	MW763 (HF-A)	G-TEMT
W2718	G-RNLI	NJ673	G-AOCR
W5856 (A2A)	G-BMGC	NJ695	G-AJXV
W9385 (YG-L)	G-ADND	NJ719	G-ANFU
Z2033 (N/275)	G-ASTL	NL750	G-AOBH
Z5140 (HA-C)	G-HURI	NL985	G-BWIK
Z5207	G-BYDL	NM181	G-AZGZ
Z5252 (GO-B)	G-BWHA	NP303	G-ANZJ
Z7015 (7-L)	G-BKTH	NX534	G-BUDL
Z7197	G-AKZN	NX611 (LE-C/DX-C)	G-ASXX
AP506	G-ACWM	PS853 (C)	G-RRGN
AP507 (KX-P)	G-ACWP	PT462 (SW-A)	G-CTIX
AR213 (PR-D)	G-AIST	PT879	G-BYDE
AR501 (NN-A)	G-AWII	RG333	G-AIEK
BB807	G-ADWO	RG333	G-AKEZ
BE417 (LK-A)	G-HURR	RH377	G-ALAH
BI-005 (RNethAF)	G-BUVN	RL962	G-AHED
BM597 (JH-C)	G-MKVB	RM221	G-ANXR
BW881	G-KAMM	RN218 (N)	G-BBJI
DE208	G-AGYU	RT486 (PF-A)	G-AJGJ
DE470	G-ANMY	RT520	G-ALYB
DE623	G-ANFI	RT610	G-AKWS
DE673	G-ADNZ	RX168	G-BWEM
DE992	G-AXXV	SG-3 (RBelAF)	G-BSKP
DF112	G-ANRM	SM845 (GZ-J)	G-BUOS
DF128 (RCO-U)	G-AOJJ	SM969 (D-A)	G-BRAF
DF155	G-ANFV	SX336	G-KASX
DF198	G-BBRB	TA634 (8K-K)	G-AWJV
DG590	G-ADMW	TA719 (6T)	G-ASKC
DR613	G-AFJB	TA805	G-PMNF
EM720	G-AXAN	TD248 (D)	G-OXVI
EN224	G-FXII	TE184 (D)	G-MXVI
EP120 (AE-A)	G-LFVB	TJ569	G-AKOW
FB226 (MT-A)	G-BDWM	TJ672	G-ANIJ
FE695 (94)	G-BTXI	TJ704 (JA)	G-ASCD
FE788	G-CTKL	TS798	G-AGNV
FR886	G-BDMS	TW439	G-ANRP
FT391	G-AZBN	TW467 (ROD-F)	G-ANIE
FX301 (FD-NQ)	G-JUDI	TW511	G-APAF
FZ626 (YS-DH)	G-AMPO	TW536 (TS-V)	G-BNGE
HB275	G-BKGM	TW591 (N)	G-ARIH
HB751	G-BCBL	TW641	G-ATDN
HD-75 (RBelAF)	G-AFDX	VF512 (PF-M)	G-ARRX
HM580	G-ACUU	VF516	G-ASMZ
JG891	G-LFVC	VF526 (T)	G-ARXU
JV579 (F)	G-RUMW	VL348	G-AVVO
KB889 (NA-I)	G-LANC	VL349	G-AWSA
KB994	G-BVBP	VM360	G-APHV
KD345 (130)	G-FGID	VP955	G-DVON
KF584 (RAI-X)	G-RAIX	VP981	G-DHDV
KZ321	G-HURY	VR192	G-APIT
LB264	G-AIXA	VR249 (FA-EL)	G-APIY
LB294	G-AHWJ	VR259 (M)	G-APJB
LB312	G-AHXE	VS356	G-AOLU
LB367	G-AHGZ	VS610 (K-L)	G-AOKL
LB375	G-AHGW	VS623	G-AOKZ
LF858	G-BLUZ	VT871	G-DHXX
LZ766	G-ALCK	VX118	G-ASNB
MD497	G-ANLW	VX147	G-AVIL
MH434 (ZD-B)	G-ASJV	VZ638 (HF)	G-JETM

Serial carried	Civil identity	Serial carried	Civil identity
VZ728	G-AGOS	WK640 (C)	G-BWUV
WA576	G-ALSS	WK642 (94)	G-BXDP
WA577	G-ALST	WL505	G-FBIX
WA591 (W)	G-BWMF	WL505	G-MKVI
WB188	G-BZPB (green)	WL626	G-BHDD
WB188	G-BZPC (red)	WM167	G-LOSM
WB531	G-BLRN	WP788	G-BCHL
WB565 (X)	G-PVET	WP790 (T)	G-BBNC
WB569	G-BYSJ	WP795 (901)	G-BVZZ
WB571 (34)	G-AOSF	WP800 (2)	G-BCXN
WB585 (RCU-X)	G-AOSY	WP803	G-HAPY
WB588 (D)	G-AOTD	WP805 (D)	G-MAJR
WB615 (E)	G-BXIA	WP808	G-BDEU
WB652	G-CHPY	WP809 (78 RN)	G-BVTX
WB654	G-BXGO	WP840 (9)	G-BXDM
WB660	G-ARMB	WP844 (85)	G-BWOX
WB671 (910)	G-BWTG	WP856 (904 RN)	G-BVWP
WB697 (95)	G-BXCT	WP857 (24)	G-BDRJ
WB702	G-AOFE	WP859 (E)	G-BXCP
WB703	G-ARMC	WP860 (6)	G-BXDA
WB711	G-APPM	WP896 (M)	G-BWVY
WB726 (E)	G-AOSK	WP901 (B)	G-BWNT
WD286 (J)	G-BBND	WP903	G-BCGC
WD292	G-BCRX	WP925 (C)	G-BXHA
WD297	G-ARMD	WP928	G-BXGM
WD305	G-ARGG	WP929 (F)	G-BXCV
WD310 (B)	G-BWUN	WP930	G-BXHF
WD331 (J)	G-BXDH	WP971	G-ATHD
WD347	G-BBRV	WP983	G-BXNN
WD363 (5)	G-BCIH	WP984 (H)	G-BWTO
WD373 (12)	G-BXDI	WR410 (N)	G-BLKA
WD379 (K)	G-APLO	WR410	G-DHUU
WD390 (68)	G-BWNK	WR421	G-DHTT
WD413	G-VROE	WT333	G-BVXC
WE402	G-VENI	WT722 (878/VL)	G-BWGN
WE569	G-ASAJ	WT933	G-ALSW
WE724 (062)	G-BUCM	WV198 (K)	G-BJWY
WF118	G-DACA	WV318	G-FFOX
WF877	G-BPOA	WV372 (R)	G-BXFI
WG307	G-BCYJ	WV493 (29)	G-BDYG
WG308 (8)	G-BYHL	WV740	G-BNPH
WG316	G-BCAH	WV783	G-ALSP
WG348	G-BBMV	WW453 (W-S)	G-TMKI
WG350	G-BPAL	WZ507	G-VTII
WG407	G-BWMX	WZ589	G-DHZZ
WG422 (16)	G-BFAX	WZ662	G-BKVK
WG465	G-BCEY	WZ711	G-AVHT
WG469 (72)	G-BWJY	WZ847 (F)	G-CPMK
WG472	G-AOTY	WZ868 (H CUAS)	G-ARMF
WG719	G-BRMA	WZ879 (73)	G-BWUT
WJ358	G-ARYD	WZ882 (K)	G-BXGP
WJ945 (21)	G-BEDV	XA880	G-BVXR
WK163	G-BVWC	XD693 (Z-Q)	G-AOBU
WK436	G-VENM	XE665 (876/VL)	G-BWGM
WK512 (A)	G-BXIM	XE685 (861/VL)	G-GAII
WK514	G-BBMO	XE689 (864/VL)	G-BWGK
WK517 (84)	G-ULAS	XE856	G-DUSK
WK522	G-BCOU	XE897	G-DHVV
WK549 (Y)	G-BTWF	XE920 (A)	G-VMPR
WK586	G-BXGX	XE956	G-OBLN
WK590 (69)	G-BWVZ	XF114	G-SWIF
WK609 (93)	G-BXDN	XF303 (105-A)	G-BWOU
WK611	G-ARWB	XF515 (R)	G-KAXF
WK622	G-BCZH	XF597 (AH)	G-BKFW
WK624 (M)	G-BWHI	XF603	G-KAPW
WK628	G-BBMW	XF690	G-MOOS
WK630 (11)	G-BXDG	XF766	G-SPDR
WK633 (A)	G-BXEC	XF785	G-ALBN

Serial carried	Civil identity	Serial carried	Civil identity
XF836 (J-G)	G-AWRY	XR724	G-BTSY
XF877 (JX)	G-AWVF	XR944	G-ATTB
XG160	G-BWAF	XR991	G-MOUR
XG452	G-BRMB	XR993	G-BVPP
XG547	G-HAPR	XS101	G-GNAT
XG775	G-DHWW	XS104	G-FRCE
XH568	G-BVIC	XS165 (37)	G-ASAZ
XH558	G-VLCN	XS587 (252/V)	G-VIXN
XJ389	G-AJJP	XS765	G-BSET
XJ615	G-BWGL	XS770	G-HRHI
XJ729	G-BVGE	XT223	G-XTUN
XJ771	G-HELV	XT435 (430)	G-RIMM
XK416	G-AYUA	XT634	G-BYRX
XK417	G-AVXY	XT671	G-BYRC
XK482	G-BJWC	XT781 (426)	G-KAWW
XK895 (19/CU)	G-SDEV	XT788 (316)	G-BMIR
XK896	G-RNAS	XV126 (X)	G-SCTA
XK940	G-AYXT	XV130 (R)	G-BWJW
XL426	G-VJET	XV134 (P)	G-BWLX
XL502	G-BMYP	XV140 (K)	G-KAXL
XL571	G-HNTR	XV268	G-BVER
XL573	G-BVGH	XW289 (73)	G-JPVA
XL577 (V)	G-BXKF	XW293	G-BWCS
XL602	G-BWFT	XW324 (K)	G-BWSG
XL621	G-BNCX	XW325 (E)	G-BWGF
XL714	G-AOGR	XW333 (79)	G-BVTC
XL716	G-AOIL	XW423 (14)	G-BWUW
XL809	G-BLIX	XW433 (63)	G-JPRO
XL812	G-SARO	XW635	G-AWSW
XL929	G-BNPU	XW784 (VL Royal Navy)	G-BBRN
XL954	G-BXES	XW866 (E)	G-BXTH
XM223 (J)	G-BWWC	XW895 (51/CU)	G-BXZD
XM370 (10)	G-BVSP	XW910 (K)	G-BXZE
XM376 (27)	G-BWDR	XX467	G-TVII
XM424	G-BWDS	XX469	G-BNCL
XM478 (33)	G-BXDL	XX525 (8)	G-CBJJ
XM479 (54)	G-BVEZ	XX622 (B)	G-CBGX
XM553	G-AWSV	XX630 (5)	G-SIJW
XM575	G-BLMC	XX713 (2)	G-CBJK
XM655	G-VULC	XX885	G-HHAA
XM685 (513/PO)	G-AYZJ	ZA250	G-VTOL
XM693	G-TIMM	ZA634 (C)	G-BUHA
XM819	G-APXW	ZB500	G-LYNX
XN351	G-BKSC	2+1 (7334 Luftwaffe)	G-SYFW
XN435	G-BGBU	3+ (Luftwaffe)	G-BAYV
XN437	G-AXWA	4+ (Luftwaffe)	G-BSLX
XN441	G-BGKT	4-97/MM52801 (Italian)	G-BBII
XN459 (N)	G-BWOT	07 (Russian AF)	G-BMJY
XN498 (16)	G-BWSH	8+ (Luftwaffe)	G-WULF
XN629	G-KNOT	F+IS (Luftwaffe)	G-BIRW
XN637 (03)	G-BKOU	BU+CC (Luftwaffe)	G-BUCC
XP242	G-BUCI	BU+CK (Luftwaffe)	G-BUCK
XP254	G-ASCC	CC+43 (Luftwaffe)	G-CJCI
XP279	G-BWKK	CF+HF (Luftwaffe)	EI-AUY
XP282	G-BGTC	LG+01 (Luftwaffe)	G-AYSJ
XP355	G-BEBC	LG+03 (Luftwaffe)	G-AEZX
XP672 (27)	G-RAFI	KG+EM (Luftwaffe)	G-ETME
XP772	G-BUCJ	NJ+C11 (Luftwaffe)	G-ATBG
XP907	G-SROE	S4+A07 (Luftwaffe)	G-BWHP
XR240	G-BDFH	S5-B06 (Luftwaffe)	G-BSFB
XR241	G-AXRR	TA+RC (Luftwaffe)	G-BPHZ
XR246	G-AZBU	6J+PR (Luftwaffe)	G-AWHB
XR486	G-RWWW	57-H (USAAC)	G-AKAZ
XR537 (T)	G-NATY	97+04 (Luftwaffe)	G-APVF
XR538 (01)	G-RORI	+114 (Luftwaffe)	G-BSMD
XR595 (M)	G-BWHU	146-11083 (5)	G-BNAI
XR673 (L)	G-BXLO		

G-BVXR, D.H.104 Devon C.2, still carries its former military identity. *DP*

G-CCHY, Bucker Bu 131 Jungmann flies in Swiss Air Force markings. *DP*

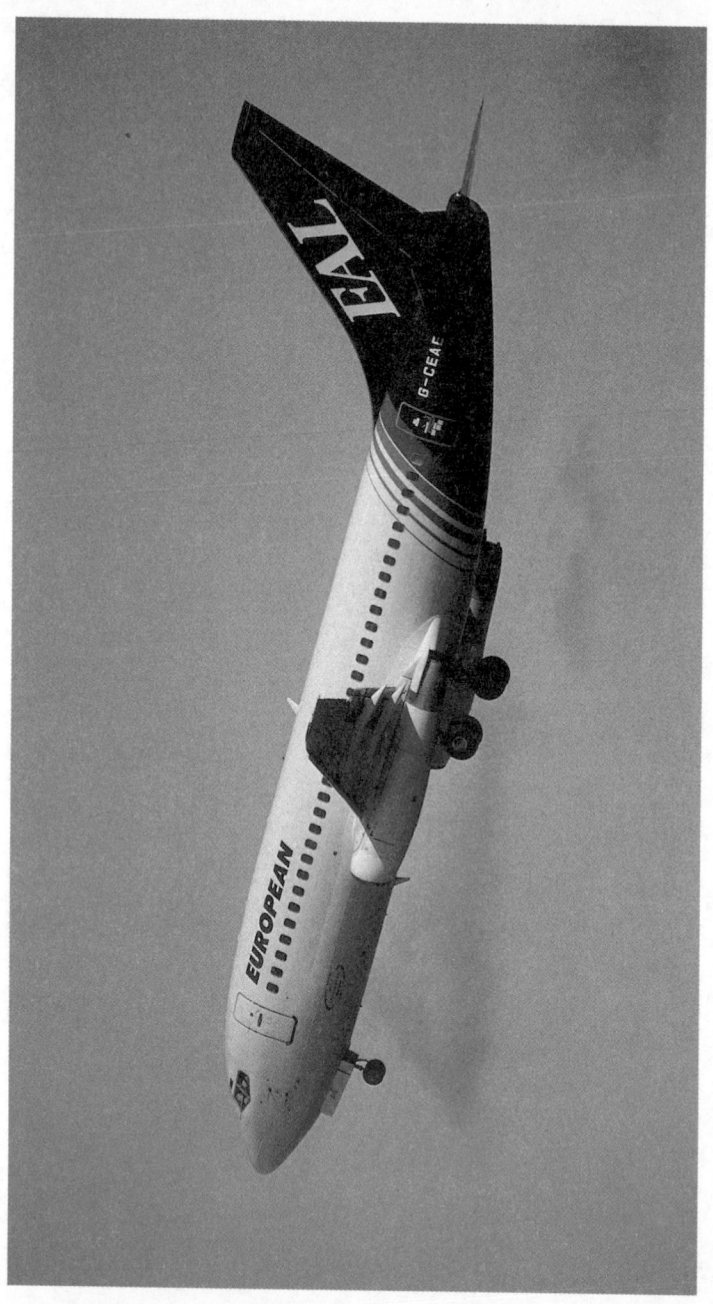

G-CEAE, Boeing 737-229 of European Aviation is often leased to other operators. *DP*

G-BXZO, Pietenpol Air Camper. *DP*

G-OROD, Piper PA-18-150 Super Cub. *DP*

G-ENCE, Partenavia P68B. *DP*

G-BMDP, Partenavia P64B Oscar is the only one of its type on the UK register. *DP*

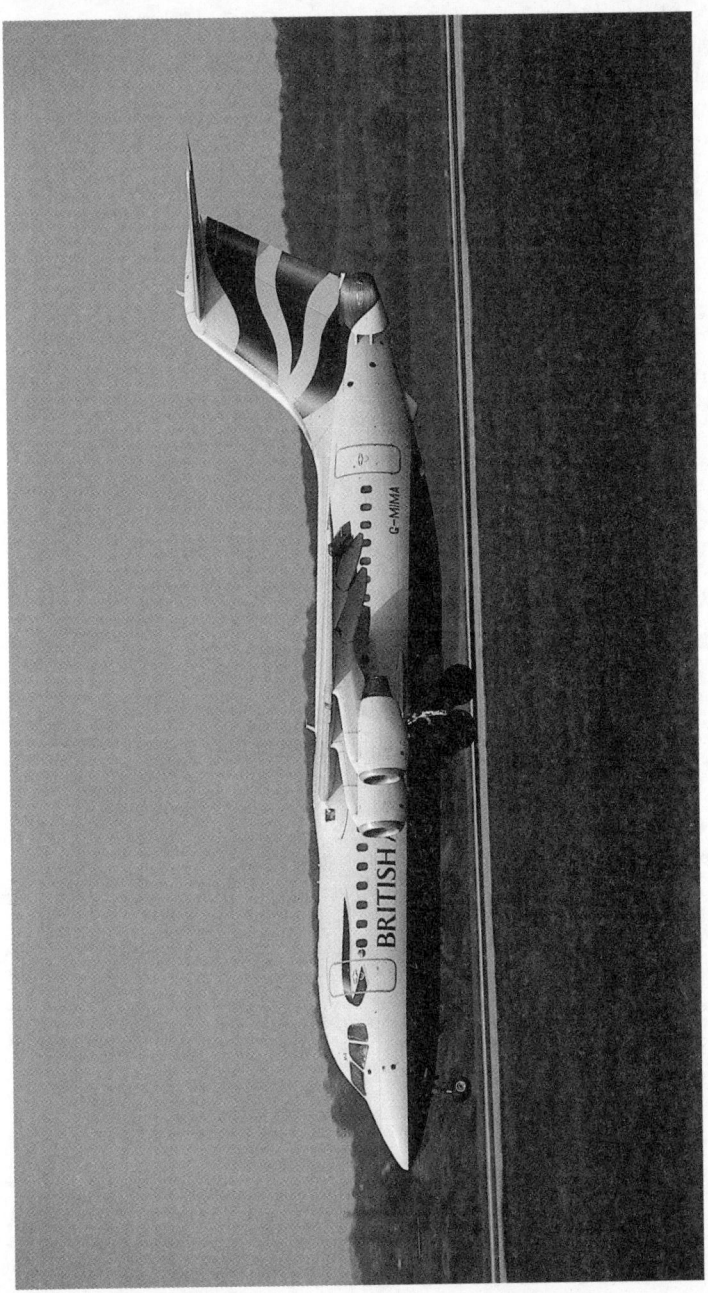

G-MIMA, BAe-146 of British Regional. *DP*

G-MAJF, BAe Jetsream 4102 of Eastern Airways. AJW

G-ZAAZ, Van's RV-8. AJW

G-BTDZ, CASA 1.131E Jungmann 2000. AJW

G-EVRO, Aerotechnik EV-97 Eurostar. *AJW*

D-ABAV, Boeing 737-8AS of Air Berlin operates low-cost services to the UK. *DP*

CS-TGP, Boeing 737-3Q8 of SATA International. AJW

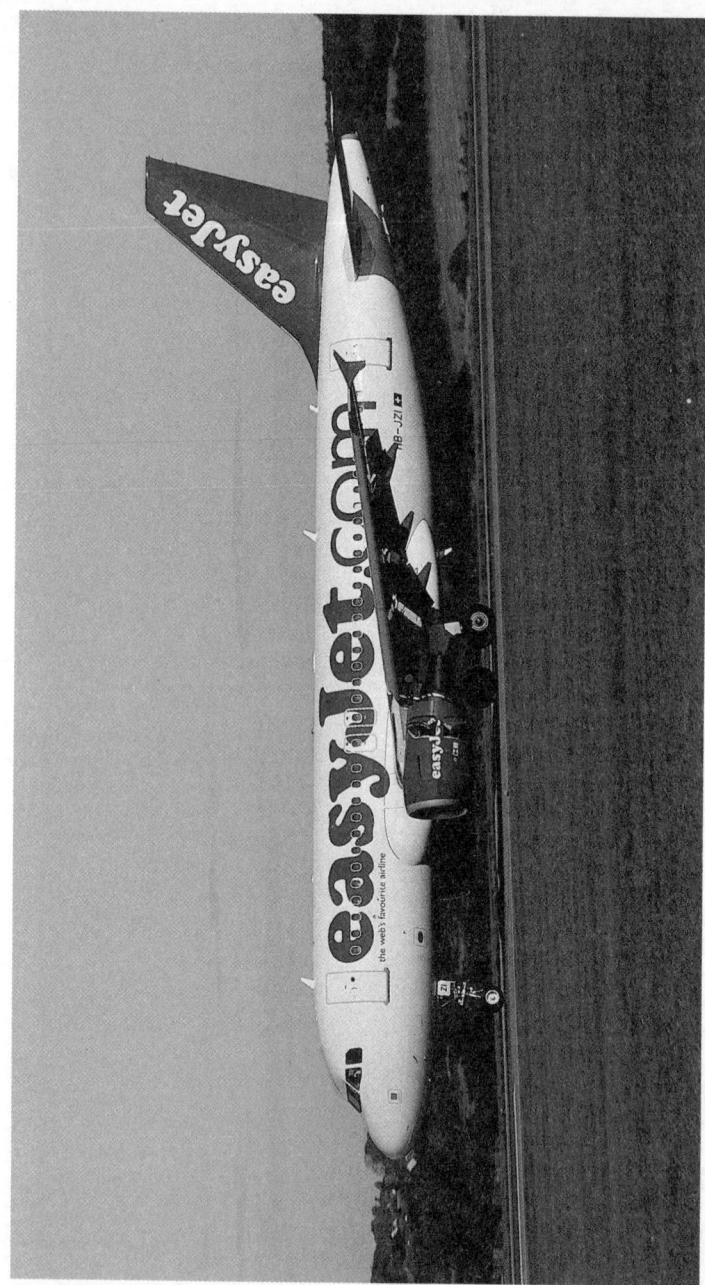

HB-JZI, Airbus A.319-111 of easyJet Switzerland. *DP*

N606FE, McDonnell-Douglas MD-11F, operates transatlantic services for FedEx. *DP*

EC-HKP, McDonnall-Douglas MD-83 of Spanair visits many UK airports on holiday charters. *DP*

Republic of Ireland Civil Registrations

Reg.	Type	Owner or Operator	Notes
EI-ABI	D.H.84 Dragon	Aer Lingus Teo *Iolar* (EI-AFK)	
EI-ADV	PA-12 Super Cruiser	R. E. Levis	
EI-AFE	Piper J3C-65 Cub	J. Conlon	
EI-AFF	B.A. Swallow 2	J. J. Sullivan & ptnrs	
EI-AGD	Taylorcraft Plus D	B. & K. O'Sullivan	
EI-AGJ	J/1 Autocrat	T. G. Rafter	
EI-AHA	D.H.82A Tiger Moth	J. H. Maher	
EI-AHI	D.H.82A Tiger Moth	High Fidelity Flyers	
EI-AKM	Piper J-3C-65 Cub	Setanta Flying Group	
EI-ALH	Taylorcraft Plus D	N. Reilly	
EI-AMK	J/1 Autocrat	J. J. Sullivan	
EI-AMY	J/1N Alpha	T. Lennon	
EI-ANT	Champion 7ECA Citabria	T. Croke & ptnrs	
EI-ANY	PA-18 Super Cub 95	Bogavia Group	
EI-AOB	PA-28 Cherokee 140	Knock Flying Group	
EI-AOS	Cessna 310B	Southair	
EI-APF	Cessna F.150F	Sligo Aero Club	
EI-APS	Schleicher ASK.14	E. Shiel & ptnrs	
EI-ARH	Currie Wot/S.E.5 Replica	L. Garrison	
EI-ARM	Currie Wot/S.E.5 Replica	L. Garrison	
EI-ARW	Jodel D.R.1050	J. Davy	
EI-ASR	Mc Candles Gyroplane MK4	J. J. Fasenfeld	
EI-AST	Cessna F.150H	Ormond Flying Club	
EI-ATJ	B.121 Srs 2	L. O'Leary	
EI-ATK	PA28-1 Cherokee	Mayo Flying Club	
EI-ATL	Aeronca 7AC	Kildare Flying Club	
EI-ATP	LA-4A Minor	Hanging in the Terminal of Miami International	
EI-ATS	M.S.880B Rallye Club	ATS Group	
EI-AUC	Cessna FA.150K	Garda Aviation Club	
EI-AUE	M.S.880B Rallye Club	Kilkenny Flying Club	
EI-AUG	M.S.894 Rallye Minerva 220	K. O'Leary	
EI-AUJ	M.S.880B Rallye Club	Ormond Flying Club	
EI-AUM	J/1 Autocrat	T. G. Rafter	
EI-AUO	Cessna FA.150K Aerobat	Kerry Aero Club	
EI-AUS	J/5F Aiglet Trainer	T. Stevens	
EI-AUT	Forney F-1A Aircoupe	Southair	
EI-AUY	Morane-Saulnier M.S.502 (CF+HF)	Historical Aircraft Preservation Group	
EI-AVB	Aeronca 7AC Champion	T. Brett	
EI-AVM	Cessna F.150L	Tojo Air Leasing	
EI-AWD	PA22 Colt-160	J. P. Montcalm	
EI-AWH	Cessna 210J	Rathcoole Flying Club	
EI-AWP	D.H.82A Tiger Moth	A. P. Bruton	
EI-AWR	Malmö MFI-9 Junior	M. Whyte & J. Brennen	
EI-AWU	M.S.880B Rallye Club	Longford Aviation	
EI-AYA	M.S.880B Rallye Club	Limerick Flying Club (Coonagh)	
EI-AYB	GY-80 Horizon 180	J. B. Smith	
EI-AYD	AA-5 Traveler	V. O'Rourke & ptnrs	
EI-AYF	Cessna FRA.150L	Limerick Flying Club (Coonagh)	
EI-AYI	M.S.880B Rallye Club	J. McNamara	
EI-AYK	Cessna F.172M	D. Gallagher	
EI-AYN	BN-2A-8 Islander	Aer Arann	
EI-AYR	Schleicher ASK-16	B. O'Broin & ptnrs	
EI-AYT	M.S.894A Rallye Minerva	K. A. O'Connor	
EI-AYV	M.S.892A Rallye Commodore 150	P. Murtagh	
EI-AYY	Evans VP-1	G. Dowd & P. O'Rourke	
EI-BAJ	Stampe SV-4C	Dublin Tiger Group	
EI-BAR	Thunder Ax8-105 balloon	J. Burke & ptnrs	
EI-BAT	Cessna F.150M	K. A. O'Connor	
EI-BAV	PA-22 Colt 108	E. Finnamore & J. Deegan	
EI-BBC	PA-28 Cherokee 180C	Vero Beach	
EI-BBD	Evans VP-1	Volksplane Group	
EI-BBE	Champion 7FC Tri-Traveler (tailwheel)	R. McNally & C. Carey	

Notes	Reg.	Type	Owner or Operator
	EI-BBG	M.S.880B Rallye Club ★	Weston (stored)
	EI-BBI	M.S.892 Rallye Commodore	Ossory Flying & Gliding Club
	EI-BBJ	M.S.880B Rallye Club	Weston
	EI-BBO	M.S.893E Rallye 180GT	G. P. Moorhead
	EI-BBV	Piper J-3C-65 Cub	F. Cronin
	EI-BCE	BN-2A-26 Islander	Aer Arann
	EI-BCF	Bensen B.8M	P. Flanagan
	EI-BCJ	Aeromere F.8L Falco 1 Srs 3	D. Kelly
	EI-BCK	Cessna F.172N II	K. A. O'Connor
	EI-BCL	Cessna 182P	L. Burke
	EI-BCM	Piper J-3C-65 Cub	Kilmoon Flying Group
	EI-BCN	Piper J-3C-65 Cub	H. Diver
	EI-BCO	Piper J-3C-65 Cub	J. Molloy
	EI-BCP	D.62B Condor	T. Delaney
	EI-BCS	M.S.880B Rallye Club	Organic Fruit & Vegetables
	EI-BCU	M.S.880B Rallye Club	Weston
	EI-BCW	M.S.880B Rallye Club	Kilkenny Flying Club
	EI-BDH	M.S.880B Rallye Club	Munster Wings
	EI-BDK	M.S.880B Rallye Club	Limerick Flying Club (Coonagh)
	EI-BDL	Evans VP-2	P. Buggle
	EI-BDM	PA-23 Aztec 250D	G. A. Costello
	EI-BDR	PA-28 Cherokee 180	Cherokee Group
	EI-BEA	M.S.880B Rallye 100ST	Weston (stored)
	EI-BEN	Piper J-3C-65 Cub	Capt. J. J. Sullivan
	EI-BEP	M.S.892A Rallye 150	H. Lynch & J. O'Leary
	EI-BFE	Cessna F.150G	Southair
	EI-BFF	Beech A.23 Musketeer	P. McCoole
	EI-BFI	M.S.880B Rallye 100ST	J. O'Neill
	EI-BFO	Piper J-3C-90 Cub	D. Gordon
	EI-BFP	M.S.800B Rallye 100ST	Limerick Flying Club (Coonagh)
	EI-BFR	M.S.880B Rallye 100ST	Wexford Flying Group
	EI-BFV	M.S.880B Rallye 100T	Ormond Flying Club
	EI-BGA	SOCATA Rallye 100ST	J. J. Frew
	EI-BGB	M.S.880B Rallye Club	Limerick Flying Club (Coonagh)
	EI-BGC	M.S.880B Rallye Club	P. Moran
	EI-BGD	M.S.880B Rallye Club	N. Kavanagh
	EI-BGF	PA-28R Cherokee	Arrow Group
	EI-BGG	M.S.892E Rallye 150	M. Martin
	EI-BGJ	Cessna F.152	Sligo Aero Club
	EI-BGS	M.S.893E Rallye	M. Farrelly
	EI-BGT	Colt 77A balloon	M. J. Mills
	EI-BGU	M.S.880B Rallye Club	M. F. Neary
	EI-BHC	Cessna F.177RG	L. Gavin
	EI-BHF	M.S.892A Rallye Commodore 150	B. Mullen
	EI-BHI	Bell 206B JetRanger 2	G. Tracy
	EI-BHM	Cessna F.337E	Clity of Dublin VEC
	EI-BHN	M.S.893A Rallye Commodore 180T	T. Garvan
	EI-BHP	M.S.893A Rallye Commodore 180T	Spanish Point Flying Club
	EI-BHT	Beech 77 Skipper	Waterford Aero Club
	EI-BHV	Champion 7EC Traveler	P. O'Donnell & ptnrs
	EI-BHW	Cessna F.150F	R. Sharpe
	EI-BHY	SOCATA Rallye 150ST	Limerick Flying Club (Coonagh)
	EI-BIB	Cessna F.152	Galway Flying Club
	EI-BID	PA-18 Super Cub 95	S. Coghlan & P. Ryan
	EI-BIG	Zlin 526	P. von Lonkhuyzen
	EI-BIJ	AB-206B JetRanger 2	Medava Properties
	EI-BIK	PA-18 Super Cub 180	Dublin Gliding Club
	EI-BIM	M.S.880B Rallye Club	D. Millar
	EI-BIO	Piper J-3C-65 Cub	Monasterevan Flying Group
	EI-BIR	Cessna F.172M	B. Harrison & ptnrs
	EI-BIS	Robin R.1180TD	Robin Aiglon Group
	EI-BIT	M.S.887 Rallye 125	Spanish Point Flying Club
	EI-BIV	Bellanca 8KCAB Citabria	Aerocrats Flying Group
	EI-BIW	M.S.880B Rallye Club	E. J. Barr
	EI-BJB	Aeronca 7AC Champion	A. W. Kennedy
	EI-BJC	Aeronca 7AC Champion	A. E. Griffin
	EI-BJI	Cessna FR.172E	Irish Parachute Club
	EI-BJJ	Aeronca 15AC Sedan	O. Bruton
	EI-BJK	M.S.880B Rallye 110ST	A. M. Keenen
	EI-BJM	Cessna A.152	Leinster Aero Club
	EI-BJO	Cessna R.172K	A. P. Hogan & ptnrs
	EI-BJT	PA-38-112 Tomahawk	S. Corrigan & W. Lennon

Reg.	Type	Owner or Operator	Notes
EI-BKC	Aeronca 15AC Sedan	G. Hendrick & M. Farrell	
EI-BKF	Cessna F.172H	D. Darby	
EI-BKK	Taylor JT.1 Monoplane	Waterford Aero Club	
EI-BKN	M.S.880B Rallye 100ST	Weston	
EI-BKU	M.S.892A Rallye Commodore 150	Limerick Flying Club (Coonagh)	
EI-BLB	SNCAN Stampe SV-4C	J. Hutchinson & R. A. Stafford	
EI-BLD	Bolkow Bo 105C	Irish Helicopters	
EI-BLE	Eipper Microlight	R. Smith & P. St George	
EI-BLN	Eipper Quicksilver MX	O. J. Conway & ptnrs	
EI-BMA	M.S.880B Rallye Club	W. Rankin & M. Kelleher	
EI-BMB	M.S.880B Rallye 100T	Glyde Court Developments	
EI-BMF	Laverda F.8L Falco	M. Slazenger	
EI-BMH	M.S.880B Rallye Club	N. S. Bracken	
EI-BMI	SOCATA TB.9 Tampico	Ashford Flying Group	
EI-BMJ	M.S.880B Rallye 100T	Limerick Flying Club (Coonagh)	
EI-BMM	Cessna F.152 II	P. Redmond	
EI-BMN	Cessna F.152 II	Sligo Light Aviation Club	
EI-BMU	Monnet Sonerai IIL	A. Fenton	
EI-BMV	Grumman AA-5B	E. Tierney & K. A. Harold	
EI-BMW	Vulcan Air Trike	L. Maddock & ptnrs	
EI-BNF	Goldwing Canard	T. Morelli	
EI-BNH	Hiway Skytrike	M. Martin	
EI-BNJ	Evans VP-2	G. Cashman	
EI-BNK	Cessna U.206F	Irish Parachute Club	
EI-BNL	Rand KR-2	K. Hayes	
EI-BNP	Rotorway 133	R. L. Renfroe	
EI-BNT	Cvjetkovic CA-65	B. Tobin & ptnrs	
EI-BNU	M.S.880B Rallye Club	P. A. Doyle	
EI-BOA	Pterodactyl Ptraveller	A. Murphy	
EI-BOE	SOCATA TB.10 Tobago	Tobago Group	
EI-BOH	Eipper Quicksilver	J. Leech	
EI-BOV	Rand KR-2	G. O'Hara & G. Callan	
EI-BOX	Jordan Duet	Dr. K. Riccius	
EI-BPE	Viking Dragonfly	G. G. Bracken	
EI-BPL	Cessna F.172K	Phoenix Flying	
EI-BPN	Flexiform Striker	P. H. Collins	
EI-BPO	Southdown Sailwings	A. Channing	
EI-BPP	Quicksilver MX	J. A. Smith	
EI-BPT	Skyhook Sabre	T. McGrath	
EI-BPU	Hiway Demon	A. Channing	
EI-BRK	Flexiform Trike	L. Maddock & ptnrs	
EI-BRS	Cessna P.172D	P. Mattews	
EI-BRU	Evans VP-1	Home Bru Flying Group	
EI-BRV	Hiway Demon	M. Garvey & C. Tully	
EI-BRW	Ultralight Deltabird	A & E Aerosport	
EI-BSB	Jodel D.112	Estartit	
EI-BSC	Cessna F.172N	S. Phelan	
EI-BSG	Bensen B.80	J. Todd	
EI-BSK	SOCATA TB.9 Tampico	Weston	
EI-BSL	PA-34-220T Seneca	P. Sreenan	
EI-BSN	Cameron O-65 balloon	C. O'Neill & T. Hooper	
EI-BSO	PA-28 Cherokee 140B	H. N. Hanley	
EI-BSV	SOCATA TB.20 Trinidad	J. Condron	
EI-BSW	Solar Wings Pegasus XL-R	E. Fitzgerald	
EI-BSX	Piper J-3C-65 Cub	J. & T. O'Dwyer	
EI-BTX	McD Douglas MD-82	Airplanes Holdings/AeroMexico	
EI-BTY	McD Douglas MD-82	Airplanes Holdings/AeroMexico	
EI-BUA	Cessna 172M	Skyhawks Flying Club	
EI-BUC	Jodel D.9 Bebe	B. Lyons	
EI-BUF	Cessna 210N	210 Group	
EI-BUG	SOCATA ST.10 Diplomate	J. Cooke	
EI-BUH	Lake LA.4-200 Buccaneer	P. Redden	
EI-BUJ	M.S.892A Rallye Commodore 150	T. Cunniffe	
EI-BUL	Whittaker MW-5 Sorcerer	J. Culleton	
EI-BUN	Beech 76 Duchess	K. A. O'Connor	
EI-BUT	M.S.893A Commodore 180	T. Keating	
EI-BVB	Whittaker MW.6 Merlin	R. England	
EI-BVJ	AMF Chevvron 232	A. Dunn	
EI-BVK	PA-38-112 Tomahawk	M. Martin	
EI-BVT	Evans VP-2	P. Morrison	
EI-BVY	Zenith 200AA-RW	J. Matthews & M. Skelly	
EI-BWH	Partenavia P.68C	K. Buckley	

Notes	Reg.	Type	Owner or Operator
	EI-BXI	Boeing 737-448	Aer Lingus Teo *St Finnian*
	EI-BXK	Boeing 737-448	Aer Lingus Teo *St Caimin*
	EI-BXL	Polaris F1B-OK350	M. McKeon
	EI-BXO	Fouga CM.170 Magister	G. Connolly
	EI-BXT	D.62B Condor	The Condor Group
	EI-BYA	Thruster TST Mk 1	E. Fagan
	EI-BYF	Cessna 150M	D. Cashin
	EI-BYG	SOCATA TB.9 Tampico	Weston
	EI-BYJ	Bell 206B JetRanger	Medeva Properties
	EI-BYL	Zenith CH.250	M. McLoughlin
	EI-BYO	Aerospatiale ATR-42-300	Aer Arann
	EI-BYR	Bell 206L-3 LongRanger 3	H. S. S.
	EI-BYX	Champion 7GCAA	P. J. Gallagher
	EI-BYY	Piper J-3C-85 Cub	The Cub Club
	EI-BZE	Boeing 737-3Y0	Paloma Developments/Philippine Airlines
	EI-BZF	Boeing 737-3Y0	Pergola/Philippine Airlines
	EI-BZJ	Boeing 737-3Y0	Pergola/Philippine Airlines
	EI-BZN	Boeing 737-3Y0	Airplanes Finance/Philippine Airlines
	EI-CAC	Grob G.115A	G. Tracey
	EI-CAD	Grob G.115A	Flightwise Training
	EI-CAE	Grob G.115A	K. A. O'Connor
	EI-CAN	Aerotech MW.5 Sorcerer	V. A. Vaughan
	EI-CAP	Cessna R.182RG	M. J. Hanlon
	EI-CAU	AMF Chevvron 232	A. J. Farrant
	EI-CAW	Bell 206B JetRanger	Celtic Helicopters
	EI-CAX	Cessna P.210N	K. A. O'Connor
	EI-CAY	Mooney M.20C	Ranger Flights
	EI-CBK	Aérospatiale ATR-42-310	GPA-ATR/Aer Arann
	EI-CBR	McD Douglas MD-83	Airplanes 111/Avianca
	EI-CBS	McD Douglas MD-83	GECAS Technical Services/Avianca
	EI-CBY	McD Douglas MD-83	GECAS Technical Services/Avianca
	EI-CBZ	McD Douglas MD-83	GECAS Technical Services/Avianca
	EI-CCC	McD Douglas MD-83	Airplanes 111/Avianca
	EI-CCD	Grob G.115A	Kal Aviation
	EI-CCE	McD Douglas MD-83	GECAS Technical Services/Avianca
	EI-CCF	Aeronca 11AC Chief	L. Murray & ptnrs
	EI-CCJ	Cessna 152 II	P. Cahill
	EI-CCK	Cessna 152 II	P. Cahill
	EI-CCL	Cessna 152 II	P. Cahill
	EI-CCM	Cessna 152 II	E. Hopkins
	EI-CCV	Cessna R.172K-XP	Kerry Aero Club
	EI-CDD	Boeing 737-548	Aer Lingus Teo *St Macartan*
	EI-CDE	Boeing 737-548	Aer Lingus Teo *St Jarlath*
	EI-CDF	Boeing 737-548	Aer Lingus Teo *St Cronan*
	EI-CDG	Boeing 737-548	Aer Lingus Teo *St Moling*
	EI-CDH	Boeing 737-548	Aer Lingus Teo *St Ronan*
	EI-CDP	Cessna 182L	Irish Parachute Club
	EI-CDV	Cessna 150G	K. A. O'Connor
	EI-CDX	Cessna 210K	Falcon Aviation
	EI-CDY	McD Douglas MD-83	GECAS Technical Services/Avianca
	EI-CEG	M.S.893A Rallye 180GT	M. Farrely
	EI-CEK	McD Douglas MD-83	Airplanes Finance/Nouvelair
	EI-CEN	Thruster T.300	P. A. J. Murphy
	EI-CEP	McD Douglas MD-83	GECAS Technical Services/Avianca
	EI-CEQ	McD Douglas MD-83	GECAS Technical Services/Avianca
	EI-CER	McD Douglas MD-83	Airplanes 111/Avianca
	EI-CES	Taylorcraft BC-65	N. O'Brien
	EI-CEX	Lake LA-4-200	C. L. Cargill
	EI-CEY	Boeing 757-2Y0	Pergola/Avianca
	EI-CEZ	Boeing 757-2Y0	Airplanes Holdings/Avianca
	EI-CFE	Robinson R-22B	ILH Enterprises
	EI-CFF	PA-12 Super Cruiser	J. & T. O'Dwyer
	EI-CFG	CP.301B Emeraude	F. Doyle
	EI-CFH	PA-12 Super Cruiser	T. Gabriel
	EI-CFN	Cessna 172P	B. Fitzmaurice & G. O'Connell
	EI-CFO	Piper J-3C-65 Cub	J. Matthews & ptnrs
	EI-CFP	Cessna 172P (floatplane)	K. A. O'Connor
	EI-CFX	Robinson R-22B	Ballaugh Motors
	EI-CFY	Cessna 172N	K. A. O'Connor
	EI-CFZ	McD Douglas MD-83	Airplanes 111/Avianca
	EI-CGB	Team Minimax	M. Garvey

Reg.	Type	Owner or Operator	Notes
EI-CGC	Stinson 108-3	A. D. Weldon	
EI-CGD	Cessna 172M	J. Murray	
EI-CGF	Luton LA-5 Major	J. Duggan	
EI-CGG	Erco Ercoupe 415C	Irish Ercoupe Group	
EI-CGH	Cessna 210N	J. Smith	
EI-CGJ	Solar Wings Pegasus XL-R	A. P. Hearty	
EI-CGM	Solar Wings Pegasus XL-R	Microflight	
EI-CGN	Solar Wings Pegasus XL-R	V. Power	
EI-CGP	PA-28 Cherokee 140C	G. Cashman	
EI-CGQ	AS.350B Ecureuil	Executive Helicopters (Cork)	
EI-CGT	Cessna 152 II	J. Rafter	
EI-CGV	Piper J-5A Cub Cruiser	J5 Group	
EI-CHH	Boeing 737-317	Airplanes Finance/Frontier Airlines	
EI-CHK	Piper J-3C-65 Cub	N. Higgins	
EI-CHM	Cessna 150M	K. A. O'Connor	
EI-CHP	D.H.C.8-103 Dash Eight	Airplanes Jetprop Finance/US Airways Express	
EI-CHR	CFM Shadow Srs BD	F. Maughan	
EI-CHS	Cessna 172M	Kerry Aero Club	
EI-CHT	Solar Wings Pegasus XL-R	J. Gratten	
EI-CHV	Agusta A.109A-II	Celtic Helicopters	
EI-CIA	M.S.880B Rallye Club	G. Hackett & C. Mason	
EI-CIF	PA-28 Cherokee 180C	AA Flying Group	
EI-CIG	PA-18 Super Cub 150	K. A. O'Connor	
EI-CIJ	Cessna 340	Airlink Airways	
EI-CIM	Avid Flyer Mk IV	P. Swan	
EI-CIN	Cessna 150K	K. A. O'Connor	
EI-CIR	Cessna 551 Citation II	Air Group Finance	
EI-CIV	PA-28 Cherokee 140	G. Cashman & E. Callanan	
EI-CIW	McD Douglas MD-83	Carotene/Meridiana	
EI-CIZ	Steen Skybolt	J. Keane	
EI-CJC	Boeing 737-204ADV	Ryanair (Hertz)	
EI-CJE	Boeing 737-204ADV	Ryanair (Jaguar) (stored)	
EI-CJF	Boeing 737-204ADV	Ryanair	
EI-CJG	Boeing 737-204ADV	Ryanair	
EI-CJH	Boeing 737-204ADV	Ryanair	
EI-CJI	Boeing 737-2E7ADV	Ryanair	
EI-CJR	SNCAN Stampe SV-4A	P. McKenna	
EI-CJS	Jodel D.120A	A. Flood	
EI-CJT	Slingsby Motor Cadet III	A. J. Tarrant	
EI-CJV	Moskito 2	M. Peril & ptnrs	
EI-CJZ	Whittaker MW.6 Merlin	M. McCarthy	
EI-CKG	Avon Hunt Weightlift	B. Kenny	
EI-CKH	PA-18 Super Cub 95	G. Brady & C. Keenan	
EI-CKI	Thruster TST Mk 1	N. Furlong	
EI-CKJ	Cameron N-77 balloon	A. F. Meldon	
EI-CKM	McD Douglas MD-83	Airplanes Finance/Meridiana	
EI-CKN	Whittaker MW.6-S Fatboy Flyer	F. Byrne & M. O.'Carroll	
EI-CKP	Boeing 737-2K2	Ryanair	
EI-CKS	Boeing 737-2T5	Ryanair (stored)	
EI-CKT	Mainair Gemini Flash	A. C. Burke	
EI-CKU	Solar Wings Pegasus SLR	M. O'Regan	
EI-CKZ	Jodel D.18	J. O'Brien	
EI-CLA	HOAC Katana DV.20	Weston	
EI-CLB	Aérospatiale ATR-72-212	Tarquin/Alitalia Express	
EI-CLC	Aérospatiale ATR-72-212	Tarquin/Alitalia Express	
EI-CLD	Aérospatiale ATR-72-212	Tarquin/Alitalia Express	
EI-CLL	Whittaker MW.6-S Fat Boy Flyer	F. Stack	
EI-CLQ	Cessna F.172N	K. Dardis & ptnrs	
EI-CLW	Boeing 737-3Y0	Airplanes Finance/Air One	
EI-CLZ	Boeing 737-3Y0	Airplanes Finance/Air One	
EI-CMB	PA-28 Cherokee 140	Dublin Flyers	
EI-CMJ	Aérospatiale ATR-72-212	Tarquin/Alitalia Express	
EI-CMK	Goldwing ST	M. Gavigan	
EI-CML	Cessna 150M	K. O'Connor	
EI-CMN	PA-12 Super Cruiser	D. Graham & ptnrs	
EI-CMR	Rutan LongEz	F. & C. O'Caoimh	
EI-CMS	BAe 146-200	CityJet	
EI-CMT	PA-34-200T Seneca II	Atlantic Air	
EI-CMU	Mainair Mercury	L. Langan & L. Laffan	
EI-CMV	Cessna 150L	K. A. O'Connor	
EI-CMW	Rotorway Executive	B. McNamee	
EI-CMY	BAe 146-200	CityJet	

Notes	Reg.	Type	Owner or Operator
	EI-CMZ	McD Douglas MD-83	Airplanes Finance/Eurofly
	EI-CNA	Letov LK-2M Sluka	G. Doody
	EI-CNB	BAe 146-200	CityJet
	EI-CNC	Team Minimax	A. M. S. Allen
	EI-CNG	Air & Space 18A gyroplane	P. Joyce
	EI-CNI	Avro 146 RJ 85	Peregrine Aviation Leasing Co (stored)
	EI-CNJ	Avro 146 RJ 85	Peregrine Aviation Leasing Co (stored)
	EI-CNL	Sikorsky S-61N	CHC Helicopters (Ireland)
	EI-CNM	PA-31-350 Navajo Chieftain	M. Goss
	EI-CNN	L.1011-385 TriStar 1	Aer Turas Teoranta
	EI-CNO	McD Douglas MD-83	Airplanes Finance (stored)
	EI-CNQ	BAe 146-200	CityJet
	EI-CNT	Boeing 737-230ADV	Ryanair (The Sun/News of the World)
	EI-CNU	Pegasus Quantum 15-912	M. Ffrench
	EI-CNV	Boeing 737-230ADV	Ryanair
	EI-CNW	Boeing 737-230ADV	Ryanair
	EI-CNX	Boeing 737-230ADV	Ryanair (Tipperary Crystal)
	EI-CNZ	Boeing 737-230ADV	Ryanair
	EI-COA	Boeing 737-230ADV	Ryanair
	EI-COB	Boeing 737-230ADV	Ryanair
	EI-COE	Shaw Europa	F. Flynn
	EI-COG	Gyroscopic Rotorcraft gyroplane	R. C. Fidler & D. Bracken
	EI-COH	Boeing 737-430	Flightlease (Ireland)/Air One
	EI-COI	Boeing 737-430	Challey/Air One
	EI-COJ	Boeing 737-430	Challey/Air One
	EI-COK	Boeing 737-430	Flightlease (Ireland)/Air One
	EI-COM	Whittaker MW.6-S Fatboy Flyer	M. Watson
	EI-CON	Boeing 737-2T5	Ryanair
	EI-COO	Carlson Sparrow II	D. Logue
	EI-COP	Cessna F.150L	High Kings Flying Group
	EI-COQ	Avro 146 RJ 70	Peregrine Aviation Leasing Co
	EI-COT	Cessna F.172N	Tojo Air Leasing
	EI-COX	Boeing 737-230	Ryanair
	EI-COY	Piper J-3C-65 Cub	D. Bruton & W. Flood
	EI-COZ	PA-28 Cherokee 140C	G. Cashman
	EI-CPC	Airbus A.321-211	Aer Lingus St Fergus
	EI-CPD	Airbus A.321-211	Aer Lingus St Davnet
	EI-CPE	Airbus A.321-211	Aer Lingus St Enda
	EI-CPF	Airbus A.321-211	Aer Lingus St Ide
	EI-CPG	Airbus A.321-211	Aer Lingus St. Aidan
	EI-CPH	Airbus A.321-211	Aer Lingus St Dervilla
	EI-CPI	Rutan LongEz	D. J. Ryan
	EI-CPJ	Avro 146 RJ 85	Peregrine Aviation Leasing Co (stored)
	EI-CPK	Avro 146 RJ 70	Peregrine Aviation Leasing Co (stored)
	EI-CPL	Avro 146 RJ 70	Peregrine Aviation Leasing Co (stored)
	EI-CPN	Auster J/4	E. Fagan
	EI-CPO	Robinson R-22B-2	Santail
	EI-CPP	Piper J-3C-65 Cub	E. Fitzgerald
	EI-CPT	Aérospatiale ATR-42-320	GPA-ATR/Aer Arran
	EI-CPX	I.I.I.Sky Arrow 650T	M. McCarthy
	EI-CRB	Lindstrand LBL-90A balloon	J. & C. Concannon
	EI-CRD	Boeing 767-31BER	ILFC Ireland/Alitalia
	EI-CRE	McD Douglas MD-83	AAR Ireland/Meridiana
	EI-CRF	Boeing 767-31BER	ILFC Ireland/Alitalia
	EI-CRG	Robin DR.400/180R	D. & B. Lodge
	EI-CRH	McD Douglas MD-83	Airplanes 111/Meridiana
	EI-CRJ	McD Douglas MD-83	C. A. Aviation/Meridiana
	EI-CRK	Airbus A.330-301	Aer Lingus St Brigid
	EI-CRL	Boeing 767-343ER	Aircraft Finance Trust Ireland
	EI-CRM	Boeing 767-343ER	GECAS Technical Services/Alitalia
	EI-CRO	Boeing 767-3Q8ER	ILFC Ireland/Alitalia
	EI-CRR	Aeronca 11AC Chief	L. Maddock & ptnrs
	EI-CRU	Cessna 152	W. Reilly
	EI-CRV	Hoffmann H-36 Dimona	The Dimona Group
	EI-CRW	McD Douglas MD-83	Airplanes IAL/Meridiana
	EI-CRX	SOCATA TB-9 Tampico	Hotel Bravo Flying Club
	EI-CRY	Medway Eclipse	G. A. Murphy
	EI-CRZ	Boeing 737-36E	ILFC Ireland/Air One
	EI-CSA	Boeing 737-8AS	Ryanair
	EI-CSB	Boeing 737-8AS	Ryanair
	EI-CSC	Boeing 737-8AS	Ryanair
	EI-CSD	Boeing 737-8AS	Ryanair

Reg.	Type	Owner or Operator	Notes
EI-CSE	Boeing 737-8AS	Ryanair	
EI-CSF	Boeing 737-8AS	Ryanair	
EI-CSG	Boeing 737-8AS	Ryanair	
EI-CSH	Boeing 737-8AS	Ryanair	
EI-CSI	Boeing 737-8AS	Ryanair	
EI-CSJ	Boeing 737-8AS	Ryanair	
EI-CSK	BAe 146-200	CityJet	
EI-CSL	BAe 146-200	CityJet	
EI-CSM	Boeing 737-8AS	Ryanair	
EI-CSN	Boeing 737-8AS	Ryanair	
EI-CSO	Boeing 737-8AS	Ryanair	
EI-CSP	Boeing 737-8AS	Ryanair	
EI-CSQ	Boeing 737-8AS	Ryanair	
EI-CSR	Boeing 737-8AS	Ryanair	
EI-CSS	Boeing 737-8AS	Ryanair	
EI-CST	Boeing 737-8AS	Ryanair	
EI-CSU	Boeing 737-36E	ILFC Ireland/Air One	
EI-CSV	Boeing 737-8AS	Ryanair	
EI-CSW	Boeing 737-8AS	Ryanair	
EI-CSX	Boeing 737-8AS	Ryanair	
EI-CSY	Boeing 737-8AS	Ryanair	
EI-CSZ	Boeing 737-8AS	Ryanair	
EI-CTA	Boeing 737-8AS	Ryanair	
EI-CTB	Boeing 737-8AS	Ryanair	
EI-CTC	Medway Eclipse	P. A. McMahon	
EI-CTD	Airbus A.320-211	Aerco Ireland/Air Europe Italy	
EI-CTG	Stoddard-Hamilton Glasair RG	K. Higgins	
EI-CTI	Cessna FRA.150L	J. Logan & T. Bradford	
EI-CTL	Aerotech MW-5B Sorcerer	M. Wade	
EI-CTT	PA-28-161 Warrior II	Conair Group	
EI-CUA	Boeing 737-4K5	Gustav Leasing XI/Blue Panorama	
EI-CUD	Boeing 737-4Q8	Castle 2003-2 Ireland/Blue Panorama	
EI-CUE	Cameron balloon	Eircom/Blue Panorama	
EI-CUG	Bell 206B Jet Ranger	Avatar Aviation	
EI-CUI	Robinson R-44	Santail	
EI-CUJ	Cessna 172N	M. Nally	
EI-CUL	Boeing 737-36N	Aircraft Finance Trust Ireland/Philippine Airlines	
EI-CUN	Boeing 737-4K5	Gustav Leasing XIV/Blue Panorama	
EI-CUP	Cessna 335	J. Greany	
EI-CUQ	Airbus A.320-214	Chamonix Aircraft Leasing *(stored)*	
EI-CUS	AB-206B JetRanger 3	Doherty Quarries and Waste Management	
EI-CUT	Maule MX-7-180A	Cosair	
EI-CUW	BN-2B-20 Islander	Aer Arann	
EI-CVA	Airbus A.320-214	Aer Lingus *St.Schira*	
EI-CVB	Airbus A.320-214	Aer Lingus *St Mobhi*	
EI-CVC	Airbus A.320-214	Aer Lingus *St Kealin*	
EI-CVD	Airbus A.320-214	Aer Lingus *St Kevin*	
EI-CVL	Ercoupe 415CD	B. Lyons	
EI-CVM	Schweizer S.269C	B. Moloney	
EI-CVN	Boeing 737-4Y0	Airplanes Finance/Philippine Airlines	
EI-CVO	Boeing 737-4S3	Aerco Ireland/Philippine Airlines	
EI-CVP	Boeing 737-4Y0	Airplanes Finance/Philippine Airlines	
EI-CVR	Aérospatiale ATR-42-300	Aer Arann	
EI-CVS	Aérospatiale ATR-42-300	Aer Arann	
EI-CVY	Brock KB-2 Gyro	G. Smyth	
EI-CWA	BAe 146-200	CityJet	
EI-CWB	BAe 146-200	CityJet	
EI-CWC	BAe 146-200	CityJet	
EI-CWD	BAe 146-200	CityJet	
EI-CWE	Boeing 737-42C	Rockshaw/Air One	
EI-CWF	Boeing 737-42C	Rockshaw/Air One	
EI-CWH	Agusta A.109E	Lochbrea Aircraft	
EI-CWL	Robinson R-22B	J. McLoughlin	
EI-CWP	Robinson R-22B	D. J. Crowley	
EI-CWR	Robinson R-22B	Coates Aviation	
EI-CWW	Boeing 737-4Y0	Airplanes Holdings/Air One	
EI-CWX	Boeing 737-4Y0	Airplanes Holdings/Air One	
EI-CXC	Raj Hamsa X'Air 502T	T. McDevitt	
EI-CXI	Boeing 737-46Q	Bellevue Aircraft Leasing/Air One	
EI-CXJ	Boeing 737-4Q8	Castle 2003-1 Ireland/Air One	
EI-CXK	Boeing 737-4S3	Bravo Aircraft Management/Transaero	
EI-CXL	Boeing 737-46N	Monroe Aircraft Ireland/Air One	

Notes	Reg.	Type	Owner or Operator
	EI-CXM	Boeing 737-4Q8	ILFC Ireland/Air One
	EI-CXN	Boeing 737-329	Embarcerado Aircraft Ireland/Transaero
	EI-CXO	Boeing 767-3G5ER	ILFC Ireland/Blue Panorama
	EI-CXR	Boeing 737-329	Embarcerado Aircraft Ireland/Transaero
	EI-CXS	Sikorsky S-61N	CHC Ireland
	EI-CXW	Boeing 737-883	Challey
	EI-CXY	Evektor EV-97 Eurostar	G. Doody & ptnrs
	EI-CXZ	Boeing 767-216ER	Embarcerado Aircraft Ireland/Transaero
	EI-CZA	ATEC Zephyr 2000	M. Higgins
	EI-CZC	CFM Streak Shadow Srs II	M. Culhane & D. Burrows
	EI-CZD	Boeing 767-216ER	ILFC Ireland/Transaero
	EI-CZG	Boeing 737-4Q8	ILFC Ireland/Air One
	EI-CZH	Boeing 767-3G5ER	ILTU Ireland/Blue Panorama
	EI-CZK	Boeing 737-4Y0	Carotene/Transaero
	EI-CZL	Schweizer 269C-1	Venturecopters
	EI-CZM	Robinson R-44	Wellingford Construction
	EI-CZN	Sikorsky S-61N	CHC Ireland
	EI-CZO	BAe 146-200	Cityjet
	EI-CZP	Schweizer 269C-1	T. N. G. Kam
	EI-DAA	Airbus A.330—202	Aer Lingus *St Keeva*
	EI-DAC	Boeing 737-8AS	Ryanair
	EI-DAD	Boeing 737-8AS	Ryanair
	EI-DAE	Boeing 737-8AS	Ryanair
	EI-DAF	Boeing 737-8AS	Ryanair
	EI-DAG	Boeing 737-8AS	Ryanair
	EI-DAH	Boeing 737-8AS	Ryanair
	EI-DAI	Boeing 737-8AS	Ryanair
	EI-DAJ	Boeing 737-8AS	Ryanair
	EI-DAK	Boeing 737-8AS	Ryanair
	EI-DAL	Boeing 737-8AS	Ryanair
	EI-DAM	Boeing 737-8AS	Ryanair
	EI-DAN	Boeing 737-8AS	Ryanair
	EI-DAO	Boeing 737-8AS	Ryanair
	EI-DAP	Boeing 737-8AS	Ryanair
	EI-DAR	Boeing 737-8AS	Ryanair
	EI-DAS	Boeing 737-8AS	Ryanair
	EI-DAT	Boeing 737-8AS	Ryanair
	EI-DAV	Boeing 737-8AS	Ryanair
	EI-DAW	Boeing 737-8AS	Ryanair
	EI-DAX	Boeing 737-8AS	Ryanair
	EI-DAY	Boeing 737-8AS	Ryanair
	EI-DAZ	Boeing 737-8AS	Ryanair
	EI-DBE	Fokker 100	EU Jet
	EI-DBF	Boeing 767-3Q8ER	ACG Acquisition Ireland/Transaero
	EI-DBG	Boeing 767-3Q8ER	Charlie Aircraft Management/Transaero
	EI-DBH	CFM Streak Shadow SA-11	M. O'Mahony
	EI-DBI	Raj Hamsa X'Air Mk.2 Falcon	E. Hamilton
	EI-DBJ	Huntwing Pegasus XL Classic	P. A. McMahon
	EI-DBK	Boeing 777-243ER	GECAS Technical Services/Alitalia
	EI-DBL	Boeing 777-243ER	GECAS Technical Services/Alitalia
	EI-DBM	Boeing 777-243ER	GECAS Technical Services/Alitalia
	EI-DBN	Bell 407	Bradbourne
	EI-DBO	Air Creation Kiss 400	E. Spain
	EI-DBP	Boeing 767-35H	CIT Ireland Leasing/Alitalia
	EI-DBR	Fokker 100	EU Jet
	EI-DBT	Canada Benjamin Co. Twinstarr	C. O'Shea
	EI-DBU	Boeing 767-37E	Pegasus Aviation Ireland/Transaero
	EI-DBV	Rand Kar X' Air 602T	S. Scanlon
	EI-DBW	Boeing 767-201	BA Finance (Ireland)/Transaero
	EI-DBX	Magni M.18 Spartan	S. Brennan
	EI-DCA	Raj Hamsa X'Air	M. O'Connell
	EI-DCB	Boeing 737-8AS	Ryanair
	EI-DCC	Boeing 737-8AS	Ryanair
	EI-DCD	Boeing 737-8AS	Ryanair
	EI-DCE	Boeing 737-8AS	Ryanair
	EI-DCF	Boeing 737-8AS	Ryanair
	EI-DCG	Boeing 737-8AS	Ryanair
	EI-DCH	Boeing 737-8AS	Ryanair
	EI-DCI	Boeing 737-8AS	Ryanair
	EI-DCJ	Boeing 737-8AS	Ryanair
	EI-DCK	Boeing 737-8AS	Ryanair
	EI-DCL	Boeing 737-8AS	Ryanair

Reg.	Type	Owner or Operator	Notes
EI-DCM	Boeing 737-8AS	Ryanair	
EI-DCN	Boeing 737-8AS	Ryanair	
EI-DCO	Boeing 737-8AS	Ryanair	
EI-DCP	Boeing 737-8AS	Ryanair	
EI-DCR	Boeing 737-8AS	Ryanair	
EI-DCS	Boeing 737-8AS	Ryanair	
EI-DCT	Boeing 737-8AS	Ryanair	
EI-DCV	Boeing 737-8AS	Ryanair	
EI-DCW	Boeing 737-8AS	Ryanair	
EI-DCX	Boeing 737-8AS	Ryanair	
EI-DCY	Boeing 737-8AS	Ryanair	
EI-DCZ	Boeing 737-8AS	Ryanair	
EI-DDA	Robinson R-44 II	J. O'R Security	
EI-DDB	Eurocopter EC.120B	J. Cudy	
EI-DDC	Cessna F.172M	Trim Flying Club	
EI-DDD	Aeronca 7AC	J. Sullivan & M. Quinn	
EI-DDE	BAe 146-200	City Jet	
EI-DDH	Boeing 777-243ER	GECAS Technical Services/Alitalia	
EI-DDI	Schweizer S.269C-1	B. Hade	
EI-DDJ	Raj Hamsa X'Air 582	J. P. McHugh	
EI-DDK	Boeing 737-4S3	Boeing Capital Leasing/Transaero	
EI-DDN	CFM Metal Fax Shadow Srs CD	F. Lynch	
EI-DDO	Montgomerie Merlin	C. Condell	
EI-DDP	Southdown International microlight	P. O'Reilly	
EI-DDW	Boeing 767-3S1ER	Pegasus Aviation Ireland/Alitalia	
EI-DDX	Cessna 172S	Atlantic Flight Training	
EI-DDY	Boeing 737-4Y0	Aerco Ireland/Transaero	
EI-DDZ	Piper PA-28-181 Archer II	Ardnari	
EI-DEA	Airbus A.320-214	Aer Lingus St Fidelma	
EI-DEB	Airbus A.320-214	Aer Lingus St Nathy	
EI-DEC	Airbus A.320-214	Aer Lingus St Fergal	
EI-DEE	Airbus A.320-214	Aer Lingus St Fintan	
EI-DEF	Airbus A.320-214	Aer Lingus St Declan	
EI-DEG	Airbus A.320-214	Aer Lingus St Fachtna	
EI-DEH	Airbus A.320-214	Aer Lingus St Malachy	
EI-DEI	Airbus A.320-214	Aer Lingus St Kilian	
EI-DEJ	Airbus A.320-214	Aer Lingus St Oliver Plunkett	
EI-DEK	Airbus A.320-214	Aer Lingus St Eunan	
EI-DEL	Airbus A.320-214	Aer Lingus St Ibar	
EI-DEM	Airbus A.320-214	Aer Lingus St Canice	
EI-DEN	Airbus A.320-214	Aer Lingus St Kieran	
EI-DEO	Airbus A.320-214	Aer Lingus St Senan	
EI-DEP	Airbus A.320-214	Aer Lingus St Eugene	
EI-DER	Airbus A.320-214	Aer Lingus St Mel	
EI-DES	Airbus A.320-214	Aer Lingus St Pappin	
EI-DEV	BAe 146-300	City Jet	
EI-DEW	BAe 146-300	City Jet	
EI-DEX	BAe 146-300	City Jet	
EI-DEY	Airbus A.319-112	Permeke Aircraft Leasing/Meridiana	
EI-DEZ	Airbus A.319-112	Permeke Aircraft Leasing/Meridiana	
EI-DFA	Airbus A.319-112	Permeke Aircraft Leasing/Meridiana	
EI-DFB	Fokker 100	EU Jet	
EI-DFC	Fokker 100	EU Jet	
EI-DFD	Boeing 737-4S3	Orix Aircraft Management/Air One	
EI-DFE	Boeing 737-4S3	Orix Aircraft Management/Air One	
EI-DFF	Boeing 737-4S3	Orix Aircraft Management/Air One	
EI-DFG	Embraer ERJ-170-100LR	GECAS Technical Services/Alitalia Express	
EI-DFH	Embraer ERJ-170-100LR	GECAS Technical Services/Alitalia Express	
EI-DFJ	Embraer ERJ-170-100LR	GECAS Technical Services/Alitalia Express	
EI-DFK	Embraer ERJ-170-100LR	GECAS Technical Services/Alitalia Express	
EI-DFL	Embraer ERJ-170-100LR	GECAS Technical Services/Alitalia Express	
EI-DFM	Evektor EV-97 Eurostar	G. Doody	
EI-DFN	Airbus A.320-211	AerFi Group/Wind Jet	
EI-DFO	Airbus A.320-211	Triton Aviation Ireland/Wind Jet	
EI-DFP	Airbus A.319-112	Delvaux Aircraft Leasing/Meridiana	
EI-DFS	Boeing 767-33AER	Jeritt Ltd/Transaero	
EI-DFW	Robinson R44	Blue Star Helicopters Ltd	
EI-DFX	Air Creation Kiss 400	L. Daly	
EI-DFY	Raj Hamsa R100 (2)	P. McGirr & R Gillespie	
EI-DFZ	Fokker 100	EU Jet	
EI-DGA	Urban Air Lambada UFM-11UK	Dr. P. & D. Durkin	
EI-DGD	Boeing 737-430	Challey Ltd/Air One	

Notes	Reg.	Type	Owner or Operator
	EI-DGE	Fokker 100	EU Jet
	EI-DGG	Raj Hamsa X'Air 582	N. Geh
	EI-DGH	Raj Hamsa X'Air 582	M. Garvey & T. McGowan
	EI-DGI	MXP-740 Savannah	N. Farrell
	EI-DGJ	Raj Hamsa X'Air 582	R. Morelli
	EI-DGK	Raj Hamsa X'Air 133	B. Chambers
	EI-DGL	Boeing 737-46J	Gelston Ltd/Air One
	EI-DGM	Boeing 737-4C9	Lux Aircraft Leasing/Blue Panorama
	EI-DGN	Boeing 737-4C9	Lux Aircraft Leasing/Blue Panorama
	EI-DGR	UFM-11UK Lambada	M. Tormeey
	EI-DGS	ATEC Zepher 2000	K. Higgins
	EI-DGT	UFM-11UK Lambada	A. & P. Aviation
	EI-DGU	Airbus A.300B4-622R	Aer Lucht
	EI-DGV	ATEC Zephyr 2000	-
	EI-DGW	Cameron Z-90 balloon	J. Leahy
	EI-DGX	Cessna 152	K. O'Connor
	EI-DGY	UFM-11 Lambada	J. Keena
	EI-DGZ	Boeing 737-86N	GECAS/Ryan International
	EI-DHA	Boeing 737-8AS	Ryanair
	EI-DHB	Boeing 737-8AS	Ryanair
	EI-DHC	Boeing 737-8AS	Ryanair
	EI-DHD	Boeing 737-8AS	Ryanair
	EI-DHE	Boeing 737-8AS	Ryanair
	EI-DHF	Boeing 737-8AS	Ryanair
	EI-DHG	Boeing 737-8AS	Ryanair
	EI-DHH	Boeing 737-8AS	Ryanair
	EI-DHI	Boeing 737-8AS	Ryanair
	EI-DHJ	Boeing 737-8AS	Ryanair
	EI-DHK	Boeing 737-8AS	Ryanair
	EI-DIA	Solar Wings Pegasus XL-Q	P. Byrne
	EI-DIB	Air Creation Kiss 400	E. Redmond
	EI-DIG	Airbus A.320-214	GECAS/Ryan International
	EI-DIH	Airbus A.320-214	GECAS/Ryan International
	EI-DIJ	Airbus A.320-212	Eir Jet
	EI-DIL	Boeing 737-83N	GECAS/Jet Airways
	EI-DIP	Airbus A.330-202	Calliope *(stored)*
	EI-DIR	Airbus A.330-202	Calliope(stored)
	EI-DIS	Boeing 737-86N	Futura Gael
	EI-DIT	Boeing 737-86N	Futura Gael
	EI-DIU	Airbus A.320-232	SALE Ireland/Flyniki
	EI-DIV	Airbus A.320-232	SALE Ireland (stored)
	EI-DIW	Airbus A.320-232	SALE Ireland (stored)
	EI-DIX	Airbus A.320-232	SALE Ireland (stored)
	EI-DIZ	Robinson R22 Beta	Blue Star Helicopters Ltd
	EI-DJH	Airbus A.320-212	ILFC Ireland (stored)
	EI-DJI	Airbus A.320-214	ILFC Ireland (stored)
	EI-DLP	Agusta A.109C	DLP Helicopters
	EI-DMG	Cessna 441	Dawn Meats Group
	EI-DOC	Robinson R-44	Donville Helicopters
	EI-DUB	Airbus A.330-301	Aer Lingus *St Patrick*
	EI-DUN	Agusta A.109E	Barkisland (Developments)
	EI-EBJ	Robinson R44	Billy Jet Ltd
	EI-ECA	Agusta A.109A-II	Beckdrive
	EI-EDR	PA-28R Cherokee Arrow 200	Kestrel Flying Group
	EI-EEC	PA-23 Aztec 250	Westair Aviation
	EI-EGG	Robinson R-44 Raven	In-Flight Aviation
	EI-EHC	Robinson R-22B	Executive Helicopters (Cork)
	EI-EHD	Robinson R-22B-2	South Coast Helicopters
	EI-EHF	Robinson R-44	Executive Helicopters Maintenance
	EI-EHG	Robinson R-22B	Executive Helicopters Maintenance
	EI-ELL	Medway Eclipse	Microflex
	EI-EUR	Eurocopter EC.120B	Atlantic Helicopters
	EI-EWR	Airbus A.330-202	Aer Lingus Lawrence O' Toole
	EI-EXC	Robinson R-44	Executive Helicopters (Cork)
	EI-EXE	Robinson R-22B-2	Executive Helicopters (Cork)
	EI-EXG	Robinson R22 Beta	21st Century Aviation
	EI-EHB	Robinson R22 Beta	Blue Star Helicopters
	EI-FBG	Cessna F.182Q	Messrs Tunney, Helly & Spelman
	EI-GAA	Boeing 767-266ER	Arbor Finance/Kras Air
	EI-GAN	Bell 407	Robswall Property
	EI-GAV	Robinson R-22B	Airo Helicopters
	EI-GBA	Boeing 767-266ER	Arbor Finance/Kras Air

Reg.	Type	Owner or Operator	Notes
EI-GER	Maule MX7-180A	L. Ryan & P. Joseph	
EI-GFC	SOCATA TB.9 Tampico	B. McGrath & ptnrs	
EI-GKL	Robinson R22 Beta	Gerry Keyes Ltd	
EI-GPT	Robinson R-22B	G & P Transport	
EI-GPZ	Robinson R-44-II	G & P Transport	
EI-GSE	Cessna F.172M	F. Doherty	
EI-GSM	Cessna 182S	Westpoint Flying Group	
EI-GWY	Cessna 172R	Galway Flying Club	
EI-HAM	Light Aero Avid Flyer	H. Goulding	
EI-HCS	Grob G.109B	H. Sydner	
EI-HER	Bell 206B JetRanger 3	Irish Helicopters	
EI-HXM	Bell 206B JetRanger	Spectra Distribution	
EI-IAW	Learjet 60	Irish Air Transport	
EI-IHL	Aerospatiale AS.350B1	Irish Helicopters	
EI-IPC	BN-2A-26 Islander	Irish Parachute Club	
EI-IRE	Canadair CL.600-2B16	Starair (Ireland)	
EI-IRV	AS.350B Ecureuil	Harrcops	
EI-IZO	Eurocopter EC.120B	Cloud 9 Helicopters	
EI-JAL	Robinson R-44-II	Heliwest	
EI-JBC	Agusta A.109A	Medeva Properties	
EI-JFD	Robinson R-44	New World Plant	
EI-JFK	Airbus A.330-301	Aer Lingus St Colmcille	
EI-JIV	L.382G-44K-30 Hercules	Air Contractors (Ireland)	
EI-JWM	Robinson R-22B	C. Shiel	
EI-KEV	Raj Hamsa R133 (1)	Kevin Glynn	
EI-LAF	Bell 206B JetRanger	Shamrock Helicopters	
EI-LAX	Airbus A.330-202	Aer Lingus St Mella	
EI-LHD	Bell 206L-III Long Ranger	Quarry & Mining Equipment Ltd	
EI-LIT	MBB Bo 105S	Irish Helicopters	
EI-LKS	Eurocopter EC-130B4	Wigaf Leasing Company	
EI-LNX	Eurocopter EC.130	Wigaf Leasing Company	
EI-LTE	Airbus A.320-214	Celestial Aviation Trading/Myair	
EI-MAG	Robinson R-22B-2	Airo Helicopters	
EI-MAX	Learjet 31A	Airlink Airways	
EI-MCC	Robinson R44-II	Coates Aviation Ltd	
EI-MCF	Cessna 172R	Galway Flying Club	
EI-MEL	Agusta A.109C	Kildare Aviation	
EI-MER	Bell 206B JetRanger	Gaelic Helicopters	
EI-MES	Sikorsky S-61N	CHC Helicopters (Ireland)	
EI-MET	Eurocopter EC.130B-4	Skyheli	
EI-MIK	Eurocopter EC.120B	Milltown Engineering	
EI-MIP	SA.365N Dauphin 2	CHC Helicopters (Ireland)	
EI-MIT	Agusta A.109E	Mercury Engineering	
EI-MJR	Robinson R-44-II	M. Melville	
EI-MMO	Robinson R-44	Morrissey Fencing	
EI-MOR	Robinson R-44	Ultimate Flight Ops	
EI-MUL	Robinson R-44	Cotton Box Design Group	
EI-MUR	Robinson R-22B	B. Murphy	
EI-NPG	Agusta A.109E	A. Logue & W. Moffet	
EI-NVL	Jora	N. Van Lonkhuyzen	
EI-NZO	Eurocopter EC.120B	Metroheli	
EI-OBJ	Robinson R-22B2	Billy Jet	
EI-ORD	Airbus A.330-301	Aer Lingus St Maeve	
EI-OZA	Airbus A.300B4-103F	Air Contractors (Ireland)	
EI-OZB	Airbus A.300B4-103F	Air Contractors (Ireland)	
EI-OZC	Airbus A.300B4-103F	Air Contractors (Ireland)	
EI-PAT	BAe 146-200	City Jet	
EI-PCI	Bell 206B JetRanger	Malcove	
EI-PEC	Robinson R44-II	P. Sexton	
EI-PJD	AS.350B-2 Twin Squirrel	New World Plant	
EI-PJW	Eurocopter EC.120B	P. White	
EI-PKS	Bell 206B JetRanger	Mountainway Builders	
EI-PMI	AB-206B JetRanger 3	Ping Golf Equipment	
EI-POD	Cessna 177B	Trim Flying Club	
EI-PRI	Bell 206B JetRanger	Brentwood Properties	
EI-RCG	Sikorsky S-61N	CHC Ireland	
EI-REA	Aérospatiale ATR-72-201	Aer Arann	
EI-REB	Aérospatiale ATR-72-201	Aer Arann	
EI-RED	Aérospatiale ATR-72-202	Aer Arann	
EI-REE	Aérospatiale ATR-72-202	Aer Arann	
EI-REF	Aerospatiale ATR-72-202	Aer Arann	
EI-RMC	Bell 206B	Westair Aviation	

Notes	Reg.	Type	Owner or Operator
	EI-SAC	Cessna 172P	Sligo Aero Club
	EI-SAM	Extra EA.300/200	D. Bruton
	EI-SAR	Sikorsky S-61N	CHC Ireland
	EI-SAT	Steen Skybolt	Capt. B. O'Sullivan
	EI-SBM	Agusta A.109E	Ballymore Management Services
	EI-SBP	Cessna T.206H	P. Morrissey
	EI-SGF	Robinson R-44	M. Reilly & S. Filan
	EI-SKS	Robin R.2160	Shemburn
	EI-SKT	PA-44-180 Seminole	Shemburn
	EI-SKU	PA-28RT-201 Arrow IV	Shemburn
	EI-SKV	Robin 2160	Shemburn
	EI-SKW	PA-28-161 Warrior II	Shemburn
	EI-SLB	Aerospatiale ATR-42-310	Air Contractors (Ireland)
	EI-SLC	Aerospatiale ATR-42-310	Air Contractors (Ireland)
	EI-SLD	Aerospatiale ATR-42-310	Air Contractors (Ireland)
	EI-SLE	Aérospatiale ATR-42-310	Air Contractors (Ireland)
	EI-SLF	Aérospatiale ATR-72-201F	Air Contractors (Ireland)
	EI-SMA	Short SD3-60 Varianr 100	Air Contractors (Ireland)
	EI-SMB	Short SD3-60 Variant 100	Air Contractors (Ireland)
	EI-SMF	F.27 Friendship MK 500	Air Contractors (Ireland)
	EI-SMK	Zenair CH-701	S. King
	EI-SNJ	Bell 407	Thornridge Services
	EI-SQG	Agusta A.109E	Quinn Group
	EI-STR	Bell 407	Barkisland (Developments)
	EI-STT	Cessna 172M	Garda Aviation Club
	EI-TAB	Airbus A.320-233	CIT Ireland Leasing/TACA
	EI-TAC	Airbus A.320-233	CIT Ireland Leasing/TACA
	EI-TAE	Airbus A.320-233	Pegasus Aviation Ireland/TACA
	EI-TBM	SOCATA TBM-700	Folens Management Services
	EI-TIP	Bell 430	Starair (Ireland)
	EI-TKI	Robinson R-22B	J. McDaid
	EI-TMH	Robinson R44	T. Maybury
	EI-TOY	Robinson R-44	Metroheli
	EI-UFO	PA-22 Tri-Pacer 150 (tailwheel)	T. William
	EI-VNE	Eurocopter EC.120B	Seafield Demesne Management
	EI-WAC	PA-23 Aztec 250E	Westair Aviation
	EI-WAV	Bell 430	Westair Aviation
	EI-WGV	G.1159 Gulfstream 5	Westair Aviation
	EI-WJN	H.S.125 Srs 700A	Westair Aviation
	EI-WMN	PA-23 Aztec 250F	Westair Aviation
	EI-WRN	PA-28-151 Warrior	Waterford Aero Club

EI-RED, Aérospatiale ATR-72-202 of Aer Arann. *DP*

Overseas Airliner Registrations

(Aircraft included in this section are those most likely to be seen at UK and major European airports on scheduled or charter services.)

Reg.	Type	Owner or Operator	Notes

A4O (Oman)

Reg.	Type	Owner or Operator	Notes
A4O-GI	Boeing 767-3P6ER (604)	Gulf Traveller	
A4O-GJ	Boeing 767-3P6ER (605)	Gulf Air *Al Muharraq*	
A4O-GK	Boeing 767-3P6ER (606)	Gulf Air *Al Burami*	
A4O-GS	Boeing 767-3P6ER (613)	Gulf Traveller	
A4O-GT	Boeing 767-3P6ER (614)	Gulf Traveller	
A4O-GU	Boeing 767-3P6ER (615)	Gulf Traveller	
A4O-GV	Boeing 767-3P6ER (616)	Gulf Air *Dukhan*	
A4O-GY	Boeing 767-3P6ER (619)	Gulf Traveller	
A4O-GZ	Boeing 767-3P6ER (620)	Gulf Traveller	
A4O-KA	Airbus A.330-243 (501)	Gulf Air	
A4O-KB	Airbus A.330-243 (502)	Gulf Air	
A4O-KC	Airbus A.330-243 (503)	Gulf Air	
A4O-KD	Airbus A.330-243 (504)	Gulf Air	
A4O-KE	Airbus A.330-243 (505)	Gulf Air	
A4O-KF	Airbus A.330-243 (506)	Gulf Air *Aldafra*	
A4O-LB	Airbus A.340-312 (402)	Gulf Air *Al Fateh*	
A4O-LC	Airbus A.340-312 (403)	Gulf Air *Doha*	
A4O-LD	Airbus A.340-312 (404)	Gulf Air *Abu Dhabi*	
A4O-LE	Airbus A.340-312 (405)	Gulf Air	
A4O-LF	Airbus A.340-312 (406)	Gulf Air	
A4O-LG	Airbus A.340-313X (407)	Gulf Air	
A4O-LH	Airbus A.340-313X (408)	Gulf Air	
A4O-LI	Airbus A.340-313X (409)	Gulf Air	
A4O-LJ	Airbus A.340-313X (410)	Gulf Air	
A4O-OMN	Boeing 747-430	Oman Government	
A4O-SO	Boeing 747SP-27	Oman Government	
A4O-SP	Boeing 747SP-27	Oman Government	

A6 (Arab Emirates)

Reg.	Type	Owner or Operator	Notes
A6-EAA	Airbus A.330-243	Emirates Airlines	
A6-EAB	Airbus A.330-243	Emirates Airlines	
A6-EAC	Airbus A.330-243	Emirates Airlines	
A6-EAD	Airbus A.330-243	Emirates Airlines	
A6-EAE	Airbus A.330-243	Emirates Airlines	
A6-EAF	Airbus A.330-243	Emirates Airlines	
A6-EAG	Airbus A.330-243	Emirates Airlines	
A6-EAH	Airbus A.330-243	Emirates Airlines	
A6-EAI	Airbus A.330-243	Emirates Airlines	
A6-EAJ	Airbus A.330-243	Emirates Airlines	
A6-EAK	Airbus A.330-243	Emirates Airlines	
A6-EAL	Airbus A.330-243	Emirates Airlines	
A6-EAM	Airbus A.330-243	Emirates Airlines	
A6-EAN	Airbus A.330-243	Emirates Airlines	
A6-EAO	Airbus A.330.243	Emirates Airlines	
A6-EAP	Airbus A.330-243	Emirates Airlines	
A6-EAQ	Airbus A.330-243	Emirates Airlines	
A6-EAR	Airbus A.330-243	Emirates Airlines	
A6-EAS	Airbus A.330-243	Emirates Airlines	
A6-EKQ	Airbus A.330-243	Emirates Airlines	
A6-EKR	Airbus A.330-243	Emirates Airlines	
A6-EKS	Airbus A.330-243	Emirates Airlines	
A6-EKT	Airbus A.330-243	Emirates Airlines	
A6-EKU	Airbus A.330-243	Emirates Airlines	
A6-EKV	Airbus A.330-243	Emirates Airlines	
A6-EKW	Airbus A.330-243	Emirates Airlines	
A6-EKX	Airbus A.330-243	Emirates Airlines	
A6-EKY	Airbus A.330-243	Emirates Airlines	
A6-EKZ	Airbus A.330-243	Emirates Airlines	
A6-EMD	Boeing 777-21H	Emirates Airlines	

Notes	Reg.	Type	Owner or Operator
	A6-EME	Boeing 777-21H	Emirates Airlines
	A6-EMF	Boeing 777-21H	Emirates Airlines
	A6-EMG	Boeing 777-21HER	Emirates Airlines
	A6-EMH	Boeing 777-21HER	Emirates Airlines
	A6-EMI	Boeing 777-21HER	Emirates Airlines
	A6-EMJ	Boeing 777-21HER	Emirates Airlines
	A6-EMK	Boeing 777-21HER	Emirates Airlines
	A6-EML	Boeing 777-21HER	Emirates Airlines
	A6-EMM	Boeing 777-31H	Emirates Airlines
	A6-EMN	Boeing 777-31H	Emirates Airlines
	A6-EMO	Boeing 777-31H	Emirates Airlines
	A6-EMP	Boeing 777-31H	Emirates Airlines
	A6-EMQ	Boeing 777-31H	Emirates Airlines
	A6-EMR	Boeing 777-31H	Emirates Airlines
	A6-EMS	Boeing 777-31H	Emirates Airlines
	A6-EMT	Boeing 777-31H	Emirates Airlines
	A6-EMU	Boeing 777-31H	Emirates Airlines
	A6-EMV	Boeing 777-31H	Emirates Airlines
	A6-EMW	Boeing 777-31H	Emirates Airlines
	A6-EMX	Boeing 777-31H	Emirates Airlines
	A6-ERA	Airbus A.340-541	Emirates Airlines
	A6-ERB	Airbus A.340-541	Emirates Airlines
	A6-ERC	Airbus A.340-541	Emirates Airlines
	A6-ERD	Airbus A.340-541	Emirates Airlines
	A6-ERE	Airbus A.340-541	Emirates Airlines
	A6-ERF	Airbus A.340-541	Emirates Airlines
	A6-ERG	Airbus A.340-541	Emirates Airlines
	A6-ERH	Airbus A.340-541	Emirates Airlines
	A6-ERI	Airbus A.340-541	Emirates Airlines
	A6-ERM	Airbus A.340-313X	Emirates Airlines
	A6-ERN	Airbus A.340-313X	Emirates Airlines
	A6-ERO	Airbus A.340-313X	Emirates Airlines
	A6-ERP	Airbus A.340-313X	Emirates Airlines
	A6-ERQ	Airbus A.340-313X	Emirates Airlines
	A6-ERR	Airbus A.340-313X	Emirates Airlines
	A6-ERS	Airbus A.340-313X	Emirates Airlines
	A6-ERT	Airbus A.340-313X	Emirates Airlines
	A6-EYA	Airbus A.330-223	Etihad Airways
	A6-EYB	Airbus A.330-223	Etihad Airways
	A6-EYC	Airbus A.340-313X	Etihad Airways
	A6-EYX	Airbus A.330-223	Etihad Airways
	A6-EYY	Airbus A.330-223	Etihad Airways
	A6-	Airbus A.330-223	Etihad Airways
	A6-	Airbus A.330-223	Etihad Airways
	A6-GDP	Boeing 747-2B4BF	Dubai Government
	A6-HRM	Boeing 747-422	Dubai Government
	A6-MMM	Boeing 747-422	Dubai Government
	A6-SMM	Boeing 747SP-31	United Arab Emirates
	A6-SMR	Boeing 747SP-31	United Arab Emirates
	A6-YAS	Boeing 747-4F6	United Arab Emirates
	A6-ZSN	Boeing 747SP-P6	United Arab Emirates

Note: United Arab Emirates also operates An-124 UR-ZYD. Emirates SkyCargo operates B.747Fs N408MC and N415MC.

A7 (Qatar)

	A7-AAF	Airbus A.310-304	Qatar Government
	A7-AAG	Airbus A.320-232	Qatar Government
	A7-ABN	Airbus A.300B4-622R	Qatar Airways Al Shaab
	A7-ABO	Airbus A.300B4-622R	Qatar Airways
	A7-ABV	Airbus A.300B4-622R	Qatar Airways
	A7-ABW	Airbus A.300B4-622R	Qatar Airways
	A7-ABX	Airbus A.300B4-622R	Qatar Airways Al Dawha
	A7-ABY	Airbus A.300B4-622R	Qatar Airways
	A7-ACA	Airbus A.330-203	Qatar Airways Alwajba
	A7-ACB	Airbus A.330-203	Qatar Airways
	A7-ACC	Airbus A.330-203	Qatar Airways Al Shahaniya
	A7-ACD	Airbus A.330-203	Qatar Airways
	A7-ACE	Airbus A.330-203	Qatar Airways
	A7-ACH	Airbus A-330-203	Qatar Airways

Reg.	Type	Owner or Operator	Notes
A7-AEA	Airbus A.330-303	Qatar Airways	
A7-AEB	Airbus A.330-303	Qatar Airways	
A7-AFA	Airbus A.300B4-622R	Qatar Airways	
A7-AFB	Airbus A.300B4-622RF	Qatar Airways	
A7-AFC	Airbus A.300B4-622R	Qatar Airways	
A7-AFD	Airbus A.300B4-622R	Qatar Airways	
A7-AFL	Airbus A.330-203	Qatar Airways	
A7-AFM	Airbus A.330-203	Qatar Airways	
A7-AFN	Airbus A.330-203	Qatar Airways	
A7-AFO	Airbus A.330-203	Qatar Airways	
A7-HHH	Airbus A.340-541	Qatar Government	
A7-HHK	Airbus A.340-211	Qatar Government	
A7-HHM	Airbus A.330-203	Qatar Government	
A7-HJJ	Airbus A.330-203	Qatar Government	

AP (Pakistan)

AP-AYV	Boeing 747-282B	Pakistan International Airlines	
AP-AYW	Boeing 747-282B	Pakistan International Airlines	
AP-BAK	Boeing 747-240B (SCD)	Pakistan International Airlines	
AP-BAT	Boeing 747-240B (SCD)	Pakistan International Airlines	
AP-BCL	Boeing 747-217B	Pakistan International Airlines	
AP-BCM	Boeing 747-217B	Pakistan International Airlines	
AP-BDZ	Airbus A.310-308	Pakistan International Airlines	
AP-BEB	Airbus A.310-308	Pakistan International Airlines	
AP-BEC	Airbus A.310-308	Pakistan International Airlines	
AP-BEG	Airbus A.310-308	Pakistan International Airlines	
AP-BEQ	Airbus A.310-308	Pakistan International Airlines	
AP-BEU	Airbus A.310-308	Pakistan International Airlines	
AP-BFU	Boeing 747-367	Pakistan International Airlines	
AP-BFV	Boeing 747-367	Pakistan International Airlines	
AP-BFW	Boeing 747-367	Pakistan International Airlines	
AP-BFX	Boeing 747-367	Pakistan International Airlines	
AP-BFY	Boeing 747-367	Pakistan International Airlines	
AP-BGG	Boeing 747-367	Pakistan International Airlines	
AP-BGJ	Boeing 777-240ER	Pakistan International Airlines	
AP-BGK	Boeing 777-240ER	Pakistan International Airlines	
AP-BGL	Boeing 777-240ER	Pakistan International Airlines	
AP-BGN	Airbus A.310-324	Pakistan International Airlines	
AP-BGO	Airbus A.310-324	Pakistan International Airlines	
AP-BGP	Airbus A.310-324	Pakistan International Airlines	
AP-BGQ	Airbus A.310-324	Pakistan International Airlines	
AP-BGR	Airbus A.310-324	Pakistan International Airlines	
AP-BGS	Airbus A.310-324	Pakistan International Airlines	

B (China/Taiwan/Hong Kong)

B-HIH	Boeing 747-267F (SCD)	Cathay Pacific Airways	
B-HKD	Boeing 747-412	Cathay Pacific Airways	
B-HKE	Boeing 747-412F	Cathay Pacific Airways	
B-HMD	Boeing 747-2L5F (SCD)	Cathay Pacific Airways	
B-HME	Boeing 747-2L5F (SCD)	Cathay Pacific Airways	
B-HMF	Boeing 747-2L5F (SCD)	Cathay Pacific Airways	
B-HOO	Boeing 747-467	Cathay Pacific Airways	
B-HOP	Boeing 747-467	Cathay Pacific Airways	
B-HOR	Boeing 747-467	Cathay Pacific Airways	
B-HOS	Boeing 747-467	Cathay Pacific Airways	
B-HOT	Boeing 747-467	Cathay Pacific Airways	
B-HOU	Boeing 747-467	Cathay Pacific Airways	
B-HOV	Boeing 747-467	Cathay Pacific Airways	
B-HOW	Boeing 747-467	Cathay Pacific Airways	
B-HOX	Boeing 747-467	Cathay Pacific Airways	
B-HOY	Boeing 747-467	Cathay Pacific Airways	
B-HOZ	Boeing 747-467	Cathay Pacific Airways	
B-HQA	Airbus A.340-642	Cathay Pacific Airways	
B-HQB	Airbus A.340-642	Cathay Pacific Airways	
B-HQC	Airbus A.340-642	Cathay Pacific Airways	
B-HQD	Airbus A.340-642	Cathay Pacific Airways	
B-	Airbus A.340-642	Cathay Pacific Airways	
B-HUA	Boeing 747-467	Cathay Pacific Airways	

Notes	Reg.	Type	Owner or Operator
	B-HUB	Boeing 747-467	Cathay Pacific Airways
	B-HUD	Boeing 747-467	Cathay Pacific Airways
	B-HUE	Boeing 747-467	Cathay Pacific Airways
	B-HUF	Boeing 747-467	Cathay Pacific Airways
	B-HUG	Boeing 747-467	Cathay Pacific Airways
	B-HUH	Boeing 747-467F (SCD)	Cathay Pacific Airways
	B-HUI	Boeing 747-467	Cathay Pacific Airways
	B-HUJ	Boeing 747-467	Cathay Pacific Airways
	B-HUK	Boeing 747-467F (SCD)	Cathay Pacific Airways
	B-HUL	Boeing 747-467F (SCD)	Cathay Pacific Airways
	B-HUO	Boeing 747-467F (SCD)	Cathay Pacific Airways
	B-HUP	Boeing 747-467F (SCD)	Cathay Pacific Airways
	B-HVX	Boeing 747-267F (SCD)	Cathay Pacific Airways
	B-HVY	Boeing 747-236F (SCD)	Cathay Pacific Airways
	B-HVZ	Boeing 747-267F (SCD)	Cathay Pacific Airways
	B-HXA	Airbus A.340-313X	Cathay Pacific Airways
	B-HXB	Airbus A.340-313X	Cathay Pacific Airways
	B-HXC	Airbus A.340-313X	Cathay Pacific Airways
	B-HXD	Airbus A.340-313X	Cathay Pacific Airways
	B-HXE	Airbus A.340-313X	Cathay Pacific Airways
	B-HXF	Airbus A.340-313X	Cathay Pacific Airways
	B-HXG	Airbus A.340-313X	Cathay Pacific Airways
	B-HXH	Airbus A.340-313X	Cathay Pacific Airways
	B-HXI	Airbus A.340-313X	Cathay Pacific Airways
	B-HXJ	Airbus A.340-313X	Cathay Pacific Airways
	B-HXK	Airbus A.340-313X	Cathay Pacific Airways
	B-HXL	Airbus A.340-313X	Cathay Pacific Airways
	B-HXM	Airbus A.340-313X	Cathay Pacific Airways
	B-HXN	Airbus A.340-313X	Cathay Pacific Airways
	B-HXO	Airbus A.340-313X	Cathay Pacific Airways
	B-KAA	Boeing 747-312F (SCD)	Dragonair Cargo
	B-KAB	Boeing 747-312F (SCD)	Dragonair Cargo
	B-KAC	Boeing 747-3H6F (SCD)	Dragonair Cargo
	B-KAD	Boeing 747-209F (SCD)	Dragonair Cargo
	B-2380	Airbus A.340-313X	China Eastern Airlines
	B-2381	Airbus A.340-313X	China Eastern Airlines
	B-2382	Airbus A.340-313X	China Eastern Airlines
	B-2383	Airbus A.340-313X	China Eastern Airlines
	B-2384	Airbus A.340-313X	China Eastern Airlines
	B-2385	Airbus A.340-313X	Air China
	B-2386	Airbus A.340-313X	Air China
	B-2387	Airbus A.340-313X	Air China
	B-2389	Airbus A.340-313X	Air China
	B-2390	Airbus A.340-313X	Air China
	B-2409	Boeing 747-412F (SCD)	Air China
	B-2443	Boeing 747-4J6	Air China
	B-2445	Boeing 747-4J6	Air China
	B-2446	Boeing 747-2J6B (SF)	Air China
	B-2447	Boeing 747-4J6	Air China
	B-2448	Boeing 747-2J6B (SF)	Air China
	B-2450	Boeing 747-2J6B (SF)	Air China
	B-2456	Boeing 747-4J6	Air China
	B-2458	Boeing 747-4J6	Air China
	B-2460	Boeing 747-4J6	Air China
	B-2462	Boeing 747-2J6F (SCD)	Air China
	B-2467	Boeing 747-4J6	Air China
	B-2468	Boeing 747-4J6	Air China
	B-2469	Boeing 747-4J6	Air China
	B-2470	Boeing 747-4J6	Air China
	B-2471	Boeing 747-4J6	Air China
	B-2472	Boeing 747-4J6	Air China
	B-6050	Airbus A.340-642	China Eastern Airlines
	B-6051	Airbus A.340-642	China Eastern Airlines
	B-6052	Airbus A.340-642	China Eastern Airlines
	B-6053	Airbus A.340-642	China Eastern Airlines
	B-6055	Airbus A.340-642	China Eastern Airlines
	B-16101	McD Douglas MD-11F	EVA Air Cargo
	B-16102	McD Douglas MD-11F	EVA Air Cargo
	B-16103	McD Douglas MD-11F	EVA Air Cargo
	B-16106	McD Douglas MD-11F	EVA Air Cargo
	B-16107	McD Douglas MD-11F	EVA Air Cargo
	B-16108	McD Douglas MD-11F	EVA Air Cargo

Reg.	Type	Owner or Operator	Notes
B-16109	McD Douglas MD-11F	EVA Air Cargo	
B-16110	McD Douglas MD-11F	EVA Air Cargo	
B-16111	McD Douglas MD-11F	EVA Air Cargo	
B-16112	McD Douglas MD-11F	EVA Air Cargo	
B-16113	McD Douglas MD-11F	EVA Air Cargo	
B-16401	Boeing 747-45E	EVA Airways	
B-16402	Boeing 747-45E	EVA Airways	
B-16403	Boeing 747-45E	EVA Airways	
B-16405	Boeing 747-45E	EVA Airways	
B-16406	Boeing 747-45E	EVA Airways	
B-16407	Boeing 747-45E	EVA Airways	
B-16408	Boeing 747-45E	EVA Airways	
B-16409	Boeing 747-45E	EVA Airways	
B-16410	Boeing 747-45E	EVA Airways	
B-16411	Boeing 747-45E	EVA Airways	
B-16412	Boeing 747-45E	EVA Airways	
B-16461	Boeing 747-45E (SCD)	EVA Airways	
B-16462	Boeing 747-45E (SCD)	EVA Airways	
B-16463	Boeing 747-45E (SCD)	EVA Airways	
B-16465	Boeing 747-45E (SCD)	EVA Airways	
B-16481	Boeing 747-45EF (SCD)	EVA Air Cargo	
B-16482	Boeing 747-45EF (SCD)	EVA Air Cargo	
B-16483	Boeing 747-45EF (SCD)	EVA Air Cargo	
B-18701	Boeing 747-409F (SCD)	China Airlines	
B-18702	Boeing 747-409F (SCD)	China Airlines	
B-18703	Boeing 747-409F (SCD)	China Airlines	
B-18705	Boeing 747-409F (SCD)	China Airlines	
B-18706	Boeing 747-409F (SCD)	China Airlines	
B-18707	Boeing 747-409F (SCD)	China Airlines	
B-18708	Boeing 747-409F (SCD)	China Airlines	
B-18709	Boeing 747-409F (SCD)	China Airlines	
B-18710	Boeing 747-409F (SCD)	China Airlines	
B-18711	Boeing 747-409F (SCD)	China Airlines	
B-18712	Boeing 747-409F (SCD)	China Airlines	
B-18715	Boeing 747-409F (SCD)	China Airlines	
B-18716	Boeing 747-409F (SCD)	China Airlines	
B-18717	Boeing 747-409F (SCD)	China Airlines	
B-18718	Boeing 747-409F (SCD)	China Airlines	
B-18719	Boeing 747-409F (SCD)	China Airlines	

Note: EVA Air Cargo operates the MD-11F N105EV together with Boeing 747-47UF N416MC.

C-F and C-G (Canada)

C-FBEF	Boeing 767-233ER (617)	Air Canada	
C-FBEG	Boeing 767-233ER (618)	Air Canada	
C-FBEM	Boeing 767-233ER (619)	Air Canada	
C-FCAB	Boeing 767-375ER (681)	Air Canada	
C-FCAE	Boeing 767-375ER (682)	Air Canada	
C-FCAF	Boeing 767-375ER (683)	Air Canada	
C-FCAG	Boeing 767-375ER (684)	Air Canada	
C-FDAT	Airbus A.310-308 (305)	Air Transat	
C-FMWP	Boeing 767-333ER (631)	Air Canada	
C-FMWQ	Boeing 767-333ER (632)	Air Canada	
C-FMWU	Boeing 767-333ER (633)	Air Canada	
C-FMWV	Boeing 767-333ER (634)	Air Canada	
C-FMWY	Boeing 767-333ER (635)	Air Canada	
C-FMXC	Boeing 767-333ER (636)	Air Canada	
C-FOCA	Boeing 767-375ER (640)	Air Canada	
C-FPCA	Boeing 767-375ER (637)	Air Canada	
C-FRAV	Airbus A.330-322	Skyservice Airlines	
C-FTCA	Boeing 767-375ER (638)	Air Canada	
C-FTNQ	Airbus A.340-313 (981)	Air Canada	
C-FUCL	Boeing 767-209ER (622)	Air Canada	
C-FVNM	Boeing 767-209ER (621)	Air Canada	
C-FXCA	Boeing 767-375ER (639)	Air Canada	
C-FYKX	Airbus A.340-313X (901)	Air Canada	
C-FYKZ	Airbus A.340-313X (902)	Air Canada	
C-FYLC	Airbus A.340-313X (903)	Air Canada	
C-FYLD	Airbus A.340-313X (904)	Air Canada	
C-FYLG	Airbus A.340-313X (905)	Air Canada	

Notes	Reg.	Type	Owner or Operator
	C-FYLU	Airbus A.340-313X (906)	Air Canada
	C-GBZR	Boeing 767-38EER (645)	Air Canada
	C-GDSP	Boeing 767-233ER (613)	Air Canada
	C-GDSS	Boeing 767-233ER (614)	Air Canada
	C-GDSU	Boeing 767-233ER (615)	Air Canada
	C-GDSY	Boeing 767-233ER (616)	Air Canada
	C-GDUZ	Boeing 767-38EER (646)	Air Canada
	C-GDVW	Airbus A.340-313X (909)	Air Canada
	C-GDVZ	Airbus A.340-313X (910)	Air Canada
	C-GEOQ	Boeing 767-375ER (647)	Air Canada
	C-GEOU	Boeing 767-375ER (648)	Air Canada
	C-GFAF	Airbus A.330-343X (931)	Air Canada
	C-GFAH	Airbus A.330-343X (932)	Air Canada
	C-GFAJ	Airbus A.330-343X (933)	Air Canada
	C-GFAT	Airbus A.310-304 (301)	Air Transat
	C-GFUR	Airbus A.330-343X (934)	Air Canada
	C-GGFJ	Boeing 767-3Y0ER (652)	Air Canada
	C-GGMX	Boeing 767-3Y0ER (653)	Air Canada
	C-GGTS	Airbus A.330-243 (101)	Air Transat
	C-GHKR	Airbus A.330-343X (935)	Air Canada
	C-GHKW	Airbus A.330-343X (936)	Air Canada
	C-GHKX	Airbus A.330-343X (937)	Air Canada
	C-GHLA	Boeing 767-35HER (656)	Air Canada
	C-GHLK	Boeing 767-35HER (657)	Air Canada
	C-GHLM	Airbus A.330-343X (938)	Air Canada
	C-GHLQ	Boeing 767-333ER (658)	Air Canada
	C-GHLT	Boeing 767-333ER (659)	Air Canada
	C-GHLU	Boeing 767-333ER (660)	Air Canada
	C-GHLV	Boeing 767-333ER (661)	Air Canada
	C-GHPA	Boeing 767-3Y0ER (686)	Air Canada
	C-GHPD	Boeing 767-3Y0ER (687)	Air Canada
	C-GITS	Airbus A.330-243 (102)	Air Transat
	C-GKOL	Airbus A.340-541 (445)	Air Canada
	C-GKOM	Airbus A.340-541 (464)	Air Canada
	C-GKTS	Airbus A.330-342 (100)	Air Transat
	C-GLAT	Airbus A.310-308 (302)	Air Transat
	C-GLCA	Boeing 767-375ER (641)	Air Canada
	C-GPAT	Airbus A.310-308 (303)	Air Transat
	C-GPTS	Airbus A.330-243 (103)	Air Transat
	C-GSAT	Airbus A.310-308 (304)	Air Transat
	C-GSCA	Boeing 767-375ER (642)	Air Canada
	C-GTSD	Airbus A.310-304 (341)	Air Transat
	C-GTSH	Airbus A.310-304 (343)	Air Transat
	C-GTSI	Airbus A.310-304 (342)	Air Transat
	C-GTSY	Airbus A.310-304	Air Transat
	C-GTSZ	L.1011-385 TriStar 100 (548)	Air Transat
	C-GVAT	Airbus A.310-304 (321)	Air Transat
	C-GZMM	Boeing 767-328ER (102)	Zoom Airlines *City of Halifax*
	C-GZUM	Boeing 767-328ER (101)	Zoom Airlines *City of Ottawa*

Note: Airline fleet number when carried on aircraft is shown in parentheses.
Air Canada lease MD-11F N702GC for transatlantic freight services.

CC (Chile)

CC-CZY	Boeing 767-316F	LAN-Chile
CC-CZZ	Boeing 767-38EF	LAN-Chile

Note: LAN-Chile also operate 767-316F N312LA.

CN (Morocco)

CN-RGA	Boeing 747-428	Royal Air Maroc
CN-RME	Boeing 747-3B6B (SCD)	Royal Air Maroc
CN-RMF	Boeing 737-4B6	Atlas Blue
CN-RMG	Boeing 737-4B6	Atlas Blue
CN-RMM	Boeing 737-2B6C	Royal Air Maroc
CN-RMN	Boeing 737-2B6C	Royal Air Maroc
CN-RMT	Boeing 757-2B6	Royal Air Maroc
CN-RMV	Boeing 737-5B6	Royal Air Maroc

Reg.	Type	Owner or Operator	Notes
CN-RMW	Boeing 737-5B6	Royal Air Maroc	
CN-RMX	Boeing 737-4B6	Atlas Blue	
CN-RMY	Boeing 737-5B6	Royal Air Maroc	
CN-RMZ	Boeing 757-2B6	Royal Air Maroc	
CN-RNA	Boeing 737-4B6	Atlas Blue	
CN-RNB	Boeing 737-5B6	Royal Air Maroc	
CN-RNC	Boeing 737-4B6	Atlas Blue	
CN-RND	Boeing 737-4B6	Atlas Blue	
CN-RNG	Boeing 737-5B6	Royal Air Maroc	
CN-RNH	Boeing 737-5B6	Royal Air Maroc	
CN-RNJ	Boeing 737-8B6	Royal Air Maroc	
CN-RNK	Boeing 737-8B6	Royal Air Maroc	
CN-RNL	Boeing 737-7B6	Royal Air Maroc	
CN-RNM	Boeing 737-7B6	Royal Air Maroc	
CN-RNP	Boeing 737-8B6	Royal Air Maroc	
CN-RNQ	Boeing 737-7B6	Royal Air Maroc	
CN-RNR	Boeing 737-7B6	Royal Air Maroc	
CN-RNS	Boeing 767-3B6ER	Royal Air Maroc	
CN-RNT	Boeing 767-3B6ER	Royal Air Maroc	
CN-RNU	Boeing 737-8B6	Royal Air Maroc	
CN-RNV	Boeing 737-7B6	Royal Air Maroc	
CN-RNW	Boeing 737-8B6	Royal Air Maroc	
CN-RNX	Airbus A.321-211	Royal Air Maroc	
CN-RNY	Airbus A.321-211	Royal Air Maroc	
CN-RNZ	Boeing 737-8B6	Royal Air Maroc	
CN-ROA	Boeing 737-8B6	Royal Air Maroc	
CN-ROB	Boeing 737-8B6	Royal Air Maroc	
CN-ROC	Boeing 737-8B6	Royal Air Maroc	

CS (Portugal)

Reg.	Type	Owner or Operator
CS-TEB	L.1011-385 TriStar 500	Euro Atlantic Airways *Naughton Simao*
CS-TEH	Airbus A.310-304	TAP - Air Portugal *Bartolomeu Dias*
CS-TEI	Airbus A.310-304	TAP - Air Portugal *Fernao de Magalhaes*
CS-TEJ	Airbus A.310-304	TAP - Air Portugal *Pedro Nunes*
CS-TEW	Airbus A.310-304	TAP - Air Portugal *Vasco da Gama*
CS-TEX	Airbus A.310-304	TAP - Air Portugal *Joao XXI*
CS-TEZ	Airbus A.310-304	TAP - Air Portugal *Viana da Mota*
CS-TGP	Boeing 737-3Q8	SATA International *Corvo*
CS-TGU	Airbus A.310-304	SATA International *Terceira*
CS-TGV	Airbus A.310-304	SATA International *Sao Miguel*
CS-TGZ	Boeing 737-43Q	SATA International *Sao Jorge*
CS-TJE	Airbus A.321-211	TAP - Air Portugal *Pero Vaz de Caminha*
CS-TJF	Airbus A.321-211	TAP - Air Portugal *Luis Vaz de Camoes*
CS-TJG	Airbus A.321-211	TAP - Air Portugal *Amelia Rodrigues*
CS-TKI	Airbus A.310-304	SATA International *Autonomia*
CS-TKJ	Airbus A.320-212	SATA International *Pico*
CS-TLM	Boeing 767-33AER	Euro Atlantic Airways
CS-TMP	L.1011-385 TriStar 500	Air Luxor
CS-TMR	L.1011-385 TriStar 500	Air Luxor
CS-TMT	Airbus A.330-322	Air Luxor
CS-TMW	Airbus A.320-214	TAP - Air Portugal *Luisa Todi*
CS-TMX	L.1011-385 TriStar 500	Yes
CS-TNA	Airbus A.320-211	TAP - Air Portugal *Grao Vasco*
CS-TNB	Airbus A.320-211	Air Luxor
CS-TNE	Airbus A.320-212	TAP - Air Portugal *Sa de Miranda*
CS-TNG	Airbus A.320-214	TAP - Air Portugal *Mouzinho da Silveira*
CS-TNH	Airbus A.320-214	TAP - Air Portugal *Almada Negreiros*
CS-TNI	Airbus A.320-214	TAP - Air Portugal *Aquilino Ribiera*
CS-TNJ	Airbus A.320-214	TAP - Air Portugal *Florbela Espanca*
CS-TNK	Airbus A.320-214	TAP - Air Portugal *Teofilo Braga*
CS-TNL	Airbus A.320-214	TAP - Air Portugal *Vitorino Nermesio*
CS-TNM	Airbus A.320-214	TAP - Air Portugal *Natalia Correia*
CS-TNN	Airbus A.320-214	TAP – Air Portugal *Gil Vicente*
CS-TOA	Airbus A.340-312	TAP - Air Portugal *Fernao Mendes Pinto*
CS-TOB	Airbus A.340-312	TAP - Air Portugal *D Joao de Castro*
CS-TOC	Airbus A.340-312	TAP - Air Portugal *Wenceslau de Moraes*
CS-TOD	Airbus A.340-312	TAP - Air Portugal *D Francisco de Almeida*
CS-TPA	Fokker 100	Portugalia *Albatroz*
CS-TPB	Fokker 100	Portugalia *Pelicano*

Notes	Reg.	Type	Owner or Operator
	CS-TPC	Fokker 100	Portugalia *Flamingo*
	CS-TPD	Fokker 100	Portugalia *Condor*
	CS-TPE	Fokker 100	Portugalia *Gaviao*
	CS-TPF	Fokker 100	Portugalia *Grifo*
	CS-TPG	Embraer RJ145EP	Portugalia *Melro*
	CS-TPH	Embraer RJ145EP	Portugalia *Pardal*
	CS-TPI	Embraer RJ145EP	Portugalia *Cuca*
	CS-TPJ	Embraer RJ145EP	Portugalia *Chapim*
	CS-TPK	Embraer RJ145EP	Portugalia *Gaio*
	CS-TPL	Embraer RJ145EP	Portugalia *Pisco*
	CS-TPM	Embraer RJ145EP	Portugalia *Rola*
	CS-TPN	Embraer RJ145EP	Portugalia *Brigao*
	CS-TQA	Airbus A.320-214	Air Luxor
	CS-TQB	Airbus A.320-214	Air Luxor
	CS-TQD	Airbus A.320-214	Air Luxor
	CS-TQE	Airbus A.320-212	Air Luxor
	CS-TQF	Airbus A.330-322	Air Luxor
	CS-TQG	Airbus A.320-211	Air Luxor
	CS-TQH	Airbus A.320-211	Air Luxor
	CS-TTA	Airbus A.319-111	TAP - Air Portugal *Vieira da Silva*
	CS-TTB	Airbus A.319-111	TAP - Air Portugal *Gago Coutinho*
	CS-TTC	Airbus A.319-111	TAP - Air Portugal *Fernando Pessoa*
	CS-TTD	Airbus A.319-111	TAP - Air Portugal *Amadeo de Souza-Cardoso*
	CS-TTE	Airbus A.319-111	TAP - Air Portugal *Francisco d'Ollanda*
	CS-TTF	Airbus A.319-111	TAP - Air Portugal *Calouste Gulbenkian*
	CS-TTG	Airbus A.319-111	TAP - Air Portugal *Humberto Delgado*
	CS-TTH	Airbus A.319-111	TAP - Air Portugal *Antonio Sergio*
	CS-TTI	Airbus A.319-111	TAP - Air Portugal *Eca de Queiros*
	CS-TTJ	Airbus A.319-111	TAP - Air Portugal *Eusebio*
	CS-TTK	Airbus A.319-111	TAP - Air Portugal *Miguel Torga*
	CS-TTL	Airbus A.319-111	TAP - Air Portugal *Almeida Garrett*
	CS-TTM	Airbus A.319-111	TAP - Air Portugal *Alexandre Herculano*
	CS-TTN	Airbus A.319-111	TAP - Air Portugal *Camilo Castelo Branco*
	CS-TTO	Airbus A.319-111	TAP - Air Portugal *Antero de Quental*
	CS-TTP	Airbus A.319-111	TAP - Air Portugal *Josefa d'Obidos*

Note: Euro Atlantic Airways also operate 737-33A PR-BRY. Aviajet also operate 737s TF-ELP and TF-ELZ.

CU (Cuba)

Note: Novair A.330s SE-RBF and SE-RBG operate European flights for Cubana.

D (Germany)

	D-ABAA	Boeing 737-76Q	Air Berlin
	D-ABAB	Boeing 737-76Q	Air Berlin
	D-ABAC	Boeing 737-86J	Air Berlin
	D-ABAD	Boeing 737-86J	Air Berlin
	D-ABAE	Boeing 737-86J	Air Berlin
	D-ABAF	Boeing 737-86J	Air Berlin
	D-ABAG	Boeing 737-86J	Air Berlin
	D-ABAH	Boeing 737-46J	Air Berlin
	D-ABAI	Boeing 737-46J	Air Berlin
	D-ABAK	Boeing 737-46J	Air Berlin
	D-ABAL	Boeing 737-46J	Air Berlin
	D-ABAM	Boeing 737-46J	Air Berlin
	D-ABAN	Boeing 737-86J	Air Berlin
	D-ABAO	Boeing 737-86J	Air Berlin
	D-ABAP	Boeing 737-86J	Air Berlin
	D-ABAQ	Boeing 737-86J	Air Berlin
	D-ABAR	Boeing 737-86J	Air Berlin
	D-ABAS	Boeing 737-86J	Air Berlin
	D-ABAT	Boeing 737-86J	Air Berlin
	D-ABAU	Boeing 737-86J	Air Berlin
	D-ABAV	Boeing 737-86J	Air Berlin
	D-ABAW	Boeing 737-86J	Air Berlin
	D-ABAX	Boeing 737-86J	Air Berlin
	D-ABAY	Boeing 737-86J	Air Berlin
	D-ABAZ	Boeing 737-86J	Air Berlin
	D-ABBA	Boeing 737-86J	Air Berlin

Reg.	Type	Owner or Operator	Notes
D-ABBB	Boeing 737-86J	Air Berlin	
D-ABBC	Boeing 737-86J	Air Berlin	
D-ABBD	Boeing 737-86J	Air Berlin	
D-ABBE	Boeing 737-86J	Air Berlin	
D-ABBF	Boeing 737-86J	Air Berlin	
D-ABBG	Boeing 737-86J	Air Berlin	
D-ABBH	Boeing 737-86J	Air Berlin	
D-ABBI	Boeing 737-86J	Air Berlin	
D-ABBJ	Boeing 737-86Q	Air Berlin	
D-ABBK	Boeing 737-8BK	Air Berlin	
D-ABBL	Boeing 737-85F	Air Berlin	
D-ABBM	Boeing 737-85F	Air Berlin	
D-ABBN	Boeing 737-76Q	Air Berlin	
D-ABBQ	Boeing 737-86N	Air Berlin	
D-ABEA	Boeing 737-330	Lufthansa *Saarbrücken*	
D-ABEB	Boeing 737-330	Lufthansa *Xanten*	
D-ABEC	Boeing 737-330	Lufthansa *Karlsrühe*	
D-ABED	Boeing 737-330	Lufthansa *Hagen*	
D-ABEE	Boeing 737-330	Lufthansa *Ulm*	
D-ABEF	Boeing 737-330	Lufthansa *Weiden i.d.Obf.*	
D-ABEH	Boeing 737-330	Lufthansa *Bad Kissingen*	
D-ABEI	Boeing 737-330	Lufthansa *Bamberg*	
D-ABEK	Boeing 737-330	Lufthansa *Wuppertal*	
D-ABEL	Boeing 737-330	Lufthansa *Pforzheim*	
D-ABEM	Boeing 737-330	Lufthansa *Eberswalde*	
D-ABEN	Boeing 737-330	Lufthansa *Neubrandenburg*	
D-ABEO	Boeing 737-330	Lufthansa *Plauen*	
D-ABEP	Boeing 737-330	Lufthansa *Naumburg (Saale)*	
D-ABER	Boeing 737-330	Lufthansa *Merseburg*	
D-ABES	Boeing 737-330	Lufthansa *Koethen/Anhalt*	
D-ABET	Boeing 737-330	Lufthansa *Gelsenkirchen*	
D-ABEU	Boeing 737-330	Lufthansa *Goslar*	
D-ABEW	Boeing 737-330	Lufthansa *Detmold*	
D-ABIC	Boeing 737-530	Lufthansa *Krefeld*	
D-ABID	Boeing 737-530	Lufthansa *Aachen*	
D-ABIE	Boeing 737-530	Lufthansa *Hildesheim*	
D-ABIF	Boeing 737-530	Lufthansa *Landau*	
D-ABIH	Boeing 737-530	Lufthansa *Bruchsal*	
D-ABII	Boeing 737-530	Lufthansa *Lörrach*	
D-ABIK	Boeing 737-530	Lufthansa *Rastatt*	
D-ABIL	Boeing 737-530	Lufthansa *Memmingen*	
D-ABIM	Boeing 737-530	Lufthansa *Salzgitter*	
D-ABIN	Boeing 737-530	Lufthansa *Langenhagen*	
D-ABIO	Boeing 737-530	Lufthansa *Wesel*	
D-ABIP	Boeing 737-530	Lufthansa *Oberhausen*	
D-ABIR	Boeing 737-530	Lufthansa *Anklam*	
D-ABIS	Boeing 737-530	Lufthansa *Rendsburg*	
D-ABIT	Boeing 737-530	Lufthansa *Neumünster*	
D-ABIU	Boeing 737-530	Lufthansa *Limburg a.d. Lahn*	
D-ABIW	Boeing 737-530	Lufthansa *Bad Nauheim*	
D-ABIX	Boeing 737-530	Lufthansa *Iserlohn*	
D-ABIY	Boeing 737-530	Lufthansa *Lingen*	
D-ABIZ	Boeing 737-530	Lufthansa *Kirchheim unter Teck*	
D-ABJA	Boeing 737-530	Lufthansa *Bad Segeberg*	
D-ABJB	Boeing 737-530	Lufthansa *Rheine*	
D-ABJC	Boeing 737-530	Lufthansa *Erding*	
D-ABJD	Boeing 737-530	Lufthansa *Freising*	
D-ABJE	Boeing 737-530	Lufthansa *Ingelheim am Rhein*	
D-ABJH	Boeing 737-530	Lufthansa *Heppenheim/Bergstrasse*	
D-ABJI	Boeing 737-530	Lufthansa *Siegburg*	
D-ABNF	Boeing 757-230	Thomas Cook Airlines/Condor	
D-ABNK	Boeing 757-230	Thomas Cook Airlines/Condor	
D-ABNN	Boeing 757-230	Thomas Cook Airlines/Condor	
D-ABNO	Boeing 757-230	Thomas Cook Airlines/Condor	
D-ABNS	Boeing 757-230	Thomas Cook Airlines/Condor	
D-ABNT	Boeing 757-230	Thomas Cook Airlines	
D-ABOA	Boeing 757-330	Thomas Cook Airlines	
D-ABOB	Boeing 757-330	Thomas Cook Airlines/Condor	
D-ABOC	Boeing 757-330	Thomas Cook Airlines/Condor	
D-ABOE	Boeing 757-330	Thomas Cook Airlines/Condor	
D-ABOF	Boeing 757-330	Thomas Cook Airlines/Condor	
D-ABOG	Boeing 757-330	Thomas Cook Airlines/Condor	

Notes	Reg.	Type	Owner or Operator
	D-ABOH	Boeing 757-330	Thomas Cook Airlines/Condor
	D-ABOI	Boeing 757-330	Thomas Cook Airlines/Condor
	D-ABOJ	Boeing 757-330	Thomas Cook Airlines/Condor
	D-ABOK	Boeing 757-330	Thomas Cook Airlines/Condor
	D-ABOL	Boeing 757-330	Thomas Cook Airlines/Condor
	D-ABOM	Boeing 757-330	Thomas Cook Airlines/Condor
	D-ABON	Boeing 757-330	Thomas Cook Airlines/Condor
	D-ABTA	Boeing 747-430 (SCD)	Lufthansa *Sachsen*
	D-ABTB	Boeing 747-430 (SCD)	Lufthansa *Brandenburg*
	D-ABTC	Boeing 747-430 (SCD)	Lufthansa *Mecklenburg-Verpommern*
	D-ABTD	Boeing 747-430 (SCD)	Lufthansa *Hamburg*
	D-ABTE	Boeing 747-430 (SCD)	Lufthansa *Sachsen-Anhalt*
	D-ABTF	Boeing 747-430 (SCD)	Lufthansa *Thüringen*
	D-ABTH	Boeing 747-430 (SCD)	Lufthansa *Duisburg*
	D-ABTK	Boeing 747-430 (SCD)	Lufthansa *Kiel*
	D-ABTL	Boeing 747-430 (SCD)	Lufthansa
	D-ABUA	Boeing 767-330ER	Thomas Cook Airlines/Condor
	D-ABUB	Boeing 767-330ER	Thomas Cook Airlines/Condor
	D-ABUC	Boeing 767-330ER	Thomas Cook Airlines/Condor
	D-ABUD	Boeing 767-330ER	Thomas Cook Airlines/Condor
	D-ABUE	Boeing 767-330ER	Thomas Cook Airlines/Condor
	D-ABUF	Boeing 767-330ER	Thomas Cook Airlines/Condor
	D-ABUH	Boeing 767-330ER	Thomas Cook Airlines/Condor
	D-ABUI	Boeing 767-330ER	Thomas Cook Airlines/Condor
	D-ABUZ	Boeing 767-330ER	Thomas Cook Airlines/Condor
	D-ABVA	Boeing 747-430	Lufthansa *Berlin*
	D-ABVB	Boeing 747-430	Lufthansa *Bonn*
	D-ABVC	Boeing 747-430	Lufthansa *Baden-Württemberg*
	D-ABVD	Boeing 747-430	Lufthansa *Bochum*
	D-ABVE	Boeing 747-430	Lufthansa *Potsdam*
	D-ABVF	Boeing 747-430	Lufthansa *Frankfurt am Main*
	D-ABVH	Boeing 747-430	Lufthansa *Düsseldorf*
	D-ABVK	Boeing 747-430	Lufthansa *Hannover*
	D-ABVL	Boeing 747-430	Lufthansa *Muenchen*
	D-ABVM	Boeing 747-430	Lufthansa *Hessen*
	D-ABVN	Boeing 747-430	Lufthansa *Dortmund*
	D-ABVO	Boeing 747-430	Lufthansa *Mulheim a.d.Ruhr*
	D-ABVP	Boeing 747-430	Lufthansa *Bremen*
	D-ABVR	Boeing 747-430	Lufthansa *Koln*
	D-ABVS	Boeing 747-430	Lufthansa *Saarland*
	D-ABVT	Boeing 747-430	Lufthansa *Rheinland Pfalz*
	D-ABVU	Boeing 747-430	Lufthansa *Bayern*
	D-ABVW	Boeing 747-430	Lufthansa *Wolfsburg*
	D-ABVX	Boeing 747-430	Lufthansa *Schleswig-Holstein*
	D-ABVY	Boeing 747-430	Lufthansa *Nordrhein Westfalen*
	D-ABVZ	Boeing 747-430	Lufthansa *Niedersachsen*
	D-ABWH	Boeing 737-330	Lufthansa *Rothenburg o. d. Taube*
	D-ABXL	Boeing 737-330	Lufthansa *Neuss*
	D-ABXM	Boeing 737-330	Lufthansa *Herford*
	D-ABXN	Boeing 737-330	Lufthansa *Böblingen*
	D-ABXO	Boeing 737-330	Lufthansa *Schwäbisch-Gmünd*
	D-ABXP	Boeing 737-330	Lufthansa *Fulda*
	D-ABXR	Boeing 737-330	Lufthansa *Celle*
	D-ABXS	Boeing 737-330	Lufthansa *Sindelfingen*
	D-ABXT	Boeing 737-330	Lufthansa *Reutlingen*
	D-ABXU	Boeing 737-330	Lufthansa *Seeheim-Jugenheim*
	D-ABXW	Boeing 737-330	Lufthansa *Hanau*
	D-ABXX	Boeing 737-330	Lufthansa *Bad Homburg v.d. Höhe*
	D-ABXZ	Boeing 737-330	Lufthansa *Bad Mergentheim*
	D-ABYZ	Boeing 747-230F (SCD)	Lufthansa Cargo
	D-ABZA	Boeing 747-230B (SCD)	Lufthansa Cargo
	D-ABZC	Boeing 747-230F (SCD)	Lufthansa Cargo
	D-ACFA	BAe 146-200	Lufthansa Regional
	D-ACHA	Canadair CL.600-2B19 RJ	Lufthansa Regional *Murrhardt*
	D-ACHB	Canadair CL.600-2B19 RJ	Lufthansa Regional *Meersburg*
	D-ACHC	Canadair CL.600-2B19 RJ	Lufthansa Regional *Füssen*
	D-ACHD	Canadair CL.600-2B19 RJ	Lufthansa Regional *Lutherstadt Eisleben*
	D-ACHE	Canadair CL.600-2B19 RJ	Lufthansa Regional *Meissen*
	D-ACHF	Canadair CL.600-2B19 RJ	Lufthansa Regional *Montabaur*
	D-ACHG	Canadair CL.600-2B19 RJ	Lufthansa Regional *Weil am Rhein*
	D-ACHH	Canadair CL.600-2B19 RJ	Lufthansa Regional *Kronach*
	D-ACHI	Canadair CL.600-2B19 RJ	Lufthansa Regional *Deidesheim*

Reg.	Type	Owner or Operator	Notes
D-ACHK	Canadair CL.600-2B19 RJ	Lufthansa Regional *Schkeuditz*	
D-ACIR	Embraer RJ145MP	Cirrus Airlines *Saarbrucken*	
D-ACJA	Canadair CL.600-2B19 RJ	Lufthansa Regional	
D-ACJB	Canadair CL.600-2B19 RJ	Lufthansa Regional	
D-ACJC	Canadair CL.600-2B19 RJ	Lufthansa Regional	
D-ACJD	Canadair CL.600-2B19 RJ	Lufthansa Regional	
D-ACJE	Canadair CL.600-2B19 RJ	Lufthansa Regional	
D-ACJF	Canadair CL.600-2B19 RJ	Lufthansa Regional	
D-ACJG	Canadair CL.600-2B19 RJ	Lufthansa Regional	
D-ACJH	Canadair CL.600-2B19 RJ	Lufthansa Regional	
D-ACJI	Canadair CL.600-2B19 RJ	Lufthansa Regional	
D-ACJJ	Canadair CL.600-2B19 RJ	Lufthansa Regional	
D-ACLA	Canadair CL.600-2B19 RJ	Lufthansa Regional	
D-ACLB	Canadair CL.600-2B19 RJ	Lufthansa Regional	
D-ACLC	Canadair CL.600-2B19 RJ	Lufthansa Regional	
D-ACLD	Canadair CL.600-2B19 RJ	Lufthansa Regional	
D-ACLE	Canadair CL.600-2B19 RJ	Lufthansa Regional	
D-ACLF	Canadair CL.600-2B19 RJ	Lufthansa Regional *Helmstedt*	
D-ACLG	Canadair CL.600-2B19 RJ	Lufthansa Regional	
D-ACLH	Canadair CL.600-2B19 RJ	Lufthansa Regional	
D-ACLI	Canadair CL.600-2B19 RJ	Lufthansa Regional	
D-ACLJ	Canadair CL.600-2B19 RJ	Lufthansa Regional	
D-ACLK	Canadair CL.600-2B19 RJ	Lufthansa Regional	
D-ACLL	Canadair CL.600-2B19 RJ	Lufthansa Regional	
D-ACLM	Canadair CL.600-2B19 RJ	Lufthansa Regional	
D-ACLP	Canadair CL.600-2B19 RJ	Lufthansa Regional	
D-ACLQ	Canadair CL.600-2B19 RJ	Lufthansa Regional	
D-ACLR	Canadair CL.600-2B19 RJ	Lufthansa Regional	
D-ACLS	Canadair CL.600-2B19 RJ	Lufthansa Regional	
D-ACLT	Canadair CL.600-2B19 RJ	Lufthansa Regional	
D-ACLU	Canadair CL.600-2B19 RJ	Lufthansa Regional	
D-ACLV	Canadair CL.600-2B19 RJ	Lufthansa Regional	
D-ACLW	Canadair CL.600-2B19 RJ	Lufthansa Regional	
D-ACLY	Canadair CL.600-2B19 RJ	Lufthansa Regional	
D-ACLZ	Canadair CL.600-2B19 RJ	Lufthansa Regional	
D-ACPA	Canadair CL.600-2C10 RJ	Lufthansa Regional *Westerland/Sylt*	
D-ACPB	Canadair CL.600-2C10 RJ	Lufthansa Regional *Rudesheim a. Rhein*	
D-ACPC	Canadair CL.600-2C10 RJ	Lufthansa Regional *Espelkamp*	
D-ACPD	Canadair CL.600-2C10 RJ	Lufthansa Regional *Vilshofen*	
D-ACPE	Canadair CL.600-2C10 RJ	Lufthansa Regional *Belzig*	
D-ACPF	Canadair CL.600-2C10 RJ	Lufthansa Regional *Uhingen*	
D-ACPG	Canadair CL.600-2C10 RJ	Lufthansa Regional *Leinfelden-Echterdingen*	
D-ACPH	Canadair CL.600-2C10 RJ	Lufthansa Regional *Eschwege*	
D-ACPI	Canadair CL.600-2C10 RJ	Lufthansa Regional *Viernheim*	
D-ACPJ	Canadair CL.600-2C10 RJ	Lufthansa Regional *Neumarkt i. d. Oberfalz*	
D-ACPK	Canadair CL.600-2C10 RJ	Lufthansa Regional *Besigheim*	
D-ACPL	Canadair CL.600-2C10 RJ	Lufthansa Regional	
D-ACPM	Canadair CL.600-2C10 RJ	Lufthansa Regional	
D-ACPN	Canadair CL.600-2C10 RJ	Lufthansa Regional	
D-ACPO	Canadair CL.600-2C10 RJ	Lufthansa Regional	
D-ACPP	Canadair CL.600-2C10 RJ	Lufthansa Regional	
D-ACPQ	Canadair CL.600-3C10 RJ	Lufthansa Regional	
D-ACPR	Canadair CL.600-3C10 RJ	Lufthansa Regional	
D-ACPS	Canadair CL.600-3C10 RJ	Lufthansa Regional	
D-ACPT	Canadair CL.600-3C10 RJ	Lufthansa Regional	
D-ACRA	Canadair CL.600-2B19 RJ	Lufthansa Regional	
D-ACRB	Canadair CL.600-2B19 RJ	Lufthansa Regional	
D-ACRC	Canadair CL.600-2B19 RJ	Lufthansa Regional	
D-ACRD	Canadair CL.600-2B19 RJ	Lufthansa Regional	
D-ACRE	Canadair CL.600-2B19 RJ	Lufthansa Regional	
D-ACRF	Canadair CL.600-2B19 RJ	Lufthansa Regional	
D-ACRG	Canadair CL.600-2B19 RJ	Lufthansa Regional	
D-ACRH	Canadair CL.600-2B19 RJ	Lufthansa Regional	
D-ACRI	Canadair CL.600-2B19 RJ	Lufthansa Regional	
D-ACRJ	Canadair CL.600-2B19 RJ	Lufthansa Regional	
D-ACRK	Canadair CL.600-2B19 RJ	Lufthansa Regional	
D-ACRL	Canadair CL.600-2B19 RJ	Lufthansa Regional	
D-ADBM	Boeing 737-31S	DBA *Schrifttanz*	
D-ADBN	Boeing 737-31S	DBA *Wolkenschreiber*	
D-ADBS	Boeing 737-31S	DBA	
D-ADEP	F.27 Friendship Mk 600	WDL	
D-ADHA	D.H.C.8Q-402 Dash Eight	Lufthansa Regional	

Notes	Reg.	Type	Owner or Operator
	D-ADHB	D.H.C.8Q-402 Dash Eight	Lufthansa Regional
	D-ADHC	D.H.C.8Q-402 Dash Eight	Lufthansa Regional
	D-ADHD	D.H.C.8Q-402 Dash Eight	Lufthansa Regional
	D-ADHE	D.H.C.8Q-402 Dash Eight	Lufthansa Regional
	D-ADIA	Boeing 737-36Q	DBA *Edelweiss*
	D-ADIB	Boeing 737-36Q	DBA *Enzian*
	D-ADIC	Boeing 737-36Q	DBA
	D-ADIF	Boeing 737-3L9	DBA
	D-ADIG	Boeing 737-3L9	DBA
	D-ADIH	Boeing 737-3Y0	DBA
	D-ADII	Boeing 737-329	DBA
	D-ADIK	Boeing 737-382	DBA
	D-AELE	F.27 Friendship Mk 600	WDL
	D-AELG	F.27 Friendship Mk 600	WDL
	D-AELH	F.27 Friendship Mk 600	WDL
	D-AELJ	F.27 Friendship Mk 600	WDL
	D-AELK	F.27 Friendship Mk 600	WDL
	D-AELM	F.27 Friendship Mk 600	WDL
	D-AERF	Airbus A.330-322	LTU
	D-AERK	Airbus A.330-322	LTU
	D-AERQ	Airbus A.330-322	LTU
	D-AEWA	BAe 146-300	Lufthansa Regional
	D-AEWB	BAe 146-300	Lufthansa Regional
	D-AEWD	BAe 146-200	Lufthansa Regional
	D-AEWE	BAe 146-200	Lufthansa Regional
	D-AEWG	Aérospatiale ATR-72-212	Lufthansa Regional
	D-AEWH	Aérospatiale ATR-72-212	Lufthansa Regional
	D-AEWI	Aérospatiale ATR-72-212	Lufthansa Regional
	D-AEWK	Aérospatiale ATR-72-212	Lufthansa Regional
	D-AGEG	Boeing 737-35B	DBA
	D-AGEJ	Boeing 737-3L9	DBA
	D-AGEK	Boeing 737-3M8	DBA
	D-AGEL	Boeing 737-75B	Hapag-Lloyd Express
	D-AGEP	Boeing 737-75B	Hapag-Lloyd Express
	D-AGEQ	Boeing 737-75B	Hapag-Lloyd Express
	D-AGER	Boeing 737-75B	Hapag-Lloyd Express
	D-AGES	Boeing 737-75B	Hapag-Lloyd Express
	D-AGET	Boeing 737-75B	Hapag-Lloyd Express
	D-AGEU	Boeing 737-75B	Hapag Lloyd Express
	D-AGEY	Boeing 737-73S	Maersk Air
	D-AGEZ	Boeing 737-73S	Maersk Air
	D-AGMR	Boeing 737-430	Air One
	D-AGPA	Fokker 100	Germania
	D-AGPB	Fokker 100	Air Berlin
	D-AGPC	Fokker 100	Germania
	D-AGPD	Fokker 100	Germania/Air Berlin
	D-AGPE	Fokker 100	Air Berlin
	D-AGPF	Fokker 100	Germania/Air Berlin
	D-AGPG	Fokker 100	Germania/Air Berlin
	D-AGPH	Fokker 100	Germania
	D-AGPI	Fokker 100	Germania/Air Berlin
	D-AGPJ	Fokker 100	Germania
	D-AGPK	Fokker 100	Air Berlin
	D-AGPL	Fokker 100	Germania
	D-AGPM	Fokker 100	Germania
	D-AGPN	Fokker 100	Germania
	D-AGPO	Fokker 100	Germania
	D-AGPP	Fokker 100	Germania
	D-AGPQ	Fokker 100	Germania/Air Berlin
	D-AGPR	Fokker 100	Germania/Air Berlin
	D-AGPS	Fokker 100	Germania/Air Berlin
	D-AHFA	Boeing 737-8K5	Hapag-Lloyd
	D-AHFB	Boeing 737-8K5	Hapag-Lloyd
	D-AHFC	Boeing 737-8K5	Hapag-Lloyd
	D-AHFD	Boeing 737-8K5	Hapag-Lloyd
	D-AHFE	Boeing 737-8K5	Hapag-Lloyd
	D-AHFF	Boeing 737-8K5	Hapag-Lloyd
	D-AHFG	Boeing 737-8K5	Hapag-Lloyd
	D-AHFH	Boeing 737-8K5	Hapag-Lloyd
	D-AHFI	Boeing 737-8K5	Hapag-Lloyd
	D-AHFJ	Boeing 737-8K5	Hapag-Lloyd
	D-AHFK	Boeing 737-8K5	Hapag-Lloyd

Reg.	Type	Owner or Operator	Notes
D-AHFL	Boeing 737-8K5	Hapag-Lloyd	
D-AHFM	Boeing 737-8K5	Hapag-Lloyd	
D-AHFN	Boeing 737-8K5	Hapag-Lloyd	
D-AHFO	Boeing 737-8K5	Hapag-Lloyd/Air Berlin	
D-AHFP	Boeing 737-8K5	Hapag-Lloyd	
D-AHFQ	Boeing 737-8K5	Hapag-Lloyd	
D-AHFR	Boeing 737-8K5	Hapag-Lloyd	
D-AHFS	Boeing 737-86N	Hapag-Lloyd/Air Berlin	
D-AHFT	Boeing 737-8K5	Hapag-Lloyd	
D-AHFU	Boeing 737-8K5	Hapag-Lloyd	
D-AHFV	Boeing 737-8K5	Hapag-Lloyd	
D-AHFW	Boeing 737-8K5	Hapag-Lloyd	
D-AHFX	Boeing 737-8K5	Hapag-Lloyd	
D-AHFY	Boeing 737-8K5	Hapag-Lloyd	
D-AHFZ	Boeing 737-8K5	Hapag-Lloyd	
D-AHIA	Boeing 737-73S	Hamburg International Airlines	
D-AHIB	Boeing 737-73S	Hamburg International Airlines	
D-AHIC	Boeing 737-7BK	Hamburg International Airlines	
D-AHID	Boeing 737-73S	Hamburg International Airlines	
D-AHIE	Boeing 737-73S	Hamburg International Airlines	
D-AHIF	Boeing 737-73S	Hamburg International Airlines	
D-AHLA	Airbus A.310-304	Hapag-Lloyd	
D-AHLD	Boeing 737-5K5	Germania/Hapag-Lloyd Express	
D-AHLG	Boeing 737-5K5	Germania/Hapag-Lloyd Express	
D-AHLH	Boeing 737-8K5	Hapag-Lloyd	
D-AHLN	Boeing 737-5K5	Germania/Hapag-Lloyd Express	
D-AHLP	Boeing 737-8K5	Hapag-Lloyd	
D-AHLQ	Boeing 737-8K5	Hapag Lloyd	
D-AHLR	Boeing 737-8K5	Hapag Lloyd	
D-AHLV	Airbus A.310-204	Hapag-Lloyd	
D-AHLW	Airbus A.310-204	Hapag-Lloyd	
D-AHLX	Airbus A.310-204	Hapag-Lloyd	
D-AHLZ	Airbus A.310-204	Hapag-Lloyd	
D-AHOI	BAe 146-300	Lufthansa Regional	
D-AIAH	Airbus A.300B4-603	Lufthansa *Lindau/Bodensee*	
D-AIAI	Airbus A.300B4-603	Lufthansa *Erbach/Odenwald*	
D-AIAK	Airbus A.300B4-603	Lufthansa *Kronberg/Taunus*	
D-AIAL	Airbus A.300B4-603	Lufthansa *Stade*	
D-AIAM	Airbus A.300B4-603	Lufthansa *Rosenheim*	
D-AIAN	Airbus A.300B4-603	Lufthansa *Nördlingen*	
D-AIAP	Airbus A.300B4-603	Lufthansa *Donauwörth*	
D-AIAR	Airbus A.300B4-603	Lufthansa *Bingen am Rhein*	
D-AIAS	Airbus A.300B4-603	Lufthansa *Mönchengladbach*	
D-AIAT	Airbus A.300B4-603	Lufthansa *Bottrop*	
D-AIAU	Airbus A.300B4-603	Lufthansa *Bocholt*	
D-AIAW	Airbus A.300B4-605R	Lufthansa *Witten*	
D-AIAX	Airbus A.300B4-605R	Lufthansa *Fürth*	
D-AIAY	Airbus A.300B4-605R	Lufthansa	
D-AIAZ	Airbus A.300B4-605R	Lufthansa	
D-AICA	Airbus A.320-212	Thomas Cook Airlines Belgium	
D-AICB	Airbus A.320-212	Thomas Cook Airlines Belgium	
D-AICC	Airbus A.320-212	Thomas Cook Airlines/Condor	
D-AICD	Airbus A.320-212	Thomas Cook Airlines/Condor	
D-AICE	Airbus A.320-212	Thomas Cook Airlines/Condor	
D-AICF	Airbus A.320-212	Thomas Cook Airlines/Condor	
D-AICG	Airbus A.320-212	Thomas Cook Airlines/Condor	
D-AICH	Airbus A.320-212	Thomas Cook Airlines/Condor	
D-AICI	Airbus A.320-212	Thomas Cook Airlines/Condor	
D-AICJ	Airbus A.320-212	Thomas Cook Airlines/Condor	
D-AICK	Airbus A.320-212	Thomas Cook Airlines/Condor	
D-AICL	Airbus A.320-212	Thomas Cook Airlines/Condor	
D-AIDD	Airbus A.310-304	Hapag-Lloyd	
D-AIDf	Airbus A.310-304	Hapag-Lloyd	
D-AIFA	Airbus A.340-313X	Lufthansa *Dorsten*	
D-AIFB	Airbus A.340-313X	Lufthansa *Gummersbach*	
D-AIFC	Airbus A.340-313X	Lufthansa *Gander/Halifax*	
D-AIFD	Airbus A.340-313X	Lufthansa *Giessen*	
D-AIFE	Airbus A.340-313X	Lufthansa *Passau*	
D-AIFF	Airbus A.340-313X	Lufthansa *Delmenhorst*	
D-AIGA	Airbus A.340-311	Lufthansa *Oldenburg*	
D-AIGB	Airbus A.340-311	Lufthansa *Recklinghausen*	
D-AIGC	Airbus A.340-311	Lufthansa *Wilhelmshaven*	

Notes	Reg.	Type	Owner or Operator
	D-AIGD	Airbus A.340-311	Lufthansa *Remscheid*
	D-AIGF	Airbus A.340-311	Lufthansa *Gottingen*
	D-AIGH	Airbus A.340-311	Lufthansa *Koblenz*
	D-AIGI	Airbus A.340-311	Lufthansa *Worms*
	D-AIGK	Airbus A.340-311	Lufthansa *Bayreuth*
	D-AIGL	Airbus A.340-313X	Lufthansa *Herne*
	D-AIGM	Airbus A.340-313X	Lufthansa *Wolfsburg*
	D-AIGN	Airbus A.340-313X	Lufthansa *Solingen*
	D-AIGO	Airbus A.340-313X	Lufthansa *Offenbach*
	D-AIGP	Airbus A.340-313X	Lufthansa
	D-AIGR	Airbus A.340-313X	Lufthansa *Leipzig*
	D-AIGS	Airbus A.340-313X	Lufthansa *Bergisch-Gladbach*
	D-AIGT	Airbus A.340-313X	Lufthansa *Viersen*
	D-AIGU	Airbus A.340-313X	Lufthansa *Castrop-Rauxei*
	D-AIGV	Airbus A.340-313X	Lufthansa *Dinslaken*
	D-AIGW	Airbus A.340-313X	Lufthansa *Gladbeck*
	D-AIGX	Airbus A.340-313X	Lufthansa *Duren*
	D-AIGY	Airbus A.340-313X	Lufthansa *Lünen*
	D-AIGZ	Airbus A.340-313X	Lufthansa *Villingen-Schwenningen*
	D-AIHA	Airbus A.340-642	Lufthansa *Nurnberg*
	D-AIHB	Airbus A.340-642	Lufthansa *Bremerhaven*
	D-AIHC	Airbus A.340-642	Lufthansa *Essen*
	D-AIHD	Airbus A.340-642	Lufthansa *Stuttgart*
	D-AIHE	Airbus A.340-642	Lufthansa *Leverkusen*
	D-AIHF	Airbus A.340-642	Lufthansa *Lubeck*
	D-AIHH	Airbus A.340-642	Lufthansa
	D-AIHI	Airbus A.340-642	Lufthansa
	D-AIHK	Airbus A.340-642	Lufthansa
	D-AIHL	Airbus A.340-642	Lufthansa
	D-AIKA	Airbus A.330-343X	Lufthansa *Minden*
	D-AIKB	Airbus A.330-343X	Lufthansa *Cuxhaven*
	D-AIKC	Airbus A.330-343X	Lufthansa
	D-	Airbus A.330-343X	Lufthansa
	D-	Airbus A.330-343X	Lufthansa
	D-	Airbus A.330-343X	Lufthansa
	D-	Airbus A.330-343X	Lufthansa
	D-	Airbus A.330-343X	Lufthansa
	D-	Airbus A.330-343X	Lufthansa
	D-AILA	Airbus A.319-114	Lufthansa *Frankfurt (Oder)*
	D-AILB	Airbus A.319-114	Lufthansa *Lutherstadt Wittenburg*
	D-AILC	Airbus A.319-114	Lufthansa *Russelsheim*
	D-AILD	Airbus A.319-114	Lufthansa *Dinkelsbühl*
	D-AILE	Airbus A.319-114	Lufthansa *Kelsterbach*
	D-AILF	Airbus A.319-114	Germanwings
	D-AILH	Airbus A.319-114	Croatia Airlines
	D-AILI	Airbus A.319-114	Germanwings
	D-AILK	Airbus A.319-114	Germanwings
	D-AILL	Airbus A.319-114	Germanwings
	D-AILM	Airbus A.319-114	Lufthansa *Friedrichshafen*
	D-AILN	Airbus A.319-114	Germanwings
	D-AILP	Airbus A.319-114	Lufthansa *Tubingen*
	D-AILR	Airbus A.319-114	Lufthansa *Tegernsee*
	D-AILS	Airbus A.319-114	Lufthansa *Heide*
	D-AILT	Airbus A.319-114	Lufthansa *Straubing*
	D-AILU	Airbus A.319-114	Lufthansa *Verden*
	D-AILW	Airbus A.319-114	Lufthansa *Donaueschingen*
	D-AILX	Airbus A.319-114	Germanwings
	D-AILY	Airbus A.319-114	Lufthansa *Schweinfurt*
	D-AIMB	Airbus A.330-223	Lufthansa
	D-AIMC	Airbus A.330-223	Lufthansa
	D-AIMD	Airbus A.330-223	Lufthansa
	D-AIME	Airbus A.330-223	Lufthansa
	D-AIMF	Airbus A.340-311	Lufthansa
	D-AIMG	Airbus A.340-311	Lufthansa
	D-AIPA	Airbus A.320-211	Lufthansa *Buxtehude*
	D-AIPB	Airbus A.320-211	Lufthansa *Heidelberg*
	D-AIPC	Airbus A.320-211	Germanwings
	D-AIPD	Airbus A.320-211	Germanwings
	D-AIPE	Airbus A.320-211	Lufthansa *Kassel*
	D-AIPF	Airbus A.320-211	Lufthansa *Deggendorf*
	D-AIPH	Airbus A.320-211	Germanwings *Spirit of Cologne/Bonn*
	D-AIPK	Airbus A.320-211	Lufthansa *Wiesbaden*

Reg.	Type	Owner or Operator	Notes
D-AIPL	Airbus A.320-211	Lufthansa *Ludwigshafen am Rhein*	
D-AIPM	Airbus A.320-211	Lufthansa *Troisdorf*	
D-AIPP	Airbus A.320-211	Lufthansa *Starnberg*	
D-AIPR	Airbus A.320-211	Lufthansa *Kaufbeuren*	
D-AIPS	Airbus A.320-211	Lufthansa *Augsburg*	
D-AIPT	Airbus A.320-211	Lufthansa *Cottbus*	
D-AIPU	Airbus A.320-211	Lufthansa *Dresden*	
D-AIPW	Airbus A.320-211	Lufthansa *Schwerin*	
D-AIPX	Airbus A.320-211	Lufthansa *Mannheim*	
D-AIPY	Airbus A.320-211	Lufthansa *Magdeburg*	
D-AIPZ	Airbus A.320-211	Lufthansa *Erfurt*	
D-AIQA	Airbus A.320-211	Lufthansa *Mainz*	
D-AIQB	Airbus A.320-211	Lufthansa *Bielefeld*	
D-AIQC	Airbus A.320-211	Lufthansa *Zwickau*	
D-AIQD	Airbus A.320-211	Lufthansa *Jena*	
D-AIQE	Airbus A.320-211	Lufthansa *Gera*	
D-AIQF	Airbus A.320-211	Lufthansa *Halle (Saale)*	
D-AIQH	Airbus A.320-211	Lufthansa *Dessau*	
D-AIQK	Airbus A.320-211	Lufthansa *Rostock*	
D-AIQL	Airbus A.320-211	Lufthansa *Stralsund*	
D-AIQM	Airbus A.320-211	Lufthansa *Nordenham*	
D-AIQN	Airbus A.320-211	Lufthansa *Laupheim*	
D-AIQP	Airbus A.320-211	Lufthansa *Suhl*	
D-AIQR	Airbus A.320-211	Germanwings	
D-AIQS	Airbus A.320-211	Lufthansa *Eisenach*	
D-AIQT	Airbus A.320-211	Lufthansa *Gotha*	
D-AIQU	Airbus A.320-211	Lufthansa *Backnang*	
D-AIQW	Airbus A.320-211	Lufthansa *Kleve*	
D-AIRA	Airbus A.321-131	Lufthansa *Finkenwerder*	
D-AIRB	Airbus A.321-131	Lufthansa *Baden-Baden*	
D-AIRC	Airbus A.321-131	Lufthansa *Erlangen*	
D-AIRD	Airbus A.321-131	Lufthansa *Coburg*	
D-AIRE	Airbus A.321-131	Lufthansa *Osnabrueck*	
D-AIRF	Airbus A.321-131	Lufthansa *Kempten*	
D-AIRH	Airbus A.321-131	Lufthansa *Garmisch-Partenkirchen*	
D-AIRK	Airbus A.321-131	Lufthansa *Freudenstadt/Schwarzwald*	
D-AIRL	Airbus A.321-131	Lufthansa *Kulmbach*	
D-AIRM	Airbus A.321-131	Lufthansa *Darmstadt*	
D-AIRN	Airbus A.321-131	Lufthansa *Kaiserslautern*	
D-AIRO	Airbus A.321-131	Lufthansa *Konstanz*	
D-AIRP	Airbus A.321-131	Lufthansa *Lüneburg*	
D-AIRR	Airbus A.321-131	Lufthansa *Wismar*	
D-AIRS	Airbus A.321-131	Lufthansa *Husum*	
D-AIRT	Airbus A.321-131	Lufthansa *Regensburg*	
D-AIRU	Airbus A.321-131	Lufthansa *Würzburg*	
D-AIRW	Airbus A.321-131	Lufthansa *Heilbronn*	
D-AIRX	Airbus A.321-131	Lufthansa *Weimar*	
D-AIRY	Airbus A.321-131	Lufthansa *Flensburg*	
D-AISB	Airbus A.321-231	Lufthansa *Hameln*	
D-AISC	Airbus A.321-231	Lufthansa *Speyer*	
D-AISD	Airbus A.321-231	Lufthansa *Chemnitz*	
D-AISE	Airbus A.321-231	Lufthansa *Neustadt an der Weinstrasse*	
D-AISF	Airbus A.321-231	Lufthansa *Lippstadt*	
D-AISG	Airbus A.321-231	Lufthansa *Dormagen*	
D-AISY	F.27 Friendship Mk 600	WDL	
D-AJET	BAe 146-200	Lufthansa Regional	
D-AKNF	Airbus A.319-112	Germanwings	
D-AKNG	Airbus A.319-112	Germanwings	
D-AKNH	Airbus A.319-112	Germanwings	
D-AKNI	Airbus A.319-112	Germanwings	
D-AKNJ	Airbus A.319-112	Germanwings	
D-AKNX	Airbus A.320-232	Germanwings	
D-AKNY	Airbus A.320-212	Germanwings	
D-AKNZ	Airbus A.320-212	Germanwings	
D-ALAQ	Airbus A.321-231	Flyniki	
D-ALCA	McD Douglas MD-11F	Lufthansa Cargo	
D-ALCB	McD Douglas MD-11F	Lufthansa Cargo	
D-ALCC	McD Douglas MD-11F	Lufthansa Cargo	
D-ALCD	McD Douglas MD-11F	Lufthansa Cargo	
D-ALCE	McD Douglas MD-11F	Lufthansa Cargo	
D-ALCF	McD Douglas MD-11F	Lufthansa Cargo	
D-ALCG	McD Douglas MD-11F	Lufthansa Cargo	

Notes	Reg.	Type	Owner or Operator
	D-ALCH	McD Douglas MD-11F	Lufthansa Cargo
	D-ALCI	McD Douglas MD-11F	Lufthansa Cargo
	D-ALCJ	McD Douglas MD-11F	Lufthansa Cargo
	D-ALCK	McD Douglas MD-11F	Lufthansa Cargo
	D-ALCL	McD Douglas MD-11F	Lufthansa Cargo
	D-ALCM	McD Douglas MD-11F	Lufthansa Cargo
	D-ALCN	McD Douglas MD-11F	Lufthansa Cargo
	D-ALCQ	McD Douglas MD-11F	Lufthansa Cargo
	D-ALCR	McD Douglas MD-11F	Lufthansa Cargo
	D-ALCS	McD Douglas MD-11F	Lufthansa Cargo
	D-ALIE	Embraer RJ170-100LR	Cirrus Airlines
	D-ALPA	Airbus A.330-223	LTU
	D-ALPB	Airbus A.330-223	LTU
	D-ALPC	Airbus A.330-223	LTU
	D-ALPD	Airbus A.330-223	LTU
	D-ALPE	Airbus A.330-223	LTU
	D-ALPF	Airbus A.330-223	LTU
	D-ALPG	Airbus A.330-223	LTU
	D-ALSA	Airbus A.321-211	LTU
	D-ALSB	Airbus A.321-211	LTU
	D-ALSC	Airbus A.321-211	LTU
	D-ALSD	Airbus A.321-211	LTU
	D-ALTB	Airbus A.320-214	LTU
	D-ALTC	Airbus A.320-214	LTU
	D-ALTD	Airbus A.320-214	LTU
	D-ALTF	Airbus A.320-214	LTU
	D-ALTG	Airbus A.320-214	LTU
	D-ALTH	Airbus A.320-214	LTU
	D-ALTI	Airbus A.320-214	LTU
	D-ALTJ	Airbus A.320-214	LTU
	D-ALTK	Airbus A.320-214	LTU
	D-ALTL	Airbus A.320-214	LTU
	D-AMAJ	BAe 146-200	WDL
	D-ANFA	Aérospatiale ATR-72-202	Lufthansa Regional
	D-ANFB	Aérospatiale ATR-72-202	Lufthansa Regional
	D-ANFC	Aérospatiale ATR-72-202	Lufthansa Regional
	D-ANFD	Aérospatiale ATR-72-202	Lufthansa Regional
	D-ANFE	Aérospatiale ATR-72-202	Lufthansa Regional
	D-ANFF	Aérospatiale ATR-72-202	Lufthansa Regional
	D-ANFG	Aérospatiale ATR-72-212A	Lufthansa Regional
	D-ANFH	Aérospatiale ATR-72-212A	Lufthansa Regional
	D-ANFI	Aérospatiale ATR-72-212A	Lufthansa Regional
	D-ANFJ	Aérospatiale ATR-72-212A	Lufthansa Regional
	D-ANFK	Aérospatiale ATR-72-212A	Lufthansa Regional
	D-ANFL	Aérospatiale ATR-72-212A	Lufthansa Regional
	D-ANJA	Airbus A.321-111	Blue Wings
	D-ANNE	Airbus A.320-232	Blue Wings
	D-AOLA	SAAB 2000	OLT
	D-AOLB	SAAB 2000	OLT
	D-AOLT	SAAB 2000	OLT Emden
	D-APAA	Airbus A.319-132	Privatair/Airbus
	D-APAB	Airbus A.319-132	Privatair/Airbus
	D-APAC	Airbus A.319-132LR	Privatair/Lufthansa
	D-APAD	Airbus A.319-132LR	Privatair/Lufthansa
	D-AQUA	BAe 146-300	Lufthansa Regional
	D-AQUI	Junkers Ju.52/3m	Lufthansa Traditionsflug
	D-ARFA	Airbus A.321-231	Aero Flight
	D-ARFB	Airbus A.321-231	Aero Flight
	D-ARFC	Airbus A.320-232	Aero Flight
	D-ARFD	Airbus A.320-232	Aero Flight
	D-ARFE	Airbus A.320-232	Aero Flight
	D-ATUI	Boeing 737-86Q	Hapag-Lloyd
	D-AVRA	Avro RJ85	Lufthansa Regional
	D-AVRB	Avro RJ85	Lufthansa Regional
	D-AVRC	Avro RJ85	Lufthansa Regional
	D-AVRD	Avro RJ85	Lufthansa Regional
	D-AVRE	Avro RJ85	Lufthansa Regional
	D-AVRF	Avro RJ85	Lufthansa Regional
	D-AVRG	Avro RJ85	Lufthansa Regional
	D-AVRH	Avro RJ85	Lufthansa Regional
	D-AVRI	Avro RJ85	Lufthansa Regional
	D-AVRJ	Avro RJ85	Lufthansa Regional

Reg.	Type	Owner or Operator	Notes
D-AVRK	Avro RJ85	Lufthansa Regional	
D-AVRL	Avro RJ85	Lufthansa Regional	
D-AVRM	Avro RJ85	Lufthansa Regional	
D-AVRN	Avro RJ85	Lufthansa Regional	
D-AVRO	Avro RJ85	Lufthansa Regional	
D-AVRP	Avro RJ85	Lufthansa Regional	
D-AVRQ	Avro RJ85	Lufthansa Regional	
D-AVRR	Avro RJ85	Lufthansa Regional	
D-AWBA	BAe 146-300	WDL/Austrian Arrows	
D-AWDL	BAe 146-100	WDL	
D-AXEL	Airbus A.330-223	Blue Wings	
D-BACH	D.H.C.8-311 Dash Eight	Lufthansa Regional	
D-BAKB	F.27 Friendship Mk 600	WDL	
D-BCRQ	Aérospatiale ATR-42-320	European Air Express	
D-BCRR	Aérospatiale ATR-42-320	European Air Express	
D-BCRS	Aérospatiale ATR-42-320	European Air Express	
D-BCRT	Aérospatiale ATR-42-320	European Air Express	
D-BDTM	D.H.C.8Q-314 Dash Eight	Lufthansa Regional	
D-BEBA	D.H.C.8Q-314 Dash Eight	Lufthansa Regional	
D-BGAL	Dornier Do.328JET	Cirrus Air	
D-BGAQ	Dornier Do.328JET	Cirrus Air	
D-BHAM	D.H.C.8-311 Dash Eight	Cirrus Air	
D-BHAT	D.H.C.8Q-311 Dash Eight	Lufthansa Regional	
D-BHOQ	D.H.C.8Q-314 Dash Eight	Lufthansa Regional	
D-BIER	D.H.C.8-102 Dash Eight	Cirrus Air	
D-BJJJ	Aérospatiale ATR-42-300	European Air Express	
D-BKIM	D.H.C.8-311 Dash Eight	Cirrus Air	
D-BKKK	Aérospatiale ATR-42-512	City-Air	
D-BLEJ	D.H.C.8Q-314 Dash Eight	Intersky	
D-BLLL	Aérospatiale ATR-42-512	Lufthansa Regional	
D-BMMM	Aérospatiale ATR-42-512	Lufthansa Regional	
D-BMUC	D.H.C.8-314 Dash Eight	Cirrus Air	
D-BNNN	Aérospatiale ATR-42-512	Lufthansa Regional	
D-BOBL	D.H.C.8-102A Dash Eight	Cirrus Air	
D-BOBO	D.H.C.8-102 Dash Eight	Cirrus Air	
D-BOBU	D.H.C.8-311 Dash Eight	Cirrus Air	
D-BOBY	D.H.C.8-102 Dash Eight	Cirrus Air	
D-BOOO	Aérospatiale ATR-42-512	Lufthansa Regional	
D-BPAD	D.H.C.8Q-314 Dash Eight	Lufthansa Regional	
D-BPPP	Aérospatiale ATR-42-512	Lufthansa Regional	
D-BQQQ	Aérospatiale ATR-42-512	Lufthansa Regional	
D-BRRR	Aérospatiale ATR-42-512	Lufthansa Regional	
D-BSSS	Aérospatiale ATR-42-512	Lufthansa Regional	
D-BTTT	Aérospatiale ATR-42-512	Lufthansa Regional	
D-CASB	SAAB SF.340B	OLT *Birdie*	
D-CBIN	Swearingen SA227AT Expediter IVC	Binair	
D-CCCC	Swearingen SA227AT Merlin IVC	Binair	
D-CCIR	Dornier Do.328-130	Cirrus Airlines	
D-CIRA	Dornier Do.328-120	Cirrus Airlines	
D-CIRB	Dornier Do.328-110	Cirrus Airlines	
D-CIRC	Dornier Do.328-110	Cirrus Airlines	
D-CIRD	Dornier Do.328-110	Cirrus Airlines	
D-CNRY	BAe Jetstream 3103	European Air Express	
D-CNRZ	BAe Jetstream 3107	European Air Express	
D-COLB	Swearingen SA227AC Metro III	OLT	
D-COLD	Swearingen SA227AC Metro III	OLT	
D-COLE	SAAB SF.340A	OLT *Bremen*	
D-COLT	Swearingen SA226AC Metro III	OLT	
D-COSA	Dornier Do.328-110	Cirrus Air	
D-CPRU	Dornier Do.328-110	City Star Airlines	
D-CUTT	Dornier Do.328-212	Cirrus Air	
D-ICRK	Swearingen SA226TC Metro II	Bin Air	

Note: Binair also operates Metro II PH-RAZ.

E3 (Eritrea)

E3-AAO	Boeing 767-366ER	Eritrean Airlines	
E3-AAQ	Boeing 767-238ER	Eritrean Airlines	

Notes	Reg.	Type	Owner or Operator

EC (Spain)

	EC-DIA	Boeing 747-256B	Iberia *Tirso de Molina*
	EC-DIB	Boeing 747-256B	Iberia *Cervantes*
	EC-DNP	Boeing 747-256B	Iberia *Juan Ramon Jimenez*
	EC-ELT	BAe 146-200QT	Pan Air/TNT Airways
	EC-EMD	Douglas DC-8-62F	Cygnus Air
	EC-EMX	Douglas DC-8-62F	Cygnus Air
	EC-EXF	McD Douglas MD-87	Iberia *Ciudad de Pamplona*
	EC-EXG	McD Douglas MD-87	Iberia *Ciudad de Almeria*
	EC-EXM	McD Douglas MD-87	Iberia *Ciudad de Zaragoza*
	EC-EXN	McD Douglas MD-87	Iberia *Ciudad de Badajoz*
	EC-EXR	McD Douglas MD-87	Iberia *Ciudad de Oviedo*
	EC-EXT	McD Douglas MD-87	Iberia *Ciudad de Albacete*
	EC-EYB	McD Douglas MD-87	Iberia *Cangas de Onis*
	EC-EYX	McD Douglas MD-87	Iberia *Ciudad de Caceres*
	EC-EYY	McD Douglas MD-87	Iberia *Ciudad de Barcelona*
	EC-EYZ	McD Douglas MD-87	Iberia *Ciudad de Las Palmas*
	EC-EZA	McD Douglas MD-87	Iberia *Ciudad de Segovia*
	EC-EZS	McD Douglas MD-87	Iberia *Ciudad de Mahon*
	EC-FCB	Airbus A.320-211	Iberia *Montana de Covadonga*
	EC-FDA	Airbus A.320-211	Iberia *Lagunas de Ruidera*
	EC-FDB	Airbus A.320-211	Iberia *Lago de Sanabria*
	EC-FEY	McD Douglas MD-87	Iberia *Ciudad de Jaen*
	EC-FEZ	McD Douglas MD-87	Iberia *Ciudad de Malaga*
	EC-FFA	McD Douglas MD-87	Iberia *Ciudad de Avila*
	EC-FFH	McD Douglas MD-87	Iberia *Ciudad de Logrono*
	EC-FFI	McD Douglas MD-87	Iberia *Ciudad de Cuenca*
	EC-FGH	Airbus A.320-211	Iberia *Caldera de Taburiente*
	EC-FGM	McD Douglas MD-88	Iberia *Torre de Hercules*
	EC-FGR	Airbus A.320-211	Iberia *Dehesa de Moncayo*
	EC-FGV	Airbus A.320-211	Iberia *Monfrague*
	EC-FHD	McD Douglas MD-87	Iberia *Ciudad de Leon*
	EC-FHG	McD Douglas MD-88	Iberia *La Almudiana*
	EC-FHK	McD Douglas MD-87	Iberia *Ciudad de Tarragona*
	EC-FIG	McD Douglas MD-88	Iberia *Penon de Ifach*
	EC-FIH	McD Douglas MD-88	Iberia *Albaicin*
	EC-FJE	McD Douglas MD-88	Iberia *Gibralfaro*
	EC-FLK	McD Douglas MD-88	Iberia *Palacio de la Magdalena*
	EC-FLN	McD Douglas MD-88	Iberia *Puerta de Tierra*
	EC-FLP	Airbus A.320-211	Iberia *Torcal de Antequera*
	EC-FLQ	Airbus A.320-211	Iberia *Dunas de Liencres*
	EC-FND	McD Douglas MD-88	Iberia *Playa de la Concha*
	EC-FNR	Airbus A.320-211	Iberia *Monte el Valle*
	EC-FOF	McD Douglas MD-88	Iberia *Puerta de Alcala*
	EC-FOG	McD Douglas MD-88	Iberia *La Giralda*
	EC-FOZ	McD Douglas MD-88	Iberia *Montjuic*
	EC-FPD	McD Douglas MD-88	Iberia *Lagos de Coradonga*
	EC-FPJ	McD Douglas MD-88	Iberia *Ria de Vigo*
	EC-FQY	Airbus A.320-211	Iberia *Joan Miro*
	EC-FTR	Boeing 757-256	Iberia *Sierra de Guadarrama*
	EC-FTS	McD Douglas MD-83	Spanair *Sunbird*
	EC-FVY	BAe 146-200QT	Pan Air/TNT Airways
	EC-FXA	McD Douglas MD-83	Spanair *Sunstar*
	EC-FXP	Boeing 737-4Q8	Air Europa *Villanueva del Conde*
	EC-FXY	McD Douglas MD-83	Spanair *Sunbeam*
	EC-FZC	McD Douglas MD-83	Spanair *Sunflower*
	EC-FZE	BAe 146-200QT	Pan Air/TNT Airways
	EC-FZZ	Boeing 737-4Y0	Air Europa *Baleares*
	EC-GAT	McD Douglas MD-83	Spanair *Sunmyth*
	EC-GBA	McD Douglas MD-83	Spanair *Sungod*
	EC-GCV	McD Douglas MD-82	Spanair *Sunburst*
	EC-GDY	Convair 580	Swiftair/DHL
	EC-GGS	Airbus A.340-313	Iberia *Concha Espina*
	EC-GGV	McD Douglas MD-83	Spanair *Sunbow*
	EC-GHE	McD Douglas MD-83	Spanair *Sunset*
	EC-GHH	McD Douglas MD-83	Spanair *Sundance*
	EC-GHN	Convair 580	Swiftair/DHL
	EC-GHX	Airbus A.340-313	Iberia *Rosalia de Castro*
	EC-GJT	Airbus A.340-313	Iberia *Rosa Chacel*
	EC-GLE	Airbus A.340-313	Iberia *Concepcion Arenal*

Reg.	Type	Owner or Operator	Notes
EC-GMU	Airbus A.310-324	Air Plus Comet	
EC-GNY	McD Douglas MD-83	Spanair *Sunflash*	
EC-GNZ	Boeing 737-4Y0	Futura International Airways	
EC-GOM	McD Douglas MD-83	Spanair *Sunlight*	
EC-GOT	Airbus A.310-324	Air Plus Comet	
EC-GOU	McD Douglas MD-83	Spanair *Sunlover*	
EC-GPB	Airbus A.340-313X	Iberia *Teresa de Avila*	
EC-GQO	BAe 146-200QT	Pan Air/TNT Airways	
EC-GQG	McD Douglas MD-83	Spanair *Sunrise*	
EC-GQK	Airbus A.340-313X	Iberia *Emilia Pardo Bazan*	
EC-GRE	Airbus A.320-211	Iberia *Sierra de Cazorla*	
EC-GRF	Airbus A.320-211	Iberia *Montseny*	
EC-GRG	Airbus A.320-211	Iberia *Timanfaya*	
EC-GRH	Airbus A.320-211	Iberia *Sierra de Segura*	
EC-GRI	Airbus A.320-211	Iberia *Delta del Ebro*	
EC-GRJ	Airbus A.320-211	Iberia *Canon del Rio Lobos*	
EC-GRK	McD Douglas MD-87	Iberia *Ciudad de Savilla*	
EC-GRL	McD Douglas MD-87	Iberia *Ciudad de Toledo*	
EC-GRM	McD Douglas MD-87	Iberia *Ciudad de Burgos*	
EC-GRN	McD Douglas MD-87	Iberia *Ciudad de Cadiz*	
EC-GRO	McD Douglas MD-87	Iberia *Arrecife de Lanzarote*	
EC-GSJ	Convair 580	Swiftair/DHL	
EC-GTO	McD Douglas MD-82	Spanair *Sunjet*	
EC-GUO	Boeing 737-4Q8	Air Europa	
EC-GUP	Airbus A.340-313X	Iberia *Agustina De Aragon*	
EC-GUQ	Airbus A.340-313X	Iberia *Beatriz Galindo*	
EC-GVE	Swearingen SA227AC Metro III	Aeronova	
EC-GVI	McD Douglas MD-83	Spanair *Sunup*	
EC-GVO	McD Douglas MD-83	Spanair *Sunspot*	
EC-GXU	McD Douglas MD-83	Spanair *Sunray*	
EC-GYI	Canadair Regional Jet 100ER	Air Nostrum/Iberia *Pinazo*	
EC-GZA	Canadair Regional Jet 100ER	Air Nostrum/Iberia *Benlliure*	
EC-GZD	Airbus A.320-214	Iberworld	
EC-HAB	Airbus A.320-214	Iberia *Cabaneros*	
EC-HAF	Airbus A.320-214	Iberia *Santiago de Compostela*	
EC-HAG	Airbus A.320-214	Iberia *Senorio de Bertiz*	
EC-HBL	Boeing 737-85P	Air Europa *Travelplan*	
EC-HBM	Boeing 737-85P	Air Europa *La Gaceta*	
EC-HBN	Boeing 737-85P	Air Europa *Llucmajor*	
EC-HBP	McD Douglas MD-83	Spanair *Sunglow*	
EC-HCH	Swearingen SA227AC Metro III	Aeronova	
EC-HDH	BAe 146-200	Pan Air/TNT Airways	
EC-HDK	Airbus A.320-214	Iberia *Mar Ortigola*	
EC-HDM	Boeing 757-256	Iberia *Brasil*	
EC-HDN	Airbus A.320-214	Iberia *Parque National de Omiedo*	
EC-HDO	Airbus A.320-214	Iberia *Formentera*	
EC-HDP	Airbus A.320-214	Iberia *Parque de Cabarceno*	
EC-HDQ	Airbus A.340-313X	Iberia *Sor Juana Ines de la Cruz*	
EC-HDR	Boeing 757-256	Iberia *Pont Aeri Madrid-Barcelona*	
EC-HDS	Boeing 757-256	Iberia *Paraguay*	
EC-HDT	Airbus A.320-214	Iberia *Museo Guggenheim Bilbao*	
EC-HDU	Boeing 757-256	Iberia *Uruguay*	
EC-HDV	Boeing 757-256	Iberia *Nicaragua*	
EC-HEK	Canadair Regional Jet 200ER	Air Nostrum/Iberia *Cecilio Pla*	
EC-HFP	McD Douglas MD-83	Spanair *Sunbreeze*	
EC-HFS	McD Douglas MD-82	Spanair *Sunbeach*	
EC-HFT	McD Douglas MD-82	Spanair *Sunspirit*	
EC-HGA	McD Douglas MD-83	Spanair *Sunisland*	
EC-HGJ	McD Douglas MD-82	Spanair *Sunworld*	
EC-HGO	Boeing 737-85P	Air Europa	
EC-HGP	Boeing 737-85P	Air Europa *Marbella*	
EC-HGQ	Boeing 737-85P	Air Europa *El Mundo-El Dia de Balaeres*	
EC-HGR	Airbus A.319-111	Iberia *Ribeira Sacra*	
EC-HGS	Airbus A.319-111	Iberia *Bardenas Reales*	
EC-HGT	Airbus A.319-111	Iberia *Icnitas de Enciso*	
EC-HGU	Airbus A.340-313X	Iberia *Maria de Molina*	
EC-HGV	Airbus A.340-313X	Iberia *Maria Guerrero*	
EC-HGX	Airbus A.340-313X	Iberia *Maria Pita*	
EC-HGY	Airbus A.320-214	Iberia *Albarracin*	
EC-HGZ	Airbus A.320-214	Iberia *Boi Taull*	
EC-HHA	Airbus A.320-214	Iberia *Serrania de Ronda*	
EC-HHF	McD Douglas MD-82	Spanair *Sunward*	

Notes	Reg.	Type	Owner or Operator
	EC-HHI	Canadair Regional Jet 200ER	Air Nostrum/Iberia *Genaro Lahuerta Lopez*
	EC-HHP	McD Douglas MD-82	Spanair *Sunshiny*
	EC-HHU	Boeing 727-277F	Swiftair/DHL
	EC-HHV	Canadair Regional Jet 200ER	Air Nostrum/Iberia *Sorolla*
	EC-HIP	Boeing 757-256	Iberia *Panama*
	EC-HIQ	Boeing 757-256	Iberia *Honduras*
	EC-HIR	Boeing 757-256	Iberia *El Salvador*
	EC-HIS	Boeing 757-256	Iberia *Bolivia*
	EC-HIT	Boeing 757-256	Iberia *Guatemala*
	EC-HIU	Boeing 757-256	Iberia *Colombia*
	EC-HIV	Boeing 757-256	Iberia *Villa de Bilbao*
	EC-HIX	Boeing 757-256	Iberia *Cuba*
	EC-HJB	McD Douglas MD-82	Spanair *Suntrek*
	EC-HJH	BAe 146-200QT	Pan Air/TNT Airways
	EC-HJP	Boeing 737-85P	Air Europa
	EC-HJQ	Boeing 737-85P	Air Europa
	EC-HJU	Convair Cv.580	Swiftair/DHL
	EC-HKO	Airbus A.319-111	Iberia *Gorbia*
	EC-HKP	McD Douglas MD-83	Spanair *Suntrail*
	EC-HKQ	Boeing 737-85P	Air Europa
	EC-HKR	Boeing 737-85P	Air Europa
	EC-HKS	Boeing 767-3Q8ER	Air Europa
	EC-HLA	Airbus A.310-324	Air Plus Comet
	EC-HMI	McD Douglas MD-87	Spanair *Sunblessed*
	EC-HMS	Convair 580	Swiftair/DHL
	EC-HNC	McD Douglas MD-83	Spanair *Sunplace*
	EC-HNY	Boeing 717-2CM	AeBal *Formentor*/Germanwings
	EC-HNZ	Boeing 717-2CM	AeBal *Espalmador*/Germanwings
	EC-HOV	McD Douglas MD-82	Spanair *Sunspeed*
	EC-HPM	Airbus A.321-231	Spanair *Camillo Jose Cela*
	EC-HPR	Canadair Regional Jet 200ER	Air Nostrum/Iberia *Mompo*
	EC-HPU	Boeing 767-3Q8ER	Air Europa
	EC-HQF	Airbus A.340-313X	Iberia *Maria de Zayas y Sotomayor*
	EC-HQG	Airbus A.320-214	Iberia *Las Hurdes*
	EC-HQH	Airbus A.340-313X	Iberia *Mariana de Silva*
	EC-HQI	Airbus A.320-214	Iberia *La Albufera*
	EC-HQJ	Airbus A.320-214	Iberia *Bosque de Muniellos*
	EC-HQK	Airbus A.320-214	Iberia *Macarella*
	EC-HQL	Airbus A.320-214	Iberia *Liebana*
	EC-HQM	Airbus A.320-214	Iberia *Rio Jucar*
	EC-HQN	Airbus A.340-313X	Iberia *Luisa Carvajal y Mendoza*
	EC-HQT	Airbus A.300B4-103F	Pan Air/TNT Airways
	EC-HQZ	Airbus A.321-231	Spanair
	EC-HRG	Airbus A.321-231	Spanair *Placido Domingo*
	EC-HRP	Airbus A.320-232	Spanair *Juan de Avalos*
	EC-HSE	Airbus A.320-214	Iberia *Hoces de Cabriel*
	EC-HSF	Airbus A.320-214	Iberia *Mar Menor*
	EC-HSH	Canadair Regional Jet 200ER	Air Nostrum/Iberia *J Michavila*
	EC-HSV	Boeing 767-3Q8ER	Air Europa
	EC-HTA	Airbus A.320-214	Iberia *Cadaques*
	EC-HTB	Airbus A.320-214	Iberia *Playa de las Americas*
	EC-HTC	Airbus A.320-214	Iberia *Alpujarra*
	EC-HTD	Airbus A.320-214	Iberia *Calblanque*
	EC-HTZ	Canadair Regional Jet 200ER	Air Nostrum/Iberia *Ricardo Verde*
	EC-HUH	Airbus A.321-211	Iberia *Benidorm*
	EC-HUI	Airbus A.321-211	Iberia *Comunidad Autonoma de la Rioja*
	EC-HUJ	Airbus A.320-214	Iberia *Getaria*
	EC-HUK	Airbus A.320-214	Iberia *Laguna Negra*
	EC-HUL	Airbus A.320-214	Iberia *Monasterio de Rueda*
	EC-HVD	Boeing 747-256B (SCD)	Iberia *Francisco de Quevedo*
	EC-HVZ	Airbus A.300B4-203F	Pan Air/TNT Airways
	EC-HXA	Airbus A.320-232	Spanair
	EC-HXM	Canadair Regional Jet 200ER	Air Nostrum/Iberia *Francisco Dominquez*
	EC-HXY	Swearingen SA227AC Metro III	BKS Air
	EC-HYC	Airbus A.320-214	Iberia *Cuidad de Ceuta*
	EC-HYD	Airbus A.320-214	Iberia *Maspalomas*
	EC-HYG	Canadair Regional Jet 200ER	Air Nostrum/Iberia *E Sales Frances*
	EC-HZH	Swearingen SA227AC Metro III	Aeronova
	EC-HZR	Canadair Regional Jet 200ER	Air Nostrum/Iberia *Jose Ribera*
	EC-HZS	Boeing 737-86Q	Air Europa
	EC-HZU	Airbus A.320-214	Iberworld
	EC-IAA	Canadair Regional Jet 200ER	Air Nostrum/Iberia *A Munoz Degrain*

Reg.	Type	Owner or Operator	Notes
EC-IAF	Boeing 747-256B	Iberia *Lope de Vega*	
EC-IAZ	Airbus A.320-232	Spanair	
EC-IBM	Canadair Regional Jet 200ER	Air Nostrum/Iberia *J Navarro Llorens*	
EC-ICD	Boeing 737-81Q	Air Europa	
EC-ICF	Airbus A.340-313X	Iberia *Maria Zambrano*	
EC-ICL	Airbus A.320-232	Spanair	
EC-ICN	Airbus A.320-214	Volar	
EC-ICQ	Airbus A.320-211	Iberia *Sierra Espuna*	
EC-ICR	Airbus A.320-211	Iberia *Isla de la Cartuja*	
EC-ICS	Airbus A.320-211	Iberia *Sierra de Grazalema*	
EC-ICT	Airbus A.320-211	Iberia *Monte Alhoya*	
EC-ICU	Airbus A.320-211	Iberia *Hayedo de Tejera Negra*	
EC-ICV	Airbus A.320-211	Iberia *Cadi Moixeroi*	
EC-IDA	Boeing 737-86Q	Air Europa	
EC-IDB	Airbus A.330-243	Iberworld *Sabine Thienemann*	
EC-IDC	Canadair Regional Jet 200ER	Air Nostrum/Iberia *Francisco Ribalta*	
EC-IDF	Airbus A.340-313X	Iberia *Mariana Pineda*	
EC-IDT	Boeing 737-86Q	Air Europa	
EC-IEF	Airbus A.320-214	Iberia *Castillo de Loarre*	
EC-IEG	Airbus A.320-214	Iberia *Costa Brava*	
EC-IEI	Airbus A.320-214	Iberia *Monasterio de Valldigna*	
EC-IEJ	Airbus A.320-232	Spanair	
EC-IEQ	Airbus A.320-214	Iberworld	
EC-IEZ	Boeing 737-33A	Hola Airlines *Trives*	
EC-IFC	Boeing 727-277F	Swiftair/DHL	
EC-IFV	Boeing 737-33A	Hola Airlines *Albacete*	
EC-IGK	Airbus A.321-211	Iberia *Costa Calida*	
EC-IGO	Canadair Regional Jet 200ER	Air Nostrum/Iberia *Juan de Juanes*	
EC-IGZ	Douglas DC-8-73F	Cygnus Air	
EC-III	Boeing 737-86Q	Air Europa	
EC-IIG	Airbus A.321-211	Iberia *Ciudad de Siguenza*	
EC-IIH	Airbus A.340-313X	Iberia *Maria Barbara de Braganza*	
EC-IIZ	Airbus A.320-232	Spanair	
EC-IJE	Canadair Regional Jet 200ER	Air Nostrum/Iberia *Joaquin Agrasot*	
EC-IJF	Canadair Regional Jet 200ER	Air Nostrum/Iberia *Vicente Magip*	
EC-IJH	Airbus A.330-322	Iberworld *Gloria Fluxa*	
EC-IJN	Airbus A.321-211	Iberia *Merida*	
EC-IJS	Canadair Regional Jet 200ER	Air Nostrum/Iberia *Manuel Benedito*	
EC-IJU	Airbus A.321-231	Spanair	
EC-IKZ	Canadair Regional Jet 200ER	Air Nostrum/Iberia *Tomas Yepes*	
EC-ILF	Canadair Regional Jet 200ER	Air Nostrum/Iberia *Vicente Lopez*	
EC-ILG	Airbus A.321-211	Volar	
EC-ILH	Airbus A.320-232	Spanair/Star Alliance	
EC-ILO	Airbus A.321-211	Iberia *Cueva de Nerja*	
EC-ILP	Airbus A.321-211	Iberia *Peniscola*	
EC-ILQ	Airbus A.320-214	Iberia *La Pedrera*	
EC-ILR	Airbus A.320-214	Iberia *San Juan de la Pena*	
EC-ILS	Airbus A.320-214	Iberia *Sierra de Cameros*	
EC-ILX	Boeing 737-86N	Travel Service Espagna *Vitava*	
EC-IMA	Airbus A.321-211	Volar	
EC-IMB	Airbus A.320-232	Spanair	
EC-IMU	Airbus A.320-214	Iberworld	
EC-INF	Canadair Regional Jet 200ER	Air Nostrum/Iberia *Pedro de Valencia*	
EC-INM	Airbus A.320-232	Spanair	
EC-INO	Airbus A.340-642	Iberia *Gaudi*	
EC-INP	Boeing 737-804	Futura International Airways	
EC-INQ	Boeing 737-4Q8	Futura International Airways	
EC-INZ	Airbus A.320-214	Iberworld	
EC-IOB	Airbus A.340-642	Iberia *Julio Romanes de Torres*	
EC-IOH	Airbus A.320-232	Spanair	
EC-IOO	Boeing 747-341	Air Pullmantur *Illusions*	
EC-IOR	Boeing 737-33A	Hola/Olympic Airways	
EC-IOU	Boeing 737-4Y0	Futura International Airways	
EC-IPI	Airbus A.320-232	Spanair	
EC-IPN	Boeing 747-212B	Air Plus Comet	
EC-IPS	Boeing 737-33A	Air Plus Comet	
EC-IPT	Airbus A.310-325ET	Air Plus Comet	
EC-IPV	Fokker 100	GirJet	
EC-IQA	Boeing 767-33AER	Air Europa	
EC-IQR	Airbus A.340-642	Iberia *Salvador Dali*	
EC-IRA	Boeing 737-4Y0	Air Europa	
EC-IRI	Canadair Regional Jet 200ER	Air Nostrum/Iberia *Benjamin Palencia*	

Notes	Reg.	Type	Owner or Operator
	EC-ISE	Boeing 737-86Q	Air Europa
	EC-ISI	Airbus A.320-214	Volar
	EC-ISN	Boeing 737-86Q	Air Europa
	EC-ISY	Boeing 757-256	Hola Airlines Privilege
	EC-ITN	Airbus A.321-211	Iberia Empuries
	EC-ITU	Canadair Regional Jet 200ER	Air Nostrum/Iberia Punta Arena
	EC-IUA	Boeing 747-230B (SCD)	Air Pullmantur
	EC-IUC	Boeing 737-86N	Futura International Airways
	EC-IVE	Boeing 727-223F	Swiftair/DHL
	EC-IVF	Boeing 727-230F	Swiftair/DHL
	EC-IVG	Airbus A.320-232	Spanair
	EC-IVH	Canadair Regional Jet 200ER	Air Nostrum/Iberia
	EC-IVO	Fokker 100	GirJet
	EC-IVR	Boeing 737-408	Futura International Airways
	EC-IVV	Boeing 737-883	Air Europa Camino de Santiago
	EC-IXD	Airbus A.321-211	Iberia Vall d'Aran
	EC-IXE	Boeing 737-883	Air Europa Disneyland Resort Paris
	EC-IXO	Boeing 737-883	Air Europa
	EC-IXY	Airbus A.321-211	Volar
	EC-IYB	Airbus A.330-202	Air Madrid Puerta del Sol
	EC-IYG	Airbus A.320-232	Spanair
	EC-IYI	Boeing 737-883	Air Europa Disneyland Resort Paris
	EC-IYN	Airbus A.330-202	Air Madrid Atocha
	EC-IYS	Boeing 737-4Y0	Futura International Airways
	EC-IZG	Boeing 737-46J	Futura International Airways
	EC-IZH	Airbus A.320-214	Iberia San Pere de Roda
	EC-IZK	Airbus A.320-232	Spanair
	EC-IZL	Boeing 747-287B	Air Plus Comet
	EC-IZM	Boeing 737-33A	Air Plus Comet
	EC-IZP	Canadair Regional Jet 200ER	Air Nostrum/Iberia
	EC-IZR	Airbus A.320-214	Iberia Urkiola
	EC-IZX	Airbus A.340-642	Iberia Mariano Benlliure
	EC-IZY	Airbus A.340-642	Iberia
	EC-JAB	Airbus A.320-214	Veuling
	EC-JAP	Boeing 737-85P	Air Europa
	EC-JAZ	Airbus A.319-111	Iberia
	EC-JBA	Airbus A.340-642	Iberia Joaquin Rodrigo
	EC-JBJ	Boeing 737-85P	Air Europa
	EC-JBK	Boeing 737-85P	Air Europa
	EC-JBL	Boeing 737-85P	Air Europa
	EC-JCG	Canadair Regional Jet 200ER	Air Nostrum/Iberia
	EC-JCL	Canadair Regional Jet 200ER	Air Nostrum/Iberia
	EC-JCM	Canadair Regional Jet 200ER	Air Nostrum/Iberia
	EC-JCO	Canadair Regional Jet 200ER	Air Nostrum/Iberia
	EC-JCU	Swearingen SA227TT Merlin IIIC	Aeronova
	EC-JCY	Airbus A.340-642	Iberia
	EC-JCZ	Airbus A.340-642	Iberia
	EC-JDH	Boeing 747-287B	Air Plus Comet
	EC-JDK	Airbus A.320-214	Veuling
	EC-JDL	Airbus A.319-111	Iberia
	EC-JDM	Airbus A.321-211	Iberia Cantabria
	EC-JDN	Fokker 100	GirJet
	EC-	Airbus A.330-202	Air Europa
	EC-	Airbus A.330-202	Air Europa

Note: Iberia also employs Boeing 747s TF-AMA, TF-AMB, TF-ATI and TF-ATJ on lease from Air Atlanta. Spanair operates MD-82s SE-DIK, SE-DIL and SE-DIN. Futura also operates 737s N251RY, N254RY, N255RY, N974RY, N975RY and N977RY.

EP (Iran)

	EP-IAA	Boeing 747SP-86	Iran Air Kurdistan
	EP-IAB	Boeing 747SP-86	Iran Air Khorasan
	EP-IAC	Boeing 747SP-86	Iran Air Fars
	EP-IAD	Boeing 747SP-86	Iran Air
	EP-IAG	Boeing 747-286B (SCD)	Iran Air Azarabadegan
	EP-IAH	Boeing 747-286B (SCD)	Iran Air Khuzestan
	EP-IAM	Boeing 747-186B	Iran Air
	EP-IBA	Airbus A.300B4-605R	Iran Air
	EP-IBB	Airbus A.300B4-605R	Iran Air

Reg.	Type	Owner or Operator	Notes
EP-MHI	Airbus A.310-304ET	Mahan Air	
EP-MHJ	Airbus A.320-232	Mahan Air	

Note: Mahan Air also operates A.310-304ET F-OJHH.

ER (Moldova)

ER-AFN	Antonov An-26B	Jet Line International	
ER-AXF	Antonov An-26	Jet Line International	
ER-AZF	Antonov An-26	Jet Line International	
ER-AZR	Antonov An-26	Jet Line International	

ES (Estonia)

ES-ABC	Boeing 737-5Q8	Estonian Air *Koit*	
ES-ABD	Boeing 737-5Q8	Estonian Air *Hamarik*	
ES-ABG	Boeing 737-505	Estonian Air *Virmaline*	
ES-ABH	Boeing 737-53S	Estonian Air *Sinilind*	
ES-ABI	Boeing 737-5L9	Estonian Air *Tuuslar*	
ES-NOB	Antonov An-72 100	Enimex	
ES-NOC	Antonov An-72 100	Enimex	
ES-NOG	Antonov An-72 100	Enimex	
ES-NOH	Antonov An-72	Enimex	
ES-NOI	Antonov An-72 100D	Enimex/Atlantic Airlines	
ES-NOK	Antonov An-72 100	Enimex	

ET (Ethiopia)

ET-AIF	Boeing 767-260ER	Ethiopian Airlines	
ET-AJS	Boeing 757-260PF	Ethiopian Airlines	
ET-AJX	Boeing 757-260	Ethiopian Airlines	
ET-AKC	Boeing 757-260	Ethiopian Airlines	
ET-AKE	Boeing 757-260	Ethiopian Airlines	
ET-AKF	Boeing 757-260	Ethiopian Airlines	
ET-ALC	Boeing 767-33AER	Ethiopian Airlines	
ET-ALH	Boeing 767-3BGER	Ethiopian Airlines	
ET-ALJ	Boeing 767-360ER	Ethiopian Airlines	
ET-ALL	Boeing 767-3BGER	Ethiopian Airlines	
ET-ALO	Boeing 767-360ER	Ethiopian Airlines	
ET-	Boeing 767-360ER	Ethiopian Airlines	

EW (Belarus)

EW-250PA	Boeing 737-524	Belavia	
EW-251PA	Boeing 737-5Q8	Belavia	
EW-85703	Tu-154M	Belavia	
EW-85706	Tu-154M	Belavia	
EW-85748	Tu-154M	Belavia	
EW-85815	Tu-154M	Belarus Government	

EZ (Turkmenistan)

EZ-A010	Boeing 757-23A	Turkmenistan Airlines	
EZ-A011	Boeing 757-22K	Turkmenistan Airlines	
EZ-A012	Boeing 757-22K	Turkmenistan Airlines	
EZ-A014	Boeing 757-22K	Turkmenistan Airlines	
EZ-A700	Boeing 767-32KER	Turkmenistan Airlines	

F (France)

F-BPVY	Boeing 747-228B	Air France	
F-BPVZ	Boeing 747-228F (SCD)	Air France Cargo	
F-BTDG	Boeing 747-2B3B (SCD)	Air France	
F-BTDH	Boeing 747-2B3B (SCD)	Air France	
F-GBOX	Boeing 747-2B3F (SCD)	Air France Cargo	
F-GCBA	Boeing 747-228B	Air France	

F

Notes	Reg.	Type	Owner or Operator
	F-GCBB	Boeing 747-228B	Air France
	F-GCBD	Boeing 747-228B (SF)	Air France
	F-GCBF	Boeing 747-228B (SCD)	Air France
	F-GCBG	Boeing 747-228F (SCD)	Air France Cargo
	F-GCBH	Boeing 747-228F (SCD)	Air France Asia Cargo
	F-GCBI	Boeing 747-228B (SCD)	Air France
	F-GCBJ	Boeing 747-228B (SCD)	Air France
	F-GCBK	Boeing 747-228F (SCD)	Air France Cargo
	F-GCBL	Boeing 747-228F (SCD)	Air France Cargo
	F-GCBM	Boeing 747-228F	Air France Asia Cargo
	F-GCJL	Boeing 737-222	Air Mediterranée
	F-GCSL	Boeing 737-222	Air Mediterranée
	F-GEMO	Airbus A.310-304	Eagle Aviation
	F-GETA	Boeing 747-3B3 (SCD)	Air France *Big Boss*
	F-GETB	Boeing 747-3B3 (SCD)	Air France
	F-GEXA	Boeing 747-4B3	Air France
	F-GEXB	Boeing 747-4B3	Air France
	F-GFEO	EMB-120RT Brasilia	Regional Airlines/Air France
	F-GFKA	Airbus A.320-111	Air France *Ville de Paris*
	F-GFKB	Airbus A.320-111	Air France *Ville de Rome*
	F-GFKD	Airbus A.320-111	Air France *Ville de Londres*
	F-GFKE	Airbus A.320-111	Air France *Ville de Bonn*
	F-GFKF	Airbus A.320-111	Air France *Ville de Madrid*
	F-GFKG	Airbus A.320-111	Air France *Ville d'Amsterdam*
	F-GFKH	Airbus A.320-211	Air France *Ville de Bruxelles*
	F-GFKI	Airbus A.320-211	Air France *Ville de Lisbonne*
	F-GFKJ	Airbus A.320-211	Air France *Ville de Copenhague*
	F-GFKK	Airbus A.320-211	Air France *Ville d'Athenes*
	F-GFKL	Airbus A.320-211	Air France *Ville de Dublin*
	F-GFKM	Airbus A.320-211	Air France *Ville de Luxembourg*
	F-GFKN	Airbus A.320-211	Air France *Ville de Strasbourg*
	F-GFKO	Airbus A.320-211	Air France *Ville de Milan*
	F-GFKP	Airbus A.320-211	Air France *Ville de Nice*
	F-GFKQ	Airbus A.320-111	Air France *Ville de Berlin*
	F-GFKR	Airbus A.320-211	Air France *Ville de Barceloune*
	F-GFKS	Airbus A.320-211	Air France *Ville de Marseilles*
	F-GFKT	Airbus A.320-211	Air France *Ville de Lyon*
	F-GFKU	Airbus A.320-211	Air France *Ville de Manchester*
	F-GFKV	Airbus A.320-211	Air France *Ville de Bordeaux*
	F-GFKX	Airbus A.320-211	Air France *Ville de Francfort*
	F-GFKY	Airbus A.320-211	Air France *Ville de Toulouse*
	F-GFKZ	Airbus A.320-211	Air France *Ville de Turin*
	F-GFUE	Boeing 737-3B3QC	Europe Air Post
	F-GFUF	Boeing 737-3B3QC	Europe Air Post *Toulouse-Montaudran*
	F-GFUG	Boeing 737-4B3	Corsair
	F-GFUH	Boeing 737-4B3	Corsair *World of TUI*
	F-GGEA	Airbus A.320-111	Air France
	F-GGEB	Airbus A.320-111	Air France
	F-GGEC	Airbus A.320-111	Air France
	F-GGEE	Airbus A.320-111	Air France
	F-GGEF	Airbus A.320-111	Air France
	F-GGEG	Airbus A.320-111	Air France
	F-GHIA	EMB-120ER Brasilia	Regional Airlines/Air France
	F-GHIB	EMB-120ER Brasilia	Regional Airlines/Air France
	F-GHPI	Aérospatiale ATR-42-300	Airlinair
	F-GHPS	Aérospatiale ATR-42-300	Brit Air/Air France Express
	F-GHPU	Aérospatiale ATR-72-101	Danish Air Transport
	F-GHPV	Aérospatiale ATR-72-101	Brit Air/Air France Express
	F-GHQA	Airbus A.320-211	Air France
	F-GHQB	Airbus A.320-211	Air France
	F-GHQC	Airbus A.320-211	Air France
	F-GHQD	Airbus A.320-211	Air France
	F-GHQE	Airbus A.320-211	Air France
	F-GHQF	Airbus A.320-211	Air France
	F-GHQG	Airbus A.320-211	Air France
	F-GHQH	Airbus A.320-211	Air France
	F-GHQI	Airbus A.320-211	Air France
	F-GHQJ	Airbus A.320-211	Air France
	F-GHQK	Airbus A.320-211	Air France
	F-GHQL	Airbus A.320-211	Air France
	F-GHQM	Airbus A.320-211	Air France
	F-GHQO	Airbus A.320-211	Air France

Reg.	Type	Owner or Operator	Notes
F-GHQP	Airbus A.320-211	Air France	
F-GHQQ	Airbus A.320-211	Air France	
F-GHQR	Airbus A.320-211	Air France	
F-GHXM	Boeing 737-53A	Air France	
F-GIDK	Douglas DC-3C	Dakota Air Legend	
F-GIOG	Fokker 100	Regional Airlines *L'Esprit Liberte*	
F-GISA	Boeing 747-428 (SCD)	Air France	
F-GISB	Boeing 747-428 (SCD)	Air France	
F-GISC	Boeing 747-428 (SCD)	Air France	
F-GISD	Boeing 747-428 (SCD)	Air France	
F-GISE	Boeing 747-428 (SCD)	Air France	
F-GITA	Boeing 747-428	Air France	
F-GITB	Boeing 747-428	Air France	
F-GITC	Boeing 747-428	Air France	
F-GITD	Boeing 747-428	Air France	
F-GITE	Boeing 747-428	Air France	
F-GITF	Boeing 747-428	Air France	
F-GITH	Boeing 747-428	Air France	
F-GITI	Boeing 747-428	Air France	
F-GITJ	Boeing 747-428	Air France	
F-GIUA	Boeing 747-428ERF (SCD)	Air France Cargo	
F-GIUB	Boeing 747-428ERF (SCD)	Air France Cargo	
F-GIUC	Boeing 747-428ERF (SCD)	Air France Cargo	
F-GIUD	Boeing 747-428ERF (SCD)	Air France Cargo	
F-GIVK	EMB-120ER Brasilia	Regional Airlines	
F-GIXA	Boeing 737-2K2C	Europe Air Post/Star Up	
F-GIXB	Boeing 737-33AQC	Europe Air Post	
F-GIXC	Boeing 737-38BQC	Europe Air Post *Saint-Louis*	
F-GIXD	Boeing 737-33AQC	Europe Air Post	
F-GIXE	Boeing 737-3B3QC	Europe Air Post	
F-GIXF	Boeing 737-3B3QC	Europe Air Post	
F-GIXG	Boeing 737-382QC	Axis Airways	
F-GIXH	Boeing 737-3S3QC	Aigle Azur	
F-GIXI	Boeing 737-348QC	Europe Air Post	
F-GIXJ	Boeing 737-3Y0QC	Europe Air Post	
F-GIXL	Boeing 737-348QC	Europe Air Post	
F-GIXM	Boeing 737-36E	Axis Airways	
F-GIXO	Boeing 737-3Q8QC	Europe Air Post	
F-GIXR	Boeing 737-3H6F	Europe Air Post	
F-GIXS	Boeing 737-3H6F	Europe Air Post	
F-GJAK	EMB-120ER Brasilia	Regional Airlines/Air France	
F-GJNA	Boeing 737-528	Air France	
F-GJNB	Boeing 737-528	Air France	
F-GJNC	Boeing 737-528	Air France	
F-GJND	Boeing 737-528	Air France	
F-GJNE	Boeing 737-528	Air France	
F-GJNF	Boeing 737-528	Air France	
F-GJNG	Boeing 737-528	Air France	
F-GJNH	Boeing 737-528	Air France	
F-GJNI	Boeing 737-528	Air France	
F-GJNJ	Boeing 737-528	Air France	
F-GJNK	Boeing 737-528	Air France	
F-GJNL	Boeing 737-5H6	Air France	
F-GJNM	Boeing 737-528	Air France	
F-GJNN	Boeing 737-528	Air France	
F-GJNO	Boeing 737-528	Air France	
F-GJNP	Boeing 737-5H6	Air France	
F-GJNX	Boeing 737-5H6	Air France	
F-GJNZ	Boeing 737-5H6	Air France	
F-GJUA	Boeing 737-548	Air France	
F-GJVA	Airbus A.320-211	Air France	
F-GJVB	Airbus A.320-211	Air France	
F-GJVD	Airbus A.320-211	Air France	
F-GJVE	Airbus A.320-211	Air France	
F-GJVF	Airbus A.320-211	Air France	
F-GJVG	Airbus A.320-211	Air France	
F-GJVW	Airbus A.320-211	Air France	
F-GKNB	Aérospatiale ATR-42-310	Airlinair	
F-GKNC	Aérospatiale ATR-42-310	Airlinair	
F-GKOB	Aérospatiale ATR-72-212	Airlinair	
F-GKTA	Boeing 737-3M8	Air One	
F-GKTB	Boeing 737-3M8	Air One	

F

OVERSEAS AIRLINERS

Notes	Reg.	Type	Owner or Operator
	F-GKXA	Airbus A.320-211	Air France *Ville de Nantes*
	F-GKXB	Airbus A.320-212	Air France
	F-GKXC	Airbus A.320-214	Air France
	F-GKXD	Airbus A.320-214	Air France
	F-GKXE	Airbus A.320-214	Air France
	F-GKXF	Airbus A.320-214	Air France
	F-GKXG	Airbus A.320-214	Air France
	F-GKXH	Airbus A.320-214	Air France
	F-GKXI	Airbus A.320-214	Air France
	F-GKXJ	Airbus A.320-214	Air France
	F-GKXK	Airbus A.320-214	Air France
	F-GKYN	Aérospatiale ATR-42-312	Airlinair
	F-GLGG	Airbus A.320-212	Air France
	F-GLGH	Airbus A.320-212	Air France
	F-GLGM	Airbus A.320-212	Air France
	F-GLIJ	Canadair CL.600-2B19 RJ	Air Littoral/Team Lufthansa
	F-GLIK	Canadair CL.600-2B19 RJ	Air Littoral/Team Lufthansa
	F-GLIR	Fokker 100	Regional Airlines/Air France
	F-GLIS	Fokker 70	Regional Airlines/Air France
	F-GLIV	Fokker 70	Regional Airlines
	F-GLIX	Fokker 70	Regional Airlines/Air France
	F-GLIY	Canadair CL.600-2B19 RJ	Air Littoral
	F-GLIZ	Canadair CL.600-2B19 RJ	Air Littoral
	F-GLND	Beech 1900D	Airlinair
	F-GLRG	EMB-120RT Brasilia	Regional Airlines/Air France
	F-GLXI	Boeing 737-4Y0	Aigle Azur
	F-GLXJ	Boeing 737-4Y0	Aigle Azur
	F-GLXQ	Boeing 737-4Y0	Axis Airways
	F-GLZA	Airbus A.340-312	Air France
	F-GLZB	Airbus A.340-312	Air France
	F-GLZC	Airbus A.340-312	Air France
	F-GLZG	Airbus A.340-312	Air France
	F-GLZH	Airbus A.340-312	Air France
	F-GLZI	Airbus A.340-312	Air France
	F-GLZJ	Airbus A.340-313X	Air France
	F-GLZK	Airbus A.340-313X	Air France
	F-GLZL	Airbus A.340-313X	Air France
	F-GLZM	Airbus A.340-313X	Air France
	F-GLZN	Airbus A.340-313X	Air France
	F-GLZO	Airbus A.340-313X	Air France
	F-GLZP	Airbus A.340-313X	Air France
	F-GLZQ	Airbus A.340-313X	Air France
	F-GLZR	Airbus A.340-313X	Air France
	F-GLZS	Airbus A.340-313X	Air France
	F-GLZT	Airbus A.340-313X	Air France
	F-GLZU	Airbus A.340-313X	Air France
	F-GMLI	McD Douglas MD-83	Blue Line
	F-GMVB	SAAB 2000	Regional Airlines/Air France
	F-GMVC	SAAB 2000	Regional Airlines
	F-GMVD	SAAB 2000	Regional Airlines/Air France
	F-GMVE	SAAB 2000	Regional Airlines/Air France
	F-GMVG	SAAB 2000	Regional Airlines/Air France
	F-GMVU	SAAB 2000	Regional Airlines/Air France
	F-GMZA	Airbus A.321-111	Air France
	F-GMZB	Airbus A.321-111	Air France
	F-GMZC	Airbus A.321-111	Air France
	F-GMZD	Airbus A.321-111	Air France
	F-GMZE	Airbus A.321-111	Air France
	F-GNIF	Airbus A.340-313X	Air France
	F-GNIG	Airbus A.340-313X	Air France
	F-GNIH	Airbus A.340-313X	Air France
	F-GNII	Airbus A.340-313X	Air France
	F-GNLG	Fokker 100	Blue Line
	F-GNLH	Fokker 100	Blue Line
	F-GNLI	Fokker 100	Regional Airlines/Air France
	F-GNLK	Fokker 100	Regional Airlines/Air France
	F-GOAF	Boeing 737-242C	Air Mediterranée
	F-GOHA	Embraer RJ135ER	Regional Airlines/Air France
	F-GOHB	Embraer RJ135ER	Regional Airlines/Air France
	F-GOHC	Embraer RJ135ER	Regional Airlines/Air France
	F-GOHD	Embraer RJ135ER	Regional Airlines/Air France
	F-GOHE	Embraer RJ135ER	Regional Airlines/Air France

Reg.	Type	Owner or Operator	Notes
F-GOHF	Embraer RJ135ER	Regional Airlines/Air France	
F-GOMA	BAe 146-200QC	Axis Airways	
F-GPAN	Boeing 747-2B3F (SCD)	Air France Cargo	
F-GPMA	Airbus A.319-113	Air France	
F-GPMB	Airbus A.319-113	Air France	
F-GPMC	Airbus A.319-113	Air France	
F-GPMD	Airbus A.319-113	Air France	
F-GPME	Airbus A.319-113	Air France	
F-GPMF	Airbus A.319-113	Air France	
F-GPMG	Airbus A.319-113	Air France	
F-GPMH	Airbus A.319-113	Air France	
F-GPMI	Airbus A.319-113	Air France	
F-GPNK	Fokker 100	Regional Airlines/Air France	
F-GPNL	Fokker 100	Regional Airlines/Air France	
F-GPOC	Aérospatiale ATR-72-212	Europe Air Post	
F-GPOD	Aérospatiale ATR-72-212	Europe Air Post	
F-GPTG	Canadair CL.600-2B19 RJ	Air Littoral	
F-GPXA	Fokker 100	Brit Air/Air France	
F-GPXB	Fokker 100	Brit Air/Air France	
F-GPXC	Fokker 100	Brit Air/Air France	
F-GPXD	Fokker 100	Brit Air/Air France	
F-GPXE	Fokker 100	Brit Air/Air France	
F-GPXF	Fokker 100	Brit Air/Air France	
F-GPXG	Fokker 100	Brit Air/Air France	
F-GPXH	Fokker 100	Brit Air/Air France	
F-GPYA	Aérospatiale ATR-42-512	Airlinair	
F-GPYB	Aérospatiale ATR-42-512	Airlinair	
F-GPYC	Aérospatiale ATR-42-512	Airlinair	
F-GPYD	Aérospatiale ATR-42-512	Airlinair	
F-GPYF	Aérospatiale ATR 42-512	Airlinair/Air France	
F-GPYK	Aérospatiale ATR-42-512	Airlinair	
F-GPYL	Aérospatiale ATR-42-512	Airlinair	
F-GPYN	Aérospatiale ATR-42-512	Air Littoral	
F-GPYO	Aérospatiale ATR-42-512	Airlinair/Air France	
F-GRGA	Embraer RJ145EU	Regional Airlines/Air France	
F-GRGB	Embraer RJ145EU	Regional Airlines/Air France	
F-GRGC	Embraer RJ145EU	Regional Airlines/Air France	
F-GRGD	Embraer RJ145EU	Regional Airlines/Air France	
F-GRGE	Embraer RJ145EU	Regional Airlines/Air France	
F-GRGF	Embraer RJ145EU	Regional Airlines/Air France	
F-GRGG	Embraer RJ145EU	Regional Airlines/Air France	
F-GRGH	Embraer RJ145EU	Regional Airlines/Air France	
F-GRGI	Embraer RJ145EU	Regional Airlines/Air France	
F-GRGJ	Embraer RJ145EU	Regional Airlines/Air France	
F-GRGK	Embraer RJ145EU	Regional Airlines/Air France	
F-GRGL	Embraer RJ145EU	Regional Airlines/Air France	
F-GRGM	Embraer RJ145EU	Regional Airlines/Air France	
F-GRGP	Embraer RJ135ER	Regional Airlines/Air France	
F-GRGQ	Embraer RJ135ER	Regional Airlines/Air France	
F-GRGR	Embraer RJ135ER	Regional Airlines/Air France	
F-GRHA	Airbus A.319-111	Air France	
F-GRHB	Airbus A.319-111	Air France	
F-GRHC	Airbus A.319-111	Air France	
F-GRHD	Airbus A.319-111	Air France	
F-GRHE	Airbus A.319-111	Air France	
F-GRHF	Airbus A.319-111	Air France	
F-GRHG	Airbus A.319-111	Air France	
F-GRHH	Airbus A.319-111	Air France	
F-GRHI	Airbus A.319-111	Air France	
F-GRHJ	Airbus A.319-111	Air France	
F-GRHK	Airbus A.319-111	Air France	
F-GRHL	Airbus A.319-111	Air France	
F-GRHM	Airbus A.319-111	Air France	
F-GRHN	Airbus A.319-111	Air France	
F-GRHO	Airbus A.319-111	Air France	
F-GRHP	Airbus A.319-111	Air France	
F-GRHQ	Airbus A.319-111	Air France	
F-GRHR	Airbus A.319-111	Air France	
F-GRHS	Airbus A.319-111	Air France	
F-GRHT	Airbus A.319-111	Air France	
F-GRHU	Airbus A.319-111	Air France	
F-GRHV	Airbus A.319-111	Air France	

Notes	Reg.	Type	Owner or Operator
	F-GRHX	Airbus A.319-111	Air France
	F-GRHY	Airbus A.319-111	Air France
	F-GRHZ	Airbus A.319-111	Air France
	F-GRJA	Canadair CL.600-2B19 RJ	Brit Air/Air France
	F-GRJB	Canadair CL.600-2B19 RJ	Brit Air/Air France
	F-GRJC	Canadair CL.600-2B19 RJ	Brit Air/Air France
	F-GRJD	Canadair CL.600-2B19 RJ	Brit Air/Air France
	F-GRJE	Canadair CL.600-2B19 RJ	Brit Air/Air France
	F-GRJF	Canadair CL.600-2B19 RJ	Brit Air/Air France
	F-GRJG	Canadair CL.600-2B19 RJ	Brit Air/Air France
	F-GRJH	Canadair CL.600-2B19 RJ	Brit Air/Air France
	F-GRJI	Canadair CL.600-2B19 RJ	Brit Air/Air France
	F-GRJJ	Canadair CL.600-2B19 RJ	Brit Air/Air France
	F-GRJK	Canadair CL.600-2B19 RJ	Brit Air/Air France
	F-GRJL	Canadair CL.600-2B19 RJ	Brit Air/Air France
	F-GRJM	Canadair CL.600-2B19 RJ	Brit Air/Air France
	F-GRJN	Canadair CL.600-2B19 RJ	Brit Air/Air France
	F-GRJO	Canadair CL.600-2B19 RJ	Brit Air/Air France
	F-GRJP	Canadair CL.600-2B19 RJ	Brit Air/Air France
	F-GRJQ	Canadair CL.600-2B19 RJ	Brit Air/Air France
	F-GRJR	Canadair CL.600-2B19 RJ	Brit Air/Air France
	F-GRJT	Canadair CL.600-2B19 RJ	Brit Air/Air France
	F-GRNB	Boeing 737-85	Air Horizons
	F-GRNE	Boeing 737-85F	Air Horizons
	F-GRNF	Boeing 737-329	Air Horizons
	F-GRNV	Boeing 737-329	Air Horizons
	F-GRSD	Airbus A.320-214	Star Airlines
	F-GRSE	Airbus A.320-214	Star Airlines
	F-GRSI	Airbus A.320-214	Star Airlines
	F-GRSN	Airbus A.320-214	Star Airlines
	F-GRSQ	Airbus A.330-243	Star Airlines
	F-GRXA	Airbus A.319-111	Air France
	F-GRXB	Airbus A.319-111	Air France
	F-GRXC	Airbus A.319-111	Air France
	F-GRXD	Airbus A.319-111	Air France
	F-GRXE	Airbus A.319-111	Air France
	F-GRXF	Airbus A.319-111	Air France
	F-GRXG	Airbus A.319-115LR	Air France
	F-GRXH	Airbus A.319-115LR	Air France
	F-GRZA	Canadair CL.600-2C10 RJ	Brit Air/Air France Express
	F-GRZB	Canadair CL.600-2C10 RJ	Brit Air/Air France Express
	F-GRZC	Canadair CL.600-2C10 RJ	Brit Air/Air France Express
	F-GRZD	Canadair CL.600-2C10 RJ	Brit Air/Air France Express
	F-GRZE	Canadair CL.600-2C10 RJ	Brit Air/Air France Express
	F-GRZF	Canadair CL.600-2C10 RJ	Brit Air/Air France Express
	F-GRZG	Canadair CL.600-2C10 RJ	Brit Air/Air France Express
	F-GRZH	Canadair CL.600-2C10 RJ	Brit Air/Air France Express
	F-GRZI	Canadair CL.600-2C10 RJ	Brit Air/Air France Express
	F-GRZJ	Canadair CL.600-2C10 RJ	Brit Air/Air France Express
	F-GSEU	Airbus A.330-243	Star Airlines
	F-GSEX	Boeing 747-312	Corsair
	F-GSKY	Boeing 747-312	Corsair
	F-GSPA	Boeing 777-228ER	Air France
	F-GSPB	Boeing 777-228ER	Air France
	F-GSPC	Boeing 777-228ER	Air France
	F-GSPD	Boeing 777-228ER	Air France
	F-GSPE	Boeing 777-228ER	Air France
	F-GSPF	Boeing 777-228ER	Air France
	F-GSPG	Boeing 777-228ER	Air France
	F-GSPH	Boeing 777-228ER	Air France
	F-GSPI	Boeing 777-228ER	Air France
	F-GSPJ	Boeing 777-228ER	Air France
	F-GSPK	Boeing 777-228ER	Air France
	F-GSPL	Boeing 777-228ER	Air France
	F-GSPM	Boeing 777-228ER	Air France
	F-GSPN	Boeing 777-228ER	Air France
	F-GSPO	Boeing 777-228ER	Air France
	F-GSPP	Boeing 777-228ER	Air France
	F-GSPQ	Boeing 777-228ER	Air France
	F-GSPR	Boeing 777-228ER	Air France
	F-GSPS	Boeing 777-228ER	Air France
	F-GSPT	Boeing 777-228ER	Air France

Reg.	Type	Owner or Operator	Notes
F-GSPU	Boeing 777-228ER	Air France	
F-GSPV	Boeing 777-228ER	Air France	
F-GSPX	Boeing 777-228ER	Air France	
F-GSPY	Boeing 777-228ER	Air France	
F-GSPZ	Boeing 777-228ER	Air France	
F-GSQA	Boeing 777-328ER	Air France	
F-GSQB	Boeing 777-328ER	Air France	
F-GSQC	Boeing 777-328ER	Air France	
F-GSQD	Boeing 777-328ER	Air France	
F-GSQE	Boeing 777-328ER	Air France	
F-GSQF	Boeing 777-328ER	Air France	
F-GSQG	Boeing 777-328ER	Air France	
F-GSQH	Boeing 777-328ER	Air France	
F-GSTA	Airbus A.300-608ST Beluga (1)	Airbus Inter Transport	
F-GSTB	Airbus A.300-608ST Beluga (2)	Airbus Inter Transport	
F-GSTC	Airbus A.300-608ST Beluga (3)	Airbus Inter Transport	
F-GSTD	Airbus A.300-608ST Beluga (4)	Airbus Inter Transport	
F-GSTF	Airbus A.300-608ST Beluga (5)	Airbus Inter Transport	
F-GSUN	Boeing 747-312	Corsair	
F-GTAD	Airbus A.321-211	Air France	
F-GTAE	Airbus A.321-211	Air France	
F-GTAF	Airbus A.321-211	Air France/Royal Jordanian Airlines	
F-GTAH	Airbus A.321-211	Air France	
F-GTAI	Airbus A.321-211	Air France	
F-GTAJ	Airbus A.321-211	Air France	
F-GTAK	Airbus A.321-211	Air France	
F-GTAL	Airbus A.321-211	Air France	
F-GTAM	Airbus A.321-211	Air France	
F-GTSG	EMB-120ER Brasilia	Regional Airlines/Air France	
F-GTSH	EMB-120ER Brasilia	Regional Airlines/Air France	
F-GTSI	EMB-120ER Brasilia	Regional Airlines/Air France	
F-GTSJ	EMB-120ER Brasilia	Regional Airlines/Air France	
F-GTSK	EMB-120ER Brasilia	Regional Airlines/Air France	
F-GTSN	EMB-120ER Brasilia	Regional Airlines/Air France	
F-GTSO	EMB-120RT Brasilia	Regional Airlines/Air France	
F-GUAA	Airbus A.321-211	Aigle Azur	
F-GUAM	Embraer RJ145MP	Regional Airlines/Air France	
F-GUBA	Embraer RJ145MP	Regional Airlines/Air France	
F-GUBB	Embraer RJ145MP	Regional Airlines/Air France	
F-GUBC	Embraer RJ145MP	Regional Airlines/Air France	
F-GUBD	Embraer RJ145MP	Regional Airlines/Air France	
F-GUBE	Embraer RJ145MP	Regional Airlines/Air France	
F-GUBF	Embraer RJ145MP	Regional Airlines/Air France	
F-GUBG	Embraer RJ145MP	Regional Airlines/Air France	
F-GUEA	Embraer RJ145MP	Regional Airlines /Air France	
F-GUFD	Embraer RJ145EU	Regional Airlines /Air France	
F-GUGA	Airbus A.318-111	Air France	
F-GUGB	Airbus A.318-111	Air France	
F-GUGC	Airbus A.318-111	Air France	
F-GUGD	Airbus A.318-111	Air France	
F-GUGE	Airbus A.318-111	Air France	
F-GUGF	Airbus A.318-111	Air France	
F-GUGG	Airbus A.318-111	Air France	
F-GUGH	Airbus A.318-111	Air France	
F-GUGI	Airbus A.318-111	Air France	
F-GUGJ	Airbus A.318-111	Air France	
F-GUGK	Airbus A.318-111	Air France	
F-GUGL	Airbus A.318-111	Air France	
F-GUGM	Airbus A.318-111	Air France	
F-GUGN	Airbus A.318-111	Air France	
F-GUGO	Airbus A.318-111	Air France	
F-GUJA	Embraer RJ145MP	Regional Airlines/Air France	
F-GUMA	Embraer RJ145MP	Regional Airlines/Air France	
F-GUPT	Embraer RJ145MP	Regional Airlines/Air France	
F-GUYH	Boeing 737-329	Aigle Azur	
F-GVBR	Embraer EMB-120RT Brasilia	Occitania Airlines	
F-GVGS	Embraer RJ145MP	Regional Airlines/Air France	
F-GVHD	Embraer RJ145EU	Regional Airlines/Air France	
F-GVZA	Aérospatiale ATR-42-512	Airlinair	
F-GVZF	Aérospatiale ATR-72-212	Airlinair/Air France	
F-GVZX	Aérospatiale ATR-42-312	Airlinair	
F-GVZY	Aérospatiale ATR-42-312	Airlinair	

Notes	Reg.	Type	Owner or Operator
	F-GVZZ	Aérospatiale ATR-42-312	Airlinair
	F-GYAM	Boeing 737-505	Air Mediterranée
	F-GYAN	Airbus A.321-111	Air Mediterranée
	F-GYAO	Airbus A.321-111	Air Mediterranée
	F-GYAP	Airbus A.321-111	Air Mediterranée
	F-GYAQ	Airbus A.321-211	Air Mediterranée
	F-GYAR	Airbus A.321-211	Air Mediterranée
	F-GYAS	Airbus A.319-115LR	Aero Services Executive/Air France
	F-GYYY	Airbus A.310-204	Eagle Aviation
	F-GZCA	Airbus A.330-203	Air France
	F-GZCB	Airbus A.330-203	Air France
	F-GZCC	Airbus A.330-203	Air France
	F-GZCD	Airbus A.330-203	Air France
	F-GZCE	Airbus A.330-203	Air France
	F-GZCF	Airbus A.330-203	Air France
	F-GZCG	Airbus A.330-203	Air France
	F-GZCH	Airbus A.330-203	Air France
	F-GZCI	Airbus A.330-203	Air France
	F-GZCJ	Airbus A.330-203	Air France
	F-GZCK	Airbus A.330-203	Air France
	F-GZCL	Airbus A.330-203	Air France
	F-GZCM	Airbus A.330-203	Air France
	F-GZCN	Airbus A.330-203	Air France
	F-GZZZ	Airbus A.320-211	Eagle Aviation
	F-HAXY	Boeing 757-2K2	Axis Air
	F-HBAB	Airbus A.321-211	Aigle Azur
	F-HBAC	Airbus A.320-214	Aigle Azur
	F-HBIL	Airbus A.330-243	Corsair
	F-HCAT	Airbus A.330-243	Corsair
	F-HJAC	Boeing 747-312	Corsair
	F-HSUN	Boeing 747-422	Corsair
	F-	Boeing 747-422	Corsair
	F-	Boeing 747-422	Corsair
	F-ODJG	Boeing 747-2Q2B	Air Gabon *President Leon Mba*
	F-ODTK	Airbus A.300B4-621	Sudan Airways
	F-ODVF	Airbus A.310-304F	Royal Jordanian *Princess Raiyah*
	F-ODVG	Airbus A.310-304F	Royal Jordanian *Prince Faisal*
	F-OFDF	Airbus A.310-223	Air Caraibes
	F-OGQR	Airbus A.310-308	Aeroflot Russian International *Rachmaninov*
	F-OGYC	Airbus A.320-212	Royal Jordanian
	F-OHGB	Airbus A.320-211	Royal Jordanian
	F-OHGC	Airbus A.320-211	Royal Jordanian
	F-OHLP	Airbus A.340-211	Royal Jordanian *Salma Bint Abdullah*
	F-OHLQ	Airbus A.340-211	Royal Jordanian
	F-OHPR	Airbus A.310-325	Yemenia
	F-OHPS	Airbus A.310-325	Yemenia
	F-OJGF	Airbus A.340-313X	Air Tahiti Nui *Mangareva*
	F-OJHH	Airbus A.310-304ET	Mahan Air
	F-OJTN	Airbus A.340-313X	Air Tahiti Nui *Bora Bora*
	F-OMAY	Boeing 777-2Q8ER	Air Austral
	F-OMEA	Airbus A.330-243	Middle East Airlines
	F-OMEB	Airbus A.330-243	Middle East Airlines
	F-OMEC	Airbus A.330-243	Middle East Airlines
	F-OMSA	Airbus A.330-301	Air Algerie
	F-OPAR	Boeing 777-2Q8ER	Air Austral
	F-OPTP	Airbus A.330-223	Air Caraibes
	F-ORME	Airbus A.321-231	Middle East Airlines
	F-ORMF	Airbus A.321-231	Middle East Airlines
	F-ORMG	Airbus A.321-231	Middle East Airlines
	F-ORMH	Airbus A.321-231	Middle East Airlines
	F-ORMI	Airbus A.321-231	Middle East Airlines
	F-ORMJ	Airbus A.321-231	Middle East Airlines
	F-ORUN	Boeing 777-2Q8ER	Air Austral
	F-OSEA	Airbus A.340-313X	Air Tahiti Nui *Rangiroa*
	F-OSUN	Airbus A.340-313X	Air Tahiti Nui *Moorea*

HA (Hungary)

	HA-FAB	F.27 Friendship Mk 500	Farnair Hungary
	HA-FAC	F.27 Friendship Mk 500	Farnair Hungary
	HA-FAD	F.27 Friendship Mk 500	Farnair Hungary

Reg.	Type	Owner or Operator	Notes
HA-FAE	F.27 Friendship Mk 500	Farnair Hungary	
HA-FAF	F.27 Friendship Mk 500	Farnair Hungary	
HA-FAH	F.27 Friendship Mk 500	Farnair Hungary	
HA-LEY	Boeing 737-4Y0	Malev	
HA-LHA	Boeing 767-27GER	Malev *Albert Szent-Gyorgyi*	
HA-LHB	Boeing 767-27GER	Malev	
HA-LKA	Boeing 737-4Y0	Travel Service Hungary	
HA-LKB	Boeing 737-86Q	Travel Service Hungary	
HA-	Boeing 737-832	Travel Service Hungary	
HA-LKO	Boeing 737-5Y0	SkyEurope Hungary	
HA-LKP	Boeing 737-5Y0	SkyEurope Hungary	
HA-LMA	Fokker 70	Malev	
HA-LMB	Fokker 70	Malev	
HA-LMC	Fokker 70	Malev	
HA-LME	Fokker 70	Malev	
HA-LMF	Fokker 70	Malev	
HA-LNA	Canadair RJ.200ER	Malev Express	
HA-LNB	Canadair RJ.200ER	Malev Express	
HA-LNC	Canadair RJ.200ER	Malev Express	
HA-LND	Canadair RJ.200ER	Malev Express	
HA-LOA	Boeing 737-7Q8	Malev	
HA-LOB	Boeing 737-7Q8	Malev	
HA-LOC	Boeing 737-8Q8	Malev	
HA-LOD	Boeing 737-6Q8	Malev	
HA-LOE	Boeing 737-6Q8	Malev	
HA-LOF	Boeing 737-6Q8	Malev	
HA-LOG	Boeing 737-6Q8	Malev	
HA-LOH	Boeing 737-6Q8	Malev	
HA-LOI	Boeing 737-8Q8	Malev	
HA-LOJ	Boeing 737-7Q8	Malev	
HA-LOK	Boeing 737-7Q8	Malev	
HA-LOL	Boeing 737-8Q8	Malev	
HA-LOM	Boeing 737-6Q8	Malev	
HA-LON	Boeing 737-6Q8	Malev	
HA-LOP	Boeing 737-7Q8	Malev	
HA-LOR	Boeing 737-7Q8	Malev	
HA-LPA	Airbus A.320-233	Wizz	
HA-LPB	Airbus A.320-233	Wizz	
HA-LPC	Airbus A.320-233	Wizz	
HA-LPD	Airbus A.320-233	Wizz	
HA-LPE	Airbus A.320-233	Wizz	
HA-LPF	Airbus A.320-233	Wizz	

HB (Switzerland)

Reg.	Type	Owner or Operator	Notes
HB-IEE	Boeing 757-23A	PrivatAir	
HB-IHR	Boeing 757-2G5	Belair *Solemar*	
HB-IHS	Boeing 757-2G5	Belair *Horizonte*	
HB-IHX	Airbus A.320-214	Edelweiss Air *Calvaro*	
HB-IHY	Airbus A.320-214	Edelweiss Air *Upali*	
HB-IHZ	Airbus A.320-214	Edelweiss Air *Viktoria*	
HB-IIO	Boeing 737-7AK	PrivatAir	
HB-IIP	Boeing 737-7AK	PrivatAir	
HB-IIQ	Boeing 737-7CN	PrivatAir/Lufthansa	
HB-IIR	Boeing 737-86Q	PrivatAir/ Swiss International	
HB-IJI	Airbus A.320-214	Swiss International *Basodino*	
HB-IJJ	Airbus A.320-214	Swiss International *Les Diablerets*	
HB-IJK	Airbus A.320-214	Swiss International *Wissigstock*	
HB-IJL	Airbus A.320-214	Swiss International *Pizol*	
HB-IJM	Airbus A.320-214	Swiss International *Schilthorn*	
HB-IJN	Airbus A.320-214	Swiss International *Vanil Noir*	
HB-IJO	Airbus A.320-214	Swiss International *Lissengrat*	
HB-IJP	Airbus A.320-214	Swiss International *Nollen*	
HB-IJQ	Airbus A.320-214	Swiss International *Agassizhorn*	
HB-IJR	Airbus A.320-214	Swiss International *Dammastock*	
HB-IJS	Airbus A.320-214	Swiss International *Creux du Van*	
HB-IJU	Airbus A.320-214	Swiss Sun *Bietschhorn*	
HB-IJV	Airbus A.320-214	Swiss Sun *Wildspitz*	
HB-IJW	Airbus A.320-214	Swiss Sun *Bachtel*	
HB-IJZ	Airbus A.320-214	MyAir	
HB-ILQ	F.27 Friendship Mk 500	Farner Air Transport	

Notes	Reg.	Type	Owner or Operator
	HB-IOC	Airbus A.321-111	Swiss International *Eiger*
	HB-IOH	Airbus A.321-111	Swiss International *Piz Palu*
	HB-IOK	Airbus A.321-111	Swiss International *Biefertenstock*
	HB-IOL	Airbus A.321-111	Swiss International *Kaiseregg*
	HB-IPR	Airbus A.319-112	Swiss International *Commune de Champagne*
	HB-IPS	Airbus A.319-112	Swiss International *Weiach*
	HB-IPT	Airbus A.319-112	Swiss International *Stadel*
	HB-IPU	Airbus A.319-112	Swiss International *Hochfelden*
	HB-IPV	Airbus A.319-112	Swiss International *Rumlang*
	HB-IPX	Airbus A.319-112	Swiss International *Steinmaur*
	HB-IPY	Airbus A.319-112	Swiss International *Hori*
	HB-IQA	Airbus A.330-223	Swiss International *Lauteraarhorn*
	HB-IQC	Airbus A.330-223	Swiss International *Breithorn*
	HB-IQG	Airbus A.330-223	Swiss International *Jungfrau*
	HB-IQH	Airbus A.330-223	Swiss International *Allalinhorn*
	HB-IQI	Airbus A.330-223	Swiss International *Piz Bernina*
	HB-IQJ	Airbus A.330-223	Swiss International *Aletschorn*
	HB-IQK	Airbus A.330-223	Swiss International *Strahlhorn*
	HB-IQO	Airbus A.330-223	Swiss International *Weissmies*
	HB-IQP	Airbus A.330-223	Swiss International *Monch*
	HB-IQZ	Airbus A.330-243	Edelweiss Air *Bahari*
	HB-ISE	Boeing 767-3Q8ER	Belair *RondoMondo*
	HB-IWE	McD Douglas MD-11	Swiss International *Nidwalden*
	HB-IXF	Avro RJ85	Swiss International *Karpf*
	HB-IXG	Avro RJ85	Swiss International *Lindenberg*
	HB-IXH	Avro RJ85	Swiss International *Montchalbeux*
	HB-IXK	Avro RJ85	Swiss International *Piz Julien*
	HB-IXN	Avro RJ100	Swiss International *Balmhorn*
	HB-IXO	Avro RJ100	Swiss International *Brisen*
	HB-IXP	Avro RJ100	Swiss International *Chestenberg*
	HB-IXQ	Avro RJ100	Swiss International *Corno Gries*
	HB-IXR	Avro RJ100	Swiss International *Hohe Winde*
	HB-IXS	Avro RJ100	Swiss International *Mont Velan*
	HB-IXT	Avro RJ100	Swiss International *Ottenberg*
	HB-IXU	Avro RJ100	Swiss International *Pfannenstiel*
	HB-IXV	Avro RJ100	Swiss International *Saxer First*
	HB-IXW	Avro RJ100	Swiss International *Schafarnisch*
	HB-IXX	Avro RJ100	Swiss International *Silberen*
	HB-IYB	SAAB 2000	Swiss International
	HB-IYC	SAAB 2000	Swiss International
	HB-IYD	SAAB 2000	Swiss International
	HB-IYE	SAAB 2000	Swiss International
	HB-IYF	SAAB 2000	Swiss International
	HB-IYW	Avro RJ100	Swiss International *Spitzmeilen*
	HB-IYX	Avro RJ100	Swiss International *Stucklistock*
	HB-IYY	Avro RJ100	Swiss International *Titlis*
	HB-IYZ	Avro RJ100	Swiss International *Tour d'Ai*
	HB-IZJ	SAAB 2000	Swiss International
	HB-IZK	SAAB 2000	Swiss International
	HB-IZL	SAAB 2000	Swiss International
	HB-IZM	SAAB 2000	Swiss International
	HB-IZO	SAAB 2000	Swiss International
	HB-IZR	SAAB 2000	Swiss International
	HB-IZZ	SAAB 2000	Swiss International
	HB-JAK	Embraer RJ145LU	Swiss International *Christalina*
	HB-JAL	Embraer RJ145LU	Swiss International
	HB-JAM	Embraer RJ145LU	Swiss International
	HB-JAN	Embraer RJ145LU	Swiss International
	HB-JAO	Embraer RJ145LU	Swiss International
	HB-JAS	Embraer RJ145LU	Swiss International
	HB-JAT	Embraer RJ145LU	Swiss International
	HB-JAU	Embraer RJ145LU	Swiss International *Schibengutsch*
	HB-JAV	Embraer RJ145LU	Swiss International *Nadienspitz*
	HB-JAW	Embraer RJ145LU	Swiss International *Geissflue*
	HB-JAX	Embraer RJ145LU	Swiss International *Corbassiere*
	HB-JAY	Embraer RJ145LU	Swiss International *Silberhorn*
	HB-JIA	McD Douglas MD-90-30	Hello/Blue Line
	HB-JIB	McD Douglas MD-90-30	Hello
	HB-JIC	McD Douglas MD-90-30	Hello
	HB-JMA	Airbus A.340-313X	Swiss International *Matterhorn*
	HB-JMB	Airbus A.340-313X	Swiss International *Dufourspitze*
	HB-JMC	Airbus A.340-313X	Swiss International *Zumsteinspitze*

Reg.	Type	Owner or Operator	Notes
HB-JMD	Airbus A.340-313X	Swiss International *Signalkuppe*	
HB-JME	Airbus A.340-313X	Swiss International *Dom*	
HB-JMF	Airbus A.340-313X	Swiss International *Liskamm*	
HB-JMG	Airbus A.340-313X	Swiss International *Weisshorn*	
HB-JMH	Airbus A.340-313X	Swiss International *Parrotspitze*	
HB-JMI	Airbus A.340-313X	Swiss International *Dent Blanche*	
HB-JVA	Fokker 100	Helvetic	
HB-JVB	Fokker 100	Helvetic	
HB-JVC	Fokker 100	Helvetic	
HB-JVD	Fokker 100	Helvetic	
HB-JVE	Fokker 100	Helvetic	
HB-JVF	Fokker 100	Helvetic	
HB-JVG	Fokker 100	Helvetic	
HB-JZF	Airbus A.319-111	easyJet Switzerland	
HB-JZG	Airbus A.319-111	easyJet Switzerland	
HB-JZH	Airbus A.319-111	easyJet Switzerland	
HB-JZI	Airbus A.319-111	easyJet Switzerland	
HB-JZJ	Airbus A.319-111	easyJet Switzerland	

HL (Korea)

HL7400	Boeing 747-4B5F	Korean Air Cargo	
HL7402	Boeing 747-4B5	Korean Air	
HL7403	Boeing 747-4B5F	Korean Air Cargo	
HL7404	Boeing 747-4B5	Korean Air	
HL7412	Boeing 747-4B5	Korean Air	
HL7420	Boeing 747-48EF (SCD)	Asiana Airlines	
HL7422	Boeing 747-48EF (SCD)	Asiana Airlines	
HL7426	Boeing 747-48EF (SCD)	Asiana Airlines	
HL7428	Boeing 747-48EF (SCD)	Asiana Airlines	
HL7434	Boeing 747-4B5F	Korean Air Cargo	
HL7436	Boeing 747-48EF (SCD)	Asiana Airlines	
HL7437	Boeing 747-4B5F	Korean Air Cargo	
HL7438	Boeing 747-4B5ERF	Korean Air Cargo	
HL7439	Boeing 747-4B5ERF	Korean Air Cargo	
HL7448	Boeing 747-4B5F (SCD)	Korean Air Cargo	
HL7449	Boeing 747-4B5F (SCD)	Korean Air Cargo	
HL7460	Boeing 747-4B5	Korean Air	
HL7461	Boeing 747-4B5	Korean Air	
HL7462	Boeing 747-4B5F	Korean Air Cargo	
HL7465	Boeing 747-4B5F	Korean Air	
HL7466	Boeing 747-4B5F	Korean Air Cargo	
HL7467	Boeing 747-4B5F	Korean Air Cargo	
HL7469	Boeing 747-3B5	Korean Air	
HL7470	Boeing 747-3B5 (SCD)	Korean Air Cargo	
HL7472	Boeing 747-4B5	Korean Air	
HL7473	Boeing 747-4B5	Korean Air	
HL7480	Boeing 747-4B5 (SCD)	Korean Air	
HL7482	Boeing 747-4B5	Korean Air	
HL7483	Boeing 747-4B5	Korean Air	
HL7484	Boeing 747-4B5	Korean Air	
HL7485	Boeing 747-4B5	Korean Air	
HL7486	Boeing 747-4B5	Korean Air	
HL7487	Boeing 747-4B5	Korean Air	
HL7488	Boeing 747-4B5	Korean Air	
HL7489	Boeing 747-4B5	Korean Air	
HL7490	Boeing 747-4B5	Korean Air	
HL7491	Boeing 747-4B5	Korean Air	
HL7492	Boeing 747-4B5	Korean Air	
HL7493	Boeing 747-4B5	Korean Air	
HL7494	Boeing 747-4B5	Korean Air	
HL7495	Boeing 747-4B5	Korean Air	
HL7497	Boeing 747-4B5F	Korean Air Cargo	
HL7498	Boeing 747-4B5	Korean Air	
HL7499	Boeing 747-4B5ERF	Korean Air Cargo	
HL7500	Boeing 777-28EER	Asiana Airlines	
HL7526	Boeing 777-2B5ER	Korean Air	
HL7530	Boeing 777-2B5ER	Korean Air	
HL7531	Boeing 777-2B5ER	Korean Air	
HL7530	Boeing 777-2B5ER	Korean Air	
HL7532	Boeing 777-3B5	Korean Air	

Notes	Reg.	Type	Owner or Operator
	HL7533	Boeing 777-3B5	Korean Air
	HL7534	Boeing 777-3B5	Korean Air
	HL7573	Boeing 777-3B5	Korean Air
	HL7575	Boeing 777-2B5ER	Korean Air
	HL7596	Boeing 777-28EER	Asiana Airlines
	HL7597	Boeing 777-28EER	Asiana Airlines
	HL7598	Boeing 777-2B5ER	Korean Air
	HL7600	Boeing 747-4B5ERF	Korean Air Cargo
	HL7601	Boeing 747-4B5ERF	Korean Air Cargo
	HL7700	Boeing 777-28EER	Asiana Airlines
	HL7714	Boeing 777-2B5ER	Korean Air
	HL7715	Boeing 777-2B5ER	Korean Air
	HL7721	Boeing 777-2B5ER	Korean Air
	HL7732	Boeing 777-28EER	Asiana Airlines

HS (Thailand)

HS-TGA	Boeing 747-4D7	Thai Airways International *Srisuriyothai*
HS-TGB	Boeing 747-4D7	Thai Airways International *Si Satchanulai*
HS-TGD	Boeing 747-3D7	Thai Airways International *Suchada*
HS-TGE	Boeing 747-3D7	Thai Airways International *Chutamat*
HS-TGF	Boeing 747-4D7	Thai Airways International *Sri Ubon*
HS-TGG	Boeing 747-4D7	Thai Airways International *Pathoomawadi*
HS-TGH	Boeing 747-4D7	Thai Airways International *Chaiprakarn*
HS-TGJ	Boeing 747-4D7	Thai Airways International *Hariphunchai*
HS-TGK	Boeing 747-4D7	Thai Airways International *Alongkorn*
HS-TGL	Boeing 747-4D7	Thai Airways International *Theparat*
HS-TGM	Boeing 747-4D7	Thai Airways International *Chao Phraya*
HS-TGN	Boeing 747-4D7	Thai Airways International *Simongkhon*
HS-TGO	Boeing 747-4D7	Thai Airways International *Bowonrangsi*
HS-TGP	Boeing 747-4D7	Thai Airways International *Thepprasit*
HS-TGR	Boeing 747-4D7	Thai Airways International *Siriwatthana*
HS-TGT	Boeing 747-4D7	Thai Airways International *Watthanothai*
HS-TGW	Boeing 747-4D7	Thai Airways International *Visuthakasatriya*
HS-TGX	Boeing 747-4D7	Thai Airways International *Sirisobhakya*
HS-TGY	Boeing 747-4D7	Thai Airways International *Dararasmi*
HS-TGZ	Boeing 747-4D7	Thai Airways International *Phimara*
HS-TJA	Boeing 777-2D7	Thai Airways International *Lamphun*
HS-TJB	Boeing 777-2D7	Thai Airways International
HS-TJC	Boeing 777-2D7	Thai Airways International
HS-TJD	Boeing 777-2D7	Thai Airways International
HS-TJE	Boeing 777-2D7	Thai Airways International
HS-TJF	Boeing 777-2D7	Thai Airways International
HS-TJG	Boeing 777-2D7	Thai Airways International
HS-TJH	Boeing 777-2D7	Thai Airways International
HS-TKA	Boeing 777-3D7	Thai Airways International *Sriwanna*
HS-TKB	Boeing 777-3D7	Thai Airways International *Chainarai*
HS-TKC	Boeing 777-3D7	Thai Airways International
HS-TKD	Boeing 777-3D7	Thai Airways International
HS-TKE	Boeing 777-3D7	Thai Airways International
HS-TKF	Boeing 777-3D7	Thai Airways International
HS-TMD	McD Douglas MD-11	Thai Airways International *Phra Nakhon*
HS-TME	McD Douglas MD-11	Thai Airways International *Pathumwan*
HS-TMF	McD Douglas MD-11	Thai Airways International *Phichit*
HS-TMG	McD Douglas MD-11	Thai Airways International *Nakhon Sawan*
HS-UTB	Boeing 747-246B	Orient Thai Airlines
HS-UTC	Boeing 747-238B	Orient Thai Airlines
HS-UTI	Boeing 747-246B	Orient Thai Airlines
HS-UTJ	Boeing 747-246B	Orient Thai Airlines
HS-UTK	Boeing 747-306M	Orient Thai Airlines
HS-UTP	Boeing 747-246B	Orient Thai Airlines
HS-VAA	Boeing 747-206B (SUD)	Phuket Airlines
HS-VAN	Boeing 747-312	Phuket Airlines
HS-VAO	Boeing 747-2U3B	Phuket Airlines
HS-VAU	Boeing 747-2U3B	Phuket Airlines
HS-VAV	Boeing 747-206B (SUD)	Phuket Airlines

HZ (Saudi Arabia)

HZ-AIF	Boeing 747SP-68	Saudi Arabian Airlines

Reg.	Type	Owner or Operator	Notes
HZ-AIJ	Boeing 747SP-68	Saudi Royal Flight	
HZ-AIV	Boeing 747-468	Saudi Arabian Airlines	
HZ-AIW	Boeing 747-468	Saudi Arabian Airlines	
HZ-AIX	Boeing 747-468	Saudi Arabian Airlines	
HZ-AIY	Boeing 747-468	Saudi Arabian Airlines	
HZ-AIZ	Boeing 747-468	Saudi Arabian Airlines	
HZ-AKA	Boeing 777-268ER	Saudi Arabian Airlines	
HZ-AKB	Boeing 777-268ER	Saudi Arabian Airlines	
HZ-AKC	Boeing 777-268ER	Saudi Arabian Airlines	
HZ-AKD	Boeing 777-268ER	Saudi Arabian Airlines	
HZ-AKE	Boeing 777-268ER	Saudi Arabian Airlines	
HZ-AKF	Boeing 777-268ER	Saudi Arabian Airlines	
HZ-AKG	Boeing 777-268ER	Saudi Arabian Airlines	
HZ-AKH	Boeing 777-268ER	Saudi Arabian Airlines	
HZ-AKI	Boeing 777-268ER	Saudi Arabian Airlines	
HZ-AKJ	Boeing 777-268ER	Saudi Arabian Airlines	
HZ-AKK	Boeing 777-268ER	Saudi Arabian Airlines	
HZ-AKL	Boeing 777-268ER	Saudi Arabian Airlines	
HZ-AKM	Boeing 777-268ER	Saudi Arabian Airlines	
HZ-AKN	Boeing 777-268ER	Saudi Arabian Airlines	
HZ-AKO	Boeing 777-268ER	Saudi Arabian Airlines	
HZ-AKP	Boeing 777-268ER	Saudi Arabian Airlines	
HZ-AKQ	Boeing 777-268ER	Saudi Arabian Airlines	
HZ-AKR	Boeing 777-268ER	Saudi Arabian Airlines	
HZ-AKS	Boeing 777-268ER	Saudi Arabian Airlines	
HZ-AKT	Boeing 777-268ER	Saudi Arabian Airlines	
HZ-ANA	McD Douglas MD-11F	Saudi Arabian Airlines	
HZ-ANB	McD Douglas MD-11F	Saudi Arabian Airlines	
HZ-ANC	McD Douglas MD-11F	Saudi Arabian Airlines	
HZ-AND	McD Douglas MD-11F	Saudi Arabian Airlines	
HZ-HM1A	Boeing 747-3G1	Saudi Royal Flight	
HZ-HM1B	Boeing 747SP-68	Saudi Royal Flight	
HZ-HM5	L.1011-385 TriStar 500	Saudi Royal Flight	
HZ-HM6	L.1011-385 TriStar 500	Saudi Royal Flight	

I (Italy)

Reg.	Type	Owner or Operator	Notes
I-ALPK	Fokker 100	Alpi Eagles San Antonio	
I-ALPL	Fokker 100	Alpi Eagles San Marco	
I-ALPQ	Fokker 100	Alpi Eagles	
I-ALPS	Fokker 100	Alpi Eagles San Zeno	
I-ALPW	Fokker 100	Alpi Eagles	
I-ALPX	Fokker 100	Alpi Eagles	
I-ALPZ	Fokker 100	Alpi Eagles San Pietro	
I-BIKA	Airbus A.320-214	Alitalia Johann Sebastian Bac	
I-BIKB	Airbus A.320-214	Alitalia Wolfgang Amadeus Mozart	
I-BIKC	Airbus A.320-214	Alitalia Zefiro	
I-BIKD	Airbus A.320-214	Alitalia Maestrale	
I-BIKE	Airbus A.320-214	Alitalia Franz Liszt	
I-BIKF	Airbus A.320-214	Alitalia Grecale	
I-BIKG	Airbus A.320-214	Alitalia Scirocco	
I-BIKI	Airbus A.320-214	Alitalia Girolamo Frescobaldi	
I-BIKL	Airbus A.320-214	Alitalia Libeccio	
I-BIKO	Airbus A.320-214	Alitalia	
I-BIKU	Airbus A.320-214	Alitalia Frederyk Chopin	
I-BIMA	Airbus A.319-112	Alitalia	
I-BIMB	Airbus A.319-112	Alitalia	
I-BIMC	Airbus A.319-112	Alitalia	
I-BIMD	Airbus A.319-112	Alitalia	
I-BIME	Airbus A.319-112	Alitalia	
I-BIMF	Airbus A.319-112	Alitalia	
I-BIMG	Airbus A.319-112	Alitalia	
I-BIMH	Airbus A.319-112	Alitalia	
I-BIMI	Airbus A.319-112	Alitalia	
I-BIMJ	Airbus A.319-112	Alitalia	
I-BIML	Airbus A.319-112	Alitalia	
I-BIMO	Airbus A.319-112	Alitalia	
I-BIXA	Airbus A.321-112	Alitalia Piazza del Duomo Milano	
I-BIXB	Airbus A.321-112	Alitalia Piazza Castello Torino	
I-BIXC	Airbus A.321-112	Alitalia Piazza del Campo Siena	
I-BIXD	Airbus A.321-112	Alitalia Piazza Pretoria Palermo	

Reg.	Type	Owner or Operator
I-BIXE	Airbus A.321-112	Alitalia *Piazza di Spagna Roma*
I-BIXF	Airbus A.321-112	Alitalia *Piazza Maggiore Bologna*
I-BIXG	Airbus A.321-112	Alitalia *Piazza dei Miracoli Pisa*
I-BIXH	Airbus A.321-112	Alitalia
I-BIXI	Airbus A.321-112	Alitalia *Piazza San Marco-Venezia*
I-BIXJ	Airbus A.321-112	Alitalia
I-BIXK	Airbus A.321-112	Alitalia *Piazza Ducale Vigevano*
I-BIXL	Airbus A.321-112	Alitalia *Piazza del Duomo Lecce*
I-BIXM	Airbus A.321-112	Alitalia *Piazza di San Franceso Assisi*
I-BIXN	Airbus A.321-112	Alitalia *Piazza del Duomo Catania*
I-BIXO	Airbus A.321-112	Alitalia *Piazza Plebiscito Napoli*
I-BIXP	Airbus A.321-112	Alitalia *Carlo Morelli*
I-BIXQ	Airbus A.321-112	Alitalia *Domenico Colapietro*
I-BIXR	Airbus A.321-112	Alitalia *Piazza dell Campidoglio-Roma*
I-BIXS	Airbus A.321-112	Alitalia *Piazza San Martino-Lucca*
I-BIXT	Airbus A.321-112	Alitalia *Piazza dei Miracoli Pisa*
I-BIXU	Airbus A.321-112	Alitalia *Piazza dell Signori Firenze*
I-BIXV	Airbus A.321-112	Alitalia *Piazza dell Rinaccimento-Urbino*
I-BIXZ	Airbus A.321-112	Alitalia *Piazza dell Duomo Orvieto*
I-DACM	McD Douglas MD-82	Alitalia *La Spezia*
I-DACN	McD Douglas MD-82	Alitalia *Rieti*
I-DACP	McD Douglas MD-82	Alitalia *Padova*
I-DACQ	McD Douglas MD-82	Alitalia *Taranto*
I-DACR	McD Douglas MD-82	Alitalia *Carrara*
I-DACS	McD Douglas MD-82	Alitalia *Maratea*
I-DACT	McD Douglas MD-82	Alitalia *Valtellina*
I-DACU	McD Douglas MD-82	Alitalia *Fabriano*
I-DACV	McD Douglas MD-82	Alitalia *Riccione*
I-DACW	McD Douglas MD-82	Alitalia *Vieste*
I-DACX	McD Douglas MD-82	Alitalia *Piacenza*
I-DACY	McD Douglas MD-82	Alitalia *Novara*
I-DACZ	McD Douglas MD-82	Alitalia *Castelfidardo*
I-DAND	McD Douglas MD-82	Alitalia *Bolzano*
I-DANF	McD Douglas MD-82	Alitalia *Vicenza*
I-DANG	McD Douglas MD-82	Alitalia *Benevento*
I-DANH	McD Douglas MD-82	Alitalia *Messina*
I-DANL	McD Douglas MD-82	Alitalia *Cosenza*
I-DANM	McD Douglas MD-82	Alitalia *Vicenza*
I-DANP	McD Douglas MD-82	Alitalia *Fabriano*
I-DANQ	McD Douglas MD-82	Alitalia *Lecce*
I-DANR	McD Douglas MD-82	Alitalia *Matera*
I-DANU	McD Douglas MD-82	Alitalia *Trapani*
I-DANV	McD Douglas MD-82	Alitalia *Forte dei Marmi*
I-DANW	McD Douglas MD-82	Alitalia *Siena*
I-DATA	McD Douglas MD-82	Alitalia *Gubbio*
I-DATB	McD Douglas MD-82	Alitalia *Bergamo*
I-DATC	McD Douglas MD-82	Alitalia *Foggia*
I-DATD	McD Douglas MD-82	Alitalia *Savona*
I-DATE	McD Douglas MD-82	Alitalia *Grosseto*
I-DATF	McD Douglas MD-82	Alitalia *Vittorio Veneto*
I-DATG	McD Douglas MD-82	Alitalia *Arezzo*
I-DATH	McD Douglas MD-82	Alitalia *Pescara*
I-DATI	McD Douglas MD-82	Alitalia *Siracusa*
I-DATJ	McD Douglas MD-82	Alitalia *Lunigiana*
I-DATK	McD Douglas MD-82	Alitalia *Ravenna*
I-DATL	McD Douglas MD-82	Alitalia *Alghero*
I-DATM	McD Douglas MD-82	Alitalia *Cividale del Friuli*
I-DATN	McD Douglas MD-82	Alitalia *Sondrio*
I-DATO	McD Douglas MD-82	Alitalia *Reggio Emilia*
I-DATP	McD Douglas MD-82	Alitalia *Latina*
I-DATQ	McD Douglas MD-82	Alitalia *Modena*
I-DATR	McD Douglas MD-82	Alitalia *Livorno*
I-DATS	McD Douglas MD-82	Alitalia *Foligno*
I-DATU	McD Douglas MD-82	Alitalia *Verona*
I-DAVA	McD Douglas MD-82	Eurofly
I-DAVB	McD Douglas MD-82	Alitalia *Ferrara*
I-DAVD	McD Douglas MD-82	Alitalia *Mantova*
I-DAVF	McD Douglas MD-82	Eurofly
I-DAVG	McD Douglas MD-82	Alitalia *Pesaro*
I-DAVH	McD Douglas MD-82	Alitalia *Salerno*
I-DAVI	McD Douglas MD-82	Alitalia *Assisi*
I-DAVJ	McD Douglas MD-82	Alitalia *Parma*

Reg.	Type	Owner or Operator	Notes
I-DAVL	McD Douglas MD-82	Alitalia *Reggio Calabria*	
I-DAVM	McD Douglas MD-82	Alitalia *Caserta*	
I-DAVP	McD Douglas MD-82	Alitalia *Gorizia*	
I-DAVR	McD Douglas MD-82	Alitalia *Pisa*	
I-DAVS	McD Douglas MD-82	Alitalia *Catania*	
I-DAVT	McD Douglas MD-82	Alitalia *Como*	
I-DAVU	McD Douglas MD-82	Alitalia *Udine*	
I-DAVV	McD Douglas MD-82	Alitalia *Pavia*	
I-DAVW	McD Douglas MD-82	Alitalia *Camerino*	
I-DAVX	McD Douglas MD-82	Alitalia *Asti*	
I-DAVZ	McD Douglas MD-82	Alitalia *Brescia*	
I-DAWA	McD Douglas MD-82	Alitalia *Roma*	
I-DAWB	McD Douglas MD-82	Alitalia *Cagliari*	
I-DAWC	McD Douglas MD-82	Alitalia *Campobasso*	
I-DAWD	McD Douglas MD-82	Alitalia *Catanzaro*	
I-DAWE	McD Douglas MD-82	Alitalia *Milano*	
I-DAWF	McD Douglas MD-82	Alitalia *Firenze*	
I-DAWG	McD Douglas MD-82	Alitalia *L'Aquila*	
I-DAWH	McD Douglas MD-82	Alitalia *Palermo*	
I-DAWI	McD Douglas MD-82	Alitalia *Ancona*	
I-DAWJ	McD Douglas MD-82	Alitalia *Genova*	
I-DAWL	McD Douglas MD-82	Alitalia *Perugia*	
I-DAWM	McD Douglas MD-82	Alitalia *Potenza*	
I-DAWO	McD Douglas MD-82	Alitalia *Bari*	
I-DAWP	McD Douglas MD-82	Alitalia *Torino*	
I-DAWQ	McD Douglas MD-82	Alitalia *Trieste*	
I-DAWR	McD Douglas MD-82	Alitalia *Venezia*	
I-DAWS	McD Douglas MD-82	Alitalia *Aosta*	
I-DAWT	McD Douglas MD-82	Alitalia *Napoli*	
I-DAWU	McD Douglas MD-82	Alitalia *Bologna*	
I-DAWV	McD Douglas MD-82	Alitalia *Trento*	
I-DAWW	McD Douglas MD-82	Eurofly	
I-DAWY	McD Douglas MD-82	Alitalia *Agrigento*	
I-DEIB	Boeing 767-33AER	Alitalia *Pier Paolo Racchetti*	
I-DEIC	Boeing 767-33AER	Alitalia *Alberto Nassetti*	
I-DEID	Boeing 767-33AER	Alitalia *Marco Polo*	
I-DEIF	Boeing 767-33AER	Alitalia *Cristoforo Colombo*	
I-DEIG	Boeing 767-33AER	Alitalia *Francesco Agello*	
I-DEIL	Boeing 767-33AER	Alitalia *Arturo Ferrarin*	
I-DIKR	Douglas DC-9-32	Alitalia *Piemonte*	
I-DISA	Boeing 777-243ER	Alitalia	
I-DISB	Boeing 777-243ER	Alitalia	
I-DISD	Boeing 777-243ER	Alitalia	
I-DISE	Boeing 777-243ER	Alitalia	
I-DISO	Boeing 777-243ER	Alitalia	
I-DISU	Boeing 777-243ER	Alitalia	
I-DIZE	Douglas DC-9-32	Alitalia *Isola della Meloria*	
I-EEZA	Airbus A.330-223	Eurofly	
I-EEZB	Airbus A.330-223	Eurofly	
I-EEZC	Airbus A.320-214	Eurofly	
I-EEZD	Airbus A.320-214	Eurofly	
I-EEZE	Airbus A.320-214	Eurofly	
I-EEZF	Airbus A.320-214	Eurofly	
I-EEZG	Airbus A.320-214	Eurofly	
I-EEZH	Airbus A.320-214	Eurofly	
I-EEZI	Airbus A.320-214	Eurofly	
I-EEZK	Airbus A.320-214	Eurofly	
I-EXMA	Embraer RJ145LR	Alitalia Express *Giosue Carducci*	
I-EXMB	Embraer RJ145LR	Alitalia Express *Salvatori Quasidomo*	
I-EXMC	Embraer RJ145LR	Alitalia Express	
I-EXMD	Embraer RJ145LR	Alitalia Express	
I-EXME	Embraer RJ145LR	Alitalia Express *Guglialmo Marconi*	
I-EXMF	Embraer RJ145LR	Alitalia Express	
I-EXMG	Embraer RJ145LR	Alitalia Express	
I-EXMH	Embraer RJ145LR	Alitalia Express	
I-EXMI	Embraer RJ145LR	Alitalia Express *Grazia Deledda*	
I-EXML	Embraer RJ145LR	Alitalia Express	
I-EXMM	Embraer RJ145LR	Alitalia Express	
I-EXMN	Embraer RJ145LR	Alitalia Express	
I-EXMO	Embraer RJ145LR	Alitalia Express *Luigi Pirandello*	
I-EXMU	Embraer RJ145LR	Alitalia Express *Enrico Fermi*	
I-JETA	Boeing 737-229	Air One	

Notes	Reg.	Type	Owner or Operator
	I-JETC	Boeing 737-230	Air One
	I-JETD	Boeing 737-230	Air One
	I-LIVD	Airbus A.321-231	Livingston
	I-LLAG	Boeing 767-330ER	Blue Panorama
	I-MSAA	BAe 146-200QT	Mistral Air/TNT
	I-OCEA	Boeing 747-230F (SCD)	Ocean Airlines
	I-OCEU	Boeing 747-230F (SCD)	Ocean Airlines
	I-PEKG	Airbus A.320-214	Air Europe Italy
	I-PEKI	Airbus A.320-214	Air Europe Italy
	I-PEKL	Airbus A.320-214	Air Europe Italy
	I-SMEB	McD Douglas MD-82	Meridiana
	I-SMEC	McD Douglas MD-83	Meridiana
	I-SMED	McD Douglas MD-83	Meridiana
	I-SMEL	McD Douglas MD-82	Meridiana
	I-SMEM	McD Douglas MD-82	Meridiana
	I-SMEP	McD Douglas MD-82	Meridiana
	I-SMER	McD Douglas MD-82	Meridiana
	I-SMES	McD Douglas MD-82	Meridiana
	I-SMET	McD Douglas MD-82	Meridiana
	I-SMEV	McD Douglas MD-82	Meridiana
	I-SMEZ	McD Douglas MD-83	Meridiana
	I-TNTC	BAe 146-200QT	Mistral Air/TNT

Note: Meridiana also operates A.319s EI-DEY, EI-DEZ, EI-DFA and EI-DFP plus MD-83s EI-CIW, EI-CKM, EI-CRE, EI-CRH, EI-CRJ and EI-CRW. Air One employs Boeing 737s D-AGMR, EI-CLW, EI-CLZ, EI-COH, EI-COI, EI-COJ, EI-COK, EI-CRZ, EI-CSU, EI-CWE, EI-CWF, EI-CWW, EI-CWX, EI-CXI, EI-CXJ, EI-CXL, EI-CXM, EI-CZG, EI-DFD, EI-DFE, EI-DFF, EI-DGD, EI-DGL, F-GKTA and F-GKTB.
Boeing 767s EI-CRD, EI-CRF, EI-CRM, EI-CRO, EI-DBP and EI-DDW plus Boeing 777s EI-DBK, EI-DBL, EI-DBM and EI-DDH are operated by Alitalia. The ATR-72s EI-CLB, EI-CLC, EI-CLD and EI-CMJ plus ERJ-170s EI-DFG, EI-DFH, EI-DFJ, EI-DFK and EI-DFL fly for Alitalia Express. Blue Panorama operates leased Boeing 737s EI-CUA, EI-CUD, EI-CUN, EI-DGM and EI-DGN together with Boeing 767s EI-CXO and EI-CZH. Myair operates A.320 EI-LTE and Wind Jet operates A.320s EI-DFN and EI-DFO.

JA (Japan)

	JA401J	Boeing 747-446F	Japan Airlines
	JA402J	Boeing 747-446F	Japan Airlines
	JA403A	Boeing 747-481	All Nippon Airways
	JA404A	Boeing 747-481	All Nippon Airways
	JA405A	Boeing 747-481	All Nippon Airways
	JA704J	Boeing 777-246ER	Japan Airlines
	JA705J	Boeing 777-246ER	Japan Airlines
	JA706J	Boeing 777-246ER	Japan Airlines
	JA707J	Boeing 777-246ER	Japan Airlines
	JA708J	Boeing 777-246ER	Japan Airlines
	JA709J	Boeing 777-246ER	Japan Airlines
	JA731J	Boeing 777-346ER	Japan Airlines
	JA732J	Boeing 777-346ER	Japan Airlines
	JA733J	Boeing 777-346ER	Japan Airlines
	JA734J	Boeing 777-346ER	Japan Airlines
	JA811J	Boeing 747-246F	Japan Airlines
	JA812J	Boeing 747-346	Japan Airlines
	JA813J	Boeing 747-346	Japan Airlines
	JA892D	Boeing 747-446	Japan Airlines
	JA8071	Boeing 747-446	Japan Airlines
	JA8072	Boeing 747-446	Japan Airlines
	JA8073	Boeing 747-446	Japan Airlines
	JA8074	Boeing 747-446	Japan Airlines
	JA8075	Boeing 747-446	Japan Airlines
	JA8076	Boeing 747-446	Japan Airlines
	JA8077	Boeing 747-446	Japan Airlines
	JA8078	Boeing 747-446	Japan Airlines
	JA8079	Boeing 747-446	Japan Airlines
	JA8080	Boeing 747-446	Japan Airlines
	JA8081	Boeing 747-446	Japan Airlines
	JA8082	Boeing 747-446	Japan Airlines
	JA8085	Boeing 747-446	Japan Airlines
	JA8086	Boeing 747-446	Japan Airlines
	JA8087	Boeing 747-446	Japan Airlines
	JA8088	Boeing 747-446	Japan Airlines
	JA8089	Boeing 747-446	Japan Airlines

Reg.	Type	Owner or Operator
JA8094	Boeing 747-481	All Nippon Airways
JA8095	Boeing 747-481	All Nippon Airways
JA8096	Boeing 747-481	All Nippon Airways
JA8097	Boeing 747-481	All Nippon Airways
JA8098	Boeing 747-481	All Nippon Airways
JA8122	Boeing 747-246B	Japan Airlines
JA8123	Boeing 747-246F (SCD)	Japan Airlines
JA8130	Boeing 747-246B	Japan Airlines
JA8131	Boeing 747-246B	Japan Airlines
JA8132	Boeing 747-246F	Japan Airlines
JA8140	Boeing 747-246B	Japan Airlines
JA8141	Boeing 747-246B	Japan Airlines
JA8151	Boeing 747-246F	Japan Airlines
JA8154	Boeing 747-246B	Japan Airlines
JA8158	Boeing 747SR-81	Nippon Cargo Airlines
JA8160	Boeing 747-221F (SCD)	Japan Airlines
JA8161	Boeing 747-246B	Japan Airlines
JA8162	Boeing 747-246B	Japan Airlines
JA8163	Boeing 747-346	Japan Airlines
JA8165	Boeing 747-221F (SCD)	Japan Airlines
JA8166	Boeing 747-346	Japan Airlines
JA8167	Boeing 747-281F (SCD)	Nippon Cargo Airlines
JA8168	Boeing 747-281F (SCD)	Nippon Cargo Airlines
JA8169	Boeing 747-246B	Japan Airlines
JA8171	Boeing 747-246F (SCD)	Japan Airlines
JA8172	Boeing 747-281F (SCD)	Nippon Cargo Airlines
JA8173	Boeing 747-346	Japan Airlines
JA8174	Boeing 747-281B	All Nippon Airways
JA8175	Boeing 747-281B	All Nippon Airways
JA8177	Boeing 747-346	Japan Airlines
JA8178	Boeing 747-346	Japan Airlines
JA8179	Boeing 747-346	Japan Airlines
JA8180	Boeing 747-246F (SCD)	Japan Airlines
JA8181	Boeing 747-281F	All Nippon Airways
JA8182	Boeing 747-281B	All Nippon Airways
JA8184	Boeing 747-346	Japan Airlines
JA8185	Boeing 747-346	Japan Airlines
JA8186	Boeing 747-346	Japan Airlines
JA8188	Boeing 747-281F (SCD)	Nippon Cargo Airlines
JA8190	Boeing 747-281B	All Nippon Airways
JA8191	Boeing 747-281F (SCD)	Nippon Cargo Airlines
JA8192	Boeing 747-2D3F(SCD)	Nippon Cargo Airlines
JA8193	Boeing 747-212F (SCD)	Japan Airlines
JA8194	Boeing 747-281F (SCD)	Nippon Cargo Airlines
JA8901	Boeing 747-446	Japan Airlines
JA8902	Boeing 747-446	Japan Airlines
JA8906	Boeing 747-446	Japan Airlines
JA8909	Boeing 747-446	Japan Airlines
JA8910	Boeing 747-446	Japan Airlines
JA8911	Boeing 747-446	Japan Airlines
JA8912	Boeing 747-446	Japan Airlines
JA8913	Boeing 747-446	Japan Airlines
JA8914	Boeing 747-446	Japan Airlines
JA8915	Boeing 747-446	Japan Airlines
JA8916	Boeing 747-446	Japan Airlines
JA8917	Boeing 747-446	Japan Airlines
JA8918	Boeing 747-446	Japan Airlines
JA8919	Boeing 747-446	Japan Airlines
JA8920	Boeing 747-446	Japan Airlines
JA8921	Boeing 747-446	Japan Airlines
JA8922	Boeing 747-446	Japan Airlines
JA8937	Boeing 747-246F	Japan Airlines
JA8958	Boeing 747-481	All Nippon Airways
JA8962	Boeing 747-481	All Nippon Airways

JY (Jordan)

JY-ABH	Airbus A.340-211	Kingdom of Jordan
JY-AGK	Airbus A.310-308	Royal Jordanian
JY-AGL	Airbus A.310-304	Royal Jordanian
JY-AGM	Airbus A.310-304	Royal Jordanian *Prince Hamzeh*

Notes	Reg.	Type	Owner or Operator
	JY-AGN	Airbus A.310-304	Royal Jordanian *Princess Haya*
	JY-AGP	Airbus A.310-304	Royal Jordanian
	JY-AGV	Airbus A.310-203	Libyan Arab Airlines
	JY-AIA	Airbus A.340-211	Royal Jordanian *Hussein Bin Abdullah*
	JY-AIB	Airbus A.340-211	Royal Jordanian
	JY-AYA	Airbus A.320-211	Royal Jordanian *Cairo*
	JY-AYB	Airbus A.320-211	Royal Jordanian *Baghdad*
	JY-JAV	Airbus A.310-222	Sudan Airways

Note: Royal Jordanian also operates A.310s registered F-ODVF and F-ODVG, A.320 F-OGYC plus F-OHGB and F-OHGC and A.340s F-OHLP and F-OHLQ.

LN (Norway)

	LN-BRD	Boeing 737-505	Braathens *Harald Gille*
	LN-BRE	Boeing 737-405	Braathens *Haakon V Magnusson*
	LN-BRH	Boeing 737-505	Braathens *Haakon den Gode*
	LN-BRI	Boeing 737-405	Braathens *Harald Haarfagre*
	LN-BRK	Boeing 737-505	Braathens *Olav Tryggvason*
	LN-BRM	Boeing 737-505	Braathens *Olav den Hellige*
	LN-BRO	Boeing 737-505	Braathens *Magnus Haraldsson*
	LN-BRQ	Boeing 737-405	Braathens *Harald Graafell*
	LN-BRR	Boeing 737-505	Braathens *Halvdan Svarte*
	LN-BRS	Boeing 737-505	Braathens *Olav Kyrre*
	LN-BRV	Boeing 737-505	Braathens *Haakon Sverresson*
	LN-BRX	Boeing 737-505	Braathens *Sigurd Munn*
	LN-BUC	Boeing 737-505	Braathens *Magnus Erlingsson*
	LN-BUD	Boeing 737-505	Braathens *Inge Krokrygg*
	LN-BUE	Boeing 737-505	Braathens *Erling Skjalgsson*
	LN-BUF	Boeing 737-405	Braathens *Magnus den Gode*
	LN-BUG	Boeing 737-505	Braathens *Oystein Haraldsson*
	LN-FAJ	BAe Jetstream 3100	Coast Air
	LN-FAM	BAe Jetstream 3100	Coast Air
	LN-FAO	Aérospatiale ATR-42-300	Coast Air
	LN-FAP	Aérospatiale ATR-42-300	Coast Air
	LN-FAV	BAe Jetstream 3100	Coast Air
	LN-FAZ	BAe Jetstream 3100	Coast Air
	LN-KKF	Boeing 737-3K2	Norwegian Air Shuttle *Fridtjof Nansen*
	LN-KKG	Boeing 737-3K2	Norwegian Air Shuttle *Gidsken Jakobse*
	LN-KKH	Boeing 737-3K2	Norwegian Air Shuttle *Otto Sverdrup*
	LN-KKI	Boeing 737-3K2	Norwegian Air Shuttle *Helge Ingstad*
	LN-KKJ	Boeing 737-36N	Norwegian Air Shuttle *Sonja Henie*
	LN-KKL	Boeing 737-36N	Norwegian Air Shuttle *Roald Amundsen*
	LN-KKM	Boeing 737-3Y0	Norwegian Air Shuttle *Thor Heyerdahl*
	LN-KKN	Boeing 737-3Y0	Norwegian Air Shuttle *Sigrid Undset*
	LN-KKO	Boeing 737-3Y0	Norwegian Air Shuttle *Henrik Ibsen*
	LN-KKP	Boeing 737-3MQ	Norwegian Air Shuttle *Kirsten Flagstad*
	LN-KKQ	Boeing 737-36Q	Norwegian Air Shuttle *Alf Proysen*
	LN-KKR	Boeing 737-3Y0	Norwegian Air Shuttle
	LN-RCN	Boeing 737-883	S.A.S. *Hedrun Viking*
	LN-RCT	Boeing 737-683	S.A.S. *Fridlev Viking*
	LN-RCU	Boeing 737-683	S.A.S. *Sigfrid Viking*
	LN-RCW	Boeing 737-683	S.A.S. *Yngvar Viking*
	LN-RCX	Boeing 737-883	S.A.S. *Hottur Viking*
	LN-RCY	Boeing 737-883	S.A.S. *Eylime Viking*
	LN-RCZ	Boeing 737-883	S.A.S. *Glitne Viking*
	LN-RDA	D.H.C.8Q-402 Dash Eight	S.A.S. Commuter *Frej Viking*
	LN-RDB	D.H.C.8Q-402 Dash Eight	S.A.S. Commuter *Kari Viking*
	LN-RDC	D.H.C.8Q-402 Dash Eight	S.A.S. Commuter *Hader Viking*
	LN-RDD	D.H.C.8Q-402 Dash Eight	S.A.S. Commuter *Loge Viking*
	LN-RDE	D.H.C.8Q-402 Dash Eight	S.A.S. Commuter *Dore Viking*
	LN-RDF	D.H.C.8Q-402 Dash Eight	S.A.S. Commuter *Fenja Viking*
	LN-RDG	D.H.C.8Q-402 Dash Eight	S.A.S. Commuter *Greip Viking*
	LN-RDH	D.H.C.8Q-402 Dash Eight	S.A.S. Commuter *Gloe Viking*
	LN-RDI	D.H.C.8Q-402 Dash Eight	S.A.S. Commuter *Asta Viking*
	LN-RDJ	D.H.C.8Q-402 Dash Eight	S.A.S. Commuter *Toke Viking*
	LN-RDK	D.H.C.8Q-402 Dash Eight	S.A.S. Commuter *Ingrid Viking*
	LN-RDL	D.H.C.8Q-402 Dash Eight	S.A.S. Commuter *Ulv Viking*
	LN-RDM	D.H.C.8Q-402 Dash Eight	S.A.S. Commuter *Banke Viking*
	LN-RDN	D.H.C.8Q-402 Dash Eight	S.A.S. Commuter *Gnupa Viking*
	LN-RDO	D.H.C.8Q-402 Dash Eight	S.A.S. Commuter *Frid Viking*

Reg.	Type	Owner or Operator	Notes
LN-RDP	D.H.C.8Q-402 Dash Eight	S.A.S. Commuter *Huge Viking*	
LN-RDQ	D.H.C.8Q-402 Dash Eight	S.A.S. Commuter *Herta Viking*	
LN-RDR	D.H.C.8Q-402 Dash Eight	S.A.S. Commuter *Terje Viking*	
LN-RDS	D.H.C.8Q-402 Dash Eight	S.A.S. Commuter *Gote Viking*	
LN-RDT	D.H.C.8Q-402 Dash Eight	S.A.S. Commuter *Kile Viking*	
LN-RKF	Airbus A.340-313X	S.A.S. *Godfred Viking*	
LN-RKG	Airbus A.340-313X	S.A.S. *Gudrod Viking*	
LN-RKH	Airbus A.330-343X	S.A.S. *Emund Viking*	
LN-RKI	Airbus A.321-231	S.A.S. *Gunnhild Viking*	
LN-RLE	McD Douglas MD-82	S.A.S. *Ketiil Viking*	
LN-RLF	McD Douglas MD-82	S.A.S. *Finn Viking*	
LN-RLG	McD Douglas MD-82	S.A.S. *Trond Viking*	
LN-RLR	McD Douglas MD-82	S.A.S. *Vegard Viking*	
LN-RMG	McD Douglas MD-87	S.A.S. *Snorre Viking*	
LN-RMH	McD Douglas MD-87	S.A.S. *Solmund Viking*	
LN-RMK	McD Douglas MD-87	S.A.S. *Ragnhild Viking*	
LN-RML	McD Douglas MD-82	S.A.S. *Aud Viking*	
LN-RMM	McD Douglas MD-82	S.A.S. *Blenda Viking*	
LN-RMN	McD Douglas MD-82	S.A.S. *Ivar Viking*	
LN-RMO	McD Douglas MD-82	S.A.S. *Bergljot Viking*	
LN-RMP	McD Douglas MD-87	S.A.S. *Reidun Viking*	
LN-RMR	McD Douglas MD-81	S.A.S. *Olav Viking*	
LN-RMS	McD Douglas MD-81	S.A.S. *Nial Viking*	
LN-RMT	McD Douglas MD-81	S.A.S. *Jarl Viking*	
LN-RMU	McD Douglas MD-87	S.A.S. *Grim Viking*	
LN-ROA	McD Douglas MD-90-30	S.A.S. *Sigurd Viking*	
LN-ROB	McD Douglas MD-90-30	S.A.S. *Isrid Viking*	
LN-ROM	McD Douglas MD-81	S.A.S. *Albin Viking*	
LN-RON	McD Douglas MD-81	S.A.S. *Holmfrid Viking*	
LN-ROO	McD Douglas MD-81	S.A.S. *Kristin Viking*	
LN-ROP	McD Douglas MD-82	S.A.S. *Bjoern Viking*	
LN-ROR	McD Douglas MD-82	S.A.S. *Assur Viking*	
LN-ROS	McD Douglas MD-82	S.A.S. *Isulv Viking*	
LN-ROT	McD Douglas MD-82	S.A.S. *Ingjaid Viking*	
LN-ROU	McD Douglas MD-82	S.A.S. *Ring Viking*	
LN-ROW	McD Douglas MD-82	S.A.S. *Ottar Viking*	
LN-ROX	McD Douglas MD-82	S.A.S. *Ulvrik Viking*	
LN-ROY	McD Douglas MD-82	S.A.S. *Spjute Viking*	
LN-ROZ	McD Douglas MD-87	S.A.S. *Slagfinn Viking*	
LN-RPA	Boeing 737-683	S.A.S. *Arnljot Viking*	
LN-RPB	Boeing 737-683	S.A.S. *Bure Viking*	
LN-RPE	Boeing 737-683	S.A.S. *Edla Viking*	
LN-RPF	Boeing 737-683	S.A.S. *Frede Viking*	
LN-RPG	Boeing 737-683	S.A.S. *Geirmund Viking*	
LN-RPH	Boeing 737-683	S.A.S. *Hamder Viking*	
LN-RPJ	Boeing 737-783	S.A.S. *Grimhild Viking*	
LN-RPK	Boeing 737-783	S.A.S. *Heimer Viking*	
LN-RPL	Boeing 737-883	S.A.S. *Svanevit Viking*	
LN-RPM	Boeing 737-883	S.A.S. *Frigg Viking*	
LN-RPN	Boeing 737-883	S.A.S. *Bergfora Viking*	
LN-RPS	Boeing 737-683	S.A.S. *Gautrek Viking*	
LN-RPT	Boeing 737-683	S.A.S. *Ellida Viking*	
LN-RPU	Boeing 737-683	S.A.S. *Ragna Viking*	
LN-RPW	Boeing 737-683	S.A.S. *Alvid Viking*	
LN-RPX	Boeing 737-683	S.A.S. *Nanna Viking*	
LN-RPY	Boeing 737-683	S.A.S. *Olof Viking*	
LN-RPZ	Boeing 737-683	S.A.S. *Bera Viking*	
LN-RRK	Boeing 737-883	S.A.S. *Gerud Viking*	
LN-RRL	Boeing 737-883	S.A.S. *Jarlabanke Viking*	
LN-RRM	Boeing 737-783	S.A.S. *Erland Viking*	
LN-RRN	Boeing 737-783	S.A.S. *Solveig Viking*	
LN-RRO	Boeing 737-683	S.A.S. *Bernt Viking*	
LN-RRP	Boeing 737-683	S.A.S. *Vilborg Viking*	
LN-RRR	Boeing 737-683	S.A.S. *Torbjorn Viking*	
LN-RRS	Boeing 737-883	S.A.S. *Ymir Viking*	
LN-RRT	Boeing 737-883	S.A.S. *Lodyn Viking*	
LN-RRU	Boeing 737-883	S.A.S. *Vingolf Viking*	
LN-RRW	Boeing 737-883	S.A.S. *Saga Viking*	
LN-RRX	Boeing 737-683	S.A.S. *Ragnfast Viking*	
LN-RRY	Boeing 737-683	S.A.S. *Signe Viking*	
LN-RRZ	Boeing 737-683	S.A.S. *Gisla Viking*	
LN-TUA	Boeing 737-705	Braathens *Ingeborg Eriksdatter*	

Notes	Reg.	Type	Owner or Operator
	LN-TUD	Boeing 737-705	Braathens *Margrete Skulesdatter*
	LN-TUF	Boeing 737-705	Braathens *Tyra Haraldsdatter*
	LN-TUH	Boeing 737-705	Braathens *Margrete Ingesdatter*
	LN-TUI	Boeing 737-705	Braathens *Kristin Knudsdatter*
	LN-TUJ	Boeing 737-705	Braathens *Eirik Blodoks*
	LN-TUK	Boeing 737-705	Braathens *Inge Bardsson*
	LN-TUL	Boeing 737-705	Braathens *Haakon IV Haakonson*
	LN-TUM	Boeing 737-705	Braathens *Oystein Magnusson*
	LN-WDA	D.H.C.8Q-402 Dash Eight	Wideroe's Flyveselskap
	LN-WDB	D.H.C.8Q-402 Dash Eight	Wideroe's Flyveselskap *Stavanger*
	LN-WDC	D.H.C.8Q-402 Dash Eight	Wideroe's Flyveselskap *Trondheim*
	LN-WFA	D.H.C.8-311 Dash Eight	Wideroe's Flyveselskap
	LN-WFB	D.H.C.8-311 Dash Eight	Wideroe's Flyveselskap
	LN-WFC	D.H.C.8-311 Dash Eight	Wideroe's Flyveselskap
	LN-WFE	D.H.C.8Q-311 Dash Eight	Wideroe's Flyveselskap *Sandefjord*
	LN-WFH	D.H.C.8-311 Dash Eight	Wideroe's Flyveselskap
	LN-WFO	D.H.C.8Q-311 Dash Eight	Wideroe's Flyveselskap
	LN-WFP	D.H.C.8Q-311 Dash Eight	Wideroe's Flyveselskap
	LN-WFR	D.H.C.8-314 Dash Eight	Wideroe's Flyveselskap
	LN-WFS	D.H.C.8Q-311 Dash Eight	Wideroe's Flyveselskap
	LN-WFT	D.H.C.8Q-311 Dash Eight	Wideroe's Flyveselskap

LV (Argentina)

Note: Aerolineas Argentinas serves London-Gatwick with a connecting service from Madrid operated by B.737s of Air Plus Comet

LX (Luxembourg)

	LX-FCV	Boeing 747-4R7F (SCD)	Cargolux *City of Luxembourg*
	LX-GCV	Boeing 747-4R7F (SCD)	Cargolux *City of Esch/Alzette*
	LX-ICV	Boeing 747-428F (SCD)	Cargolux *City of Ettelbruck*
	LX-KCV	Boeing 747-4R7F (SCD)	Cargolux *City of Dudelange*
	LX-LCV	Boeing 747-4R7F (SCD)	Cargolux *City of Grevenmacher*
	LX-LGC	Fokker 50	Luxair *Prince Guillaume*
	LX-LGD	Fokker 50	Luxair *Prince Felix*
	LX-LGE	Fokker 50	Luxair *Prince Louis*
	LX-LGI	Embraer RJ145LU	Luxair
	LX-LGJ	Embraer RJ145LU	Luxair
	LX-LGK	Embraer RJ145LR	Luxair
	LX-LGN	Boeing 737-59D	Luxair
	LX-LGO	Boeing 737-5C9	Luxair *Chateau de Clervaux*
	LX-LGP	Boeing 737-5C9	Luxair *Chateau de Bourglinster*
	LX-LGQ	Boeing 737-7C9	Luxair *Chateau de Burg*
	LX-LGR	Boeing 737-7C9	Luxair *Chateau de Fischbach*
	LX-LGS	Boeing 737-7C9	Luxair
	LX-LGU	Embraer RJ145EP	Luxair *Prince Sebastian*
	LX-LGV	Embraer RJ145LU	Luxair
	LX-LGW	Embraer RJ145LU	Luxair
	LX-LGX	Embraer RJ145LU	Luxair
	LX-LGY	Embraer RJ145LU	Luxair
	LX-LGZ	Embraer RJ145LU	Luxair
	LX-MCV	Boeing 747-4R7F (SCD)	Cargolux *City of Echternach*
	LX-NCV	Boeing 747-4R7F (SCD)	Cargolux *City of Vianden*
	LX-OCV	Boeing 747-4R7F (SCD)	Cargolux *City of Differdange*
	LX-PCV	Boeing 747-4R7F (SCD)	Cargolux *City of Diekirch*
	LX-RCV	Boeing 747-4R7F (SCD)	Cargolux *City of Schengen*
	LX-SCV	Boeing 747-4R7F (SCD)	Cargolux *City of Niederanven*
	LX-TCV	Boeing 747-4R7F (SCD)	Cargolux *City of Sandweiler*

LY (Lithuania)

	LY-AGQ	Boeing 737-524	Lithuanian Airlines
	LY-AGT	Tupolev Tu-204C	Aviapaslauga
	LY-AGZ	Boeing 737-524	Lithuanian Airlines
	LY-AZY	Boeing 737-548	Lithuanian Airlines
	LY-BSD	Boeing 737-2T4	Lithuanian Airlines *Steponas Darius*
	LY-BSG	Boeing 737-2T2	Lithuanian Airlines *Stasys Girenas*

Reg.	Type	Owner or Operator	Notes

LZ (Bulgaria)

LZ-BHA	Airbus A.320-211	Balkan Holidays	
LZ-BHB	Airbus A.320-212	Balkan Holidays	
LZ-BHC	Airbus A.320-212	Balkan Holidays	
LZ-BOG	Boeing 737-330	Bulgaria Air	
LZ-BOH	Boeing 737-330	Bulgaria Air	
LZ-BOI	Boeing 737-530	Bulgaria Air	
LZ-BOJ	Boeing 737-3L9	Bulgaria Air	
LZ-BOK	Boeing 737-3M8	Bulgaria Air	
LZ-BOL	Boeing 737-3M8	Bulgaria Air	
LZ-BOM	Boeing 737-31S	Bulgaria Air	
LZ-BON	Boeing 737-31S	Bulgaria Air	
LZ-HMF	Tupolev Tu-154M	Hemus Air	
LZ-HMH	Tupolev Tu-154M	Hemus Air	
LZ-HMI	Tupolev Tu-154M	Balkan Holidays	
LZ-HMN	Tupolev Tu-154M	Balkan Holidays	
LZ-HMQ	Tupolev Tu-154M	Balkan Holidays	
LZ-HMS	Tupolev Tu-154M	Hemus Air	
LZ-HMW	Tupolev Tu-154M	Balkan Holidays	
LZ-HMY	Tupolev Tu-154M	Hemus Air	
LZ-LCA	Tupolev Tu-154M	Bulgarian Air Charter	
LZ-LCQ	Tupolev Tu-154M	Bulgarian Air Charter	
LZ-LCS	Tupolev Tu-154M	Bulgarian Air Charter	
LZ-LCT	Tupolev Tu-154M	Bulgarian Air Charter	
LZ-LCV	Tupolev Tu-154M	Bulgarian Air Charter	
LZ-LCX	Tupolev Tu-154M	Bulgarian Air Charter	
LZ-LDK	McD Douglas MD-82	Bulgarian Air Charter/Albanian Airlines	
LZ-LDX	McD Douglas MD-83	Bulgarian Air Charter	
LZ-LDZ	McD Douglas MD-83	Bulgarian Air Charter	

N (USA)

N100UN	Boeing 737-7K9	Transaero *Bavaria*	
N101UN	Boeing 737-7K9	Transaero *Seattle*	
N104UA	Boeing 747-422	United Airlines	
N105EV	McD Douglas MD-11F	EVA Air Cargo	
N105UA	Boeing 747-451	United Airlines	
N106UA	Boeing 747-451	United Airlines	
N107UA	Boeing 747-422	United Airlines *William A Patterson*	
N116UA	Boeing 747-422	United Airlines	
N117UA	Boeing 747-422	United Airlines	
N118UA	Boeing 747-422	United Airlines	
N119UA	Boeing 747-422	United Airlines	
N120UA	Boeing 747-422	United Airlines	
N121UA	Boeing 747-422	United Airlines	
N122UA	Boeing 747-422	United Airlines	
N127UA	Boeing 747-422	United Airlines	
N128UA	Boeing 747-422	United Airlines	
N154DL	Boeing 767-3P6ER	Delta Air Lines	
N155DL	Boeing 767-3P6ER	Delta Air Lines	
N156DL	Boeing 767-3P6ER	Delta Air Lines	
N161AT	L.1011-385 TriStar 500	ATA Airlines	
N162AT	L.1011-385 TriStar 500	ATA Airlines	
N163AT	L.1011-385 TriStar 500	ATA Airlines	
N164AT	L.1011-385 TriStar 500	ATA Airlines	
N169DZ	Boeing 767-332ER	Delta Air Lines	
N171DN	Boeing 767-332ER	Delta Air Lines	
N171UA	Boeing 747-422	United Airlines *Spirit of Seattle II*	
N172DN	Boeing 767-332ER	Delta Air Lines	
N172UA	Boeing 747-422	United Airlines	
N173DN	Boeing 767-332ER	Delta Air Lines	
N173UA	Boeing 747-422	United Airlines	
N174DN	Boeing 767-332ER	Delta Air Lines	
N174UA	Boeing 747-422	United Airlines	
N175DN	Boeing 767-332ER	Delta Air Lines	
N175UA	Boeing 747-422	United Airlines	
N176DN	Boeing 767-332ER	Delta Air Lines	
N176UA	Boeing 747-422	United Airlines	
N177DN	Boeing 767-332ER	Delta Air Lines	

Reg.	Type	Owner or Operator
N177UA	Boeing 747-422	United Airlines
N178DN	Boeing 767-332ER	Delta Air Lines
N178UA	Boeing 747-422	United Airlines
N179DN	Boeing 767-332ER	Delta Air Lines
N179UA	Boeing 747-422	United Airlines
N180DN	Boeing 767-332ER	Delta Air Lines
N180UA	Boeing 747-422	United Airlines
N181DN	Boeing 767-332ER	Delta Air Lines
N181UA	Boeing 747-422	United Airlines
N182DN	Boeing 767-332ER	Delta Air Lines
N182UA	Boeing 747-422	United Airlines
N183AN	Boeing 757-223ET	American Airlines
N183DN	Boeing 767-332ER	Delta Air Lines
N183UA	Boeing 747-422	United Airlines
N184AN	Boeing 757-223ET	American Airlines
N184DN	Boeing 767-332ER	Delta Air Lines
N184UA	Boeing 747-422	United Airlines
N185AN	Boeing 757-223ET	American Airlines
N185DN	Boeing 767-332ER	Delta Air Lines
N185UA	Boeing 747-422	United Airlines
N186AN	Boeing 757-223ET	American Airlines
N186DN	Boeing 767-332ER	Delta Air Lines
N187AN	Boeing 757-223ET	American Airlines
N187DN	Boeing 767-332ER	Delta Air Lines
N187UA	Boeing 747-422	United Airlines
N188AN	Boeing 757-223ET	American Airlines
N188DN	Boeing 767-332ER	Delta Air Lines
N189AN	Boeing 757-223ET	American Airlines
N189DN	Boeing 767-332ER	Delta Air Lines
N190AA	Boeing 757-223ET	American Airlines
N190DN	Boeing 767-332ER	Delta Air Lines
N190UA	Boeing 747-422	United Airlines
N191AN	Boeing 757-223ET	American Airlines
N191DN	Boeing 767-332ER	Delta Air Lines
N192AN	Boeing 757-223ET	American Airlines
N192DN	Boeing 767-332ER	Delta Air Lines
N192UA	Boeing 747-422	United Airlines
N193AN	Boeing 757-223ET	American Airlines
N193DN	Boeing 767-332ER	Delta Air Lines
N193UA	Boeing 747-422	United Airlines
N194AA	Boeing 757-223ET	American Airlines
N194AT	L.1011-385 TriStar 100	ATA Airlines
N194DN	Boeing 767-332ER	Delta Air Lines
N194UA	Boeing 747-422	United Airlines
N195DN	Boeing 767-332ER	Delta Air Lines
N195UA	Boeing 747-422	United Airlines
N196DN	Boeing 767-332ER	Delta Air Lines
N196UA	Boeing 747-422	United Airlines
N197DN	Boeing 767-332ER	Delta Air Lines
N197UA	Boeing 747-422	United Airlines
N198DN	Boeing 767-332ER	Delta Air Lines
N198UA	Boeing 747-422	United Airlines
N199DN	Boeing 767-332ER	Delta Air Lines
N199UA	Boeing 747-422	United Airlines
N204UA	Boeing 777-222ER	United Airlines
N206UA	Boeing 777-222ER	United Airlines
N209UA	Boeing 777-222ER	United Airlines
N210UA	Boeing 777-222ER	United Airlines
N211NW	Douglas DC-10-30	Northwest Airlines
N211UA	Boeing 777-222ER	United Airlines
N212UA	Boeing 777-222ER	United Airlines
N213UA	Boeing 777-222ER	United Airlines
N214UA	Boeing 777-222ER	United Airlines
N215UA	Boeing 777-222ER	United Airlines
N216UA	Boeing 777-222ER	United Airlines
N217UA	Boeing 777-222ER	United Airlines
N218UA	Boeing 777-222ER	United Airlines
N219UA	Boeing 777-222ER	United Airlines
N220UA	Boeing 777-222ER	United Airlines
N221NW	Douglas DC-10-30	Northwest Airlines
N221UA	Boeing 777-222ER	United Airlines
N222UA	Boeing 777-222ER	United Airlines

OVERSEAS AIRLINERS

Reg.	Type	Owner or Operator	Notes
N223NW	Douglas DC-10-30	Northwest Airlines	
N223UA	Boeing 777-222ER	United Airlines	
N224NW	Douglas DC-10-30	Northwest Airlines	
N224UA	Boeing 777-222ER	United Airlines	
N225NW	Douglas DC-10-30	Northwest Airlines	
N225UA	Boeing 777-222ER	United Airlines	
N226NW	Douglas DC-10-30	Northwest Airlines	
N226UA	Boeing 777-222ER	United Airlines	
N227NW	Douglas DC-10-30	Northwest Airlines	
N227UA	Boeing 777-222ER	United Airlines	
N228NW	Douglas DC-10-30	Northwest Airlines	
N228UA	Boeing 777-222ER	United Airlines	
N229NW	Douglas DC-10-30	Northwest Airlines	
N229UA	Boeing 777-222ER	United Airlines	
N230NW	Douglas DC-10-30	Northwest Airlines	
N232NW	Douglas DC-10-30	Northwest Airlines	
N234NW	Douglas DC-10-30	Northwest Airlines	
N235NW	Douglas DC-10-30	Northwest Airlines	
N236NW	Douglas DC-10-30	Northwest Airlines	
N237NW	Douglas DC-10-30	Northwest Airlines	
N238NW	Douglas DC-10-30	Northwest Airlines	
N239NW	Douglas DC-10-30	Northwest Airlines	
N240NW	Douglas DC-10-30	Northwest Airlines	
N241NW	Douglas DC-10-30	Northwest Airlines	
N242NW	Douglas DC-10-30	Northwest Airlines	
N243NW	Douglas DC-10-30	Northwest Airlines	
N244NW	Douglas DC-10-30	Northwest Airlines	
N250UP	McD Douglas MD-11F	United Parcel Service	
N251RY	Boeing 737-4Y0	Futura International Airways	
N251UP	McD Douglas MD-11F	United Parcel Service	
N252UP	McD Douglas MD-11F	United Parcel Service	
N253UP	McD Douglas MD-11F	United Parcel Service	
N254RY	Boeing 737-4Y0	Futura International Airways	
N254UP	McD Douglas MD-11F	United Parcel Service	
N255RY	Boeing 737-86N	Futura International Airways	
N270UP	McD Douglas MD-11F	United Parcel Service	
N271UP	McD Douglas MD-11F	United Parcel Service	
N271WA	McD Douglas MD-11	World Airways	
N272UP	McD Douglas MD-11F	United Parcel Service	
N272WA	McD Douglas MD-11	World Airways	
N273UP	McD Douglas MD-11F	United Parcel Service	
N273WA	McD Douglas MD-11	World Airways	
N274UP	McD Douglas MD-11F	United Parcel Service	
N274WA	McD Douglas MD-11F	World Airways	
N275UP	McD Douglas MD-11F	United Parcel Service	
N275WA	McD Douglas MD-11CF	World Airways	
N276UP	McD Douglas MD-11F	United Parcel Service	
N276WA	McD Douglas MD-11CF	World Airways	
N277UP	McD Douglas MD-11F	United Parcel Service	
N277WA	McD Douglas MD-11	World Airways	
N279AX	Douglas DC-10-30F	Centurion Air Cargo *Wings of Miami*	
N279WA	McD Douglas MD-11	World Airways	
N301UP	Boeing 767-34AFER	United Parcel Service	
N302UP	Boeing 767-34AFER	United Parcel Service	
N303UP	Boeing 767-34AFER	United Parcel Service	
N303WL	Douglas DC-10-30	World Airways	
N304UP	Boeing 767-34AFER	United Parcel Service	
N304WL	Douglas DC-10-30	World Airways	
N305UP	Boeing 767-34AFER	United Parcel Service	
N306UP	Boeing 767-34AFER	United Parcel Service	
N307UP	Boeing 767-34AFER	United Parcel Service	
N308UP	Boeing 767-34AFER	United Parcel Service	
N309UP	Boeing 767-34AFER	United Parcel Service	
N310UP	Boeing 767-34AFER	United Parcel Service	
N311UP	Boeing 767-34AFER	United Parcel Service	
N312LA	Boeing 767-316F	LAN-Chile	
N312UP	Boeing 767-34AFER	United Parcel Service	
N313UP	Boeing 767-34AFER	United Parcel Service	
N314UP	Boeing 767-34AFER	United Parcel Service	
N315UP	Boeing 767-34AFER	United Parcel Service	
N316UP	Boeing 767-34AFER	United Parcel Service	
N317UP	Boeing 767-34AFER	United Parcel Service	

Reg.	Type	Owner or Operator
N318UP	Boeing 767-34AFER	United Parcel Service
N319UP	Boeing 767-34AFER	United Parcel Service
N320UP	Boeing 767-34AFER	United Parcel Service
N322UP	Boeing 767-34AFER	United Parcel Service
N323MC	Boeing 747-2D7B	Atlas Air
N323UP	Boeing 767-34AFER	United Parcel Service
N324UP	Boeing 767-34AFER	United Parcel Service
N325UP	Boeing 767-34AFER	United Parcel Service
N326UP	Boeing 767-34AFER	United Parcel Service
N327UP	Boeing 767-34AFER	United Parcel Service
N328UP	Boeing 767-34AFER	United Parcel Service
N329UP	Boeing 767-34AER	United Parcel Service
N330UP	Boeing 767-34AER	United Parcel Service
N331UP	Boeing 767-34AER	United Parcel Service
N332UP	Boeing 767-34AER	United Parcel Service
N334UP	Boeing 767-34AER	United Parcel Service
N342AN	Boeing 767-223ER	American Airlines
N343AN	Boeing 767-223ER	American Airlines
N344AN	Boeing 767-223ER	American Airlines
N345AN	Boeing 767-223ER	American Airlines
N346AN	Boeing 767-223ER	American Airlines
N347AN	Boeing 767-223ER	American Airlines
N348AN	Boeing 767-223ER	American Airlines
N349AN	Boeing 767-223ER	American Airlines
N350AN	Boeing 767-223ER	American Airlines
N351AA	Boeing 767-323ER	American Airlines
N352AA	Boeing 767-323ER	American Airlines
N352WL	Douglas DC-10-30	World Airways
N353AA	Boeing 767-323ER	American Airlines
N353WL	Douglas DC-10-30	World Airways
N354AA	Boeing 767-323ER	American Airlines
N355AA	Boeing 767-323ER	American Airlines
N355MC	Boeing 747-341F	Polar Air Cargo
N357AA	Boeing 767-323ER	American Airlines
N358AA	Boeing 767-323ER	American Airlines
N359AA	Boeing 767-323ER	American Airlines
N360AA	Boeing 767-323ER	American Airlines
N361AA	Boeing 767-323ER	American Airlines
N362AA	Boeing 767-323ER	American Airlines
N363AA	Boeing 767-323ER	American Airlines
N366AA	Boeing 767-323ER	American Airlines
N368AA	Boeing 767-323ER	American Airlines
N369AA	Boeing 767-323ER	American Airlines
N369AX	Boeing 757-28A	Omni Air International
N370AA	Boeing 767-323ER	American Airlines
N371AA	Boeing 767-323ER	American Airlines
N372AA	Boeing 767-323ER	American Airlines
N373AA	Boeing 767-323ER	American Airlines
N374AA	Boeing 767-323ER	American Airlines
N376AN	Boeing 767-323ER	American Airlines
N377AN	Boeing 767-323ER	American Airlines
N378AN	Boeing 767-323ER	American Airlines
N379AA	Boeing 767-323ER	American Airlines
N380AN	Boeing 767-323ER	American Airlines
N381AN	Boeing 767-323ER	American Airlines
N382AN	Boeing 767-323ER	American Airlines
N383AN	Boeing 767-323ER	American Airlines
N384AA	Boeing 767-323ER	American Airlines
N385AM	Boeing 767-323ER	American Airlines
N386AA	Boeing 767-323ER	American Airlines
N387AM	Boeing 767-323ER	American Airlines
N388AA	Boeing 767-323ER	American Airlines
N389AA	Boeing 767-323ER	American Airlines
N390AA	Boeing 767-323ER	American Airlines
N391AA	Boeing 767-323ER	American Airlines
N392AN	Boeing 767-323ER	American Airlines
N393AN	Boeing 767-323ER	American Airlines
N394AN	Boeing 767-323ER	American Airlines
N394DL	Boeing 767-324ER	Delta Air Lines
N395AN	Boeing 767-323ER	American Airlines
N396AN	Boeing 767-323ER	American Airlines
N397AN	Boeing 767-323ER	American Airlines

Reg.	Type	Owner or Operator	Notes
N398AN	Boeing 767-323ER	American Airlines	
N399AN	Boeing 767-323ER	American Airlines	
N401JR	Douglas DC-10-30F	DAS Air Cargo	
N408MC	Boeing 747-47UF	Atlas Air/Emirates SkyCargo	
N409MC	Boeing 747-47UF	Atlas Air	
N412MC	Boeing 747-47UF	Atlas Air	
N415MC	Boeing 747-47UF	Atlas Air/ Emirates SkyCargo	
N416MC	Boeing 747-47UF	Atlas Air/EVA Air Cargo	
N418MC	Boeing 747-47UF	Atlas Air	
N450PA	Boeing 747-46NF	Polar Air Cargo	
N451PA	Boeing 747-46NF	Polar Air Cargo	
N452PA	Boeing 747-46NF	Polar Air Cargo	
N453PA	Boeing 747-46NF	Polar Air Cargo	
N454PA	Boeing 747-46NF	Polar Air Cargo	
N459AX	Boeing 757-28A	Omni Air International	
N471EV	Boeing 747-273C	Evergreen International Airlines	
N478EV	Boeing SR-46 (SCD)	Evergreen International Airlines	
N479EV	Boeing 747-132 (SCD)	Evergreen International Airlines	
N480EV	Boeing 747-121F	Evergreen International Airlines	
N481EV	Boeing 747-132 (SCD)	Evergreen International Airlines	
N482EV	Boeing 747-212B (SCD)	Evergreen International Airlines	
N485EV	Boeing 747-212B (SCD)	Evergreen International Airlines	
N486EV	Boeing 747-212B (SCD)	Evergreen International Airlines	
N492MC	Boeing 747-47UF	Atlas Air *Spirit of Panalpina*	
N493MC	Boeing 747-47UF	Atlas Air	
N496MC	Boeing 747-47UF	Polar Air Cargo	
N497MC	Boeing 747-47UF	Emirates SkyCargo	
N498MC	Boeing 747-47UF	Emirates SkyCargo	
N499MC	Boeing 747-47UF	Atlas Air	
N505MC	Boeing 747-2D3B (SCD)	Atlas Air	
N506MC	Boeing 747-2D3B (SCD)	Atlas Air	
N508MC	Boeing 747-230B (SCD)	Atlas Air	
N509MC	Boeing 747-230B (SCD)	Polar Air Cargo	
N512MC	Boeing 747-230B (SCD)	Atlas Air	
N513AT	Boeing 757-28A	ATA Airlines	
N514AT	Boeing 757-23N	ATA Airlines	
N515AT	Boeing 757-23N	ATA Airlines	
N516AT	Boeing 757-23N	ATA Airlines	
N516MC	Boeing 747-243F (SCD)	Atlas Air	
N517AT	Boeing 757-23N	ATA Airlines	
N517MC	Boeing 747-243F (SCD)	Atlas Air	
N518AT	Boeing 767-23N	ATA Airlines	
N518MC	Boeing 747-243F (SCD)	Atlas Air	
N519AT	Boeing 757-23N	ATA Airlines	
N520AT	Boeing 757-23N	ATA Airlines	
N520UP	Boeing 747-212B (SF)	United Parcel Service	
N521AT	Boeing 757-28A	ATA Airlines	
N521UP	Boeing 747-212B (SF)	United Parcel Service	
N522AT	Boeing 757-23N	ATA Airlines	
N522UP	Boeing 747-212B (SF)	United Parcel Service	
N523AT	Boeing 757-23N	ATA Airlines	
N523UP	Boeing 747-283B (SF)	United Parcel Service	
N523MC	Boeing 747-2D7BF	Atlas Air	
N524AT	Boeing 757-23N	ATA Airlines	
N524MC	Boeing 747-2D7BF	Atlas Air	
N525AT	Boeing 757-23N	ATA Airlines	
N526AT	Boeing 757-23N	ATA Airlines	
N526MC	Boeing 747-2D7BF	Atlas Air	
N526NA	Boeing 757-236	Ryan International	
N526UP	Boeing 747-212B (SF)	United Parcel Service	
N527AT	Boeing 757-23N	ATA Airlines	
N527MC	Boeing 747-2D7BF	Atlas Air	
N528AT	Boeing 757-23N	ATA Airlines	
N528MC	Boeing 747-2D7BF	Atlas Air/British Airways	
N534MC	Boeing 747-2F6B	Polar Air Cargo	
N536MC	Boeing 747-228F	Atlas Air/Alitalia Cargo System	
N537MC	Boeing 747-271C (SCD)	Atlas Air	
N540MC	Boeing 747-243B (SCD)	Atlas Air	
N549AX	Boeing 757-23A	Omni Air International	
N550TZ	Boeing 757-33N	ATA Airlines	
N551TZ	Boeing 757-33N	ATA Airlines	
N552TZ	Boeing 757-33N	ATA Airlines	

Notes	Reg.	Type	Owner or Operator
	N553TZ	Boeing 757-33N	ATA Airlines
	N554TZ	Boeing 757-33N	ATA Airlines
	N555TZ	Boeing 757-33N	ATA Airlines
	N556TZ	Boeing 757-33N	ATA Airlines
	N557TZ	Boeing 757-33N	ATA Airlines
	N558TZ	Boeing 757-33N	ATA Airlines
	N559TZ	Boeing 757-33N	ATA Airlines
	N560TZ	Boeing 757-33N	ATA Airlines
	N561TZ	Boeing 757-33N	ATA Airlines
	N578FE	McD Douglas MD-11F	Federal Express *Stephen*
	N579FE	McD Douglas MD-11F	Federal Express *Nash*
	N580FE	McD Douglas MD-11F	Federal Express *Ashton*
	N582FE	McD Douglas MD-11F	Federal Express *Jamie*
	N583FE	McD Douglas MD-11F	Federal Express *Nancy*
	N584FE	McD Douglas MD-11F	Federal Express *Jeffrey Wellington*
	N585FE	McD Douglas MD-11F	Federal Express *Katherine*
	N586FE	McD Douglas MD-11F	Federal Express *Dylan*
	N587FE	McD Douglas MD-11F	Federal Express *Jeanna*
	N588FE	McD Douglas MD-11F	Federal Express *Kendra*
	N589FE	McD Douglas MD-11F	Federal Express
	N590FE	McD Douglas MD-11F	Federal Express
	N591FE	McD Douglas MD-11F	Federal Express *Giovanni*
	N592FE	McD Douglas MD-11F	Federal Express *Joshua*
	N593FE	McD Douglas MD-11F	Federal Express *Harrison*
	N594FE	McD Douglas MD-11F	Federal Express
	N595FE	McD Douglas MD-11F	Federal Express *Avery*
	N596FE	McD Douglas MD-11F	Federal Express
	N597FE	McD Douglas MD-11F	Federal Express
	N598FE	McD Douglas MD-11F	Federal Express
	N599FE	McD Douglas MD-11F	Federal Express *Mariana*
	N600GC	Douglas DC-10-30F	Gemini Air Cargo *Kristiana*
	N601FE	McD Douglas MD-11F	Federal Express *Jim Riedmeyer*
	N601GC	Douglas DC-10-30F	Gemini Air Cargo *Molly*
	N602AL	Douglas DC-8-73CF	Air Transport International
	N602FE	McD Douglas MD-11F	Federal Express *Malcolm Baldridge 1990*
	N602GC	Douglas DC-10-30F	Gemini Air Cargo *Helai*
	N603AL	Douglas DC-8-73AF	Air Transport International
	N603FE	McD Douglas MD-11F	Federal Express *Elizabeth*
	N604FE	McD Douglas MD-11F	Federal Express *Hollis*
	N604GC	Douglas DC-10-30F	Gemini Air Cargo *Daman*
	N605AL	Douglas DC-8-73CF	Air Transport International
	N605FE	McD Douglas MD-11F	Federal Express *April Star*
	N605GC	Douglas DC-10-30F	Gemini Air Cargo *Barbara*
	N606AL	Douglas DC-8-73AF	Air Transport International
	N606FE	McD Douglas MD-11F	Federal Express *Charles & Theresa*
	N606GC	Douglas DC-10-30F	Gemini Air Cargo *Laine*
	N607FE	McD Douglas MD-11F	Federal Express *Christina*
	N608AA	Boeing 757-223ET	American Airlines
	N608FE	McD Douglas MD-11F	Federal Express *Karen*
	N609AA	Boeing 757-223ET	American Airlines
	N609FE	McD Douglas MD-11F	Federal Express *Scott*
	N610FE	McD Douglas MD-11F	Federal Express *Marisa*
	N612FE	McD Douglas MD-11F	Federal Express *Alyssa*
	N612GC	Douglas DC-10-30F	Centurion Air Cargo
	N613FE	McD Douglas MD-11F	Federal Express *Krista*
	N613US	Boeing 747-251B	Northwest Airlines
	N614FE	McD Douglas MD-11F	Federal Express *Christy Allison*
	N615FE	McD Douglas MD-11F	Federal Express *Max*
	N616FE	McD Douglas MD-11F	Federal Express *Shanita*
	N616US	Boeing 747-251F (SCD)	Northwest Airlines
	N617FE	McD Douglas MD-11F	Federal Express *Travis*
	N617US	Boeing 747-251F (SCD)	Northwest Airlines
	N618FE	McD Douglas MD-11F	Federal Express *Justin*
	N618US	Boeing 747-251F (SCD)	Northwest Airlines
	N619FE	McD Douglas MD-11F	Federal Express *Lyndon*
	N619US	Boeing 747-251F (SCD)	Northwest Airlines
	N620FE	McD Douglas MD-11F	Federal Express
	N621FE	McD Douglas MD-11F	Federal Express
	N623FE	McD Douglas MD-11F	Federal Express *Meghan*
	N624US	Boeing 747-251B	Northwest Airlines
	N625FE	McD Douglas MD-11F	Federal Express
	N627FE	McD Douglas MD-11F	Federal Express

Reg.	Type	Owner or Operator	Notes
N627US	Boeing 747-251B	Northwest Airlines	
N629US	Boeing 747-251F (SCD)	Northwest Airlines	
N630AX	Douglas DC-10-30	Omni Air International	
N630US	Boeing 747-2J9F (SCD)	Northwest Airlines	
N631US	Boeing 747-251B	Northwest Airlines	
N632US	Boeing 747-251B	Northwest Airlines	
N636FE	Boeing 747-245F (SCD)	Federal Express	
N637US	Boeing 747-251B	Northwest Airlines	
N638US	Boeing 747-251B	Northwest Airlines	
N639FE	Boeing 747-2R7F (SCD)	Federal Express	
N639US	Boeing 747-251F (SCD)	Northwest Airlines	
N640US	Boeing 747-251F (SCD)	Northwest Airlines	
N641UA	Boeing 767-322ER	United Airlines	
N642UA	Boeing 767-322ER	United Airlines	
N643NW	Boeing 747-249F (SCD)	Northwest Airlines	
N643UA	Boeing 767-322ER	United Airlines	
N644NW	Boeing 747-212F (SCD)	Northwest Airlines	
N644UA	Boeing 767-322ER	United Airlines	
N645NW	Boeing 747-222SF	Northwest Cargo	
N645UA	Boeing 767-322ER	United Airlines	
N645US	Boeing 767-201ER	US Airways	
N646NW	Boeing 747-222SF	Northwest Cargo	
N646UA	Boeing 767-322ER	United Airlines	
N646US	Boeing 767-201ER	US Airways	
N647UA	Boeing 767-322ER	United Airlines	
N648UA	Boeing 767-322ER	United Airlines	
N648US	Boeing 767-201ER	US Airways	
N649UA	Boeing 767-322ER	United Airlines	
N649US	Boeing 767-201ER	US Airways	
N650UA	Boeing 767-322ER	United Airlines	
N650US	Boeing 767-201ER	US Airways	
N651UA	Boeing 767-322ER	United Airlines	
N651US	Boeing 767-2B7ER	US Airways	
N652UA	Boeing 767-322ER	United Airlines	
N652US	Boeing 767-2B7ER	US Airways	
N653UA	Boeing 767-322ER	United Airlines	
N653US	Boeing 767-2B7ER	US Airways	
N654UA	Boeing 767-322ER	United Airlines	
N655UA	Boeing 767-322ER	United Airlines	
N655US	Boeing 767-2B7ER	US Airways	
N656UA	Boeing 767-322ER	United Airlines	
N656US	Boeing 767-2B7ER	US Airways	
N657UA	Boeing 767-322ER	United Airlines	
N658UA	Boeing 767-322ER	United Airlines	
N659UA	Boeing 767-322ER	United Airlines	
N660UA	Boeing 767-322ER	United Airlines	
N661UA	Boeing 767-322ER	United Airlines	
N661US	Boeing 747-451	Northwest Airlines	
N662UA	Boeing 767-322ER	United Airlines	
N662US	Boeing 747-451	Northwest Airlines	
N663UA	Boeing 767-322ER	United Airlines	
N663US	Boeing 747-451	Northwest Airlines	
N664US	Boeing 747-451	Northwest Airlines *The Spirit of Beijing*	
N665US	Boeing 747-451	Northwest Airlines	
N666US	Boeing 747-451	Northwest Airlines	
N667US	Boeing 747-451	Northwest Airlines	
N668US	Boeing 747-451	Northwest Airlines	
N669US	Boeing 747-451	Northwest Airlines	
N670US	Boeing 747-451	Northwest Airlines *The Alliance-Spirit*	
N670UW	Airbus A.330-323X	US Airways	
N671US	Boeing 747-451	Northwest Airlines *City of Detroit*	
N671UW	Airbus A.330-323X	US Airways	
N672UP	Boeing 747-123F (SCD)	United Parcel Service	
N672US	Boeing 747-451	Northwest Airlines *Spirit of Asia*	
N672UW	Airbus A.330-323X	US Airways	
N673US	Boeing 747-451	Northwest Airlines *Spirit of Tokyo*	
N673UW	Airbus A.330-323X	US Airways	
N674US	Boeing 747-451	Northwest Airlines *City of Shanghai*	
N674UW	Airbus A.330-323X	US Airways	
N675NW	Boeing 747-451	Northwest Airlines *Spirit of the Northwest People*	
N675UP	Boeing 747-123F (SCD)	United Parcel Service	
N675US	Airbus A.330-323X	US Airways	

Notes	Reg.	Type	Owner or Operator
	N676NW	Boeing 747-451	Northwest Airlines
	N676UP	Boeing 747-123F (SCD)	United Parcel Service
	N676UW	Airbus A.330-323X	US Airways
	N677UP	Boeing 747-123F (SCD)	United Parcel Service
	N677UW	Airbus A.330-323X	US Airways
	N678US	Airbus A.330-323X	US Airways
	N680UP	Boeing 747SR-46F (SF)	United Parcel Service
	N681UP	Boeing 747-121F (SF)	United Parcel Service
	N682UP	Boeing 747-121F (SF)	United Parcel Service
	N683UP	Boeing 747-121F (SF)	United Parcel Service
	N687AA	Boeing 757-223ET	American Airlines
	N688AA	Boeing 757-223ET	American Airlines
	N689AA	Boeing 757-223ET	American Airlines
	N690AA	Boeing 757-223ET	American Airlines
	N691AA	Boeing 757-223ET	American Airlines
	N692AA	Boeing 757-223ET	American Airlines
	N701GC	McD Douglas MD-11F	Gemini Air Cargo
	N702CK	Boeing 747-146 (SF)	Kalitta Air
	N702GC	McD Douglas MD-11F	Gemini Air Cargo/Air Canada
	N703GC	McD Douglas MD-11F	Gemini Air Cargo
	N704CK	Boeing 747-209F (SCD)	Kalitta Air
	N705CK	Boeing 747-246F (SCD)	Kalitta Air
	N705GC	McD Douglas MD-11F	Gemini Air Cargo
	N706CK	Boeing 747-249F (SCD)	Kalitta Air
	N709CK	Boeing 747-132 (SF)	Kalitta Air
	N710CK	Boeing 747-2B4B (SF)	Kalitta Air
	N712CK	Boeing 747-122 (SF)	Kalitta Air
	N713CK	Boeing 747-2B4B (SF)	Kalitta Air
	N714CK	Boeing 747-209B (SF)	Kalitta Air
	N715CK	Boeing 747-209B (SF)	Kalitta Air
	N716CK	Boeing 747-122 (SF)	Kalitta Air
	N717CK	Boeing 747-123 (SF)	Kalitta Air
	N720AX	Douglas DC-10-30	Omni Air International
	N727CK	Boeing 747-246B	Kalitta Air
	N740SA	Boeing 747-230B (SCD)	Southern Air
	N746SA	Boeing 747-206F (SCD)	Southern Air
	N750AN	Boeing 777-223ER	American Airlines
	N750NA	Boeing 757-28A	North American Airlines
	N751AN	Boeing 777-223ER	American Airlines
	N751SA	Boeing 747-228F	Southern Air
	N752AN	Boeing 777-223ER	American Airlines
	N752NA	Boeing 757-28A	North American Airlines *Alisa Ferrera*
	N753NA	Boeing 777-223ER	American Airlines
	N753NA	Boeing 757-28A	North American Airlines
	N754AN	Boeing 777-223ER	American Airlines
	N754NA	Boeing 757-28A	North American Airlines
	N755AN	Boeing 777-223ER	American Airlines
	N755NA	Boeing 757-28A	North American Airlines *John Plueger*
	N756AM	Boeing 777-223ER	American Airlines
	N756NA	Boeing 757-28A	North American Airlines *Claudette Abrahams*
	N757AN	Boeing 777-223ER	American Airlines
	N758AN	Boeing 777-223ER	American Airlines
	N759AN	Boeing 777-223ER	American Airlines
	N760AN	Boeing 777-223ER	American Airlines
	N760NA	Boeing 767-39HER	North American Airlines
	N761AJ	Boeing 777-223ER	American Airlines
	N762AN	Boeing 777-223ER	American Airlines
	N765AN	Boeing 777-223ER	American Airlines
	N766AN	Boeing 777-223ER	American Airlines
	N766UA	Boeing 777-222	United Airlines
	N767NA	Boeing 767-324ER	North American Airlines *Janice M.*
	N767UA	Boeing 777-222	United Airlines
	N768NA	Boeing 767-36NER	North American Airlines *Lisa Caroline*
	N768UA	Boeing 777-222	United Airlines
	N769UA	Boeing 777-222	United Airlines
	N770AN	Boeing 777-223ER	American Airlines
	N770UA	Boeing 777-222	United Airlines
	N771AN	Boeing 777-223ER	American Airlines
	N771UA	Boeing 777-222	United Airlines
	N772AN	Boeing 777-223ER	American Airlines
	N772UA	Boeing 777-222	United Airlines
	N773AN	Boeing 777-223ER	American Airlines

Reg.	Type	Owner or Operator	Notes
N773UA	Boeing 777-222	United Airlines	
N774AN	Boeing 777-223ER	American Airlines	
N774UA	Boeing 777-222	United Airlines	
N775AN	Boeing 777-223ER	American Airlines	
N775UA	Boeing 777-222	United Airlines	
N776AN	Boeing 777-223ER	American Airlines	
N776UA	Boeing 777-222	United Airlines	
N777AN	Boeing 777-223ER	American Airlines	
N777UA	Boeing 777-222	United Airlines	
N778AN	Boeing 777-223ER	American Airlines	
N778UA	Boeing 777-222	United Airlines	
N779AN	Boeing 777-223ER	American Airlines	
N779UA	Boeing 777-222	United Airlines	
N780AN	Boeing 777-223ER	American Airlines	
N780UA	Boeing 777-222	United Airlines	
N781AN	Boeing 777-223ER	American Airlines	
N781UA	Boeing 777-222	United Airlines	
N782AN	Boeing 777-223ER	American Airlines	
N782UA	Boeing 777-222ER	United Airlines	
N783AN	Boeing 777-223ER	American Airlines	
N783UA	Boeing 777-222ER	United Airlines	
N784AN	Boeing 777-223ER	American Airlines	
N784UA	Boeing 777-222ER	United Airlines	
N785AN	Boeing 777-223ER	American Airlines	
N785UA	Boeing 777-222ER	United Airlines	
N786AN	Boeing 777-223ER	American Airlines	
N786UA	Boeing 777-222ER	United Airlines	
N787AL	Boeing 777-223ER	American Airlines	
N787UA	Boeing 777-222ER	United Airlines	
N788AN	Boeing 777-223ER	American Airlines	
N788UA	Boeing 777-222ER	United Airlines	
N789AN	Boeing 777-223ER	American Airlines	
N790AN	Boeing 777-223ER	American Airlines	
N791AN	Boeing 777-223ER	American Airlines	
N791UA	Boeing 777-222ER	United Airlines	
N792AN	Boeing 777-223ER	American Airlines	
N792UA	Boeing 777-222ER	United Airlines	
N793AN	Boeing 777-223ER	American Airlines	
N793UA	Boeing 777-222ER	United Airlines	
N794AN	Boeing 777-223ER	American Airlines	
N794UA	Boeing 777-222ER	United Airlines	
N795AN	Boeing 777-223ER	American Airlines	
N795UA	Boeing 777-222ER	United Airlines	
N796AN	Boeing 777-223ER	American Airlines	
N796UA	Boeing 777-222ER	United Airlines	
N797AN	Boeing 777-223ER	American Airlines	
N797UA	Boeing 777-222ER	United Airlines	
N798AN	Boeing 777-223ER	American Airlines	
N798UA	Boeing 777-222ER	United Airlines	
N799AN	Boeing 777-223ER	American Airlines	
N799UA	Boeing 777-222ER	United Airlines	
N801DE	McD Douglas MD-11	World Airways	
N801DH	Douglas DC-8-73AF	DHL Air Cargo	
N801NW	Airbus A.330-323X	Northwest Airlines	
N802DH	Douglas DC-8-73AF	DHL Air Cargo	
N802NW	Airbus A.330-323X	Northwest Airlines	
N803DE	McD Douglas MD-11	World Airways	
N803DH	Douglas DC-8-73AF	DHL Air Cargo	
N803NW	Airbus A.330-323X	Northwest Airlines	
N804DE	McD Douglas MD-11	World Airways	
N804DH	Douglas DC-8-73AF	DHL Air Cargo	
N804NW	Airbus A.330-323X	Northwest Airlines	
N805DH	Douglas DC-8-73AF	DHL Air Cargo	
N805NW	Airbus A.330-323X	Northwest Airlines	
N806DH	Douglas DC-8-73AF	DHL Air Cargo	
N806NW	Airbus A.330-323X	Northwest Airlines	
N807DH	Douglas DC-8-73AF	DHL Air Cargo	
N807NW	Airbus A.330-323X	Northwest Airlines	
N808MC	Boeing 747-212B (SCD)	Atlas Air	
N808NW	Airbus A.330-323X	Northwest Airlines	
N809MC	Boeing 747-228F (SCD)	Atlas Air	
N809NW	Airbus A.330-323X	Northwest Airlines	

Notes	Reg.	Type	Owner or Operator
	N810AX	Douglas DC-10-30	Omni Air International
	N810NW	Airbus A.330-323X	Northwest Airlines
	N811NW	Airbus A.330-323X	Northwest Airlines
	N812NW	Airbus A.330-323X	Northwest Airlines
	N820BX	Douglas DC-8-71AF	Air Transport International
	N821BX	Douglas DC-8-71AF	Air Transport International
	N822BX	Douglas DC-8-71AF	Air Transport International
	N825BX	Douglas DC-8-71AF	Air Transport International
	N828BX	Douglas DC-8-71AF	Air Transport International
	N829BX	Douglas DC-8-71AF	BAX Global
	N830BX	Douglas DC-8-71AF	Air Transport International
	N860DA	Boeing 777-232ER (7001)	Delta Air Lines
	N861DA	Boeing 777-232ER (7002)	Delta Air Lines
	N862DA	Boeing 777-232ER (7003)	Delta Air Lines
	N863DA	Boeing 777-232ER (7004)	Delta Air Lines
	N864DA	Boeing 777-232ER (7005)	Delta Air Lines
	N865DA	Boeing 777-232ER (7006)	Delta Air Lines
	N866DA	Boeing 777-232ER (7007)	Delta Air Lines
	N867DA	Boeing 777-232ER (7008)	Delta Air Lines
	N868DA	Boeing 777-232ER (7009)	Delta Air Lines
	N869DA	Boeing 777-232ER (7010)	Delta Air Lines
	N870DA	Boeing 777-232ER (7011)	Delta Air Lines
	N871DA	Boeing 777-232ER (7012)	Delta Air Lines
	N872DA	Boeing 777-232ER (7013)	Delta Air Lines
	N920FT	Boeing 747-249F (SCD)	Polar Air Cargo *Thomas G Taaffe*
	N921FT	Boeing 747-283F	Polar Air Cargo
	N922FT	Boeing 747-2U3BF	Tradewinds Cargo
	N923FT	Boeing 747-2U3BF	Tradewinds Cargo
	N929RD	Boeing 757-2G5	Ryan International
	N974RY	Boeing 737-86N	Futura International Airways
	N975RY	Boeing 737-86N	Futura International Airways
	N977RY	Boeing 737-86N	Futura International Airways
	N1200K	Boeing 767-332ER (200)	Delta Air Lines
	N1201P	Boeing 767-332ER (201)	Delta Air Lines
	N1501P	Boeing 767-3P6ER (1501)	Delta Air Lines
	N1602	Boeing 767-332ER (1602)	Delta Air Lines
	N1603	Boeing 767-332ER (1603)	Delta Air Lines
	N1604R	Boeing 767-332ER (1604)	Delta Air Lines
	N1605	Boeing 767-332ER (1605)	Delta Air Lines
	N1606P	Boeing 767-332ER (1606)	Delta Air Lines
	N1607B	Boeing 767-332ER (1607)	Delta Air Lines
	N1608B	Boeing 767-332ER (1608)	Delta Air Lines
	N1609B	Boeing 767-332ER (1609)	Delta Air Lines
	N1610D	Boeing 767-332ER (1610)	Delta Air Lines
	N1611B	Boeing 767-332ER (1611)	Delta Air Lines
	N1612T	Boeing 767-332ER (1612)	Delta Air Lines
	N1613B	Boeing 767-332ER (1613)	Delta Air Lines
	N7375A	Boeing 767-323ER	American Airlines
	N12114	Boeing 757-224 (114)	Continental Airlines
	N12116	Boeing 757-224 (116)	Continental Airlines
	N14075	Douglas DC-10-30	World Airways
	N14102	Boeing 757-224 (102)	Continental Airlines
	N14115	Boeing 757-224 (115)	Continental Airlines
	N16065	Boeing 767-332ER	Delta Air Lines
	N16078	Boeing 767-332ER	Delta Air Lines
	N17085	Douglas DC-10-30	Omni Air International
	N17128	Boeing 757-224 (128)	Continental Airlines
	N17133	Boeing 757-224 (133)	Continental Airlines
	N19117	Boeing 757-224 (117)	Continental Airlines
	N19130	Boeing 757-224 (130)	Continental Airlines
	N19136	Boeing 757-224 (136)	Continental Airlines
	N27015	Boeing 777-224ER (015)	Continental Airlines
	N29129	Boeing 757-224 (129)	Continental Airlines
	N33103	Boeing 757-224 (103)	Continental Airlines
	N39356	Boeing 767-323ER	American Airlines
	N39364	Boeing 767-323ER	American Airlines
	N39365	Boeing 767-323ER	American Airlines
	N39367	Boeing 767-323ER	American Airlines
	N41135	Boeing 757-224 (135)	Continental Airlines
	N41140	Boeing 757-224 (140)	Continental Airlines
	N47888	Douglas DC-10-30F	Centurion Air Cargo *Captain Mike*
	N49082	Douglas DC-10-30	Omni Air International

Reg.	Type	Owner or Operator	Notes
N57016	Boeing 777-224ER	Continental Airlines	
N57111	Boeing 757-224	Continental Airlines	
N58101	Boeing 757-224 (101)	Continental Airlines	
N59053	Boeing 767-424ER (053)	Continental Airlines	
N59083	Douglas DC-10-30	Omni Air International	
N66051	Boeing 767-424ER (051)	Continental Airlines	
N66056	Boeing 767-424ER (056)	Continental Airlines	
N66057	Boeing 767-424ER (057)	Continental Airlines	
N67052	Boeing 767-424ER (052)	Continental Airlines	
N67058	Boeing 767-424ER (058)	Continental Airlines	
N67134	Boeing 757-224 (134)	Continental Airlines	
N67157	Boeing 767-224ER (157)	Continental Airlines	
N67158	Boeing 767-224ER (158)	Continental Airlines	
N68061	Boeing 767-424ER (061)	Continental Airlines	
N68155	Boeing 767-224ER (155)	Continental Airlines	
N68159	Boeing 767-224ER (159)	Continental Airlines	
N68160	Boeing 767-224ER (160)	Continental Airlines	
N69059	Boeing 767-424ER (059)	Continental Airlines	
N69154	Boeing 767-224ER (154)	Continental Airlines	
N73152	Boeing 767-224ER (152)	Continental Airlines	
N74007	Boeing 777-224ER (007)	Continental Airlines	
N76010	Boeing 777-224ER (010)	Continental Airlines	
N76054	Boeing 767-424ER (054)	Continental Airlines	
N76055	Boeing 767-424ER (055)	Continental Airlines	
N76151	Boeing 767-224ER (151)	Continental Airlines	
N76153	Boeing 767-224ER (153)	Continental Airlines	
N76156	Boeing 767-224ER (156)	Continental Airlines	
N77006	Boeing 777-224ER (006)	Continental Airlines	
N77012	Boeing 777-224ER (012)	Continental Airlines	
N77014	Boeing 777-224ER (014))	Continental Airlines	
N78001	Boeing 777-224ER (001)	Continental Airlines	
N78002	Boeing 777-224ER (002)	Continental Airlines	
N78003	Boeing 777-224ER (003)	Continental Airlines	
N78004	Boeing 777-224ER (004)	Continental Airlines	
N78005	Boeing 777-224ER (005)	Continental Airlines	
N78008	Boeing 777-224ER (008)	Continental Airlines	
N78009	Boeing 777-224ER (009)	Continental Airlines	
N78013	Boeing 777-224ER (013)	Continental Airlines	
N78017	Boeing 777-224ER (017)	Continental Airlines	
N78060	Boeing 767-424ER (060)	Continental Airlines	
N79011	Boeing 777-224ER (011)	Continental Airlines	

OD (Lebanon)

Middle East Airlines operate Airbus A.321s F-ORME to F-ORMJ together with Airbus A330s F-OMEA, F-OMEB and F-OMEC.

OE (Austria)

OE-ILF	Boeing 737-3Z9	Lauda Air *Bob Marley*
OE-LAE	Boeing 767-3Z9ER	Lauda Air *Louis Armstrong*
OE-LAG	Airbus A.340-212	Austrian Airlines *Europe*
OE-LAH	Airbus A.340-212	Austrian Airlines *Asia*
OE-LAK	Airbus A.340-313X	Austrian Airlines *Afrika*
OE-LAL	Airbus A.340-313X	Austrian Airlines *America*
OE-LAM	Airbus A.330-223	Austrian Airlines *Daschstein*
OE-LAN	Airbus A.330-223	Austrian Airlines *Arlberg*
OE-LAO	Airbus A.330-223	Austrian Airlines *Grossglockner*
OE-LAP	Airbus A.330-223	Austrian Airlines *Semmering*
OE-LAT	Boeing 767-31AER	Lauda Air *Enzo Ferrari*
OE-LAW	Boeing 767-3Z9ER	Lauda Air *Ayrton Senna*
OE-LAX	Boeing 767-3Z9ER	Lauda Air *James Dean*
OE-LAY	Boeing 767-3Z9ER	Lauda Air *Steve McQueen*
OE-LAZ	Boeing 767-3Z9ER	Lauda Air *Frank Sinatra*
OE-LBA	Airbus A.321-111	Austrian Airlines *Salzkammergut*
OE-LBB	Airbus A.321-111	Austrian Airlines *Pinzgau*
OE-LBC	Airbus A.321-111	Austrian Airlines *Sudtirol*
OE-LBD	Airbus A.321-111	Austrian Airlines *Steirisches Weinland*
OE-LBE	Airbus A.321-111	Austrian Airlines *Wachau*
OE-LBF	Airbus A.321-111	Austrian Airlines *Wien*

Notes	Reg.	Type	Owner or Operator
	OE-LBN	Airbus A.320-214	Austrian Airlines *Osttirol*
	OE-LBO	Airbus A.320-214	Austrian Airlines *Pyhrn-Eisenwurzen*
	OE-LBP	Airbus A.320-214	Austrian Airlines *Neusiedler See*
	OE-LBQ	Airbus A.320-214	Austrian Airlines *Wienerwald*
	OE-LBR	Airbus A.320-214	Austrian Airlines *Bregenzer Wald*
	OE-LBS	Airbus A.320-214	Austrian Airlines *Waldviertel*
	OE-LBT	Airbus A.320-214	Austrian Airlines *Worthersee*
	OE-LBU	Airbus A.320-214	Austrian Airlines *Muhlviertel*
	OE-LCF	Canadair CL.600-2B19 RJ	Tyrolean Airways *Stadt Dusseldorf*
	OE-LCG	Canadair CL.600-2B19 RJ	Tyrolean Airways *Stadt Köln*
	OE-LCH	Canadair CL.600-2B19 RJ	Tyrolean Airways *Stadt Amsterdam*
	OE-LCI	Canadair CL.600-2B19 RJ	Tyrolean Airways *Stadt Zürich*
	OE-LCJ	Canadair CL.600-2B19 RJ	Austrian Arrows *Stadt Hannover*
	OE-LCK	Canadair CL.600-2B19 RJ	Tyrolean Airways *Stadt Brussel*
	OE-LCL	Canadair CL.600-2B19 RJ	Austrian Arrows *Stadt Oslo*
	OE-LCM	Canadair CL.600-2B19 RJ	Austrian Arrows *Stadt Bologna*
	OE-LCN	Canadair CL.600-2B19 RJ	Tyrolean Airways *Stadt Bremen*
	OE-LCO	Canadair CL.600-2B19 RJ	Tyrolean Airways
	OE-LCP	Canadair CL.600-2B19 RJ	Tyrolean Airways *Stadt Hamburg*
	OE-LCQ	Canadair CL.600-2B19 RJ	Tyrolean Airways *Stadt Strassburg*
	OE-LCR	Canadair CL.600-2B19 RJ	Austrian Arrows
	OE-LDA	Airbus A.319-112	Austrian Airlines *Sofia*
	OE-LDB	Airbus A.319-112	Austrian Airlines *Bucharest*
	OE-LDC	Airbus A.319-112	Austrian Airlines *Kiev*
	OE-LFG	Fokker 70	Tyrolean Airways *Stadt Innsbruck*
	OE-LFH	Fokker 70	Austrian Arrows *Stadt Salzburg*
	OE-LFI	Fokker 70	Austrian Arrows *Stadt Klagenfurt*
	OE-LFJ	Fokker 70	Austrian Arrows *Stadt Graz*
	OE-LFK	Fokker 70	Tyrolean Airways *Stadt Wien*
	OE-LFL	Fokker 70	Austrian Arrows *Stadt Linz*
	OE-LFP	Fokker 70	Austrian Airlines *Wels*
	OE-LFQ	Fokker 70	Austrian Airlines *Dornbirn*
	OE-LFR	Fokker 70	Austrian Airlines *Steyr*
	OE-LGA	D.H.C.8Q-402 Dash Eight	Tyrolean Airways *Land Karnten*
	OE-LGB	D.H.C.8Q-402 Dash Eight	Tyrolean Airways *Land Tirol*
	OE-LGC	D.H.C.8Q-402 Dash Eight	Tyrolean Airways *Land Salzburg*
	OE-LGD	D.H.C.8Q-402 Dash Eight	Tyrolean Airways *Land Steiermark*
	OE-LGE	D.H.C.8Q-402 Dash Eight	Tyrolean Airways *Land Oberosterreich*
	OE-LGF	D.H.C.8Q-402 Dash Eight	Tyrolean Airways *Land Niederosterreich*
	OE-LMD	McD Douglas MD-83	Austrian Airlines *Villach*
	OE-LME	McD Douglas MD-83	Austrian Airlines *Krems*
	OE-LMK	McD Douglas MD-87	Austrian Airlines *St Pölten*
	OE-LML	McD Douglas MD-87	Austrian Airlines *Salzburgstadt*
	OE-LMN	McD Douglas MD-87	Austrian Airlines *Klagenfurt*
	OE-LMO	McD Douglas MD-87	Austrian Airlines *Bregenz*
	OE-LNI	Boeing 737-4Z9	Lauda Air *Janis Joplin*
	OE-LNJ	Boeing 737-8Z9	Lauda Air *Falco*
	OE-LNK	Boeing 737-8Z9	Lauda Air *Freddie Mercury*
	OE-LNM	Boeing 737-6Z9	Lauda Air *Innsbruck*
	OE-LNN	Boeing 737-7Z9	Lauda Air *Maria Callas*
	OE-LNO	Boeing 737-7Z9	Lauda Air *Greta Garbo*
	OE-LNP	Boeing 737-8Z9	Lauda Air *George Harrison*
	OE-LNQ	Boeing 737-8Z9	Lauda Air *Gregory Peck*
	OE-LOE	Airbus A.320-232	Flyniki
	OE-LOF	Airbus A.320-232	Flyniki *Kimi Eberl*
	OE-LOR	Airbus A.321-231	Flyniki
	OE-LOS	Airbus A.321-231	Flyniki *Kurt Hofmeister*
	OE-LPA	Boeing 777-2Z9	Lauda Air *Pablo Picasso*
	OE-LPB	Boeing 777-2Z9	Lauda Air *Ernest Hemingway*
	OE-LPC	Boeing 777-2Z9	Lauda Air *Donald Bradman*
	OE-LRE	Canadair CL.600-2B19 RJ	Austrian Arrows
	OE-LRF	Canadair CL.600-2B19 RJ	Austrian Arrows
	OE-LRG	Canadair CL.600-2B19 RJ	Austrian Arrows
	OE-LRH	Canadair CL.600-2B19 RJ	Austrian Arrows
	OE-LSB	D.H.C.8-311 Dash Eight	Intersky *Espace Mittelland*
	OE-LTD	D.H.C.8-314 Dash Eight	Austrian Arrows
	OE-LTF	D.H.C.8-314 Dash Eight	Austrian Arrows
	OE-LTG	D.H.C.8-314 Dash Eight	Austrian Arrows
	OE-LTH	D.H.C.8-314 Dash Eight	Austrian Arrows *Stadt Kitzbuhel*
	OE-LTI	D.H.C.8-314 Dash Eight	Austrian Arrows
	OE-LTJ	D.H.C.8-314 Dash Eight	Austrian Arrows *Seefeld*
	OE-LTK	D.H.C.8-314 Dash Eight	Austrian Arrows *Oetztal*

Reg.	Type	Owner or Operator	Notes
OE-LTL	D.H.C.8-314 Dash Eight	Austrian Arrows	
OE-LTM	D.H.C.8-314 Dash Eight	Austrian Arrows *Achensee*	
OE-LTN	D.H.C.8-314 Dash Eight	Austrian Arrows *St Anton*	
OE-LTO	D.H.C.8-314 Dash Eight	Tyrolean Airways *Stadt Kufstein*	
OE-LTP	D.H.C 8-314 Dash Eight	Tyrolean Airways *Stadt Lienz*	
OE-LTU	Airbus A.320-214	LTU Austria	
OE-LVA	Fokker 100	Austrian Arrows *Riga*	
OE-LVB	Fokker 100	Austrian Arrows	
OE-LVC	Fokker 100	Austrian Arrows	
OE-LVD	Fokker 100	Austrian Arrows *Belgrade*	
OE-LVE	Fokker 100	Austrian Arrows *Zagreb*	
OE-LVF	Fokker 100	Austrian Arrows	

OH (Finland)

OH-AFI	Boeing 757-2K2	Air Finland	
OH-AFJ	Boeing 757-2Q8	Air Finland	
OH-AFK	Boeing 757-28A	Air Finland	
OH-LBO	Boeing 757-2Q8	Finnair	
OH-LBR	Boeing 757-2Q8	Finnair	
OH-LBS	Boeing 757-2Q8	Finnair	
OH-LBT	Boeing 757-2Q8	Finnair	
OH-LBU	Boeing 757-2Q8	Finnair	
OH-LBV	Boeing 757-2Q8	Finnair	
OH-LBX	Boeing 757-2Q8	Finnair	
OH-LGA	McD Douglas MD-11	Finnair	
OH-LGB	McD Douglas MD-11	Finnair	
OH-LGC	McD Douglas MD-11	Finnair	
OH-LGD	McD Douglas MD-11	Finnair	
OH-LGE	McD Douglas MD-11	Finnair	
OH-LGF	McD Douglas MD-11	Finnair	
OH-LMG	McD Douglas MD-83	Finnair	
OH-LMH	McD Douglas MD-82	Finnair	
OH-LMW	McD Douglas MD-82	Finnair	
OH-LMX	McD Douglas MD-82	Finnair	
OH-LMY	McD Douglas MD-82	Finnair	
OH-LMZ	McD Douglas MD-82	Finnair	
OH-LPA	McD Douglas MD-82	Finnair	
OH-LPB	McD Douglas MD-83	Finnair	
OH-LPC	McD Douglas MD-83	Finnair	
OH-LPD	McD Douglas MD-83	Finnair	
OH-LPG	McD Douglas MD-83	Finnair	
OH-LVA	Airbus A.319-112	Finnair	
OH-LVB	Airbus A.319-112	Finnair	
OH-LVC	Airbus A.319-112	Finnair	
OH-LVD	Airbus A.319-112	Finnair	
OH-LVE	Airbus A.319-112	Finnair	
OH-LVF	Airbus A.319-112	Finnair	
OH-LVG	Airbus A.319-112	Finnair	
OH-LVH	Airbus A.319-112	Finnair	
OH-LVI	Airbus A.319-112	Finnair	
OH-LVK	Airbus A.319-112	Finnair	
OH-LVL	Airbus A.319-112	Finnair	
OH-LXA	Airbus A.320-214	Finnair	
OH-LXB	Airbus A.320-214	Finnair	
OH-LXC	Airbus A.320-214	Finnair	
OH-LXD	Airbus A.320-214	Finnair	
OH-LXE	Airbus A.320-214	Finnair	
OH-LXF	Airbus A.320-214	Finnair	
OH-LXG	Airbus A.320-214	Finnair	
OH-LXH	Airbus A.320-214	Finnair	
OH-LXI	Airbus A.320-214	Finnair	
OH-LXK	Airbus A.320-214	Finnair	
OH-LXL	Airbus A.320-214	Finnair	
OH-LXM	Airbus A.320-214	Finnair	
OH-LYP	Douglas DC-9-51	Finnair	
OH-LYU	Douglas DC-9-51	Finnair	
OH-LZA	Airbus A.321-211	Finnair	
OH-LZB	Airbus A.321-211	Finnair	
OH-LZC	Airbus A.321-211	Finnair	
OH-LZD	Airbus A.321-211	Finnair	

Notes	Reg.	Type	Owner or Operator
	OH-LZE	Airbus A.321-211	Finnair
	OH-LZF	Airbus A.321-211	Finnair
	OH-SAH	Avro RJ85	Blue1
	OH-SAI	Avro RJ85	Blue1
	OH-SAJ	Avro RJ85	Blue1
	OH-SAK	Avro RJ85	Blue1
	OH-SAL	Avro RJ85	Blue1
	OH-SAM	Avro RJ100	Blue1
	OH-SAN	Avro RJ100	Blue1
	OH-SAO	Avro RJ85	Blue1
	OH-SAP	Avro RJ85	Blue1
	OH-SAS	SAAB 2000	Blue1
	OH-SAT	SAAB 2000	Blue1
	OH-SAU	SAAB 2000	Blue1
	OH-SAW	SAAB 2000	Blue1
	OH-SAX	SAAB 2000	Blue1

OK (Czech Republic)

	OK-BGQ	Boeing 737-43Q	CSA Czech Airlines *Karlovy Vary*
	OK-CGH	Boeing 737-55S	CSA Czech Airlines *Usti n. Labem*
	OK-CGI	Boeing 737-49R	CSA Czech Airlines *Prostejov*
	OK-CGJ	Boeing 737-55S	CSA Czech Airlines *Hradec Kralove*
	OK-CGK	Boeing 737-55S	CSA Czech Airlines *Pardubice*
	OK-CGT	Boeing 737-46M	CSA Czech Airlines *Pisek*
	OK-DGB	Boeing 737-5L9	CSA Czech Airlines *Beroun*
	OK-DGC	Boeing 737-5L9	CSA Czech Airlines
	OK-DGL	Boeing 737-55S	CSA Czech Airlines *Tabor*
	OK-DGM	Boeing 737-45S	CSA Czech Airlines *Trebon*
	OK-DGN	Boeing 737-45S	CSA Czech Airlines *Trebic*
	OK-EGO	Boeing 737-55S	CSA Czech Airlines *Jindrichuv Hradec*
	OK-EGP	Boeing 737-45S	CSA Czech Airlines *Kladno*
	OK-FAN	Boeing 737-33A	Fischer Air
	OK-FGR	Boeing 737-45S	CSA Czech Airlines *Ostrava*
	OK-FGS	Boeing 737-45S	CSA Czech Airlines *Brno*
	OK-FIT	Boeing 737-36N	Fischer Air
	OK-FUN	Boeing 737-33A	Fischer Air
	OK-SWY	Boeing 737-530	Smart Wings
	OK-SWZ	Boeing 737-530	Smart Wings
	OK-TVA	Boeing 737-86N	Travel Service Airlines
	OK-TVB	Boeing 737-8CX	Travel Service Airlines
	OK-TVC	Boeing 737-86Q	Travel Service Airlines
	OK-TVD	Boeing 737-86N	Travel Service Airlines
	OK-TVR	Boeing 737-4Y0	Travel Service Airlines/Flyglobespan
	OK-VGZ	Boeing 737-4K5	CSA Czech Airlines
	OK-WAA	Airbus A.310-304	CSA Czech Airlines *Praha*
	OK-WAB	Airbus A.310-304	CSA Czech Airlines *Bratislava*
	OK-WGF	Boeing 737-4Y0	CSA Czech Airlines *Jihlava*
	OK-WGG	Boeing 737-4Y0	CSA Czech Airlines *Liberec*
	OK-WGX	Boeing 737-436	CSA Czech Airlines
	OK-WGY	Boeing 737-436	CSA Czech Airlines *Roudnice*
	OK-XGA	Boeing 737-55S	CSA Czech Airlines *Plzen*
	OK-XGB	Boeing 737-55S	CSA Czech Airlines *Olomouc*
	OK-XGC	Boeing 737-55S	CSA Czech Airlines *Ceske Budejovice*
	OK-XGD	Boeing 737-55S	CSA Czech Airlines *Poprad*
	OK-XGE	Boeing 737-55S	CSA Czech Airlines *Kosice*
	OK-XGV	Boeing 737-5H6	CSA Czech Airlines *Frantisovky lazne*
	OK-XGW	Boeing 737-5H6	CSA Czech Airlines *Marianske lazne*
	OK-YAC	Airbus A.310-325ET	CSA Czech Airlines *Zlin*
	OK-YGA	Boeing 737-4Q8	CSA Czech Airlines *Bechyne*
	OK-YGU	Boeing 737-4Q8	CSA Czech Airlines *Melnik*

OM (Slovakia)

	OM-AAD	Boeing 737-33A	Slovak Airlines
	OM-BYO	Tupolev Tu-154M	Slovak Government.
	OM-BYR	Tupolev Tu-154M	Slovak Government
	OM-DGK	Boeing 757-236	Air Slovakia *City of Golden Temple*
	OM-NSH	Boeing 767-231	Slovak Airlines
	OM-SEA	Boeing 737-5Y0	SkyEurope Airlines

Reg.	Type	Owner or Operator	Notes
OM-SEB	Boeing 737-5Y0	SkyEurope Airlines	
OM-SEC	Boeing 737-5Y0	SkyEurope Airlines	
OM-SED	Boeing 737-53C	SkyEurope Airlines	
OM-SEE	Boeing 737-53C	SkyEurope Airlines	
OM-SNA	Boeing 757-27B	Air Slovakia	

OO (Belgium)

Reg.	Type	Owner or Operator	Notes
OO-DJE	BAe 146-200	SN Brussels Airlines	
OO-DJF	BAe 146-200	SN Brussels Airlines	
OO-DJG	BAe 146-200	SN Brussels Airlines	
OO-DJH	BAe 146-200	SN Brussels Airlines	
OO-DJK	Avro RJ85	SN Brussels Airlines	
OO-DJL	Avro RJ85	SN Brussels Airlines	
OO-DJN	Avro RJ85	SN Brussels Airlines	
OO-DJO	Avro RJ85	SN Brussels Airlines	
OO-DJP	Avro RJ85	SN Brussels Airlines	
OO-DJQ	Avro RJ85	SN Brussels Airlines	
OO-DJR	Avro RJ85	SN Brussels Airlines	
OO-DJS	Avro RJ85	SN Brussels Airlines	
OO-DJT	Avro RJ85	SN Brussels Airlines	
OO-DJV	Avro RJ85	SN Brussels Airlines	
OO-DJW	Avro RJ85	SN Brussels Airlines	
OO-DJX	Avro RJ85	SN Brussels Airlines	
OO-DJY	Avro RJ85	SN Brussels Airlines	
OO-DJZ	Avro RJ85	SN Brussels Airlines	
OO-DIB	Airbus A.300B4-203F	European Air Transport (DHL)	
OO-DIC	Airbus A.300B4-203F	European Air Transport (DHL)	
OO-DLC	Airbus A.300B4-203F	European Air Transport (DHL)	
OO-DLD	Airbus A.300B4-203F	European Air Transport (DHL)	
OO-DLE	Airbus A.300B4-203F	European Air Transport (DHL)	
OO-DLG	Airbus A.300B4-203F	European Air Transport (DHL)	
OO-DLI	Airbus A.300B4-203F	European Air Transport (DHL)	
OO-DLJ	Boeing 757-23APF	European Air Transport (DHL)	
OO-DLN	Boeing 757-236F	European Air Transport (DHL)	
OO-DLP	Boeing 757-236F	European Air Transport (DHL)	
OO-DLQ	Boeing 757-236F	European Air Transport (DHL)	
OO-DLR	Airbus A.300B4-203F	European Air Transport (DHL)	
OO-DLT	Airbus A.300B4-203F	European Air Transport (DHL)	
OO-DLU	Airbus A.300B4-203F	European Air Transport (DHL)	
OO-DLV	Airbus A.300B4-203F	European Air Transport (DHL)	
OO-DLW	Airbus A.300B4-203F	European Air Transport (DHL)	
OO-DLY	Airbus A.300B4-203F	European Air Transport (DHL)	
OO-DPB	Boeing 757-236F	European Air Transport (DHL)	
OO-DPF	Boeing 757-236F	European Air Transport (DHL)	
OO-DPI	Boeing 757-236F	European Air Transport (DHL)	
OO-DPJ	Boeing 757-236F	European Air Transport (DHL)	
OO-DPK	Boeing 757-236F	European Air Transport (DHL)	
OO-DPL	Boeing 757-236F	European Air Transport (DHL)	
OO-DPM	Boeing 757-236F	European Air Transport (DHL)	
OO-DPN	Boeing 757-236F	European Air Transport (DHL)	
OO-DPO	Boeing 757-236F	European Air Transport (DHL)	
OO-DWA	Avro RJ100	SN Brussels Airlines	
OO-DWB	Avro RJ100	SN Brussels Airlines	
OO-DWC	Avro RJ100	SN Brussels Airlines	
OO-DWD	Avro RJ100	SN Brussels Airlines	
OO-DWE	Avro RJ100	SN Brussels Airlines	
OO-DWF	Avro RJ100	SN Brussels Airlines	
OO-DWG	Avro RJ100	SN Brussels Airlines	
OO-DWH	Avro RJ100	SN Brussels Airlines	
OO-DWI	Avro RJ100	SN Brussels Airlines	
OO-DWJ	Avro RJ100	SN Brussels Airlines	
OO-DWK	Avro RJ100	SN Brussels Airlines	
OO-DWL	Avro RJ100	SN Brussels Airlines	
OO-MJE	BAe 146-200	SN Brussels Airlines	
OO-SFM	Airbus A.330-301	SN Brussels Airlines	
OO-SFN	Airbus A.330-301	SN Brussels Airlines	
OO-SFO	Airbus A.330-301	SN Brussels Airlines	
OO-SSG	Airbus A.319-112	SN Brussels Airlines	
OO-SSK	Airbus A.319-112	SN Brussels Airlines	
OO-SSM	Airbus A.319-112	SN Brussels Airlines	

Notes	Reg.	Type	Owner or Operator
	OO-TAA	BAe 145-300QT	TNT Airways
	OO-TAD	BAe 146-300QT	TNT Airways
	OO-TAE	BAe 146-300QT	TNT Airways
	OO-TAF	BAe 146-300QT	TNT Airways
	OO-TAH	BAe 146-300QT	TNT Airways
	OO-TAJ	BAe 146-300QT	TNT Airways
	OO-TAK	BAe 146-300QT	TNT Airways
	OO-TAR	BAe 146-200QT	TNT Airways
	OO-TAS	BAe 146-300QT	TNT Airways
	OO-TAU	BAe 146-200QT	TNT Airways
	OO-TAW	BAe 146-200QT	TNT Airways
	OO-TCC	Airbus A.320-231	Thomas Cook Airlines Belgium
	OO-TCF	Airbus A.320-231	Thomas Cook Airlines Belgium
	OO-TCH	Airbus A.320-214	Thomas Cook Airlines Belgium
	OO-TCI	Airbus A.320-214	Thomas Cook Airlines Belgium
	OO-TCJ	Airbus A.320-214	Thomas Cook Airlines Belgium
	OO-TNA	Boeing 737-3T0F	TNT Airways
	OO-TNB	Boeing 737-3T0F	TNT Airways
	OO-TNC	Boeing 737-3T0F	TNT Airways
	OO-TNE	Boeing 737-3Q8	TNT Airways
	OO-TUA	Boeing 737-4K5	TUI Airlines Belgium
	OO-TUB	Boeing 737-4K5	TUI Airlines Belgium/Jetair *Devotion*
	OO-TUI	Boeing 737-4K5	TUI Airlines Belgium/Jetair
	OO-TZA	Airbus A.300B4-203F	TNT Airways
	OO-TZB	Airbus A.300B4-203F	TNT Airways
	OO-TZC	Airbus A.300B4-203F	TNT Airways
	OO-TZD	Airbus A.300B4-203F	TNT Airways
	OO-VAC	Boeing 737-8BK	TUI Airlines Belgium/Jetair *Rising Sun*
	OO-VLE	Fokker 50	V.L.M. Airlines *City of Dusseldorf*
	OO-VLJ	Fokker 50	V.L.M. Airlines *City of London*
	OO-VLK	Fokker 50	V.L.M. Airlines *City of Monchengladbach*
	OO-VLL	Fokker 50	V.L.M. Airlines *Island of Jersey*
	OO-VLM	Fokker 50	V.L.M. Airlines *City of Rotterdam*
	OO-VLN	Fokker 50	V.L.M. Airlines *City of Antwerp*
	OO-VLO	Fokker 50	V.L.M. Airlines *City of Geneva*
	OO-VLQ	Fokker 50	V.L.M. Airlines *City of Brussels*
	OO-VLR	Fokker 50	V.L.M. Airlines *Grand Duchy of Luxembourg*
	OO-VLS	Fokker 50	V.L.M. Airlines *City of Manchester*
	OO-VLV	Fokker 50	V.L.M. Airlines *States of Guernsey*
	OO-VLX	Fokker 50	V.L.M. Airlines *City of Luxembourg*
	OO-VLY	Fokker 50	V.L.M. Airlines *City of Liverpool*

Note: DHL Air also operates a number of Boeing 757s which retain their UK registrations.
Thomas Cook Airlines Belgium also operate A.320s D-AICA and D-AICB.

OY (Denmark)

Reg.	Type	Owner or Operator
OY-APB	Boeing 737-5L9	Maersk Air
OY-APH	Boeing 737-5L9	Maersk Air
OY-API	Boeing 737-5L9	Maersk Air
OY-APK	Boeing 737-5L9	Maersk Air/DBA
OY-APL	Boeing 737-5L9	Maersk Air/DBA
OY-BJP	Swearingen SA227AC Metro III	Ben Air
OY-CIG	Aérospatiale ATR-42-300	Cimber Air
OY-CIH	Aérospatiale ATR-42-300	Cimber Air
OY-CIL	Aérospatiale ATR-42-512	Cimber Air
OY-CIM	Aérospatiale ATR-72-212A	Cimber Air
OY-CIN	Aérospatiale ATR-72-212A	Cimber Air
OY-CIO	Aérospatiale ATR-72-212A	Cimber Air/Team Lufthansa
OY-CIP	Aérospatiale ATR-72-201	Cimber Air
OY-CIR	Aérospatiale ATR-42-310	Danish Air Transport
OY-CIU	Aérospatiale ATR-42-320	Danish Air Transport
OY-CIV	Aérospatiale ATR-72-212	Cimber Air
OY-CIW	Aérospatiale ATR-72-212A	Cimber Air
OY-CRG	BAe 146-200	Atlantic Airways
OY-JRF	Beech 1900C	Danish Air Transport
OY-JRI	Beech 1900C-1	Danish Air Transport
OY-JRJ	Aérospatiale ATR-42-320	Danish Air Transport
OY-JRK	SC.7 Skyvan	Danish Air Transport
OY-JRV	Beech 1900D	Danish Air Transport
OY-JRY	Aérospatiale ATR-42-320	Danish Air Transport

Reg.	Type	Owner or Operator	Notes
OY-KBA	Airbus A.340-313X	S.A.S. *Adalstein Viking*	
OY-KBB	Airbus A.321-231	S.A.S. *Hjorulf Viking*	
OY-KBC	Airbus A.340-313X	S.A.S. *Fredis Viking*	
OY-KBD	Airbus A.340-313X	S.A.S. *Toste Viking*	
OY-KBE	Airbus A.321-231	S.A.S *Emma Viking*	
OY-KBF	Airbus A.321-231	S.A.S. *Skapti Viking*	
OY-KBH	Airbus A.321-231	S.A.S. *Sulke Viking*	
OY-KBI	Airbus A.340-313X	S.A.S. *Rurik Viking*	
OY-KBK	Airbus A.321-231	S.A.S. *Arne Viking*	
OY-KBL	Airbus A.321-231	S.A.S. *Gynnbjorn Viking*	
OY-KBM	Airbus A.340-313X	S.A.S. *Astrid Viking*	
OY-KBN	Airbus A.330-343X	S.A.S. *Eystein Viking*	
OY-KCD	D.H.C.8Q-402 Dash Eight	S.A.S. Commuter *Bjarke Viking*	
OY-KCE	D.H.C.8Q-402 Dash Eight	S.A.S. Commuter *Alf Viking*	
OY-KCF	D.H.C.8Q-402 Dash Eight	S.A.S. Commuter *Asa Viking*	
OY-KCG	D.H.C.8Q-402 Dash Eight	S.A.S. Commuter *Sote Viking*	
OY-KCH	D.H.C.8Q-402 Dash Eight	S.A.S. Commuter *Igle Viking*	
OY-KGT	McD Douglas MD-82	S.A.S. *Hake Viking*	
OY-KGY	McD Douglas MD-81	S.A.S. *Rollo Viking*	
OY-KGZ	McD Douglas MD-81	S.A.S. *Hagbard Viking*	
OY-KHC	McD Douglas MD-82	S.A.S. *Faste Viking*	
OY-KHE	McD Douglas MD-82	S.A.S. *Saxo Viking*	
OY-KHF	McD Douglas MD-87	S.A.S. *Ragnar Viking*	
OY-KHG	McD Douglas MD-82	S.A.S. *Alle Viking*	
OY-KHI	McD Douglas MD-87	S.A.S. *Torkel Viking*	
OY-KHM	McD Douglas MD-82	S.A.S. *Mette Viking*	
OY-KHN	McD Douglas MD-81	S.A.S. *Dan Viking*	
OY-KHP	McD Douglas MD-81	S.A.S. *Harild Viking*	
OY-KHR	McD Douglas MD-81	S.A.S. *Torkild Viking*	
OY-KHT	McD Douglas MD-82	S.A.S. *Gorm Viking*	
OY-KHU	McD Douglas MD-87	S.A.S. *Ravn Viking*	
OY-KHW	McD Douglas MD-87	S.A.S. *Ingemund Viking*	
OY-KIL	McD Douglas MD-90-30	S.A.S. *Kaare Viking*	
OY-KIM	McD Douglas MD-90-30	S.A.S. *Jon Viking*	
OY-KIN	McD Douglas MD-90-30	S.A.S. *Tormod Viking*	
OY-KKG	Boeing 737-683	S.A.S. *Sindre Viking*	
OY-KKH	Boeing 737-683	S.A.S *Embla Viking*	
OY-KKI	Boeing 737-783	S.A.S. *Borgny Viking*	
OY-KKR	Boeing 737-783	S.A.S. *Gjuke Viking*	
OY-KKS	Boeing 737-683	S.A.S. *Ramveig Viking*	
OY-MAA	Boeing 737-5L9	Maersk Air	
OY-MAE	Boeing 737-5L9	Maersk Air	
OY-MAV	Canadair CL.600-2B19 RJ	Cimber Air	
OY-MBI	Canadair CL.600-2B19 RJ	Maersk Air	
OY-MBJ	Canadair CL.600-2B19 RJ	Cimber Air/S.A.S.	
OY-MBT	Canadair CL.600-2B19 RJ	Maersk Air	
OY-MBU	Canadair CL.600-2B19 RJ	Cimber Air/S.A.S.	
OY-MLW	Boeing 737-73S	Maersk Air	
OY-MRC	Boeing 737-7L9	Maersk Air	
OY-MRD	Boeing 737-7L9	Maersk Air	
OY-MRE	Boeing 737-7L9	Maersk Air	
OY-MRF	Boeing 737-7L9	Maersk Air	
OY-MRG	Boeing 737-7L9	Maersk Air	
OY-MRH	Boeing 737-7L9	Maersk Air	
OY-MRI	Boeing 737-7L9	Maersk Air	
OY-MRJ	Boeing 737-7L9	Maersk Air	
OY-MRK	Boeing 737-7L9	Maersk Air/Sterling European	
OY-MRL	Boeing 737-7L9	Maersk Air	
OY-MUE	BAe Jetstream 3100	Sun-Air/British Airways	
OY-NCS	Dornier 328-100	Sun-Air/British Airways	
OY-	Dornier 328-100	Sun-Air/British Airways	
OY-PBH	Let L410UVP-E20	Ben Air	
OY-PBI	Let L410UVP-E20	Air Centre West	
OY-RCA	BAe 146-200	Atlantic Airways	
OY-RCB	BAe 146-200	Atlantic Airways	
OY-RJA	Canadair CL.600-2B19 RJ	Cimber Air	
OY-RJB	Canadair CL.600-2B19 RJ	Cimber Air	
OY-RTA	Aérospatiale ATR-72-201	Cimber Air	
OY-RTD	Aérospatiale ATR-72-202	Cimber Air	
OY-RTE	Aérospatiale ATR-72-202	Cimber Air	
OY-RUB	Aérospatiale ATR-72-201	Danish Air Transport	
OY-RUM	Aérospatiale ATR-42-312	Danish Air Transport	

Notes	Reg.	Type	Owner or Operator
	OY-SEA	Boeing 737-8Q8	Sterling European Airlines
	OY-SEB	Boeing 737-8Q8	Sterling European Airlines
	OY-SEC	Boeing 737-8Q8	Sterling European Airlines
	OY-SED	Boeing 737-8Q8	Sterling European Airlines
	OY-SEH	Boeing 737-85H	Sterling European Airlines
	OY-SEI	Boeing 737-85H	Sterling European Airlines
	OY-SEJ	Boeing 737-85H	Sterling European Airlines
	OY-SEK	Boeing 737-86Q	Sterling European Airlines
	OY-SEL	Boeing 737-8BK	Sterling European Airlines
	OY-SEM	Boeing 737-8BK	Sterling European Airlines
	OY-SVB	BAe Jetstream 3202	Sun-Air/British Airways
	OY-SVF	BAe Jetstream 3102	Sun-Air/British Airways
	OY-SVI	BAe ATP	Sun-Air/British Airways
	OY-SVJ	BAe Jetstream 3102	Sun-Air/British Airways
	OY-SVR	BAe Jetstream 3202	Sun-Air/British Airways
	OY-SVT	BAe ATP	Sun-Air/British Airways
	OY-SVU	BAe ATP	Sun-Air/British Airways
	OY-SVW	BAe Jetstream 4102	Sun-Air/British Airways
	OY-SVY	BAe Jetstream 3202	Sun-Air/British Airways
	OY-SVZ	BAe Jetstream 3102	Sun-Air/British Airways
	OY-USA	Boeing 757-24APF	Starair/UPS
	OY-USB	Boeing 757-24APF	Starair/UPS
	OY-USC	Boeing 757-24APF	Starair/UPS
	OY-USD	Boeing 757-24APF	Starair/UPS
	OY-VKA	Airbus A.321-211	My Travel Airways
	OY-VKB	Airbys A.321-211	My Travel Airways
	OY-VKC	Airbus A.321-211	My Travel Airways
	OY-VKF	Airbus A.330-243	My Travel Airways
	OY-VKG	Airbus A.330-343X	My Travel Airways
	OY-VKH	Airbus A.330-343X	My Travel Airways
	OY-VKI	Airbus A.330-343X	My Travel Airways
	OY-VKM	Airbus A.320-214	My Travel Airways
	OY-VKN	Airbus A.320-214	My Travel Airways
	OY-VKO	Airbus A.320-214	My Travel Airways
	OY-VKP	Airbus A.320-214	My Travel Airways
	OY-VKR	Airbus A.320-214	My Travel Airways
	OY-VKS	Airbus A.320-214	My Travel Airways

P4 (Aruba)

	P4-EAS	Boeing 757-2G5	Air Astana
	P4-FAS	Boeing 757-2G5	Air Astana
	P4-GAS	Boeing 757-2G5	Air Astana

PH (Netherlands)

	PH-AHQ	Boeing 767-383ER	Dutch Caribbean Exel
	PH-AHX	Boeing 767-383ER	Holland Exel
	PH-AHY	Boeing 767-383ER	Belgium Exel
	PH-AJU	Douglas DC-2	Dutch Dakota Association *Uiver* (N39165)
	PH-BDA	Boeing 737-306	K.L.M. *Willem Barentsz*
	PH-BDC	Boeing 737-306	K.L.M. *Cornelis De Houteman*
	PH-BDD	Boeing 737-306	K.L.M. *Anthony van Diemen*
	PH-BDE	Boeing 737-306	K.L.M. *Abel J. Tasman*
	PH-BDG	Boeing 737-306	K.L.M. *Michiel A. de Ruyter*
	PH-BDI	Boeing 737-306	K.L.M. *Maarten H. Tromp*
	PH-BDK	Boeing 737-306	K.L.M. *Jan H. van Linschoten*
	PH-BDN	Boeing 737-306	K.L.M. *Willem van Ruysbroeck*
	PH-BDO	Boeing 737-306	K.L.M. *Jacob van Heemskerck*
	PH-BDP	Boeing 737-306	K.L.M. *Jacob Roggeveen*
	PH-BDR	Boeing 737-406	K.L.M. *Willem C. Schouten*
	PH-BDS	Boeing 737-406	K.L.M. *Jorris van Spilbergen*
	PH-BDT	Boeing 737-406	K.L.M. *Gerrit de Veer*
	PH-BDU	Boeing 737-406	K.L.M. *Marco Polo*
	PH-BDW	Boeing 737-406	K.L.M. *Leifur Eiriksson*
	PH-BDY	Boeing 737-406	K.L.M. *Vasco da Gama*
	PH-BDZ	Boeing 737-406	K.L.M. *Christophorus Columbus*
	PH-BFA	Boeing 747-406	K.L.M. *City of Atlanta*
	PH-BFB	Boeing 747-406	K.L.M. *City of Bangkok*
	PH-BFC	Boeing 747-406 (SCD)	K.L.M. *City of Calgary*

Reg.	Type	Owner or Operator	Notes
PH-BFD	Boeing 747-406 (SCD)	K.L.M. *City of Dubai*	
PH-BFE	Boeing 747-406 (SCD)	K.L.M. *City of Melbourne*	
PH-BFF	Boeing 747-406 (SCD)	K.L.M. *City of Freetown*	
PH-BFG	Boeing 747-406	K.L.M. *City of Guayaquil*	
PH-BFH	Boeing 747-406 (SCD)	K.L.M. *City of Hong Kong*	
PH-BFI	Boeing 747-406 (SCD)	K.L.M. *City of Jakarta*	
PH-BFK	Boeing 747-406 (SCD)	K.L.M. *City of Karachi*	
PH-BFL	Boeing 747-406	K.L.M. *City of Lima*	
PH-BFM	Boeing 747-406 (SCD)	K.L.M. *City of Mexico*	
PH-BFN	Boeing 747-406	K.L.M. *City of Nairobi*	
PH-BFO	Boeing 747-406 (SCD)	K.L.M. *City of Orlando*	
PH-BFP	Boeing 747-406 (SCD)	K.L.M. *City of Paramaribo*	
PH-BFR	Boeing 747-406 (SCD)	K.L.M. *City of Rio de Janeiro*	
PH-BFS	Boeing 747-406 (SCD)	K.L.M. *City of Seoul*	
PH-BFT	Boeing 747-406 (SCD)	K.L.M. *City of Tokyo*	
PH-BFU	Boeing 747-406 (SCD)	K.L.M. *City of Beijing*	
PH-BFV	Boeing 747-406	K.L.M. *City of Vancouver*	
PH-BFW	Boeing 747-406	K.L.M. *City of Shanghai*	
PH-BFY	Boeing 747-406	K.L.M. *City of Johannesburg*	
PH-BPB	Boeing 737-4Y0	K.L.M. *Jan Tinbergen*	
PH-BPC	Boeing 737-4Y0	K.L.M. *Ernest Hemingway*	
PH-BQA	Boeing 777-206ER	K.L.M. *Dr Albert Plesman*	
PH-BQB	Boeing 777-206ER	K.L.M. *Borobudur*	
PH-BQC	Boeing 777-206ER	K.L.M. *Chichen-Itza*	
PH-BQD	Boeing 777-206ER	K.L.M. *Darjeeling Highway*	
PH-BQE	Boeing 777-206ER	K.L.M. *Epidaurus*	
PH-BQF	Boeing 777-206ER	K.L.M. *Ferrara City*	
PH-BQG	Boeing 777-206ER	K.L.M. *Galapagos Islands*	
PH-BQH	Boeing 777-206ER	K.L.M. *Hadrian's Wall*	
PH-BQI	Boeing 777-206ER	K.L.M. *Iguazu Falls*	
PH-BQK	Boeing 777-206ER	K.L.M. *Mount Kilimanjaro*	
PH-BTA	Boeing 737-406	K.L.M. *Fernao Magalhaes*	
PH-BTB	Boeing 737-406	K.L.M. *Henry Hudson*	
PH-BTC	Boeing 737-406	K.L.M. *David Livingstone*	
PH-BTD	Boeing 737-306	K.L.M. *James Cook*	
PH-BTE	Boeing 737-306	K.L.M. *Roald Amundsen*	
PH-BTF	Boeing 737-406	K.L.M. *Alexander von Humboldt*	
PH-BTG	Boeing 737-406	K.L.M. *Henry Morton Stanley*	
PH-BTH	Boeing 737-306	K.L.M. *Heike Kamerlingh-Onnes*	
PH-BTI	Boeing 737-306	K.L.M. *Niels Bohr*	
PH-BUH	Boeing 747-206F (SCD)	Martinair Cargo	
PH-BXA	Boeing 737-8K2	K.L.M. *Zwaan/Swan*	
PH-BXB	Boeing 737-8K2	K.L.M. *Valk/Falcon*	
PH-BXC	Boeing 737-8K2	K.L.M. *Korhoen/Grouse*	
PH-BXD	Boeing 737-8K2	K.L.M. *Arend/Eagle*	
PH-BXE	Boeing 737-8K2	K.L.M. *Harvik/Hawk*	
PH-BXF	Boeing 737-8K2	K.L.M. *Zwallou/Swallow*	
PH-BXG	Boeing 737-8K2	K.L.M./Transavia *Kraanvogel/Crane*	
PH-BXH	Boeing 737-8K2	K.L.M. *Gans/Goose*	
PH-BXI	Boeing 737-8K2	K.L.M. *Zilvermeeuw*	
PH-BXK	Boeing 737-8K2	K.L.M. *Gierzwallou/Swift*	
PH-BXL	Boeing 737-8K2	K.L.M. *Sperwer/Sparrow*	
PH-BXM	Boeing 737-8K2	K.L.M. *Kluut/Avocet*	
PH-BXN	Boeing 737-8K2	K.L.M. *Merel/Blackbird*	
PH-BXO	Boeing 737-9K2	K.L.M. *Plevier/Plover*	
PH-BXP	Boeing 737-9K2	K.L.M. *Meerkoet/Crested Coot*	
PH-BXR	Boeing 737-9K2	K.L.M. *Nachtegaal/Nightingale*	
PH-BXS	Boeing 737-9K2	K.L.M. *Buizerd/Buzzard*	
PH-BXT	Boeing 737-9K2	K.L.M. *Zeestern/Sea Tern*	
PH-BZA	Boeing 767-306ER	K.L.M. *Blue Bridge*	
PH-BZB	Boeing 767-306ER	K.L.M. *Pont Neuf*	
PH-BZC	Boeing 767-306ER	K.L.M. *Brooklyn Bridge*	
PH-BZD	Boeing 767-306ER	K.L.M. *King Hussain Bridge*	
PH-BZE	Boeing 767-306ER	K.L.M. *Ponte Rialto*	
PH-BZF	Boeing 767-306ER	K.L.M. *Golden Gate Bridge*	
PH-BZG	Boeing 767-306ER	K.L.M. *Erasmus Bridge*	
PH-BZH	Boeing 767-306ER	K.L.M. *Tower Bridge*	
PH-BZI	Boeing 767-306ER	K.L.M. *Bosporus Bridge*	
PH-BZK	Boeing 767-306ER	K.L.M. *Zeeland Bridge*	
PH-BZM	Boeing 767-306ER	K.L.M. *Garibaldi Bridge*	
PH-BZO	Boeing 767-306ER	K.L.M. *Karmsund Bridge*	
PH-CKA	Boeing 747-406ERF	K.L.M. *Eendracht*	

Notes	Reg.	Type	Owner or Operator
	PH-CKB	Boeing 747-406ERF	K.L.M.*Leeuwin*
	PH-CKC	Boeing 747-406ERF	K.L.M. *Oranje*
	PH-DDZ	Douglas DC-3	Dutch Dakota Association
	PH-DMS	Fokker 50	Denim Air
	PH-DMT	Fokker 50	Denim Air
	PH-DYM	Swearingen SA.227AC Metro III	Dynamic Airlines
	PH-HZA	Boeing 737-8K2	Transavia
	PH-HZB	Boeing 737-8K2	Transavia
	PH-HZC	Boeing 737-8K2	Transavia
	PH-HZD	Boeing 737-8K2	Transavia
	PH-HZE	Boeing 737-8K2	Transavia *City of Rhodos*
	PH-HZF	Boeing 737-8K2	Transavia
	PH-HZG	Boeing 737-8K2	Transavia
	PH-HZI	Boeing 737-8K2	Transavia
	PH-HZJ	Boeing 737-8K2	Transavia
	PH-HZK	Boeing 737-8K2	Transavia
	PH-HZL	Boeing 737-8K2	Transavia/Air Berlin
	PH-HZM	Boeing 737-8K2	Transavia
	PH-HZN	Boeing 737-8K2	Transavia
	PH-HZS	Boeing 737-8K2	Transavia
	PH-HZV	Boeing 737-8K2	Transavia
	PH-HZW	Boeing 737-8K2	Transavia
	PH-HZX	Boeing 737-8K2	Transavia
	PH-HZY	Boeing 737-8K2	Transavia
	PH-JCE	Fokker 50	Denim Air
	PH-JCH	Fokker 70	K.L.M. CityHopper
	PH-JCT	Fokker 70	K.L.M. CityHopper
	PH-JXJ	Fokker 50	Denim Air
	PH-JXK	Fokker 50	Denim Air
	PH-JXN	Fokker 50	Denim Air/Air Nostrum
	PH-KBX	Fokker 70	Netherlands Government
	PH-KCA	McD Douglas MD-11	K.L.M. *Amy Johnson*
	PH-KCB	McD Douglas MD-11	K.L.M. *Maria Montessori*
	PH-KCC	McD Douglas MD-11	K.L.M. *Marie Curie*
	PH-KCD	McD Douglas MD-11	K.L.M. *Florence Nightingale*
	PH-KCE	McD Douglas MD-11	K.L.M. *Audrey Hepburn*
	PH-KCF	McD Douglas MD-11	K.L.M. *Annie Romein*
	PH-KCG	McD Douglas MD-11	K.L.M. *Maria Callas*
	PH-KCH	McD Douglas MD-11	K.L.M. *Anna Pavlova*
	PH-KCI	McD Douglas MD-11	K.L.M. *Ingrid Bergman*
	PH-KCK	McD Douglas MD-11	K.L.M. *Marie Servaes*
	PH-KLD	Fokker 100	K.L.M. CityHopper
	PH-KLE	Fokker 100	K.L.M. CityHopper
	PH-KLG	Fokker 100	K.L.M. CityHopper
	PH-KLI	Fokker 100	K.L.M. CityHopper
	PH-KVA	Fokker 50	K.L.M. CityHopper *Bremen*
	PH-KVB	Fokker 50	K.L.M. CityHopper *Brussels*
	PH-KVC	Fokker 50	K.L.M. CityHopper *Stavanger*
	PH-KVD	Fokker 50	K.L.M. CityHopper *Dusseldorf*
	PH-KVE	Fokker 50	K.L.M. CityHopper *Amsterdam*
	PH-KVF	Fokker 50	K.L.M. CityHopper *Paris/Paris*
	PH-KVG	Fokker 50	K.L.M. CityHopper *Stuttgart*
	PH-KVH	Fokker 50	K.L.M. CityHopper *Hannover*
	PH-KVI	Fokker 50	K.L.M. CityHopper *Bordeaux*
	PH-KVK	Fokker 50	K.L.M. CityHopper *London*
	PH-KXH	Fokker 50	K.L.M. CityHopper *City of Bradford*
	PH-KXM	Fokker 50	Denim Air
	PH-KZA	Fokker 70	K.L.M. CityHopper
	PH-KZB	Fokker 70	K.L.M. CityHopper
	PH-KZC	Fokker 70	K.L.M. CityHopper
	PH-KZD	Fokker 70	K.L.M. CityHopper
	PH-KZE	Fokker 70	K.L.M. CityHopper
	PH-KZF	Fokker 70	K.L.M. CityHopper
	PH-KZG	Fokker 70	K.L.M. CityHopper
	PH-KZH	Fokker 70	K.L.M. CityHopper
	PH-KZI	Fokker 70	K.L.M. CityHopper
	PH-KZK	Fokker 70	K.L.M. CityHopper
	PH-KZL	Fokker 70	K.L.M. CityHopper
	PH-KZM	Fokker 70	K.L.M. CityHopper
	PH-KZN	Fokker 70	K.L.M. CityHopper
	PH-KZO	Fokker 70	K.L.M. CityHopper
	PH-KZP	Fokker 70	K.L.M. CityHopper

Reg.	Type	Owner or Operator	Notes
PH-KZR	Fokker 70	K.L.M. CityHopper	
PH-LMT	Fokker 50	Denim Air/Air Nostrum	
PH-LXJ	Fokker 50	K.L.M. CityHopper *City of Hull*	
PH-LXK	Fokker 50	K.L.M. CityHopper *City of York*	
PH-LXP	Fokker 50	K.L.M. CityHopper *City of Durham*	
PH-LXR	Fokker 50	K.L.M. CityHopper *City of Amsterdam*	
PH-LXT	Fokker 50	K.L.M. CityHopper *City of Stavanger*	
PH-MCE	Boeing 747-21AC (SCD)	Martinair *Prins van Oranje*	
PH-MCF	Boeing 747-21AC (SCD)	Martinair *Prins Claus*	
PH-MCG	Boeing 767-31AER	Martinair *Prins Johan Friso*	
PH-MCH	Boeing 767-31AER	Martinair *Prins Constantijn*	
PH-MCI	Boeing 767-31AER	Martinair *Prins Pieter-Christiaan*	
PH-MCL	Boeing 767-31AER	Martinair *Koningin Beatrix*	
PH-MCM	Boeing 767-31AER	Martinair *Prins Floris*	
PH-MCN	Boeing 747-228F	Martinair Cargo *Prins Bernhard Junior*	
PH-MCP	McD Douglas MD-11CF	Martinair Cargo	
PH-MCR	McD Douglas MD-11CF	Martinair Cargo	
PH-MCS	McD Douglas MD-11CF	Martinair Cargo	
PH-MCT	McD Douglas MD-11CF	Martinair Cargo	
PH-MCU	McD Douglas MD-11F	Martinair Cargo *Prinses Maxima*	
PH-MCV	Boeing 767-31AER	Holland Exel	
PH-MCW	McD Douglas MD-11CF	Martinair Cargo	
PH-MCY	McD Douglas MD-11F	Martinair Cargo	
PH-MPD	Airbus A.320-232	Martinair	
PH-MPE	Airbus A.320-232	Martinair	
PH-MPF	Airbus A.320-232	Martinair	
PH-OFA	Fokker 100	K.L.M. CityHopper	
PH-OFB	Fokker 100	K.L.M. CityHopper	
PH-OFC	Fokker 100	K.L.M. CityHopper	
PH-OFD	Fokker 100	K.L.M. CityHopper	
PH-OFE	Fokker 100	K.L.M. CityHopper	
PH-OFF	Fokker 100	K.L.M. CityHopper	
PH-OFG	Fokker 100	K.L.M. CityHopper	
PH-OFH	Fokker 100	K.L.M. CityHopper	
PH-OFI	Fokker 100	K.L.M. CityHopper	
PH-OFJ	Fokker 100	K.L.M. CityHopper	
PH-OFK	Fokker 100	K.L.M. CityHopper	
PH-PBA	Douglas DC-3C	Dutch Dakota Association	
PH-PRH	Fokker 50	Denim Air	
PH-PRJ	Fokker 50	Denim Air	
PH-RAH	Beech 1900D	Rossair Europe	
PH-RAR	Beech 1900D	Rossair Europe	
PH-RAZ	Swearingen SA226TC Metro II	Binair *The Deerhunter*	
PH-RXA	Embraer RJ145MP	Air Exel	
PH-RXB	Embraer RJ145MP	Air Exel	
PH-RXC	Embraer RJ145LR	Air Exel	
PH-WXA	Fokker 70	K.L.M. CityHopper	
PH-WXC	Fokker 70	K.L.M. CityHopper	
PH-WXD	Fokker 70	K.L.M. CityHopper	
PH-XLE	Aérospatiale ATR-42-320	Air Exel	
PH-XLK	Aérospatiale ATR-42-320	Air Exel	
PH-XLO	Aérospatiale ATR-42-512	Air Exel	
PH-XLP	Aérospatiale ATR-42-512	Air Exel	
PH-XLQ	Aérospatiale ATR-42-320	Air Exel	
PH-XLS	Aérospatiale ATR-42-512	Air Exel	
PH-XRA	Boeing 737-7K2	Transavia *Leontien van Moorsel*	
PH-XRB	Boeing 737-7K2	Transavia	
PH-XRC	Boeing 737-7K2	Transavia	
PH-XRD	Boeing 737-7K2	Transavia	
PH-XRE	Boeing 737-7K2	Transavia	
PH-XRW	Boeing 737-7K2	Transavia	
PH-XRX	Boeing 737-7K2	Transavia	
PH-XRY	Boeing 737-7K2	Transavia	
PH-XRZ	Boeing 737-7K2	Transavia	

PP/PR/PT (Brazil)

PP-VMT	Douglas DC-10-30F	VARIG Log	
PP-VMU	Douglas DC-10-30F	VARIG Log	
PP-VOI	Boeing 767-341ER	VARIG	
PP-VOJ	Boeing 767-341ER	VARIG	

Notes	Reg.	Type	Owner or Operator
	PP-VOK	Boeing 767-341ER	VARIG
	PP-VOL	Boeing 767-341ER	VARIG
	PP-VPJ	McD Douglas MD-11	VARIG
	PP-VPK	McD Douglas MD-11	VARIG
	PP-VPV	Boeing 767-375ER	VARIG
	PP-VPW	Boeing 767-375ER	VARIG
	PP-VQF	McD Douglas MD-11	VARIG
	PP-VQG	McD Douglas MD-11	VARIG
	PP-VQH	McD Douglas MD-11	VARIG
	PP-VQI	McD Douglas MD-11	VARIG
	PP-VQJ	McD Douglas MD-11	VARIG
	PP-VQK	McD Douglas MD-11	VARIG
	PP-VQL	McD Douglas MD-11	VARIG
	PP-VQX	McD Douglas MD-11	VARIG
	PP-VQY	Douglas DC-10-30F	VARIG Log
	PP-VRA	Boeing 777-2Q8ER	VARIG
	PP-VRB	Boeing 777-2Q8ER	VARIG
	PP-VRC	Boeing 777-236	VARIG
	PP-VRD	Boeing 777-236	VARIG
	PP-VRE	Boeing 777-222ER	VARIG
	PP-VRF	Boeing 777-222ER	VARIG
	PP-VRG	Boeing 777-228ER	VARIG
	PP-VTC	Boeing 767-3Y0ER	VARIG
	PP-VTE	Boeing 767-3Y0ER	VARIG
	PP-VTH	McD Douglas MD-11	VARIG
	PP-VTI	McD Douglas MD-11	VARIG
	PP-VTJ	McD Douglas MD-11	VARIG
	PP-VTK	McD Douglas MD-11	VARIG
	PP-VTP	McD Douglas MD-11	VARIG
	PP-VTU	McD Douglas MD-11F	VARIG
	PT-MVC	Airbus A.330-223	TAM Brasil *The Magic Red Carpet*
	PT-MVF	Airbus A.330-203	TAM Brasil
	PT-MVG	Airbus A.330-203	TAM Brasil
	PT-MVH	Airbus A.330-203	TAM Brasil
	PT-MVK	Airbus A.330-203	TAM Brasil

PZ (Surinam)

PZ-TCM	Boeing 747-306 (SCD)	Surinam Airways *Ronald Elwin Kappel*

RA (Russia)

RA-74016	An-74	Gazpromavia/Instone Air
RA-82010	An-124	Polet
RA-82014	An-124	Polet
RA-82024	An-124	Polet
RA-82026	An-124	Polet
RA-82042	An-124	Volga Dnepr
RA-82043	An-124	Volga Dnepr
RA-82044	An-124	Volga Dnepr
RA-82045	An-124	Volga Dnepr
RA-82046	An-124	Volga Dnepr
RA-82047	An-124	Volga Dnepr
RA-82074	An-124	Volga Dnepr
RA-82075	An-124	Polet
RA-82077	An-124	Polet
RA-82078	An-124	Volga Dnepr
RA-82079	An-124	Volga Dnepr
RA-82080	An-124	Polet
RA-82081	An-124	Volga Dnepr
RA-85171	Tu-154M	Pulkovo Aviation
RA-85185	Tu-154M	Pulkovo Aviation
RA-85187	Tu-154M	Pulkovo Aviation
RA-85204	Tu-154M	Pulkovo Aviation
RA-85658	Tu-154M	Pulkovo Aviation
RA-85695	Tu-154M	Pulkovo Aviation
RA-85739	Tu-154M	Pulkovo Aviation
RA-85753	Tu-154M	Pulkovo Aviation
RA-85767	Tu-154M	Pulkovo Aviation
RA-85769	Tu-154M	Pulkovo Aviation

Reg.	Type	Owner or Operator	Notes
RA-85770	Tu-154M	Pulkovo Aviation	
RA-85771	Tu-154M	Pulkovo Aviation	
RA-85779	Tu-154M	Pulkovo Aviation	
RA-85785	Tu-154M	Pulkovo Aviation	
RA-85800	Tu-154M	Pulkovo Aviation	
RA-85832	Tu-154M	Pulkovo Aviation	
RA-85834	Tu-154M	Pulkovo Aviation	
RA-85835	Tu-154M	Pulkovo Aviation	
RA-85836	Tu-154M	Pulkovo Aviation	
RA-86466	IL-62M	Russia State Transport	
RA-86467	IL-62M	Russia State Transport	
RA-86468	IL-62M	Russia State Transport	
RA-86536	IL-62M	Russia State Transport	
RA-86537	IL-62M	Russia State Transport	
RA-86540	IL-62M	Russia State Transport	
RA-86559	IL-62M	Russia State Transport	
RA-86561	IL-62M	Russia State Transport	
RA-86710	IL-62M	Russia State Transport	
RA-86712	IL-62M	Russia State Transport	
RA-96005	IL-96	Aeroflot Russian International Airlines *V. Chkalov*	
RA-96007	IL-96	Aeroflot Russian International Airlines	
RA-96008	IL-96	Aeroflot Russian International Airlines	
RA-96010	IL-96	Aeroflot Russian International Airlines *Nikolaj Karpajev*	
RA-96011	IL-96	Aeroflot Russian International Airlines	
RA-96012	IL-96	Russia State Transport	
RA-96015	IL-96	Aeroflot Russian International Airlines	

Note: Aeroflot Russian International also operates DC-10-40Fs VP-BDE, VP-BDF, VP-BDG and VP-BDH, Airbus A.310 F-OGQR, A319s registered VP-BDM, VP-BDN, VP-BDO, VP-BWA, VP-BWG, VP-BWJ, VP-BWK AND VP-BWL plus A320s VP-BDK, VP-BWD, VP-BWE, VP-BWF, VP-BWH, VP-BWI and VP-BWM and Boeing 767-300s VP-BAV, VP-BAY, VP-BAZ and VP-BDI. Transaero employs Boeing 737s EI-CXK, EI-CXN, EI-CXR, EI-CZK, EI-DDK, EI-DDY, N100UN and N101UN as well as Boeing 767s EI-CXZ, EI-DBF, EI-DBG, EI-DBU and EI-DBW.

S2 (Bangladesh)

S2-ACO	Douglas DC-10-30	Bangladesh Biman *City of Shah Makhdum (R.A.)*	
S2-ACP	Douglas DC-10-30	Bangladesh Biman *The City of Dhaka*	
S2-ACQ	Douglas DC-10-30	Bangladesh Biman *The City of Hazarat-Shah Jalal (R.A.)*	
S2-ACR	Douglas DC-10-30F	Bangladesh Biman *The New Era*	
S2-ACS	Douglas DC-10-30	Bangladesh Biman	
S2-ADE	Airbus A.310-325	Bangladesh Biman *City of Hazrat Khan Jahan Ali (R.A.)*	
S2-ADF	Airbus A.310-325	Bangladesh Biman *City of Chittagong*	
S2-ADH	Airbus A.310-324	Bangladesh Biman	
S2-ADK	Airbus A.310-324	Bangladesh Biman	
S2-ADN	Douglas DC-10-30	Bangladesh Biman	

S5 (Slovenia)

S5-AAA	Airbus A.320-231	Adria Airways	
S5-AAB	Airbus A.320-231	Adria Airways	
S5-AAC	Airbus A.320-231	Adria Airways	
S5-AAD	Canadair CL.600-2B19 RJ	Adria Airways	
S5-AAE	Canadair CL.600-2B19 RJ	Adria Airways	
S5-AAF	Canadair CL.600-2B19 RJ	Adria Airways	
S5-AAG	Canadair CL.600-2B19 RJ	Adria Airways	
S5-AAH	Canadair CL.600-2B19 RJ	Adria Airways	
S5-AAI	Canadair CL.600-2B19 RJ	Adria Airways	
S5-AAJ	Canadair CL.600-2B19 RJ	Adria Airways	

S7 (Seychelles)

S7-AHM	Boeing 767-37DER	Air Seychelles *Vallee de Mai*	
S7-ASY	Boeing 767-3Q8ER	Air Seychelles *Aldabra*	

SE (Sweden)

Reg.	Type	Owner or Operator
SE-CFP	Douglas DC-3	Flygande Veteraner *Fridtjof Viking*
SE-DIB	McD Douglas MD-87	S.A.S. *Varin Viking*
SE-DIC	McD Douglas MD-87	S.A.S. *Grane Viking*
SE-DIF	McD Douglas MD-87	S.A.S. *Hjorulf Viking*
SE-DIK	McD Douglas MD-82	S.A.S. *Stenkil Viking* /Spanair
SE-DIL	McD Douglas MD-82	S.A.S. *Tord Viking* /Spanair
SE-DIN	McD Douglas MD-82	S.A.S. *Eskil Viking*/Spanair
SE-DIP	McD Douglas MD-87	S.A.S. *Margret Viking*
SE-DIR	McD Douglas MD-81	S.A.S. *Nora Viking*
SE-DIS	McD Douglas MD-81	S.A.S. *Sigmund Viking*
SE-DIU	McD Douglas MD-87	S.A.S. *Torsten Viking*
SE-DIZ	McD Douglas MD-82	S.A.S. *Sigyn Viking*
SE-DMB	McD Douglas MD-81	S.A.S. *Bjarne Viking*/Scanair
SE-DMF	McD Douglas MD-90-30	S.A.S. *Heidrek Viking*
SE-DMG	McD Douglas MD-90-30	S.A.S. *Hervor Viking*
SE-DMH	McD Douglas MD-90-30	S.A.S. *Torolf Viking*
SE-DMT	McD Douglas MD-81	Nordic Leisure
SE-DNT	Boeing 737-683	S.A.S. *Snefrid Viking*
SE-DNU	Boeing 737-683	S.A.S. *Unn Viking*
SE-DNX	Boeing 737-683	S.A.S. *Torvald Vikingi*
SE-DOR	Boeing 737-683	S.A.S. *Elisabeth Viking*
SE-DOT	Boeing 737-683	S.A.S. *Edla Viking*
SE-DPB	Boeing 737-33A (QC)	Falcon Air/FlyMe *Pilgrimsfalken*
SE-DPC	Boeing 737-33A (QC)	Falcon Air/FlyMe *Tornfalken*
SE-DRA	BAe 146-200	Malmö Aviation
SE-DRM	BAe 146-200	Malmö Aviation
SE-DSO	Avro RJ100	Malmö Aviation
SE-DSP	Avro RJ100	Malmö Aviation
SE-DSR	Avro RJ100	Malmö Aviation
SE-DSS	Avro RJ100	Malmö Aviation
SE-DST	Avro RJ100	Malmö Aviation
SE-DSU	Avro RJ100	Malmö Aviation
SE-DSV	Avro RJ100	Malmö Aviation
SE-DSX	Avro RJ100	Malmö Aviation
SE-DSY	Avro RL100	Malmö Aviation
SE-DTH	Boeing 737-683	S.A.S. *Vile Viking*
SE-DVU	Boeing 737-85F	Novair Airlines
SE-DZA	Embraer RJ145EP	Skyways Express
SE-DZB	Embraer RJ145EP	Skyways Express
SE-DZC	Embraer RJ145EP	Skyways Express
SE-DZD	Embraer RJ145EP	Skyways Express
SE-DZH	Boeing 737-804	Britannia Airways AB
SE-DZI	Boeing 737-804	Britannia Airways AB
SE-DZK	Boeing 737-804	Britannia Airways AB
SE-DZL	Boeing 737-804	Britannia Airways AB
SE-DZM	Boeing 737-804	Britannia Airways AB
SE-DZV	Boeing 737-804	Britannia Airways AB
SE-LGU	BAe ATP	West Air Sweden
SE-LGV	BAe ATP	West Air Sweden
SE-LGX	BAe ATP	West Air Sweden
SE-LGY	BAe ATP	West Air Sweden
SE-LGZ	BAe ATP	West Air Sweden
SE-LOG	SAAB 2000	Golden Air
SE-LOM	SAAB 2000	Golden Air
SE-LOT	SAAB 2000	Golden Air
SE-LOX	SAAB 2000	Golden Air
SE-LPR	BAe ATP	West Air Sweden
SE-LPS	BAe ATP	West Air Sweden
SE-LPT	BAe ATP	West Air Sweden
SE-LPU	BAe ATP	West Air Europe
SE-LPV	BAe ATP	West Air Sweden
SE-LTR	Fokker 50	Swe Fly
SE-LTS	Fokker 50	Swe Fly
SE-RAA	Embraer RJ135ER	City Airline *City of Gothenburg*
SE-RAB	Embraer RJ135LR	City Airline *City of Linkoping*
SE-RAC	Embraer RJ145LR	City Airline
SE-RBE	McD Douglas MD-82	Nordic Airlink
SE-RBF	Airbus A.330-223	Novair Airlines
SE-RBG	Airbus A.330-223	Novair Airlines

Reg.	Type	Owner or Operator	Notes
SE-RCO	Boeing 737-33A	FlyMe	
SE-RCP	Boeing 737-33A	FlyMe	
SE-RCR	Boeing 737-33A	FlyMe	
SE-RCS	Boeing 737-3Q8	FlyMe	
SE-RDE	McD Douglas MD-83	Viking Airlines	
SE-RDF	McD Douglas MD-83	Viking Airlines	
SE-RDG	McD Douglas MD-83	Viking Airlines	
SE-RDM	McD Douglas MD-83	Nordic Leisure	
SE-RDN	Airbus A.321-231	Novair Airlines	
SE-RDO	Airbus A.321-231	Novair Airlines	
SE-RDR	McD Douglas MD-82	FlyNordic	
SE-RDS	McD Douglas MD-83	FlyNordic	
SE-RDT	McD Douglas MD-82	FlyNordic	
SE-RDU	McD Douglas MD-82	FlyNordic	
SE-RDV	McD Douglas MD-83	FlyNordic	
SE-REE	Airbus A.340-313X	S.A.S. *Freydis Viking*	
SE-REF	Airbus A.340-313X	S.A.S. *Toste Viking*	
SE-REG	Airbus A.321-231	S.A.S. *Viger Viking*	
SE-RFA	McD Douglas MD-83	FlyNordic	

SP (Poland)

Reg.	Type	Owner or Operator	Notes
SP-FDO	An-26B	Exin Air	
SP-FDP	An-26B	Exin Air/DHL	
SP-FDR	An-26B	Exin Air/DHL	
SP-FDS	An-26B	Exin Air	
SP-FDT	An-26B	Exin Air	
SP-LDA	Embraer RJ170 100ST	Polskie Linie Lotnicze (LOT)	
SP-LDB	Embraer RJ170 100ST	LOT	
SP-LDC	Embraer RJ170 100ST	LOT	
SP-LDD	Embraer RJ170 100ST	LOT	
SP-LDE	Embraer RJ170 100LR	LOT	
SP-LDF	Embraer RJ170 100LR	LOT	
SP-LGA	Embraer RJ145EP	LOT	
SP-LGB	Embraer RJ145EP	LOT	
SP-LGC	Embraer RJ145EP	LOT	
SP-LGD	Embraer RJ145EP	LOT	
SP-LGE	Embraer RJ145LR	LOT	
SP-LGF	Embraer RJ145MP	LOT	
SP-LGG	Embraer RJ145MP	LOT	
SP-LGH	Embraer RJ145MP	LOT	
SP-LGI	Embraer RJ145MP	LOT	
SP-LGK	Embraer RJ145MP	LOT	
SP-LGL	Embraer RJ145MP	LOT	
SP-LGM	Embraer RJ145MP	LOT	
SP-LGN	Embraer RJ145MP	LOT	
SP-LGO	Embraer RJ145MP	LOT	
SP-LKA	Boeing 737-55D	LOT	
SP-LKB	Boeing 737-55D	LOT	
SP-LKC	Boeing 737-55D	LOT	
SP-LKD	Boeing 737-55D	LOT	
SP-LKE	Boeing 737-55D	LOT	
SP-LKF	Boeing 737-55D	LOT	
SP-LKG	Boeing 737-53C	Cirrus Airlines	
SP-LLA	Boeing 737-45D	LOT	
SP-LLB	Boeing 737-45D	LOT	
SP-LLC	Boeing 737-45D	LOT	
SP-LLD	Boeing 737-45D	LOT	
SP-LLE	Boeing 737-45D	LOT	
SP-LLF	Boeing 737-45D	LOT	
SP-LMC	Boeing 737-36N	LOT	
SP-LMD	Boeing 737-36N	LOT	
SP-LOA	Boeing 767-25DER	LOT *Gniezno*	
SP-LOB	Boeing 767-25DER	LOT *Krakow*	
SP-LPA	Boeing 767-35DER	LOT *Warszawa*	
SP-LPB	Boeing 767-35DER	LOT *Gdansk*	
SP-LPC	Boeing 767-35DER	LOT *Poznan*	

Note: Some LOT 737s have been transferred to low-cost subsidiary Central Wings.

ST (Sudan)

| ST-UAA | Douglas DC-8-62F | United Arabian Airlines |

Note: Sudan Airways operates A.300-622R F-ODTK and A.310-222 JY-JAV plus other aircraft leased on an *ad hoc* basis.

SU (Egypt)

SU-BDG	Airbus A.300B4-203F	EgyptAir Cargo *Toshki*
SU-BMM	Airbus A.300B4-203	AMC Airlines *Zeaid*
SU-BOW	Airbus A.310-322	AMC Airlines *Jomana*
SU-GAC	Airbus A.300B4-203F	EgyptAir Cargo *New Valley*
SU-GAL	Boeing 747-366 (SCD)	EgyptAir *Hatshepsut*
SU-GAM	Boeing 747-366 (SCD)	EgyptAir *Cleopatra*
SU-GAU	Airbus A.300B4-622R	EgyptAir *Mycerinus*
SU-GAV	Airbus A.300B4-622R	EgyptAir *Menes*
SU-GAW	Airbus A.300B4-622R	EgyptAir *Ahmuse*
SU-GBA	Airbus A.320-231	EgyptAir *Aswan*
SU-GBB	Airbus A.320-231	EgyptAir *Luxor*
SU-GBC	Airbus A.320-231	EgyptAir *Hurghada*
SU-GBD	Airbus A.320-231	EgyptAir *Taba*
SU-GBE	Airbus A.320-231	EgyptAir *El Alamein*
SU-GBF	Airbus A.320-231	EgyptAir *Sharm El Sheikh*
SU-GBG	Airbus A.320-231	EgyptAir *Saint Catherine*
SU-GBM	Airbus A.340-212	EgyptAir *Osiris Express*
SU-GBN	Airbus A.340-212	EgyptAir *Cleo Express*
SU-GBO	Airbus A.340-212	EgyptAir *Hathor Express*
SU-GBP	Boeing 777-266	EgyptAir *Nefertiti*
SU-GBR	Boeing 777-266	EgyptAir *Nefertari*
SU-GBS	Boeing 777-266	EgyptAir *Tyie*
SU-GBT	Airbus A.321-231	EgyptAir *Red Sea*
SU-GBU	Airbus A.321-231	EgyptAir *Sinai*
SU-GBV	Airbus A.321-231	EgyptAir *Mediterranean*
SU-GBW	Airbus A.321-231	EgyptAir/Air Cairo *The Nile*
SU-GBX	Boeing 777-266ER	EgyptAir *Neit*
SU-GBY	Boeing 777-266ER	EgyptAir *Titi*
SU-GBZ	Airbus A.320-232	EgyptAir
SU-GCA	Airbus A.320-232	EgyptAir
SU-GCB	Airbus A.320-232	EgyptAir
SU-GCC	Airbus A.320-232	EgyptAir
SU-GCD	Airbus A.320-232	EgyptAir
SU-GCE	Airbus A.330-243	EgyptAir
SU-GCF	Airbus A.330-243	EgyptAir

SX (Greece)

SX-BBT	Boeing 737-33A	Aegean Airlines *Kastalia*
SX-BBU	Boeing 737-33A	Aegean Airlines *Joanna*
SX-BEK	Airbus A.300B4-605R	Olympic Airlines *Macedonia*
SX-BEL	Airbus A.300B4-605R	Olympic Airlines *Athena*
SX-BEM	Airbus A.300B4-605R	Olympic Airlines *Creta*
SX-BGH	Boeing 737-4Y0	Aegean Airlines *Iniochos*
SX-BGI	Boeing 737-3L9	Aegean Airlines
SX-BGJ	Boeing 737-4S3	Aegean Airlines
SX-BGK	Boeing 737-3Y0	Aegean Airlines *Thessaloniki*
SX-BGN	Boeing 737-45D	Aegean Airlines
SX-BGR	Boeing 737-4Q8	Aegean Airlines
SX-BGS	Boeing 737-4Q8	Aegean Airlines
SX-BGV	Boeing 737-4Q8	Aegean Airlines
SX-BGW	Boeing 737-31S	Aegean Airlines
SX-BGY	Boeing 737-31S	Aegean Airlines
SX-BGZ	Boeing 737-31S	Aegean Airlines
SX-BKA	Boeing 737-484	Olympic Airlines *Vergina*
SX-BKB	Boeing 737-484	Olympic Airlines *Olynthos*
SX-BKC	Boeing 737-484	Olympic Airlines *Philipoli*
SX-BKD	Boeing 737-484	Olympic Airlines *Amphipoli*
SX-BKE	Boeing 737-484	Olympic Airlines *Stagira*
SX-BKF	Boeing 737-484	Olympic Airlines *Dion*

Reg.	Type	Owner or Operator	Notes
SX-BKG	Boeing 737-484	Olympic Airlines *Pella*	
SX-BKH	Boeing 737-4Q8	Olympic Airlines	
SX-BKI	Boeing 737-4Q8	Olympic Airlines	
SX-BKK	Boeing 737-4Q8	Olympic Airlines	
SX-BKL	Boeing 737-4Y0	Olympic Airlines	
SX-BKM	Boeing 737-4Q8	Olympic Airlines	
SX-BKN	Boeing 737-4Q8	Olympic Airlines	
SX-BLA	Boeing 737-33R	Olympic Airlines	
SX-BLC	Boeing 737-3Q8	Olympic Airlines	
SX-BLM	Boeing 737-42C	Aegean Airlines	
SX-BLW	Boeing 757-236 (ER)	Greece Airways/Air Scotland	
SX-BMC	Boeing 737-42J	Olympic Airlines *City of Alexandroupoli*	
SX-BOA	Boeing 717-2K9	Olympic Aviation *Andromeda*	
SX-BOB	Boeing 717-2K9	Olympic Aviation *Kassiopi*	
SX-BOC	Boeing 717-23S	Olympic Aviation *Iridanos*	
SX-BVA	Airbus A.320-232	Hellas Jet *Pegasus*	
SX-BVB	Airbus A.320-232	Hellas Jet *Hermes*	
SX-BVC	Airbus A.320-232	Hellas Jet *Orion*	
SX-CVP	Douglas DC-10-15	Electra Airlines *Gerhard Mercator*	
SX-DFA	Airbus A.340-313X	Olympic Airlines *Olympia*	
SX-DFB	Airbus A.340-313X	Olympic Airlines *Delphi*	
SX-DFC	Airbus A.340-313X	Olympic Airlines *Marathon*	
SX-DFD	Airbus A.340-313X	Olympic Airlines *Epidaurus*	

Note: Olympic Airlines also operates B.737 EC-IOR.

TC (Turkey)

TC-AAB	Boeing 737-86N	Pegasus Airlines
TC-AAP	Boeing 737-86N	Pegasus Airlines
TC-ABK	Airbus A.300B4-203F	Kuzu Cargo
TC-AKP	Airbus A.310-203	Turkish Airlines
TC-APD	Boeing 737-42R	Pegasus Airlines/Atlas Jet
TC-APF	Boeing 737-86N	Pegasus Airlines
TC-APH	Boeing 737-8S3	Pegasus Airlines/Air Algerie
TC-API	Boeing 737-86N	Pegasus Airlines
TC-APJ	Boeing 737-86N	Pegasus Airlines
TC-APM	Boeing 737-809	Pegasus Airlines
TC-APN	Boeing 737-86N	Pegasus Airlines/Fly Air
TC-APR	Boeing 737-4Y0	Pegasus Airlines
TC-APU	Boeing 737-82R	Pegasus Airlines
TC-APY	Boeing 737-86N	Pegasus Airlines
TC-APZ	Boeing 737-809	Pegasus Airlines
TC-FBB	McD Douglas MD-83	Freebird Airlines
TC-FBD	McD Douglas MD-83	Freebird Airlines
TC-FBE	Airbus A.320-212	Freebird Airlines/Air Algerie
TC-FBF	Airbus A.320-212	Freebird Airlines
TC-FBG	McD Douglas MD-83	Freebird Airlines
TC-FBT	McD Douglas MD-83	Freebird Airlines
TC-FLA	Airbus A.300B4-203	Fly Air
TC-FLE	Airbus A.300B2K-3C	Fly Air
TC-FLF	Airbus A.300B4-2C	Fly Air
TC-FLG	Airbus A.300B4-2C	Fly Air
TC-FLH	Boeing 737-3Q8	Fly Air
TC-FLI	Boeing 737-4Q8	Fly Air
TC-FLJ	Airbus A.300B2K-3C	Fly Air
TC-FLK	Airbus A.300B4-203	Fly Air
TC-FLL	Airbus A.300B4-2C	Fly Air
TC-FLM	Airbus A.300B2K-3C	Fly Air
TC-IEA	Boeing 737-8CX	Inter Express Airlines *Berke*
TC-IEB	Boeing 737-8CX	Inter Express Airlines *Cem & Emre*
TC-JCO	Airbus A.310-203	Kibris Turkish Airlines *Lerkosa*
TC-JCV	Airbus A.310-304	Turkish Airlines *Aras*
TC-JCY	Airbus A.310-304	Turkish Airlines *Coruh*
TC-JCZ	Airbus A.310-304	Turkish Airlines *Ergene*
TC-JDA	Airbus A.310-304	Turkish Airlines *Aksu*
TC-JDB	Airbus A.310-304	Turkish Airlines *Göksu*
TC-JDF	Boeing 737-4Y0	Turkish Airlines *Ayvalik*
TC-JDG	Boeing 737-4Y0	Turkish Airlines *Marmaris*
TC-JDH	Boeing 737-4Y0	Turkish Airlines *Amasra*
TC-JDJ	Airbus A.340-311	Turkish Airlines *Istanbul*

Notes	Reg.	Type	Owner or Operator
	TC-JDK	Airbus A.340-311	Turkish Airlines *Isparta*
	TC-JDL	Airbus A.340-311	Turkish Airlines *Ankara*
	TC-JDM	Airbus A.340-311	Turkish Airlines *Izmir*
	TC-JDN	Airbus A.340-313	Turkish Airlines *Adana*
	TC-JDT	Boeing 737-4Y0	Turkish Airlines *Alanya*
	TC-JDY	Boeing 737-4Y0	Turkish Airlines *Antalya*
	TC-JEN	Boeing 737-4Q8	Turkish Airlines *Gelibolu*
	TC-JEO	Boeing 737-4Q8	Turkish Airlines *Anadolu*
	TC-JER	Boeing 737-4Y0	Turkish Airlines *Mugla*
	TC-JET	Boeing 737-4Y0	Turkish Airlines *Canakkale*
	TC-JEU	Boeing 737-4Y0	Turkish Airlines *Kayseri*
	TC-JEV	Boeing 737-4Y0	Turkish Airlines *Efes*
	TC-JEY	Boeing 737-4Y0	Turkish Airlines *Side*
	TC-JEZ	Boeing 737-4Y0	Turkish Airlines *Bergama*
	TC-JFC	Boeing 737-8F2	Turkish Airlines *Diyarbakir*
	TC-JFD	Boeing 737-8F2	Turkish Airlines *Rize*
	TC-JFE	Boeing 737-8F2	Turkish Airlines *Hatay*
	TC-JFF	Boeing 737-8F2	Turkish Airlines *Afyon*
	TC-JFG	Boeing 737-8F2	Turkish Airlines *Mardi*
	TC-JFH	Boeing 737-8F2	Turkish Airlines *Igdir*
	TC-JFI	Boeing 737-8F2	Turkish Airlines *Sivas*
	TC-JFJ	Boeing 737-8F2	Turkish Airlines *Agri*
	TC-JFK	Boeing 737-8F2	Turkish Airlines *Zonguldak*
	TC-JFL	Boeing 737-8F2	Turkish Airlines *Ordu*
	TC-JFM	Boeing 737-8F2	Turkish Airlines *Nigde*
	TC-JFN	Boeing 737-8F2	Turkish Airlines *Bitlis*
	TC-JFO	Boeing 737-8F2	Turkish Airlines *Batman*
	TC-JFP	Boeing 737-8F2	Turkish Airlines *Amasya*
	TC-JFR	Boeing 737-8F2	Turkish Airlines *Giresun*
	TC-JFT	Boeing 737-8F2	Turkish Airlines *Kastamonu*
	TC-JFU	Boeing 737-8F2	Turkish Airlines *Elazig*
	TC-JFV	Boeing 737-8F2	Turkish Airlines *Tunceli*
	TC-JFY	Boeing 737-8F2	Turkish Airlines *Manisa*
	TC-JFZ	Boeing 737-8F2	Turkish Airlines *Bolu*
	TC-JGA	Boeing 737-8F2	Turkish Airlines *Malatya*
	TC-JGB	Boeing 737-8F2	Turkish Airlines *Eskisehir*
	TC-JGC	Boeing 737-8F2	Turkish Airlines *Kocaeli*
	TC-JGD	Boeing 737-8F2	Turkish Airlines *Nevsehir*
	TC-JGE	Boeing 737-8F2	Turkish Airlines *Tekirdag*
	TC-JGF	Boeing 737-8F2	Turkish Airlines *Ardahan*
	TC-JIH	Airbus A.340-313X	Turkish Airlines *Hakkari*
	TC-JII	Airbus A.340-313X	Turkish Airlines *Aydin*
	TC-JKA	Boeing 737-4Q8	Turkish Airlines *Kars*
	TC-JKB	Boeing 737-4Y0	Turkish Airlines *Konya*
	TC-JKC	Boeing 737-4Y0	Turkish Airlines *Bursa*
	TC-JKD	Boeing 737-4Y0	Turkish Airlines *Trabzon*
	TC-JLA	Airbus A.320-214	Turkish Airlines *Sanijurfa*
	TC-JLB	Airbus A.320-214	Turkish Airlines *Balikesir*
	TC-JLC	Airbus A.320-214	Turkish Airlines *Kahramanmaras*
	TC-JLD	Airbus A.320-214	Turkish Airlines *Mersin*
	TC-JLE	Airbus A.320-214	Turkish Airlines *Erzurum*
	TC-JLF	Airbus A.320-214	Turkish Airlines *Sakarya*
	TC-JLG	Airbus A.320-214	Turkish Airlines *Adiyman*
	TC-JMA	Airbus A.321-111	Turkish Airlines *Yozgat*
	TC-KTC	Airbus A.321-211	Kibris Turkish Airlines *Girne*
	TC-KTD	Airbus A.321-211	Kibris Turkish Airlines *Iskele*
	TC-KTY	Airbus A.321-211	Kibris Turkish Airlines *Lefke*
	TC-KZU	Airbus A.300B4-203F	Kuzu Cargo
	TC-MAO	Boeing 737-86N	Kibris Turkish Airlines *Karpaz*
	TC-MNA	Airbus A.300B4-203F	MNG Cargo
	TC-MNB	Airbus A.300B4-203F	MNG Cargo
	TC-MNC	Airbus A.300B4-203F	MNG Cargo
	TC-MND	Airbus A.300B4-203F	MNG Cargo
	TC-MNJ	Airbus A.300B4-203F	MNG Cargo
	TC-MNL	Boeing 737-4Q8	MNG Pax/Air Algerie
	TC-MNM	Boeing 737-4Q8	MNG Pax/Air Algerie
	TC-MNY	Airbus A.300B4-203	MNG Pax
	TC-	Airbus A.300B4-203	MNG Pax
	TC-MSO	Boeing 737-8S3	Kibris Turkish Airlines *Magusa*
	TC-MZZ	Boeing 737-8S3	Kibris Turkish Airlines *Guzelyurt*
	TC-OAA	Airbus A.300B4-605R	Onur Air
	TC-OAB	Airbus A.300B4-605R	Onur Air

Reg.	Type	Owner or Operator	Notes
TC-OAC	Airbus A.320-212	Onur Air	
TC-OAD	Airbus A.320-212	Onur Air	
TC-OAE	Airbus A.321-231	Onur Air	
TC-OAF	Airbus A.321-231	Onur Air	
TC-OAG	Airbus A.300B4-605R	Onur Air	
TC-OAH	Airbus A.300B4-605R	Onur Air	
TC-OAI	Airbus A.321-231	Onur Air	
TC-OAK	Airbus A.321-231	Onur Air	
TC-OAL	Airbus A.321-231	Onur Air	
TC-OAN	Airbus A.321-231	Onur Air	
TC-OGB	Boeing 757-225	Atlas Jet International *Perihan*	
TC-OGD	Boeing 757-2G5	Atlas Jet International/Air Astana	
TC-OGE	Airbus A.320-214	Atlas Jet International	
TC-OGF	Airbus A.320-214	Atlas Jet International	
TC-OGG	Boeing 757-2G5	Atlas Jet International *Sait*	
TC-OGH	Boeing 757-225	Atlas Jet International *Yanikali*	
TC-ONJ	Airbus A.321-131	Onur Air *Kaptan Soray Sahin*	
TC-ONK	Airbus A.300B4-103	Onur Air *Pinar*	
TC-ONL	Airbus A.300B4-103	Onur Air *Selin*	
TC-ONM	McD Douglas MD-88	Onur Air *Yasemin*	
TC-ONN	McD Douglas MD-88	Onur Air *Ece*	
TC-ONO	McD Douglas MD-88	Onur Air *Yonca*	
TC-ONP	McD Douglas MD-88	Onur Air *Esra*	
TC-ONR	McD Douglas MD-88	Onur Air *Evren*	
TC-ONS	Airbus A.321-131	Onur Air *Funda*	
TC-ONT	Airbus A.300B4-203	Onur Air *B. Basar*	
TC-ONU	Airbus A.300B4-203	Onur Air	
TC-ONY	Airbus A.300B2K-3C	Onur Air	
TC-ORH	Airbus A.300C4-203	Orbit Express *Aida*	
TC-ORI	Airbus A.300-203F	Orbit Express	
TC-SKA	Boeing 737-4Y0	Sky Airlines *Sun*	
TC-SKB	Boeing 737-430	Sky Airlines *Star*	
TC-SKC	Boeing 737-85F	Sky Airlines *Moon*	
TC-SKD	Boeing 737-4Q8	Sky Airlines *Black Eagle*	
TC-SKE	Boeing 737-4Q8	Sky Airlines *Milky Way*	
TC-SNS	Boeing 737-4Y0	SunExpress	
TC-SUC	Boeing 737-86N	SunExpress	
TC-SUG	Boeing 737-8CX	SunExpress	
TC-SUH	Boeing 737-8CX	SunExpress	
TC-SUI	Boeing 737-8CX	SunExpress	
TC-SUJ	Boeing 737-8CX	SunExpress	
TC-SUL	Boeing 737-85F	SunExpress	
TC-SUM	Boeing 737-85F	SunExpress	

TF (Iceland)

TF-ABA	Boeing 747-267B	Air Atlanta Icelandic/Malaysian Airlines	
TF-ABO	Boeing 747-1D1	Air Atlanta Icelandic/Saudi Arabian Airlines	
TF-ABP	Boeing 747-267B	Air Atlanta Icelandic/Malaysian Airlines	
TF-AMA	Boeing 747-412	Air Atlanta Icelandic/Iberia	
TF-AMB	Boeing 747-412	Air Atlanta Icelandic/Iberia	
TF-AME	Boeing 747-312	Air Atlanta Icelandic	
TF-ARD	Boeing 757-225	Air Atlanta Icelandic/Excel Airways	
TF-ARE	Boeing 757-225	Air Atlanta Icelandic/Excel Airways	
TF-ARG	Boeing 747-236B	Air Atlanta Europe	
TF-ARH	Boeing 747-230B (SF)	Air Atlanta Cargo	
TF-ARI	Boeing 757-2Y0	Air Atlanta Icelandic/Excel Airways	
TF-ARJ	Boeing 747-236B (M)	Air Atlanta Cargo	
TF-ARK	Boeing 757-225	Air Atlanta Icelandic/Excel Airways	
TF-ARL	Boeing 747-230F (SCD)	Air Atlanta Icelandic/MAS Kargo	
TF-ARM	Boeing 747-230BF (SCD)	Air Atlanta Icelandic/MAS Kargo	
TF-ARN	Boeing 747-2F6B (SCD)	Air Atlanta Icelandic/MAS Kargo	
TF-ARO	Boeing 747-243B	Air Atlanta Icelandic	
TF-ARP	Boeing 747-230F (SCD)	Air Atlanta Cargo	
TF-ARS	Boeing 747-357	Air Atlanta Icelandic/Air Algerie	
TF-ARU	Boeing 747-344	Air Atlanta Icelandic	
TF-ARV	Boeing 747-230F (SCD)	Air Atlanta Cargo	
TF-ARW	Boeing 747-256B (SF)	Air Atlanta Icelandic/MAS Kargo	
TF-ARY	Boeing 747-329 (SF)	Air Atlanta Cargo	
TF-ATD	Boeing 747-267B	Air Atlanta Icelandic	
TF-ATI	Boeing 747-341	Air Atlanta Icelandic/Iberia *Gonzalo de Berceo*	

Notes	Reg.	Type	Owner or Operator
	TF-ATJ	Boeing 747-341	Air Atlanta Icelandic/Iberia *Jose Zorrilla*
	TF-ATT	Boeing 767-383ER	Air Atlanta Icelandic
	TF-ATU	Boeing 767-3Y0ER	Air Atlanta Icelandic/Excel Airways
	TF-ATX	Boeing 747-236B (SF)	Air Atlanta Icelandic/MAS Kargo
	TF-ATY	Boeing 767-204ER	Air Atlanta Icelandic/Air Algerie
	TF-ATZ	Boeing 747-236B (SF)	Air Atlanta Icelandic/MAS Kargo
	TF-BBC	Boeing 737-3Q4F	Bluebird Cargo
	TF-BBD	Boeing 737-3Y0F	Bluebird Cargo
	TF-BBE	Boeing 737-36EF	Bluebird Cargo
	TF-BBF	Boeing 737-36EF	Bluebird Cargo
	TF-ELA	Boeing 737-33AQC	Islandsflug
	TF-ELB	Airbus A.300B4-622RF	Islandsflug/DHL
	TF-ELC	Boeing 737-3M8	Islandsflug/flyglobescan
	TF-ELD	Boeing 737-46B	Islandsflug
	TF-ELE	Airbus A.310-304F	Islandsflug
	TF-ELG	Airbus A.300C4-605RF	Islandsflug/Air Hong Kong
	TF-ELJ	Boeing 737-46B	Islandsflug
	TF-ELL	Boeing 737-210C	Islandsflug/ATA Brasil (std Southend)
	TF-ELM	Boeing 737-3M8QC	Islandsflug
	TF-ELN	Boeing 737-3Q8QC	Islandsflug/flyglobescan
	TF-ELO	Boeing 737-33AQC	Islandsflug Cargo
	TF-ELP	Boeing 737-429	Islandsflug/Aviajet
	TF-ELR	Airbus A.310-324	Islandsflug/Corsair
	TF-ELS	Airbus A.310-304F	Islandsflug Cargo
	TF-ELU	Airbus A.300B4-622RF	Islandsflug Cargo
	TF-ELV	Boeing 737-4S3	Islandsflug
	TF-ELW	Airbus A.300C4-605RF	Islandsflug Cargo
	TF-ELY	Boeing 737-4Q8	Islandsflug
	TF-ELZ	Boeing 737-4Y0	Islandsflug/Aviajet
	TF-FIA	Boeing 767-3Y0ER	Icelandair/Loftleidir Icelandic
	TF-FIB	Boeing 767-383ER	Icelandair/Loftleidir Icelandic
	TF-FIG	Boeing 757-23APF	Icelandair Cargo
	TF-FIH	Boeing 757-208	Icelandair *Hafdis*
	TF-FII	Boeing 757-208	Icelandair *Fanndis*
	TF-FIJ	Boeing 757-208	Icelandair *Svandis*
	TF-FIK	Boeing 757-28A	Icelandair *Soldis*
	TF-FIN	Boeing 757-208	Icelandair *Bryndis*
	TF-FIO	Boeing 757-208	Icelandair *Valdis*
	TF-FIP	Boeing 757-208	Icelandair *Leifur Eiriksson*
	TF-FIR	Boeing 757-256	Icelandair/Loftleidir Icelandic
	TF-FIU	Boeing 757-256	Icelandair/Loftleidir Icelandic
	TF-FIV	Boeing 757-208	Icelandair *Gudridur Porbjarnardottir*
	TF-FIW	Boeing 757-27B	Icelandair/Loftleidir Icelandic
	TF-FIX	Boeing 757-308	Icelandair *Snorri Porfinnsson*
	TF-JXA	McD Douglas MD-82	JetX
	TF-JXB	McD Douglas MD-82	JetX/MyAir/Blue Line

Note: Iceland Express operates Boeing 737 G-STRE. Air Atlanta, Islandsflug and Loftleidir Icelandic aircraft are frequently leased to other airlines on a short-term basis.

TJ (Cameroon)

TJ-CAC	Boeing 767-33AER	Cameroon Airlines *Le Dja*
TJ-CAE	Boeing 747-312	Cameroon Airlines *Big Boss*

TR (Gabon)

TR-LFH	Boeing 767-266ER	Air Gabon

Note: Air Gabon also operates Boeing 747-2Q2B F-ODJG *President Leon Mba*

TS (Tunisia)

TS-IAX	Airbus A.300-622R	Fly International/Libyan Arab Airlines
TS-IAY	Airbus A.300-622R	Fly International/Afriquiyah Airways
TS-IAZ	Airbus A.300-622R	Fly International/Libyan Arab Airlines
TS-IEC	Boeing 737-33A	Karthago Airlines
TS-IED	Boeing 737-33A	Karthago Airlines
TS-IEE	Boeing 737-33A	Karthago Airlines

Reg.	Type	Owner or Operator	Notes
TS-IEF	Boeing 737-3Q8	Karthago Airlines	
TS-IGU	Airbus A.310-203	Fly International/Libyan Arab Airlines	
TS-IMB	Airbus A.320-211	Tunis Air *Fahrat Hached*	
TS-IMC	Airbus A.320-211	Tunis Air *7 Novembre*	
TS-IMD	Airbus A.320-211	Tunis Air *Khereddine*	
TS-IME	Airbus A.320-211	Tunis Air *Tabarka*	
TS-IMF	Airbus A.320-211	Tunis Air *Djerba*	
TS-IMG	Airbus A.320-211	Tunis Air *Abou el Kacem Chebbi*	
TS-IMH	Airbus A.320-211	Tunis Air *Ali Belhaouane*	
TS-IMI	Airbus A.320-211	Tunis Air *Jughurta*	
TS-IMJ	Airbus A.319-114	Tunis Air *El Kantaoui*	
TS-IMK	Airbus A.319-114	Tunis Air *Kerkenah*	
TS-IML	Airbus A.320-211	Tunis Air *Gafsa el Ksar*	
TS-IMM	Airbus A.320-211	Tunis Air *Le Bardo*	
TS-IMN	Airbus A.320-211	Tunis Air *Ibn Khaldoun*	
TS-IMO	Airbus A.319-114	Tunis Air *Hannibal*	
TS-IMP	Airbus A.320-211	Tunis Air *La Galite*	
TS-INA	Airbus A.320-214	Nouvelair/Afriquiyah Airways	
TS-INB	Airbus A.320-214	Nouvelair	
TS-INC	Airbus A.320-214	Nouvelair *Youssef*	
TS-IND	Airbus A.320-212	Nouvelair	
TS-INE	Airbus A.320-212	Nouvelair	
TS-INF	Airbus A.320-212	Nouvelair	
TS-ING	Airbus A.320-211	Nouvelair/Afriquiyah Airways	
TS-INH	Airbus A.320-211	Nouvelair/Afriquiyah Airways	
TS-INI	Airbus A.320-212	Nouvelair	
TS-INJ	Airbus A.320-211	Libyan Arab Airlines	
TS-IOG	Boeing 737-5H3	Tunis Air *Sfax*	
TS-IOH	Boeing 737-5H3	Tunis Air *Hammamet*	
TS-IOI	Boeing 737-5H3	Tunis Air *Mahida*	
TS-IOJ	Boeing 737-5H3	Tunis Air *Monastir*	
TS-IOK	Boeing 737-6H3	Tunis Air *Kairouan*	
TS-IOL	Boeing 737-6H3	Tunis Air *Tozeur-Nefta*	
TS-IOM	Boeing 737-6H3	Tunis Air *Carthage*	
TS-ION	Boeing 737-6H3	Tunis Air *Utique*	
TS-IOP	Boeing 737-6H3	Tunis Air *El Jem*	
TS-IOQ	Boeing 737-6H3	Tunis Air *Bizerte*	
TS-IOR	Boeing 737-6H3	Tunis Air *Tahar Haddad*	
TS-IPA	Airbus A.300B4-605R	Tunis Air *Sidi Bou Said*	
TS-IPB	Airbus A.300B4-605R	Tunis Air *Tunis*	
TS-IPC	Airbus A.300B4-605R	Tunis Air *Amilcar*	

Note: Nouvelair also operates MD-83s EI-CEK and EI-CNO.

UK (Uzbekistan)

UK-31001	Airbus A.310-324	Uzbekistan Airways *Tashkent*
UK-31002	Airbus A.310-324	Uzbekistan Airways *Fergana*
UK-31003	Airbus A.310-324	Uzbekistan Airways *Bukhara*
UK-75700	Boeing 757-23P	Uzbekistan Airways

Note: Uzbekistan Airways also operates Boeing 767-3CBER VP-BUE, 767-33PERs VP-BUA and VP-BUZ, 757-23Ps VP-BUB and VP-BUD, and 757-231s VP-BUH, VP-BUI and VP-BUJ.

UN (Kazakhstan)

Note: Air Astana operates Boeing 757-2G5s P4-EAS, P4-FAS and P4-GAS.

UR (Ukraine)

UR-GAC	Boeing 737-247	Ukraine International
UR-GAH	Boeing 737-32Q	Ukraine International *Mayrni*
UR-GAJ	Boeing 737-5Y0	Ukraine International
UR-GAK	Boeing 737-5Y0	Ukraine International
UR-GAL	Boeing 737-341	Ukraine International
UR-GAM	Boeing 737-4Y0	Ukraine International
UR-GAN	Boeing 737-36N	Ukraine International
UR-GAO	Boeing 737-4Z9	Ukraine International
UR-VVF	Boeing 767-383ER	Aerosvit Airlines

Notes	Reg.	Type	Owner or Operator
	UR-VVG	Boeing 767-383ER	Aerosvit Airlines
	UR-ZYD	Antonov An-124	United Arab Emirates
	UR-09307	Antonov An-22A	Antonov Design Bureau
	UR-74057	Antonov An-74-200	Atlantic Airlines
	UR-82007	Antonov An-124	Antonov Design Bureau
	UR-82008	Antonov An-124	Antonov Design Bureau
	UR-82009	Antonov An-124	Antonov Design Bureau
	UR-82027	Antonov An-124	Antonov Design Bureau
	UR-82029	Antonov An-124	Antonov Design Bureau
	UR-82060	Antonov An-225	Antonov Design Bureau
	UR-82072	Antonov An-124	Antonov Design Bureau
	UR-82073	Antonov An-124	Antonov Design Bureau

V5 (Namibia)

	V5-NMA	Boeing 747-48E	Air Namibia *Welwitschia*
	V5-NMC	McD Douglas MD-11	Air Namibia

V8 (Brunei)

	V8-ALI	Boeing 747-430	Brunei Sultan's Flight
	V8-BKH	Airbus A.340-212	Brunei Sultan's Flight
	V8-MHB	Boeing 767-27GER	Brunei Sultan's Flight
	V8-RBF	Boeing 767-33AER	Royal Brunei Airlines
	V8-RBG	Boeing 767-33AER	Royal Brunei Airlines
	V8-RBH	Boeing 767-33AER	Royal Brunei Airlines
	V8-RBJ	Boeing 767-33AER	Royal Brunei Airlines
	V8-RBK	Boeing 767-33AER	Royal Brunei Airlines
	V8-RBL	Boeing 767-33AER	Royal Brunei Airlines
	V8-RBM	Boeing 767-328ER	Royal Brunei Airlines/Air Europa
	V8-RBN	Boeing 767-328ER	Royal Brunei Airlines

VH (Australia)

	VH-OEB	Boeing 747-48E	QANTAS *Phillip Island*
	VH-OEC	Boeing 747-4H6	QANTAS *King Island*
	VH-OED	Boeing 747-4H6	QANTAS *Kangaroo Island*
	VH-OEE	Boeing 747-438ER	QANTAS
	VH-OEF	Boeing 747-438ER	QANTAS
	VH-OEG	Boeing 747-438ER	QANTAS
	VH-OEH	Boeing 747-438ER	QANTAS
	VH-OEI	Boeing 747-438ER	QANTAS
	VH-OEJ	Boeing 747-438ER	QANTAS
	VH-OJA	Boeing 747-438	QANTAS *City of Canberra*
	VH-OJB	Boeing 747-438	QANTAS *City of Sydney*
	VH-OJC	Boeing 747-438	QANTAS *City of Melbourne*
	VH-OJD	Boeing 747-438	QANTAS *City of Brisbane*
	VH-OJE	Boeing 747-438	QANTAS *City of Adelaide*
	VH-OJF	Boeing 747-438	QANTAS *City of Perth*
	VH-OJG	Boeing 747-438	QANTAS *City of Hobart*
	VH-OJH	Boeing 747-438	QANTAS *City of Darwin*
	VH-OJI	Boeing 747-438	QANTAS *Longreach*
	VH-OJJ	Boeing 747-438	QANTAS *Winton*
	VH-OJK	Boeing 747-438	QANTAS *City of Newcastle*
	VH-OJL	Boeing 747-438	QANTAS *City of Ballaarat*
	VH-OJM	Boeing 747-438	QANTAS *City of Gosford*
	VH-OJN	Boeing 747-438	QANTAS *City of Dubbo*
	VH-OJO	Boeing 747-438	QANTAS *City of Toowoomba*
	VH-OJP	Boeing 747-438	QANTAS *City of Albury*
	VH-OJQ	Boeing 747-438	QANTAS *City of Mandurah*
	VH-OJR	Boeing 747-438	QANTAS *City of Bathurst*
	VH-OJS	Boeing 747-438	QANTAS
	VH-OJT	Boeing 747-438	QANTAS
	VH-OJU	Boeing 747-438	QANTAS

Reg.	Type	Owner or Operator	Notes

VN (Vietnam)

VN-A141	Boeing 777-2Q8ER	Vietnam Airlines
VN-A142	Boeing 777-2Q8ER	Vietnam Airlines
VN-A143	Boeing 777-26KER	Vietnam Airlines
VN-A144	Boeing 777-26KER	Vietnam Airlines
VN-A145	Boeing 777-26KER	Vietnam Airlines
VN-A146	Boeing 777-26KER	Vietnam Airlines

Note: Vietnam Airlines hope to commence a London-Hanoi service during 2005.

VP-B (Bermuda)

VP-BAT	Boeing 747SP-21	Worldwide Aviation
VP-BAV	Boeing 767-36NER	Aeroflot Russian International *L.Tolstoy*
VP-BAX	Boeing 767-36NER	Aeroflot Russian International *F. Dostoevsky*
VP-BAY	Boeing 767-36NER	Aeroflot Russian International *I.Turgenev*
VP-BAZ	Boeing 767-36NER	Aeroflot Russian International *N. Nekrasov*
VP-BBR	Boeing 757-22L	Azerbaijan Airlines
VP-BBS	Boeing 757-22L	Azerbaijan Airlines
VP-BBT	Boeing 737-705	Ford Motor Company/Stansted
VP-BBU	Boeing 737-705	Ford Motor Company/Stansted
VP-BDE	Douglas DC-10-40F	Aeroflot Russian International
VP-BDF	Douglas DC-10-40F	Aeroflot Russian International
VP-BDG	Douglas DC-10-40F	Aeroflot Russian International
VP-BDH	Douglas DC-10-40F	Aeroflot Russian International
VP-BDI	Boeing 767-38AER	Aeroflot Russian International
VP-BDK	Airbus A.320-214	Aeroflot Russian International *G. Sviridov*
VP-BDM	Airbus A.319-111	Aeroflot Russian International
VP-BDN	Airbus A.319-111	Aeroflot Russian International
VP-BDO	Airbus A.319-111	Aeroflot Russian International *I. Stravinsky*
VP-BIA	Boeing 747-243F (SCD)	Volga Dnepr
VP-BIB	Boeing 747-243M	Volga Dnepr
VP-BUA	Boeing 767-33PER	Uzbekistan Airlines *Samarkand*
VP-BUB	Boeing 757-23P	Uzbekistan Airlines *Urgench*
VP-BUD	Boeing 757-23P	Uzbekistan Airlines *Shahrisabz*
VP-BUE	Boeing 767-3CBER	Uzbekistan Airlines
VP-BUF	Boeing 767-33PER	Uzbekistan Airlines
VP-BUI	Boeing 757-23H	Uzbekistan Airlines
VP-BUI	Boeing 757-231	Uzbekistan Airlines
VP-BUJ	Boeing 757-231	Uzbekistan Airlines
VP-BUZ	Boeing 767-33PER	Uzbekistan Airlines *Khiva*
VP-BWA	Airbus A.319-111	Aeroflot Russian International
VP-BWD	Airbus A.320-214	Aeroflot Russian International *D. Aliabiev*
VP-BWE	Airbus A.320-214	Aeroflot Russian International *H.Rimsky-Korsakov*
VP-BWF	Airbus A.320-214	Aeroflot Russian International *D. Shostakovich*
VP-BWG	Airbus A.319-111	Aeroflot Russian International
VP-BWH	Airbus A.320-214	Aeroflot Russian International *M. Balakirev*
VP-BWI	Airbus A.320-214	Aeroflot Russian International *A. Glazunov*
VP-BWJ	Airbus A.319-111	Aeroflot Russian International *A. Shnitke*
VP-BWK	Airbus A.319-111	Aeroflot Russian International *S. Taneyev*
VP-BWL	Airbus A.319-111	Aeroflot Russian International *A. Grechaninov*
VP-BWM	Airbus A.320-214	Aeroflot Russian International *S. Rakhmaninov*
VP-BWN	Airbus A.321-211	Aeroflot Russian International *A. Skriabin*
VP-BWO	Airbus A.321-211	Aeroflot Russian International *P. Chaikovsky*
VP-BWP	Airbus A.321-211	Aeroflot Russian International *M. Musorgsky*
VP-BWQ	Boeing 767-341ER	Aeroflot Russian International *M. Lermontov*

VT (India)

VT-AIC	Boeing 747-4B5	Air-India *Sanchi*
VT-AID	Boeing 747-4B5	Air-India *Kaziranga*
VT-AIE	Boeing 747-412	Air-India *Mamallapuram*
VT-AIF	Boeing 747-412	Air-India *Ellora*
VT-AII	Boeing 777-222ER	Air-India
VT-AIJ	Boeing 777-222ER	Air-India
VT-AIK	Boeing 777-222ER	Air-India *Megh Malhaar*
VT-EDU	Boeing 747-237B	Air-India *Akbar*

Notes	Reg.	Type	Owner or Operator
	VT-EFU	Boeing 747-237B	Air-India Krishna Deva Raya
	VT-EGA	Boeing 747-237B	Air-India Samudra Gupta
	VT-EGB	Boeing 747-237B	Air-India Mahendra Varman
	VT-EGC	Boeing 747-237B	Air-India Harsha Vardhana
	VT-EPW	Boeing 747-337 (SCD)	Air-India Shivaji
	VT-EPX	Boeing 747-337 (SCD)	Air-India Narasimha Varman
	VT-ESM	Boeing 747-437	Air-India Konark
	VT-ESN	Boeing 747-437	Air-India Tanjore
	VT-ESO	Boeing 747-437	Air-India Khajuraho
	VT-ESP	Boeing 747-437	Air-India Ajanta
	VT-EVA	Boeing 747-437	Air-India Agra
	VT-EVB	Boeing 747-437	Air-India Velha Goa
	VT-EVJ	Boeing 747-4B5	Air-India Fatehpur Sikri

XA (Mexico)

	XA-APB	Boeing 767-3Q8ER	Aeromexico
	XA-JBC	Boeing 767-284ER	Aeromexico
	XA-OAM	Boeing 767-2B1ER	Aeromexico
	XA-RVZ	Boeing 767-284ER	Aeromexico
	XA-TNS	Boeing 767-283ER	Aeromexico
	XA-TOJ	Boeing 767-283ER	Aeromexico

YA (Afghanistan)

	YA-BAB	Airbus A.300B4-203	Ariana Afghan Airlines
	YA-BAD	Airbus A.300B4-203	Ariana Afghan Airlines

YI (Iraq)

Note: Limited operations by Iraqi Airways to regional destinations have been resumed.

YK (Syria)

	YK-AHA	Boeing 747SP-94	Syrianair November 16
	YK-AHB	Boeing 747SP-94	Syrianair Arab Solidarity
	YK-AKA	Airbus A.320-232	Syrianair Ugarit
	YK-AKB	Airbus A.320-232	Syrianair Ebla
	YK-AKC	Airbus A.320-232	Syrianair Afamia
	YK-AKD	Airbus A.320-232	Syrianair Mari
	YK-AKE	Airbus A.320-232	Syrianair Bosra
	YK-AKF	Airbus A.320-232	Syrianair Amrit

YL (Latvia)

	YL-BBA	Boeing 737-505	Air Baltic
	YL-BBB	Boeing 737-505	Air Baltic
	YL-BBD	Boeing 737-53S	Air Baltic
	YL-BBE	Boeing 737-53S	Air Baltic
	YL-BBF	Boeing 737-548	Air Baltic
	YL-BBG	Boeing 737-548	Air Baltic
	YL-	Boeing 737-548	Air Baltic
	YL-KSA	Antonov An-74-200	KS-Avia
	YL-KSB	Antonov An-74	KS-Avia
	YL-RAA	Antonov An-26	RAF-Avia
	YL-RAB	Antonov An-26	RAF-Avia
	YL-RAC	Antonov An-26B	RAF-Avia
	YL-RAD	Antonov An-26B	RAF-Avia
	YL-RAE	Antonov An-26	RAF-Avia
	YL-RAF	Antonov An-74TK-100	RAF-Avia

YR (Romania)

	YR-BGA	Boeing 737-38J	TAROM Alba Iulia
	YR-BGB	Boeing 737-38J	TAROM Bucuresti
	YR-BGC	Boeing 737-38J	TAROM Constanta

Reg.	Type	Owner or Operator	Notes
YR-BGD	Boeing 737-38J	TAROM *Deva*	
YR-BGE	Boeing 737-38J	TAROM *Timisoara*	
YR-BGF	Boeing 737-78J	TAROM *Braila*	
YR-BGG	Boeing 737-78J	TAROM *Craiova*	
YR-BGH	Boeing 737-78J	TAROM *Hunedoara*	
YR-BGI	Boeing 737-78J	TAROM *Iasi*	

YU (Yugoslavia)

YU-AMB	Douglas DC-10-30	JAT Airways *City of Belgrade*	
YU-AND	Boeing 737-3H9	JAT Airways *City of Krusevac*	
YU-ANF	Boeing 737-3H9	JAT Airways	
YU-ANH	Boeing 737-3H9	JAT Airways	
YU-ANJ	Boeing 737-3H9	JAT Airways	
YU-ANK	Boeing 737-3H9	JAT Airways	
YU-ANW	Boeing 737-3H9	JAT Airways	
YU-AON	Boeing 737-3Q4	JAT Airways	
YU-AOO	Boeing 737-4Q8	JAT Airways	
YU-AOR	Boeing 737-4B7	JAT Airways	
YU-AOS	Boeing 737-4B7	JAT Airways	

Z (Zimbabwe)

Z-ARL	Douglas DC-10-30CF	Avient	
Z-WPE	Boeing 767-2N0ER	Air Zimbabwe *Victoria Falls*	
Z-WPF	Boeing 767-2N0ER	Air Zimbabwe *Chimanimani*	

Z3 (Macedonia)

Z3-AAA	Boeing 737-3H9	Macedonian Airlines	
Z3-AAF	Boeing 737-3B7	Macedonian Airlines	
Z3-ARF	Boeing 737-3H9	Macedonian Airlines	

ZK (New Zealand)

ZK-NBS	Boeing 747-419	Air New Zealand	
ZK-NBT	Boeing 747-419	Air New Zealand	
ZK-NBU	Boeing 747-419	Air New Zealand	
ZK-NBV	Boeing 747-419	Air New Zealand	
ZK-NBW	Boeing 747-419	Air New Zealand	
ZK-SUH	Boeing 747-475	Air New Zealand	
ZK-SUI	Boeing 747-441	Air New Zealand	
ZK-SUJ	Boeing 747-4F6	Air New Zealand	

ZS (South Africa)

ZS-OSI	Douglas DC-8-62F	African International Airways	
ZS-OZV	Douglas DC-8-62F	African International Airways	
ZS-PAE	Douglas DC-8-54F	African International Airways	
ZS-PBI	Boeing 767-3Y0ER	Nationwide Air	
ZS-SAC	Boeing 747-312	South African Airways *Shosholoza*	
ZS-SAJ	Boeing 747-312	South African Airways *Ndizani*	
ZS-SAK	Boeing 747-444	South African Airways *Ibhayi*	
ZS-SAV	Boeing 747-444	South African Airways *Durban*	
ZS-SAW	Boeing 747-444	South African Airways *Bloemfontein*	
ZS-SAX	Boeing 747-444	South African Airways *Kempton Park*	
ZS-SAY	Boeing 747-444	South African Airways *Vulindlela*	
ZS-SAZ	Boeing 747-444	South African Airways *Imonti*	
ZS-SBK	Boeing 747-4F6	South African Airways *The Great North*	
ZS-SBS	Boeing 747-4F6	South African Airways	
ZS-SLA	Airbus A.340-211	South African Airways	
ZS-SLB	Airbus A.340-211	South African Airways	
ZS-SLC	Airbus A.340-211	South African Airways	
ZS-SLD	Airbus A.340-211	South African Airways	
ZS-SLE	Airbus A.340-211	South African Airways	
ZS-SLF	Airbus A.340-211	South African Airways	
ZS-SNA	Airbus A.340-642	South African Airways	

Notes	Reg.	Type	Owner or Operator
	ZS-SNB	Airbus A.340-642	South African Airways
	ZS-SNC	Airbus A.340-642	South African Airways
	ZS-SND	Airbus A.340-642	South African Airways
	ZS-SNE	Airbus A.340-642	South African Airways
	ZS-SNF	Airbus A.340-642	South African Airways
	ZS-SNG	Airbus A.340-642	South African Airways
	ZS-SNH	Airbus A.340-642	South African Airways
	ZS-SNI	Airbus A.340-642	South African Airways
	ZS-SXA	Airbus A.340-313X	South African Airways
	ZS-SXB	Airbus A.340-313X	South African Airways
	ZS-SXC	Airbus A.340-313X	South African Airways
	ZS-SXD	Airbus A.340-313X	South African Airways
	ZS-SXE	Airbus A.340-313X	South African Airways
	ZS-SXF	Airbus A.340-313X	South African Airways

3B (Mauritius)

3B-NAU	Airbus A.340-312	Air Mauritius *Pink Pigeon*
3B-NAV	Airbus A.340-312	Air Mauritius *Kestrel*
3B-NAY	Airbus A.340-313X	Air Mauritius *Cardinal*
3B-NBD	Airbus A.340-313X	Air Mauritius *Parakeet*
3B-NBE	Airbus A.340-313X	Air Mauritius *Paille en Queue*

3D (Swaziland)

3D-AFR	Douglas DC-8-54F	African International Airways

Note: African International Airways also operate DC-8s ZS-OSI, ZS-OZV & ZS-PAE.

4K (Azerbaijan)

4K-AZ38	Boeing 757-256	Azerbaijan Airlines
4K-AZ43	Boeing 757-256	Azerbaijan Airlines

Note: Azerbaijan Airways also operate Boeing 757s VP-BBR & VP-BBS.

4R (Sri Lanka)

4R-ADA	Airbus A.340-311	SriLankan Airlines
4R-ADB	Airbus A.340-311	SriLankan Airlines
4R-ADC	Airbus A.340-311	SriLankan Airlines
4R-ADE	Airbus A.340-313X	SriLankan Airlines
4R-ADF	Airbus A.340-313X	SriLankan Airlines
4R-ALA	Airbus A.330-243	SriLankan Airlines
4R-ALB	Airbus A.330-243	SriLankan Airlines
4R-ALC	Airbus A.330-243	SriLankan Airlines
4R-ALD	Airbus A.330-243	SriLankan Airlines

4X (Israel)

4X-AXF	Boeing 747-258C	El Al Cargo
4X-AXH	Boeing 747-258B (SCD)	El Al Cargo
4X-AXK	Boeing 747-245F (SCD)	El Al Cargo
4X-AXL	Boeing 747-245F (SCD)	El Al Cargo
4X-AXM	Boeing 747-2B5BF	El Al Cargo
4X-AXQ	Boeing 747-238B	El Al
4X-BAU	Boeing 757-3E7	Arkia
4X-BAW	Boeing 757-3E7	Arkia
4X-BAZ	Boeing 757-236	Arkia
4X-EAA	Boeing 767-258ER	El Al
4X-EAB	Boeing 767-258ER	El Al
4X-EAC	Boeing 767-258ER	El Al
4X-EAD	Boeing 767-258ER	El Al
4X-EAE	Boeing 767-27EER	El Al
4X-EAF	Boeing 767-27EER	El Al
4X-EAJ	Boeing 767-330ER	El Al
4X-EBI	Boeing 757-258	Israir

Reg.	Type	Owner or Operator
4X-EBM	Boeing 757-258	Israir
4X-EBO	Boeing 757-258	El Al/Sun d'Or
4X-EBS	Boeing 757-258	El Al
4X-EBT	Boeing 757-258	El Al
4X-EBU	Boeing 757-258	El Al
4X-EBV	Boeing 757-258	El Al
4X-EBY	Boeing 757-27B	El Al/Sun d'Or
4X-ECA	Boeing 777-258ER	El Al *Galilee*
4X-ECB	Boeing 777-258ER	El Al *Negev*
4X-ECC	Boeing 777-258ER	El Al *Hasharon*
4X-ECD	Boeing 777-258ER	El Al *Carmel*
4X-EKA	Boeing 737-858	El Al *Tiberias*
4X-EKB	Boeing 737-858	El Al *Eilat*
4X-EKC	Boeing 737-858	El Al *Beit Shean*
4X-EKD	Boeing 737-758	El Al *Ashkelon*
4X-EKE	Boeing 737-758	El Al *Nazareth*
4X-EKI	Boeing 737-86N	El Al
4X-ELA	Boeing 747-458	El Al *Tel Aviv-Jaffa*
4X-ELB	Boeing 747-458	El Al *Haifa*
4X-ELC	Boeing 747-458	El Al *Beer Sheva*
4X-ELD	Boeing 747-458	El Al *Jerusalem*
4X-ICL	Boeing 747-271C (SCD)	Cargo Air Lines
4X-ICM	Boeing 747-271C (SCD)	Cargo Air Lines
4X-ICN	Boeing 747-2B5B (SF)	Cargo Air Lines

5A (Libya)

5A-DKL	An-124	Libyan Arab Air Cargo
5A-DKN	An-124	Libyan Arab Air Cargo

Note: Libyan Arab Airlines operates Airbus A.320 TS-INJ, A.310 JY-AGV and TS-IGU, and A300-622Rs
TS+IAX and TS-IAZ.
Afriquiyah Airways operates Airbus A.320 TS-ING and TS-INH, and A300-622R TS-IAY.

5B (Cyprus)

5B-DAT	Airbus A.320-231	Cyprus Airways *Praxandros*
5B-DAU	Airbus A.320-231	Cyprus Airways *Evelthon*
5B-DAV	Airbus A.320-231	Cyprus Airways *Kinyras*
5B-DAW	Airbus A.320-231	Cyprus Airways *Agapinor*
5B-DBA	Airbus A.320-231	Cyprus Airways *Evagoras*
5B-DBB	Airbus A.320-231	Cyprus Airways *Akamas*
5B-DBC	Airbus A.320-231	Cyprus Airways *Tefkros*
5B-DBD	Airbus A.320-231	Cyprus Airways *Onisillos*
5B-DBH	Boeing 737-86N	Helios Airways *Zela*
5B-DBI	Boeing 737-86N	Helios Airways *Beni*
5B-DBO	Airbus A.319-132	Cyprus Airways *Nikoklis*
5B-DBP	Airbus A.319-132	Cyprus Airways *Chalkanor*
5B-DBS	Airbus A.330-223	Cyprus Airways *Ammochostos*
5B-DBT	Airbus A.330-223	Cyprus Airways *Keryneia*
5B-DBU	Boeing 737-8Q8	Eurocypria Airlines *Zephyros*
5B-DBV	Boeing 737-8Q8	Eurocypria Airlines *Levantes*
5B-DBW	Boeing 737-8Q8	Eurocypria Airlines *Maistros*
5B-DBX	Boeing 737-8Q8	Eurocypria Airlines *Grekos*
5B-DBY	Boeing 737-31S	Helios Airways *Olympia*

5N (Nigeria)

Note: Nigerian Airways ceased operations during 2004.

5R (Madagascar)

5R-MFF	Boeing 767-3S1ER	Air Madagascar
5R-MFG	Boeing 767-383ER	Air Madagascar
5R-MFT	Boeing 747-2B2B (SCD)	Air Madagascar/stored Filton

5X (Uganda)

5X-BON	Douglas DC-10-30F	DAS Air Cargo
5X-JCR	Douglas DC-10-30F	DAS Air Cargo
5X-JOE	Douglas DC-10-30F	DAS Air Cargo
5X-JOS	Douglas DC-10-30F	DAS Air Cargo
5X-ROY	Douglas DC-10-30F	DAS Air Cargo

Note: DAS Air Cargo also operates a Douglas DC-10-30F registered N401JR.

5Y (Kenya)

5Y-CCC	Boeing 767-3Y0ER	East African Safari Air
5Y-KQR	Boeing 767-3P6ER	Kenya Airways
5Y-KQU	Boeing 777-2U8ER	Kenya Airways
5Y-KQV	Boeing 767-3Y0ER	Kenya Airways
5Y-KQW	Boeing 767-3Y0ER	Kenya Airways
5Y-KQX	Boeing 767-36NER	Kenya Airways
5Y-KQY	Boeing 767-36NER	Kenya Airways
5Y-KQZ	Boeing 767-36NER	Kenya Airways
5Y-VIP	Airbus A.310-308	African Safari Airways

6Y (Jamaica)

6Y-JMC	Airbus A.340-312	Air Jamaica *Spirit of New York*
6Y-JMM	Airbus A.340-313X	Air Jamaica *Spirit of Jamaica – Atlantic Limousine*
6Y-JMP	Airbus A.340-313X	Air Jamaica *Spirit of Jamaica – Atlantic Limousine II*

7O (Yemen)

7O-ADJ	Airbus A.310-324	Yemenia
7O-ADL	Boeing 737-8Q8	Yemenia
7O-ADM	Boeing 737-8Q8	Yemenia
7O-ADN	Boeing 737-8Q8	Yemenia
7O-ADP	Airbus A.330-243	Yemenia *Sana'a*
7O-ADT	Airbus A.330-243	Yemenia *Aden*
7O-YMN	Boeing 747SP-27	Yemenia

Note: Yemenia also operates Airbus A.310s registered F-OHPR and F-OHPS.

7T (Algeria)

7T-VHG	L.100-30 Hercules	Air Algerie
7T-VHL	L.100-30 Hercules	Air Algerie
7T-VJG	Boeing 767-3D6ER	Air Algerie
7T-VJH	Boeing 767-3D6ER	Air Algerie
7T-VJI	Boeing 767-3D6ER	Air Algerie
7T-VJJ	Boeing 737-8D6	Air Algerie *Jugurtha*
7T-VJK	Boeing 737-8D6	Air Algerie *Mansourah*
7T-VJL	Boeing 737-8D6	Air Algerie *Allizi*
7T-VJM	Boeing 737-8D6	Air Algerie
7T-VJN	Boeing 737-8D6	Air Algerie
7T-VJO	Boeing 737-8D6	Air Algerie
7T-VJP	Boeing 737-8D6	Air Algerie
7T-VJQ	Boeing 737-6D6	Air Algerie
7 T-VJR	Boeing 737-6D6	Air Algerie
7T-VJS	Boeing 737-6D6	Air Algerie
7T-VJT	Boeing 737-6D6	Air Algerie
7T-VJU	Boeing 737-6D6	Air Algerie

9A (Croatia)

9A-CBC	McD Douglas MD-82	Air Adriatic *My Dream*
9A-CBD	McD Douglas MD-81	Air Adriatic *I Never Give Up*
9A-CBE	McD Douglas MD-81	Air Adriatic *No Guts, No Glory*
9A-CTF	Airbus A.320-211	Croatia Airlines *Rijeka*

Reg.	Type	Owner or Operator	Notes
9A-CTG	Airbus A.319-112	Croatia Airlines *Zadar*	
9A-CTH	Airbus A.319-112	Croatia Airlines *Zagreb*	
9A-CTI	Airbus A.319-112	Croatia Airlines *Vukovar*	
9A-CTJ	Airbus A.320-214	Croatia Airlines *Dubrovnik*	
9A-CTK	Airbus A.320-214	Croatia Airlines *Split*	
9A-CTL	Airbus A.319-112	Croatia Airlines *Pula*	
9A-	Airbus A.320-212	Croatia Airlines	

Note: Croatia Airlines also operates A.319 D-AILH.

9G (Ghana)

9G-AND	Douglas DC-10-30	Ghana Airways	
9G-ANE	Douglas DC-10-30	Ghana Airways	
9G-FAB	Douglas DC-8-63F	Johnsons Air	
9G-LCA	Canadair CL-44-0	Johnsons Air/stored Bournemouth	
9G-MKG	Douglas DC-8-62AF	MK Airlines	
9G-MKH	Douglas DC-8-62AF	MK Airlines	
9G-MKK	Douglas DC-8-62AF	MK Airlines	
9G-MKL	Boeing 747-2R7F	MK Airlines	
9G-MKM	Boeing 747-2B5F	MK Airlines	
9G-MKN	Douglas DC-8-63CF	MK Airlines	
9G-MKO	Douglas DC-8-63CF	MK Airlines	
9G-MKP	Boeing 747-245F	MK Airlines	
9G-MKQ	Boeing 747-2S4F	MK Airlines	
9G-MKR	Boeing 747-2B5F	MK Airlines	
9G-MKS	Boeing 747-2B5F	MK Airlines	
9G-MKT	Douglas DC-8-55F	MK Airlines	
9G-PEL	Douglas DC-8-62F	Johnsons Air	
9G-RMF	Douglas DC-8-62F	Johnsons Air	

Note: Ghana Airways' services were suspended at the time of publication.

9H (Malta)

9H-ABQ	Airbus A.320-211	Air Malta/Air Mediterranée	
9H-ABR	Boeing 737-3Y5	Air Malta *Birkirkara*	
9H-ABS	Boeing 737-3Y5	Air Malta *Sliema*	
9H-ABT	Boeing 737-3Y5	Air Malta *Hamrun*	
9H-ADH	Boeing 737-33A	Air Malta	
9H-ADI	Boeing 737-33A	Air Malta	
9H-ADM	Boeing 737-382	Air Malta	
9H-AEF	Airbus A.320-214	Air Malta *Valetta*	
9H-AEG	Airbus A.319-112	Air Malta *Mdina*	
9H-AEH	Airbus A.319-112	Air Malta	
9H-AEI	Airbus A.320-214	Air Malta	
9H-AEJ	Airbus A.319-112	Air Malta	
9H-AEK	Airbus A.320-214	Air Malta	
9H-AEL	Airbus A.319-112	Air Malta	
9H-AER	Airbus A.320-214	Air Malta/Veuling	

9K (Kuwait)

9K-ADE	Boeing 747-469 (SCD)	Kuwait Airways *Al-Jabariya*	
9K-ALA	Airbus A.310-308	Kuwait Airways *Al-Jahra*	
9K-ALB	Airbus A.310-308	Kuwait Airways *Gharnada*	
9K-ALC	Airbus A.310-308	Kuwait Airways *Al-Salmiya*	
9K-ALD	Airbus A.310-308	State of Kuwait *Al-Sabahiya*	
9K-AMA	Airbus A.300B4-605R	Kuwait Airways *Failaka*	
9K-AMB	Airbus A.300B4-605R	Kuwait Airways *Burghan*	
9K-AMC	Airbus A.300B4-605R	Kuwait Airways *Wafra*	
9K-AMD	Airbus A.300B4-605R	Kuwait Airways *Wara*	
9K-AME	Airbus A.300B4-605R	Kuwait Airways *Al-Rawdhatain*	
9K-ANA	Airbus A.340-313	Kuwait Airways *Warba*	
9K-ANB	Airbus A.340-313	Kuwait Airways *Al-Sabahiya*	
9K-ANC	Airbus A.340-313	Kuwait Airways *Al-Mobarakiya*	
9K-AND	Airbus A.340-313	Kuwait Airways *Al-Riggah*	
9K-AOA	Boeing 777-269ER	Kuwait Airways *Al-Grain*	
9K-AOB	Boeing 777-269ER	Kuwait Airways *Garouh*	

9M (Malaysia)

Reg.	Type	Owner or Operator
9M-MHI	Boeing 747-236F (SF)	MAS Kargo *Kuching*
9M-MHJ	Boeing 747-236F (SF)	MAS Kargo *Kota Kinabalu*
9M-MPA	Boeing 747-4H6	Malaysian Airlines *Ipoh*
9M-MPB	Boeing 747-4H6	Malaysian Airlines *Shah Alam*
9M-MPC	Boeing 747-4H6	Malaysian Airlines *Kuantan*
9M-MPD	Boeing 747-4H6	Malaysian Airlines *Serembam*
9M-MPE	Boeing 747-4H6	Malaysian Airlines *Kangar*
9M-MPF	Boeing 747-4H6	Malaysian Airlines *Kota Bharu*
9M-MPG	Boeing 747-4H6	Malaysian Airlines *Kuala Terengganu*
9M-MPH	Boeing 747-4H6	Malaysian Airlines *Langkawi*
9M-MPI	Boeing 747-4H6	Malaysian Airlines *Tioman*
9M-MPJ	Boeing 747-4H6	Malaysian Airlines *Labuan*
9M-MPK	Boeing 747-4H6	Malaysian Airlines *Johor Bahru*
9M-MPL	Boeing 747-4H6	Malaysian Airlines *Penang*
9M-MPM	Boeing 747-4H6	Malaysian Airlines *Melaka*
9M-MPN	Boeing 747-4H6	Malaysian Airlines *Pangkor*
9M-MPO	Boeing 747-4H6	Malaysian Airlines *Alor Setar*
9M-MPP	Boeing 747-4H6	Malaysian Airlines *Putrajaya*
9M-MPQ	Boeing 747-4H6	Malaysian Airlines *Kuala Lumpur*
9M-MRA	Boeing 777-2H6ER	Malaysian Airlines
9M-MRB	Boeing 777-2H6ER	Malaysian Airlines
9M-MRC	Boeing 777-2H6ER	Malaysian Airlines
9M-MRD	Boeing 777-2H6ER	Malaysian Airlines
9M-MRE	Boeing 777-2H6ER	Malaysian Airlines
9M-MRF	Boeing 777-2H6ER	Malaysian Airlines
9M-MRG	Boeing 777-2H6ER	Malaysian Airlines
9M-MRH	Boeing 777-2H6ER	Malaysian Airlines
9M-MRI	Boeing 777-2H6ER	Malaysian Airlines
9M-MRJ	Boeing 777-2H6ER	Malaysian Airlines
9M-MRK	Boeing 777-2H6ER	Malaysian Airlines
9M-MRL	Boeing 777-2H6ER	Malaysian Airlines
9M-MRM	Boeing 777-2H6ER	Malaysian Airlines
9M-MRN	Boeing 777-2H6ER	Malaysian Airlines
9M-MRO	Boeing 777-2H6ER	Malaysian Airlines
9M-MRP	Boeing 777-2H6ER	Malaysian Airlines
9M-MRQ	Boeing 777-2H6ER	Malaysian Airlines

Note: MAS Kargo also operate B.747Fs leased from Air Atlanta Icelandic.

9V (Singapore)

Reg.	Type	Owner or Operator
9V-SFA	Boeing 747-412F	Singapore Airlines Cargo
9V-SFB	Boeing 747-412F	Singapore Airlines Cargo
9V-SFD	Boeing 747-412F	Singapore Airlines Cargo
9V-SFE	Boeing 747-412F	Singapore Airlines Cargo
9V-SFF	Boeing 747-412F	Singapore Airlines Cargo
9V-SFG	Boeing 747-412F	Singapore Airlines Cargo
9V-SFH	Boeing 747-412F	Singapore Airlines Cargo
9V-SFI	Boeing 747-412F	Singapore Airlines Cargo
9V-SFJ	Boeing 747-412F	Singapore Airlines Cargo
9V-SFK	Boeing 747-412F	Singapore Airlines Cargo
9V-SFL	Boeing 747-412F	Singapore Airlines Cargo
9V-SFM	Boeing 747-412F	Singapore Airlines Cargo
9V-SFN	Boeing 747-412F	Singapore Airlines Cargo
9V-SFO	Boeing 747-412F	Singapore Airlines Cargo
9V-SFP	Boeing 747-412F	Singapore Airlines Cargo
9V-SFQ	Boeing 747-412F	Singapore Airlines Cargo
9V-SGA	Airbus A.340-541	Singapore Airlines
9V-SGB	Airbus A.340-541	Singapore Airlines
9V-SGC	Airbus A.340-541	Singapore Airlines
9V-SGD	Airbus A.340-541	Singapore Airlines
9V-SGE	Airbus A.340-541	Singapore Airlines
9V-SMJ	Boeing 747-412	Singapore Airlines
9V-SMM	Boeing 747-412	Singapore Airlines
9V-SMO	Boeing 747-412	Singapore Airlines
9V-SMP	Boeing 747-412	Singapore Airlines
9V-SMQ	Boeing 747-412	Singapore Airlines
9V-SMR	Boeing 747-412	Singapore Airlines

Reg.	Type	Owner or Operator	Notes
9V-SMS	Boeing 747-412	Singapore Airlines	
9V-SMT	Boeing 747-412	Singapore Airlines	
9V-SMU	Boeing 747-412	Singapore Airlines	
9V-SMV	Boeing 747-412	Singapore Airlines	
9V-SMW	Boeing 747-412	Singapore Airlines	
9V-SMY	Boeing 747-412	Singapore Airlines	
9V-SMZ	Boeing 747-412	Singapore Airlines	
9V-SPA	Boeing 747-412	Singapore Airlines	
9V-SPB	Boeing 747-412	Singapore Airlines	
9V-SPC	Boeing 747-412	Singapore Airlines	
9V-SPD	Boeing 747-412	Singapore Airlines	
9V-SPE	Boeing 747-412	Singapore Airlines	
9V-SPF	Boeing 747-412	Singapore Airlines	
9V-SPG	Boeing 747-412	Singapore Airlines	
9V-SPH	Boeing 747-412	Singapore Airlines	
9V-SPI	Boeing 747-412	Singapore Airlines	
9V-SPJ	Boeing 747-412	Singapore Airlines	
9V-SPL	Boeing 747-412	Singapore Airlines	
9V-SPM	Boeing 747-412	Singapore Airlines	
9V-SPN	Boeing 747-412	Singapore Airlines	
9V-SPO	Boeing 747-412	Singapore Airlines	
9V-SPP	Boeing 747-412	Singapore Airlines	
9V-SPQ	Boeing 747-412	Singapore Airlines	
9V-SVA	Boeing 777-212ER	Singapore Airlines	
9V-SVB	Boeing 777-212ER	Singapore Airlines	
9V-SVC	Boeing 777-212ER	Singapore Airlines	
9V-SVD	Boeing 777-212ER	Singapore Airlines	
9V-SVE	Boeing 777-212ER	Singapore Airlines	
9V-SVG	Boeing 777-212ER	Singapore Airlines	
9V-SVH	Boeing 777-212ER	Singapore Airlines	
9V-SVI	Boeing 777-212ER	Singapore Airlines	
9V-SVJ	Boeing 777-212ER	Singapore Airlines	
9V-SVK	Boeing 777-212ER	Singapore Airlines	
9V-SVL	Boeing 777-212ER	Singapore Airlines	
9V-SVM	Boeing 777-212ER	Singapore Airlines	
9V-SVN	Boeing 777-212ER	Singapore Airlines	
9V-SVO	Boeing 777-212ER	Singapore Airlines	

9Y (Trinidad and Tobago)

9Y-JIL	Airbus A.340-311	B.W.I.A. West Indies Airways	
9Y-TJN	Airbus A.340-313X	B.W.I.A. West Indies Airways	

EC-INF, Canadair RJ200ER of Air Nostrum. *AJW*

Radio Frequencies

The frequencies used by the larger airfields/airports are listed below. Abbreviations used: TWR — Tower, APP - Approach, A/G - Air-Ground advisory. It is possible that changes will be made from time to time with the frequencies allocated, all of which are quoted in Megahertz (MHz)

Airfield	TWR	APP	A/G	Airfield	TWR	APP	A/G
Aberdeen	118.1	119.05		Inverness	122.6		
Alderney	125.35	128.65		Jersey	119.45	120.3	
Andrewsfield			130.55	Kemble	118.9		
Audley End			122.35	Kidlington	133.425	125.325	
Barton			122.7	Land's End	130.7		
Barrow			123.2	Leeds Bradford	120.3	123.75	
Beccles	134.6			Leicester	122.125		
Belfast Intl	118.3	120.9		Liverpool	126.35	119.85	
Belfast City	130.75	130.85		London City	118.075	132.7	
Bembridge		123.25		Luton	132.55	129.55	
Biggin Hill	134.8	129.4		Lydd			120.7
Birmingham	118.3	118.05		Manchester	118.625	119.525	
Blackbushe			122.3	Manston	119.925	129.45	
Blackpool	118.4	119.95		Netherthorpe			123.275
Bodmin			122.7	Newcastle	119.7	124.375	
Booker			126.55	Newquay	123.4		
Bourn			129.8	North Denes	123.4		
Bournemouth	125.6	119.475		North Weald	123.525		
Breighton	129.80			Norwich	124.25	119.35	
Bristol/Filton	132.35	122.725		Old Warden	123.05		
Bristol/Lulsgate	133.85	136.075		Penzance	118.1		
Bruntingthorpe	122.825			Perth	119.8		
Caernarfon	122.25			Plymouth	118.15	133.55	
Cambridge	122.2	123.6		Popham			129.8
Cardiff	125.0	125.85		Prestwick	118.15	120.55	
Carlisle		123.6		Redhill	119.6		
Clacton	135.40			Rochester			122.25
Compton Abbas			122.7	Ronaldsway	118.9	120.85	
Conington			129.725	Sandown			123.5
Cosford	128.65	135.875		Sandtoft			130.425
Coventry	124.8	119.25		Scilly Isles			123.15
Cranfield	134.925	122.85		Seething			122.6
Denham			130.725	Sheffield City	128.525		
Dundee	122.9			Sheffield Robin Hood	128.7775	126.225	
Dunkeswell			123.475	Sherburn			122.6
Durham Tees Valley	119.8	118.85		Shipdham			119.55
Duxford			122.075	Shobdon			123.5
Earls Colne			122.425	Shoreham	125.4	123.15	
East Midlands	124.0	134.175		Sibson			122.3
Edinburgh	118.7	121.2		Sleap			122.45
Elstree			122.4	Southampton	118.2	128.85	
Exeter	119.8	128.975		Southend	127.725	130.775	
Fairoaks			123.425	Stansted	123.8	126.95	
Fenland			122.925	Stapleford			122.8
Fowlmere			120.925	Sumburgh	118.25	123.15	
Gamston			130.475	Swansea	119.7		
Gatwick	124.225	126.825		Swanton Morley			123.5
Glasgow	118.8	119.1		Sywell			122.7
Gloucester/				Tatenhill	124.075		
Staverton	122.9	128.55		Thruxton			130.45
Goodwood	135.5			Tollerton			134.875
Guernsey	119.95	128.65		Wellesbourne			124.02
Haverfordwest			122.2	Welshpool	123.25		
Hawarden	124.95	123.35		White Waltham			122.6
Headcorn			122.0	Wick	119.7		
Heathrow	118.7	119.725		Wickenby			122.45
	118.5	134.975		Wolverhampton	123.0		
Hethel			122.35	Woodford	120.7	130.75	
Hucknall			130.8	Woodvale	119.75	121.0	
Humberside	118.55	119.125		Yeovil	125.4	130.8	

Airline Flight Codes

Those listed below identify both UK and overseas carriers appearing in the book.

Code	Airline	
AAE	Atlantic Airlines	G
AAF	Aigle Azur	F
AAL	American A/L	N
AAR	Asiana A/L	HL
AAW	Afriqiyah Airways	?????
ABD	Air Atlanta Iceland	TF
ABR	Air Contractors	EI
ACA	Air Canada	C
ADB	Air One	I
ADR	Adria A/W	S5
AEA	Air Europa	EC
AEL	Air Europe Italy	I
AEU	Astraeus	G
AEW	Aerosvit	UR
AFG	Ariana	YA
AFL	Aeroflot	RA
AFR	Air France	F
AHK	Air Hong Kong	B
AIC	Air-India	VT
AIN	African International	3D
AJM	Air Jamaica	6Y
AKL	Air Kilroe	G
ALK	SriLankan A/L	4R
AMC	Air Malta	9H
AMM	Air 2000	G
AMT	ATA Airlines	N
AMV	AMC Airlines	SU
AMX	Aeromexico	XA
ANA	All Nippon A/W	JA
ANS	Air Nostrum	EC
ANZ	Air New Zealand	ZK
ARF	Airlinair	F
ARG	Aerdineas Argentinas	LV
ATN	Air Transport International	N
AUA	Austrian A/L	OE
AUI	Ukraine Intl	UR
AUR	Aurigny A/S	G
AVA	Avianca	HK
AWC	Titan A/W	G
AWW	Air Wales	G
AXL	Air exel	PH
AXY	Axis Airways	F
AZA	Alitalia	I
AZW	Air Zimbabwe	Z
BAG	Deutsche BA	D
BAL	Britannia A/L	G
BAW	British Airways	G
BBC	Bangladesh Biman	S2
BBD	Bluebird CArgo	TF
BCS	European A/T	OO
BCY	CityJet	EI
BDI	Benair A/L	OY
BEE	Flybe	G
BEH	Balkan Holidays	LZ
BER	Air Berlin	D
BIE	Air Mediterranee	F
BIH	CHC Scotia	G
BLE	Blue Line	F
BMA	bmi british midland	G
BMI	bmi Baby	G
BPA	§Blue Panorama	I
BRA	Braathens	LN
BRT	BA Citiexpress	G
BRU	Belavia	EW
BTI	Air Baltic	YL
BWA	BWIA	9Y
BZH	Brit Air	F
CAJ	Air Caraibes	F
CAL	China Airlines	B
CCA	Air China	B
CES	China Eastern	B
CFG	Thomas Cook/Condor	D
CIM	Cimber Air	OY
CKS	Kalitta Air	N
CLH	Lufthansa CityLine	D
CLX	Cargolux	LX
COA	Continental A/L	N
CPA	Cathay Pacific	B
CRL	Corsair	F
CSA	Czech A/L	OK
CTN	Croatia A/L	9A
CUB	Cubana	CU
CWC	Centurian Air Cargo	N
CYP	Cyprus A/W	5B
DAH	Air Algerie	7T
DAL	Delta A/L	N
DAN	Maersk Air	OY
DAT	SN Brussels A/L	OO
DHL	DHL Express	N/OO
DLH	Lufthansa	D
DNM	Denim Air	PH
DSR	DAS Air Cargo	5X
DTR	Danish Air Transport	OY
EAF	European A/Ch	G
ECA	Eurofly	I
EDW	Edelweiss Air	HB
EIA	Evergreen Intl	N
EIN	Aer Lingus	EI
ELY	El Al	4X
ERT	Eritrean Airlines	E3
ETD	Etihad Airways	A6
ETH	Ethiopian A/L	ET
EUH	Air Horizons	F
EUJ	EuJet	EI
EUK	Air Atlanta Europe	TF
EVA	EVA Airways	B
EWG	Eurowings	D
EXS	Channel Express	G
EZE	Eastern Airways	G
EZS	easyJet Switzerland	HB
EZY	easyJet	G
FCA	First Choice A/L	G
FDX	Federal Express	N
FHY	Freebird Airlines	TC
FIF	Air Finland	OH
FIN	Finnair	OH
FJE	Fly Jet	G
FLI	Atlantic A/W	OY
FLT	Flightline	G
FLY	Fly Me	SE
FOB	Ford	G
FUA	Futura	EC
GBL	GB Airways	G
GCO	Gemini Air Cargo	N
GEC	Lufthansa Cargo	D
GFA	Gulf Air	A40
GHA	Ghana A/W	9G
GIA	Garuda	PK
GJJ	Gir Jet	EC
GMI	Germania	D
GSM	Flyglobespan	G
GTI	Atlas Air	N
GWI	German Wings	D
HCY	Helios A/L	5B
HEJ	Hellas Jet	SX
HHI	Hamburg Intl	D
HLF	Hapag-Lloyd	D
HLX	Hapag Lloyd Express	D
HMS	Hemus Air	LZ
HOA	Hola A/L	EC
HSK	Skyeurope Hungary	HA
HWY	Highland A/W	G
HXL	Holland Exel	PH
IBE	Iberia	EC
ICB	Islands Flag	TF
ICE	Icelandair	TF
ICL	Cargo Airlines	4X
IOS	Isles of Scilly Skybus	G
IRA	Iran Air	EP
IRM	Mahan Air	EP
ISK	Intersky	D
ISS	Meridiana	I
IWD	Iberworld	EC
IYE	Yemenia	7O
JAL	Japan A/L	JA
JAT	JAT Airways	YU
JEM	Emerald A/W	G
JET	Wind Jet	I
JKK	Spanair	EC
KAC	Kuwait A/W	9K
KAJ	Karthago A/L	TS
KAL	Korean Air	HL
KLC	KLM CityHopper	PH
KLM	KLM	PH
KQA	Kenya A/W	5Y
KYV	Kibris Turkish	TC
KZR	Air Astana	UN
LAA	Libyan Arch	5A
LAJ	British Meditrn	G
LAN	Lan Chile	CC
LAZ	Bulgarian A/L	LZ
LBT	Nouvelair	TS
LDA	Lauda Air	OE
LGL	Luxair	LX
LIL	Lithuanian A/L	LY
LOG	Loganair	G
LOT	Polish A/L (LOT)	SP
LTE	Volar	EC
LTU	LTU	D
MAH	Malev	HA
MAK	Macedonian A/L	Z3
MAS	Malaysian A/L	9M

403

MAU	Air Mauritius	3B	QSC	African Safaris	5Y	SXS	Sun Express	TC
MEA	Middle East A/L	OD	QTR	Qatar Airways	AT	SYR	Syrian Arab	YK
MKA	MKA Airlines	9G	RAM	Royal Air Maroc	CN	TAP	Air Portugal	CS
MMZ	Euro Atlantic	CS	RBA	Royal Brunei	V8	TAR	Tunis Air	TS
MNB	MNG Airlines	TC	REA	Aer Arann	EI	TAY	TNT A/W	OO
MON	Monarch A/L	G	REU	Air Austral	F	TCW	Thomas Cook	
MPD	Air Plus Comet	EC	RJA	Royal Jordanian	JY		Belgium	OO
MPH	Martinair	PH	ROT	Tarom	YR	TCX	Thomas Cook A/L	G
MSR	Egyptair	SU	RPX	BAC Express A/L	G	THA	Thai A/W Intl	HS
MTL	RAF-Avia	YL	RUS	Cirrus Airlines	D	THT	Air Tahiti Nui	F
MYT	My Travel A/W	G	RWD	Air Rwanda	9XR	THY	Turkish A/L	TC
NAO	North American A/L	N	RZO	SATA International	CS	TRA	Transavia	PH
NAX	Norweigian Air Shuttle	LN	SAA	South African A/W	ZS	TSC	Air Transat	C
NCA	Nippon Cargo	JA	SAS	SAS	SE OY LN	TVS	Smart Wings	N
NEX	Northern Executive	G	SAY	Scot Airways	G	TYR	Tyrolean	OE
NLY	Flyniks	OE	SCW	Malmo Aviation	SE	UAE	Emirates A/L	A6
NMB	Air Namibia	V5	SCY	Air Scandic	G	UAL	United A/L	N
NVR	Novair	SE	SDR	City Airline	Se	UPS	United Parcels	N
NWA	Northwest A/L	N	SEU	Star Airlines	F	USA	US Airways	N
OAE	Omni Air Intnl	N	SEY	Air Seychelles	S7	UYC	Cameroon A/L	TJ
OAL	Olympic A/L	SX	SIA	Singapore A/L	9V	UZB	Uzbekistan A/W	UK
OAW	Helvetic Airways	HB	SLL	Slovak Airlines	OM	VDA	Volga-Dnepr	UR
OGE	Atlasjet	TC	SNB	Sterling European	OY	VIR	Virgin Atlantic	G
OHY	Onur Air	TC	SQC	Singapore Cargo	9V	VLM	VLM	OO
OOM	Zoom Airways	C	SRR	Starair	OY	VRG	Varig	PP
PAC	Polar Air Cargo	N	SUD	Sudan A/W	ST	WDL	WDL	D
PAL	Philippine A/L	RP	SUS	Sun-Air	OY	WIF	Wideroe's	LN
PGA	Portugalia	CS	SVA	Saudia	HZ	WOA	World A/W	N
PGT	Pegasus A/L	TC	SVK	Air Slovak	OM	WZZ	Wizz Air	HA
PIA	Pakistan Intl	AP	SWE	Swedair	SE	XLA	Excel A/W	G
POT	Polet	RA	SWN	Western Air Sweden	SE			
QFA	Qantas	VH	SWR	Swiss	HB			

SP-FDR, Antonov AN-26 of Polish operator EXIN on a freight charter. *DP*

British Aircraft Preservation Council Register

The British Aircraft Preservation Council was formed in 1967 to co-ordinate the works of all bodies involved in the preservation, restoration and display of historical aircraft. Membership covers the whole spectrum of national, Service, commercial and voluntary groups, and meetings are held regularly at the bases of member organisations. The Council is able to provide a means of communication, helping to resolve any misunderstandings or duplication of effort. Every effort is taken to encourage the raising of standards of both organisation and technical capacity amongst the member groups to the benefit of everyone interested in aviation. To assist historians, the B.A.P.C. register has been set up and provides an identity for those aircraft which do not qualify for a Service serial or inclusion in the UK Civil Register.

Aircraft on the current B.A.P.C. Register are as follows:

Reg.	Type	Owner or Operator	Notes
1	Roe Triplane Type 4 (replica)	Shuttleworth Collection as G-ARSG (not carried)	
2	Bristol Boxkite (replica)	Shuttleworth Collection as G-ASPP (not carried)	
6	Roe Triplane Type IV (replica)	Manchester Museum of Science & Industry	
7	Southampton University MPA	Southampton Hall of Aviation	
8	Dixon ornithopter	The Shuttleworth Collection	
9	Humber Monoplane (replica)	Midland Air Museum/Coventry	
10	Hafner R.II Revoplane	Museum of Army Flying/Middle Wallop	
12	Mignet HM.14	Museum of Flight/E. Fortune	
13	Mignet HM.14	Brimpex Metal Treatments	
14	Addyman Standard Training Glider	A. Lindsay & N. H. Ponsford	
15	Addyman Standard Training Glider	The Aeroplane Collection	
16	Addyman ultra-light aircraft	N. H. Ponsford	
17	Woodhams Sprite	The Aeroplane Collection	
18	Killick MP Gyroplane	A. Lindsay & N. H. Ponsford	
20	Lee-Richards annular biplane (replica)	Visitor Centre Shoreham Airport	
21	Thruxton Jackaroo	M. J. Brett	
22	Mignet HM.14 (G-AEOF)	Aviodome/Netherlands	
23	SE-5A Scale Model	Newark Air Museum	
24	Currie Wot (replica)	Newark Air Museum	
25	Nyborg TGN-III glider	Midland Air Museum	
26	Auster AOP.6 (fuselage frame)	Now scrapped	
27	Mignet HM.14	M. J. Abbey	
28	Wright Flyer (replica)	Corn Exchange/Leeds	
29	Mignet HM.14 (replica) (G-ADRY)	Brooklands Museum of Aviation/Weybridge	
32	Crossley Tom Thumb	Midland Air Museum	
33	DFS.108-49 Grunau Baby IIb	-	
34	DFS.108-49 Grunau Baby IIb	D. Elsdon	
35	EoN primary glider	-	
36	Fieseler Fi 103 (V-1) (replica)	Kent Battle of Britain Museum/Hawkinge	
37	Blake Bluetit (G-BXIY)	The Shuttleworth Collection/O. Warden	
38	Bristol Scout replica (A1742)	K. Williams & M. Thorn	
39	Addyman Zephyr sailplane	A. Lindsay & N. H. Ponsford	
40	Bristol Boxkite (replica)	Bristol City Museum	
41	B.E.2C (replica) (6232)	Yorkshire Air Museum/Elvington	
42	Avro 504 (replica) (H1968)	Yorkshire Air Museum/Elvington	
43	Mignet HM.14	Newark Air Museum/Winthorpe	
44	Miles Magister (L6906)	Museum of Berkshire Aviation (G-AKKY)/ Woodley	
45	Pilcher Hawk (replica)	Stanford Hall Museum	
46	Mignet HM.14	Stored	
47	Watkins Monoplane	National Museum of Wales	
48	Pilcher Hawk (replica)	Glasgow Museum of Transport	
49	Pilcher Hawk	Royal Scottish Museum/East Fortune	
50	Roe Triplane Type 1	Science Museum/S. Kensington	
51	Vickers Vimy IV	Science Museum/S. Kensington	
52	Lilienthal glider	Science Museum Store/Hayes	
53	Wright Flyer (replica)	Science Museum/S. Kensington	
54	JAP-Harding monoplane	Science Museum/S. Kensington	
55	Levavasseur Antoinette VII	Science Museum/S. Kensington	
56	Fokker E.III (210/16)	Science Museum/S. Kensington	
57	Pilcher Hawk (replica)	Science Museum/S. Kensington	
58	Yokosuka MXY7 Ohka II (15-1585)	F.A.A. Museum/Yeovilton	

Notes	Reg.	Type	Owner or Operator
	59	Sopwith Camel (replica) (D3419)	Aerospace Museum/Cosford
	60	Murray M.1 helicopter	The Aeroplane Collection Ltd
	61	Stewart man-powered ornithopter	Lincolnshire Aviation Museum
	62	Cody Biplane (304)	Science Museum/S. Kensington
	63	Hurricane (replica) (P3208)	Kent Battle of Britain Museum/Hawkinge
	64	Hurricane (replica) (P3059)	Kent Battle of Britain Museum/Hawkinge
	65	Spitfire (replica) (N3289)	Kent Battle of Britain Museum/Hawkinge
	66	Bf 109 (replica) (1480)	Kent Battle of Britain Museum/Hawkinge
	67	Bf 109 (replica) (14)	Kent Battle of Britain Museum/Hawkinge
	68	Hurricane (replica) (H3426)	Midland Air Museum
	69	Spitfire (replica) (N3313)	Kent Battle of Britain Museum/Hawkinge
	70	Auster AOP.5 (TJ398)	Museum of Flight/E. Fortune
	71	Spitfire (replica) (P8140)	Norfolk & Suffolk Aviation Museum
	72	Hurricane (model) (V6779)	Gloucestershire Aviation Collection
	73	Hurricane (replica)	-
	74	Bf 109 (replica) (6357)	Kent Battle of Britain Museum/Hawkinge
	75	Mignet HM.14 (G-AEFG)	N. H. Ponsford
	76	Mignet HM.14 (G-AFFI)	Yorkshire Air Museum/Elvington
	77	Mignet HM.14 (replica) (G-ADRG)	Lower Stondon Transport Museum
	78	Hawker Hind (K5414) (G-AENP)	The Shuttleworth Collection/O. Warden
	79	Fiat G.46-4B (MM53211)	British Air Reserve/France
	80	Airspeed Horsa (KJ351)	Museum of Army Flying/Middle Wallop
	81	Hawkridge Dagling	Russavia Collection
	82	Hawker Hind (Afghan)	RAF Museum/Hendon
	83	Kawasaki Ki-100-1b (24)	Aerospace Museum/Cosford
	84	Nakajima Ki-46 (Dinah III)(5439)	Aerospace Museum/Cosford
	85	Weir W-2 autogyro	Museum of Flight/E. Fortune
	86	de Havilland Tiger Moth (replica)	Yorkshire Aircraft Preservation Soc
	87	Bristol Babe (replica) (G-EASQ)	Bristol Aero Collection/Kemble
	88	Fokker Dr 1 (replica) (102/17)	F.A.A. Museum/Yeovilton
	89	Cayley glider (replica)	Manchester Museum of Science & Industry
	90	Colditz Cock (replica)	Imperial War Museum/Duxford
	91	Fieseler Fi 103 (V-1)	Lashenden Air Warfare Museum
	92	Fieseler Fi 103 (V-1)	RAF Museum/Hendon
	93	Fieseler Fi 103 (V-1)	Imperial War Museum/Duxford
	94	Fieseler Fi 103 (V-1)	Aerospace Museum/Cosford
	95	Gizmer autogyro	F. Fewsdale
	96	Brown helicopter	N.E. Aircraft Museum
	97	Luton L.A.4A Minor	N.E. Aircraft Museum
	98	Yokosuka MXY7 Ohka II (997)	Manchester Museum of Science & Industry
	99	Yokosuka MXY7 Ohka II (8486M)	Aerospace Museum/Cosford
	100	Clarke Chanute biplane gliderr	RAF Museum/Hendon
	101	Mignet HM.14	Newark Air Museum/Winthorpe
	102	Mignet HM.14	Not completed
	103	Hulton hang glider (replica)	Personal Plane Services Ltd
	104	Bleriot XI	Sold in France
	105	Blériot XI (replica)	Arango Collection/Los Angeles
	106	Blériot XI (164)	RAF Museum/Hendon
	107	Blériot XXVII	RAF Museum/Hendon
	108	Fairey Swordfish IV (HS503)	RAF Restoration Centre/Wyton
	109	Slingsby Kirby Cadet TX.1	RAF Museum/Henlow store
	110	Fokker D.VII replica (static) (5125)	Stored
	111	Sopwith Triplane replica (static) (N5492)	F.A.A. Museum/Yeovilton
	112	D.H.2 replica (static) (5964)	Museum of Army Flying/Middle Wallop
	113	S.E.5A replica (static) (B4863)	Stored
	114	Vickers Type 60 Viking (static) (G-EBED)	Brooklands Museum of Aviation/ Weybridge
	115	Mignet HM.14	Norfolk & Suffolk Aviation Museum/Flixton
	116	Santos-Dumont Demoiselle (replica)	Cornwall Aero Park/Helston
	117	B.E.2C (replica)(1701)	RAF Manston History Museum
	118	Albatros D.V (replica) (C19/18)	North Weald Aircraft Restoration Flight
	119	Bensen B.7	N.E. Aircraft Museum
	120	Mignet HM.14 (G-AEJZ)	Bomber County Museum/Hemswell
	121	Mignet HM.14 (G-AEKR)	S. Yorks Aviation Soc
	122	Avro 504 (replica) (1881)	Stored
	123	Vickers FB.5 Gunbus (replica)	A. Topen (stored)/Cranfield
	124	Lilienthal Glider Type XI (replica)	Science Museum/S. Kensington

Reg.	Type	Owner or Operator	Notes
125	Clay Cherub	*ground trainer*/Coventry	
126	D.31 Turbulent (static)	Midland Air Museum/Coventry	
127	Halton Jupiter MPA	The Shuttleworth Collection	
128	Watkinson Cyclogyroplane Mk IV	IHM/Weston-s-Mare	
129	Blackburn 1911 Monoplane (replica)	Cornwall Aero Park/Helston store	
130	Blackburn 1912 Monoplane (replica)	Yorkshire Air Museum	
131	Pilcher Hawk (replica)	C. Paton	
132	Blériot XI (G-BLXI)	*Stored*	
133	Fokker Dr 1 (replica) (425/17)	Kent Battle of Britain Museum/Hawkinge	
134	Pitts S-2A static (G-CARS)	Toyota Ltd/Sywell	
135	Bristol M.1C (replica) (C4912)	*Stored*	
136	Deperdussin Seaplane (replica)	National Air Race Museum/USA	
137	Sopwith Baby Floatplane (replica) (8151)	*Stored*	
138	Hansa Brandenburg W.29 Floatplane (replica) (2292)	*Stored*	
139	Fokker Dr 1 (replica) 150/17	*Stored*	
140	Curtiss 42A (replica)	Stored	
141	Macchi M39 (replica)	Switzerland	
142	SE-5A (replica) (F5459)	*Stored*	
143	Paxton MPA	R. A. Paxton/Gloucestershire	
144	Weybridge Mercury MPA	Cranwell Gliding Club	
145	Oliver MPA	*Stored*	
146	Pedal Aeronauts Toucan MPA	*Stored*	
147	Bensen B.7	Norfolk & Suffolk Aviation Museum/Flixton	
148	Hawker Fury II (replica) (K7271)	High Ercall Aviation Museum	
149	Short S.27 (replica)	F.A.A. Museum (stored)/Yeovilton	
150	SEPECAT Jaguar GR.1 (replica) (XX728)	RAF M & R Unit/St Athan RAF Marketing & Recruitment Unit/St Athan	
151	SEPECAT Jaguar GR.1 (replica) (XZ363)	RAF M & R Unit/St Athan	
152	BAe Hawk T.1 (replica) (XX226)	RAF M & R Unit/St Athan	
153	Westland WG.33	IHM/Weston-s-Mare	
154	D.31 Turbulent	Lincolnshire Aviation Museum/E. Kirkby	
155	Panavia Tornado GR.1 (model) (ZA556)	RAF M & R Unit/St Athan	
156	Supermarine S-6B (replica)	National Air Race Museum/USA	
157	Waco CG-4A(237123)	Yorkshire Air Museum/Elvington	
158	Fieseler Fi 103 (V-1)	Defence Ordnance Disposal School/Chattenden	
159	Yokosuka MXY7 Ohka II	Defence Ordnance Disposal School/Chattenden	
160	Chargus 18/50 hang glider	Museum of Flight/E. Fortune	
161	Stewart Ornithopter Coppelia	Bomber County Museum	
162	Goodhart MPA	Science Museum/Wroughton	
163	AFEE 10/42 Rotabuggy (replica)	Museum of Army Flying/Middle Wallop	
164	Wight Quadruplane Type 1 (replica)	Hall of Aviation/Southampton	
165	Bristol F.2b (E2466)	RAF Museum/Hendon	
166	Bristol F.2b (D7889)	*Stored*	
167	Bristol SE-5A	*Stored*	
168	D.H.60G Moth (static replica)	Hilton Hotel/Gatwick (G-AAAH)	
169	BAC/Sepecat Jaguar GR.1 (XX110)	RAF Training School/Cosford	
170	Pilcher Hawk (replica)	A. Gourlay/Strathallan	
171	BAe Hawk T.1 (model) (XX253)	RAF Marketing & Recruitment Unit/St Athan	
172	Chargus Midas Super 8 hang glider	Science Museum/Wroughton	
173	Birdman Promotions Grasshopper	Science Museum/Wroughton	
174	Bensen B.7	Science Museum/Wroughton	
175	Volmer VJ-23 Swingwing	Manchester Museum of Science & Industry	
176	SE-5A (replica) (A4850)	S. Yorks Aviation Soc/Firbeck	
177	Avro 504K (replica) (G-AACA)	Brooklands Museum of Aviation/Weybridge	
178	Avro 504K (replica) (E373)	Bygone Times Antique Warehouse/ Eccleston, Lancs	
179	Sopwith Pup (replica) (A7317)	Midland Air Museum/Coventry	
180	McCurdy Silver Dart (replica)	Reynolds Pioneer Museum/Canada	
181	RAF B.E.2b (replica) (687)	RAF Museum/Hendon	
182	Wood Ornithopter	Manchester Museum of Science & Industry	
183	Zurowski ZP.1 helicopter	Newark Air Museum/Winthorpe	
184	Spitfire IX (replica) (EN398)	Fighter Wing Display Team/North Weald	
185	Waco CG-4A (243809)	Museum of Army Flying/Middle Wallop	

Notes	Reg.	Type	Owner or Operator
	186	D.H.82B Queen Bee (LF789)	de Havilland Heritage Museum
	187	Roe Type 1 biplane (replica)	Brooklands Museum of Aviation/Weybridge
	188	McBroom Cobra 88	Science Museum/Wroughton
	189	Bleriot XI (replica)	Stored
	190	Spitfire (replica) (K5054)	P. Smith/Hawkinge
	191	BAe Harrier GR.7 (model) (ZH139)	RAF M & R Unit/St Athan
	192	Weedhopper JC-24	The Aeroplane Collection
	193	Hovey WD-11 Whing Ding	The Aeroplane Collection
	194	Santos Dumont Demoiselle (replica)	Brooklands Museum of Aviation/Weybridge
	195	Moonraker 77 hang glider	Museum of Flight/E. Fortune
	196	Sigma 2M hang glider	Museum of Flight/E. Fortune
	197	Scotkites Cirrus III hang glider	Museum of Flight/E. Fortune
	198	Fieseler Fi 103 (V-1)	Imperial War Museum/Lambeth
	199	Fieseler Fi 103 (V-1)	Science Museum/S. Kensington
	200	Bensen B.7	K. Fern Collection/Stoke
	201	Mignet HM.14	Caernarfon Air Museum
	202	Spitfire V (model) (MAV467)	Maes Artro Craft Centre
	203	Chrislea LC.1 Airguard (G-AFIN)	The Aeroplane Collection
	204	McBroom hang glider	Newark Air Museum
	205	Hurricane (replica) (BE421)	RAF Museum/Hendon
	206	Spitfire (replica) (MH486)	RAF Museum/Hendon
	207	Austin Whippet (replica) (K.158)	South Yorkshire Aviation MuseumDoncaster
	208	SE-5A (replica) (D276)	Prince's Mead Shopping Precinct/Farnborough
	209	Spitfire IX (replica) (MJ751)	Museum of D-Day Aviation/Shoreham
	210	Avro 504J (replica) (C4451)	Southampton Hall of Aviation
	211	Mignet HM.14 (replica) (G-ADVU)	N.E. Aircraft Museum
	212	Bensen B.8	IHM/Weston-s-Mare
	213	Vertigo MPA	IHM/Weston-s-Mare
	214	Spitfire prototype (replica) (K5054)	Tangmere Military Aviation Museum
	215	Airwave hang-glider prototype	Southampton Hall of Aviation
	216	D.H.88 Comet (replica) (G-ACSS)	de Havilland Heritage Museum
	217	Spitfire (replica) (K9926)	RAF Museum/Bentley Priory
	218	Hurricane (replica) (P3386)	RAF Museum/Bentley Priory
	219	Hurricane (replica) (L1710)	RAF Memorial Chapel/Biggin Hill
	220	Spitfire 1 (replica) (N3194)	RAF Memorial Chapel/Biggin Hill
	221	Spitfire LF.IX (replica) (MH777)	RAF Museum/Northolt
	222	Spitfire IX (replica) (BR600)	RAF Museum/Uxbridge
	223	Hurricane 1 (replica) (V7467)	RAF Museum/Coltishall
	224	Spitfire V (replica) (BR600)	Ambassador Hotel/Norwich
	225	Spitfire IX (replica) (P8448)	RAF Museum/Cranwell
	226	Spitfire XI (replica) (EN343)	RAF Museum/Benson
	227	Spitfire 1A (replica) (L1070)	RAF Museum/Turnhouse
	228	Olympus hang-glider	N.E. Aircraft Museum/Usworth
	229	Spitfire IX (replica) (MJ832)	RAF Museum/Digby
	230	Spitfire (replica) (AA550)	Eden Camp/Malton
	231	Mignet HM.14 (G-ADRX)	South Copeland Aviation Group
	232	AS.58 Horsa I/II	de Havilland Heritage Museum
	233	Broburn Wanderlust sailplane	Museum of Berkshire Aviation/Woodley
	234	Vickers FB.5 Gunbus (replica)	RAF Manston Museum
	235	Fieseler Fi 103 (V-1) (replica)	Eden Camp Wartime Museum
	236	Hurricane (replica) (P2793)	Eden Camp Wartime Museum
	237	Fieseler Fi 103 (V-1)	RAF Museum/Wyton
	238	Waxflatter ornithopter	Personal Plane Services Ltd
	239	Fokker D.VIII 5/8 scale replica	Norfolk & Suffolk Aviation Museum/Flixton
	240	Messerschmitt Bf.109G (replica)	Yorkshire Air Museum/Elvington
	241	Hurricane 1 (replica) (L1679)	Tangmere Military Aviation Museum
	242	Spitfire Vb (replica) (BL924)	Tangmere Military Aviation Museum
	243	Mignet HM.14 (replica) (G-ADYV)	P. Ward
	244	Solar Wings Typhoon	Museum of Flight/E. Fortune
	245	Electraflyer Floater hang glider	Museum of Flight/E. Fortune
	246	Hiway Cloudbase hang glider	Museum of Flight/E. Fortune
	247	Albatross ASG.21 hang glider	Museum of Flight/E. Fortune
	248	McBroom hang glider	Museum of Berkshire Aviation/Woodley
	249	Hawker Fury 1 (replica) (K5673)	Brooklands Museum of Aviation/Weybridge
	250	RAF SE-5A (replica) (F5475)	Brooklands Museum of Aviation/Weybridge
	251	Hiway Spectrum hang glider (replica)	Manchester Museum of Science & Industry
	252	Flexiform Wing hang glider	Manchester Museum of Science & Industry

Reg.	Type	Owner or Operator	Notes
253	Mignet HM.14 (G-ADZW)	H. Shore/Sandown	
254	Hawker Hurricane (P3873)	Yorkshire Air Museum/Elvington	
255	NA P-51D Mustang (replica) (463209)	American Air Museum/Duxford	
256	Santos Dumont Type 20 (replica)	Brooklands Museum of Aviation/Weybridge	
257	D.H.88 Comet (G-ACSS)	The Galleria/Hatfield	
258	Adams balloon	British Balloon Museum	
259	Gloster Gamecock (replica)	Jet Age Museum Gloucestershire	
260	Not yet allocated		
261	GAL Hotspur (replica)	Museum of Army Flying/ Middle Wallop	
262	Catto CP-16	Museum of Flight/East Fortune	
263	Chargus Cyclone	Ulster Aviation Heritage/Langford Lodge	
264	Bensen B.8M	IHM/Weston-super-Mare	
265	Spitfire 1 (P3873)	Yorkshire Air Museum/Elvington	
266	Rogallo hang glider	Ulster Aviation Heritage	
267	Hurricane (model)	Duxford	
268	Spifire (model)	-	
269	Spitfire (model) USAF	Lakenheath	
270	D.H.60 Moth (model)	Yorkshire Air Museum	
271	Messerschmitt Me 163B	Shuttleworth Collection/Old Warden	
272	Hurricane (model)	Kent Battle of Britain Museum	
273	Hurricane (model)	Kent Battle of Britain Museum	
274	Boulton & Paul P.6 (model)	Boulton & Paul Aircraft Heritage Project	
275	Bensen B.7 gyroglider	Doncaster Museum	
276	Hartman Ornithopter	Science Museum/Wroughton	
277	Mignet HM.14	Visitor Centre Shoreham Airport	
278	Hurricane (model)	Kent Battle of Britain Mussum	
279	Airspeed Horsa	Shawbury	
280	D.H.89A Dragon Rapide (model)	-	
281	Boulton & Paul Defian (model)t	-	
282	Manx Elder Duck	Isle of Man Airport Terminal	
283	Spitfire (model	Jurby Isle of Man	
284	Gloster E.28/39 (model)	Lutterworth Leics	
285	Gloster E.28/39 (model)	Farnborough	
286	Mignet HM.14	Caernarfon Air Museum	
287	Blackburn F.2 Lincock (model)	Street Life Museum/Hull	
288	Hurricane (model)	Wonderland Pleasure Park, Mansfield	
289	Gyro Boat	IHM Weston-super-Mare	
290	Fieseler Fi 103 (V1) (model)	Dover Museum	
291	Hurricane (model)	National Battle of Britain Memorial, Capel le Ferne, Kent	
292	Eurofighter Typhoon (model)	RAF Museum/Hendon	
293	Spitfire (model)	RAF Museum/Hendon	
294	Fairchild Argus (model)	Visitor Centre, Thorpe Camp, Woodall Spar	
295	Da Vinci hang glider (replica)	Skysport Engineering	
296	Army Balloon Factory NuIII (replica)	-	
297	Spitfire (replica)	Kent Battle of Britain Museum/Hawkinge	
298	Spitfire IX (Model)	Kent Battle of Britain Museum/ Hawkinge	
299	Spitfire 1 (model).	National Battle of Britain Memorial. Capel le Ferne, Kent	
300	Hummingbird (replica)	Shoreham Airport Historical Association	

Note: Registrations/Serials carried are mostly false identities.
MPA = Man Powered Aircraft, IHM = International Helicopter Museum The aircraft, listed as 'models' are
generally intended for exhibition purposes and are not airworthy although they are full scale replicas.
However, in a few cases the machines have the ability to taxi when used for film work

Future Allocations Log

The grid provides the facility to record future in-sequence registrations as they are issued or seen. To trace a particular code, refer to the left hand column which contains the three letters following the G prefix. The final letter can be found by reading across the columns headed A to Z. For example, the box for G-CEBT is located 27 rows down (CET) and then 19 across to the T column.

G-	A	B	C	D	E	F	G	H	I	J	K	L	M	N	O	P	R	S	T	U	V	W	X	Y	Z
CDA																									
CDB																									
CDC																									
CDD																									
CDE																									
CDF																									
CDG																									
CDH																									
CDI																									
CDJ																									
CDK																									
CDL																									
CDM																									
CDN																									
CDO																									
CDP																									
CDR																									
CDS																									
CDT																									
CDU																									
CDV																									
CDW																									
CDX																									
CDY																									
CDZ																									
CEA																									
CEB																									
CEC																									
CED																									
CEE																									
CEF																									
CEG																									
CEH																									
CEI																									
	A	B	C	D	E	F	G	H	I	J	K	L	M	N	O	P	R	S	T	U	V	W	X	Y	Z

Credit: *Wal Gandy*

Future Allocation Groups

This grid can be used to record registrations as they are issued or seen. The first column is provided for the ranges prefixed with G-C, ie from G-CYxx to G-CZxx. The remaining columns cover the sequences from G-Dxxx to G-Zxxx and in this case it is necessary to insert the last three letters in the appropriate section.

G-C	G-D	G-F	G-H	G-J	G-L	G-N	G-O	G-P	G-S	G-U
										G-V
	G-E	G-G			G-M	G-O				
				G-K						
										G-W
			G-I							
								G-R		
									G-T	
										G-X
G-D	G-F			G-L	G-N					
										G-Y
										G-Z

Overseas Airliners Registration Log

This grid may be used to record airliner registrations not included in the main section.

Reg.	Type	Operator

Reg.	Type	Owner or Operator	Notes
Additions and changes of ownership			
G-AGYH	Alpha Auster J1N	I. M. Staves	
G-AWGR	Cessna F172H	R. G. Hallam	
G-AYRG	Cessna F172K	Comed Schedule Services Ltd	
G-BADV	Brochet MB50	W. B. Cooper	
G-BAMT	CEA DR400/160	S. G. Jones	
G-BBBK	Piper PA-28 Cherokee 140	Comed Schedule Services Ltd	
G-BBEV	Piper PA-28 Cherokee 140	Comed Schedule Services Ltd	
G-BCMJ	Squarecraft Cavalier SA.102-5	N. F. Andrews	
G-BDGM	Piper PA-28 Cherokee 151	Comed Schedule Services Ltd	
G-BDWY	Piper PA-28 Cherokee 140	Comed Schedule Services Ltd	
G-BMKK	Piper PA-28 Cherokee 200	GBMKK Ltd	
G-BRJC	Cessna 120	A. L. Hall-Carpenter	
G-BRPL	Piper PA-28 Cherokee 140 Arrow	Comed Schedule Services Ltd	
G-BULH	Cessna 172N	Comed Schedule Services Ltd	
G-BVCX	Sikorsky S76A (modified)	CHC Scotia Ltd	
G-CDBH	Pitts S-2B	J. W. Sullivan	
G-CDBU	Ikarus C42 FB100	S. E. Meehan	
G-CDCI	Pegasus Quik	S. G. Murray	
G-CDCR	Savannah Jabiru(1)	T. Davidson	
G-CDDM	Lindstrand LBL 90A	Lindstrand Hot Air Balloons Ltd	
G-CDDN	Lindstrand LBL 90A	Lindstrand Hot Air Balloons Ltd	
G-CDEW	Pegasus Quik	M. M. Chittenden	
G-CDEA	Saab 2000	Air Kilroe Ltd	
G-CDEC	Pagasus Quik	S. Bradie	
G-CDED	Robinson R22 Beta	Heli Air Ltd	
G-CDEZ	Robinson R44 II	Heli Air Ltd	
G-CDFE	Yak-52	D. R. Farley	
G-CDFO	Pegasus Quik	C. J. Gordon	
G-DGC	Pegasus Quik	A. T. K. Crozier	
G-CDGI	Thruster T600N 450	Thruster Air Services Ltd	
G-CDGP	Zenair CH 601XL	T. J. Bax	
G-CDGR	Zenair CH 701UL	M. Morris	
G-CDGT	Montgomerie-Parsons Two Place	A. A. Craig	
G-CDGU	Spitfire I	A. J. E. Smith	
G-CDGV	Bell 206B	Multiflight Ltd	
G-CDGW	Piper PA-28-181	Barry Fielding Aviation Ltd	
G-CDGX	Pegasus Quantum 15-912	T. Griffiths	
G-CDGY	Spitfire Mk VC	Aero Vintage Ltd	
G-DHA	Skyranger 912S(1)	K. J. Gay	
G-CDHC	Slingsby T67C	N. J. Morgan	
G-CDHE	Skyranger 912(2)	S. Owen	
G-CDHM	Pegasus Quantum 15	M. K. Morgan	
G-CDII	Schleicher ASW 22BLE 50R	R. A. Cheetham	
G-CDIP	Skyranger 912S(1)	M. S. McCrudden	
G-CDIZ	Escapade 912(1)	E. G. Bishop	
G-CDJB	Vans RV-4	J. M. Keane	
G-CDJP	Skyranger 912(2)	J. S. Potts	
G-CDKM	Pegasus Quik	P. Barrow	
G-CDPL	EV-97 Teameurostar UK	C. I. D. H Garrison	
G-CFRY	Zenair CH 601UL	C. K. Fry	
G-DSPK	Cameron Z-140	Bailey Balloons Ltd	
G-EZEY	Airbus A.319-111	easyJet Airline Co Ltd/Luton	
G-EZMK	Airbus A.319-111	easyJet Airline Co Ltd/Luton	
G-EZMS	Airbus A.319-111	easyJet Airline Co Ltd/Luton	
G-FORM	Learjet 45	Broomco 3598 Ltd	
G-FRNK	Skyranger 912(2)	F. Tumelty	
G-GATT	Robinson R44 II	N. R. Gatt	
G-GILI	Robinson R44	Twylight Management Ltd	
G-GOGB	Lindstrand LBL ,90A	J. Dyer (G-CDFX)	
G-IIVI	Cap 232	Skylane Aviation Ltd	
G-INIT	SOCATA TB7	P. R. Shakeshaft	
G-JAME	Zenair CH 601UL	J. P. Harris	
G-JECG	DHC-8-402	Flybe.com British European	
G-JMJR	Cameron Z-90	J. M. Reck	

Notes	Reg.	Type	Owner or Operator
	G-MGPA	Ikarus C42 FB100	P. D. Ashley
	G-MNID	Gemini Flash	D. Sykes
	G-MUCK	Lindstrand LBL 77A	C. J. Wootton
	G-MUMY	Vans RV-4	S. D. Howes
	G-OMPW	Pegasus Quik	MPW Decorators Ltd
	G-OMSS	Skyranger 912(2)	M. S. Schoield
	G-OPEN	Bell 206B	Gazelle Aviation LLP
	G-PLAD	Kolb Twinstar Mk 3 Extra	P. J. Ladd
	G-RGNT	Robinson R44 II	P. R. Nott (G-DMCG)
	G-RMMT	Europa XS	N. Schmitt
	G-RPAF	Europa XS	R. P. Frost
	G-ROMP	Extra 230H	J. S. Allison
	G-RTWO	Robinson R44 II	Heli Air Ltd
	G-RVDG	Vans RV-9A	D. M. Gill
	G-SAJP	Canadair CL600-2B16	Jetlinks (CI) Ltd
	G-SAUK	Rans S6-ES	D. A. Smith
	G-SHAN	Robinson R44 II	Heli Air Ltd
	G-SKYW	AS355F1	Skywalker Aviation Ltd (G-BTIS)
	G-SSJP	Robinson R44 II	P. Cripps
	G-SVIP	Cessna 421B	T. Stone-Brown (G-BNYJ)
	G-TAFC	Maule M-7-235B	The Amphibious Flying Club Ltd
	G-TAME	Schweizer 269D	Total Air Management Services Ltd
	G-TEDB	Cessna F150L	E. L. Bamford (G-AZLZ)
	G-TREX	Pioneer 300	R. K. King
	G-TUTU	Cameron O-105	A. C. K. Rawson
	G-TVEE	Hughes 369HS	M. J. Gee
	G-WCIL	Agusta A109E	Castle Air Charters Ltd
	G-WOCO	Waco YMF-5C	Airpark Flight Centre Ltd
	G-XIOO	X'Air 133(1)	J. Campbell

Cancellations

G-AZLZ, G-AWGR, G-BAAU, G-BDGY, G-BGED, G-BLKZ, G-BPKZ, G-BPNL, G-BRPO, G-BTIS, G-BTRE, G-BUOC, G-BVAA, G-BYFP, G-BYIW, G-BZDF, G-CBFC, G-CBXX, G-CCTJ, G-CDEZ, G-CDFX, G-CDFZ, G-DMCG, G-DRIC, G-EZYF, G-GATT, G-MMAN, G-MMOW, G-MMUT, G-MNGL, G-MNGN, G-MNRF, G-MTCX, G-MTLR, G-MVOA, G-MWUF, G-MWVZ, G-PFKD

VX147 (G-AVIL) Alon A-2 Aircoupe. *AJW*

OK-WGG, Boeing 373-4Y0 of CSA Czech Airlines. AJW